A Survey of Numerical Mathematics

by David M. Young
Center for Numerical Analysis
The University of Texas at Austin

and Robert Todd Gregory
Formerly at the Center for Numerical Analysis
The University of Texas at Austin

In Two Volumes

Volume II

Dover Publications, Inc., *New York*

Published in Canada by General Publishing Company, Ltd., 30 Lesmill
Road, Don Mills, Toronto, Ontario.
Published in the United Kingdom by Constable and Company, Ltd.

This Dover edition, first published in 1988, is an unabridged, corrected
republication of the work originally published by the Addison-Wesley Publishing Company (Addison-Wesley Series in Mathematics), Reading, Mass., 1972
(Vol. I) and 1973 (Vol. II). In the Dover edition the Bibliography and Index of
Volume I, given in incomplete form in the original edition, have been replaced
by the complete versions that first appeared in Volume II. Appendix B and
Appendix C, newly added to the Dover edition, include revisions of two
passages in Volume I.

Manufactured in the United States of America
Dover Publications, Inc., 31 East 2nd Street, Mineola, N.Y. 11501

Library of Congress Cataloging-in-Publication Data

Young, David M., 1923–
    A survey of numerical mathematics.

    Reprint. Originally published: Reading, Mass. : Addison-Wesley,
c1972–c1973. (Addison-Wesley series in mathematics)
    Includes bibliographies and index.
    1. Numerical analysis—Data processing. I. Gregory, Robert Todd,
1920–1984. II. Title. III. Series: Addison-Wesley series in mathematics.
QA297.Y63   1988   519.4   88-3630
ISBN 0-486-65691-8 (pbk. : v. 1)
ISBN 0-486-65692-6 (pbk. : v. 2)

To the memory of
GEORGE E. FORSYTHE

PREFACE

Since the advent of automatic digital computers there has been a rapid development of the branch of applied mathematics known as *numerical analysis*. Numerical analysis is concerned with the development, analysis, and evaluation of numerical algorithms which can be carried out on a computer (usually an automatic digital computer) for obtaining numerical solutions to mathematical problems. Although the methods and results of classical analysis can often be helpful, they usually provide only background and/or a starting point for the numerical analyst. For example, the pure mathematician may be fully satisfied if he can prove that a unique solution to a given problem exists, but it is up to the numerical analyst to devise a procedure for actually computing a solution to within a specified accuracy and within a reasonable time. In developing an algorithm to use in solving a given problem, the numerical analyst must be concerned not only with the number of arithmetic operations and the theoretical accuracy but also with the cumulative effects of rounding errors which are made when the algorithm is implemented on a computer.

The purpose of this book (Volume I and Volume II) is to study computer-oriented numerical algorithms for solving various types of mathematical problems. An attempt has been made to provide a judicious mixture of mathematics, numerical analysis, and computation. It is intended that the reader should not only acquire a working knowledge of practical techniques for solving real problems, but that he should also be prepared for deeper studies of particular topics at the graduate level.

In Chapter 1, we try to define numerical analysis as a subject area and to distinguish it from classical mathematical analysis. We attempt to give an idea of some of the kinds of problems encountered by the numerical analyst and the kinds of techniques used to solve these problems. In Chapter 2, we describe some of the characteristics of automatic digital computers which are of concern to the numerical analyst, and we show, by way of illustration, how accurate and reliable procedures can be developed for evaluating many of the elementary mathematical functions. Chapter 3 presents methods for monitoring and controlling the size of numbers involved in certain types of problems. Chapter 4 is concerned with the solution of the single nonlinear equation $f(x) = 0$ and systems of such equations. The special case of polynomial equations is treated in Chapter 5. Interpolation and approximation are examined in Chapter 6 and numerical differentiation and numerical quadrature in Chapter 7.

Chapters 8, 9, and 10 deal with methods for solving ordinary differential equations. Chapter 8 describes some of the basic methods for solving initial-value problems for a single equation or for systems of equations, and this portion of the material is included in Volume I. Chapter 9, which begins Volume II, is concerned with stability, convergence, and accuracy of the various methods. Two-point boundary-value problems involving second-order differential equations are treated in Chapter 10. Much of the analysis of Chapter 10 is applicable to the study of boundary-value problems involving elliptic partial differential equations in Chapter 15.

Chapter 11 contains a review of the elements of linear algebra and matrix theory, along with an introduction to vector and matrix norms (with emphasis on those results which are used in this book). Arbitrary systems of linear algebraic equations are discussed in Chapter 12. The discussion is restricted to Gaussian elimination with pivoting (and the equivalent triangular decomposition) followed by iterative improvement. Difficulties associated with numerical instability and with ill-conditioned systems are discussed.

We survey the elements of residue arithmetic in Chapter 13 and describe an algorithm (a modified version of Gauss-Jordan elimination) for solving systems of linear algebraic equations *exactly*, using residue arithmetic. This chapter is a modified version of a University of Texas Computation Center report, TNN 82 (revised) by Howell and Gregory [1969a]. Moreover, the report itself is a revised version of a Master's thesis written by Mrs. Howell under the direction of the second author.

The algebraic eigenvalue–eigenvector problem is presented in Chapter 14, and various algorithms are described for handling Hermitian matrices as well as non-Hermitian matrices. Again, the difficulties associated with numerical instability and with ill-conditioned matrices are discussed.

In Chapters 15, 16, and 17 we deal with the numerical solution of partial differential equations. Finite difference methods for solving elliptic equations are studied in Chapter 15. It is shown that the problem can be reduced to that of solving a large system of linear algebraic equations. Iterative methods, such as the successive overrelaxation method for solving these systems, are described in Chapter 16. Initial-value problems, such as the heat equation in two and three (space) dimensions, are considered in the final chapter.

The mathematical and computational background required of the reader varies over different parts of the book. He should have had a course on the elementary aspects of real analysis as covered in a good course in "advanced calculus" or "elementary analysis," and he should also be familiar with elementary complex analysis to some extent. In Appendix A, we have included the basic theorems which we use. In addition to the mathematical background required, the reader should have some knowledge of automatic digital computers and be able to program in an algebraic compiler language such as Fortran or Algol.

For the study of Chapters 8, 9, and 10 the reader should have some knowledge of the elementary theory of ordinary differential equations. Similarly, although the treatment in Chapters 11, 12, 13, and 14 is largely self-contained, a beginning course in linear algebra is almost essential. For Chapters 15, 16, and 17 some background in partial differential equations would be helpful.

The book is designed as a text for a sequence of three one-semester courses beginning at the senior year and continuing into the first year of graduate work. The first course would cover Volume I. The second course dealing with computational methods in linear algebra would cover Chapters 11–14 included in Volume II. The third course would cover ordinary differential equations (Chapters 9 and 10 along with a review of Chapter 8) and partial differential equations (Chapters 15, 16, and 17).

Exercises are given for most of the sections. Some of them are designed to extend the theory, but the primary objective of the exercises is to enhance the student's understanding of the material. In some cases he is asked to work numerical problems, whereas in other cases he is required to fill in gaps in the discussion in the text. A few problems can be worked with the aid of a slide rule or a desk calculator and tables, but many have to be solved on an automatic digital computer.

Decimal notation is used for the numbering of the sections and chapters. For example, the third section of Chapter 6 is numbered 6.3 and referenced as Section 6.3. Equations, theorems, lemmas, figures, and tables are numbered consecutively as items within sections, e.g., the tenth item in Section 6.3 is numbered 3.10. This item might be referenced in several ways. For instance, if it is an *equation*, it is referred to as (3.10), but if it is any other kind of item (such as a theorem), it is called Theorem 3.10. If it is referenced outside Chapter 6, then the notation is 6-(3.10) or Theorem 6-3.10. At this point we should add that A-(3.10) refers to the tenth item in the third section of Appendix A.

We decided to divide the material into two volumes because of its length. The most logical point of division appears to be between Chapters 8 and 9, primarily because the material in the first eight chapters seems suitable for a one-semester introductory course. Appendix A and the complete bibliography are included in both volumes.

Austin, Texas   D.M.Y.
September 1972   R.T.G.

ACKNOWLEDGMENTS

We wish to express our appreciation to the many people who helped make this book a reality. We are especially grateful to Richard Varga who reviewed the entire manuscript. We are also grateful to A. E. McDonald, Y. Ikebe, J. Dauwalder, T. Lyche, Jo Ann Howell, and V. Benokraitis, each of whom reviewed substantial portions of the material. All of the reviewers made excellent suggestions for improving the book. We benefited greatly from our discussion with G. W. Stewart, who generously allowed us to read the manuscript of his forthcoming book on numerical linear algebra. Jo Ann Howell's greatest contribution —over and above her contribution in reviewing some of the material—stems from the fact that Chapter 13 is essentially a revised version of her Master's thesis.

An early draft of portions of this book was prepared in the form of class notes taken by Sylvia Goodrich from lectures given by the first author.

Finally, we must thank Dorothy Baker for her complete dedication to this effort. Her superb job of preparing the manuscript enabled us to bring this project to completion. She was aided at various stages by Barbara Allen, Marge Dragoo, and Linda Brothers.

We wish to dedicate this book to the memory of Professor George E. Forsythe, who was a pioneer and an inspiration in the development of the field of numerical mathematics.

CONTENTS

Chapter 12 The Solution of Systems of Linear Algebraic Equations by Direct Methods

VOLUME I

CHAPTER 9

ORDINARY DIFFERENTIAL EQUATIONS: STABILITY, CONVERGENCE, AND ACCURACY

9.1 INTRODUCTION

In Chapter 8 we considered various standard methods for solving ordinary differential equations. We gave some consideration to the questions of convergence and accuracy but none to questions of stability. In this chapter we consider stability, convergence, and accuracy for a more general class of methods.

In Section 9.2 we consider the class of linear multistep methods which includes as special cases methods based on numerical quadrature as well as methods based on numerical differentiation. We consider the convergence, consistency, and stability of linear multistep methods in Section 9.3. In Section 9.4 we show that the attainable accuracy is limited by stability considerations. The behavior of several linear multistep methods when applied to the special case

1.1
$$y' = Ay,$$

where A is a constant, is studied in detail in Sections 9.5 and 9.6. The concepts of weak, strong, and conditional stability for more general methods are introduced and studied in Section 9.7. A numerical example is presented in Section 9.8. The stability of methods for solving systems of first-order equations is treated in Section 9.9.

9.2 LINEAR MULTISTEP METHODS

We now consider a class of methods defined by

2.1
$$\sum_{i=0}^{N} \alpha_i y_{n+1-i} = h \sum_{i=0}^{N} \beta_i f(x_{n+1-i}, y_{n+1-i})$$

where the α_i and β_i are constants independent of n and h and where

2.2
$$\begin{cases} \alpha_0 \neq 0 \\ |\alpha_N| + |\beta_N| > 0. \end{cases}$$

Here

$$h = x_{n+1} - x_n = x_n - x_{n-1} = \cdots = x_{n+2-N} - x_{n+1-N}.$$

Methods of this type are known as *linear multistep methods*. The method defined by (2.1) is a linear *N-step method*. If $\beta_0 = 0$ the method is *open*; otherwise, it is *closed*.

The *order* of a linear multistep method is the largest integer p such that

2.3
$$(R[y])_n = \sum_{i=0}^{N} \alpha_i y_{n+1-i} - h \sum_{i=0}^{N} \beta_i y'_{n+1-i} = 0$$

for any polynomial $y(x)$ of degree p or less. It is evident that if the order of (2.1) is at least zero, then we must have

2.4
$$\sum_{i=0}^{N} \alpha_i = 0.$$

If the order is at least one, then we must have (2.4) and

2.5
$$\sum_{i=0}^{N} (N - i)\alpha_i = \sum_{i=0}^{N} \beta_i.$$

This follows from requiring that (2.3) hold for $y = x - x_{n+1-N}$.

The general method based on numerical quadrature 8-(6.25) is a linear N-step method where $N = \max(q, M)$. Conversely, given a linear N-step method with order at least zero such that all α_i vanish except α_0 and one other α_q, then dividing by α_0 we have a method based on numerical quadrature. This follows since by (2.4) we have $\alpha_q = -\alpha_0$ and thus (2.1) becomes

2.6
$$y_{n+1} - y_{n+1-q} = h \sum_{i=0}^{N} \frac{\beta_i}{\alpha_0} f(x_{n+1-i}, y_{n+1-i})$$

which has the form 8-(6.25) with β_i replaced by β_i/α_0.

A method based on numerical differentiation of the form 8-(10.3) is a linear multistep method. For we can multiply both sides of 8-(10.3) by h and obtain the form (2.1).* Conversely, given a linear N-step method such that all β_i vanish except one, say β_q, then upon dividing both sides of (2.1) by $h\beta_q$ we get the form 8-(10.3).

If the method is open, then we can determine y_{n+1} by solving (2.1) for y_{n+1} obtaining

2.7
$$y_{n+1} = - \sum_{i=1}^{N} \frac{\alpha_i}{\alpha_0} y_{n+1-i} + h \sum_{i=1}^{N} \frac{\beta_i}{\alpha_0} f(x_{n+1-i}, y_{n+1-i}).$$

If the method is closed, then in general (2.1) defines a nonlinear equation for y_{n+1}. If $f(x, y)$ satisfies a Lipschitz condition in y, then one can solve for y_{n+1} provided h is sufficiently small, i.e., if

---

* It follows from a result of Section 7.3 that $\alpha_0 \neq 0$.

2.8
$$\frac{hL|\beta_0|}{|\alpha_0|} < 1$$

where L is the Lipschitz constant. One can obtain y_{n+1} using an iterative method as in the case of methods based on numerical quadrature. If a sufficiently good predictor is used, it may be sufficient to use one iteration. The predictor may be a linear multistep method of the same order as (2.1). In general, the predictor will correspond to a linear N^*-step method with $N^* > N$. As in the case of predictor-corrector methods considered in Chapter 8, it is necessary to use a special starting procedure to get $y_1, y_2, \ldots, y_{N^*-1}$.

While there are $2N + 2$ coefficients in (2.1), there are really only $2N + 1$ "free" coefficients which can be chosen to make (2.3) hold for any polynomial of degree p or less. We thus expect to be able to make the order of the method $2N$. The following theorem shows that this is indeed possible.

2.9 Theorem. There exists a unique open linear N-step method of order at least $2N - 1$ and a unique closed linear N-step method of order at least $2N$ such that $\alpha_0 = 1$.

Proof. For the open case the proof is a direct application of Hermite interpolation. Given $y_n, y_n', y_{n-1}, y_{n-1}', \ldots, y_{n+1-N}, y_{n+1-N}'$ there is a unique polynomial $F(x)$ of degree $2N - 1$ or less such that

2.10 $F(x_i) = y_{n+1-i}$ and $F'(x_i) = y_{n+1-i}'$, $i = 1, 2, \ldots, N$.

(See Section 6.7.) In fact $F(x)$ has the form

2.11 $$F(x) = \sum_{i=1}^{N} [v_i(x)y_{n+1-i} + w_i(x)y_{n+1-i}']$$

where the $v_i(x)$ and $w_i(x)$ are polynomials of degree $2N - 1$ or less. Therefore, if $y(x)$ is a polynomial of degree $2N - 1$ or less it follows that

2.12 $$y_{n+1} = \sum_{i=1}^{N} [v_i(x_{n+1})y_{n+1-i} + w_i(x_{n+1})y_{n+1-i}'].$$

The reader should verify from 6-(7.5) that the $v_i(x_{n+1})$ and $w_i(x_{n+1})/h$ are independent of h and x_{n+1}. Thus we have a linear N-step method where $\alpha_0 = 1$ and

2.13 $$\begin{cases} \alpha_i = -v_i \\ \beta_i = \dfrac{w_i}{h}, \quad i = 1, 2, \ldots, N. \end{cases}$$

In order to treat the closed case we now prove the following lemma concerning a slight generalization of the Hermite interpolation problem.

2.14 Lemma. Given $m + 2$ points $x_0, x_1, \ldots, x_{m+1}$ such that $x_0 < x_1 < \cdots$ $< x_m < x_{m+1}$ and given any $2m + 3$ numbers $\rho_0, \rho_1, \ldots, \rho_m$, $\sigma_0, \sigma_1, \ldots, \sigma_{m+1}$ there exists a unique polynomial $F(x)$ of degree $2m + 2$ or less such that

2.15
$$\begin{cases} F(x_i) = \rho_i, & i = 0, 1, \ldots, m \\ F'(x_i) = \sigma_i, & i = 0, 1, \ldots, m + 1. \end{cases}$$

Proof. We seek $2m + 3$ coefficients $a_0, a_1, \ldots, a_{2m+2}$ such that

2.16
$$F(x) = \sum_{j=0}^{2m+2} a_j x^{2m+2-j}$$

and such that (2.15) is satisfied. This leads to the system of $2m + 3$ equations in $2m + 3$ unknowns

2.17
$$\begin{cases} \displaystyle\sum_{j=0}^{2m+2} a_j x_i^{2m+2-j} = \rho_i, & i = 0, 1, \ldots, m \\ \displaystyle\sum_{j=0}^{2m+1} (2m + 2 - j)a_j x_i^{2m+1-j} = \sigma_i, & i = 0, 1, \ldots, m + 1. \end{cases}$$

A unique solution exists if and only if the determinant of the system (2.17) does not vanish. Moreover, the determinant does not vanish if and only if the homogeneous system formed by replacing the right-hand sides of the equations (2.17) by zeros has only the trivial solution. Suppose that there exist coefficients $\hat{a}_0, \hat{a}_1, \ldots, \hat{a}_{2m+2}$ not all zero such that

2.18
$$\begin{cases} \hat{F}(x_i) = 0, & i = 0, 1, 2, \ldots, m \\ \hat{F}'(x_i) = 0, & i = 0, 1, 2, \ldots, m + 1 \end{cases}$$

where

2.19
$$\hat{F}(x) = \sum_{i=0}^{2m+2} \hat{a}_i x^{2m+2-i}.$$

Evidently $\hat{F}'(x)$ vanishes for $x_0, x_1, \ldots, x_{m+1}$. Also, by Rolle's theorem, $\hat{F}'(x)$ vanishes for $\xi_0, \xi_1, \ldots, \xi_{m-1}$ where

2.20
$$\xi_i \in (x_i, x_{i+1}), \qquad i = 0, 1, \ldots, m - 1.$$

Therefore $\hat{F}'(x)$ vanishes at $2m + 2$ distinct points. Since $\hat{F}(x)$ is a polynomial of degree $2m + 2$ or less, $\hat{F}'(x)$ is a polynomial of degree $2m + 1$ or less. Hence $\hat{F}'(x) \equiv 0$ and $\hat{F}(x)$ is a constant which by (2.18) must be zero. Therefore $\hat{F}(x) \equiv 0$ and $\hat{a}_0 = \hat{a}_1 = \cdots = \hat{a}_{2m+2} = 0$. Thus the determinant of the system (2.17) does not vanish and a unique polynomial $F(x)$ exists. This completes the proof of Lemma 2.14.

We now seek to determine $\alpha_0, \alpha_1, \ldots, \alpha_N, \beta_0, \beta_1, \ldots, \beta_N$ such that (2.3) holds for the functions

2.21
$$(x - x_{n+1-N})^k, \qquad k = 0, 1, \ldots, 2N.$$

We obtain the system of $2N + 1$ linear algebraic equations

2.22
$$
\begin{cases}
\displaystyle\sum_{i=0}^{N} \alpha_i = 0 \\[2ex]
\displaystyle\sum_{i=0}^{N} \alpha_i(N - i) = \sum_{i=0}^{N} \beta_i \\[2ex]
\displaystyle\sum_{i=0}^{N} \alpha_i(N - i)^k = \sum_{i=0}^{N} k\beta_i(N - i)^{k-1}, \qquad k = 2, 3, \ldots, 2N.
\end{cases}
$$

The system (2.22) can be written in the form

2.23
$$
\begin{cases}
\displaystyle\sum_{i=0}^{N} \left(\frac{\alpha_i}{\alpha_0}\right) = 0 \\[2ex]
\displaystyle\sum_{i=1}^{N} \left(\frac{\alpha_i}{\alpha_0}\right)(N - i) - \sum_{i=0}^{N} \left(\frac{\beta_i}{\alpha_0}\right) = -N \\[2ex]
\displaystyle\sum_{i=1}^{N} \left(\frac{\alpha_i}{\alpha_0}\right)(N - i)^k - \sum_{i=0}^{N} k\left(\frac{\beta_i}{\alpha_0}\right)(N - i)^{k-1} = -N^k, \qquad k = 2, 3, \ldots, 2N.
\end{cases}
$$

Thus we have $2N + 1$ equations in the $2N + 1$ unknowns

2.24
$$\frac{\alpha_1}{\alpha_0}, \frac{\alpha_2}{\alpha_0}, \ldots, \frac{\alpha_N}{\alpha_0}, \frac{\beta_0}{\alpha_0}, \frac{\beta_1}{\alpha_0}, \ldots, \frac{\beta_N}{\alpha_0}.$$

A unique solution exists if and only if the determinant of the system does not vanish.

Consider the problem of finding coefficients $a_0, a_1, \ldots, a_{2N}$ such that

2.25
$$
\begin{cases}
F(x_{n+1-i}) = y_{n+1-i}, & i = 1, 2, \ldots, N \\
F'(x_{n+1-i}) = y'_{n+1-i}, & i = 0, 1, 2, \ldots, N
\end{cases}
$$

where

2.26
$$F(x) = \sum_{i=0}^{2N} a_i(x - x_{n+1-N})^{2N-i}.$$

The determinant of (2.25) considered as a linear system in the a_i is the same to within a factor of $\pm h^s$, for some integer s, as the determinant of (2.23). Since the former does not vanish, by Lemma 2.14, it follows that the latter also does not vanish. Thus the system (2.23) has a unique solution where the α_i/α_0 and β_i/α_0 are clearly independent of h. This completes the proof of Theorem 2.9.

We define the *interpolation error* of a linear multistep method of order p as

2.27
$$(IE[y])_n = \frac{1}{\alpha_0} (R[(x - \alpha)^{p+1}])_n \frac{y^{(p+1)}(x_n)}{(p+1)!}.$$

Here we let

2.28
$$(R[y])_n = \sum_{i=0}^{N} \alpha_i y_{n+1-i} - h \sum_{i=0}^{N} \beta_i y'_{n+1-i}.$$

The constant α can be chosen for convenience and the interpolation error is independent of α.

If the operator $(R[y])_n$ is a definite representation of the operator zero and if $y(x) \in C^{(p+1)}[x_{n+1-N}, x_{n+1}]$, then by 7-(2.28) we have

2.29
$$(R[y])_n = (R[(x - \alpha)^{p+1}])_n \frac{y^{(p+1)}(\xi)}{(p+1)!}$$

for some $\xi \in (x_{n+1-N}, x_{n+1})$. In particular, it can be shown that for the closed methods based on the generalized Hermite interpolation method, as well as for the open methods based on the regular Hermite interpolation method, the operator $(R[y])_n$ is definite. (See Exercise 16.) Even if $(R[y])_n$ is not definite one can still show that

2.30
$$|(R[y])_n| \leq h^{p+1} G M_{p+1}$$

for some constant G independent of y. Here M_{p+1} is a bound for $|y^{(p+1)}(x)|$ in $[x_{n+1-N}, x_{n+1}]$. (See, for instance, Henrici [1962].)

For methods based on numerical differentiation and for methods based on numerical quadrature, the interpolation error is the same as the differentiation error (see Section 8.10) and the quadrature error (see Section 8.6), respectively. We have already seen that many of these methods correspond to definite formulas, and hence (2.29) holds in such cases.

We define the *maximum-order open linear N-step method* as the unique* open linear N-step method of order at least $2N - 1$. Similarly, we define the *maximum-order closed linear N-step method* as the unique closed linear N-step method of order at least $2N$. In each case we assume that $\alpha_0 = 1$. We designate the maximum-order methods by

        IO.N   (maximum-order open method)

        IC.N   (maximum-order closed method).

In Table 2.33 we give the coefficients for methods *IO.N* and *IC.N* for

---

*The method is unique except for multiplication of all $\{\alpha_i\}$ and $\{\beta_i\}$ by a nonzero constant.

$N = 1, 2, 3, 4$ as well as the corresponding interpolation errors. On the basis of the interpolation errors it would appear that very high accuracy could be obtained using a linear N-step method even for a relatively small value of N. For example, instead of using the Adams-Moulton method ($QC.1.3$) which has quadrature error

2.31
$$-\tfrac{19}{720}h^5 y^{(5)}$$

one might be tempted to use $IC.3$ which has interpolation error

2.32
$$-\tfrac{1}{770}h^7 y^{(7)}.$$

Unfortunately, as we shall see in Section 9.3, the method $IC.3$ is unstable and totally unsuited for numerical calculation.

As we have seen, there is no loss of generality in assuming that $\alpha_0 = 1$. According to Theorem 2.9, if the remaining $2N + 1$ coefficients are "free," we can obtain a method which is of order $2N$. On the other hand, if all but r of the α_i and

2.33 Table   Table of Coefficients of Maximum Order Linear N-step Methods of Order $2N - 1$ or $2N$.

		α_0	α_1	α_2	α_3	α_4	β_0	β_1	β_2	β_3	β_4	Error	Name	Symbol
$N=1$	open	1	-1	—	—	—	1	—	—	—	—	$\tfrac{1}{2}h^2 y^{(2)}$	Euler	IO.1
	closed	1	-1	—	—	—	$\tfrac{1}{2}$	$\tfrac{1}{2}$	—	—	—	$-\tfrac{1}{12}h^3 y^{(3)}$	Modified Euler	IC.1
$N=2$	open	1	4	-5	—	—	—	4	2	—	—	$\tfrac{1}{6}h^4 y^{(4)}$	—	IO.2*
	closed	1	0	-1	—	—	$\tfrac{1}{3}$	$\tfrac{4}{3}$	$\tfrac{1}{3}$	—	—	$-\tfrac{1}{90}h^5 y^{(5)}$	Milne-Simpson Corrector	IC.2
$N=3$	open	1	18	-9	-10	—	—	9	18	3	—	$\tfrac{1}{20}h^6 y^{(6)}$	—	IO.3*
	closed	1	$\tfrac{27}{11}$	$-\tfrac{27}{11}$	-1	—	$\tfrac{3}{11}$	$\tfrac{27}{11}$	$\tfrac{27}{11}$	$\tfrac{3}{11}$	—	$-\tfrac{1}{770}h^7 y^{(7)}$	—	IC.3*
$N=4$	open	1	$\tfrac{128}{3}$	36	-64	$-\tfrac{47}{3}$	—	16	72	48	4	$\tfrac{1}{70}h^8 y^{(8)}$	—	IO.4*
	closed	1	$\tfrac{32}{5}$	0	$-\tfrac{32}{5}$	-1	$\tfrac{96}{25}$	$\tfrac{6}{25}$	$\tfrac{216}{25}$	$\tfrac{6}{25}$	$\tfrac{96}{25}$	$-\tfrac{551}{12,000}h^9 y^{(9)}$	—	IC.4*

* These methods are unstable and should not be used for numerical computation.

With the coefficients indicated, the linear N-step method

2.34
$$\sum_{i=0}^{N} \alpha_i y_{n+1-i} = h \sum_{i=0}^{N} \beta_0 f(x_{n+1-i}, y_{n+1-i})$$

is of order $2N - 1$ for the *open* case (where $\beta_0 = 0$) and is of order $2N$ in the *closed* case (where $\beta_0 \neq 0$). The methods are denoted by

         IO.N   open case
         IC.N   closed case

all but s of the β_i are specified, then we would hope that we could obtain a method of order $r + s - 1$. We shall show that this is true in the following cases:

a) $r = 1$

b) $r = N$

c) $1 \leqq r \leqq N$ and $s = 0$.

We remark that for the order of the method to be at least zero it is necessary that (2.4) hold. Hence one cannot arbitrarily specify all of the α_i (in addition to letting $\alpha_0 = 1$). If one specifies all but one of the α_i the remaining α_i will be determined by (2.4). Thus if we specify all the values of the α_i and require that (2.4) hold, this is equivalent to letting $r = 1$.

We now prove

2.35 Theorem. Let $\alpha_0 = 1$ and let $\alpha_1, \alpha_2, \ldots, \alpha_N$ be given so that (2.4) holds. Let $N + 1 - s$ of the numbers $\beta_0, \beta_1, \ldots, \beta_N$ be given where $0 \leqq s \leqq N + 1$. There exists a unique set of the s values of the β_i which are not specified so that the order of (2.1) is at least s.

Proof. Let us consider the first $s + 1$ equations of (2.22). The first equation is satisfied because of (2.4). The remaining s equations uniquely determine the s unspecified β_i. For, if we divide the kth such equation by k, for $k = 1, 2, \ldots, s$, the determinant of the resulting system is a Vandermonde determinant and therefore does not vanish. Thus the equations can be satisfied by a unique set of the previously unspecified β_i and the theorem follows.

Suppose now that all α_i except α_0 and α_q are assumed to vanish and that all of the $N + 1 - s$ values of the β_i are also assumed to vanish. If we let $\alpha_0 = 1$, then by (2.4) we have

2.36 $\alpha_q = -\alpha_0 = -1$

and the method has the form

2.37 $y_{n+1} - y_{n+1-q} = h \sum_{i=0}^{N} \beta_i f(x_{n+1-i}, y_{n+1-i})$

where $N + 1 - s$ of the β_i are assumed to vanish. This method is related to a numerical quadrature formula with, in general, unequally spaced points. It is easy to show that the optimum associated numerical quadrature formula, with $N + 1 - s$ of the β_i required to vanish, has order at least $s - 1$ and that the order of the associated method based on numerical quadrature is at least s. Hence the latter is the method given in Theorem 2.35. In particular, if the values of the β_i which are assumed to vanish are $\beta_s, \beta_{s+1}, \ldots, \beta_N$, then we get method $QC.q.s - 1$. If $\beta_0, \beta_{s+1}, \beta_{s+2}, \ldots, \beta_N$ are assumed to vanish we get method $QO.q.s$.

2.38 Theorem. Let $\alpha_0 = 1$ and let $N + 1 - s$ of the numbers $\beta_0, \beta_1, \ldots, \beta_N$ be given where $0 \leqq s \leqq N + 1$. There exists a unique set of the numbers $\alpha_1, \alpha_2, \ldots, \alpha_N$ and s values of β_i which are not specified so that the order of (2.1) is at least $N + s - 1$.

Proof. The proof is similar to that of Theorem 2.9. It is simply necessary to extend Lemma 2.14 to allow for the possibility that $F'(x_i)$ is specified only for s of the values $x_0, x_1, \ldots, x_{m+1}$. The details are left to the reader. (See Exercise 20.)

Suppose that all of the β_i except one, say β_q, vanish. If we divide both sides of (2.1) by $h\beta_q$ we get a method based on numerical differentiation of the form 8-(10.3). As shown in Section 8.10, the method $D.q.N$, has order at least N; hence the corresponding linear multistep method has order at least N. Therefore, the method of Theorem 2.38 of order N corresponding to the prescribed (and vanishing) values of the β_i is $D.q.N$.

2.39 Theorem. Let $\alpha_0 = 1$ and let $N - r$ of the numbers $\alpha_1, \alpha_2, \ldots, \alpha_N$ be given where $1 \leqq r \leqq N$. Given $\beta_0, \beta_1, \ldots, \beta_N$ there exists a unique set of the r values of the α_i which are not specified so that the order of (2.1) is at least $r - 1$.

Proof. The first r equations (2.22) can be satisfied by proper choice of the unspecified α_i since the associated determinant is a Vandermonde determinant and hence does not vanish.

If one prescribes $N - r$ of the α_i and $N + 1 - s$ of the β_i, where $1 < r < N$ and $s > 0$ then there may not exist a linear N-step method with the prescribed α_i and β_i with order $r + s - 1$. Thus, for example, consider the case $N = 3$ where we prescribe $\alpha_0 = 1$, $\alpha_2 = 0$, $\beta_0 = \beta_1 = \beta_3 = 0$. Here $r = 2$ and $s = 1$ and yet we cannot choose α_1, α_3, and β_2 to obtain order 2. For by (2.22) if the order were 2 we would have

2.40
$$\begin{cases} \alpha_1 + \alpha_3 = -1 \\ 2\alpha_1 - \beta_2 = -3 \\ 4\alpha_1 - 2\beta_2 = -9. \end{cases}$$

But the last two conditions are inconsistent. Hence there does not exist a linear 3-step method of order 2 such that $\alpha_0 = 1$, $\alpha_2 = 0$, $\beta_0 = \beta_1 = \beta_3 = 0$.

EXERCISES 9.2

1. Find N for each of the following linear N-step methods. Which methods are open and which are closed?

$$y_{n+1} - y_{n-1} = 2hf(x_n, y_n)$$

$$y_{n+1} - y_{n-1} = 2hf(x_{n-2}, y_{n-2})$$

$$y_{n+1} - y_{n-1} = 2hf(x_{n+1}, y_{n+1})$$

$$y_{n+1} - y_n = \tfrac{1}{2}hf(x_{n+1}, y_{n+1}) + \tfrac{1}{2}hf(x_{n-1}, y_{n-1}).$$

What is the order of each method?

2. How small must h be so that the equation corresponding to the closed methods of the preceding exercise can be solved for y_{n+1} at each step for the problem $y' = 3y$, $y(0) = 1$?

3. Verify that the corrector for the Adams-Moulton method is a linear 3-step method of order 4. What about the predictor?

4. By the method of undetermined weights, find $F(x_2)$ where $F(x)$ is a polynomial of degree three or less such that

$$F(x_0) = f(x_0), \qquad F(x_1) = f(x_1)$$
$$F'(x_0) = f'(x_0), \qquad F'(x_1) = f'(x_1)$$

where $x_2 - x_1 = x_1 - x_0 = h$. What is the corresponding linear multistep method? Give the "interpolation error." Find the method in Table 2.33.

5. Using the method of undetermined weights, find $F(x_2)$ where $F(x)$ is a polynomial of degree four or less such that

$$F(x_0) = f(x_0), \qquad F(x_1) = f(x_1)$$
$$F'(x_0) = f'(x_0), \qquad F'(x_1) = f'(x_1), \qquad F'(x_2) = f'(x_2)$$

where $x_2 - x_1 = x_1 - x_0 = h$. What is the order of accuracy of the formula thus obtained? What is the corresponding linear multistep method? Give the "interpolation error." Find the method in Table 2.33.

6. Find the polynomial $F(x)$ of degree 4 or less such that

$$F(0) = 0, \qquad F'(0) = 0, \qquad F(1) = 0, \qquad F'(1) = 0, \qquad F'(2) = 1.$$

7. Apply the methods of Exercises 4 and 5 to the initial-value problem $y' = -y$, $y(0) = 1$. Obtain $y(0.1)$ by the Euler method and carry out the calculations until $x = 0.6$. For the closed method solve for y_{n+1} explicitly.

8. What can be said concerning a polynomial of degree 3 or less such that $F(0) = F'(0) = F'(1) = F''(2) = 0$?

9. Prove that there is a unique polynomial $F(x)$ of degree $n + 1$ or less such that

$$F'(x_k) = f'(x_k), \qquad k = 0, 1, \dots, n$$
$$F(x_s) = f(x_s)$$

for some integer s such that $0 \le s \le n$. Assume $x_0 < x_1 < \cdots < x_n$. (Hint: The proof is similar to the proof of Lemma 2.14.)

10. Prove the second part of Theorem 2.9 as follows. Find by Hermite interpolation the unique polynomial $F(x)$ of degree $2N + 1$ such that for $i = 0, 1, \dots, N$

$$F(x_{n+1-i}) = y_{n+1-i}$$
$$F'(x_{n+1-i}) = y'_{n+1-i}.$$

Then find a condition on $y_{n+1}, y_n, \dots, y_{n+1-N}, y'_{n+1}, y'_m, \dots, y'_{n+1-N}$ such that $F(x)$ is of degree $2N$. Justify this procedure and use it to determine the linear 1-step method of order 2 and the linear 2-step method of order 4 (Dahlquist [1956]).

11. Verify that the determinant of (2.23) is the same to within a factor of $\pm h^s$ for some integer s as the determinant of (2.25) for the case $N = 3$.

12. Give an explicit formula analogous to 6-(7.5) for the polynomial $F(x)$ of Lemma 2.14. (Hint: Assume

$$F(x) = \sum_{k=0}^{m} (A_k x^2 + B_k x + C_k) Q_k(x) \rho_k$$

$$+ \sum_{k=0}^{m} (D_k x^2 + E_k x + F_k) Q_k(x) \sigma_k$$

$$+ \frac{v(x)}{v'(x_{m+1})} \sigma_{m+1}$$

where

$$v(x) = \prod_{j=0}^{m} (x - x_j)^2$$

$$Q_k(x) = \prod_{\substack{j=0 \\ j \neq k}}^{m} (x - x_j)^2, \qquad k = 0, 1, \ldots, m.$$

Determine $A_k, B_k, C_k, D_k, E_k,$ and F_k so that

$$u_k(x_k) = 1, \qquad u_k'(x_k) = 0, \qquad u_k'(x_{m+1}) = 0,$$

$$v_k(x_k) = 0, \qquad v_k'(x_k) = 1, \qquad v_k'(x_{m+1}) = 0,$$

$$k = 0, 1, \ldots, m.$$

Here

$$u_k(x) = (A_k x^2 + B_k x + C_k) Q_k(x),$$

and

$$v_k(x) = (D_k x^2 + E_k x + F_k) Q_k(x).$$

Apply the formula to the case $m = 1$.

13. Solve the generalized Hermite interpolation problem of Lemma 2.14 for the cases $m = 0$ and $m = 1$. Assume evenly spaced interpolation points. What is the corresponding linear multistep method and the interpolation error in each case?

14. Prove (2.31).

15. Verify the expressions for the interpolation errors given for methods IO.4 and IC.4 in Table 2.33.

16. Show that the generalized Hermite interpolation method (see Lemma 2.14) is definite, as defined in Section 7.7, provided $x \leq x_{m+1}$.

17. Verify that the unique linear 2-step method of order 4 is method IC.2.

18. Verify the coefficients given in Table 2.33 for the closed linear N-step methods for $N = 1$,

2, 3, 4. Also, find the coefficients for the case $N = 5$. Show that for $N = 1, 2, 3$ the operator

$$(R[y])_n = \sum_{i=0}^{N} \alpha_i y_{n+1-i} - h \sum_{i=0}^{N} \beta_i y'_{n+1-i}$$

is definite. (See Section 7.7.) Then verify the interpolation errors for $N = 1, 2, 3$.

19. Show that there does not exist a polynomial $F(x)$ of degree two or less such that $F(0) = F(1) = 0, F'(\frac{1}{2}) = 1$, but that there does exist such a polynomial with $F(0) = F(1) = 0$, $F'(2) = 1$.

20. Work out all the details of the proof of Theorem 2.38 for the case $N = 3$ where β_1 and β_2 are given. Then give a proof for the general case.

21. Find the linear 2-step method of maximum order of accuracy such that $\beta_0 = \beta_1 = 0$.

22. Find a linear 2-step method of order at least 3 such that $\alpha_0 = 1, \beta_1 = 1$. Find the interpolation error.

23. Find a linear 3-step method of order at least one such that $\alpha_0 = 1, \alpha_2 = 2, \beta_0 = \beta_1 = \beta_2 = \beta_3 = 1$. Find the interpolation error.

24. Does there exist a linear 3-step method of order 3 such that $\alpha_0 = 1, \alpha_2 = 1, \beta_0 = 2$, $\beta_1 = 0$? If so, find the interpolation error.

25. Find the linear 3-step method of order at least 3 such that $\alpha_0 = 1, \alpha_1 = 0, \alpha_2 = -\frac{1}{2}$, $\alpha_3 = -\frac{1}{2}$.

26. Construct a linear 2-step method of order 2 such that $\beta_0 = 0$. Find the method with the smallest interpolation error such that $\beta_1 \leq 2$.

27. Show that for $r = 0, 1, 2, 3, 4$ there exists a linear 2-step method of order at least $4 - r$ such that $\alpha_0 = 1$ and such that r of the quantities $\alpha_1, \alpha_2, \beta_0, \beta_1, \beta_2$ vanish, provided α_1 and α_2 do not both vanish.

28. Find a linear 2-step method of order at least 2 such that $\alpha_0 = 1, \alpha_1 = -2, \alpha_2 = 1, \beta_1 = 3$. Find the interpolation error.

29. Find the linear 3-step method of order at least 4 such that $\alpha_0 = 1, \alpha_1 = 0, \alpha_2 = -3$, $\alpha_3 = 2$. Also find the interpolation error.

30. Let $\alpha_0 = 1$ and let $N - r$ of the numbers $\alpha_1, \alpha_2, \ldots, \alpha_N$ be given where $1 \leq r \leq N$. Let $N + 1 - s$ of the numbers $\beta_0, \beta_1, \ldots, \beta_N$ also be prescribed where $0 \leq s \leq N + 1$. Show that if $r = 1$ or if for any i and j such that α_i and β_j are not prescribed we have $i < j$, then there exists a unique linear N-step method of order at least $r + s - 1$ with the prescribed α_i and β_j.

9.3 STABILITY, CONSISTENCY, AND CONVERGENCE

In this discussion it is convenient to make a slight change of notation. We assume we are seeking to solve the initial-value problem

3.1
$$\begin{cases} y' = f(x, y) \\ y(x_0) = \eta. \end{cases}$$

We let $y_0, y_1, y_2, \ldots$ be the approximate numerical values produced by the procedure. We assume that $y_0, y_1, \ldots, y_{N^*-1}$ are determined by a special procedure and that $y_{N^*}, y_{N^*+1}, \ldots$ are determined by a linear N-step method, where $N \leq N^*$.

For any $h > 0$ let the function $y_h(x)$ be defined as follows. Let $y_{h,n} = y_h(x_n)$ be determined by (2.1) for $n = N^*, N^* + 1, \ldots$ and let

3.2
$$y_{h,n} = \eta_n(h), \qquad n = 0, 1, \ldots, N^* - 1$$

where $\eta_0(h), \eta_1(h), \ldots, \eta_{N^*-1}(h)$ are given functions of h. (Actually, $\eta_0(h), \eta_1(h), \ldots,$ $\eta_{N^*-1}(h)$ are the values produced by the special procedure for given h—normally $\eta_0(h) = \eta$.)

For values of x between x_n and x_{n+1} we use linear interpolation, i.e., we let

3.3
$$y_h(x) = \frac{x_{n+1} - x}{h} y_{h,n} + \frac{x - x_n}{h} y_{h,n+1}, \qquad x_n \leq x \leq x_{n+1}.$$

3.4 Definition. Let $\mathscr{I}$ be the class of initial-value problems (3.1) such that $f(x, y)$ satisfies the conditions of Theorem 8-2.31, including the Lipschitz condition. The procedure described above is *convergent* if, for some $\bar{x} > x_0$ such that a unique solution $\bar{y}(x)$ of (3.1) exists for $[x_0, \bar{x}]$, for any functions $\eta_0(h), \eta_1(h), \ldots, \eta_{N^*-1}(h)$ satisfying

3.5
$$\lim_{h \to 0} \eta_i(h) = \eta, \qquad i = 0, 1, \ldots, N^* - 1$$

we have

3.6
$$\lim_{h \to 0} y_h(x) = \bar{y}(x)$$

uniformly in $[x_0, \bar{x}]$. The convergence must hold for any initial-value problem of the class $\mathscr{I}$.

Corresponding to the linear N-step method (2.1) let us define the polynomials $\rho(\lambda)$ and $\sigma(\lambda)$ by

3.7
$$\begin{cases} \rho(\lambda) = \sum_{i=0}^{N} \alpha_i \lambda^{N-i} = \alpha_0 \lambda^N + \alpha_1 \lambda^{N-1} + \cdots + \alpha_N \\ \\ \sigma(\lambda) = \sum_{i=0}^{N} \beta_i \lambda^{N-i} = \beta_0 \lambda^N + \beta_1 \lambda^{N-1} + \cdots + \beta_N. \end{cases}$$

Evidently the conditions (2.4) and (2.5) can be expressed in the form

3.8
$$\rho(1) = 0$$

and

3.9
$$\sigma(1) = \rho'(1)$$

respectively. We now define the *condition of consistency* as follows.

3.10 Definition. The linear N-step method (2.1) satisfies the *condition of consistency* if (3.8) and (3.9) are satisfied.

We also define the *condition of stability* by

3.11 Definition. The linear N-step method satisfies the *condition of stability* if all roots of the equation

3.12 $$\rho(\lambda) = 0$$

do not exceed unity in modulus and if all roots of modulus unity are simple.

We now prove

3.13 Theorem. If the linear N-step method (2.1) is convergent, then the condition of stability is satisfied.

Proof. If the method is convergent, then (3.6) must hold for the problem

3.14 $$y' = 0, \qquad y(0) = 0$$

whose exact solution is

3.15 $$\bar{y}(x) = 0.$$

Convergence must hold for any choice of the functions $\eta_0(h), \eta_1(h), \ldots, \eta_{N^*-1}(h)$ such that (3.5) holds.

Let us first consider the case where $N = 2$ and let us assume that $N^* = 2$. The linear 2-step method becomes (with $y_n = y_{h,n}$)

3.16 $$\alpha_0 y_{n+1} + \alpha_1 y_n + \alpha_2 y_{n-1} = 0, \qquad n = 1, 2, \ldots$$

with

3.17 $$y_0 = \eta_0(h), \qquad y_1 = \eta_1(h).$$

Let λ_1 and λ_2 be the roots of

3.18 $$\alpha_0 \lambda^2 + \alpha_1 \lambda + \alpha_2 = 0.$$

One can easily verify that for any constants c_1 and c_2 a solution of (3.16) is given by

3.19 $$y_n = c_1 \lambda_1^n + c_2 \lambda_2^n.$$

If $\lambda_1 \neq \lambda_2$ and if $\lambda_1 \neq 0, \lambda_2 \neq 0$, we can satisfy (3.17) by letting

3.20 $$c_1 = \frac{\eta_1 - \eta_0 \lambda_2}{\lambda_1 - \lambda_2}, \qquad c_2 = \frac{\lambda_1 \eta_0 - \eta_1}{\lambda_1 - \lambda_2}$$

so that we have

3.21
$$y_n = \left(\frac{\eta_1 - \eta_0 \lambda_2}{\lambda_1 - \lambda_2}\right)\lambda_1^n + \left(\frac{\lambda_1 \eta_0 - \eta_1}{\lambda_1 - \lambda_2}\right)\lambda_2^n, \qquad n \geq 0.$$

The fact that (3.21) is the only solution of (3.16) and (3.17) follows from the fact that $\alpha_0 \neq 0$; hence any two solutions y_n and z_n of (3.16) for which

3.22
$$y_0 = z_0, \qquad y_1 = z_1$$

are identical. One can also verify that even if $\lambda_2 = 0$, the unique solution is given by (3.21) for $n \geq 1$ and (3.17) for $n = 0$.

If $\lambda_1 = \lambda_2 = \lambda \neq 0$, then the solution of (3.16) and (3.17) is

3.23
$$y_n = \eta_0 \lambda^n + (\eta_1 - \eta_0 \lambda)n\lambda^{n-1}, \qquad n \geq 0.$$

In order not to make the example seem too artificial we let

3.24
$$\eta_0(h) = 0, \qquad \eta_1(h) = h.$$

Thus, by (3.21), for the case $\lambda_1 \neq \lambda_2$ we have

3.25
$$y_n = h\frac{\lambda_1^n - \lambda_2^n}{\lambda_1 - \lambda_2}.$$

For any $x > 0$, we have

3.26
$$y_h(x) = h\frac{\lambda_1^n - \lambda_2^n}{\lambda_1 - \lambda_2}$$

for each value of h such that for some integer n we have

3.27
$$\frac{x}{h} = n.$$

If $|\lambda_1| > |\lambda_2|$ or if $|\lambda_2| > |\lambda_1|$ it is clear that (3.6) implies the condition

3.28
$$\max(|\lambda_1|, |\lambda_2|) \leq 1.$$

This follows from the fact that if $|\lambda| > 1$, we have

3.29
$$\lim_{h \to 0} h|\lambda|^{x/h} = \infty.$$

Now suppose that $\lambda_1 \neq \lambda_2$ but $|\lambda_1| = |\lambda_2|$. Then either λ_1 is real and

3.30
$$\lambda_1 = -\lambda_2$$

or else λ_1 is complex and

3.31
$$\begin{cases} \lambda_1 = \rho e^{i\theta} \\ \lambda_2 = \rho e^{-i\theta} \end{cases}$$

for some positive ρ and real θ such that θ is not a multiple of π. In the former case we have, by (3.26), for any $x > 0$

3.32
$$y_h(x) = h \frac{(\lambda_1)^n - (-\lambda_1)^n}{2\lambda_1} = \begin{cases} 0, & n \text{ even} \\ h\lambda_1^{n-1}, & n \text{ odd}, \end{cases}$$

for all h such that (3.27) holds for some integer n. For convergence we must clearly have $|\lambda_1| \leq 1$.

In the case λ_1 is complex we have

3.33
$$y_h(x) = h\rho^{n-1} \frac{\sin n\theta}{\sin \theta}.$$

If $\rho > 1$, then (3.6) cannot possibly hold unless

3.34
$$\lim_{n \to \infty} \sin n\theta = 0.$$

We now prove

3.35 Lemma. If $\lim_{n \to \infty} \sin n\alpha = 0$, then α is a multiple of π.

Proof. If $\lim_{n \to \infty} \sin n\alpha = 0$, then given any $\varepsilon > 0$ there exists $n_0 > 0$ such that for all $n > n_0$ we have $|\sin n\alpha| < \varepsilon$. Let $\varepsilon = \sqrt{2}/2$. We seek to show that for any n_0, there exists $n > n_0$ such that $|\sin n\alpha| > \varepsilon$ unless α is a multiple of π. If α is not a multiple of π, then $\alpha/\pi = p + f$ where p is an integer and $0 < f < 1$. Hence $\alpha = p\pi + f\pi$. Let m be an integer greater than n_0. We claim that $m\alpha$ and $(m + 1)\alpha$ cannot both be multiples of π. Suppose $m\alpha$ is a multiple of π. Then mf is an integer, say r. But if $(m + 1)\alpha$ is a multiple of π then $(m + 1)f = r + f$ also is an integer. But this is impossible since $0 < f < 1$.

Let $\beta = m'\alpha$ where $m' = m$ if $m\alpha$ is not a multiple of π; otherwise, $m' = m + 1$. Choose β' such that $0 < |\beta'| \leq \pi/2$ and $\beta' \equiv \beta \pmod{\pi}$. Suppose that $0 < \beta' \leq \pi/2$. If $\beta' \geq \pi/4$, then $|\sin m'\alpha| = |\sin \beta| = |\sin \beta'| \geq \sqrt{2}/2$. Otherwise, for some integer k we have $\pi/4 \leq k\beta' \leq \pi/2$. Evidently $|\sin k\beta'| \geq \sqrt{2}/2$ and $|\sin km'\alpha| \geq \sqrt{2}/2$. A similar argument holds when $-\pi/2 \leq \beta' < 0$. Thus, we have found an integer, say s, greater than n_0 such that $|\sin s\alpha| \geq \sqrt{2}/2$. This contradiction shows that α is a multiple of π and the lemma follows.

From Lemma 3.35 it follows that if $\lambda_1 \neq \lambda_2$, then in order for convergence to hold we must have $|\lambda_1| \leq 1$ and $|\lambda_2| \leq 1$. For otherwise θ must be a multiple of π and hence $\lambda_1 = \lambda_2$, by (3.31).

Let us now consider the case where

3.36 $$\lambda_1 = \lambda_2 = \lambda \neq 0.$$

Evidently λ is real. We again let

3.37 $$\eta_0(h) = 0, \qquad \eta_1(h) = h$$

and obtain, by (3.23)

3.38 $$\begin{cases} y_n = hn\lambda^{n-1}, & n > 0 \\ y_0 = 0. \end{cases}$$

Given $x > 0$ we have

3.39 $$y_h(x) = x\lambda^{n-1}$$

for each value of h such that (3.27) holds for some integer n. Evidently we cannot have (3.6) unless $|\lambda| < 1$. Thus we have shown that in order for convergence to hold we must either have two distinct roots of (3.18) with modulus not greater than unity or else one double root with modulus less than unity. In other words, the method must satisfy the condition of stability.

We now sketch a proof for the case $N > 2$. As before, we let $N^* = N$. Let λ_1 be any zero of $\rho(\lambda)$ of maximum modulus. We consider the following cases:

a) *λ_1 is real and simple ($\lambda_1 \neq 0$).*

Let λ_2 be any other zero of $\rho(\lambda)$. If λ_2 is real we let $\eta_0(h) = 0$ and

3.40 $$\eta_k(h) = h\frac{\lambda_1^k - \lambda_2^k}{\lambda_1 - \lambda_2}, \qquad k = 1, \ldots, N - 1.$$

We obtain

3.41 $$y_n = h\frac{\lambda_1^n - \lambda_2^n}{\lambda_1 - \lambda_2}, \qquad n = N, N + 1, \ldots .$$

The argument then proceeds as in the case $N = 2$. We consider separately the cases where $|\lambda_2| < |\lambda_1|$ and where $|\lambda_2| = |\lambda_1|$. In either case we must have $|\lambda_1| \leq 1$.

If λ_2 is complex, we let $\lambda_2 = \rho e^{i\phi}$, $\eta_0(h) = 0$, and

3.42 $$\eta_k(h) = h[\lambda_1^k - \rho^k \cos k\phi], \qquad k = 1, \ldots, N - 1.$$

We obtain

3.43 $$y_n = h[\lambda_1^n - \rho^n \cos n\phi], \qquad n = N, N + 1, \ldots .$$

If $|\lambda_2| < |\lambda_1|$, then we clearly must have $|\lambda_1| \leq 1$. If $|\lambda_2| = |\lambda_1|$, then we have for $n \geq N$

3.44
$$y_n = \begin{cases} h\lambda_1^n(1 - \cos n\phi) = 2h\lambda_1^n \sin^2 \dfrac{n\phi}{2}, & \text{if } \lambda_1 > 0 \\[3mm] h|\lambda_1|^n((-1)^n - \cos n\phi), & \text{if } \lambda_1 < 0. \end{cases}$$

In the case $\lambda_1 > 0$ we must have $|\lambda_1| \leq 1$ unless ϕ is a multiple of 2π, by Lemma 3.35. This would imply $\lambda_2 = \lambda_1$, a contradiction. In the case $\lambda_1 < 0$ we have for $n \geq N$

3.45
$$y_n = \begin{cases} 2h|\lambda_1|^n \sin^2 \dfrac{n\phi}{2}, & \text{if } n \text{ is even} \\[3mm] -2h|\lambda_1|^n \cos^2 \dfrac{n\phi}{2}, & \text{if } n \text{ is odd.} \end{cases}$$

For the sequence to converge to zero, the subsequence associated with n even must converge to zero. If $|\lambda_1| > 1$, this implies that ϕ is a multiple of 2π, by Lemma 3.35. Thus, whether or not λ_1 is positive, if $|\lambda_1| > 1$, then λ_2 is real and we have a contradiction. Therefore $|\lambda_1| \leq 1$.

b) λ_1 *is complex and simple.*

Let $\lambda_1 = \rho e^{i\theta}$, and let

3.46
$$\eta_k(h) = h\rho^k \sin k\theta, \qquad k = 0, 1, \dots, N - 1.$$

We obtain

3.47
$$y_n = h\rho^n \sin n\theta, \qquad n = N, N + 1, \dots .$$

If $\rho > 1$, then for convergence θ must be a multiple of π, by Lemma 3.35. But this contradicts the assumption that λ_1 is complex. It therefore follows that if λ_1 is simple, $|\lambda_1| \leq 1$.

c) λ_1 *is real and multiple* $(\lambda_1 \neq 0)$.

We let

3.48
$$\eta_k(h) = kh\lambda_1^{k-1}, \qquad k = 0, 1, \dots, N - 1$$

and obtain

3.49
$$y_n = nh\lambda_1^{n-1}, \qquad n = N, N + 1, \dots .$$

We can then show, as in the case $N = 2$, that for convergence we must have $|\lambda_1| < 1$.

d) λ_1 *is complex and multiple.*

We let $\lambda_1 = \rho e^{i\theta}$ and

3.50 $$\eta_k(h) = kh\rho^{k-1}\sin(k-1)\theta, \qquad k = 0, 1, 2, \dots, N-1.$$

(Note that $N \geq 4$ since $\bar{\lambda}_1$ is also a multiple root.) We obtain

3.51 $$y_n = nh\rho^{n-1}\sin(n-1)\theta, \qquad n = N, N+1, \dots .$$

For any $x > 0$, we have

3.52 $$y_h(x) = x\rho^{n-1}\sin(n-1)\theta$$

for any h such that (3.27) holds for some integer n. If $\rho \geq 1$, in order that $y_h(x) \to 0$ as $h \to 0$, we must have

3.53 $$\lim_{n\to\infty}\sin(n-1)\theta = 0.$$

But by Lemma 3.35 this implies that θ is a multiple of π and λ_1 is real. This contradicts the assumption that λ_1 is complex. Thus we have proved that if λ_1 is a multiple zero of $\rho(\lambda)$ then $|\lambda_1| < 1$. This completes the proof of Theorem 3.13.

In the proof given above we have avoided choosing $\eta_0(h)$ to be different from zero in order to avoid making the example appear too artificial. The proof can be simplified if we allow $\eta_0(h)$ to be different from zero; see Henrici [1962, pp. 218–219]. Thus, for example, one can show that each zero $\lambda = \rho e^{i\theta}$ of $\rho(\lambda)$ is not greater than unity in modulus by letting

3.54 $$\eta_k(h) = h\rho^k \cos k\theta, \qquad k = 0, 1, \dots, N-1.$$

We remark that to treat the case $N = 1$, we let $\eta_0(h) = h$ and have $y_n = h\lambda^n$, $n \geq 1$, where $\lambda = -\alpha_1/\alpha_0$. Hence, for convergence, $|\lambda| \leq 1$.

While it may appear that the failure of an unstable method to converge will only happen for a very special choice of the $\eta_k(h)$, in actual fact, the convergence will occur only in very special cases. For example, if $N = 2$ and if $|\lambda_1| > 1$ and $|\lambda_2| \leq 1$, then, by (3.21), if $\eta_0(h) = 0$ convergence cannot occur in the interval $0 \leq x \leq \bar{x}$ for $\bar{x} > 0$ unless

3.55 $$\eta_1(h) = o(|\lambda_1|^{-\bar{x}/h}).$$

We now prove

3.56 Theorem. If the linear multistep method (2.1) is convergent, then the condition of consistency is satisfied.

Proof. For simplicity we assume that $N^* = N$. We first consider the problem

3.57 $$y'(x) = 0, \qquad y(0) = 1,$$

whose solution is

3.58 $$\bar{y}(x) \equiv 1.$$

We let $\eta_0(h) = \eta_1(h) = \cdots = \eta_{N-1}(h) = 1$. The linear multistep method reduces to

3.59 $$\alpha_0 y_{n+1} + \alpha_1 y_n + \cdots + \alpha_N y_{n-(N-1)} = 0.$$

Therefore we have

3.60 $$y_N = \frac{-1}{\alpha_0}[\alpha_1 y_{N-1} + \alpha_2 y_{N-2} + \cdots + \alpha_N y_0]$$

$$= -\frac{1}{\alpha_0}[\alpha_1 + \alpha_2 + \cdots + \alpha_N].$$

Hence for each h we have

3.61 $$y_h(Nh) = -\frac{1}{\alpha_0}[\alpha_1 + \alpha_2 + \cdots + \alpha_N].$$

Thus, for each h, there exists x, namely $x = Nh$, such that

3.62 $$|y_h(x) - \bar{y}(x)| = \left|\frac{\rho(1)}{\alpha_0}\right|$$

where

3.63 $$\rho(1) = \sum_{i=0}^{N} \alpha_i = \alpha_0 + \alpha_1 + \cdots + \alpha_N.$$

If $\rho(1) \neq 0$, then for any $\bar{x} > 0$ and for h sufficiently small, there exists $x \in [0, \bar{x}]$ such that $|y_h(x) - \bar{y}(x)| \geq |\rho(1)/\alpha_0|$. This violates the uniform convergence of $y_h(x)$ in the interval $[x_0, \bar{x}]$. Hence we must have (3.8).

Next, let us consider the problem

3.64 $$y'(x) = 1, \qquad y(0) = 0$$

which has the solution

3.65 $$\bar{y}(x) \equiv x.$$

Since $\rho(1) = 0$, it follows that $\lambda = 1$ is a root of $\rho(\lambda) = 0$. Since the method is convergent, it follows from Theorem 3.13 that $\rho'(1) \neq 0$; otherwise, $\lambda = 1$ would be a double root. We now show that with the choice

3.66 $$\eta_0(h) = 0, \qquad \eta_1(h) = hK, \qquad \eta_2(h) = 2hK, \ldots, \eta_{N-1}(h) = (N-1)hK$$

where

3.67
$$K = \frac{\beta_0 + \beta_1 + \cdots + \beta_N}{\rho'(1)} = \frac{\sigma(1)}{\rho'(1)},$$

then

3.68
$$y_n = hnK$$

satisfies the equation

3.69
$$\alpha_0 y_{n+1} + \alpha_1 y_n + \cdots + \alpha_N y_{n-(N-1)} = h(\beta_0 + \beta_1 + \cdots + \beta_N)$$

which corresponds to our linear multistep method. We first note that (3.68) holds for $n = 0, 1, 2, \ldots, N - 1$ by (3.66). If y_n is given by (3.68), then the left member of (3.69) becomes

3.70
$$Kh[n\rho(1) + \alpha_0 - \alpha_2 - 2\alpha_3 - \cdots - (N - 1)\alpha_N]$$
$$= Kh[\alpha_0 - \alpha_2 - 2\alpha_3 - \cdots - (N - 1)\alpha_N + (N - 1)\rho(1)]$$
$$= Kh[N\alpha_0 + (N - 1)\alpha_1 + \cdots + \alpha_{N-1}] = Kh\rho'(1),$$

since $\rho(1) = 0$. Thus, if $K = \sigma(1)/\rho'(1)$, the left member of (3.69) equals $h\sigma(1)$, which is precisely the right member of (3.69). Since the solution of (3.69) and (3.66) is clearly unique, it is given by (3.68).

Because linear interpolation is used to determine $y_h(x)$ for values of x such that x/h is not an integer we have

3.71
$$y_h(x) = Kx$$

for all x. In order that $y_h(x)$ should converge uniformly to $\bar{y}(x) = x$, it is necessary that $K = 1$. Thus (3.9) and the condition of consistency holds. This completes the proof of Theorem 3.56.

The following theorem is essentially proved by Henrici [1962]. (See also Dahlquist [1956].)

3.72 Theorem. If the linear multistep method (2.1) satisfies the condition of consistency and the condition of stability, then it is convergent.

EXERCISES 9.3

1. Verify that the methods IO.3 and IC.3 are unstable. Show, however, that they are of orders 5 and 6 respectively.

2. Does the linear 3-step method

$$11y_{n+1} - 18y_n + 9y_{n-1} - 2y_{n-2} = 6hf(x_{n+1}, y_{n+1})$$

satisfy the condition of stability and the condition of consistency? What is the order of the method?

3. Show that any linear multistep method which satisfies the condition of consistency and for which all β_i vanish does not satisfy the condition of stability.

4. Show that there exists a linear 2-step method of order 3 such that $\beta_0 = 0$ but that there does not exist such a method of order 4. Does the method satisfy the condition of consistency?

5. Find the linear 2-step method of maximum order such that $\beta_1 = 0$. Determine the order of accuracy and the interpolation error. Is the method stable?

6. Verify that if the roots λ_1 and λ_2 of

$$a_0\lambda^2 + a_1\lambda + a_2 = 0$$

are distinct, then any solution of

(*) $a_0 y_{n+1} + a_1 y_n + a_2 y_{n-1} = 0, \qquad n = 1, 2, \ldots$

can be written in the form

$$y_n = c_1\lambda_1^n + c_2\lambda_2^n, \qquad n = 1, 2, \ldots$$

for some constants c_1 and c_2. Assume that $a_0 \neq 0$. Also show that if $\lambda_1 = \lambda_2 = \lambda$, then any solution can be written in the form

$$y_n = c_1\lambda^n + c_2 n\lambda^{n-1}, \qquad n = 2, 3, \ldots .$$

(In the case $\lambda_1 \neq \lambda_2$ be sure to consider the case $\lambda_2 = 0$; also, in the case $\lambda_1 = \lambda_2 = \lambda$ consider the case $\lambda = 0$.) Note that if y_0 and y_1 are given, then $y_2, y_3, \ldots$ are completely determined by (*).

7. Let λ_1 and λ_2 be the roots of (3.18). Show that if $|\lambda_1| = |\lambda_2| > 1$ and if $\eta_0(h) = 0$, then the method (3.16) and (3.17) will converge to the solution of (3.14) in $[0, \bar{x}]$ only if $\eta_1(h) = o(|\lambda_1|^{-\bar{x}/h})$. Also show that if $\lambda_1 = \lambda_2$ and if $|\lambda_1| \geq 1$, then convergence will occur only if $\eta_1(h) = o(h|\lambda_1|^{-\bar{x}/h})$.

8. Verify that if $\lambda_1 \neq 0$ and $\lambda_2 = 0$, then the unique solution of (3.16) and (3.17) is given by (3.21) for $n \geq 1$. (If both λ_1 and λ_2 vanish, verify that (3.23) holds for $n \geq 2$.)

9. For $h = \frac{1}{2}, \frac{1}{4}$, and $\frac{1}{8}$ carry out method IO.2 for the problem $y' = 0$, $y(0) = 0$ with $y_0 = 0$, $y_1 = h$ and carry the calculation as far as $x = 1$. In each case check the numerical solution of the difference equation with the analytic solution.

10. Evaluate $h^2 3^{1/h}$ and $(1 + h)^{1/h}$ for $h = \frac{1}{2}, \frac{1}{4}, \frac{1}{8}, \frac{1}{16}$, and $\frac{1}{32}$ and speculate on the limit as $h \to 0$ in each case.

11. Find n such that $|\sin n\alpha| > \sqrt{\frac{1}{2}}$ where

$$\alpha = \frac{17\pi}{18}.$$

Show that for any integer $n_0 > 0$ there exists $n > n_0$ such that $|\sin n\alpha| > \sqrt{\frac{1}{2}}$. If $n_0 = 1000$, find such an n.

12. In the preceding problem find n if $n_0 = 19$.

13. Given that $\alpha_0 = 1$, $\alpha_1 = -2$, find the linear 2-step method of order 3. Is the method stable?

14. Derive the linear 3-step method of highest possible order such that $\alpha_0 = 1, \beta_1 = \beta_2 = 0$. Is the method stable?

15. Does there exist a linear 3-step method of order 3 such that $\alpha_0 = 1, \alpha_2 = 0, \beta_0 = \beta_3 = 0$? If so, find it. Is it stable?

16. Let the linear multistep method

$$\alpha_0 y_{n+1} + \alpha_1 y_n + \alpha_2 y_{n-1} + \alpha_3 y_{n-2}$$
$$= h[\beta_0 f(x_{n+1}, y_{n+1}) + \beta_1 f(x_n, y_n) + \beta_2 f(x_{n-1}, y_{n-1})$$
$$+ \beta_3 f(x_{n-2}, y_{n-2})]$$

be stable and consistent. Show that if the polynomials

$$\rho(\lambda) = \alpha_0 \lambda^3 + \alpha_1 \lambda^2 + \alpha_2 \lambda + \alpha_3$$
$$\sigma(\lambda) = \beta_0 \lambda^3 + \beta_1 \lambda^2 + \beta_2 \lambda + \beta_3$$

have a common factor, say $\lambda - c$, then $c \neq 1$ and the linear multistep method corresponding to

$$\frac{\rho(\lambda)}{(\lambda - c)} \quad \text{and} \quad \frac{\sigma(\lambda)}{(\lambda - c)}$$

is also stable and consistent.

17. Does there exist a linear N-step method which does not satisfy the condition of consistency but is "weakly convergent" in the sense that uniform convergence is not required? (See Definition 3.10.)

9.4 LIMITATIONS ON THE ORDER OF A STABLE METHOD

We have already seen in Section 9.2 that there exists a linear N-step method with order $2N$. However, as we shall show, for $N > 2$ the linear N-step method of order $2N$ is not stable and hence is not convergent. Before doing so, we shall prove the following positive result.

4.1 Theorem. There exists a stable linear N-step method of order at least $N + 1$.

Proof. By Theorem 2.35 if we let $\alpha_0 = 1, \alpha_1 = -1, \alpha_2 = \alpha_3 = \cdots = \alpha_N = 0$ there exists a unique set of $\beta_0, \beta_1, \ldots, \beta_N$ such that the corresponding linear N-step method has order at least $N + 1$. As a matter of fact, from the discussion following Theorem 2.35, it follows that the method is the method QC.1.N which is based on numerical quadrature. Since $\rho(\lambda) = \lambda^N - \lambda^{N-1}$, the method clearly satisfies the condition of stability, and the theorem follows.

The basic result of this section is the following

4.2 Theorem. The order of a stable linear N-step method cannot exceed $N + 1$ if N is odd and $N + 2$ if N is even. If N is even, there exists a stable linear N-step method of order $N + 2$ and the zeros of $\rho(\lambda)$ all have modulus unity.

Proof. The verification is not difficult for the cases $N = 1$ and $N = 2$ and is left to the reader. (See Exercises 5 and 6.) We shall give here a proof for $N = 3$ and we shall prove the second statement for $N = 4$. A complete proof together with a description of an algorithm for finding stable methods of maximum order for N even is given by Henrici [1962]. (See also Dahlquist [1956].)

We already know by Theorem 4.1 that there exists a stable linear 3-step method of order 4. One such method is method QC.1.3, the Adams-Moulton corrector. We now seek to show that if the linear 3-step method

4.3 $y_{n+1} + \alpha_1 y_n + \alpha_2 y_{n-1} + \alpha_3 y_{n-2}$

$$= h[\beta_0 f(x_{n+1}, y_{n+1}) + \beta_1 f(x_n, y_n) + \beta_2 f(x_{n-1}, y_{n-1}) + \beta_3(x_{n-2}, y_{n-2})]$$

is of order 5 or more, then it is unstable. If the method is of order 5 or more then we have, by (2.22) (letting $\alpha_0 = 1$),

4.4
$$\begin{cases} 1 + \quad \alpha_1 + \alpha_2 + \alpha_3 & = 0 \\ 3 + \quad 2\alpha_1 + \alpha_2 \quad + \quad \delta_0 + \quad \delta_1 + \delta_2 + \delta_3 = 0 \\ 9 + \quad 4\alpha_1 + \alpha_2 \quad + \quad 6\delta_0 + 4\delta_1 + 2\delta_2 \quad = 0 \\ 27 + \quad 8\alpha_1 + \alpha_2 \quad + \quad 27\delta_0 + 12\delta_1 + 3\delta_2 \quad = 0 \\ 81 + 16\alpha_1 + \alpha_2 \quad + \quad 108\delta_0 + 32\delta_1 + 4\delta_2 \quad = 0 \\ 243 + 32\alpha_1 + \alpha_2 \quad + \quad 405\delta_0 + 80\delta_1 + 5\delta_2 \quad = 0 \end{cases}$$

where $\delta_i = -\beta_i, i = 0, 1, 2, 3$.

Let us write the last four equations in the form

4.5
$$\begin{cases} 4\alpha_1 + \alpha_2 + \quad 4\delta_1 + 2\delta_2 = \quad -9 - \quad 6\delta_0 \\ 8\alpha_1 + \alpha_2 + 12\delta_1 + 3\delta_2 = \quad -27 - \quad 27\delta_0 \\ 16\alpha_1 + \alpha_2 + 32\delta_1 + 4\delta_2 = \quad -81 - 108\delta_0 \\ 32\alpha_1 + \alpha_2 + 80\delta_1 + 5\delta_2 = -243 - 405\delta_0. \end{cases}$$

We consider the augmented matrix

4.6
$$\begin{pmatrix} 4 & 1 & 4 & 2 & -9 - \quad 6\delta_0 \\ 8 & 1 & 12 & 3 & -27 - \quad 27\delta_0 \\ 16 & 1 & 32 & 4 & -81 - 108\delta_0 \\ 32 & 1 & 80 & 5 & -243 - 405\delta_0 \end{pmatrix}$$

Applying a number of elementary row operations to the matrix (4.6), as in the Gaussian elimination method (see Chapter 12), we get

4.7
$$\begin{pmatrix} 4 & 1 & 4 & 2 & -9 - 6\delta_0 \\ 4 & 0 & 8 & 1 & -18 - 21\delta_0 \\ 0 & 0 & 4 & -1 & -18 - 39\delta_0 \\ 0 & 0 & 4 & 0 & -36 - 96\delta_0 \end{pmatrix}$$

Additional elementary row operations yield

4.8
$$\begin{pmatrix} 0 & 1 & 0 & 0 & -9 - 24\delta_0 \\ 4 & 0 & 0 & 0 & 72 + 228\delta_0 \\ 0 & 0 & 0 & -1 & 18 + 57\delta_0 \\ 0 & 0 & 4 & 0 & -36 - 96\delta_0 \end{pmatrix}$$

and we have

4.9
$$\begin{cases} \alpha_1 = 18 + 57\delta_0 = 18 - 57\beta_0 \\ \alpha_2 = -9 - 24\delta_0 = -9 + 24\beta_0. \end{cases}$$

Since $1 + \alpha_1 + \alpha_2 + \alpha_3 = 0$, we have

4.10
$$\begin{aligned} \rho(\lambda) &= \lambda^3 + \alpha_1\lambda^2 + \alpha_2\lambda + \alpha_3 \\ &= (\lambda - 1)[\lambda^2 + (\alpha_1 + 1)\lambda + (\alpha_2 + \alpha_1 + 1)] = 0. \end{aligned}$$

We now prove

4.11 Lemma. If b and c are real numbers and if the roots of the quadratic equation

4.12
$$x^2 + bx + c = 0$$

do not exceed unity in modulus then

4.13
$$|c| \leq 1, \qquad |b| \leq 1 + c.$$

Proof. Let r_1 and r_2 be the roots of (4.12). If $|r_1| \leq 1$ and $|r_2| \leq 1$, then since $c = r_1 r_2$ we have $|c| \leq 1$. Suppose now that $|b| > 1 + c$. Then

$$b^2 - 4c > (1 - c)^2 \geq 0$$

and both roots of (4.12) are real. Moreover,

4.14
$$\max(|r_1|, |r_2|) = \frac{|b| + \sqrt{b^2 - 4c}}{2} > \frac{1 + c + (1 - c)}{2} = 1$$

which contradicts the assumption that $|r_1| \leq 1$, $|r_2| \leq 1$. Hence $|b| \leq 1 + c$ and the lemma follows.

From (4.10) and (4.13) we thus seek to determine conditions on δ_0 such that the conditions

4.15
$$\begin{cases} |\alpha_1 + \alpha_2 + 1| \leq 1 \\ \quad |\alpha_1 + 1| \leq 2 + \alpha_1 + \alpha_2 \end{cases}$$

hold.

If $|\alpha_1 + \alpha_2 + 1| \leq 1$, then

4.16
$$-1 \leq \alpha_1 + \alpha_2 + 1 \leq 1$$

or

4.17
$$-2 \leq \alpha_1 + \alpha_2 \leq 0.$$

But by (4.9), this implies

4.18
$$-2 \leq 9 + 33\delta_0 \leq 0$$

or

4.19
$$-\tfrac{1}{3} \leq \delta_0 \leq -\tfrac{3}{11}.$$

On the other hand, if $|\alpha_1 + 1| \leq 2 + \alpha_1 + \alpha_2$, we have

4.20
$$-2 - \alpha_1 - \alpha_2 \leq \alpha_1 + 1 \leq 2 + \alpha_1 + \alpha_2$$

or

4.21
$$\alpha_2 + 1 \geq 0, \qquad 2\alpha_1 + \alpha_2 \geq -3.$$

These conditions imply

4.22
$$-8 - 24\delta_0 \geq 0 \qquad \text{or} \qquad \delta_0 \leq -\tfrac{1}{3}$$

and

4.23
$$27 + 90\delta_0 \geq -3 \qquad \text{or} \qquad \delta_0 \geq -\tfrac{1}{3}.$$

These latter conditions imply that $\delta_0 = -\tfrac{1}{3}$. In this case, we have

4.24 $\quad \alpha_1 = -1, \qquad \alpha_2 = -1, \qquad \alpha_1 + 1 = 0, \qquad \alpha_1 + \alpha_2 + 1 = -1,$

and the quadratic equation

4.25
$$\lambda^2 + (\alpha_1 + 1)\lambda + (\alpha_2 + \alpha_1 + 1) = 0$$

becomes

4.26
$$\lambda^2 - 1 = 0$$

whose roots are ± 1. But in this case, $\lambda = 1$ is a double root of $\rho(\lambda) = 0$ and hence the method does not satisfy the condition of stability.

Let us now consider the case $N = 4$. We seek to show that there exist stable linear 4-step methods of order 6 and that for each such method all zeros of $\rho(\lambda)$ have modulus unity. If the linear 4-step method has order at least 6, then the first seven equations of (2.22) are satisfied. We are led to consider the matrix*

	(α_0)	(α_1)	(α_2)	(α_3)	(α_4)	(δ_0)	(δ_1)	(δ_2)	(δ_3)	(δ_4)
	1	1	1	1	1	0	0	0	0	0
	2	1	0	-1	-2	1	1	1	1	1
	4	1	0	1	4	4	2	0	-2	-4
4.27	8	1	0	-1	-8	12	3	0	3	12
	16	1	0	1	16	32	4	0	-4	-32
	32	1	0	-1	-32	80	5	0	5	80
	64	1	0	1	64	192	6	0	-6	-192

At the top of each column we indicate in parentheses the corresponding coefficient. As before, we let $\delta_i = -\beta_i$, $i = 0, 1, 2, 3, 4$.

Let us introduce the new variables

4.28
$$\begin{cases} \theta_0 = \alpha_0 + \alpha_4, & \theta_5 = \delta_0 + \delta_4, \\ \theta_1 = \alpha_0 - \alpha_4, & \theta_6 = \delta_0 - \delta_4, \\ \theta_3 = \alpha_1 + \alpha_3, & \theta_7 = \delta_1 + \delta_3, \\ \theta_4 = \alpha_1 - \alpha_3, & \theta_8 = \delta_1 - \delta_3. \end{cases}$$

This leads to the new matrix

	(θ_0)	(θ_1)	(α_2)	(θ_3)	(θ_4)	(θ_5)	(θ_6)	(δ_2)	(θ_7)	(θ_8)
	1	0	1	1	0	0	0	0	0	0
	0	2	0	0	1	1	0	1	1	0
	4	0	0	1	0	0	4	0	0	2
4.29	0	8	0	0	1	12	0	0	3	0
	16	0	0	1	0	0	32	0	0	4
	0	32	0	0	1	80	0	0	5	0
	64	0	0	1	0	0	192	0	0	6

---

* Actually, we develop the analog of conditions (2.22) by considering the functions $(x - x_{n-1})^k$, $k = 0, 1, \ldots, 6$, rather than $(x - x_{n-3})^k$.

Using elementary row operations we get

4.30

$$
\begin{array}{ccccccccc}
(\theta_0) & (\theta_1) & (\alpha_2) & (\theta_3) & (\theta_4) & (\theta_5) & (\theta_6) & (\delta_2) & (\theta_7) & (\theta_8) \\
\end{array}
$$

$$
\left(\begin{array}{cccccccccc}
1 & 0 & \boxed{1} & 1 & 0 & 0 & 0 & 0 & 0 & 0 \\
0 & 2 & 0 & 0 & 1 & 1 & 0 & \boxed{1} & 1 & 0 \\
4 & 0 & 0 & \boxed{1} & 0 & 0 & 4 & 0 & 0 & 2 \\
0 & 8 & 0 & 0 & \boxed{1} & 12 & 0 & 0 & 3 & 0 \\
\boxed{12} & 0 & 0 & 0 & 0 & 0 & 28 & 0 & 0 & 2 \\
0 & \boxed{24} & 0 & 0 & 0 & 68 & 0 & 0 & 2 & 0 \\
0 & 0 & 0 & 0 & 0 & 0 & 48 & 0 & 0 & \boxed{-6}
\end{array}\right)
$$

Using the "pivots" indicated we get

4.31

$$
\begin{cases}
\delta_1 - \delta_3 = 8(\delta_0 - \delta_4) \\
\alpha_0 - \alpha_4 = -\tfrac{17}{6}(\delta_0 + \delta_4) - \tfrac{1}{12}(\delta_1 + \delta_3) \\
\alpha_0 + \alpha_4 = -\tfrac{11}{3}(\delta_0 - \delta_4) \\
\alpha_1 + \alpha_3 = -\tfrac{16}{3}(\delta_0 - \delta_4) \\
\alpha_1 - \alpha_3 = \tfrac{32}{3}(\delta_0 + \delta_4) - \tfrac{7}{3}(\delta_1 + \delta_3).
\end{cases}
$$

Since $\alpha_0 + \alpha_1 + \alpha_2 + \alpha_3 + \alpha_4 = 0$ it follows that

4.32

$$
\begin{cases}
\delta_0 - \delta_4 = \tfrac{1}{9}\alpha_2 \\
\alpha_0 + \alpha_4 = -\tfrac{11}{27}\alpha_2 \\
\alpha_1 + \alpha_3 = -\tfrac{16}{27}\alpha_2.
\end{cases}
$$

We now seek to determine conditions on the coefficients α_0, α_1, α_2, α_3, and α_4 such that $\rho(\lambda) = \alpha_0\lambda^4 + \alpha_1\lambda^3 + \alpha_2\lambda^2 + \alpha_3\lambda + \alpha_4$ has zeros which do not exceed one in modulus. Following Henrici [1962, pp. 229–230] we let

4.33

$$
z = \frac{\lambda - 1}{\lambda + 1}, \qquad \lambda = \frac{1 + z}{1 - z}.
$$

This transformation maps the circle $|\lambda| \leqq 1$ onto the left half-plane $\mathrm{Re}\, z \leqq 0$. We consider the polynomial

4.34 $r(z) = (1-z)^4 \rho\left(\dfrac{1+z}{1-z}\right) = (\alpha_0 - \alpha_1 + \alpha_2 - \alpha_3 + \alpha_4)z^4 + (4\alpha_0 - 2\alpha_1 + 2\alpha_3 - 4\alpha_4)z^3$

$$+ (6\alpha_0 - 2\alpha_2 + 6\alpha_4)z^2 + (4\alpha_0 + 2\alpha_1 - 2\alpha_3 - 4\alpha_4)z$$

$$+ (\alpha_0 + \alpha_1 + \alpha_2 + \alpha_3 + \alpha_4).$$

Evidently, since $\alpha_0 + \alpha_1 + \alpha_2 + \alpha_3 + \alpha_4 = 0$, the roots of $r(z) = 0$ are 0, x_1, $x_2 \pm iy_2$ where x_1, x_2, and y_2 are real; hence we can write $r(z)$ in the form

4.35 $r(z) = Az(z - x_1)(z^2 - 2x_2 z + x_2^2 + y_2^2)$

for some constant A. Since $x_1 \leq 0, x_2 \leq 0$ it follows that each coefficient of $r(z)$ either vanishes or has the same sign as A.*

It follows from the above that the following quantities either vanish or have the same sign

4.36
$$\begin{cases} c_0 = \alpha_0 - \alpha_1 + \alpha_2 - \alpha_3 + \alpha_4 = \frac{32}{27}\alpha_2 \\[4pt] c_1 = 4\alpha_0 - 2\alpha_1 + 2\alpha_3 - 4\alpha_4 \\[4pt] c_2 = 6\alpha_0 - 2\alpha_2 + 6\alpha_4 = -\frac{40}{9}\alpha_2 \\[4pt] c_3 = 4\alpha_0 + 2\alpha_1 - 2\alpha_3 - 4\alpha_4. \end{cases}$$

In order that c_0 and c_2 have the same sign, we clearly must have $\alpha_2 = 0$. But this implies

4.37 $\delta_0 = \delta_4, \qquad \alpha_1 = -\alpha_3, \qquad \alpha_0 = -\alpha_4.$

Letting $\alpha_0 = 1$, we have

4.38 $c_1 = 8 - 4\alpha_1, \qquad c_3 = 8 + 4\alpha_1.$

In order that either c_1 or c_3 vanish or else c_1 and c_3 have the same sign, we must have

$$|\alpha_1| \leq 2.$$

Therefore,

4.39 $\rho(\lambda) = \lambda^4 + \alpha_1 \lambda^3 - \alpha_1 \lambda - 1 = (\lambda - 1)(\lambda^3 + (1 + \alpha_1)\lambda^2 + (1 + \alpha_1)\lambda + 1)$

$$= (\lambda - 1)(\lambda + 1)(\lambda^2 + \alpha_1 \lambda + 1).$$

If $|\alpha_1| < 2$ there are two complex roots with modulus equal to unity. If $\alpha_1 = \pm 2$, then ∓ 1 would be a triple root of $\rho(\lambda) = 0$ and the method would not be stable.

---

* Although the condition $\gamma_1 \geq 0, \gamma_2 \geq 0, \ldots, \gamma_N \geq 0$ is *necessary* in order that the roots of $z^N + \gamma_1 z^{N-1} + \cdots + \gamma_N = 0$ have nonpositive real parts, it is not sufficient. For, the equation $z^3 + 1$ has nonnegative coefficients but the roots are $-1, \frac{1}{2} \pm i\sqrt{3}/2$.

Thus we have proved that if the method is stable and of order 6, then all roots of $\rho(\lambda) = 0$ have modulus unity. Moreover, assuming $\alpha_0 = 1$, we have

4.40
$$\begin{cases} \alpha_0 = 1, \quad \alpha_1 = -\alpha_3, \quad \alpha_4 = -1 \\ |\alpha_1| < 2. \end{cases}$$

We now seek to determine various choices of the β_i assuming $\alpha_0 = 1$. Since $\alpha_2 = 0$ and $\delta_0 = \delta_4$, we have from (4.31) that $\delta_1 = \delta_3$ and

4.41
$$\begin{cases} \delta_1 = -12 - 34\delta_0 \\ \alpha_1 = 28 + 90\delta_0. \end{cases}$$

The condition $|\alpha_1| < 2$ implies that $\beta_0 = -\delta_0$ lies in the range

4.42
$$\tfrac{13}{45} < \beta_0 < \tfrac{1}{3}.$$

Having selected β_0 in this range, we have

4.43
$$\begin{cases} \alpha_0 = 1 \\ \alpha_1 = 28 - 90\beta_0, \quad \beta_1 = 12 - 34\beta_0 \\ \alpha_2 = 0, \quad \beta_2 = 4 + 2\alpha_1 - 2\beta_0 - 2\beta_1 \\ \alpha_3 = -\alpha_1, \quad \beta_3 = \beta_1 \\ \alpha_4 = -1, \quad \beta_4 = \beta_0. \end{cases}$$

With the above choices, the method is stable since all of the roots of $\rho(\lambda) = 0$ have modulus unity and all are simple.

As an example, let $\beta_0 = \tfrac{3}{10}$. We have

4.44
$$\begin{cases} \alpha_0 = 1, \quad \beta_0 = \tfrac{3}{10} \\ \alpha_1 = 1, \quad \beta_1 = \tfrac{18}{10} \\ \alpha_2 = 0, \quad \beta_2 = \tfrac{18}{10} \\ \alpha_3 = -1, \quad \beta_3 = \tfrac{18}{10} \\ \alpha_4 = -1, \quad \beta_4 = \tfrac{3}{10}. \end{cases}$$

The interpolation error is

4.45 $(IE[y])_n = (R_h[(x - x_{n-1})^7])_n \dfrac{y^{(7)}}{7!} = (-36h^7)\dfrac{y^{(7)}}{7!} = -\tfrac{1}{140}h^7 y^{(7)}.$

We remark that with the choice of β_0 given above, the method is identical to that given by Dahlquist [1956, formula (1.10)]. If one lets $\beta_0 = \tfrac{29}{90}$, one obtains formula

(1.11) of Dahlquist's paper. According to Dahlquist, the method with $\beta_0 = \frac{3}{10}$ has a smaller error but is "less stable," while the other has a larger error but is "more stable." (The concepts of "more stable" and "less stable" relate to the notions of "weak stability" and "weak instability" which are discussed in Section 9.7.)

Dahlquist [1956] showed that if a linear N-step method with N even is stable, then the order of the method is $N + 2$ if and only if

4.46 $$\alpha_k = -\alpha_{N-k}, \qquad \beta_k = \beta_{N-k}, \qquad k = 0, 1, \dots, \frac{N}{2}.$$

For the case $N = 4$ this result follows from the proof of Theorem 4.2.

We now give values of N and p for some of the methods which we have been considering.

	N	p
Euler method	1	1
modified Euler method	1	2
midpoint method	2	2
Milne-Simpson corrector	2	4
Milne-Simpson predictor	4	4
Adams-Moulton corrector	3	4
Adams-Moulton predictor	4	4

We note that of the above methods only the Milne-Simpson corrector is of order $N + 2$ and the roots of $\rho(\lambda) = \lambda^2 - 1 = 0$, which are ± 1, do indeed have modulus 1.

EXERCISES 9.4

1. Find a stable linear 3-step method of order 4. Also, find the corresponding interpolation error.

2. Find the linear 3-step method such that $\alpha_0 = 1$, $\alpha_2 = 0$, $\beta_0 = \beta_1 = \beta_3 = 0$ which has order unity and which is most stable in the sense that the maximum of the moduli of all zeros of $\rho(\lambda)$, other than $\lambda = 1$, is minimized. Find the interpolation error of the method.

3. Find the open linear 3-step method of order 4 such that $\beta_1 = 0$. (It need not be stable.) For this method find the root radius of $\rho(\lambda)$, i.e., the maximum of the moduli of the zeros of $\rho(\lambda)$. Also find the open linear 3-step method of order 4 such that the root radius of $\rho(\lambda)$ is minimized.

4. Find a stable linear 5-step method of order 6. Determine the interpolation error.

5. Verify Theorem 4.2 for the case $N = 1$. Find the method of order 2 such that $\alpha_0 = 1$.

6. Verify Theorem 4.2 for the case $N = 2$. Find the method of order 4 such that $\alpha_0 = 1$.

7. Find the solution of (4.4) corresponding to $\beta_0 = \frac{1}{3}$ and the solution corresponding to $\beta_0 = 0.3$. In the latter case determine the zeros of $\rho(\lambda)$.

8. Show that if $|c| \leq 1$ and $|b| \leq 1 + c$, then neither root of the quadratic equation

$$x^2 + bx + c = 0,$$

where b and c are real, exceeds one in modulus. (Hint: Consider separately the cases where $b^2 - 4c \geq 0$ and $b^2 - 4c < 0$.)

9. Show that if $|\lambda| \leq 1$, then $Re\, z \leq 0$ where

$$z = \frac{\lambda - 1}{\lambda + 1}.$$

Also, show that if $Re\, z \leq 0$, then $|\lambda| \leq 1$ where

$$\lambda = \frac{1 + z}{1 - z}.$$

(Hint: Let $\lambda = \rho e^{i\theta}$ where ρ and θ are real.)

10. Find the polynomial $\rho(\lambda)$ of degree 4 with leading coefficients unity whose roots are $-2, -3 + i, -3 - i$, and -1. Verify that all of its coefficients have the same sign.

11. By considering the transformation given by (4.33), derive necessary conditions to ensure that both roots of the quadratic equation

$$\rho(\lambda) = \lambda^2 + \alpha_1 \lambda + \alpha_2 = 0$$

(where α_1 and α_2 are real) do not exceed unity in modulus. Show that the conditions are sufficient.

12. As in the previous exercise, derive necessary conditions to guarantee that all roots of the cubic equation

$$\rho(\lambda) = \lambda^3 + b\lambda^2 + c\lambda + d = 0$$

do not exceed one in modulus. (Assume b, c, and d are real.) What can be said about the roots of the following equations?

a) $\lambda^3 + 3\lambda = 0$

b) $\lambda^3 - \frac{3}{2}\lambda^2 + \lambda - \frac{1}{4} = 0$

c) $\lambda^3 - \lambda^2 - \lambda - 2 = 0$.

(In each case try to find the actual roots and verify your conclusions.)

13. Show that if α_1 is real and $|\alpha_1| < 2$ then any linear 4-step method with $\alpha_0 = 1, \alpha_3 = -\alpha_1$, $\alpha_2 = 0$, and $\alpha_4 = -1$ satisfies the condition of stability.

14. Find the stable linear 4-step method of order 6 such that $\beta_0 = \frac{29}{90}$. Find the roots of the equation $\rho(\lambda) = 0$. Also find the interpolation error.

15. Find the interpolation error for the stable linear 4-step method of order 6 with $\beta_0 = \frac{3}{10}$.

16. Find the stable linear 4-step method of order 6 such that $\alpha_1 = 0$. Also, find the interpolation error and the zeros of $\rho(\lambda)$.

17. Find the value of β_0 in the range given by (4.42) such that for a stable linear 4-step method of order 6 the coefficient of $y^{(7)}$ in the expression for the interpolation error is minimized. Give the interpolation error and the values of the α_i and β_i.

18. Show that there does not exist a stable linear 4-step method of order 7 or more.

19. What order of accuracy can be obtained with a stable linear 9-step method? With a stable linear 10-step method?

20. Show that the method D.q.N is unstable if $q = 1$ and $N \geq 3$ or if $q = 2, N \geq 2$.

21. Show that the method D.q.N is unstable if $q = 0$ and $N \geq 7$.

22. Construct the most general stable linear 6-step method of order 8.

9.5 THE SPECIAL CASE $y' = Ay$

We now study the behavior of several of the methods which we have considered above when applied to the initial-value problem

5.1
$$\begin{cases} y' = Ay \\ y(x_0) = y_0 \end{cases}$$

where A is a given constant. As we have seen, the solution of (5.1) is

5.2
$$\bar{y}(x) = y_0 e^{A(x - x_0)}.$$

As usual, we assume that we are seeking approximate values of the solution for

5.3
$$x_n = x_0 + nh, \qquad n = 0, 1, 2, \ldots$$

where h is the step-size. Having obtained such values $y_0, y_1, \ldots$, we can define a continuous function $y_h(x)$ as follows:

5.4
$$y_h(x) = \begin{cases} y_n, & \text{if } x/h = n, \text{ an integer} \\ \dfrac{x_{n+1} - x}{h} y_n + \dfrac{x - x_n}{h} y_{n+1}, & \text{if } x_n \leq x \leq x_{n+1}. \end{cases}$$

This procedure would be reasonable for such methods as the Euler, modified Euler, Heun, and midpoint methods whose accumulated error* is $0(h^2)$, $(0(h)$ for the Euler method) since the error of linear interpolation is $0(h^2)$. For a fourth-order method, however, whose accumulated error is $0(h^4)$ the use of a Hermite interpolation formula involving $y_n, y_{n+1}, y'_n, y'_{n+1}$ is recommended since the interpolation error is $0(h^4)$ and since the derivatives are readily available.

For the sake of simplicity, however, in this section we shall consider the behavior of the numerical solution $y_h(x)$ only for values of x such that x/h is an integer. As before, we define

5.5
$$e_n = \bar{y}_n - y_n.$$

---

* See Section 8.7.

By Lemma 8-5.20 we have for the Euler method

5.6
$$|e_n| \leq \begin{cases} \dfrac{h}{2} A^2 |y_0|(x_n - x_0)e^{A(x_n - x_0)}, & \text{if } A \geq 0 \\[3mm] \dfrac{h}{2} A^2 |y_0|(x_n - x_0)e^{A(x_n - x_0 - h)}, & \text{if } A < 0. \end{cases}$$

Therefore, assuming $y_0 \neq 0$, the relative error satisfies

5.7
$$\frac{|e_n|}{|\bar{y}_n|} \leq \begin{cases} \dfrac{h}{2} A^2 (x_n - x_0), & \text{if } A \geq 0 \\[3mm] \dfrac{h}{2} A^2 (x_n - x_0)e^{-Ah}, & \text{if } A < 0. \end{cases}$$

Hence the relative error, like the absolute error, $|e_n|$, is $0(h)$. Moreover, for fixed h the relative error increases only linearly with $x_n - x_0$. We now perform similar studies on some of the other methods which we have considered.

The Modified Euler Method

In studying the behavior of the modified Euler method for the problem (5.1), we first assume that the numerical solution satisfies 8-(6.8) exactly at each step. Thus we have, by (5.1),

5.8
$$y_{n+1} = y_n + \frac{h}{2}(Ay_n + Ay_{n+1})$$

or

5.9
$$y_{n+1} = \left(\frac{1 + \dfrac{Ah}{2}}{1 - \dfrac{Ah}{2}} \right) y_n$$

and

5.10
$$y_n = \left(\frac{1 + \dfrac{Ah}{2}}{1 - \dfrac{Ah}{2}} \right)^n y_0.$$

As in the case of the Euler method, y_n increases exponentially as $n \to \infty$, if $A > 0$

and $Ah < 2$. On the other hand, if $A < 0$ and $Ah > -2$, then y_n decreases exponentially.

5.11 Theorem. Let y_n be determined by the modified Euler method (5.8) where at each step, y_{n+1} is obtained exactly. Then if $|Ah| < 2$, we have

5.12
$$|\bar{y}_n - y_n| \leq \begin{cases} \frac{1}{12}A^3h^2|y_0|\dfrac{(x_n - x_0)}{\left(1 - \dfrac{hA}{2}\right)}e^{A(x_n - x_0)/(1 - hA/2)}, & A \geq 0 \\[4mm] \frac{1}{12}|A|^3h^2|y_0|(x_n - x_0)e^{A(x_n - x_0 - h)}, & A < 0. \end{cases}$$

Proof. For convenience let

5.13
$$\alpha = Ah$$

and

5.14
$$\Delta = \frac{1 + \dfrac{Ah}{2}}{1 - \dfrac{Ah}{2}} - e^{Ah} = \frac{1 + \alpha/2}{1 - \alpha/2} - e^{\alpha}.$$

If $0 \leq \alpha < 2$, then

5.15 $\Delta\left(1 - \dfrac{\alpha}{2}\right) = 1 + \dfrac{\alpha}{2} - e^{\alpha}\left(1 - \dfrac{\alpha}{2}\right) = \dfrac{\alpha}{2}e^{\alpha} - \left(e^{\alpha} - \left(1 + \dfrac{\alpha}{2}\right)\right)$

$$= \frac{\alpha}{2}\left(1 + \alpha + \frac{\alpha^2}{2!} + \cdots\right) - \left(\frac{\alpha}{2} + \frac{\alpha^2}{2!} + \frac{\alpha^3}{3!} + \cdots\right)$$

$$= \tfrac{1}{12}\alpha^3 + \tfrac{1}{24}\alpha^4 + \tfrac{1}{80}\alpha^5 + \cdots + \left(\frac{1}{2(p - 1)!} - \frac{1}{p!}\right)\alpha^p + \cdots.$$

But since

5.16
$$q_p = \frac{1}{2(p - 1)!} - \frac{1}{p!} = \frac{p - 2}{2p!} > 0$$

for $p > 2$ and

5.17
$$\frac{q_{p+1}}{q_p} = \left(\frac{p - 1}{p - 2}\right)\left(\frac{1}{p + 1}\right)$$

it follows that $\Delta \geqq 0$, and by (5.14)

5.18
$$\frac{1 + \dfrac{\alpha}{2}}{1 - \dfrac{\alpha}{2}} \geqq e^{\alpha}.$$

Moreover, q_{p+1}/q_p is a decreasing function of p for $p \geq 3$. Also, for $p \geq 3$ we have $q_{p+1}/q_p \leq \frac{1}{2}$. Hence

5.19
$$\Delta \left(1 - \frac{\alpha}{2}\right) \leq \tfrac{1}{12}\alpha^3 \left(1 + \alpha + \frac{\alpha^2}{2} + \frac{\alpha^3}{4} + \cdots\right)$$
$$= \tfrac{1}{12}\alpha^3 \left(1 + \frac{\alpha}{1 - \dfrac{\alpha}{2}}\right) = \tfrac{1}{12}\alpha^3 \frac{1 + \alpha/2}{1 - \alpha/2}$$

and we have

5.20
$$0 \leqq \Delta \leqq \left(\frac{1 + \dfrac{\alpha}{2}}{\left(1 - \dfrac{\alpha}{2}\right)^2}\right) \tfrac{1}{12}\alpha^3.$$

If we now let

5.21
$$S_n = e^{A(x_n - x_0)} - \left(\frac{1 + \dfrac{\alpha}{2}}{1 - \dfrac{\alpha}{2}}\right)^n = (e^{\alpha})^n - \left(\frac{1 + \dfrac{\alpha}{2}}{1 - \dfrac{\alpha}{2}}\right)^n$$
$$= \left(e^{\alpha} - \frac{1 + \dfrac{\alpha}{2}}{1 - \dfrac{\alpha}{2}}\right)\left[(e^{\alpha})^{n-1} + \left(\frac{1 + \dfrac{\alpha}{2}}{1 - \dfrac{\alpha}{2}}\right)(e^{\alpha})^{n-2} + \cdots + \left(\frac{1 + \dfrac{\alpha}{2}}{1 - \dfrac{\alpha}{2}}\right)^{n-1}\right].$$

By (5.20) and (5.18) we have

5.22
$$|S_n| \leqq \left(\frac{1 + \dfrac{\alpha}{2}}{\left(1 - \dfrac{\alpha}{2}\right)^2}\right)\tfrac{1}{12}\alpha^3 n \left(\frac{1 + \dfrac{\alpha}{2}}{1 - \dfrac{\alpha}{2}}\right)^{n-1} = \tfrac{1}{12}\alpha^3 \left(\frac{n}{1 - \dfrac{\alpha}{2}}\right)\left(\frac{1 + \dfrac{\alpha}{2}}{1 - \dfrac{\alpha}{2}}\right)^n.$$

But since for $0 \leqq \alpha < 2$ we have

5.23
$$\frac{1 + \dfrac{\alpha}{2}}{1 - \dfrac{\alpha}{2}} = 1 + \frac{\alpha}{1 - \dfrac{\alpha}{2}} \leqq e^{\alpha/(1 - \alpha/2)},$$

it follows that

5.24
$$|S_n| \leqq \tfrac{1}{12}\alpha^3 \left(\frac{1}{1 - \dfrac{\alpha}{2}}\right) n e^{n\alpha/(1 - \alpha/2)}$$

or, by (5.13) since $x_n - x_0 = nh$,

5.25
$$|S_n| \leqq \tfrac{1}{12}A^3 h^2 \frac{(x_n - x_0)}{1 - Ah/2} e^{A(x_n - x_0)/(1 - Ah/2)}.$$

This proves (5.12) for the case $A \geqq 0$.

To treat the case $A < 0$ we observe that for $0 > \alpha > -2$

5.26
$$\log \frac{1 + \alpha/2}{1 - \alpha/2} = -2\left(\frac{|\alpha|}{2} + \frac{1}{3}\left(\frac{|\alpha|}{2}\right)^3 + \frac{1}{5}\left(\frac{|\alpha|}{2}\right)^5 + \cdots\right)$$
$$\leqq -|\alpha|.$$

Therefore

5.27
$$e^\alpha \geqq \frac{1 + \alpha/2}{1 - \alpha/2}.$$

Moreover, we have from (5.15)

5.28
$$\Delta(1 - \alpha/2) = \tfrac{1}{12}\alpha^3 + \tfrac{1}{24}\alpha^4 + \tfrac{1}{80}\alpha^5 + \cdots,$$

which is an alternating series whose terms, after the second, have decreasing magnitude for $|\alpha| < 2$. Therefore,

5.29
$$\tfrac{1}{12}\alpha^3 \leqq \Delta\left(1 - \frac{\alpha}{2}\right) \leqq \tfrac{1}{12}\alpha^3 + \tfrac{1}{24}\alpha^4$$

or

5.30
$$|\Delta(1 - \alpha/2)| \leqq \tfrac{1}{12}|\alpha|^3.$$

By (5.27), (5.30) and (5.21) we have

5.31
$$|S_n| \le \tfrac{1}{12}|\alpha|^3 \frac{1}{1 - \alpha/2} n e^{\alpha(n-1)}$$

$$\le \tfrac{1}{12}|\alpha|^3 e^{-\alpha} n e^{\alpha n}$$

and (5.12) follows.

From (5.12) and (5.2) we have, assuming $y_0 \ne 0$,

5.32
$$\frac{|e_n|}{|\bar{y}_n|} \le \begin{cases} \dfrac{\tfrac{1}{12}A^3 h^2 \dfrac{x_n - x_0}{hA} e^{A(x_n - x_0) \, hA/(2 - hA)}}{1 - \dfrac{hA}{2}}, & \text{if } A \ge 0 \\[4mm] \tfrac{1}{12}|A|^3 h^2 (x_n - x_0) e^{|A|h}, & \text{if } A < 0. \end{cases}$$

Thus the relative error and the accumulated error are each $0(h^2)$. For $A > 0$, the relative error increases slightly faster than linearly with $x_n - x_0$ for fixed h.

A similar result holds for the Heun method. Here we have

5.33
$$y_{n+1} = y_n + \frac{h}{2}(Ay_n + A(y_n + hAy_n))$$

$$= y_n + Ahy_n + \frac{A^2 h^2}{2} y_n$$

and

5.34
$$y_n = \left(1 + Ah + \frac{A^2 h^2}{2}\right)^n y_0.$$

5.35 Theorem. If $|Ah| < 2$ and if y_n is obtained by the Heun method, then

5.36
$$|\bar{y}_n - y_n| \le \begin{cases} \tfrac{1}{6}A^3 h^2 |y_0|(x_n - x_0) e^{A(x_n - x_0)}, & A \ge 0 \\ \tfrac{1}{6}|A|^3 h^2 |y_0|(x_n - x_0) e^{A(1 + Ah/2)(x_n - x_0 - h)}, & A < 0. \end{cases}$$

Proof. The proof is similar to that given for Theorem 5.11. If $A > 0$, we note that $0 \le e^\alpha - (1 + \alpha + \alpha^2/2) \le \alpha^3 e^\alpha/6$. If $A < 0$, we have $1 + \alpha + \alpha^2/2 \le e^{\alpha(1 + \alpha/2)}$, since $\log(1 + \alpha + \alpha^2/2) = (\alpha + \alpha^2/2) - \tfrac{1}{2}(\alpha + \alpha^2/2)^2 + \tfrac{1}{3}(\alpha + \alpha^2/2)^3 - \cdots \le \alpha + \alpha^2/2$ for $-2 < \alpha < 0$. Moreover, $|e^\alpha - (1 + \alpha + \alpha^2/2)| \le |\alpha|^3/6$ since the series for e^α is an alternating series where terms, after the second, are decreasing in magnitude.

Again, the solution of the Heun method behaves very much like the true solution (5.2), though the error bound is approximately twice as large as that for the modified Euler method.

The Midpoint Method

In contrast to the Euler method and the modified Euler method, the behavior of the solution obtained from the midpoint method may in some cases differ in a qualitative way from the exact solution. Actually, if $A > 0$, the behavior is quite similar, but if $A < 0$, the numerical solution will not behave at all like the exact solution for x very large. In spite of this bad behavior, which we shall examine in more detail later, one can prove the following.

5.37 Theorem. Let y_n be determined by the midpoint method for $n \geq 2$. Then if $|Ah| < 1$

$$\mathbf{5.38} \quad |\bar{y}_n - y_n| \leq \begin{cases} |y_0|A^2h^2 \left\{ \left(\dfrac{A(x_n - x_0)}{6} + \dfrac{1}{4} \right) e^{A(x_n - x_0)} + \dfrac{1}{4} \right\}, \ A \geq 0 \\[3mm] |y_0|A^2h^2 \left\{ \dfrac{|A|(x_n - x_0)}{6} + \dfrac{1}{4} + \dfrac{1}{4}e^{|A|(x_n - x_0)} \right\}, \qquad A < 0 \end{cases}$$

if y_1 is determined by the Euler method, and

$$\mathbf{5.39} \quad |\bar{y}_n - y_n| \leq \begin{cases} \frac{1}{6}A^3h^2|y_0|(x_n - x_0)e^{A(x_n - x_0)} + \dfrac{|y_0|A^4h^4}{16}(e^{A(x - x_0)} + 1), \qquad A \geq 0 \\[3mm] \frac{1}{6}|A|^3h^2|y_0|(x_n - x_0) + \dfrac{|y_0|A^4h^4}{16}(e^{|A|(x_n - x_0)} + 1), \qquad A < 0 \end{cases}$$

if y_1 is determined by the Heun method.

We remark that using the Heun starting value, the accuracy is one-half that obtained with the modified Euler method (see (5.12)) to within terms of order h^4. Thus even though the single-step error of the midpoint method is 4 times that of the modified Euler method, the overall error is only twice as great. This is due to the fact that the midpoint method is a two-step method. See also the bounds on the accumulated error given in Section 8.7.

Proof. By 8-(6.22) we have,

$$\mathbf{5.40} \qquad\qquad y_{n+1} = y_{n-1} + 2hAy_n, \qquad n \geq 1.$$

The above difference equation is linear with constant coefficients and can be solved in a manner similar to that used for linear differential equations with constant coefficients. Indeed, let us assume a solution of the form $y_n = c\lambda^n$ with $c \neq 0, \lambda \neq 0$. Substituting in (5.40) we obtain $c\lambda^{n+1} = c\lambda^{n-1} + 2chA\lambda^n$, or

$$\mathbf{5.41} \qquad\qquad\qquad \lambda^2 - 2hA\lambda - 1 = 0.$$

The roots of this equation are given by

5.42 $$\lambda_1 = Ah + \sqrt{A^2h^2 + 1}, \qquad \lambda_2 = Ah - \sqrt{A^2h^2 + 1}.$$

For any c_1 and c_2 the following is a solution of (5.40)

5.43 $$y_n = c_1 \lambda_1^n + c_2 \lambda_2^n$$

provided h is sufficiently small that $\lambda_1 \neq 0$, $\lambda_2 \neq 0$, we now determine c_1 and c_2 in terms of the assigned values y_0 and y_1. Letting $n = 0$ and $n = 1$, we obtain

5.44 $$\begin{cases} y_0 = c_1 + c_2 \\ y_1 = c_1\lambda_1 + c_2\lambda_2. \end{cases}$$

Solving for c_1 and c_2, we have

5.45 $$c_1 = \frac{y_1 - \lambda_2 y_0}{\lambda_1 - \lambda_2}, \qquad c_2 = \frac{\lambda_1 y_0 - y_1}{\lambda_1 - \lambda_2}.$$

The desired solution of (5.40) is therefore

5.46 $$y_n = \left(\frac{y_1 - \lambda_2 y_0}{\lambda_1 - \lambda_2} \right) \lambda_1^n + \left(\frac{\lambda_1 y_0 - y_1}{\lambda_1 - \lambda_2} \right) \lambda_2^n.$$

We remark that the solution is unique as can be shown by induction. Thus y_2 is uniquely determined by (5.40) for given y_0 and y_1, then y_3 is uniquely determined, etc.

Let us now digress from the proof of Theorem 5.40 to examine the behavior of the solution (5.46). If $A > 0$, then $\lambda_1 > 1$ and $-1 < \lambda_2 < 0$. Moreover, as we shall show, the term $c_1\lambda_1^n$ behaves very much like $y_0 e^{A(x_n - x_0)}$, whereas the term $c_2\lambda_2^n$ decreases in magnitude as $(x_n - x_0)$ increases. On the other hand, if $A < 0$, the situation is quite different. We now have $0 < \lambda_1 < 1$ and $\lambda_2 < -1$. As before, the term $c_1\lambda_1^n$ behaves very much like $y_0 e^{A(x_n - x_0)}$ and decreases exponentially as $(x_n - x_0)$ increases. On the other hand, the term $c_2\lambda_2^n$, while small at first, will eventually, for sufficiently large $(x_n - x_0)$, become very large and will completely overwhelm the other term. When this occurs, the numerical solution is, of course, useless from that point on. This phenomenon is a form of instability. It is known as *weak instability* since for a given value of $\bar{x}$, we can choose h small enough so that the instability is delayed until $x > \bar{x}$. As a matter of fact, the numerical solutions converge to the exact solution (pointwise) as $h \to 0$. This is in contrast to a *strongly unstable method*, such as the method D.1.3 considered later in this section, where the condition of stability is not satisfied. We shall discuss weak and strong stability and instability in more detail in Section 9.7.

We remark that even if the *exact* value $y_0 e^{Ah}$ were assigned to y_1, the coefficient c_2 would not vanish. Thus the (weak) instability would still occur. If we let $y_1 = \lambda_1 y_0$, then c_2 would vanish. If exact calculations were used from then on, the numerical solution would indeed decrease exponentially to zero. However, if at any stage of the calculation a rounding error were introduced, then the term $c_2 \lambda_2^n$ would be introduced. Actually, suppose that $y_1 = \lambda_1 y_0$ and the calculations are carried out exactly until one reaches x_p where we obtain $y_p^* = y_p + \varepsilon$. Here ε is a rounding error and y_p is the value which would have been obtained had there been no rounding error. If all subsequent calculations are carried out exactly, the numerical solution for $n \geq p + 1$ would be given by

5.47
$$y_n^* = \left(\frac{y_p^* - \lambda_2 y_{p-1}}{\lambda_1 - \lambda_2} \right) \lambda_1^{n-(p-1)} + \left(\frac{\lambda_1 y_{p-1} - y_p^*}{\lambda_1 - \lambda_2} \right) \lambda_2^{n-(p-1)}.$$

Since $y_p = \lambda_1 y_{p-1}$, it follows that $y_p^* = (y_p + \varepsilon) \neq \lambda_1 y_{p-1}$, and hence the second term does not vanish.

Returning to the proof of Theorem 5.37, we first assume that y_1 is determined by the Euler method, i.e., $y_1 = (1 + Ah)y_0$. By (5.45) and (5.42) we have

5.48
$$c_1 = \frac{(1 + Ah - \lambda_2)}{\lambda_1 - \lambda_2} y_0 = \frac{1 + \sqrt{1 + A^2 h^2}}{2\sqrt{1 + A^2 h^2}} y_0 = \left(1 + \frac{1 - \sqrt{1 + A^2 h^2}}{2\sqrt{1 + A^2 h^2}} \right) y_0$$

$$= \left(1 - \frac{A^2 h^2}{2\sqrt{1 + A^2 h^2}(1 + \sqrt{1 + A^2 h^2})} \right) y_0$$

and

5.49
$$|c_1 - y_0| \leq \frac{A^2 h^2}{4} |y_0|.$$

By (5.44) we have $c_2 = y_0 - c_1$ and hence

5.50
$$|c_2| \leq \frac{A^2 h^2}{4} |y_0|.$$

For convenience, let $\alpha = Ah$. Then $\lambda_1 = \alpha + \sqrt{1 + \alpha^2}$. We show that

5.51
$$0 < e^\alpha - \lambda_1 < \tfrac{1}{6}\alpha^3 e^\alpha, \qquad 0 < \alpha < 1.$$

Evidently,

5.52
$$\log \lambda_1 = \log(\alpha + \sqrt{1 + \alpha^2}) = \sinh^{-1}\alpha$$

$$= \alpha - \frac{1}{2}\left(\frac{\alpha^3}{3} \right) + \frac{(1)(3)}{(2)(4)}\left(\frac{\alpha^5}{5} \right) - \frac{(1)(3)(5)}{(2)(4)(6)}\left(\frac{\alpha^7}{7} \right) + \cdots$$

which is an alternating series with terms decreasing in magnitude. Therefore,

5.53
$$\alpha - \frac{\alpha^3}{6} < \log \lambda_1 < \alpha,$$

and by the mean-value theorem,

5.54 $$e^\alpha - \lambda_1 = e^\alpha - e^{\log \lambda_1} = e^z(\alpha - \log \lambda_1)$$

where $\log \lambda_1 < z < \alpha$. Hence, (5.51) follows.
On the other hand, if $-1 < \alpha < 0$, then we have

5.55 $$-|\alpha| \le \log \lambda_1 \le -|\alpha| + \tfrac{1}{6}|\alpha|^3 \le 0$$

and

5.56 $$0 \le \lambda_1 - e^\alpha \le \tfrac{1}{6}|\alpha|^3, \qquad 0 > \alpha > -1.$$

We now show that

5.57 $$|e^{A(x_n - x_0)} - \lambda_1^n| \le \begin{cases} \dfrac{h^2 A^3}{6}(x_n - x_0)e^{A(x_n - x_0)}, & A \ge 0 \\[2ex] \dfrac{h^2|A|^3}{6}(x_n - x_0), & A < 0. \end{cases}$$

But

5.58 $$e^{A(x_n - x_0)} - \lambda_1^n = e^{n\alpha} - \lambda_1^n = (e^\alpha - \lambda_1)(e^{\alpha(n-1)} + e^{\alpha(n-2)}\lambda_1 + \cdots + \lambda_1^{n-1})$$

where $\alpha = Ah$ and

5.59 $$|e^{A(x_n - x_0)} - \lambda_1^n| \le |e^\alpha - \lambda_1|ne^{(n-1)\alpha}$$
$$\le \tfrac{1}{6}A^3h^2(x_n - x_0)e^{A(x_n - x_0)}$$

if $A > 0$ by (5.51), and

5.60 $$|e^{A(x_n - x_0)} - \lambda_1^n| \le \tfrac{1}{6}|A|^3h^2(x_n - x_0)|\lambda_1^{n-1}| \le \tfrac{1}{6}|A|^3h^2(x_n - x_0)$$

if $A < 0$. Hence (5.57) holds.
We now have

5.61 $$\bar{y}_n - y_n = y_0 e^{A(x_n - x_0)} - (c_1\lambda_1^n + c_2\lambda_2^n)$$
$$= T_1 + T_2 + T_3$$

where

5.62 $$\begin{cases} T_1 = y_0(e^{A(x_n - x_0)} - \lambda_1^n) \\ T_2 = (y_0 - c_1)\lambda_1^n \\ T_3 = -c_2\lambda_2^n. \end{cases}$$

By (5.57) we have

5.63
$$|T_1| \leq \begin{cases} |y_0| \dfrac{h^2 A^3}{6} (x_n - x_0) e^{A(x_n - x_0)}, & A \geq 0 \\ |y_0| \dfrac{h^2 |A|^3}{6} (x_n - x_0), & A < 0. \end{cases}$$

By (5.51), (5.55), and (5.49) we have

5.64
$$|T_2| \leq \begin{cases} |y_0| \dfrac{A^2 h^2}{4} e^{A(x_n - x_0)}, & A \geq 0 \\ |y_0| \dfrac{A^2 h^2}{4}, & A < 0. \end{cases}$$

Finally, by (5.50), (5.51), and (5.56) we have, since $\lambda_2 = -\lambda_1^{-1}$,

5.65
$$|T_3| \leq \begin{cases} |y_0| \dfrac{A^2 h^2}{4}, & A \geq 0 \\ |y_0| \dfrac{A^2 h^2}{4} e^{|A|(x_n - x_0)}, & A < 0. \end{cases}$$

The result (5.38) now follows.

If the Heun method is used to determine y_1 we have

5.66
$$y_1 = (1 + Ah + \tfrac{1}{2} A^2 h^2) y_0.$$

Hence, one can verify from (5.45) and (5.42)

5.67
$$|y_0 - c_1| \leq \frac{|y_0| A^4 h^4}{16},$$

5.68
$$|c_2| \leq \frac{|y_0| A^4 h^4}{16},$$

and (5.39) follows.

Method D.1.3

Let us now consider the method D.1.3, which is defined in Section 8.10. As applied to the initial-value problem (5.1), the method is given by

5.69
$$2y_{n+1} + 3y_n - 6y_{n-1} + y_{n-2} = 6hAy_n.$$

The above formula can be used for $n = 2, 3, \ldots$. However, y_1 and y_2 must be determined by another procedure.

We first seek to solve (5.69) analytically. One can show by the methods used for the midpoint method that

5.70
$$y_n = c_1 \lambda_1^n + c_2 \lambda_2^n + c_3 \lambda_3^n$$

where λ_1, λ_2, and λ_3 are the roots of

5.71
$$2\lambda^3 + (3 - 6Ah)\lambda^2 - 6\lambda + 1 = 0$$

and where c_1, c_2, and c_3 are determined by

5.72
$$\begin{cases} y_0 = c_1 \quad + c_2 \quad + c_3 \\ y_1 = c_1\lambda_1 + c_2\lambda_2 + c_3\lambda_3 \\ y_2 = c_1\lambda_1^2 + c_2\lambda_2^2 + c_3\lambda_3^2. \end{cases}$$

Solving, we get

5.73
$$\begin{cases} c_1 = \dfrac{y_0\lambda_2\lambda_3 - y_1(\lambda_2 + \lambda_3) + y_2}{(\lambda_3 - \lambda_1)(\lambda_2 - \lambda_1)} \\[2mm] c_2 = \dfrac{-y_0\lambda_1\lambda_3 + y_1(\lambda_3 + \lambda_1) - y_2}{(\lambda_3 - \lambda_2)(\lambda_2 - \lambda_1)} \\[2mm] c_3 = \dfrac{y_0\lambda_1\lambda_2 - y_1(\lambda_1 + \lambda_2) + y_2}{(\lambda_3 - \lambda_1)(\lambda_3 - \lambda_2)}. \end{cases}$$

Let us study the behavior of the roots of (5.71) for small h. We consider first the limiting equation

5.74
$$2\bar{\lambda}^3 + 3\bar{\lambda}^2 - 6\bar{\lambda} + 1 = 0$$

which has roots

5.75 $\bar{\lambda}_1 = 1, \qquad \bar{\lambda}_2 = \dfrac{-5 + \sqrt{33}}{4} \doteq 0.186, \qquad \bar{\lambda}_3 = \dfrac{-5 - \sqrt{33}}{4} \doteq -2.686.$

By the continuity of the roots of (5.71) as a function of the coefficients, it follows that for h small enough, the roots are distinct. Let us seek to determine λ_1 by the Newton method with $\lambda_1^{(0)} = 1$ as an initial approximation. Letting the left member of (5.71) be $g(\lambda)$, we have $g(1) = -6Ah$, $g'(1) = 6 - 12Ah$, and

5.76
$$\lambda_1^{(1)} = \lambda_1^{(0)} - \frac{g(\lambda_1^{(0)})}{g'(\lambda_1^{(0)})} = 1 + \frac{Ah}{1 - 2Ah}$$

$$= 1 + Ah + 0(h^2).$$

Letting $\lambda_1^{(1)} = 1 + Ah$, we get $\lambda_1^{(2)} = 1 + Ah + \frac{1}{2}A^2h^2 + 0(h^3)$. Continuing in this way we have

5.77
$$\begin{cases} \lambda_1^{(3)} = 1 + Ah + \frac{1}{2}A^2h^2 + \frac{1}{6}A^3h^3 + 0(h^4) \\ \lambda_1^{(4)} = 1 + Ah + \frac{1}{2}A^2h^2 + \frac{1}{6}A^3h^3 - \frac{1}{24}A^4h^4 + 0(h^5). \end{cases}$$

Now let

5.78
$$\lambda_1^* = 1 + Ah + \frac{1}{2}A^2h^2 + \frac{1}{6}A^3h^3 - \frac{1}{24}A^4h^4.$$

We can show that

5.79
$$|\lambda_1 - \lambda_1^*| = 0(h^5)$$

as follows. Evidently,

5.80
$$g(\lambda_1^*) = 0(h^5) \quad \text{and} \quad g'(\lambda_1^*) = 6 + 0(h).$$

Hence, by 5-(5.34) and 5-(5.37), we have

5.81
$$|\lambda_1 - \lambda_1^*| \leq 3\left| \frac{g(\lambda_1^*)}{g'(\lambda_1^*)} \right| = 0(h^5).$$

Let us now assume that y_1 and y_2 are determined by the Euler method; thus

5.82
$$y_1 = (1 + Ah)y_0, \qquad y_2 = (1 + Ah)^2 y_0.$$

By (5.73) we have

5.83
$$\begin{cases} c_1 = \frac{y_0(\lambda_2 - (1 + Ah))(\lambda_3 - (1 + Ah))}{(\lambda_3 - \lambda_1)(\lambda_2 - \lambda_1)} \\ \lim_{h \to 0} c_1 = \frac{y_0(\bar{\lambda}_2 - 1)(\bar{\lambda}_3 - 1)}{(\bar{\lambda}_3 - 1)(\bar{\lambda}_2 - 1)} = y_0 \end{cases}$$

5.84
$$\begin{cases} c_2 = -\frac{y_0(\lambda_1 - (1 + Ah))(\lambda_3 - (1 + Ah))}{(\lambda_3 - \lambda_2)(\lambda_2 - \lambda_1)} \\ \lim_{h \to 0} c_2 = 0 \end{cases}$$

5.85
$$\begin{cases} c_3 = \dfrac{y_0(\lambda_1 - (1 + Ah))(\lambda_2 - (1 + Ah))}{(\lambda_3 - \lambda_1)(\lambda_3 - \lambda_2)} \\[2ex] \lim_{h \to 0} c_3 = 0. \end{cases}$$

Let x be fixed and let h be such that $(x - x_0)/h$ is a positive integer, say n. Then

5.86
$$y_h(x) = y_{h,n} = c_1 \lambda_1^{(x - x_0)/h} + c_2 \lambda_2^{(x - x_0)/h} + c_3 \lambda_3^{(x - x_0)/h}$$

Since $\lambda_1 = 1 + Ah + 0(h^2)$, it follows that

$$\log \lambda_1^{(x - x_0)/h} = \frac{x - x_0}{h} \log \lambda_1 = \frac{(x - x_0)}{h}(Ah + 0(h^2)) = A(x - x_0) + 0(h).$$

Hence

5.87
$$\lim_{h \to 0} \lambda_1^{(x - x_0)/h} = e^{A(x - x_0)}.$$

Moreover,

5.88
$$\lim_{h \to 0} \lambda_2^{(x - x_0)/h} = 0.$$

Thus, it follows from (5.83) and (5.84) that

5.89
$$\lim_{h \to 0} y_h(x) = \lim_{h \to 0} c_3 \lambda_3^{(x - x_0)/h} + y_0 e^{A(x - x_0)}.$$

Evidently, we need a more detailed analysis of the product $c_3 \lambda_3^{(x - x_0)/h}$. By (5.78), (5.81), and (5.85) we have

5.90
$$\lim_{h \to 0} \frac{c_3}{h^2} = \frac{y_0 \frac{1}{2} A^2 (\bar{\lambda}_2 - 1)}{(\bar{\lambda}_3 - \bar{\lambda}_2)(\bar{\lambda}_3 - 1)}$$

which is a finite, nonvanishing number for $y_0 \neq 0$ and $A \neq 0$. Therefore,

5.91
$$\lim_{h \to 0} c_3 \lambda_3^{(x - x_0)/h} = \lim_{h \to 0} \left[\left(\frac{c_3}{h^2} \right) h^2 \lambda_3^{(x - x_0)/h} \right]$$

$$= \left[\lim_{h \to 0} \left(\frac{c_3}{h^2} \right) \right] \left[\lim_{h \to 0} h^2 \lambda_3^{(x - x_0)/h} \right]$$

provided both limits exist. But we now show that

5.92
$$\lim_{h \to 0} h^2 |\lambda_3|^{(x - x_0)/h} = \infty.$$

Indeed,

5.93
$$\log h^2 |\lambda_3|^{(x-x_0)/h} = \left(\frac{x-x_0}{h}\right) \log |\lambda_3| + 2 \log h$$

$$= \frac{1}{h} [(x - x_0) \log |\lambda_3| + 2h \log h]$$

which approaches ∞ as $h \to 0$ since $\log |\lambda_3| \geq 0.5$ for h small enough and since*

5.94
$$\lim_{h \to 0} h \log h = 0.*$$

Hence (5.92) holds and

5.95
$$\lim_{h \to 0} |c_3 \lambda_3^{(x-x_0)/h}| = \infty.$$

Therefore, the method does not converge.

Let us now consider the case where y_1 and y_2 are chosen to be the exact values of the solution of the differential equation, i.e.,

5.96
$$y_1 = \bar{y}_1 = y_0 e^{Ah}, \qquad y_2 = \bar{y}_2 = y_0 e^{2Ah}.$$

It is easy to show that (5.89) holds. We now proceed to show that (5.95) holds. But

5.97
$$c_3 = y_0 \frac{(\lambda_1 - e^{Ah})(\lambda_2 - e^{Ah})}{(\lambda_3 - \lambda_2)(\lambda_3 - \lambda_1)};$$

hence by (5.78) and (5.81), since $e^{Ah} = 1 + Ah + \frac{1}{2}A^2 h^2 + \frac{1}{6}A^3 h^3 + \frac{1}{24}A^4 h^4 + 0(h^5)$

5.98
$$\lambda_1 - e^{Ah} = -\frac{1}{12} A^4 h^4 + 0(h^5)$$

and

5.99
$$\lim_{h \to 0} \frac{c_3}{h^4} = \frac{-\frac{1}{12} y_0 A^4 (\bar{\lambda}_2 - 1)}{(\bar{\lambda}_3 - \bar{\lambda}_2)(\bar{\lambda}_3 - 1)},$$

which is a finite, nonvanishing number for $y_0 \neq 0$ and $A \neq 0$. Therefore, we have

5.100
$$\lim_{h \to 0} c_3 |\lambda_3|^{(x-x_0)/h} = \left(\lim_{h \to 0} \frac{c_3}{h^4}\right) \left(\lim_{h \to 0} h^4 |\lambda_3|^{(x-x_0)/h}\right)$$

and the second limit can be shown to be infinite in the same way that (5.92) was proved.

---

* This can be seen by applying L'Hospital's rule (Theorem A-3.52) to the function $(\log h)/h^{-1}$

The Runge-Kutta Method

The Runge-Kutta method applied to the initial-value problem (5.1) yields

5.101
$$y_{n+1} = (1 + Ah + \tfrac{1}{2}A^2h^2 + \tfrac{1}{6}A^3h^3 + \tfrac{1}{24}A^4h^4)y_n;$$

hence

5.102
$$y_n = (1 + Ah + \tfrac{1}{2}A^2h^2 + \tfrac{1}{6}A^3h^3 + \tfrac{1}{24}A^4h^4)^n \, y_0.$$

Moreover,

5.103
$$\bar{y}_n - y_n = y_0(e^{\alpha n} - \lambda^n)$$

where

5.104
$$\alpha = Ah, \qquad \lambda = 1 + \alpha + \frac{\alpha^2}{2} + \frac{\alpha^3}{6} + \frac{\alpha^4}{24}.$$

But

5.105
$$e^{\alpha n} - \lambda^n = (e^\alpha - \lambda)(e^{\alpha(n-1)} + e^{\alpha(n-2)}\lambda + \cdots + \lambda^{n-1}).$$

By Taylor's theorem

5.106
$$e^\alpha = \lambda + \frac{\alpha^5}{120} e^z$$

where z lies between 0 and α. Therefore,

5.107
$$\begin{cases} 0 \leqq e^\alpha - \lambda \leqq \dfrac{\alpha^5}{120} e^\alpha, & \alpha > 0 \\[2mm] 0 \leqq \lambda - e^\alpha \leqq \dfrac{|\alpha|^5}{120}, & \alpha < 0. \end{cases}$$

It thus follows that

5.108
$$|e^{\alpha n} - \lambda^n| \leqq \begin{cases} \dfrac{|\alpha|^5}{120} e^\alpha n e^{\alpha(n-1)}, & \alpha \geqq 0 \\[2mm] \dfrac{|\alpha|^5}{120} n, & -2 \leqq \alpha < 0. \end{cases}$$

Therefore, for $|Ah| \leqq 2$,

5.109
$$|\bar{y}_n - y_n| \leqq \begin{cases} |y_0| \dfrac{A^5 h^4}{120}(x_n - x_0)e^{Ah(x_n - x_0)}, & A \geqq 0 \\[2mm] |y_0| \dfrac{|A|^5 h^4}{120}(x_n - x_0), & A < 0. \end{cases}$$

EXERCISES 9.5

1. State and prove analogs of Theorem 5.11 for the corrected Heun method and for the Heun-midpoint method. For the latter method assume that y_1 is determined by the Heun method.

2. Give an analysis similar to that derived for the modified Euler method for the Adams-Moulton method where the corrector equation is solved exactly at each step. Assume that exact values are available for y_1 and y_2. Also, consider the case where the predictor-corrector procedure is used and where the starting values y_1, y_2, and y_3 are obtained using the Runge-Kutta method.

3. Carry out an analysis similar to that given for the modified Euler method for method QC.1.2. Assume that y_{n+1} is obtained exactly at each step. Consider the cases where y_1 is determined by the Euler method and by the Heun method. Also, find the coefficients c_1, c_2, c_3, c_4, and c_5 in the expansion $\lambda_1 = 1 + c_1\alpha + c_2\alpha^2 + c_3\alpha^3 + c_4\alpha^4 + c_5\alpha^5 + \cdots$ about $\alpha = 0$ where λ_1 is a root of the equation

$$\left(1 - \frac{5\alpha}{12}\right)\lambda^2 + \left(-1 - \frac{8\alpha}{12}\right)\lambda + \frac{\alpha}{12} = 0.$$

4. Give a proof of the convergence of the method IC.2 for solving $y' = Ay$, $y(0) = 1$ similar to that given for the midpoint method. Assume that for each h we determine $y(h)$ using the Euler method. (We assume that at each step y_{n+1} is determined explicitly in terms of y_n and y_{n-1}.)

5. Consider the use of method IC.2 for solving $y' = -y$, $y(0) = 1$. Assume that at each step one obtains y_{n+1} exactly. If the Heun method is used to get y_1, how small a value of h must be used so that the extraneous term in the analytic solution of the difference equation does not exceed 25 percent of the representative term for $x = 2$?

6. Work out the details of the proof of (5.67) and (5.68).

7. Carry out an analysis similar to that given for the midpoint method in Theorem 5.37 for method I.C.2 where y_{n+1} is obtained exactly at each step. Consider the case where y_1 is obtained by the Heun method and also the case where y_1 is obtained by the Runge-Kutta method.

8. Carry out an analysis similar to that given for method D.1.3 for method I.0.2. Consider the cases where y_1 is determined by the Euler method, the Heun method, and the exact value of the solution. Verify that for small α one of the roots of

$$\lambda^2 + (4 - 4\alpha)\lambda + (-5 - 2\alpha) = 0$$

is given by

$$\lambda = 1 + \alpha + \frac{\alpha^2}{2} + \frac{\alpha^3}{6} + \frac{\alpha^4}{72} + \cdots.$$

9. Consider the initial-value problem $y' = -2y$, $y(0) = 1$. Determine $e^{-2} - y_{10}$ where y_{10} is the numerical solution obtained after 10 steps with $h = 0.1$ using each of the following methods:

   a) The Euler method.

   b) The modified Euler method.

c) The Heun method.
d) Method QC.1.2 with y_{n+1} being obtained exactly at each step. Assume y_1 is obtained by the Heun method.
e) The midpoint method with y_1 obtained by the Heun method.
f) The Adams-Moulton method (assume the starting values are obtained by the Runge-Kutta method).
g) Method IC.2 with y_{n+1} being obtained exactly at each step. (Assume y_1 is obtained by the Runge-Kutta method.)

10. Consider the linear 2-step method for solving $y' = f(x, y)$, $y(x_0) = y_0$ given by

$$y_{n+1} - 4y_n + 3y_{n-1} = -2hf(x_{n-1}, y_{n-1}).$$

For the case $y' = -y$, $y(0) = 1$, compute the exact solution of the difference equation for $x = 0.6$, $h = 0.1$. Assume $y(0.1)$ is determined by the Euler method. Also carry out the numerical procedure and compare the results. Do the same if y_1 is chosen to be $e^{-0.1}$.

11. Write a computer program for solving the initial value problem $y' = -y$, $y(0) = 1$ by the midpoint method and also by method IO.2 with $h = 0.1$ in the interval $0 \le x \le 4$. For the midpoint method use the Euler method for $y(0.1)$. For method IO.2 use three values of $y(0.1)$, namely: (a) the value given by the Euler method; (b) e^{-h}; (c) λ_1, the root of the characteristic equation closest to unity. Verify that $e^{-0.1} \doteq 0.904837$, $\lambda_1 \doteq 0.904834$. In each case give an output table listing $x_n(= nh)$, y_n, e^{-nh}, and $e^{-nh} - y_n$. In any given case the calculation should terminate if $|y_{n+1}|$ becomes too large. In each case, determine $y_n = c_1\lambda_1^n + c_2\lambda_2^n$, the exact solution of the difference equation considered.

12. Verify (5.77).

13. Verify (5.89) if y_1 and y_2 are given by (5.96).

14. Verify (5.101).

9.6 ACCURACY OF LINEAR MULTISTEP METHODS

In Section 8.7 we derived a bound for the accumulated error for a predictor-corrector method where the predictor and the corrector are each methods based on numerical quadrature. We now consider the case where the predictor and the corrector are linear multistep methods, not necessarily based on numerical quadrature.

It can be shown (see Henrici [1962, pp. 242–243]) that if the linear multistep method (2.1) is stable, then the coefficients $\gamma_0, \gamma_1, \gamma_2, \dots,$ in the expansion

6.1 $$J(\lambda) = (\alpha_0 + \alpha_1\lambda + \cdots + \alpha_N\lambda^N)^{-1} = \gamma_0 + \gamma_1\lambda + \gamma_2\lambda^2 + \cdots$$

are bounded. In this case we let

6.2 $$\Gamma = \sup_{i=0,1,2,\dots} |\gamma_i|.$$

If all roots of $\rho(\lambda) = 0$ are simple, the boundedness of the γ_i can be shown as follows. Let $\lambda_1, \lambda_2, \ldots, \lambda_N$ be the roots of $\rho(\lambda) = 0$ and let

6.3
$$\mu_i = \lambda_i^{-1}.$$

Then we have, by a partial fraction decomposition,

6.4
$$J(\lambda) = \frac{1}{\alpha_N(\lambda - \mu_1)(\lambda - \mu_2)\cdots(\lambda - \mu_N)} = \sum_{i=1}^{N} \frac{c_i}{\lambda - \mu_i}$$

for some constants $c_1, c_2, \ldots, c_N$. Since

6.5
$$\frac{c_i}{\lambda - \mu_i} = \frac{c_i/\mu_i}{\lambda/\mu_i - 1} = -\frac{c_i}{\mu_i}\left(1 + \left(\frac{\lambda}{\mu_i}\right) + \left(\frac{\lambda}{\mu_i}\right)^2 + \cdots\right)$$

then we have

6.6
$$\gamma_k = -\sum_{i=1}^{N} \frac{c_i}{\mu_i^{k+1}}$$

which is bounded for all k since, by the stability, we have

6.7
$$|\mu_i| \geq 1, \qquad i = 1, 2, \ldots, N.$$

We now state without proof the following theorem. We make the same assumptions as in Section 8.7. The proof is similar but somewhat more complicated than that of Theorem 8-7.30. For details of the proof, see Henrici [1962].*

6.8 Theorem. Let $y_1, y_2, \ldots, y_{N^*-1}$ be chosen so that†

6.9
$$|\bar{y}_n - y_n| \leq \delta, \qquad n = 0, 1, 2, \ldots, N^* - 1$$

and for $n = N^* - 1, N^*, \ldots$ let y_{n+1} be determined by the predictor-corrector method

6.10
$$\alpha_0^* y_{n+1}^{(P)} + \sum_{i=1}^{N^*} \alpha_i^* y_{n+1-i} = h \sum_{i=1}^{N^*} \beta_i^* f(x_{n+1-i}, y_{n+1-i})$$

6.11 $\quad \alpha_0 y_{n+1} + \sum_{i=1}^{N} \alpha_i y_{n+1-i} = h \sum_{i=1}^{N} \beta_i f(x_{n+1-i}, y_{n+1-i}) + h\beta_0 f(x_{n+1}, y_{n+1}^{(P)})$

where the corrector is stable and where $N^* \geq N$. Let the order of the predictor

---

* Henrici actually derives a bound for the accumulated error for the case where the corrector equation is solved exactly at each step. While he does not give a specific bound for the predictor-corrector method, he does indicate how one could be obtained. (See Henrici [1962, pp. 260–261].)

† Here, as in Section 9.3, we assume that $\bar{y}(x_0) = \eta$ and that y_0 is not necessarily equal to η.

and the corrector be p where $p \geqq 1$, and let the interpolation errors for the predictor and the corrector satisfy

6.12
$$|IE_n^*| \leqq |K^*| h^{p+1} M_{p+1}$$

and

6.13
$$|IE_n| \leqq |K| h^{p+1} M_{p+1}$$

respectively. Then for $n = 0, 1, 2, \ldots, (\bar{x} - x_0)/h$ we have

6.14 $\quad |e_n| \leqq \Gamma \left[\left(\bar{A} + Lh \left| \dfrac{\beta_0}{\alpha_0} \right| \bar{A}^* \right) \delta N^* + (x_n - x_0) h^p M_{p+1} \left(|K| + Lh \left| \dfrac{\beta_0}{\alpha_0} \right| |K^*| \right) \right] \times$

$$e^{L\Gamma(x_n - x_0)} \left(\bar{B} + hL \left| \frac{\beta_0}{\alpha_0} \right| \bar{B}^* \right)$$

where

6.15
$$\bar{A} = \sum_{i=0}^{N} |\alpha_i|, \qquad \bar{A}^* = \sum_{i=0}^{N^*} |\alpha_i^*|$$

6.16
$$\bar{B} = \sum_{i=0}^{N} |\beta_i|, \qquad \bar{B}^* = \sum_{i=1}^{N^*} |\beta_i^*|$$

where Γ is defined by (6.2).

We remark that if the corrector corresponds to an open method then $N^* = N$ and, of course, $\beta_0 = 0$.

Let us now apply Theorem 6.8 to various methods for solving the initial-value problem

6.17
$$\begin{cases} y' = Ay \\ y(x_0) = \eta. \end{cases}$$

For the Euler method we have $N = 1$, $\alpha_0 = 1$, $\alpha_1 = -1$, $\beta_0 = 0$, $\beta_1 = 1$. To determine Γ we have

6.18
$$\frac{1}{\alpha_0 + \alpha_1 \lambda} = \frac{1}{1 - \lambda} = 1 + \lambda + \lambda^2 + \cdots.$$

and hence

6.19
$$\Gamma = 1.$$

This result holds, incidentally, for any method based on numerical quadrature. Moreover, we have $N^* = 1$ and

6.20
$$\bar{A} = 2, \qquad \bar{B} = 1, \qquad \bar{A}^* = 0, \qquad \bar{B}^* = 0.$$

Since $|K| = \frac{1}{2}, L = |A|, p = 1$ we have, by (6.14),

6.21 $\quad |e_n| \leqq \left[2\delta + (x_n - x_0) \dfrac{h}{2} M_2 \right] e^{|A|(x_n - x_0)}$

$$\leqq \begin{cases} \left[2\delta + (x_n - x_0) \dfrac{hA^2}{2} |\eta| e^{A(x_n - x_0)} \right] e^{A(x_n - x_0)}, & \text{if } A \geqq 0 \\[3mm] \left[2\delta + (x_n - x_0) \dfrac{hA^2}{2} |\eta| \right] e^{|A|(x_n - x_0)}, & \text{if } A < 0. \end{cases}$$

Here

6.22 $$\delta = |y_0 - \eta|.$$

Evidently the result (6.21) is somewhat less sharp than that given by Lemma 8-5.20.

Next, let us consider the modified Euler method with the method QO.1.2 as the predictor, i.e., the improved Heun method. Evidently we have $N = 1, N^* = 2$, $p = 2$, and

6.23 $\begin{cases} \alpha_0 = 1, \ \alpha_1 = -1, \ \bar{A} = 2, \ \beta_0 = \frac{1}{2}, \ \beta_1 = \frac{1}{2}, \ \bar{B} = 1, \ |K| = \frac{1}{12} \\ \alpha_0^* = 1, \ \alpha_1^* = -1, \ \alpha_2^* = 0, \ \bar{A}^* = 2, \ \beta_1^* = \frac{3}{2}, \ \beta_2^* = -\frac{1}{2}, \ \bar{B}^* = 2, \ |K^*| = \frac{5}{12}. \end{cases}$

Thus, by (6.14) we have, since $\Gamma = 1, L = |A|$

6.24 $\quad |e_n| \leqq \left[4(1 + |A|h/2)\delta + (x_n - x_0) \dfrac{h^2 M_3}{12} (1 + \frac{5}{2}h|A|) \right] e^{|A|(x_n - x_0)(1 + h|A|)}$

$$\leqq \begin{cases} \left[4(1 + Ah/2)\delta + (x_n - x_0) \dfrac{h^2 A^3}{12} |\eta|(1 + \frac{5}{2}hA)e^{A(x_n - x_0)} \right] e^{A(x_n - x_0)(1 + hA)} \\[2mm] \hspace{9cm} \text{if } A \geqq 0 \\[3mm] \left[4(1 + |A|h/2)\delta + (x_n - x_0) \dfrac{h^2 |A|^3}{12} |\eta|(1 + \frac{5}{2}h|A|) \right] e^{|A|(x_n - x_0)(1 + h|A|)} \\[2mm] \hspace{9cm} \text{if } A < 0. \end{cases}$$

Here

6.25 $$\delta = \max(|y_0 - \eta|, |y_1 - \bar{y}_1|).$$

This result is less sharp than that given in Section 6.12.

For the midpoint method we have $N = N^* = 2$ and

6.26 $\qquad\qquad \bar{A} = 2, \qquad \bar{B} = 2, \qquad \bar{A}^* = 0, \qquad \bar{B}^* = 0.$

Since $|K| = \frac{1}{3}$, $L = |A|$, $\Gamma = 1$, $p = 2$ we have

6.27 $\displaystyle |e_n| \leq \left[4\delta + (x_n - x_0)\frac{h^2 M_3}{3} \right] e^{2|A|(x_n - x_0)}$

$$\leq \begin{cases} \left[4\delta + (x_n - x_0)|\eta|\dfrac{h^2 A^3}{3} e^{A(x_n - x_0)} \right] e^{2A(x_n - x_0)}, & \text{if } A \geqq 0 \\[3mm] \left[4\delta + (x_n - x_0)|\eta|\dfrac{h^2|A|^3}{3} \right] e^{2|A|(x_n - x_0)}, & \text{if } A < 0. \end{cases}$$

For the Adams-Moulton method we have $N = 3$, $N^* = 4$, $\beta_0 = \frac{9}{24}$ and

6.28 $\qquad\qquad \bar{A} = 2, \qquad \bar{B} = \frac{17}{12}, \qquad \bar{A}^* = 2, \qquad \bar{B}^* = \frac{20}{3}.$

Since $p = 4$, $L = |A|$, $\Gamma = 1$, $|K| = \frac{19}{720}$, $|K^*| = \frac{251}{720}$, we have

6.29 $\displaystyle |e_n| \leq \left[8(1 + \frac{9}{24}|A|h)\delta + (x_n - x_0)\frac{19h^4 M_5}{720}(1 + \frac{753}{152}h|A|) \right] \times$

$$e^{|A|(x_n - x_0)(17/12 + (5/2)h|A|)}$$

$$\leq \begin{cases} \left[8(1 + \frac{9}{24}Ah)\delta + (x_n - x_0)\dfrac{19h^4 A^5}{720}|\eta|(1 + \frac{753}{152}hA)\,e^{A(x_n - x_0)} \right] \times \\[3mm] \qquad\qquad e^{A(x_n - x_0)(17/12 + (5/2)hA)}, \qquad \text{if } A \geqq 0 \\[3mm] \left[8(1 + \frac{9}{24}|A|h)\delta + (x_n - x_0)\dfrac{19h^4|A|^5}{720}|\eta|(1 + \frac{753}{152}h|A|) \right] \times \\[3mm] \qquad\qquad e^{|A|(x_n - x_0)(17/12 + (5/2)h|A|)}, \qquad \text{if } A < 0. \end{cases}$$

Here

6.30 $\qquad\qquad\qquad\qquad \delta = \max_{i = 0,1,2,3} |y_i - \bar{y}_i|.$

From Theorem 6.8 it follows that if the starting values are such that

6.31 $\qquad\qquad\qquad\qquad\qquad \delta \leqq \hat{K} h^p$

for some constant $\hat{K}$ independent of h, then

6.32 $\qquad\qquad\qquad\qquad\qquad |e_n| \leq |E| h^p$

for some constant E. Actually, as shown by Henrici [1962], we have

6.33 $\qquad\qquad\qquad\qquad e_n = e(x_n)h^p + 0(h^{p+1})$

for some function $e(x_n)$ independent of h, provided that $\delta = O(h^{p+1})$.

For the method IO.2, the condition of stability is not satisfied. For the roots of

6.34 $$\rho(\lambda) = \lambda^2 + 4\lambda - 5 = 0$$

are 1 and -5. Hence we have

6.35 $$\frac{1}{1 + 4\lambda - 5\lambda^2} = \frac{1}{(1 - \lambda)(1 + 5\lambda)} = \frac{\frac{1}{6}}{1 - \lambda} + \frac{\frac{5}{6}}{1 + 5\lambda}$$

$$= \tfrac{1}{6}(1 + \lambda + \lambda^2 + \cdots) + \tfrac{5}{6}(1 - 5\lambda + (5\lambda)^2 - (5\lambda)^3 \cdots).$$

Therefore, by (6.1) we have

6.36 $$\gamma_k = \tfrac{1}{6} + \tfrac{5}{6}(-1)^k 5^k, \qquad k = 0, 1, 2, \ldots$$

and the γ_k are clearly not bounded. Thus we cannot apply Theorem 6.8.

EXERCISES 9.6

1. Prove that the γ_i in (6.1) are bounded if the associated method is stable, assuming that each root of $\rho(\lambda) = 0$ is either simple or double.

2. Show that the γ_i of (6.1) are not bounded for method D.1.3. Find $\gamma_0, \gamma_1, \gamma_2,$ and γ_3.

3. Prove that (6.19) holds for any method based on numerical quadrature.

4. Derive an analog of Theorem 6.8 for the case where the corrector equation is solved exactly. Apply this result to the initial-value problem (6.17).

5. Find a bound on the error $|\bar{y}_5 - y_5|$ for the problem $y' = -3y$, $y(0) = 1$ where $\bar{y}_n$ is the exact solution at $x_n = nh$ and y_n is the numerical solution using

$$y_{n+1} - \tfrac{3}{2}y_n + \tfrac{1}{2}y_{n-1} = \frac{h}{2} f(x_{n+1}, y_{n+1})$$

where y_1 is obtained using the Euler method and where $h = 0.1$.

6. Consider the initial-value problem $y' = -2y$, $y(0) = 1$. Using Theorem 6.8 determine $e^{-2} - y_{10}$ where y_{10} is the numerical solution obtained after 10 steps with $h = 0.1$ using each of the following methods:

   a) The Euler method.
   b) The Heun method.
   c) The modified Euler method with QO.1.2 as the predictor. Assume y_1 is obtained by the Heun method.
   d) The midpoint method. Assume y_1 is obtained by the Heun method.
   e) The Adams-Moulton method. Assume the starting values are obtained by the Heun method.
   f) The Milne-Simpson method. Assume the starting values are obtained by the Heun method.

Compare the results with those obtained using the methods of Section 9.5.

7. Work out the analog of (6.29) for the Milne-Simpson method.

9.7 WEAK, STRONG, AND CONDITIONAL STABILITY

We now describe an alternative stability theory which is compatible with but not identical to the theory developed above for linear multistep methods. This alternative theory can be applied to any method such that, when the method is applied to the differential equation

7.1
$$y' = Ay$$

one obtains a difference equation of the form

7.2
$$A_0 y_{n+1} + A_1 y_n + \cdots + A_N y_{n-(N-1)} = 0,$$

for some $N \geq 1$ where the A_i depend on h and A and where $\lim_{h \to 0} A_i$ exists for $i = 0, 1, 2, \ldots, N$. Thus if one has available $y_0, y_1, \ldots, y_{N-1}$ one can use (7.2) to determine $y_N, y_{N+1}, \ldots$. Clearly, any linear N-step method fits into this category since we have, by (2.1),

7.3
$$A_i = \alpha_i - Ah\beta_i.$$

Moreover, for the Runge-Kutta method we have by 8-(9.17) and 8-(9.18),

7.4
$$
\begin{cases}
\Delta_1 = \alpha y_n \\[2mm]
\Delta_2 = \left(\alpha + \dfrac{\alpha^2}{2} \right) y_n \\[2mm]
\Delta_3 = \left(\alpha + \dfrac{\alpha^2}{2} + \dfrac{\alpha^3}{4} \right) y_n \\[2mm]
\Delta_4 = \left(\alpha + \alpha^2 + \dfrac{\alpha^3}{2} + \dfrac{\alpha^4}{4} \right) y_n
\end{cases}
$$

where $\alpha = Ah$, and

7.5
$$y_{n+1} = y_n + \tfrac{1}{6}(\Delta_1 + 2\Delta_2 + 2\Delta_3 + \Delta_4)$$
$$= \left(1 + \alpha + \frac{\alpha^2}{2} + \frac{\alpha^3}{6} + \frac{\alpha^4}{24} \right) y_n;$$

hence,

7.6
$$A_0 = 1, \qquad A_1 = -\left(1 + Ah + \frac{A^2 h^2}{2} + \frac{A^3 h^3}{3!} + \frac{A^4 h^4}{4!} \right).$$

Similarly, the method of Taylor's series with 5 terms also gives the same result for the special case (7.1).

Let us define the coefficients $\bar{A}_0, \bar{A}_1, \ldots, \bar{A}_N$ as

7.7
$$\bar{A}_i = \lim_{h \to 0} A_i, \qquad i = 0, 1, 2, \ldots, N.$$

Thus in the case of a linear N-step method we have, by (7.3),

7.8
$$\bar{A}_i = \alpha_i, \qquad i = 0, 1, 2, \ldots, N$$

and in the case of the Runge-Kutta method

7.9
$$\bar{A}_0 = 1, \qquad \bar{A}_1 = -1.$$

Evidently, if $A = 0$, then the exact solution of (7.1) with the initial condition $y(x_0) = y_0$ is $\bar{y}(x) \equiv y_0$. If the solution of (7.2) is to converge uniformly to y_0, then we must have, as in the proof of Theorem 3.56.

7.10
$$\bar{A}_0 + \bar{A}_1 + \cdots + \bar{A}_N = 0.$$

(This condition is, of course, satisfied in the case of (7.2) if the linear N-step method is consistent and in the case of the Runge-Kutta method.) Consequently, $\lambda = 1$ is a root of the *limiting characteristic equation*

7.11
$$\rho(\lambda) = \bar{A}_0 \lambda^N + \bar{A}_1 \lambda^{N-1} + \cdots + \bar{A}_N = 0.$$

Let the roots of (7.11) be $\bar{\lambda}_1 = 1, \bar{\lambda}_2, \ldots, \bar{\lambda}_N$ and let $\lambda_1, \lambda_2, \ldots, \lambda_N$ be the roots of the *characteristic equation*

7.12
$$g(\lambda) = A_0 \lambda^N + A_1 \lambda^{N-1} + \cdots + A_N = 0.$$

By the continuity of the roots of a polynomial equation as a function of its coefficients* it follows that as $h \to 0$, the roots of (7.12) converge to those of (7.11). Thus, in particular, one root of (7.12), say λ_1, converges to the root $\bar{\lambda}_1 = 1$ of (7.11). The root λ_1 is said to be the *representative* root; all other roots are said to be *extraneous* roots.

For a stable and consistent linear N-step method, we show that for h sufficiently small we have

7.13
$$\lambda_1 = 1 + Ah + 0(h^2).$$

But since $g(\lambda) = \rho(\lambda) - Ah\sigma(\lambda)$, it follows that

---

* See, for instance, Ostrowski [1966, p. 220].

7.14 $g(1 + Ah) = \rho(1 + Ah) - Ah\sigma(1 + Ah)$

$$= \rho(1) + Ah\rho'(1) + \frac{A^2h^2}{2}\rho''(\xi_1) - Ah\sigma(1) - A^2h^2\sigma'(\xi_2)$$

where ξ_1 and ξ_2 lie between 1 and $1 + Ah$. But by the conditions of consistency we have $\rho(1) = 0$, $\rho'(1) = \sigma(1)$ and hence

7.15 $g(1 + Ah) = 0(h^2)$.

Similarly,

7.16 $g'(\lambda) = \rho'(\lambda) - Ah\sigma'(\lambda)$

so that

7.17 $g'(1 + Ah) = \rho'(1 + Ah) - Ah\sigma'(1 + Ah)$

$$= \rho'(1) + Ah\rho''(\xi_3) - Ah\sigma'(1) - A^2h^2\sigma''(\xi_4)$$

$$= \rho'(1) + 0(h)$$

where ξ_3 and ξ_4 lie between 1 and $1 + Ah$. Consequently, since $\rho'(1) \neq 0$ (otherwise 1 would be a double root of $\rho(\lambda) = 0$ and the method would not be stable), we have

7.18 $\dfrac{g(1 + Ah)}{g'(1 + Ah)} = \dfrac{0(h^2)}{\rho'(1) + 0(h)} = 0(h^2)$

provided Ah is small enough. But by 5-(5.34) and 5-(5.37), we have

7.19 $|\lambda_1 - (1 + Ah)| \leqq N\left|\dfrac{g(1 + Ah)}{g'(1 + Ah)}\right| = 0(h^2)$

and hence (7.13) follows.

We now introduce the following definitions:

1. the method is *strongly stable* if $|\bar{\lambda}_i| < 1$ for $i \neq 1$;

2. the method is *strongly unstable* if $|\bar{\lambda}_i| > 1$ for some i or if there is a double root of (7.11) of modulus unity;

3. the method is *conditionally stable* if it is not strongly unstable and if for some $i \neq 1$ we have $|\bar{\lambda}_i| = 1$;

   3a. the method is *weakly stable* if for h sufficiently small

7.20 $|\lambda_i| \leqq |\lambda_1|$ for $i \neq 1$,

3b. the method is *weakly unstable* if for any $h_0 > 0$ there exists $h < h_0$ such that for some $i \neq 1$ we have

7.21
$$|\lambda_i| > |\lambda_1|.$$

It should be noted that the above classification is relative to the differential equation (7.1) and that the particular classification may depend on A or at least on the sign of A.

For linear multistep methods, if the condition of stability is not satisfied the method is strongly unstable. If the condition of stability is satisfied, then the method is either strongly stable or conditionally stable.

The various definitions of stability are motivated as follows: if the roots $\lambda_1, \lambda_2, \ldots, \lambda_N$ of the characteristic equation (7.12) are distinct, then we can write the solution of (7.2) in the form

7.22
$$y_n = c_1 \lambda_1^n + c_2 \lambda_2^n + \cdots + c_N \lambda_N^n$$

where $c_1, c_2, \ldots, c_N$ are determined by $y_0, y_1, \ldots, y_{N-1}$. If the method is weakly stable or strongly stable, then for h sufficiently small, the first term, which involves the representative root, λ_1, is not dominated by any other term. On the other hand, if the method is weakly unstable or strongly unstable, the extraneous terms may dominate. We have already seen that if a linear N-step method is weakly unstable, this is not fatal in the sense that for a given range in x we can choose h small enough so as to obtain convergence in the desired range. For strongly unstable methods, however, choosing a smaller value of h only makes things worse.

Evidently, according to the above definitions, method D.1.3 is strongly unstable while the Euler method and the modified Euler method are strongly stable. The midpoint method is conditionally stable, since the roots of the limiting characteristic equation are ± 1. When applied to the equation $y' = Ay$ with $A > 0$, it is weakly stable since the roots of

7.23
$$\lambda^2 - 2Ah\lambda - 1 = 0$$

are, by (5.42)

7.24
$$\lambda_1 = Ah + \sqrt{1 + A^2h^2} = 1 + Ah + 0(h^2), \qquad \lambda_2 = Ah - \sqrt{1 + A^2h^2};$$

hence λ_1 is the representative root. Thus if $A > 0$, we have $|\lambda_2| < |\lambda_1|$. On the other hand, if $A < 0$, then $|\lambda_2| > |\lambda_1|$ and the method is weakly unstable.

Let us now consider the Runge-Kutta method. Evidently, by (7.6) we have $\bar{A}_0 = 1$, $\bar{A}_1 = -1$, and the limiting characteristic equation is

7.25
$$\bar{\lambda} - 1 = 0.$$

Thus the limiting characteristic equation has the single root $\bar{\lambda} = 1$ and the Runge-Kutta method is strongly stable.

Evidently, a linear N-step method based on numerical quadrature is strongly stable if it is based on an integration from x_n to x_{n+1} since in this case the limiting characteristic equation has $N - 1$ roots at $\lambda = 0$ in addition to the root $\lambda = 1$. If the method is based on integration from x_{n-k} to x_n, where $k = 1, 2, \ldots, N - 1$, then the method is conditionally stable.

EXERCISES 9.7

1. Give the details of the proof of (7.10).

2. Analyze the stability of the Heun-midpoint method, the Milne-Simpson method, and the Adams-Moulton method using the alternative stability theory.

3. Discuss the stability of the method based on the use of

$$y_{n+1} - 4y_n + 3y_{n-1} = -2hf(x_{n-1}, y_{n-1})$$

   as a predictor and method IC.2 as a corrector for the differential equation $y' = -y$.

4. Perform the alternative stability analysis for the method based on IO.2 as a predictor followed by QC.1.2 as a corrector. Apply the method to the problem $y' = x + y^2$, $y(0) = 1$, with $h = 0.1$, and find $y(0.2)$. Use the corrected Heun method to get $y(0.1)$.

5. Find the linear 2-step method such that $\alpha_0 = 1$, $\alpha_1 = -\frac{9}{8}$, $\alpha_2 = \frac{1}{8}$ which has the highest possible order. Find the order and the interpolation error. Discuss the stability. For the equation $y' = -y$, how small must h be so that all roots of the characteristic equation are less than unity in absolute value?

6. Find the single-step error for the method based on

$$y_{n+1}^{(p)} = y_{n-3} + \frac{4h}{3}[2f(x_n, y_n) - f(x_{n-1}, y_{n-1}) + 2f(x_{n-2}, y_{n-2})]$$

   used as a predictor followed by one application of the method of the preceding exercise as the corrector. Analyze the stability of the procedure.

7. Classify the method

$$y_{n+1} - \tfrac{3}{2}y_n + \tfrac{1}{2}y_{n-1} = \tfrac{1}{2}hf(x_{n+1}, y_{n+1})$$

   according to the alternative stability theory for the equation $y' = -3y$. Find the roots of the characteristic equation for $h = 0.1$ and the roots of the limiting characteristic equation. Do the same for method IC.2.

8. Analyze the stability of the following method for solving $y' = Ay$. Given y_n, y_{n-1}, and y_{n-2}, let $y_{n+1} = (\hat{y}_{n+1} + y_{n+1}^*)/2$ where

$$\hat{y}_{n+1} = y_{n-1} + \frac{Ah}{3}[\hat{y}_{n+1} + 4y_n + y_{n-1}]$$

$$y_{n+1}^* = y_{n-2} + \frac{3Ah}{8}[y_{n+1}^* + 3y_n + 3y_{n-1} + y_{n-2}].$$

9. In the Runge-Kutta method for solving $y' = -2y$, how small must h be chosen so that all roots of the characteristic equation are less than unity in modulus?

10. Given a weakly stable linear N-step method we can measure the amount of stability for the equation $y' = Ay$ with $A < 0$ in terms of max Q_k where $Q_k = -\operatorname{Re} P_k$ and where the "growth parameters," P_k, are given by

$$P_k = \frac{\sigma(\bar\lambda_k)}{\bar\lambda_k \rho'(\bar\lambda_k)}.$$

Here $\bar\lambda_1 = 1$ and $\bar\lambda_2, \bar\lambda_3, \ldots, \bar\lambda_n$ are the zeros of $\rho(\lambda)$. The growth parameters are defined only for those k such that $|\bar\lambda_k| = 1$. Show that for the equation $y' = Ay$ the root λ_k of the characteristic equation associated with the zero $\bar\lambda_k$ of $\rho(\lambda)$ is given by

$$\lambda_k \sim \bar\lambda_k(1 + hAP_k)$$

and

$$|\lambda_k^{x/h}| \sim |\bar\lambda_k^{x/h}| \, |1 + hAP_k|^{x/h} \sim |e^{AP_k}|^x = e^{A(\operatorname{Re} P_k)x}.$$

Consider now the stable linear 4-step methods of order 6 given by methods A, B, and C where

Method	α_0	α_1	α_2	α_3	α_4	β_0	β_1	β_2	β_3	β_4
A	1	1	0	-1	-1	$\frac{3}{10}$	$\frac{18}{10}$	$\frac{18}{10}$	$\frac{18}{10}$	$\frac{3}{10}$
B	1	-1	0	1	-1	$\frac{29}{90}$	$\frac{94}{90}$	$-\frac{66}{90}$	$\frac{94}{90}$	$\frac{29}{90}$
C	1	0	0	0	-1	$\frac{28}{90}$	$\frac{128}{90}$	$\frac{48}{90}$	$\frac{128}{90}$	$\frac{28}{90}$

Determine the measure of stability in each case.

    Compute the roots of the characteristic equation in each case for the equation $y' = -y$ and $h = 0.1$.

9.8 A NUMERICAL EXAMPLE

We shall now illustrate some of the preceding results by solving the initial-value problem

8.1 $$y' = -y, \qquad y(0) = 1$$

using several different methods and several different choices of starting values. The following methods were used:

Strongly Stable	Weakly Unstable	Strongly Unstable
Adams-Moulton	Midpoint	Method D.2.2
	Milne-Simpson	Method D.1.3

The following procedures were used to determine the necessary starting values:

              Euler
              Corrected Heun
              Runge-Kutta
              Exact Values

The step sizes of $h = 0.4, 0.2, 0.1, 0.05$ were used. The corrector equation was solved exactly in each case following the procedure illustrated by 8-(6.10) for the modified Euler method. Thus we have

Adams-Moulton Method

8.2 $$y_{n+1} = \left[\left(1 - \frac{19h}{24}\right)y_n + \frac{5h}{24}y_{n-1} - \frac{h}{24}y_{n-2} \right] \bigg/ \left(1 + \frac{9h}{24}\right), \quad n = 2, 3, \ldots .$$

Milne-Simpson Method

8.3 $$y_{n+1} = [(3 - h)y_{n-1} - 4hy_n]/(3 + h), \quad n = 1, 2, \ldots .$$

Midpoint Method

8.4 $$y_{n+1} = y_{n-1} - 2hy_n, \quad n = 1, 2, \ldots .$$

Method D.2.2

8.5 $$y_{n+1} = (2h - 3)y_{n-1} + 4y_n, \quad n = 1, 2, \ldots .$$

Method D.1.3

8.6 $$y_{n+1} = (-3h - \tfrac{3}{2})y_n + 3y_{n-1} - \tfrac{1}{2}y_{n-2}, \quad n = 2, 3, \ldots .$$

Table 8.7 shows the behavior of the errors of the various methods as a function of the step size h. Errors are given both for $x = 0.8$ and $x = 1.6$. The Runge-Kutta starting values were used in all cases except for the midpoint method where the corrected Heun method was used. It can be seen that for the Adams-Moulton and Milne-Simpson methods the errors behave roughly as h^4. If this were exactly the case, then if h is divided by two, the error is divided by 16. Actually, the ratios in many cases are somewhat greater than 16, but presumably would approach 16 as $h \to 0$. We note that for the values of x and h considered, the Milne-Simpson method appears more accurate than the Adams-Moulton method by a factor of 3 or more. However, as we shall see, with the Milne-Simpson method one must be concerned with the problem of weak instability.

For the midpoint method, the error appears to approach zero somewhat faster than the theoretical rate of h^2, since as we reduce h by a factor of 2 the errors are, with one exception, reduced by a factor greater than 4. The exception is probably due to the fact that the overall error is the sum of the error caused by the inaccurate starting values together with the basic error in the midpoint method. Thus if the exact starting values were used we would have

	$h = 0.4$	$h = 0.2$	$h = 0.1$	$h = 0.05$
$x = 0.8$	-0.0144	-0.00329	-0.000729	-0.000167
$x = 1.6$	-0.0224	-0.00460	-0.000891	-0.000181

and the behavior is much more regular.

For Methods D.2.2 and D.1.3 the errors actually increase as h decreases. This is a manifestation of the strong instability, which we shall describe in more detail below.

8.7 Table   Accuracy As a Function of Step Size.

$x = 0.8$	$h = 0.4$	$h = 0.2$	$h = 0.1$	$h = 0.05$
Adams-Moulton	-1.07×10^{-4}	6.42×10^{-6}	9.04×10^{-7}	6.06×10^{-8}
Milne-Simpson	1.05×10^{-4}	5.01×10^{-6}	2.58×10^{-7}	1.43×10^{-8}
Midpoint	-0.0195	-0.00392	-0.000806	-0.000177
Method D.2.2	-0.0323	-0.0800	-1.086	$*$
Method D.1.3	-1.07×10^{-4}	-3.87×10^{-4}	-2.36×10^{-3}	-0.468

$x = 1.6$				
Adams-Moulton	2.73×10^{-4}	2.15×10^{-5}	1.14×10^{-6}	6.27×10^{-8}
Milne-Simpson	1.42×10^{-4}	6.06×10^{-6}	2.82×10^{-7}	1.44×10^{-8}
Midpoint	-0.0357	-0.00628	-0.001097	-0.000207
Method D.2.2	-0.545	-8.680	$*$	$*$
Method D.1.3	-7.36×10^{-3}	-5.13×10^{-2}	-11.807	$*$

$*$ Indicates error greater than 100.

Table 8.8 shows the effect of the choice of starting values on the accuracy of the Adams-Moulton method. Using the Euler starting values, the errors behave like h^2, the behavior of the single-step error of the Euler method. Thus as we reduce h by a factor of two, the error is reduced by a factor of four. With the corrected Heun starting values, the errors behave like h^3, as does the single-step error of the corrected Heun method. With the Runge-Kutta and the exact starting values, the behavior is at least as good as h^4, the same as that of the overall error of the Adams-Moulton method.

Figures 8.9–8.12 show the effect of the choice of starting values on the stability characteristics of the various methods considered. The more accurate starting values delay the onset of the various types of instability, but the qualitative behavior is the same in each case.

Figures 8.13–8.15 show the behavior of the solutions of the Milne-Simpson and midpoint methods and Methods D.2.2 and D.1.3 as functions of x for various values of h. The starting values were obtained by the Euler method. The graph of the solution of the Adams-Moulton method is not given since it cannot be distinguished from the exact solution.

8.8 Table   Adams-Moulton Method: Effect of Choice of Starting Values.

x	$h = 0.4$	$h = 0.2$	$h = 0.1$	$h = 0.05$
		Euler Starting Values		
0	0	0	0	0
1.0	—	0.0245	0.0058	0.0014
2.0	0.0382	0.0091	0.0021	0.0005
3.0	—	0.0033	0.0008	—
4.0	0.0055	0.0012	0.0003	—
		Corrected Heun Starting Values		
0	0	0	0	0
1.0	—	0.00099	0.000107	0.000012
2.0	0.00417	0.00038	0.000040	0.000005
3.0	—	0.00015	0.000015	—
4.0	0.00073	0.00006	0.000006	—
		Runge-Kutta Starting Values		
0	0	0	0	0
1.0	—	0.000014	0.0000011	0.00000007
2.0	0.00004₁	0.000021	0.0000010	0.00000005
3.0	—	0.000013	0.0000006	—
4.0	0.00020	0.000007	0.0000003	—
		Exact Starting Values		
0	0	0	0	0
1.0	—	0.000017	0.0000012	0.00000007
2.0	0.00046	0.000022	0.0000010	0.00000006
3.0	—	0.000014	0.0000006	—
4.0	0.00021	0.000007	0.0000003	—

For the Milne-Simpson method and the midpoint method the solutions eventually "blew up," though the phenomenon of weak instability is less severe and occurs later for the Milne-Simpson method. For a given value of $\bar{x}$, if h is chosen small enough, no serious stability problem will occur with either method for $x \leqq \bar{x}$.

For methods D.2.2 and D.1.3, on the other hand, the instability occurs very early and is extremely severe. The smaller h is chosen, the greater the errors become. Thus the numerical solutions are totally useless.

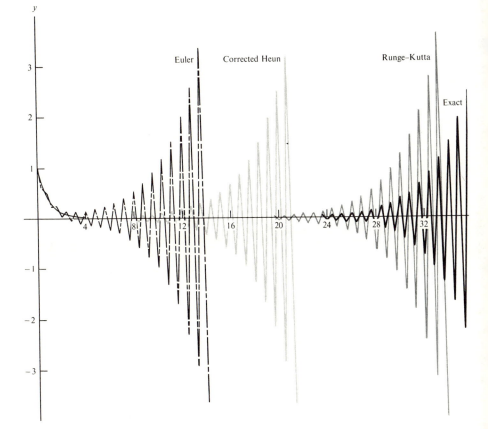

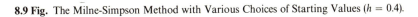

8.9 Fig. The Milne-Simpson Method with Various Choices of Starting Values ($h = 0.4$).

EXERCISES 9.8

1. Solve the problem given by (8.1) using the following methods:

Method	Starting Values
Adams-Moulton	Corrected Heun; Runge-Kutta
midpoint	Corrected Heun
method IC.2 (Milne-Simpson)	Corrected Heun; Runge-Kutta
method IO.2	Euler; Runge-Kutta; exact

Let $h = 0.4, 0.2, 0.1, 0.05$.
Integrate from $x = 0$ to $x = 3.2$.
(In the case of a closed method obtain y_{n+1} exactly at each step.) Compute the exact values of the solution and the error.

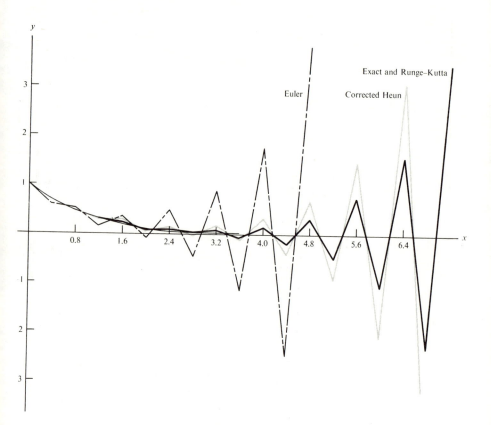

8.10 Fig. The Midpoint Method with Various Choices of Starting Values ($h = 0.4$).

9.9 SYSTEMS: MATHEMATICAL AND NUMERICAL INSTABILITY

In this section we consider methods for solving systems of first-order ordinary differential equations. For the sake of simplicity we shall consider a system of two equations with two independent variables, though most of the discussion would apply in the general case. Let us consider the system

9.1
$$\begin{cases} y' = f(x, y, z) \\ z' = g(x, y, z) \end{cases}$$

with the initial conditions

9.2
$$y(x_0) = y_0, \qquad z(x_0) = z_0.$$

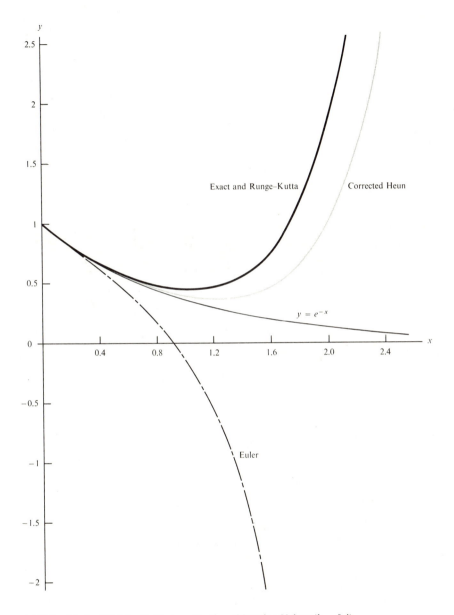

8.11 Fig. Method D.2.2 with Various Choices of Starting Values ($h = 0.4$).

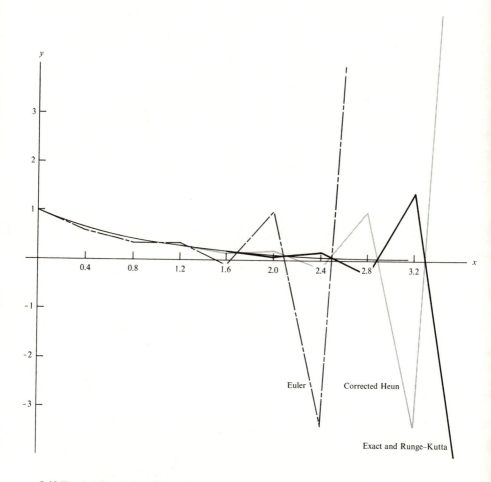

8.12 Fig. Method D.1.3 with Various Choices of Starting Values ($h = 0.4$).

Any of the methods considered so far could be used to solve (9.1)–(9.2). In particular, with a linear N-step method one would obtain

9.3
$$\begin{cases} \displaystyle\sum_{i=0}^{N} \alpha_i y_{n+1-i} = h \sum_{i=0}^{N} \beta_i f(x_{n+1-i}, y_{n+1-i}, z_{n+1-i}) \\ \displaystyle\sum_{i=0}^{N} \alpha_i z_{n+1-i} = h \sum_{i=0}^{N} \beta_i g(x_{n+1-i}, y_{n+1-i}, z_{n+1-i}). \end{cases}$$

Given $y_0, z_0, y_1, z_1, \ldots, y_{N-1}, z_{N-1}$ one could solve the above pair of equations

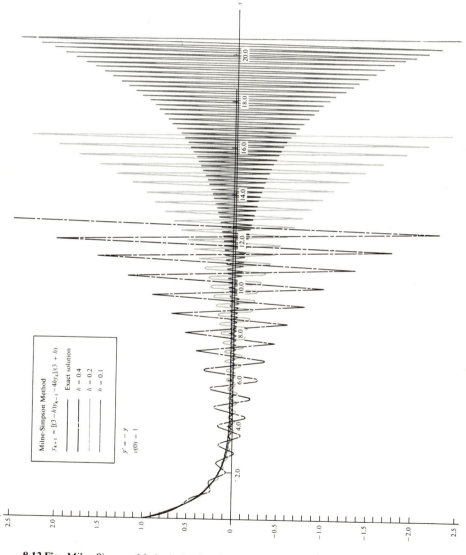

8.13 Fig. Milne-Simpson Method with Starting Values Determined by the Euler Method
$(h = 0.4, 0.2, 0.1)$.

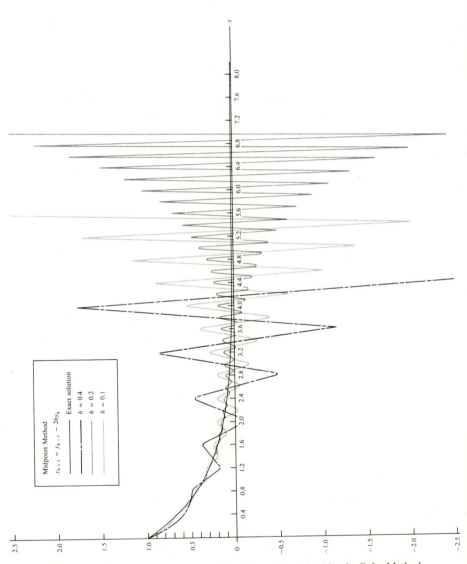

8.14 Fig. Midpoint Method with Starting Values Determined by the Euler Method
$(h = 0.4, 0.2, 0.1)$.

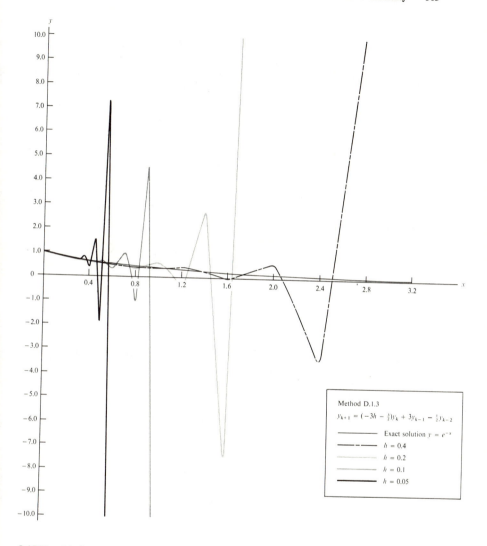

8.15 Fig. Method D.1.3 with Starting Values Determined by the Euler Method
($h = 0.4, 0.2, 0.1, 0.05$).

with $n = N - 1$ to obtain y_N and z_N and then could proceed to determine $y_{N+1}, z_{N+1}, \ldots$. As in the case of a single equation, one can define the concepts of stability and consistency in terms of the polynomials $\rho(\lambda) = \sum_{i=0}^{N} \alpha_i \lambda^{N-i}$ and $\sigma(\lambda) = \sum_{i=0}^{N} \beta_i \lambda^{N-i}$. One can also define convergence, and it can be shown (see Henrici [1962]), that a linear N-step method is convergent if and only if it is both stable and consistent.

We now describe a difficulty which may occur, namely, *mathematical instability* (also called *physical instability*, Henrici [1957a]). In the case of numerical instability, a small perturbation in the numerical solution at any stage results in a large change in the numerical solution at subsequent stages. However, with a more stable method such an effect might not occur. Thus, if we solve the single equation $y' = Ay$, by method D.1.3 and if we chose y_1 and y_2 appropriately, we would obtain a reasonable solution, provided all calculations were done exactly. However, a slight roundoff error at any stage could result in the introduction of an extraneous solution, and in this case the errors would grow rapidly. However, if we replaced this method by a strongly stable method such as the modified Euler method, the introduction of a small error such as a roundoff error at any stage would not result in large subsequent errors.* However, with a system of equations it may happen that if a small error is introduced at some stage, large subsequent errors may occur at later stages no matter what method is used.

An understanding of this phenomenon depends on a knowledge of the behavior of solutions of systems of ordinary differential equations. We now review the derivation of the analytic solution of the system

9.4
$$\begin{cases} y' = a_{1,1}y + a_{1,2}z \\ z' = a_{2,1}y + a_{2,2}z \end{cases}$$

subject to the initial conditions (9.2) where $a_{1,1}$, $a_{1,2}$, $a_{2,1}$, and $a_{2,2}$ are constants. We first seek a solution of the form

9.5
$$y = ce^{mx}, \qquad z = de^{mx}.$$

Substituting in (9.4) we obtain

9.6
$$\begin{cases} (a_{1,1} - m)c + a_{1,2}d = 0 \\ a_{2,1}c + (a_{2,2} - m)d = 0. \end{cases}$$

We can obtain a nontrivial solution (c, d) to the above system provided m satisfies

9.7 $\det \begin{pmatrix} a_{1,1} - m & a_{1,2} \\ a_{2,1} & a_{2,2} - m \end{pmatrix} = (a_{1,1} - m)(a_{2,2} - m) - a_{2,1}a_{1,2} = 0.$

This leads to the quadratic equation

9.8 $m^2 - (a_{1,1} + a_{2,2})m + (a_{1,1}a_{2,2} - a_{1,2}a_{2,1}) = 0$

---

* At least the errors would not be large relative to the exact solution. Thus for the case $y' = Ay$ and $A > 0$ the errors will indeed grow but will not be large relative to the exact solution.

whose solutions are

9.9
$$m_1, m_2 = \frac{a_{1,1} + a_{2,2} \pm \sqrt{(a_{1,1} - a_{2,2})^2 + 4a_{1,2}a_{2,1}}}{2}.$$

Let us now assume that $a_{1,2}$ and $a_{2,1}$ have the same sign and that neither vanishes. This implies that $m_1 \neq m_2$. Evidently, the general solution of (9.4) can be written in the form

9.10
$$\begin{cases} y = c_1 e^{m_1 x} + c_2 e^{m_2 x} \\ z = d_1 e^{m_1 x} + d_2 e^{m_2 x} \end{cases}$$

where for $i = 1, 2$ we have

9.11
$$\begin{cases} (a_{1,1} - m_i)c_i + a_{1,2}d_i = 0 \\ a_{2,1}c_i + (a_{2,2} - m_i)d_i = 0. \end{cases}$$

We now seek to determine c_1, c_2, d_1, d_2 so that the initial conditions (9.2) are satisfied. From the first equation of (9.11) we have, since $a_{1,2} \neq 0$,

9.12
$$d_1 = -\frac{a_{1,1} - m_1}{a_{1,2}}c_1, \qquad d_2 = -\frac{a_{1,1} - m_2}{a_{1,2}}c_2.$$

Substituting in (9.10) and using (9.2) we have

9.13
$$\begin{cases} y_0 = \hat{c}_1 + \hat{c}_2 \\ z_0 = -\hat{c}_1\left(\frac{a_{1,1} - m_1}{a_{1,2}}\right) - \hat{c}_2\left(\frac{a_{1,1} - m_2}{a_{1,2}}\right) \end{cases}$$

where

9.14
$$\hat{c}_1 = c_1 e^{m_1 x_0}, \qquad \hat{c}_2 = c_2 e^{m_2 x_0}.$$

Solving for $\hat{c}_1$ and $\hat{c}_2$ we have

9.15
$$\begin{cases} \hat{c}_1 = (m_2 - m_1)^{-1}[-a_{1,2}z_0 - (a_{1,1} - m_2)y_0] \\ \hat{c}_2 = (m_2 - m_1)^{-1}[a_{1,2}z_0 + (a_{1,1} - m_1)y_0] \end{cases}$$

so that we have, finally, using (9.10), (9.12), (9.14), and (9.8),

9.16
$$\begin{cases} y(x) = (m_2 - m_1)^{-1}\{[-a_{1,2}z_0 - (a_{1,1} - m_2)y_0]e^{m_1(x - x_0)} \\ \qquad + [a_{1,2}z_0 + (a_{1,1} - m_1)y_0]e^{m_2(x - x_0)}\} \\ z(x) = (m_2 - m_1)^{-1}\{[-a_{2,1}y_0 + (a_{1,1} - m_1)z_0]e^{m_1(x - x_0)} \\ \qquad + [a_{2,1}y_0 - (a_{1,1} - m_2)z_0]e^{m_2(x - x_0)}\}. \end{cases}$$

As an example, let us consider the system

9.17
$$\begin{cases} y' = 3y + 5z \\ z' = 5y + 3z \end{cases}$$

with $y(0) = y_0$, $z(0) = z_0$. Evidently, $m_1 = 8$, $m_2 = -2$, and we have

9.18
$$\begin{cases} y(x) = \dfrac{y_0 + z_0}{2} e^{8x} + \dfrac{y_0 - z_0}{2} e^{-2x} \\ z(x) = \dfrac{y_0 + z_0}{2} e^{8x} + \dfrac{-y_0 + z_0}{2} e^{-2x}. \end{cases}$$

If $y_0 = 1$, $z_0 = -1$, we obtain

9.19
$$y(x) = e^{-2x}, \qquad z(x) = -e^{-2x}$$

while if $y_0 = 1 + 2\varepsilon$, $z_0 = -1$ we have

9.20 $y(x) = (1 + \varepsilon)e^{-2x} + \varepsilon e^{8x}, \qquad z(x) = -(1 + \varepsilon)e^{-2x} + \varepsilon e^{8x}.$

Clearly, a small perturbation ε in the initial value of y can result in a very large change in later values of y and z. If we were to try to solve (9.17) by any numerical procedure and with the initial values $y_0 = 1$, $z_0 = -1$, we would expect that there would be at least a slight deviation from the solution (9.19) due to the inherent inaccuracy of the method and/or rounding error. When that happens, the numerical solution will attempt to imitate (9.20) and hence will differ greatly from (9.19).

The form of instability illustrated above will be referred to as *mathematical instability* and cannot be blamed on the numerical method being used. This is in contrast to numerical instability which does depend on the method being used.

Following the discussion of Section 9.7, for a single equation, we now describe an alternative stability theory for methods applied to systems. The theory can be applied to any method such that when the method is used to solve the system (9.4) one obtains a system of difference equations of the form

9.21
$$\begin{cases} \displaystyle\sum_{i=0}^{N} A_i y_{n+1-i} + \sum_{i=0}^{N} B_i z_{n+i-1} = 0 \\ \displaystyle\sum_{i=0}^{N} C_i y_{n+1-i} + \sum_{i=0}^{N} D_i z_{n+i-1} = 0 \end{cases}$$

where the coefficients A_i, B_i, C_i, and D_i depend on h as well as the $a_{i,j}$ of (9.4). We assume that the limit of each coefficient exists as $h \to 0$ and, moreover, that

9.22 $\displaystyle\lim_{h \to 0} B_i = \lim_{h \to 0} C_i = 0, \qquad i = 0, 1, 2, \ldots, N.$

Evidently, linear N-step methods and the Runge-Kutta method are of this type. Let us now seek a solution of (9.21) in the case $N = 2$. If we let

9.23
$$y_n = c\lambda^n, \qquad z_n = d\lambda^n$$

and substitute in (9.21) we obtain the conditions

9.24
$$\begin{cases} (A_0\lambda^2 + A_1\lambda + A_2)c + (B_0\lambda^2 + B_1\lambda + B_2)d = 0 \\ (C_0\lambda^2 + C_1\lambda + C_2)c + (D_0\lambda^2 + D_1\lambda + D_2)d = 0. \end{cases}$$

There exists a nontrivial solution (c, d) provided that λ satisfies the *characteristic equation*

9.25
$$g(\lambda) = \det\begin{pmatrix} A_0\lambda^2 + A_1\lambda + A_2 & B_0\lambda^2 + B_1\lambda + B_2 \\ C_0\lambda^2 + C_1\lambda + C_2 & D_0\lambda^2 + D_1\lambda + D_2 \end{pmatrix} = 0.$$

Evidently, the *limiting characteristic equation* formed from (9.25) by replacing each coefficient by its limit as $h \to 0$ has the form

9.26
$$\rho(\lambda) = \det\begin{pmatrix} \bar{A}_0\lambda^2 + \bar{A}_1\lambda + \bar{A}_2 & 0 \\ 0 & \bar{D}_0\lambda^2 + \bar{D}_1\lambda + \bar{D}_2 \end{pmatrix} = 0$$

where $\bar{A}_i = \lim_{h\to 0} A_i$, etc. As in the case of a single equation we can show that if the method is convergent, we must have $\bar{A}_0 + \bar{A}_1 + \bar{A}_2 = \bar{D}_0 + \bar{D}_1 + \bar{D}_2 = 0$; hence $\lambda = 1$ is a double root of the limiting characteristic equation. One can also show, as we shall do later for the case of the midpoint method, that two of the roots of (9.25) have the form

9.27
$$\lambda_1 = 1 + m_1 h + 0(h^2), \qquad \lambda_2 = 1 + m_2 h + 0(h^2).$$

These roots are *representative roots*. The other roots are *extraneous roots*. We now introduce the following definitions:

1. The method is *strongly stable* if $|\bar{\lambda}_i| < 1$ for $i \neq 1, 2$ where the $\bar{\lambda}_i$ are roots of (9.26).

2. The method is *strongly unstable* if $|\bar{\lambda}_i| > 1$ for some i or if there is a root of the limiting characteristic equation of modulus unity with multiplicity greater than two.

3. The method is *conditionally stable* if it is not strongly unstable and if for some $i \neq 1, 2$ we have $|\bar{\lambda}_i| = 1$.

3a. The method is *weakly stable*, if for h sufficiently small,

9.28
$$|\lambda_i| \leq \max(|\lambda_1|, |\lambda_2|) \text{ for } i \neq 1, 2,$$

where the λ_i are roots of (9.25).

3b. The method is *weakly unstable* if for any $h_0 > 0$ there exists $h < h_0$ such that for some $i \neq 1, 2$ we have

9.29
$$|\lambda_i| > \max(|\lambda_1|, |\lambda_2|).$$

As in the case of a single equation, the classification of a method is relative to the differential equation (9.4) and may also depend on the roots of (9.25).

Since the limiting characteristic equation is the product of two polynomials of degree N the above definitions of strongly stable, strongly unstable, and conditionally stable methods are natural extensions of the definitions of Section 9.7. The condition of weak stability implies that when we write the solution of (9.21) in the form

9.30
$$y_n = \sum_{i=1}^{2N} c_i \lambda_i^n, \qquad z_n = \sum_{i=1}^{2N} d_i \lambda_i^n,$$

then the representative terms, involving λ_1 and λ_2, will "dominate" for h small enough. On the other hand, with weak instability, other terms may dominate, at least for large n.

For example, for the midpoint method we have

9.31
$$\begin{cases} y_{n+1} - y_{n-1} = 2h(a_{1,1}y_n + a_{1,2}z_n) \\ z_{n+1} - z_{n-1} = 2h(a_{2,1}y_n + a_{2,2}z_n) \end{cases}$$

and

9.32
$$\begin{cases} A_0 = 1, & A_1 = -2ha_{1,1}, & A_2 = -1, & B_0 = 0, & B_1 = -2ha_{1,2}, & B_2 = 0 \\ C_0 = 0, & C_1 = -2ha_{2,1}, & C_2 = 0, & D_0 = 1, & D_1 = -2ha_{2,2}, & D_2 = -1. \end{cases}$$

The limiting characteristic equation is

9.33
$$\rho(\lambda) = \det \begin{pmatrix} \lambda^2 - 1 & 0 \\ 0 & \lambda^2 - 1 \end{pmatrix} = (\lambda^2 - 1)^2 = 0$$

whose roots are $1, 1, -1, -1$.

The characteristic equation is

9.34
$$g(\lambda) = \det \begin{pmatrix} \lambda^2 - 2ha_{1,1}\lambda - 1 & -2ha_{1,2}\lambda \\ -2ha_{2,1}\lambda & \lambda^2 - 2ha_{2,2}\lambda - 1 \end{pmatrix} = 0$$

or

9.35
$$\lambda^4 - 2h(a_{1,1} + a_{2,2})\lambda^3 - [2 - 4h^2(a_{1,1}a_{2,2} - a_{1,2}a_{2,1})]\lambda^2 + 2h(a_{1,1} + a_{2,2})\lambda + 1 = 0.$$

We now show that the roots of (9.34) are μ_1, ν_1, μ_2, ν_2 where μ_i and ν_i are the roots of

9.36
$$\lambda^2 - 2hm_i\lambda - 1 = 0.$$

Suppose that c and d are numbers which do not both vanish such that

9.37
$$\begin{pmatrix} a_{1,1} & a_{1,2} \\ a_{2,1} & a_{2,2} \end{pmatrix}\begin{pmatrix} c \\ d \end{pmatrix} = m\begin{pmatrix} c \\ d \end{pmatrix}.$$

Then $y_n = c\lambda^n$, $z = d\lambda^n$ satisfies (9.31) provided

9.38
$$\begin{cases} c(\lambda^2 - 1) = 2h(a_{1,1}c + a_{1,2}d)\lambda = 2hm\lambda c \\ d(\lambda^2 - 1) = 2h(a_{2,1}c + a_{2,2}d)\lambda = 2hm\lambda d \end{cases}$$

i.e., if

9.39
$$\lambda^2 - 2hm\lambda - 1 = 0.$$

Evidently we have

9.40 $\quad \mu_i = hm_i + \sqrt{1 + h^2m_i^2}, \qquad \nu_i = hm_i - \sqrt{1 + h^2m_i^2}, \qquad i = 1, 2.$

Thus $\mu_1 = 1 + hm_1 + 0(h^2)$ and $\mu_2 = 1 + hm_2 + 0(h^2)$ are representative roots and ν_1 and ν_2 are extraneous roots.

The midpoint method is conditionally stable. It is weakly unstable if $m_1 < 0$ and $m_2 < 0$. If $m_1 > 0$ and $m_2 > 0$, it is weakly stable. Suppose now that $m_1 > 0$ and $m_2 < 0$. The method will be weakly stable provided

9.41
$$|\mu_1| \geq |\nu_2|,$$

i.e., provided

9.42
$$hm_1 + \sqrt{1 + h^2m_1^2} \geq -hm_2 + \sqrt{1 + h^2m_2^2}.$$

But this holds provided

9.43
$$hm_1 \geq -hm_2$$

i.e.,

9.44
$$m_1 + m_2 \geq 0.$$

Thus, as shown by Henrici [1957a], the midpoint method is weakly stable provided $m_1 + m_2 \geq 0$; otherwise it is weakly unstable.

Evidently the Runge-Kutta method for solving the system (9.4) is strongly stable since the limiting characteristic equation has only the double root $\lambda = 1$.

With a system of equations, there is apt to be a more serious limitation on the step size h than in the case of a single equation. For example, let us consider the system

9.45
$$\begin{cases} y' = -2y + z \\ z' = y - 2z \end{cases}$$

with the initial conditions

9.46
$$y(0) = y_0, \qquad z(0) = z_0.$$

By (9.16) the solution is

9.47
$$\begin{cases} y(x) = \tfrac{1}{2}(z_0 + y_0)e^{-x} + \tfrac{1}{2}(-z_0 + y_0)e^{-3x} \\ z(x) = \tfrac{1}{2}(y_0 + z_0)e^{-x} + \tfrac{1}{2}(-y_0 + z_0)e^{-3x}. \end{cases}$$

Clearly, the dominant term is $\tfrac{1}{2}(y_0 + z_0)e^{-x}$ for both y and z. Thus, we compare the limitation on the step size h with that for the single equation

9.48
$$y' = -y, \qquad y(0) = 1$$

whose solution is

9.49
$$y(x) = e^{-x}.$$

For the midpoint method we have

9.50
$$\begin{cases} y_{n+1} - y_{n-1} - 2h(-2y_n + z_n) = 0 \\ z_{n+1} - z_{n-1} - 2h(y_n - 2z_n) = 0. \end{cases}$$

Let us assume that y_1 and z_1 are obtained by the Euler method, i.e.,

9.51
$$\begin{cases} y_1 = y_0 + h(-2y_0 + z_0) \\ z_1 = z_0 + h(y_0 - 2z_0). \end{cases}$$

The general solution of (9.50) is

9.52
$$\begin{cases} y_n = c_1\mu_1^n + c_2\mu_2^n + c_3v_1^n + c_4v_2^n \\ z_n = d_1\mu_1^n + d_2\mu_2^n + d_3v_1^n + d_4v_2^n \end{cases}$$

where μ_i and v_i are given by (9.40).

To determine the d_i in terms of the c_i we proceed as follows. For $i = 1, 3$ we have

9.53
$$\begin{pmatrix} -2 & 1 \\ 1 & -2 \end{pmatrix}\begin{pmatrix} c_i \\ d_i \end{pmatrix} = m_1\begin{pmatrix} c_i \\ d_i \end{pmatrix}$$

and for $i = 2, 4$, we have

9.54
$$\begin{pmatrix} -2 & 1 \\ 1 & -2 \end{pmatrix}\begin{pmatrix} c_i \\ d_i \end{pmatrix} = m_2\begin{pmatrix} c_i \\ d_i \end{pmatrix}.$$

Since $m_1 = -1$, $m_2 = -3$ we have

9.55
$$c_1 = d_1, \qquad c_3 = d_3, \qquad c_2 = -d_2, \qquad c_4 = -d_4$$

and hence

9.56
$$\begin{cases} y_n = c_1\mu_1^n + c_2\mu_2^n + c_3v_1^n + c_4v_2^n \\ z_n = c_1\mu_1^n - c_2\mu_2^n + c_3v_1^n - c_4v_2^n. \end{cases}$$

We determine c_1, c_2, c_3, c_4 by the initial conditions

9.5/
$$\begin{cases} y_0 = c_1 + c_2 + c_3 + c_4, \qquad z_0 = c_1 - c_2 + c_3 - c_4 \\ y_0 + h(-2y_0 + z_0) = c_1\mu_1 + c_2\mu_2 + c_3v_1 + c_4v_2 \\ z_0 + h(y_0 - 2z_0) = c_1\mu_1 - c_2\mu_2 + c_3v_1 - c_4v_2. \end{cases}$$

These conditions are equivalent to

9.58
$$\begin{cases} c_1 + c_3 = \tfrac{1}{2}(y_0 + z_0) \\[2mm] c_1\mu_1 + c_3v_1 = \tfrac{1}{2}(y_0 + z_0) + \dfrac{h}{2}(-y_0 - z_0) = \tfrac{1}{2}(1 - h)(y_0 + z_0) \\[2mm] c_2 + c_4 = \tfrac{1}{2}(y_0 - z_0) \\[2mm] c_2\mu_2 + c_4v_2 = \tfrac{1}{2}(y_0 - z_0) + \dfrac{3h}{2}(-y_0 + z_0) = \tfrac{1}{2}(1 - 3h)(y_0 - z_0). \end{cases}$$

Solving, we obtain

9.59
$$\begin{cases} c_1 = \dfrac{1}{2(v_1 - \mu_1)}[v_1 - (1 - h)](y_0 + z_0) \\[4mm] c_3 = \dfrac{1}{2(v_1 - \mu_1)}[1 - h - \mu_1](y_0 + z_0) \\[4mm] c_2 = \dfrac{1}{2(v_2 - \mu_2)}[v_2 - (1 - 3h)](y_0 - z_0) \\[4mm] c_4 = \dfrac{1}{2(v_2 - \mu_2)}[1 - 3h - \mu_2](y_0 - z_0). \end{cases}$$

In spite of the fact that

9.60 $c_3 = 0(h^2), \qquad c_4 = 0(h^2)$

nevertheless, the extraneous terms $c_3 v_1''$ and $c_4 v_2''$ may become large during the course of the calculation. In the case of the single equation, we only had one extraneous term which would involve v_1''. Evidently, if $n = (x - x_0)/h$, then $|v_1''|$ behaves like e^x, and this corresponds to a much less rapid growth than e^{3x}, which corresponds to $|v_2''|$. Thus, even though the term e^{-3x} does not contribute much to the solution of the differential equation itself, the associated extraneous solution may cause a more serious buildup of errors.

In view of the situation, it would appear that the use of weakly unstable methods for systems may be very unsatisfactory in some cases.

Because of the difficulties sometimes encountered in solving systems of ordinary differential equations by the standard methods, which are essentially based on numerical differentiation or quadrature using polynomials, some scheme such as that suggested below may be appropriate in some cases. Suppose we have the system

9.61 $\begin{cases} y' = a_{1,1}(x)y + a_{1,2}(x)z + \phi_1(x) \\ z' = a_{2,1}(x)y + a_{2,2}(x)z + \phi_2(x) \end{cases}$

with the initial conditions (9.2).

The first step is to solve the system

9.62 $\begin{cases} y' = a_{1,1}(x_0)y + a_{1,2}(x_0)z + \phi_1(x_0) \\ z' = a_{2,1}(x_0)y + a_{2,2}(x_0)z + \phi_2(x_0) \end{cases}$

analytically as for (9.4) and then determine y_1 and z_1. Unless the roots m_1 and m_2 of $(a_{11} - m)(a_{22} - m) - a_{12}a_{21} = 0$ are equal, then we can write

9.63 $\begin{cases} y = c_1 e^{m_1 x} + c_2 e^{m_2 x} + e_1 \\ z = d_1 e^{m_1 x} + d_2 e^{m_2 x} + e_2 \end{cases}$

where c_1, c_2, c_3, c_4 are determined as in (9.16) and e_1 and e_2 are determined by

9.64 $\begin{cases} a_{1,1}e_1 + a_{1,2}e_2 = -\phi_1(x_0) \\ a_{2,1}e_1 + a_{2,2}e_2 = -\phi_2(x_0). \end{cases}$

Alternatively, we could use $a_{1,1}((x_0 + x_1)/2)$ in place of $a_{1,1}(x_0)$, etc. to get y_1 and z_1. Presumably, more accurate schemes could be found. With such a scheme we could handle much more easily a case such as $m_1 = -1$, $m_2 = -1000$, which would be very difficult to treat by the usual methods unless a very small mesh size is used.

For the case of a system involving r equations and r unknowns for $r \gg 2$ the above procedure could be carried out. It then becomes necessary to find, at each step, the eigenvalues and eigenvectors of an $r \times r$ matrix. (See Chapter 14.)

EXERCISES 9.9

1. Find the analytic solution of the system

$$y' = -2y + z$$

$$z' = y + 2z$$

where $y(0) = 2$, $z(0) = 1$. Evaluate the solution for $x = 0.2$.

2. In the preceding exercise find $y(0.1)$ and $z(0.1)$ by the Euler method with $h = 0.1$ and then find $y(0.2)$ and $z(0.2)$ by the midpoint method.

3. For the system of Exercise 1, find the characteristic equation and the limiting characteristic equation. What can be said concerning the stability?

4. Develop analytic formulas for the values of y_n and z_n produced by using the midpoint method for the system of Exercise 1 with starting values determined by the Euler method. Evaluate these formulas for the case $n = 2$.

5. Analyze the stability of the midpoint method for the system

$$y' = 4y - z$$

$$z' = -2y - z.$$

Determine the roots of the characteristic equation for $h = 0.1$ and the roots of the limiting characteristic equation. Do the same for method IC.2.

6. Carry out four steps of the midpoint method for solving the system of Exercise 1 with $h = 0.1$. (In the first step use the Euler method.) Compare the numerical solution thus obtained with the analytic solution given by (9.56) and also with the exact solution of the system.

7. Discuss the stability of the midpoint method as applied to the following systems:

$$\begin{cases} y' = 7y + 4z \\ z' = 4y + z \end{cases} \qquad \begin{cases} y' = -7y - 4z \\ z' = -4y - z \end{cases}$$

$$\begin{cases} y' = -2y + z \\ z' = y - 2z \end{cases} \qquad \begin{cases} y' = -2y + z \\ z' = y + 2z \end{cases}$$

$$\begin{cases} y' = 3y + 5z \\ z' = 5y + 3z \end{cases} \qquad \begin{cases} y' = -y - 3z \\ z' = -3y - z \end{cases}$$

8. Show that method IC.2 is weakly stable when applied to the system given by (9.4) provided the roots m_1 and m_2 of (9.8) satisfy $m_1 + m_2 \geq 0$ (Henrici [1957a]).

9. Solve the system $y' = 3y + 5z$, $z' = 5y + 3z$ with $y(0) = 1$, $z(0) = -1$ and also with $y(0) = 1.1$ and $z(0) = -1$. Use the Euler method. Integrate from $x = 0$ to $x = 0.5$ with $h = 0.1$. Compare the results with the exact solution.

10. Consider the system $y' = -1001y + 999z$, $z' = 999y - 1001z$, with $y(0) = z(0) = 1$. How small a step size must be used in order that the numerical values obtained by the Euler method will be a reasonable representation of the solution? What if the Runge-Kutta method is used?

SUPPLEMENTARY DISCUSSION

Section 9.2

Dahlquist [1956] gives explicit formulas for the coefficients of the methods IC.N. Numerical values are given by Salzer [1956].

Section 9.3

The definitions of "condition of consistency," "conditions of stability," and "convergence" are based on those used by Henrici [1962] (see also Dahlquist [1956]).

Section 9.4

It is possible to overcome the limitation on the order of a stable linear N-step method. For example, Hansen [1969] describes a procedure involving the use of N different methods in a periodic manner. He is able to develop a stable scheme of order $2N - 1$. This scheme does not involve the use of additional computations over and above those required using a single method. An alternative scheme described by Gragg and Stetter [1964] also produces high accuracy but involves additional computation.

As stated by Henrici [1962] the methods D.q.N are unstable if $q = 0$, $N \geq 7$ or if $q = 1$, $N \geq 3$. It is also easy to show the methods D.2.N are unstable for $N \geq 1$.

Section 9.6

Bounds for the accumulated error for predictor-corrector methods and for more general methods are given by Gragg and Stetter [1964].

Section 9.7

Milne and Reynolds [1959] describe a scheme for stabilizing the Milne-Simpson method. This scheme involves replacing on every M step the value of y_n by $\hat{y}_n$ which is the average of y_n obtained by using the basic method and y_n^* obtained by using a corrector based on the use of QC.1.2. This procedure greatly reduces the effect of the extraneous solution. Timlake [1965] shows that similar schemes exist for all conditionally stable multistep methods. Thomason [1968] describes some numerical experiments based on these procedures.

The method of Milne and Reynolds can be derived as a special case of the scheme proposed by Hansen [1969]. (See Supplementary Discussion for Section 9.4.)

Section 9.8

As stated in Section 9.6, the fact that the error for the various methods is proportional to h^p in the sense that $e_n = e(x_n)h^p + O(h^{p+1})$ for some function $e(x)$, independent of h, is shown by Henrici [1962]. The starting values are assumed to be accurate to within $O(h^{p+1})$. This assumption is valid for $p = 4$, for instance for the Adams-Moulton method with Runge-Kutta starting values.

Section 9.9

If the coefficients $a_{i,j}$ appearing in (9.61) are slowly varying functions of y and z, then we can use an analog of the Heun method.

CHAPTER 10

ORDINARY DIFFERENTIAL EQUATIONS: BOUNDARY VALUE AND EIGENVALUE PROBLEMS

10.1 INTRODUCTION

In this chapter we consider problems involving a second-order differential equation where the auxiliary conditions are given at two values of the independent variable. Such a problem is often referred to as a "boundary value problem" as contrasted with "initial value problems" which we have been considering in Chapters 8 and 9 and where the auxiliary conditions are all specified at a single point, say x_0.

A typical example of a boundary value problem involves the second-order equation

1.1
$$y'' = g(x, y, y')$$

with the boundary conditions

1.2
$$y(a) = y_a, \qquad y(b) = y_b.$$

This problem is, in general, more difficult to solve than the initial-value problem involving the initial conditions

1.3
$$y(a) = y_a, \qquad y'(a) = y'_a.$$

In Section 10.2 we shall consider the "shooting method" for solving the two-point boundary value problem (1.1)–(1.2). This method involves solving the initial value problem (1.1)–(1.3) for various assumed values of $y'(a)$ with the hope of finding one such value such that $y(b) = y_b$. In Section 10.3 we show that if $g(x, y, y')$ is a linear function of y and y', then the shooting method can be applied in a more systematic manner. The resulting method is referred to as the "method of superposition." In Section 10.4 we use the method of finite differences where one replaces the differential equation (1.1) by a difference equation involving values of y at certain points in the interval $[a, b]$. In Section 10.5, we describe a procedure based on the work of Gerschgorin [1930] and Collatz [1933] for determining the accuracy of finite difference solutions. The use of higher-order finite difference methods is considered in Section 10.6 along with the difference correction method of L. Fox [1947] and an "extrapolation to zero grid size" procedure of L. F. Richardson [1910]. In Section 10.7 we consider methods for obtaining higher accuracy which involve the use of derivatives of the coefficients of the differential equation. Such methods normally involve difference equations involving fewer mesh points than the methods considered in Section 10.6. In Section 10.8 we show that in many cases an "exact" difference equation can be constructed whose solution is the exact solution of the given problem.

In Section 10.9 we shall consider eigenvalue problems involving the differential equation

1.4 $$y'' = g(x, y, y'; \lambda)$$

and the homogeneous boundary conditions

1.5 $$y(a) = y(b) = 0.$$

In many cases the above two-point boundary value problem has only the "trivial solution" $(y(x) \equiv 0)$ except for certain discrete values of λ known as "eigenvalues." We shall consider methods for finding such eigenvalues, when $g(x, y, y'; \lambda)$ is linear in y and y', based on the method of superposition and on the method of finite differences.

10.2 THE SHOOTING METHOD

In order to solve (1.1)–(1.2) by the shooting method we assume a value, say λ, for $y'(a)$ and solve* the initial-value problem

2.1 $$\begin{cases} y'' = g(x, y, y') \\ y(a) = y_a, y'(a) = \lambda. \end{cases}$$

If $g(x, y, y')$ is a sufficiently well-behaved function, there exists a solution, say $y^{(\lambda)}(x)$, of this initial-value problem in the interval $[a, \hat{x}]$ for some $\hat{x} > a$. Any of the methods of Chapter 8 can be used to obtain an approximate solution. If we can carry the solution procedure to $x = b$ we can evaluate

2.2 $$f(\lambda) = y^{(\lambda)}(b) - y_b.$$

If $f(\lambda) = 0$, then the choice of λ is correct and $y^{(\lambda)}(x)$ is a solution of (1.1)–(1.2). Otherwise, we try other values of λ. Since we are attempting to solve the nonlinear equation

2.3 $$f(\lambda) = 0,$$

---

* By Theorem 8-11.27 there exists a unique solution of (2.1) defined in the interval $[a, \hat{x}]$ for some $\hat{x} > a$ if $g(x, y, y')$ is continuous and satisfies a Lipschitz condition in both y and y' in the region $a \leqq x \leqq a + \eta$, $|y - y_a| \leqq \delta_1$, $|y' - \lambda| \leqq \delta_2$ for some positive η, δ_1, and δ_2.

and since we have a procedure (which is admittedly long) for evaluating $f(\lambda)$ for any value of λ, we can use any of the methods of Chapter 4, which involve only the values of the function $f(\lambda)$ itself, such as the method of bisection, the method of false position, the secant method, and Muller's method. Since we are seeking only real roots, the secant method should be adequate for most cases. If not, Muller's method should be tried.

For the actual evaluation of $f(\lambda)$ we first reduce (2.1) to the system of first-order equations

2.4
$$\begin{cases} \dfrac{dy^{(\lambda)}}{dx} = z^{(\lambda)} \\[2mm] \dfrac{dz^{(\lambda)}}{dx} = g(x, y^{(\lambda)}, z^{(\lambda)}) \\[2mm] y^{(\lambda)}(a) = y_a, \qquad z^{(\lambda)}(a) = \lambda. \end{cases}$$

Solving this numerically, we evaluate $f(\lambda)$ by (2.2).

As an example, let us consider the problem

2.5
$$\begin{cases} y'' = y + 1 \\ y(0) = 1, \qquad y(1) = 2. \end{cases}$$

It can be verified that the exact solution of this problem is

2.6
$$y = \left(\frac{3 - 2e^{-1}}{e - e^{-1}} \right) e^x + \left(\frac{2e - 3}{e - e^{-1}} \right) e^{-x} - 1.$$

If we were to apply the shooting method to evaluate $f(\lambda)$ we would find $y^{(\lambda)}(1)$ by solving the initial value problem

2.7
$$\begin{cases} \dfrac{dy^{(\lambda)}}{dx} = z^{(\lambda)} \\[2mm] \dfrac{dz^{(\lambda)}}{dx} = y^{(\lambda)} + 1 \\[2mm] y^{(\lambda)}(0) = 1, \qquad z^{(\lambda)}(0) = \lambda. \end{cases}$$

The functions $y^{(\lambda)}(x)$ and $z^{(\lambda)}(x)$ could be obtained numerically using any of the methods of Chapter 8. However, in this particularly simple case we can obtain an analytic solution to (2.7), namely,

2.8
$$\begin{cases} y^{(\lambda)}(x) = \dfrac{2 + \lambda}{2} e^x + \dfrac{2 - \lambda}{2} e^{-x} - 1 \\[4mm] z^{(\lambda)}(x) = \dfrac{2 + \lambda}{2} e^x - \dfrac{2 - \lambda}{2} e^{-x}. \end{cases}$$

Therefore,

2.9
$$f(\lambda) = y^{(\lambda)}(1) - y(1) = \frac{2 + \lambda}{2} e + \frac{2 - \lambda}{2} e^{-1} - 3$$

$$= 2 \cosh 1 + \lambda \sinh 1 - 3.$$

Because of the simple nature of the function $f(\lambda)$, we can solve the equation $f(\lambda) = 0$ for λ explicitly, obtaining

2.10
$$\lambda_0 = \frac{3 - 2 \cosh 1}{\sinh 1} \doteq -0.0733.$$

Therefore, upon substitution in (2.8) we get

2.11
$$y(x) = y^{(\lambda_0)}(x) = \left(\frac{3 - 2e^{-1}}{e - e^{-1}} \right) e^x + \left(\frac{2e - 3}{e - e^{-1}} \right) e^{-x} - 1,$$

which agrees with (2.6).

EXERCISES 10.2

1. Solve the two-point boundary value problem

$$y'' - y = 1, \qquad 0 < x < 1$$
$$y(0) = 1, \qquad y(1) = 2$$

by the shooting method using the Euler method with $h = 1/4$. Use the secant method with initial values $\lambda = 1, \lambda = 2$ to solve the equation $f(\lambda) = 0$. Compare the numerical results with the exact solution.

2. Solve the two-point boundary value problem

$$y'' = 1 + yy', \qquad 0 < x < 0.6$$
$$y(0) = 1, \qquad y(0.6) = 2$$

using the shooting method with the Euler method and $h = 0.2$. The solution should give $y(0.6)$ correct to within ± 0.02.

10.3 THE METHOD OF SUPERPOSITION

Let us now consider the linear differential equation

3.1
$$L[y] = y'' + Py' + Qy = R,$$

where $P = P(x)$, $Q = Q(x)$, and $R = R(x)$ are well behaved functions of x. We could solve the boundary value problem involving (3.1) and the conditions (1.2) using the shooting method described above. Here

3.2 $$g(x, y, y') = -Py' - Qy + R,$$

which is a linear function of y and y'. We shall show that because of this, $f(\lambda)$ is a linear function of λ; hence it is sufficient for the shooting method to find $y^{(\lambda)}(b)$ for just two values of λ, say 0 and 1.

We note that it follows from Theorem 8-11.31 that the auxiliary initial value problems are uniquely solvable. For, if the coefficients $P(x)$ and $Q(x)$ are continuous on $[a, b]$, the function $g(x, y, y') = -Py' - Qy$ satisfies a Lipschitz condition in y and y' on $[a, b]$ and for all y and y'.

Thus, for convenience, let $u(x)$ and $v(x)$ correspond to $\lambda = 0$ and $\lambda = 1$, respectively. Evidently we have

3.3 $$\begin{cases} L[u] = R \\ u(a) = y_a, \qquad u'(a) = 0, \end{cases}$$

and

3.4 $$\begin{cases} L[v] = R \\ v(a) = y_a, \qquad v'(a) = 1. \end{cases}$$

For any λ the function

3.5 $$w(x) = u(x) + \lambda(v(x) - u(x))$$

satisfies the differential equation $L[u] = R$. Moreover, we have

3.6 $$\begin{cases} w(a) = u(a) = y_a \\ w'(a) = u'(a) + \lambda(v'(a) - u'(a)) = \lambda. \end{cases}$$

Therefore,

3.7 $$f(\lambda) = w(b) - y_b = u(b) + \lambda(v(b) - u(b)) - y_b$$

is a linear function of λ. Hence as long as $v(b) \neq u(b)$, we can solve the equation $f(\lambda) = 0$ for λ explicitly obtaining

3.8 $$\lambda_0 = \frac{y_b - u(b)}{v(b) - u(b)}.$$

Substituting in (3.5) we get

3.9
$$w(x) = u(x) + \frac{y_b - u(b)}{v(b) - u(b)}(v(x) - u(x))$$

$$= \left(\frac{v(b) - y_b}{v(b) - u(b)}\right) u(x) + \left(\frac{y_b - u(b)}{v(b) - u(b)}\right) v(x),$$

which is the solution of our two-point boundary value problem. Thus, we have, in effect, "superimposed" the solutions of two simple initial value problems.

We now seek to show that if $Q(x) \leq 0$ in $[a, b]$ then $u(b) \neq v(b)$. This will establish the existence of a solution of the boundary value problem (3.1)–(1.2). Let

3.10
$$r(x) = v(x) - u(x).$$

Evidently, we have

3.11
$$\begin{cases} L[r] = L[v] - L[u] = 0 \\ r(a) = v(a) - u(a) = 0 \\ r(b) = v(b) - u(b). \end{cases}$$

We seek to show that $r(b) \neq 0$.

Since $L[r] = 0$ we have, by (3.1),

3.12
$$e^{\int_a^x P dt} L[r] = e^{\int_a^x P dt} r'' + P e^{\int_a^x P dt} r' + Q e^{\int_a^x P dt} r = 0,$$

or

3.13
$$\frac{d}{dx}\left[e^{\int_a^x P dt} r'\right] = -Q e^{\int_a^x P dt} r.$$

By (3.10), (3.3), and (3,4), we have

3.14
$$r'(a) = v'(a) - u'(a) = 1,$$

so that $r(x)$ must be positive for all values of x in the open interval $(a, \tilde{x})$ for some $\tilde{x} > a$.

If $r(x) = 0$ for some $x \in (a, b]$ let x^* be the smallest value† of x such that

---

† Actually, in order to make the argument rigorous, we should let x^* be the greatest lower bound of all numbers x such that $x > a$ and $r(x) = 0$. Then either $r(x^*) = 0$ or x^* is the limit of a sequence of values of x such that $r(x) = 0$. However, by continuity it follows that $r(x^*) = 0$. Moreover, one can show using the mean-value theorem that if there is a sequence of values of x approaching a with $r(x) = 0$, then $r'(a) = 0$, contrary to (3.14). We conclude, then, that there exists $x^* > a$ such that $r(x^*) = 0$ and such that $r(x)$ does not vanish in the interval (a, x^*).

3.15
$$\begin{cases} x^* > a \\ r(x^*) = 0. \end{cases}$$

If $r(b) = 0$, then $x^* \leq b$. Moreover, for $x \in (a, x^*)$, we have

3.16
$$r(x) > 0.$$

From (3.13), since $Q \leq 0$ and $r \geq 0$, we have

3.17
$$\frac{d}{dx}\left[e^{\int_a^x P dt} r' \right] \geq 0$$

For $x \in [a, x^*]$, by Rolle's theorem, since $r(a) = r(x^*) = 0$, there exists an $x^{**} \in (a, x^*)$ such that

3.18
$$r'(x^{**}) = 0.$$

However, using (3.17) and the mean-value theorem we have

3.19
$$r'(a) = e^{\int_a^a P dt} r'(a) \leq e^{\int_a^{x^{**}} P dt} r'(x^{**})$$

and therefore,

3.20
$$r'(x^{**}) \geq \frac{r'(a)}{e^{\int_a^{x^{**}} P dt}} = e^{-\int_a^{x^{**}} P dt} > 0.$$

Since we have a contradiction, our assumption that $r(b) = 0$ is untrue, and hence

3.21
$$v(b) \neq u(b).$$

The above argument can also be used to show the uniqueness of a solution of the problem (3.1)–(1.2). If $y(x)$ and $\hat{y}(x)$ are two solutions, then $\delta(x) = y(x) - \hat{y}(x)$ satisfies the conditions

3.22
$$\begin{cases} L[\delta] = 0 \\ \delta(a) = \delta(b) = 0. \end{cases}$$

If $\delta'(a) = 0$, then $\delta(x) \equiv 0$ is a solution of (3.22). By Theorem 8-11.31, it follows that this is the only solution of (3.22). If $\delta'(a) \neq 0$, then since $Q \leq 0$ it follows from the above argument on $r(x)$ that $\delta(b) \neq 0$. This contradiction implies that $\delta'(a) = 0$ and $\delta(x) \equiv y(x) - \hat{y}(x) \equiv 0$. Hence the uniqueness follows.

As an example, we now solve the problem (2.5) by the method of super-position. By (3.3) we solve the initial-value problem

3.23
$$\begin{cases} u'' - u = 1, \\ u(0) = 1, \quad u'(0) = 0, \end{cases}$$

obtaining $u(x) = e^x + e^{-x} - 1$. By (3.4) we solve

3.24
$$\begin{cases} v'' - v = 1, \\ v(0) = 1, \qquad v'(0) = 1, \end{cases}$$

and obtain $v(x) = \frac{3}{2} e^x + \frac{1}{2} e^{-x} - 1$. We now compute the quantities

3.25
$$\begin{cases} u(1) = e + e^{-1} - 1, \\ v(1) = \frac{3}{2} e + \frac{1}{2} e^{-1} - 1. \end{cases}$$

By (3.8) we choose λ_0 as

3.26
$$\lambda_0 = \frac{2 - u(1)}{v(1) - u(1)} = \frac{3 - e - e^{-1}}{\frac{1}{2}(e - e^{-1})} = \frac{6 - 2(e + e^{-1})}{e - e^{-1}}.$$

Substituting in (3.5) we obtain (2.6).

EXERCISES 10.3

1. Solve Exercise 1 of Section 10.2 using the method of superposition and the Euler method with $h = 1/4$. Compare the numerical results with the exact solution.
2. Solve the two-point boundary value problem

$$y'' - xy = 1, \qquad 0 < x < 1$$
$$y(0) = 1, \qquad y(1) = 2$$

by the method of superposition. Use the Euler method with $h = 1/3$.

10.4 FINITE DIFFERENCE METHODS

We now consider the solution of the two-point boundary value (1.1)–(1.2) by finite difference methods. Since some of the methods discussed generalize rather easily to two and higher dimensions, it is convenient to let u, rather than y, denote the dependent variable, thus reserving y to be an independent variable in the two-dimensional case. Thus, we are dealing with the differential equation

4.1
$$u'' = g(x, u, u')$$

with the boundary conditions

4.2
$$u(a) = u_a, \qquad u(b) = u_b.$$

The first step in the application of finite difference methods is the selection of a set of mesh points in $[a, b]$. A simple way of doing this is to choose an integer N and to use the mesh points

4.3 $$x_i = a + ih, \qquad i = 0, 1, 2, \ldots, N$$

where the mesh size h is given by

4.4 $$h = \frac{b-a}{N}.$$

Evidently, we have

4.5 $$x_0 = a, \qquad x_N = b.$$

The next step is to represent the differential equation (4.1) by a difference equation involving values of the unknown function u at the mesh points. One way of doing this is to represent the derivatives by difference quotients and then substitute in (4.1). Thus, for example, we can replace u' and u'' by

4.6 $$\frac{u(x+h) - u(x-h)}{2h} \sim u'(x)$$

and

4.7 $$\frac{u(x+h) + u(x-h) - 2u(x)}{h^2} \sim u''(x),$$

respectively. If this is done at each interior mesh point $x_1, x_2, \ldots, x_{N-1}$ we obtain the difference equation

4.8 $$\frac{u_{i+1} + u_{i-1} - 2u_i}{h^2} = g\left(x_i, u_i, \frac{u_{i+1} - u_{i-1}}{2h}\right), \qquad i = 1, 2, \ldots, N-1,$$

where

4.9 $$u_i = u(x_i).$$

This particular choice of difference quotients has the advantage that the same difference equation can be used at each interior mesh point. If we had represented the derivatives by difference quotients involving more mesh points, we would have had to use special formulas near the ends of the interval. This will be done later in Section 10.6. Also, in Section 10.7, we shall show that one can in many cases obtain greater accuracy with three-point difference equations other than (4.8).

Evidently, using the fact that $u_0 = u_a$ and $u_N = u_b$, we obtain from (4.8) a system of $N-1$ equations which are in general nonlinear. Some of the methods of Section 4.13 may be applicable in a given case. We shall be primarily concerned with the linear case involving the differential equation

4.10 $$L[u] = Au'' + Du' + Fu = G$$

where A, D, F, and G are well-behaved functions of x such that $A(x) > 0$ and $F(x) \leqq 0$ in $[a, b]$.

Using the difference quotients (4.6) and (4.7) we obtain

4.11　　$L_h[u] = A \dfrac{u(x + h) + u(x - h) - 2u(x)}{h^2} + D \dfrac{u(x + h) - u(x - h)}{2h}$

$$+ Fu(x) = G$$

or

4.12　　$L_h[u] = \alpha_0 u(x) + \alpha_1 u(x + h) + \alpha_2 u(x - h) = G$

where

4.13　　　　$\alpha_0 = -\dfrac{2A}{h^2} + F, \qquad \alpha_1 = \dfrac{A}{h^2} + \dfrac{D}{2h}, \qquad \alpha_2 = \dfrac{A}{h^2} - \dfrac{D}{2h}.$

If h is sufficiently small so that for all x in $[a, b]$, we have

4.14　　　　　　　　　　　　$A(x) - \dfrac{h|D(x)|}{2} > 0,$

then the coefficients α_1 and α_2 are positive. In any case, α_0 is negative, since $F \leqq 0$. Moreover, we have

4.15　　　　　　　　　　　　$-\alpha_0 \geqq \alpha_1 + \alpha_2.$

Evidently, we can write (4.12) as a system of $N - 1$ linear algebraic equations

4.16　　　　$\alpha_0 u_i + \alpha_1 u_{i+1} + \alpha_2 u_{i-1} = G_i, \qquad i = 1, 2, \ldots, N - 1$

where $u_0 = u_a$, $u_N = u_b$. We show that the system has a unique solution $u_1, u_2, \ldots, u_{N-1}$. To do this we show that the determinant of the system does not vanish. If the determinant vanishes, then there exists numbers $u_1, u_2, \ldots, u_{N-1}$ not all zero which satisfy (4.16) with $G_i = 0$ and $u_0 = u_N = 0$ (see Theorem A-7.7). This corresponds to the homogeneous system of equations. We can assume that at least one of the u_i is positive. Otherwise, we can work with $-u_1, -u_2, \ldots, -u_{N-1}$ which will also satisfy the homogeneous system. Let $u_{i_0} = M = \max_i u_i > 0$. Since

$$-\alpha_0 u_{i_0} = \alpha_1 u_{i_0+1} + \alpha_2 u_{i_0-1},$$

we have, by (4.15),

4.17　$0 = (-\alpha_0)M - [\alpha_1 u_{i_0+1} + \alpha_2 u_{i_0-1}] \geqq \alpha_1 M + \alpha_2 M - [\alpha_1 u_{i_0+1} + \alpha_2 u_{i_0-1}]$

$$= \alpha_1(M - u_{i_0+1}) + \alpha_2(M - u_{i_0-1}).$$

This is strictly positive unless $u_{i_0+1} = u_{i_0-1} = M$. Similarly, we can show that $u_{i_0+2} = u_{i_0-2} = M$, etc. But eventually, this will imply that $u_0 = M$ or $u_N = M$, which contradicts the requirement that $u_0 = u_N = 0$. This contraction shows that the determinant in question does not vanish.

As an example, let us consider the problem

4.18
$$\begin{cases} u'' - xu = x^2 \\ u(1) = 2, \quad u(2) = 1. \end{cases}$$

Letting $h = \frac{1}{4}$ and replacing the derivative u'' by the difference quotient (4.7) we get

4.19
$$\begin{cases} \dfrac{u_2 + u_0 - 2u_1}{h^2} - \frac{5}{4}u_1 = (\frac{5}{4})^2, \\[2ex] \dfrac{u_3 + u_1 - 2u_2}{h^2} - \frac{3}{2}u_2 = (\frac{3}{2})^2, \\[2ex] \dfrac{u_4 + u_2 - 2u_3}{h^2} - \frac{7}{4}u_3 = (\frac{7}{4})^2. \end{cases}$$

Using the fact that $u_0 = 2$ and $u_4 = 1$, we can write (4.19) in the form

4.20
$$\begin{cases} (2 + \frac{5}{4}h^2)u_1 - u_2 & = -h^2 \, (\frac{5}{4})^2 + 2, \\[2ex] -u_1 + (2 + \frac{3}{2}h^2)u_2 - u_3 & = -(\frac{3}{2})^2 \, h^2, \\[2ex] -u_2 + (2 + \frac{7}{4}h^2)u_3 & = -h^2 \, (\frac{7}{4})^2 + 1. \end{cases}$$

Here, and for any value of h, we have a linear system with a tri-diagonal matrix.

We now describe an algorithm for solving such a linear system. Let us write the system in the form

4.21
$$\begin{cases} B_1 T_1 + C_1 T_2 & = D_1 \\ A_i T_{i-1} + B_i T_i + C_i T_{i+1} = D_i, & i = 2, 3, \dots, M-1 \\ A_M T_{M-1} + B_M T_M & = D_M. \end{cases}$$

where $B_i > 0$, $i = 1, 2, \dots, M$. This, of course, may be written in the equivalent matrix form

$$
\mathbf{4.22}\quad
\begin{pmatrix}
B_1 & C_1 & 0 & 0 & \ldots & 0 & 0 & 0 \\
A_2 & B_2 & C_2 & 0 & \ldots & 0 & 0 & 0 \\
0 & A_3 & B_3 & C_3 & \ldots & 0 & 0 & 0 \\
\cdot & \cdot & \cdot & \cdot & \ldots & \cdot & \cdot \\
\cdot & \cdot & \cdot & \cdot & \ldots & \cdot & \cdot \\
\cdot & \cdot & \cdot & \cdot & \ldots & \cdot & \cdot \\
0 & 0 & 0 & 0 & \ldots & A_{M-1} & B_{M-1} & C_{M-1} \\
0 & 0 & 0 & 0 & \ldots & 0 & A_M & B_M
\end{pmatrix}
\begin{pmatrix}
T_1 \\ T_2 \\ T_3 \\ \cdot \\ \cdot \\ \cdot \\ T_{M-1} \\ T_M
\end{pmatrix}
=
\begin{pmatrix}
D_1 \\ D_2 \\ D_3 \\ \cdot \\ \cdot \\ \cdot \\ D_{M-1} \\ D_M
\end{pmatrix}
$$

The algorithm which we give is based on the Gaussian elimination method and was apparently first used for solving elliptic differential equations by L. H. Thomas [1949].

The T_i are determined by

$$
\mathbf{4.23}\quad
\begin{cases}
b_1 = \dfrac{C_1}{B_1}, & b_i = \dfrac{C_i}{B_i - A_i b_{i-1}}, & i = 2, 3, \ldots, M-1, \\[2ex]
q_1 = \dfrac{D_1}{B_1}, & q_i = \dfrac{D_i - A_i q_{i-1}}{B_i - A_i b_{i-1}}, & i = 2, 3, \ldots, M, \\[2ex]
T_M = q_M, & T_i = q_i - b_i T_{i+1}, & i = M-1, M-2, \ldots, 1.
\end{cases}
$$

These formulas can be derived by using the Gaussian elimination procedure (see Chapter 12). One first divides the first row by B_1 and then subtracts A_2 times the new first row from the second. One then divides the new second row by its diagonal element, namely, $B_2 - A_2 b_1$. One then subtracts A_3 times the new second row from the third. Continuing this process we obtain an upper triangular matrix. The corresponding linear system can then be solved by back substitution.

We now show that the process defined by (4.23) can be carried out if the following conditions are satisfied:

$$
\mathbf{4.24}\quad
\begin{cases}
B_i > 0, & i = 1, 2, \ldots, M, \\[1ex]
B_1 > |C_1|, \\[1ex]
B_i \geqq |C_i| + |A_i| \text{ and } A_i \neq 0,\, C_i \neq 0 \text{ for } i = 2, 3, \ldots, M-1, \\[1ex]
B_M \geqq |A_M|.
\end{cases}
$$

It is sufficient to show that $B_i - A_i b_{i-1} > 0$ for $i = 2, 3, \ldots, M$. We first show that $|b_i| < 1$ for $i = 1, 2, \ldots, M-1$. Evidently $|b_1| = |C_1/B_1| < 1$. Assume now that $|b_i| < 1$ for some $i \leqq M - 2$. To show that $|b_{i+1}| < 1$ we need only show that $|B_{i+1} - A_{i+1}b_i| > |C_{i+1}|$.   But $|A_{i+1}b_i| < |A_{i+1}|$ since $|b_i| < 1$ and $A_{i+1} \neq 0$. Therefore $B_{i+1} - A_{i+1}b_i > 0$.   Consequently,

$$B_{i+1} - A_{i+1}b_i \geqq B_{i+1} - |A_{i+1}b_i| > B_{i+1} - |A_{i+1}| \geqq |C_{i+1}| > 0$$

and hence $|b_{i+1}| < 1$. Thus, by induction we have $|b_i| < 1$, $i = 1, 2, \ldots, M - 1$. Therefore, since $|A_i| < B_i$ it follows that

$$B_i - A_i b_{i-1} \geqq B_i - |A_i| \, |b_{i-1}| > B_i - |A_i| > 0 \text{ for } i = 2, 3, \ldots, M - 1.$$

Also $B_M - b_{M-1}A_M > 0$ if $A_M = 0$; otherwise, if $A_M \neq 0$, then

$$B_M - b_{M-1}A_M \geqq B_M - |b_{M-1}| \, |A_M| > B_M - |A_M| \geqq 0.$$

Thus $B_i - b_{i-1}A_i > 0$ for $i = 2, 3, \ldots, M$.

We apply the algorithm (4.23) to (4.20) which we write in the form

4.25
$$\begin{cases} \dfrac{133}{64}u_1 - u_2 & = \dfrac{487}{256}, \\[2mm] - u_1 + \dfrac{67}{32}u_2 - u_3 & = - \dfrac{9}{64}, \\[2mm] - u_2 + \dfrac{135}{64}u_3 & = \dfrac{207}{256}. \end{cases}$$

Using (4.23) we have

4.26
$$\begin{cases} b_1 = \dfrac{-64}{133} \doteq -0.4812, & b_2 = \dfrac{-1}{(67/32) - (64/133)} \doteq -0.6201, \\[3mm] q_1 = \dfrac{487}{532} \doteq 0.9154, & q_2 = \dfrac{(-9/64) + (487/532)}{(67/32) - (64/133)} \doteq 0.4805, \\[3mm] & q_3 = \dfrac{(207/256) + (0.4805)}{(135/64) + (-0.6201)} = 0.8656, \end{cases}$$

4.27
$$\begin{cases} u_3 = T_3 = q_3 = 0.8656 \\[1mm] u_2 = T_2 = q_2 - b_2 T_3 = 1.0173 \\[1mm] u_1 = T_1 = q_1 - b_1 T_2 = 1.4049. \end{cases}$$

One can readily verify that the values of $u_1, u_2,$ and u_3 thus determined satisfy (4.25).

If the differential equation is "nearly linear" one can often use the above procedure treating the nonlinear part as known.   For example, consider the problem defined by (4.18) but with the differential equation replaced by

4.28 $$u'' - xu = x^2 + \alpha e^u$$

where α is a small positive quantity. It can be shown (see, for instance, Henrici [1962]) that a unique solution exists. If α is sufficiently small we can use an iterative procedure. Letting $u^{(0)}(x) \equiv 0$ we solve the linear problem

4.29 $$u'' - xu = x^2 + \alpha e^{u^{(0)}}$$

as above obtaining $u^{(1)}$. Then we solve

4.30 $$u'' - xu = x^2 + \alpha e^{u^{(1)}}$$

to obtain $u^{(2)}$, etc. Refinements of this scheme are often used to accelerate convergence or to achieve convergence when it would not otherwise occur.

Self-adjoint Operators

Let us now consider the case where $L[u]$ is *self-adjoint*, i.e., where

4.31 $$D = A'.$$

In this case the differential equation (4.10) can be written in the form

4.32 $$L[u] = (Au')' + Fu = G.$$

Even if (4.31) does not hold, there exists an "integrating factor" $\mu(x)$ such that $\mu L[u]$ is self-adjoint. Thus, we choose $\mu(x)$ such that

4.33 $$\mu D = (\mu A)',$$

i.e., such that

4.34 $$\mu = \frac{c}{A} e^{\int (D/A)dx}$$

for some constant c. Multiplying both sides of (4.10) by μ we get the self-adjoint equation

4.35 $$\mu L[u] = (\mu Au')' + \mu Fu = \mu G.$$

Thus, for example, consider the differential equation

4.36 $$L[u] = u'' - \frac{1}{x}u' + u = x^2.$$

Evidently (4.31) does not hold. However, by (4.34) with $c = 1$ we have

4.37 $$\mu = e^{-\int(dx/x)} = \frac{1}{x}.$$

Hence if we multiply both sides of (4.36) by μ we get the self-adjoint equation

4.38
$$\frac{1}{x} L[u] = \frac{1}{x} u'' - \frac{1}{x^2} u' + \frac{1}{x} u$$
$$= \left(\frac{1}{x} u'\right)' + \frac{1}{x} u = x.$$

There are several alternative discrete representations of the expression

4.39
$$M[u] = (Au')'$$

which are as accurate as the usual representation

4.40 $M_h[u] = A(x) \dfrac{u(x + h) + u(x - h) - 2u(x)}{h^2} + A'(x) \dfrac{u(x + h) - u(x - h)}{2h}$,

which is based on the expanded form

4.41
$$M[u] = Au'' + A'u'.$$

The following two alternative representations are as accurate as (4.40) and in many cases have more desirable properties:

4.42 $M_h^{(1)}[u] = \dfrac{1}{h}\left[A\left(x + \dfrac{h}{2}\right)\left(\dfrac{u(x + h) - u(x)}{h}\right) - A\left(x - \dfrac{h}{2}\right)\left(\dfrac{u(x) - u(x - h)}{h}\right)\right]$,

4.43 $M_h^{(2)}[u] = \dfrac{1}{h}\left[\left(\dfrac{A(x + h) + A(x)}{2}\right)\left(\dfrac{u(x + h) - u(x)}{h}\right)\right.$
$$\left. - \left(\dfrac{A(x) + A(x - h)}{2}\right)\left(\dfrac{u(x) - u(x - h)}{h}\right)\right].$$

Let us now analyze the difference between $M_h^{(1)}[u]$ and $M[u]$ under the assumption that A and u belong to $C^{(4)}$. We show that in this case

4.44
$$M_h^{(1)}[u] - M[u] = O(h^2).$$

Before verifying this in general, let us consider the case where $u = x^k$ for some integer k. Evidently

4.45
$$M_h^{(1)}[1] - M[1] = 0,$$

4.46 $M_h^{(1)}[x] - M[x] = \dfrac{1}{h}\left[A\left(x + \dfrac{h}{2}\right) - A\left(x - \dfrac{h}{2}\right)\right] - A'(x)$
$$= \dfrac{h^2}{24} A^{(3)}(\xi) = O(h^2)$$

for some $\xi \in I = [x - h/2, x + h/2]$. Moreover,

4.47 $M_h^{(1)}[x^2] - M[x^2] = 2x \left\{ \dfrac{A\left(x + \dfrac{h}{2}\right) - A\left(x - \dfrac{h}{2}\right)}{h} - A'(x) \right\}$

$$+ \left[A\left(x + \frac{h}{2}\right) + A\left(x - \frac{h}{2}\right) - 2A(x) \right]$$

$$= 2x \left(\frac{h^2}{24} A^{(3)}(\xi) \right) + \frac{h^2}{4} A''(\xi') = O(h^2)$$

where ξ and ξ' are in I.

   Also, we have

4.48 $M_h^{(1)}[x^3] - M[x^3] = 3x^2 \left\{ \dfrac{A\left(x + \dfrac{h}{2}\right) - A\left(x - \dfrac{h}{2}\right)}{h} - A'(x) \right\}$

$$+ 3x \left\{ A\left(x + \frac{h}{2}\right) + A\left(x - \frac{h}{2}\right) - 2A(x) \right\}$$

$$+ h \left\{ A\left(x + \frac{h}{2}\right) - A\left(x - \frac{h}{2}\right) \right\} = O(h^2).$$

The reader should verify that

4.49 $M_h^{(1)}[x^k] - M[x^k] = O(h^2),$

for $k \geqq 4$.

   To prove (4.44) we consider the function

4.50 $\psi(h) = A\left(x + \dfrac{h}{2}\right)\left(\dfrac{u(x + h) - u(x)}{h} \right).$

Evidently

4.51 $M_h^{(1)}[u] = \dfrac{\psi(h) - \psi(-h)}{h} = M[u] + \dfrac{h^2}{3}\psi'''(\xi)$

for some $\xi \in (-h, h)$. This follows since for $h \neq 0$ we have

4.52 $\psi'(h) = \tfrac{1}{2}A'\left(x + \dfrac{h}{2}\right)\left[\dfrac{u(x + h) - u(x)}{h} \right]$

$$+ A\left(x + \frac{h}{2}\right)\left[\frac{hu'(x + h) - [u(x + h) - u(x)]}{h^2} \right]$$

and, by L'Hospital's rule,

4.53
$$\lim_{h \to 0} \psi'(h) = \tfrac{1}{2}A'(x)u'(x) + \tfrac{1}{2}A(x)u''(x).$$

If we let

4.54
$$\Gamma(h) = A\left(x + \frac{h}{2}\right)(u(x + h) - u(x)),$$

then we have

4.55
$$\psi'''(h) = \frac{\tfrac{1}{6}h^3\Gamma'''(h) - \tfrac{1}{2}h^2\Gamma''(h) + h\Gamma'(h) - \Gamma(h)}{\tfrac{1}{6}h^4}$$

$$= \tfrac{1}{4}\Gamma^{(4)}(\zeta')$$

since by Taylor's theorem

4.56
$$0 = \Gamma(0) = \Gamma(h) - h\Gamma'(h) + \frac{h^2}{2}\Gamma''(h) - \frac{h^3}{6}\Gamma'''(h) + \frac{h^4}{24}\Gamma^{(4)}(\zeta').$$

Here ζ' lies between 0 and h. Also we have

4.57
$$\Gamma^{(4)}(h) = \tfrac{1}{16}A^{(4)}\left(x + \frac{h}{2}\right)[u(x + h) - u(x)] + \tfrac{1}{2}A^{(3)}\left(x + \frac{h}{2}\right)u'(x + h)$$

$$+ \tfrac{3}{2}A''\left(x + \frac{h}{2}\right)u''(x + h) + 2A'\left(x + \frac{h}{2}\right)u^{(3)}(x + h)$$

$$+ A\left(x + \frac{h}{2}\right)u^{(4)}(x + h).$$

Since $u(x + h) - u(x) = hu'(\xi_1)$, for some ξ_1 between x and $x + h$, we have, by (4.51), (4.55), and (4.57),

4.58
$$|M_h^{(1)}[u] - M[u]| \le \frac{h^2}{12}|\Gamma^{(4)}(\zeta')|$$

$$\le \frac{h^2}{12}[(1/16)hM_1N_4 + \tfrac{1}{2}M_1N_3$$

$$+ \tfrac{3}{2}M_2N_2 + 2M_3N_1 + M_4N_0],$$

where for $k = 0, 1, 2, \ldots$, we let

4.59
$$M_k = \max_{x\in I} |u^{(k)}(x)|, \qquad N_k = \max_{x\in I} |A^{(k)}(x)|,$$

and $I = [a, b]$. Thus (4.44) follows.

The reader should carry out a similar analysis for $M_h^{(2)}[u]$.

The difference equation corresponding to $M_h^{(1)}[u]$ is

4.60 $\qquad M_h^{(1)}[u] + Fu = \alpha_0^{(1)}u(x) + \alpha_1^{(1)}u(x + h) + \alpha_2^{(1)}u(x - h) = G,$

where

4.61
$$\begin{cases} \alpha_0^{(1)} = -\dfrac{A\left(x + \dfrac{h}{2}\right) + A\left(x - \dfrac{h}{2}\right)}{h^2} + F \\[2em] \alpha_1^{(1)} = \dfrac{A\left(x + \dfrac{h}{2}\right)}{h^2}, \qquad \alpha_2^{(1)} = \dfrac{A\left(x - \dfrac{h}{2}\right)}{h^2}. \end{cases}$$

Evidently for all h the coefficients $\alpha_1^{(1)}$ and $\alpha_2^{(1)}$ are positive and $\alpha_0^{(1)}$ is negative. Moreover, $-\alpha_0^{(1)} \geqq \alpha_1^{(1)} + \alpha_2^{(1)}$. Thus, using the argument given following (4.16) we can show that the linear system has a unique solution. We recall that for the difference equation (4.11) we are only assured of a unique solution for h sufficiently small.

The matrix of the linear system corresponding to (4.60) is symmetric, whereas this is not in general the case for the system corresponding to (4.11). The symmetry follows at once from the fact that the coefficient of $u(x + h)$ in the equation for $u(x)$, namely $h^{-2}A(x + h/2)$, is the same as the coefficient of $u(x)$ in the equation for $u(x + h)$, namely,

$$h^{-2}A(x + h - h/2) = h^{-2}A(x + h/2).$$

Whether it is actually worthwhile to reduce the differential equation to self-adjoint form depends to some extent on how easy it is to determine μ. Of course, if we cannot integrate D/A in closed form, we could perform a numerical integration in any given case.

We now show that for h sufficiently small there exists a "discrete integrating factor," i.e., a function $\mu(x)$ such that $\mu L_h[u]$ corresponds to a symmetric matrix. For some x_0 we let

4.62 $\qquad\qquad\qquad\qquad \mu(x_0) = 1.$

We choose $\mu(x)$ such that for all x the coefficient of $u(x + h)$ in the equation for $u(x)$ is equal to the coefficient of $u(x)$ in the equation for $u(x + h)$. By (4.11) this condition implies that

4.63 $\qquad \mu(x)\left[A(x) + \dfrac{h}{2}D(x)\right] = \mu(x + h)\left[A(x + h) - \dfrac{h}{2}D(x + h)\right],$

i.e.,

4.64 $\qquad \mu(x + h) = \mu(x)\dfrac{A(x) + \dfrac{h}{2}D(x)}{A(x + h) - \dfrac{h}{2}D(x + h)}.$

Therefore for $k = 1, 2, \ldots$ we have

4.65 $\qquad \mu(x_0 + kh) = \displaystyle\prod_{i=0}^{k-1} \dfrac{A(x_0 + ih) + \dfrac{h}{2}D(x_0 + ih)}{A(x_0 + (i + 1)h) - \dfrac{h}{2}D(x_0 + (i + 1)h)}.$

Thus the function $\mu(x)$ exists and is positive for all mesh points provided (4.14) holds in $[a, b]$.

With this choice of $\mu(x)$ we have, by (4.11) and (4.63),

4.66 $\qquad \mu(x)L_h[u](x) = \dfrac{1}{h}\Bigg\{ \mu(x)\left[A(x) + \dfrac{h}{2}D(x)\right]\left[\dfrac{u(x + h) - u(x)}{h}\right]$

$\qquad\qquad - \mu(x)\left[A(x - h) + \dfrac{h}{2}D(x - h)\right]\times$

$\qquad\qquad \left[\dfrac{u(x) - u(x - h)}{h}\right]\Bigg\} + \mu(x)F(x)u(x).$

EXERCISES 10.4

1. Solve Exercise 1, Section 10.2, using the method of finite differences with $h = 1/4$. Compare the numerical values with the exact solution.

2. Solve the two-point boundary value problem

$$y'' - xy = 1, \qquad 0 < x < 1$$

$$y(0) = 1, \qquad y(1) = 2$$

by the method of finite differences with $h = 1/3$ and also with $h = 1/4$. In the latter case use the special algorithm for solving a linear system with a tri-diagonal matrix.

3. Consider the two-point boundary value problem

$$u'' - x^{-1}u' + u = x^2, \qquad 1 < x < 2$$

$$u(1) = 1, \qquad u(2) = 3.$$

Find the discrete integrating factor corresponding to $h = 1/4$ and compare with the integrating factor defined by (4.37).

4. For the boundary value problem of the preceding exercise, obtain the numerical solution
   a) corresponding to the use of the standard difference equation,
   b) corresponding to the self-adjoint equation obtained by multiplying the differential equation by the integrating factor and using the symmetric difference equation.

5. Consider the two-point boundary value problem defined by

$$L[u] = 4u'' + x^2u' = 3x^2 + 1, \qquad 0 < x < 3$$

$$u(0) = 1, \qquad u(3) = 2.$$

Find an integrating factor $\mu(x)$ such that when the differential equation is multiplied by $\mu(x)$ it becomes self-adjoint. Set up and solve the symmetric difference equation derived from (4.42) with $h = 3/4$.

6. Verify that (4.44) is true under the same assumptions on A and u if we replace $M_h^{(1)}[u]$ by $M_h^{(2)}[u]$.

7. Verify (4.49) for $k \geq 4$.

8. Prove that $B_i - A_i b_{i-1}$ does not vanish for $i = 2, 3, \ldots, M$, where $b_1, b_2, \ldots, b_{M-1}$ are given by (4.23), under the following assumptions:
   a) $B_i > 0, \qquad i = 1, 2, \ldots, M$
   b) $B_1 > |C_1|$
   c) $B_i \geq |A_i| + |C_i|, \qquad i = 2, 3, \ldots, M - 1$
   d) $B_M > |A_M|$
   e) $A_i \leq 0, \qquad i = 2, 3, \ldots, M$
   f) $C_i \leq 0, \qquad i = 1, 2, \ldots, M - 1$
   g) the matrix of (4.22) is nonsingular.

9. Consider the following method for solving a linear system with a real tri-diagonal matrix which is positive definite. For the case of 4 equations and 4 unknowns the system is

$$\begin{pmatrix} B_1 & C_1 & 0 & 0 \\ C_1 & B_2 & C_2 & 0 \\ 0 & C_2 & B_3 & C_3 \\ 0 & 0 & C_3 & B_4 \end{pmatrix} \begin{pmatrix} u_1 \\ u_2 \\ u_3 \\ u_4 \end{pmatrix} = \begin{pmatrix} b_1 \\ b_2 \\ b_3 \\ b_4 \end{pmatrix}.$$

Compute

$$d_1 = \sqrt{\Delta_1}, \qquad d_2 = \sqrt{\Delta_2/\Delta_1}, \qquad d_3 = \sqrt{\Delta_3/\Delta_2}, \qquad d_4 = \sqrt{\Delta_4/\Delta_3}$$

$$e_1 = C_1/\sqrt{\Delta_2}, \qquad e_2 = C_2/\sqrt{\Delta_3/\Delta_1}, \qquad e_3 = C_3/\sqrt{\Delta_4/\Delta_2}$$

where

$$\Delta_1 = B_1, \qquad \Delta_2 = B_2\Delta_1 - C_1^2, \qquad \Delta_3 = B_3\Delta_2 - C_2^2\Delta_1, \qquad \Delta_4 = B_4\Delta_3 - C_3^2\Delta_2.$$

Let

$$D = \begin{pmatrix} d_1 & 0 & 0 & 0 \\ 0 & d_2 & 0 & 0 \\ 0 & 0 & d_3 & 0 \\ 0 & 0 & 0 & d_4 \end{pmatrix}, \qquad E = \begin{pmatrix} 1 & e_1 & 0 & 0 \\ 0 & 1 & e_2 & 0 \\ 0 & 0 & 1 & e_3 \\ 0 & 0 & 0 & 1 \end{pmatrix}.$$

Solve

$$E^T v = D^{-1} b$$

for v. Then solve

$$Ew = v$$

for w. Then compute

$$u = D^{-1} w.$$

Apply the method to solve the system

$$\begin{pmatrix} 2 & -1 & 0 & 0 \\ -1 & 2 & -1 & 0 \\ 0 & -1 & 2 & -1 \\ 0 & 0 & -1 & 2 \end{pmatrix} \begin{pmatrix} u_1 \\ u_2 \\ u_3 \\ u_4 \end{pmatrix} = \begin{pmatrix} 1 \\ 0 \\ 0 \\ 2 \end{pmatrix}.$$

(First obtain the solution of the system by using the algorithm given by (4.23).) Also, determine the formulas for the case of five equations and five unknowns.

10. Consider the linear system

$$\begin{aligned} 12u_1 - 4u_2 \qquad\qquad &= 4 \\ -u_1 + 12u_2 - \quad u_3 &= 20 \\ -9u_2 + 12u_3 &= 18. \end{aligned}$$

a) Solve the system using the algorithm given by (4.23).

b) Find μ_2 and μ_3 so that if we multiply the second equation by μ_2 and the third by μ_3 we obtain a symmetric matrix. Solve the resulting system by the algorithm described in the preceding exercise.

c) Find v_2 and v_3 so that if we introduce the new variables $\hat{u}_1 = u_1$, $\hat{u}_2 = v_2 u_2$, $\hat{u}_3 = v_3 u_3$ and then multiply the second equation by v_2 and the third by v_3 we get a system with a symmetric matrix and the same diagonal elements as the original one. Solve this system for $\hat{u}_1$, $\hat{u}_2$, and $\hat{u}_3$ by the algorithm described in the preceding exercise. Then determine u_1, u_2, u_3.

10.5 ACCURACY OF FINITE DIFFERENCE SOLUTIONS

In this section we study the accuracy of the finite difference solution of the two-point boundary value problem

5.1
$$L[u] = Au'' + Du' + Fu = G$$

5.2
$$u(a) = u_a, \qquad u(b) = u_b.$$

We assume that $A > 0$ and $F \leq 0$ on $[a, b]$. We also assume that the functions A, D, F, and G belong to $C^{(2)}[a, b]$. From the analysis of Section 10.3 and from the existence theorems of Section 8.11 it follows that a unique solution $\bar{u} = \bar{u}(x)$ exists and that $\bar{u} \in C^{(4)}[a, b]$.

We shall be primarily concerned with the difference equation

5.3 $$L_h[u] = A\frac{u(x + h) + u(x - h) - 2u(x)}{h^2} + D\frac{u(x + h) - u(x - h)}{2h}$$
$$+ Fu(x) = G.$$

We have already seen in Section 10.3 that for h sufficiently small, there exists a unique solution u of (5.3)–(5.2). We now seek a bound on the error

5.4
$$e(x) = u(x) - \bar{u}(x).$$

From 7-(3.36), 7-(3.37) and 7-(3.63) we have, for any function u in $C^{(4)}[x - h, x + h]$,

5.5
$$\frac{u(x + h) + u(x - h) - 2u(x)}{h^2} - u''(x) = \frac{h^2}{12}u^{(4)}(\xi),$$

5.6
$$\frac{u(x + h) - u(x - h)}{2h} - u'(x) = \frac{h^2}{6}u^{(3)}(\eta),$$

where ξ and η lie in $(x - h, x + h)$. Consequently, we have

5.7
$$L_h[u] - L[u] = \frac{Ah^2}{12}u^{(4)}(\xi) + \frac{Dh^2}{6}u^{(3)}(\eta)$$

and hence

5.8
$$|L_h[u] - L[u]| \leq \frac{Ah^2}{12}M_4 + \frac{|D|h^2}{6}M_3,$$

where, as usual, we let

5.9
$$M_k = \max_{a \leq x \leq b} |u^{(k)}(x)|, \qquad k = 3, 4.$$

In order that our analysis can be more conveniently extended to the two-dimensional case, we shall use the following notation. Let R be the open interval (a, b) and S be the set containing the points a and b. Let R_h be the set of mesh points $x_1, x_2, \ldots, x_{N-1}$ where $N = (b - a)/h$ and N is an integer, and $S_h = S$. We prove the following lemmas.

5.10 Lemma. Let $L_h[u]$ be a discrete operator of the form

5.11 $$L_h[u] = \alpha_1 u(x + h) + \alpha_2 u(x - h) + \alpha_0 u(x),$$

where $-\alpha_0, \alpha_1$, and α_2 are positive functions such that $-\alpha_0 \geq \alpha_1 + \alpha_2$. If $u \geq 0$ on S_h and $-L_h[u] \geq 0$ on R_h, then $u \geq 0$ in $R_h + S_h$.

Proof. If $u < 0$ for some point of R_h, then for some point $\bar{x}$ of R_h we have $u(\bar{x}) \leq u(x)$ for all $x \in R_h$ and $u(\bar{x}) < 0$. Let $M = -u(\bar{x})$. We seek to show that $u(\bar{x} + h)$ and $u(\bar{x} - h)$ both equal $-M$. But since $L_h[u](\bar{x}) \leq 0$ we have

5.12 $$-\alpha_0 u(\bar{x}) \geq \alpha_1 u(\bar{x} + h) + \alpha_2 u(\bar{x} - h).$$

Therefore,

5.13 $$0 \geq \alpha_1 u(\bar{x} + h) + \alpha_2 u(\bar{x} - h) - \alpha_0 M \geq \alpha_1 u(\bar{x} + h) + \alpha_2 u(\bar{x} - h)$$
$$+ \alpha_1 M + \alpha_2 M = \alpha_1[u(\bar{x} + h) + M] + \alpha_2[u(\bar{x} - h) + M].$$

Since $\alpha_1 > 0$, $\alpha_2 > 0$, $u(\bar{x} + h) + M \geq 0$, $u(\bar{x} - h) + M \geq 0$, the last expression can be nonpositive only if

5.14 $$u(\bar{x} + h) = u(\bar{x} - h) = -M.$$

In a similar way, we can show that $u(\bar{x} - 2h) = u(\bar{x} + 2h) = -M$. Continuing this process we can show that $u(x) = -M$ for all x in $R_h + S_h$. But since $u \geq 0$ in S_h we have a contradiction. Hence $u \geq 0$ in $R_h + S_h$.

5.15 Lemma. Let $L_h[u]$ satisfy the hypotheses of Lemma 5.10. If $|u| \leq v$ on S_h and $|L_h[u]| \leq -L_h[v]$ in R_h, then $|u| \leq v$ in $R_h + S_h$.

Proof. Evidently $L_h[v] \leq 0$ in R_h and $v \geq 0$ on S_h. Moreover, by Lemma 5.10 it follows that $v \geq 0$ in $R_h + S_h$. We now show that $v \geq u$ and that $v \geq -u$ in $R_h + S_h$, implies that $v \geq |u|$.

To prove that $v \geq u$ in $R_h + S_h$, we let $w = v - u$. Since $v \geq |u|$ on S_h, it follows that $w \geq 0$ on S_h. Moreover,

5.16 $$L_h[w] = L_h[v - u] = L_h[v] - L_h[u] \leq L_h[v] + |L_h[u]| \leq 0.$$

The last inequality follows from an initial assumption. Using Lemma 5.10 the above relations imply that $w \geq 0$ and hence $v \geq u$ in $R_h + S_h$.

To prove that $v \geq -u$, let $\hat{w} = v + u$. Again, since $v \geq |u|$ on S_h, we have $v \geq -u$ and $\hat{w} \geq 0$ on S_h. Moreover,

5.17
$$L_h[\hat{w}] = L_h[v + u] = L_h[v] + L_h[u]$$
$$\leq L_h[v] + |L_h[u]| \leq 0.$$

By Lemma 5.10 we have $\hat{w} \geq 0$ in $R_h + S_h$, which implies that $v \geq -u$ in $R_h + S_h$. The lemma follows.

5.18 Lemma. Let $L_h[u]$ satisfy the hypotheses of Lemma 5.10. Let $w(x)$ be any function such that $-L_h[w] > 0$ in R_h and $w \geq 0$ on S_h. For any function $e(x)$ we have, for all x in $R_h + S_h$,

5.19
$$|e(x)| \leq W \max_{R_h} \left[\frac{|L_h[e]|}{-L_h[w]} \right] + \max_{S_h} |e(x)|$$

where

5.20
$$W = \max_{R_h + S_h} |w(x)|.$$

Proof. We consider the "majorant" function

5.21
$$v(x) = w(x) \max_{R_h} \left[\frac{|L_h[e]|}{-L_h[w]} \right] + \max_{S_h} |e(x)|.$$

Evidently, since $-L_h(c) \geq 0$ for any nonnegative constant c, we have

5.22
$$-L_h[v] \geq -L_h[w] \max_{R_h} \left[\frac{|L_h[e]|}{-L_h[w]} \right]$$
$$\geq |L_h[e]|$$

in R_h. Moreover, since $w(x) \geq 0$ on S_h, we also have by Lemma 5.10

5.23
$$|e(x)| \leq v(x)$$

on S_h by (5.21). Therefore, by Lemma 5.15 we have $|e(x)| \leq v(x)$ in R_h and hence (5.19) follows.

We now apply Lemma 5.18 to some special cases where an appropriate $w(x)$ can be found.

5.24 Theorem. Let $\bar{u}(x)$ satisfy (5.1)–(5.2) and let $u(x)$ satisfy (5.3) on R_h. If $\bar{u} \in C^{(4)}$ in $R + S$, then for all $x \in R_h + S_h$ we have

5.25 $|u(x) - \bar{u}(x)| \leqq \dfrac{h^2 r^2}{24} \max\limits_{R+S} \left[\dfrac{A M_4 + 2|D| M_3}{(A - r|D|)} \right] + \max\limits_{x \in S_h} |u(x) - \bar{u}(x)|,$

provided that for all x in $R + S$ we have

5.26 $$A(x) - r|D(x)| > 0.$$

Moreover, whether or not (5.26) holds, we have

5.27 $|u(x) - \bar{u}(x)| \leqq \dfrac{h^2 r^2}{6} M_4 + \dfrac{h^2 r}{3} M_3 + \max\limits_{x \in S_h} |u(x) - \bar{u}(x)|,$

provided $D(x) \geqq 0$ for all $x \in R$ or $D(x) \leqq 0$ for all $x \in R$. Here

5.28 $$r = \frac{b - a}{2}.$$

If $D = 0$ in R, then

5.29 $$|u(x) - \bar{u}(x)| \leqq \frac{h^2 r^2}{24} M_4 + \max\limits_{x \in S_h} |u(x) - \bar{u}(x)|.$$

Proof. Let $e(x) = u(x) - \bar{u}(x)$. Evidently, we have

5.30 $$L_h[e] = L_h[u] - L_h[\bar{u}] = L[\bar{u}] - L_h[\bar{u}],$$

since $L_h[u] = G = L[\bar{u}]$. Moreover, by (5.8) and (5.30) we have

5.31 $$|L_h[e]| = |L_h[\bar{u}] - L[\bar{u}]| \leqq \frac{A h^2}{12} M_4 + \frac{|D| h^2}{6} M_3.$$

To prove (5.25) we apply Lemma 5.2 with the function

5.32 $$w(x) = 1 - \frac{(x - c)^2}{r^2},$$

where $c = (a + b)/2$. Evidently

5.33 $$- L_h[w] = \frac{2}{r^2} (A(x) + (x - c) D(x)) - F(x) \left(1 - \frac{(x - c)^2}{r^2} \right)$$

$$\geqq \frac{2}{r^2} (A(x) - r|D(x)|) > 0$$

since $F \leqq 0$. Moreover,

5.34 $$W = \max\limits_{R_h + S_h} |w(x)| = 1,$$

and (5.25) follows by Lemma 5.18. Clearly (5.29) follows from (5.25) by letting $D = 0$. To prove (5.27) for the case $D \geq 0$ we consider the function

5.35 $$w(x) = \frac{r^2 M_4 h^2}{24}\left[3 - \frac{2(x - c)}{r} - \frac{(x - c)^2}{r^2}\right] + \frac{rM_3 h^2}{6}\left[1 + \frac{x - c}{r}\right] + \max_{x \in S_h}|e(x)|.$$

Evidently $w(x) \geq |e(x)|$ on S_h and

5.36 $$-L_h[w] \geq \frac{M_4 h^2}{12}\left[A + D(r + (x - c)) - F\left(\frac{3r^2}{2} - r(x - c) - \frac{(x - c)^2}{2}\right)\right]$$

$$+ \frac{rM_3 h^2}{6}\left[-F\left(1 - \frac{x - c}{r}\right) + \frac{D}{r}\right]$$

$$\geq \frac{AM_4 h^2}{12} + \frac{DM_3 h^2}{6} \geq |L_h[e]|.$$

Also, we have

5.37 $$W = \max_{R_h + S_h}|w(x)| = \frac{r^2 M_4 h^2}{6} + \frac{rM_3 h^2}{3} + \max_{x \in S_h}|e(x)|.$$

(We note that $0 \leq 3 - 2(x - c)/r - (x - c)^2/r^2 = 4 - [1 + (x - c)/r]^2 \leq 4$.) The result (5.27) follows from Lemma 5.15. To handle the case where $D \leq 0$ we replace $x - c$ by $-(x - c)$ in (5.35).

We remark that Lemma 5.18 can often be used to determine a bound on the difference between an approximate solution, say $\tilde{u}$, of the difference equation

5.38 $$L_h[u] = G$$

and the exact solution, u. Thus suppose we have

5.39 $$L_h[\tilde{u}] - G = \delta(x)$$

and $\tilde{u} = u$ on S_h. Evidently

5.40 $$|L_h[u - \tilde{u}]| = |\delta(x)|.$$

If $w(x)$ is any function satisfying $-L_h[w] > 0$ in R_h and such that $w \geq 0$ in S_h, then from Lemma 5.18 we have

5.41 $$|u(x) - \tilde{u}(x)| \leq W \max_{R_h}\left[\frac{|\delta(x)|}{-L_h[w]}\right],$$

where $W = \max|w(x)|$ for $x \in R_h + S_h$. In the case where (5.26) holds, we let w be given by (5.32) and obtain by (5.33)–(5.34)

5.42
$$|u(x) - \tilde{u}(x)| \leq \frac{r^2}{2} \max_{R_h} \left[\frac{|\delta|}{A - r|D|} \right].$$

On the other hand, if $D \geq 0$ in $R + S$, then we consider the function

5.43
$$w(x) = 3 - \frac{2(x - c)}{r} - \frac{(x - c)^2}{r^2}.$$

Evidently

5.44
$$W = 4,$$

and, as in the derivation of (5.36),

5.45
$$- L_h[w] \geq \frac{2A}{r^2}.$$

Therefore, by (5.41) we have

5.46
$$|u(x) - \tilde{u}(x)| \leq 2r^2 \max_{R_h} \left(\frac{|\delta|}{A} \right).$$

This result also holds if $D \leq 0$ in $R + S$.

We remark that in the application of the results of Theorem 5.24 one can often, without serious error, estimate M_3 and M_4 by using appropriate difference quotients. Thus, for example, at $x = a + 2h, a + 3h, \ldots, b - 2h$, we could estimate $u^{(4)}(x)$ by

5.47
$$u^{(4)}(x) \sim \frac{u(x + 2h) - 4u(x + h) + 6u(x) - 4u(x - h) + u(x - 2h)}{h^4}.$$

For a point next to the boundary such as $a + h$ we could use a formula involving $u(a)$, $u(a + h)$, $u(a + 2h)$, $u(a + 3h)$, and $u(a + 4h)$; similar formulas could be developed for $u^{(3)}(x)$.

Let us now consider the case where $D(x)$ is not necessarily one-signed and does not necessarily satisfy (5.26). We apply Lemma 5.18 with the function

5.48
$$\hat{w}(x) = 1 - e^{m(x - a) - 2rm}$$

where m is a positive number to be chosen later. Since $F(x) \leq 0$ and $\hat{w}(x) \geq 0$ we have

5.49
$$- L[\hat{w}] = (m^2 A + mD)e^{m[(x - a) - 2r]} - F\hat{w}(x) \geq e^{-2rm}(m^2 A - m|D|).$$

We choose m large enough so that

5.50
$$m^2 A - m|D| \geq A$$

for all x in $R + S$, i.e., so that

5.51
$$m \geq \sigma + \sqrt{1 + \sigma^2}$$

where

5.52
$$\sigma = \max_{R+S} \frac{|D|}{2A}.$$

Evidently, if m satisfies (5.51), we have

5.53
$$-L[\hat{w}] \geq Ae^{-2rm}$$

and

5.54
$$W = \max_{R+S} |\hat{w}(x)| \leq 1.$$

Moreover, by (5.8) we have

5.55
$$-L_h[\hat{w}] \geq -L[\hat{w}] - \frac{Ah^2}{12} N_4 - \frac{|D|h^2}{6} N_3,$$

where for $k = 3, 4$ we let

5.56
$$N_k = \max_{R+S} |\hat{w}^{(k)}(x)|.$$

Therefore, by (5.31) and Lemma 5.18 we have, for h sufficiently small,

5.57
$$|u(x) - \bar{u}(x)| \leq \frac{h^2}{12} \max_{R_h} \left[\frac{AM_4 + 2M_3|D|}{Ae^{-2rm} - \frac{Ah^2}{12} N_4 - \frac{|D|h^2}{6} N_3} \right]$$
$$+ \max_{S_h} |u(x) - \bar{u}(x)|.$$

But since

5.58
$$N_3 \leq m^3, \qquad N_4 \leq m^4$$

we have

5.59 $$|u(x) - \bar{u}(x)| \leq \frac{h^2}{12} \frac{M_4 + 4\sigma M_3}{e^{-2rm} - \frac{h^2}{12}[m^4 + 4m^3\sigma]} + \max_{S_h} |u(x) - \bar{u}(x)|,$$

where σ is given by (5.52). Again, h must be sufficiently small.

Thus we see that in all cases we have

5.60 $$|u(x) - \bar{u}(x)| \leq \max_{S_h} |u(x) - \bar{u}(x)| + O(h^2)$$

for h sufficiently small, provided $\bar{u} \in C^{(4)}(R + S)$. In the next two sections we shall consider methods for improving the order of accuracy. We shall use the following result which follows from Lemma 5.18 and from the construction of the above majorant function $\hat{w}(x)$.

5.61 Theorem. If $L_h[u]$ is a discrete operator of the form (5.11) satisfying the hypotheses of Lemma 5.10, then there exists a constant β such that for any function $e(x)$ and for all h sufficiently small we have

5.62 $$|e(x)| \leq \beta \max_{R_h} |L_h[e]| + \max_{S_h} |e(x)|.$$

EXERCISES 10.5

1. Show that the existence and uniqueness of a solution of (5.1)–(5.2), where $L_h[u]$ satisfies the hypotheses of Lemma 5.10, follows from Lemma 5.10.

2. Let $L_h[u]$ be a discrete linear operator of the form given by (5.11) where $-\alpha_0, \alpha_1$, and α_2 are positive functions such that $-\alpha_0 \geq \alpha_1 + \alpha_2$. Show that if $L_h[u] = 0$ in R_h then for all x in R_h

   $$\min(u(a), u(b)) \leq u(x) \leq \max(u(a), u(b)).$$

3. Give the details of the proof of Theorem 5.24 for the case where $D(x) \leq 0$ for $x \in [a, b]$.

4. Consider the two-point boundary value problem

   $$u'' - u = x^2, \qquad 0 < x < 2$$

   $$u(0) = 1, \qquad u(2) = 2.$$

   Give a bound for $|u - \bar{u}|$ on R_h where $\bar{u}$ is the exact solution and u is the numerical solution obtained by the use of the standard finite difference equation with $h = \frac{1}{2}$. Determine $|u - \bar{u}|$ by direct computation based on exact determination of $\bar{u}$ and u and also by the use of Theorem 5.24. (Determine M_3 and M_4 using the analytic solution of the problem. Note that the general solution of the differential equation is $u = c_1 e^x + c_2 e^{-x} - x^2 - 2$ for arbitrary constants c_1 and c_2.)

5. For the boundary value problem of the preceding exercise develop formulas for $u^{(3)}(x)$ and $u^{(4)}(x)$, based on 4 and 5 points, respectively, for each interior mesh point. Use the results to estimate M_3 and M_4 and compare with the exact values.

6. For the boundary value problem of Exercise 5, Section 10.4, find a bound on $|u(x) - \bar{u}(x)|$ in terms of h and

$$M_3 = \max_{0 \leq x \leq 3} |\bar{u}^{(3)}(x)|, \qquad M_4 = \max_{0 \leq x \leq 3} |\bar{u}^{(4)}(x)|$$

where $\bar{u}(x)$ is the exact solution and $u(x)$ is the solution of the three-point difference equation obtained by direct replacement of derivatives by difference quotients. What if the interval is $[0, 1]$? Also, find a bound on $|u(x) - \bar{u}(x)|$ if the coefficient of u' is replaced by $\frac{2}{9}(x - \frac{3}{2})$. (Here the interval is $[0, 3]$.)

7. Consider the two-point boundary value problem

$$u'' + (4x - 2)u' + u = 1, \qquad 0 < x < 1$$

$$u(0) = 1, \qquad u(1) = 2.$$

Find a bound on the error in terms of M_3 and M_4 for the finite difference solution corresponding to $h = \frac{1}{10}$.

8. State and prove an analog of Theorem 5.24 with $L_h[u]$ replaced by $L_h^{(1)}[u] = M_h^{(1)}[u] + Fu$ where $M_h^{(1)}[u]$ is given by (4.42) and where the differential operator $L[u]$ is given by (4.32).

9. For the differential equation

$$L[u] = u'' - 5xu' - u = 0$$

where $x \in (0, 2)$, find a function $\hat{w}(x)$ such that $\max|\hat{w}(x)| \leq 1$ for $x \in (0, 2)$, such that $w(x) \geq 0$ for $x \in [0, 2]$, and such that, for h sufficiently small, $\min(-L_h[\hat{w}]) \geq v > 0$, where v is independent of h. Here $L_h[u]$ is given by (4.11) and $2/h$ is an integer greater than unity.

10.6 MORE GENERAL DISCRETE OPERATORS

So far in our discussion of finite difference methods we have been almost entirely concerned with the discrete operator (4.11), although we have also considered the symmetric operators based on (4.42) and (4.43). In this section and in the next section we consider the use of more general discrete operators with a view to obtaining higher accuracy.

In order to simplify the discussion we shall assume for the rest of this chapter that the coefficients A, D, F, and G are analytic in $I = [a, b]$. Hence it follows from Theorems 8-11.31 and 8-11.33 that the exact solution $\bar{u}$ of (5.1) and (5.2) is analytic on $[a, b]$.

Given the integers $m, s_1, s_2, s_3, \ldots, s_m$ we consider the class of linear discrete operators of the form

6.1
$$L_h[u] = L_h[u](x) = \sum_{k=0}^{m} \alpha_k(x)u(x + s_k h)$$

where $s_0 = 0$. Evidently the operator of (4.11) is of the above form with $m = 2$, $s_1 = 1$, $s_2 = -1$ and

6.2 $\alpha_0 = -\dfrac{2A(x)}{h^2} + F(x)$, $\alpha_1 = \dfrac{A(x)}{h^2} + \dfrac{D(x)}{2h}$, $\alpha_2 = \dfrac{A(x)}{h^2} - \dfrac{D(x)}{2h}$.

Consistency and Degree of Approximation

We say that a linear discrete operator of the form (6.1) is *consistent* with a linear differential operator $L[u]$ at the point x if for every function $u(x)$ which is analytic at x, we have

6.3 $$\lim_{h \to 0} [L_h[u](x) - L[u](x)] = 0.$$

If $L_h[u]$ is consistent with $L[u]$, then the (*absolute*) *degree of approximation* of $L_h[u]$ to $L[u]$ is the largest integer q such that

6.4 $$L_h[u](x) - L[u](x) = O(h^{q+1})$$

for every analytic function $u(x)$. The operator is *uniformly consistent* in an interval I if (6.3) holds uniformly for all x in I. If $L_h[u]$ is uniformly consistent with $L[u]$, then the (absolute) uniform degree of approximation is the largest integer q such that (6.4) holds uniformly in I. Frequently we shall omit the word "uniformly" where no confusion will arise.

If $L_h[u]$ is consistent with $L[u]$ then, in particular, (6.3) holds whenever $u(x)$ is a polynomial. The converse is also true. Thus we have

6.5 Theorem   The linear discrete operator $L_h[u]$ is consistent with $L[u]$ if (6.3) holds for any polynomial $u(x)$. Moreover, if q is the largest integer such that (6.4) holds for any polynomial $u(x)$, then q is the absolute degree of approximation of $L_h[u]$ to $L[u]$.

Proof. We consider only the differential operator $L[u]$ given by (4.10) although the proof could easily be extended to more general operators. We first prove

6.6 Lemma.   If $L[u]$ is given by (4.10), if $L_h[u]$ is given by (6.1), and if (6.3) holds for any polynomial $u(x)$, then

6.7 $$\alpha_i = O(h^{-2}), \qquad i = 0, 1, \ldots, m.$$

Proof. For each $k = 0, 1, \ldots, m$ let $u_k(x)$ be the polynomial

6.8 $$u_k(x) = \prod_{\substack{j=0 \\ j \neq k}}^{m} \left(\frac{x - x_j}{x_k - x_j} \right)$$

where $x_k = x_0 + s_k h$, $k = 0, 1, \ldots, m$. Evidently

6.9
$$u_k(x_j) = \begin{cases} 1, & \text{if } j = k \\ 0, & \text{if } j \neq k \end{cases}$$

and

6.10
$$L_h[u_k](x_0) = \alpha_k(x_0).$$

Moreover,

6.11
$$L[u_k](x_0) = L\left[\prod_{\substack{j=0 \\ j \neq k}}^{m} \left(\frac{x - x_j}{x_k - x_j} \right) \right](x_0) = O(h^{-2}).$$

The lemma now follows since, by (6.3),

6.12
$$L_h[u_k](x_0) - L[u_k](x_0) = o(1).$$

Let $\bar{a}$ and $\bar{b}$ be lower and upper bounds, respectively, for the numbers

6.13
$$x_k = x_0 + s_k h, \qquad k = 0, 1, \ldots, m.$$

By Taylor's theorem (Theorem A—3.66) we have

6.14
$$u(x) = u(\bar{a}) + (x - \bar{a})u'(\bar{a}) + \frac{(x - \bar{a})^2}{2!}u''(\bar{a}) + \ldots + \frac{(x - \bar{a})^s}{s!}u^{(s)}(\bar{a})$$
$$+ \int_{\bar{a}}^{x} (x - t)^s \frac{u^{(s+1)}}{s!} dt.$$

Therefore, if (6.4) holds for any polynomial $u(x)$ for some integer $q \geq 0$ we have, as in Section 7.7,

6.15
$$R[u](x) = L_h[u](x) - L[u](x)$$
$$= 0(h^{q+1}) + R\left[\int_{\bar{a}}^{x} (x - t)^s \frac{u^{(s+1)}(t)}{s!} dt \right](x)$$
$$= 0(h^{q+1}) + \int_{\bar{a}}^{\bar{b}} R[\phi_s(x - t)] \frac{u^{(s+1)}(t)}{s!} dt,$$

where $\phi_s(x - t)$ is defined by 7-(7.55). Since for $s \geq 2$ we have

6.16
$$L[\phi_s(x - t)](x) = As(s - 1)\phi_{s-2}(x - t) + D s \phi_{s-1}(x - t) + F\phi_s(x - t)$$

and

6.17
$$L_h[\phi_s(x - t)](x) = \sum_{k=0}^{m} \alpha_k(x)\phi_s(x - t + s_k h),$$

it follows that for $s \geq q + 3$ and from Lemma 6.6,

6.18
$$\begin{cases} L[\phi_s(x - t)](x) = O(h^{q+1}) \\ L_h[\phi_s(x - t)](x) = O(h^{q+1}) \end{cases}$$

and hence

6.19
$$R[u](x) = O(h^{q+1}).$$

This completes the proof of Theorem 6.5.

Let us now consider the discrete operator

6.20
$$L_h[u] = \alpha_0 u(x_0) + \alpha_1 u(x_0 + h) + \alpha_2 u(x_0 - h).$$

Let us define $\delta_0, \delta_1, \delta_2, \ldots$ by

6.21
$$\delta_k = L_h[(x - x_0)^k] - L[(x - x_0)^k], \qquad k = 0, 1, \ldots .$$

Evidently, we have

6.22
$$\begin{cases} \delta_0 = \alpha_0 + \alpha_1 + \alpha_2 - F(x_0) \\ \delta_1 = h(\alpha_1 - \alpha_2) - D(x_0) \\ \delta_2 = h^2(\alpha_1 + \alpha_2) - 2A(x_0) \\ \delta_3 = h^3(\alpha_1 - \alpha_2) \\ \delta_4 = h^4(\alpha_1 + \alpha_2) \end{cases}$$

etc. With the α_i given by (6.2) we have $\delta_0 = \delta_1 = \delta_2 = 0$. Moreover,

6.23
$$\delta_3 = h^2 D(x_0), \qquad \delta_4 = 2h^2 A(x_0).$$

Similarly, for $k = 5, 6, \ldots$ we have

6.24
$$\delta_k = O(h^2).$$

Therefore, by Theorem 6.5 the discrete operator $L_h[u]$ given by (4.11) is consistent with $L[u]$ and has a degree of approximation of one.

The above result follows from the considerations of Section 10.5. As a matter of fact, an exact expression for $L_h[u] - L[u]$ is given by (5.7).

In Section 10.4 we showed that

6.25 $$M_h^{(1)}[x^k] + Fx^k - L[x^k] = O(h^2)$$

for $k = 0, 1, 2, \ldots$ where $M_h^{(1)}[u]$ is defined by (4.60). Here $L[u]$ is the self-adjoint operator

6.26 $$L[u] = (Au')' + Fu.$$

From this it follows that $M_h^{(1)}[u] + Fu$ is consistent with $L[u]$ and has a degree of approximation of one.

If $L_h[u]$ is consistent with $L[u]$ then the δ_i given by (6.21) must be $o(1)$ for all i and hence by (6.22) we have*

6.27
$$
\begin{cases}
\alpha_0 = -\dfrac{2\tilde{A}(x)}{h^2} + \tilde{F}(x) \\[2ex]
\alpha_1 = \dfrac{\tilde{A}(x)}{h^2} + \dfrac{\tilde{D}(x)}{2h} \\[2ex]
\alpha_2 = \dfrac{\tilde{A}(x)}{h^2} - \dfrac{\tilde{D}(x)}{2h},
\end{cases}
$$

where we let

6.28
$$
\begin{cases}
\tilde{A}(x) = A(x) + \varepsilon_2(h) \\
\tilde{D}(x) = D(x) + \varepsilon_1(h) \\
\tilde{F}(x) = F(x) + \varepsilon_0(h)
\end{cases}
$$

and

6.29 $$\varepsilon_i(h) = o(1), \qquad i = 0, 1, 2,$$

as $h \to 0$. If $F(x) < 0$ for all x in I, then for sufficiently small h we have $\alpha_1 > 0$, $\alpha_2 > 0$, and

6.30 $$-\alpha_0 \geqq \alpha_1 + \alpha_2.$$

Furthermore, we can show that the discrete problem

6.31
$$
\begin{cases}
L_h[u] = G, & \text{on } R_h \\
u(a) = u_a, & u(b) = u_b
\end{cases}
$$

is uniquely solvable. Moreover, Lemmas 5.10, 5.15, and 5.18 are applicable. On

---

* We note that $\alpha_i = O(h^{-2})$, $i = 0, 1, 2, \ldots$ as required by Lemma 6.6.

the other hand, if $F(x)$ vanishes for some or all values of x, then no matter how small h may be, it is still possible that

6.32
$$-\alpha_0 < \alpha_1 + \alpha_2.$$

This follows since $\tilde{F}(x)$ may be negative for some x. However, as we show in Section 10.8, in spite of this, most of the analysis of Section 10.5 is applicable provided h is sufficiently small.

We note that for the discrete operator $M_h^{(1)}[u]$ given by (4.60) the coefficients $\alpha_0^{(1)}$, $\alpha_1^{(1)}$, $\alpha_2^{(1)}$ satisfy (6.27) with $\tilde{F}(x)$ replaced by $F(x)$ and with $\varepsilon_0(h) = 0$ and

6.33
$$\begin{cases} \varepsilon_1(h) = \dfrac{A(x + (h/2)) - A(x - (h/2))}{h} - A'(x) = O(h^2) \\[4mm] \varepsilon_2(h) = \dfrac{A(x + (h/2)) + A(x - (h/2)) - 2A(x)}{2} = O(h^2). \end{cases}$$

Difference Equations Involving More Than Three Points

Perhaps the most direct way to increase the degree of approximation of $L_h[u]$ to $L[u]$ is to simply increase the number of points considered. Let us consider the case $m = 4$, $s_1 = -s_2 = 1$, $s_3 = -s_4 = 2$. By the method of undetermined weights or by the use of Stirling's formula of interpolation we have

6.34
$$\frac{-u(x + 2h) + 8u(x + h) - 8u(x - h) + u(x - 2h)}{12h} = u'(x) - \frac{h^4}{30} u^{(5)}(\xi)$$

and

6.35
$$\frac{-u(x + 2h) + 16u(x + h) - 30u(x) + 16u(x - h) - u(x - 2h)}{12h^2}$$

$$= u''(x) - \frac{h^4}{90} u^{(6)}(\eta),$$

where ξ and η lie in the interval $(x - 2h, x + 2h)$. If we substitute the difference quotients for the corresponding derivatives in (4.10), we obtain the corresponding difference equation

6.36 $\tilde{L}_h[u] = A\left(\dfrac{-u(x + 2h) + 16u(x + h) - 30u(x) + 16u(x + h) - u(x - 2h)}{12h^2}\right)$

$$+ D\left(\frac{-u(x + 2h) + 8u(x + h) - 8u(x - h) + u(x - 2h)}{12h}\right) + Fu = G.$$

We remark that the formula (6.35) (without the error term) can be derived as follows. We know that

6.37
$$\frac{u(x + h) + u(x - h) - 2u(x)}{h^2} = u''(x) + \frac{h^2}{12}u^{(4)}(x) + \frac{h^4}{360}u^{(6)}(\xi)$$

for some ξ in the interval $(x - h, x + h)$. To see this, use Taylor's theorem and the method 7-(7.6)–7-(7.10). If we replace $u^{(4)}$ by

6.38 $u^{(4)}(x) \sim \dfrac{u(x + 2h) - 4u(x + h) + 6u(x) - 4u(x - h) + u(x - 2h)}{h^4}$

and solve for $u''(x)$, we get (6.35). Similarly, we can obtain (6.34) by using (5.6) and replacing $u^{(3)}(\eta)$ by

6.39 $u^{(3)}(\eta) \sim \dfrac{u(x + 2h) - 2u(x + h) + 2u(x - h) - u(x - 2h)}{2h^3}.$

Moreover, if $u(x) \in C^{(6)}$ we have

6.40 $\tilde{L}_h[u] - L[u] = -\dfrac{h^4 A}{90}u^{(6)}(\eta) - \dfrac{h^4 D}{30}u^{(5)}(\xi)$

so that the absolute degree of approximation of $\tilde{L}_h[u]$ to $L[u]$ is 3.

Let us define the sets $R_h = \{a + 2h, a + 3h, \ldots, b - 2h\}$, $S_h^* = \{a + h, b - h\}$ and $S_h = \{a, b\}$, as before. Clearly, we cannot apply (6.36) at the points of S_h^* without using points outside of the interval $R + S$. One could, of course, avoid this by using unsymmetric five-point difference representations of the derivatives. Thus, for instance, we have

6.41 $u'(x) = \dfrac{-3u(x - h) - 10u(x) + 18u(x + h) - 6u(x + 2h) + u(x + 3h)}{12h}$

$$-\frac{1}{20}h^4 u^{(5)}(\xi)$$

and

6.42 $u''(x) = \dfrac{11u(x - h) - 20u(x) + 6u(x + h) + 4u(x + 2h) - u(x + 3h)}{12h^2}$

$$+\frac{1}{12}h^3 u^{(5)}(\xi).$$

The fact that error in the approximation of $u''(x)$ is $O(h^3)$ instead of $O(h^4)$ as at points of R_h is at first disturbing. However, as we shall show, we can even use a less accurate representation and still get overall accuracy of $O(h^4)$.

Treatment of Point Near the Boundary

Before discussing various procedures for treating points of S_h^* for the higher-order difference equation (6.36), let us first consider the situation where we are working with the simple difference equation (4.11). We show that the order of the error would be affected to a surprisingly small extent by using various crude procedures at points of S_h^*. We consider the following procedures:

Procedure I: We let

6.43 $$u(a + h) = u(a), \qquad u(b - h) = u(b).$$

Procedure II: We let $u(a + h)$ and $u(b - h)$ be determined by linear interpolation, i.e., by

6.44 $$u(a + h) = \tfrac{1}{2}(u(a) + u(a + 2h)), \qquad u(b - h) = \tfrac{1}{2}(u(b) + u(b - 2h)).$$

Procedure III: We use a difference equation of less accuracy for S_h^*. In each case we require that (4.11) be satisfied on R_h.

The Gerschgorin analysis can be applied directly to determine the accuracy of Procedure I. Indeed, we apply the analysis of Section 10.5 replacing S_h by S_h^* and using the smaller set R_h defined above. Evidently, we have by the mean-value theorem

6.45 $$\bar{u}(a + h) = \bar{u}(a) + h\bar{u}'(\theta)$$

where $a < \theta < a + h$. Hence, since $u(a + h) = \bar{u}(a)$ we have

6.46 $$|u(a + h) - \bar{u}(a + h)| \leq hM_1$$

where

6.47 $$M_1 = \max_{x \in R + S} |\bar{u}'(x)|.$$

Thus we have

6.48 $$\max_{S_h^*} |u(x) - \bar{u}(x)| \leq hM_1.$$

The rest of the results of Section 10.5 go through as before.

Procedure II is a specialization for the one-dimensional case of a suggestion of Collatz [1933] for improving the Gerschgorin results for the two-dimensional case. By the theory of linear interpolation (see Section 6.3) we have

6.49 $$\frac{\bar{u}(a) + \bar{u}(a + 2h)}{2} - \bar{u}(a + h) = \frac{h^2}{2}\bar{u}''(\zeta),$$

where $a < \zeta < a + 2h$. Therefore, by (6.44) it follows that

6.50
$$|e(a + h)| \leq \tfrac{1}{2}|e(a + 2h)| + \frac{h^2}{2} M_2$$

and

6.51
$$\max_{S_h^*} |e(x)| \leq \tfrac{1}{2} \max_{R_h} |e(x)| + \frac{h^2}{2} M_2.$$

Let us consider the case where $D(x) \equiv 0$. By (5.29) we have

6.52
$$|e(x)| \leq \frac{h^2 r^2}{24} M_4 + \max_{S_h^*} |e(x)|$$

so that

6.53
$$\max_{R_h} |e(x)| \leq \frac{h^2 r^2}{24} M_4 + \tfrac{1}{2} \max_{R_h} |e(x)| + \frac{h^2}{2} M_2,$$

and hence

6.54
$$\max_{R_h} |e(x)| \leq \frac{h^2 r^2}{12} M_4 + h^2 M_2.$$

A similar analysis can be carried out for the other cases considered in Section 10.5.

As an example of Procedure III let us suppose that for the differential equation

6.55
$$Au'' + Fu = G$$

with $A(x) > 0$ and $F(x) \leq 0$ in $[a, b]$ we use the difference equation

6.56 $$\hat{L}_h[u](a + h) = A \frac{2u(a) - 3u(a + h) + u(a + 3h)}{3h^2} + Fu(a + h) = G$$

at $x = a + h$ and a similar one at $b - h$. We verify directly that

6.57
$$\hat{L}_h[u] - L[u] = \frac{h u^{(3)}(\mu) A}{3},$$

where μ lies in the interval $x - h < \mu < x + 2h$. Thus we have

6.58 $$A \frac{2\bar{u}(a) - 3\bar{u}(a + h) + \bar{u}(a + 3h)}{3h^2} + F\bar{u}(a + h) = G + \varepsilon$$

where

6.59
$$|\varepsilon| \leqq A \frac{hM_3}{3}.$$

Therefore,

6.60
$$e(a + h) = \frac{1}{3 - 3(F/A)h^2} \left\{ e(a + 3h) + 2e(a) + \varepsilon \frac{3h^2}{A} \right\},$$

and since $e(a) = 0$ and $(F/A)h^2 \leqq 0$,

6.61
$$|e(a + h)| \leqq \tfrac{1}{3} |e(a + 3h)| + O(h^3).$$

As in the case of Procedure II, we can show that

6.62
$$\max_{R_h} |e(x)| = O(h^2).$$

The above analysis suggests the use of the following procedure. We use the five-point difference equation (6.36) at points of R_h and the three-point difference equation (4.11) at points of S_h^*. An analysis of this procedure for the case $A \equiv 1$, $D \equiv 0$ was given by Bramble and Hubbard [1964], who showed that the accuracy is $O(h^4)$.

Extrapolation to Zero Grid Size

One can obtain an accuracy substantially as great as that associated with the higher-order method described above and with less computational effort by using the *difference correction method*. Before describing the difference correction method, however, we first consider the somewhat related process of "extrapolation to zero grid size," due to L. F. Richardson [1910]. This process, also known as "deferred approach to the limit," involves the solution of the difference equation for two values of the step size h. The justification for the procedure is the fact that there exists an analytic function $f(x)$ such that

6.63
$$u(x) = \bar{u}(x) + h^2 f(x) + O(h^4)$$

where u satisfies (4.11) and (4.2) while $\bar{u}(x)$ satisfies (4.10) and (4.2) (see Keller [1968], pp. 78–79). Indeed, $f(x)$ is defined by the conditions

6.64
$$\begin{cases} L[f] = Af'' + Df' + Ff = -\tfrac{1}{12} A\bar{u}^{(4)} - \tfrac{1}{6} D\bar{u}^{(3)} \\ f(a) = f(b) = 0. \end{cases}$$

The existence of an analytic function $f(x)$ satisfying (6.64) follows from the analyticity of A, D, F, and $\bar{u}$ and from the fact that $A > 0$ in $[a, b]$.

We now seek to show that

6.65
$$L_h[\bar u + h^2 f] - G = O(h^4).$$

It is easy to show that

6.66 $$L_h[\bar u] = L[\bar u] + \frac{h^2 A}{12} \bar u^{(4)} + \frac{h^2 D}{6} \bar u^{(3)} + \frac{h^4 A}{360} \bar u^{(6)}(\xi) + \frac{h^4 D}{120} \bar u^{(5)}(\eta)$$

for some ξ and η in the interval $I = (x - h, x + h)$. Moreover,

6.67 $$L_h[f] = L[f] + \frac{h^2 A}{12} f^{(4)}(\xi_1) + \frac{h^2 D}{6} f^{(3)}(\xi_2)$$

for some ξ_1 and ξ_2 in I. Hence since $L[\bar u] = G$ the result (6.65) follows. Moreover, (6.63) follows from Theorem 5.61 with

6.68
$$e(x) = u(x) - \bar u(x) - h^2 f(x).$$

Suppose we solve (4.11)–(4.2) with h and $h/2$ obtaining u and u^*, respectively. Then

6.69
$$\begin{cases} u(x) = \bar u(x) + h^2 f(x) + O(h^4) \\ u^*(x) = \bar u(x) + (h/2)^2 f(x) + O(h^4) \end{cases}$$

so that

6.70
$$\bar u(x) = \tfrac{4}{3} u^*(x) - \tfrac{1}{3} u(x) + O(h^4).$$

Thus $\tfrac{4}{3} u^*(x) - \tfrac{1}{3} u(x)$ gives an $O(h^4)$ representation of $\bar u(x)$.

The Difference Correction Method

We now describe a procedure which is a slight variant of the difference correction method of L. Fox [1947]. Here we first solve the discrete problem (4.11)–(4.2) obtaining a solution u. We then solve the following problem for w:

6.71
$$\begin{cases} L_h[w] = \frac{h^2}{12} A \mathscr{D}_4[\bar u] + \frac{h^2}{6} D \mathscr{D}_3[\bar u] \text{ on } R_h + S_h^* \\ w = 0 \qquad\qquad\qquad\qquad\qquad \text{on } S_h. \end{cases}$$

Here $\mathscr{D}_4[\bar u]$ and $\mathscr{D}_3[\bar u]$ are representations of $\bar u^{(4)}(x)$ and $\bar u^{(3)}(x)$, respectively, as defined below. Having obtained w we accept as our numerical solution $\hat u$ as defined by

6.72
$$\hat u = u + w.$$

We define $\mathscr{D}_4[u]$ and $\mathscr{D}_3[u]$ on $R_h + S_h^*$ as follows:

6.73 $\quad \mathscr{D}_4[u] = \begin{cases} \dfrac{\Delta^4 u(x - 2h)}{h^4}, & \text{if } x \in R_h \\[2mm] \dfrac{\Delta^4 u(a)}{h^4}, & \text{if } x = a + h \\[2mm] \dfrac{\Delta^4 u(b - 4h)}{h^4}, & \text{if } x = b - h \end{cases}$

6.74 $\quad \mathscr{D}_3[u] = \begin{cases} \dfrac{\Delta^3 u(x - 2h) + \Delta^3 u(x - h)}{2h^3}, & \text{if } x \in R_h \\[3mm] \dfrac{- 3u(a) + 10u(a + h) - 12u(a + 2h) + 6u(a + 3h) - u(a + 4h)}{2h^3}, \\[1mm] \qquad\qquad\qquad\qquad \text{if } x = a + h \\[3mm] \dfrac{u(b - 4h) - 6u(b - 3h) + 12u(b - 2h) - 10u(b - h) + 3u(b)}{2h^3}, \\[1mm] \qquad\qquad\qquad\qquad \text{if } x = b - h. \end{cases}$

Here $\Delta^4 u(x)$ and $\Delta^3 u(x)$ are the forward differences defined in Section 6.6.

We now show that there exists an analytic function $g(x)$ such that

6.75 $$u(x) = \bar{u}(x) + h^2 f(x) + h^4 g(x) + O(h^6).$$

The function $g(x)$ is determined by the conditions

6.76 $\quad \begin{cases} L[g] = -\dfrac{A\bar{u}^{(6)}}{360} - \dfrac{D\bar{u}^{(5)}}{120} - \dfrac{A f^{(4)}}{12} - \dfrac{D f^{(3)}}{6} \\[3mm] g(a) = g(b) = 0 \end{cases}$

Here $f(x)$ is determined by (6.64). As in the case of $f(x)$, the existence of an analytic function $g(x)$ satisfying (6.76) follows from the analyticity of A, D, F, $\bar{u}$ and f. Evidently since

6.77 $\quad L_h[\bar{u}] = L[\bar{u}] + \dfrac{A h^2}{12} \bar{u}^{(4)} + \dfrac{D h^2}{6} \bar{u}^{(3)} + \dfrac{A h^4}{360} \bar{u}^{(6)} + \dfrac{D h^4}{120} \bar{u}^{(5)} + O(h^6)$

6.78 $\quad L_h[f] = L[f] + \dfrac{A h^2}{12} f^{(4)} + \dfrac{D h^2}{6} f^{(3)} + O(h^4)$

6.79 $\quad L_h[g] = L[g] + O(h^2),$

we have

6.80 $$L_h[\bar{u} + h^2f + h^4g] - L_h[u] = L_h[\bar{u} + h^2f + h^4g] - L[\bar{u}]$$

$$= h^2\left(\frac{A}{12}\,\bar{u}^{(4)} + \frac{D}{6}\,\bar{u}^{(3)} + L[f]\right) + h^4\left(\frac{A}{360}\,\bar{u}^{(6)} + \frac{D}{120}\,\bar{u}^{(5)} + \frac{A}{12}\,f^{(4)} + \frac{D}{6}\,f^{(3)} + L[g]\right)$$

$$+ O(h^6)$$

$$= O(h^6).$$

The result (6.75) follows from Theorem 5.61 with

6.81 $$e(x) = u(x) - \bar{u}(x) - h^2f(x) - h^4g(x).$$

The reader should verify that for any function $v \in C^{(6)}$ we have, on R_h,

6.82
$$\begin{cases} \mathscr{D}_4[v] = v^{(4)} + \dfrac{h^2}{6}\,v^{(6)}(\xi) \\[2mm] \mathscr{D}_3[v] = v^{(3)} + \dfrac{h^2}{4}\,v^{(5)}(\eta), \end{cases}$$

where ξ and η lie in the interval $(x - 2h, x + 2h)$. Therefore by (6.75) we have

6.83
$$\begin{cases} \mathscr{D}_4[u] = \bar{u}^{(4)} + h^2f^{(4)} + h^4g^{(4)} + O(h^6) \\ \mathscr{D}_3[u] = \bar{u}^{(3)} + h^2f^{(3)} + h^4g^{(3)} + O(h^6). \end{cases}$$

Hence by (6.71) and (6.64)

6.84 $$L_h[w] = \frac{h^2}{12}\,A\bar{u}^{(4)} + \frac{h^2}{6}\,D\bar{u}^{(3)} + O(h^4)$$

$$= -h^2L[f] + O(h^4).$$

Moreover, since

6.85 $$L_h[f] = L[f] + O(h^2)$$

we have

6.86 $$L_h[w + h^2f] = O(h^4).$$

Now let us define $\zeta(x)$ by

6.87 $$\zeta(x) = w(x) + h^2f(x).$$

The reader should verify that on S_h^*

6.88 $$L_h(\zeta) = O(h^3)$$

and that

6.89
$$\begin{cases} |\zeta(a + h)| \le \frac{1}{2}\left(1 + \frac{h|D|}{2A}\right)|\zeta(a + 2h)| + O(h^5) \\ |\zeta(b - h)| \le \frac{1}{2}\left(1 + \frac{h|D|}{2A}\right)|\zeta(b - 2h)| + O(h^5). \end{cases}$$

Since for h sufficiently small we have

6.90
$$\frac{1}{2}\left(1 + \frac{h|D|}{2A}\right) \le \gamma < 1,$$

it follows from Theorem 5.61 with S_h replaced by S_h^* that

6.91
$$\zeta(x) = O(h^4)$$

for $x \in R_h$. Thus we have finally, using (6.63) and (6.87),

6.92
$$\hat{u} - \bar{u} = u + w - \bar{u} = O(h^4).$$

We remark that the result (6.92) would also hold if we let

6.93
$$L_h[w] = 0 \quad \text{on } S_h^*.$$

A Numerical Example

To illustrate the above methods let us consider the following problem:

6.94
$$\begin{cases} u'' - u = x^2, & 0 < x < 2 \\ u(0) = 1, & u(2) = 2. \end{cases}$$

It is easy to verify that the exact solution is

6.95
$$\bar{u}(x) = \left(\frac{8}{\sinh 2}\right)\sinh x + \left(\frac{3}{\sinh 2}\right)\sinh(2 - x) - x^2 - 2.$$

First, let us apply the simple three-point difference equation

6.96
$$\frac{u(x + h) + u(x - h) - 2u(x)}{h^2} - u(x) = x^2.$$

Letting $h = \frac{1}{2}$ and $u_i = u(ih)$, $i = 0, 1, \dots, 4$, we obtain

6.97
$$\begin{cases} \dfrac{u_2 + u_0 - 2u_1}{1/4} - u_1 = \left(\dfrac{1}{2}\right)^2 \\[3mm] \dfrac{u_3 + u_1 - 2u_2}{1/4} - u_2 = 1 \\[3mm] \dfrac{u_4 + u_2 - 2u_3}{1/4} - u_3 = \left(\dfrac{3}{2}\right)^2, \end{cases}$$

or since $u_0 = 1, u_4 = 2$,

6.98
$$\begin{cases} 9u_1 - 4u_2 \qquad\quad = \dfrac{15}{4} \\[3mm] 9u_2 - 4u_1 - 4u_3 = -1 \\[3mm] 9u_3 - 4u_2 \qquad\quad = \dfrac{23}{4}. \end{cases}$$

Solving, we obtain

6.99
$$\begin{cases} u_1 = \left(\dfrac{4}{9}\right)\left(\dfrac{29}{49}\right) + \dfrac{15}{36} \doteq 0.67971 \\[3mm] u_2 = \dfrac{29}{49} \qquad\qquad\qquad \doteq 0.59184 \\[3mm] u_3 = \left(\dfrac{4}{9}\right)\left(\dfrac{29}{49}\right) + \dfrac{23}{36} \doteq 0.90193. \end{cases}$$

Thus we have the following table.

x	u	Δu	$\Delta^2 u$	$\Delta^3 u$	$\Delta^4 u$	$\bar{u}(x)$	$u(x) - \bar{u}(x)$
0	1.00000					1.00000	0
		−0.32029					
1/2	0.67971		0.23242			0.66067	0.01904
		−0.08787		0.16554			
1	0.59184		0.39796		0.22448	0.56430	0.02754
		0.31009		0.39002			
3/2	0.90193		0.78798			0.87772	0.02421
		1.09807					
2	2.00000					2.00000	0

We can estimate the error $u(x) - \bar{u}(x)$ using (5.29) obtaining, since $r = 1$,

6.100
$$|u(x) - \bar{u}(x)| \leq \frac{h^2}{24} M_4.$$

We replace M_4 by the fourth difference quotient $\Delta^4 u(x)/h^4$ obtaining the approximate bound

6.101
$$\frac{h^2}{24} \frac{\Delta^4 u}{h^4} \sim \frac{1}{6}(0.22448) \doteq 0.03741$$

which agrees reasonably well with the true bound, namely, 0.02754.

Let us now seek to improve the accuracy using the difference correction method. Using (6.71), we seek to find w such that

6.102
$$\begin{cases} L_h[w] = \dfrac{h^2}{12} \mathscr{D}_4[\bar{u}] \\ w(0) = w(2) = 0. \end{cases}$$

By (6.73) we replace $\mathscr{D}_4[\bar{u}]$ at each of the three points of R_h by $\Delta^4 u(0)/h^4$. We thus seek to solve

6.103
$$\begin{cases} \dfrac{w_0 + w_2 - 2w_1}{1/4} - w_1 = \dfrac{h^2}{12}\left(\dfrac{0.22448}{h^4}\right) \doteq 0.07483 = \delta \\[2mm] \dfrac{w_1 + w_3 - 2w_2}{1/4} - w_2 = 0.07483 \\[2mm] \dfrac{w_2 + w_4 - 2w_3}{1/4} - w_3 = 0.07483 \end{cases}$$

or

6.104
$$\begin{cases} 9w_1 - 4w_2 = -\delta \\ 9w_2 - 4w_1 - 4w_3 = -\delta \\ 9w_3 - 4w_2 = -\delta. \end{cases}$$

Solving, we get

6.105
$$\begin{cases} w_1 = \dfrac{4}{9}\left(-\dfrac{17}{49}\delta\right) - \dfrac{\delta}{9} \doteq -0.01985 \\[2mm] w_2 = -\dfrac{17}{49}\delta \qquad\qquad \doteq -0.02596 \\[2mm] w_3 = w_1 \qquad\qquad\qquad \doteq -0.01985. \end{cases}$$

Thus we have the following table.

x	u	w	$\tilde{u} = u + w$	$\bar{u}$	$\tilde{u} - \bar{u}$
0	1.00000	0	1.00000	1.00000	0
1/2	0.67971	-0.01985	0.65986	0.66067	-0.00081
1	0.59184	-0.02596	0.56588	0.56430	0.00158
3/2	0.90193	-0.01985	0.88208	0.87772	0.00436
2	2.00000	0	2.00000	2.00000	0

Evidently $\tilde{u}$ is much closer to $\bar{u}$ than is u.

We remark that the difference correction process could be repeated. Thus we could estimate $u^{(4)}(x)$ in terms of the fourth difference quotient for $\tilde{u}$, then compute a new w and repeat. For h sufficiently small this process will converge to the solution of

6.106
$$\begin{cases} \tilde{L}_h[\hat{u}] = G \\ \hat{u}(a) = u_a, \qquad \hat{u}(b) = u_b \end{cases}$$

where $\tilde{L}_h[u]$ is the five-point difference operator (6.36) at points of R_h, and the unsymmetric five-point operator based on formulas like (6.41) and (6.42) at points of S_h^*.

We now seek to explicitly construct the function $f(x)$ which has the property that

6.107
$$u(x) = \bar{u}(x) + h^2 f(x) + O(h^4).$$

By (6.64) we must have

6.108
$$L[f] = -\frac{1}{12}\bar{u}^{(4)}(x).$$

If we write $\bar{u}(x)$ in the form $\bar{u}(x) = c_1 e^x + c_2 e^{-x} - x^2 - 2$ where

6.109
$$c_1 = \frac{8 - 3e^{-2}}{e^2 - e^{-2}}, \qquad c_2 = \frac{3e^2 - 8}{e^2 - e^{-2}},$$

then we have

6.110
$$\bar{u}^{(4)}(x) = c_1 e^x + c_2 e^{-x}$$

so that

6.111
$$L[f] = f'' - f = -\frac{1}{12}(c_1 e^x + c_2 e^{-x}).$$

The general solution of the above equation is

6.112 $$f(x) = d_1 e^x + d_2 e^{-x} - \frac{c_1}{24} x e^x + \frac{c_2}{24} x e^{-x}.$$

If we require that $f(0) = f(2) = 0$ we get

6.113
$$\begin{cases} d_1 + d_2 = 0 \\ d_1 e^2 + d_2 e^{-2} - \dfrac{c_1}{12} e^2 + \dfrac{c_2}{12} e^{-2} = 0 \end{cases}$$

so that

6.114
$$\begin{cases} d_1(e^2 - e^{-2}) = \dfrac{1}{12}(c_1 e^2 - c_2 e^{-2}) \\ d_1 = -d_2 = \dfrac{1}{12} \dfrac{c_1 e^2 - c_2 e^{-2}}{e^2 - e^{-2}}. \end{cases}$$

Thus we have

6.115 $$f(x) = \frac{1}{12} \frac{c_1 e^2 - c_2 e^{-2}}{e^2 - e^{-2}} (e^x - e^{-x}) - \frac{c_1}{24} x e^x + \frac{c_2}{24} x e^{-x},$$

where c_1 and c_2 are given by (6.109).

Let us now consider the use of extrapolation to zero grid size. Letting $h = 1$ we have

6.116 $$\frac{u(2) + u(0) - 2u(1)}{1} - u(1) = 1$$

or

6.117 $$u_1(1) = u(1) = 2/3 \doteq 0.66667.$$

Thus by (6.70) the extrapolated value is

6.118 $$u_E(1) = \tfrac{4}{3} u_{\frac{1}{2}}(1) - \tfrac{1}{3} u_1(1) \doteq \tfrac{4}{3}(0.59184) - \tfrac{1}{3}(0.66667) \doteq 0.56690$$

which is appreciably closer to the exact value, $\bar{u}(1) \doteq 0.56430$, than is $u_{\frac{1}{2}}(1) \doteq 0.59184$.

EXERCISES 10.6

1. Develop a symmetric five-point discrete operator $L_h[u]$ of the form

$L_h[u] = \sum_{j=-2}^{2} \alpha_j u(x + jh)$ which is consistent with the differential operator

$$L[u] = \frac{d}{dx}(Au') + Fu$$

and which has an absolute degree of approximation of three.

2. For the boundary value problem given by (6.94) with $h = \frac{1}{2}$, solve the difference equation based on the five-point difference formula defined by (6.36) for points of R_h and the usual three-point formula for points of S_h^*. Also, carry out the solution using the three-point formula for R_h and each of Procedures I, II, and III for points of S_h^*.

3. In the example (6.94) given in the text for the difference correction method, show that the repeated application of the method will converge to the solution of the difference equation $\tilde{L}_h[u] = G$ where $\tilde{L}_h[u]$ is defined by (6.36) for points of R_h, and for points of S_h^* the operator $\tilde{L}_h[u]$ is determined by using unsymmetric difference representations based on (6.41) and (6.42).

4. Verify (6.41) and (6.42).

5. Work out the details of the derivation of (6.62).

6. Carry out an analysis of the accuracy of Procedure II under the assumption given by (5.26) where $D(x) \geqq 0$ for all $x \in [a, b]$. Also, carry out an analysis in the more general case assuming that h is sufficiently small, as in the derivation of (5.59).

7. Show that $L_h^{(2)}[u] = M_h^{(2)}[u] + Fu$ where $M_h^{(2)}[u]$ is defined by (4.43) is consistent with $L[u]$ as defined by (4.32) and has an absolute degree of approximation of unity.

8. In the example (6.94) find $g(x)$ such that $u(x) = \bar{u}(x) + h^2 f(x) + h^4 g(x) + O(h^6)$ where $f(x)$ is determined by (6.64). Also evaluate $f(1)$, compute $u(1) - h^2 f(1) - \bar{u}(1)$ and compare with $h^4 g(1)$.

9. Supply the details of the derivation of (6.34) and (6.35) by the method of undetermined weights. Also, carry out the details of the alternative method described in the text for deriving both (6.34) and (6.35) if the corresponding error terms are neglected.

10. Consider the two-point boundary value problem

$$u'' - 2u' - 3u = 1 + x^2, \qquad 0 < x < 1$$

$$u(0) = 1, \qquad u(2) = 2.$$

a) Find the exact solution $\bar{u}(x)$ and evaluate it for $x = 0, \frac{1}{2}, 1, \frac{3}{2}, 2$.

b) Solve the above problem by the ordinary finite difference method with $h = 1$ and also with $h = \frac{1}{2}$. Use extrapolation to zero grid size to get an improved value of $u(1)$.

c) Estimate $\bar{u}^{(4)}$ by an appropriate difference quotient in $u(x)$ and estimate the error $|u(1) - \bar{u}(1)|$. Compare with the computed value of the error.

d) Apply the difference correction method to improve the accuracy of the solution.

e) Find a function $f(x)$ such that $f(0) = f(2) = 0$ and such that for all h such that $2/h$ is an integer $u(x) = \bar{u}(x) + h^2 f(x) + O(h^4)$.

f) Estimate the accuracy of the solution using the methods of Section 10.5.

11. Prove Theorem 6.5 using Taylor's Theorem with the Lagrange form of the remainder (Theorem A-3.64).

12. Is it true that if $u(x) \in C^{(6)}$ then $L_h[u](x) = L[u](x) + (h^2/12)Au^{(4)}(\xi) + (h^2/6)Du^{(3)}(\xi)$, for

some ξ, where $L[u]$ and $L_h[u]$ are given by (4.10) and (4.11), respectively? Is the result true if we replace $u^{(3)}(\xi)$ by $u^{(3)}(x)$?

13. Find $\delta_0, \delta_1, \ldots$ as defined by (6.21) where $L[u]$ is given by (4.32) and where

$$L_h^{(1)}[u] = M_h^{(1)}[u] + Fu$$

and $M_h^{(1)}[u]$ is given by (4.42).

14. Carry out the details of the proof of Theorem 6.5 if $L_h[u]$ is the discrete operator (4.11).

15. Consider the two-point boundary value problem

$$u'' - (1 + x)u = 1, \qquad 0 < x < 2$$
$$u(0) = 1, \qquad u(2) = 2.$$

First solve the problem using the method of finite differences developed in Section 10.4 with $h = \frac{1}{2}$. Then obtain an improved result using one application of the difference correction method.

16. Verify (6.88).

17. Determine the result of applying the difference correction method to the boundary value problem given by (6.94) by letting $\mathcal{D}_4[u] = 0$ on S_h^*.

10.7 REFINED HIGHER-ORDER METHODS

We now show that one can construct a three-point difference equation based on a discrete operator $L_h[u]$ of the form

7.1 $$L_h[u] = \alpha_0 u(x) + \alpha_1 u(x + h) + \alpha_2 u(x - h)$$

which leads to greater accuracy than does the simple difference equation (4.11). As in the analysis of Section 10.6 we shall assume that the functions A, D, F, and G are analytic in $I = [a, b]$ so that the solution $\bar{u}(x)$ of (4.10) and (4.2) is also analytic in I.

For motivation let us consider a modification of the difference correction method as applied to the example (6.94) considered in the previous section. We again seek to determine $w = w(x)$ such that

7.2 $$L_h[w] = \frac{h^2}{12} \bar{u}^{(4)}(x)$$

where $\bar{u}(x)$ is the exact solution of (6.94). Previously, we used a fourth difference quotient of $u(x)$ to represent $\bar{u}^{(4)}(x)$. However, we now observe that we can express $\bar{u}^{(4)}(x)$ in terms of lower-order derivatives of $\bar{u}(x)$ using the differential equation and higher-order differential equations obtained by differentiating the differential equation. Thus we have from (6.94)

7.3 $$\bar{u}^{(4)} - \bar{u}^{(2)} = 2$$

and hence

7.4 $$\bar{u}^{(4)} = 2 + \bar{u}^{(2)} = 2 + x^2 + \bar{u}.$$

Thus $w(x)$ satisfies

7.5 $$L_h[w] = \frac{h^2}{12}[2 + x^2 + \bar{u}].$$

This is equivalent to

7.6 $$L_h[u + w] = \frac{h^2}{12}[2 + x^2 + \bar{u}] + x^2.$$

If we let $\tilde{u} = u + w$ and replace $\bar{u}$ by $\tilde{u}$ in (7.6), we have

7.7 $$L_h[\tilde{u}] = \frac{h^2}{12}[2 + x^2 + \tilde{u}] + x^2$$

or

7.8 $$\frac{\tilde{u}(x + h) + \tilde{u}(x - h) - 2\tilde{u}(x)}{h^2} - \tilde{u}(x) = \frac{h^2}{12}(2 + x^2 + \tilde{u}(x)) + x^2,$$

which can be written in the form

7.9 $$\tilde{L}_h[\tilde{u}] = \frac{1}{h^2\left(1 + \frac{h^2}{12}\right)}(\tilde{u}(x + h) + \tilde{u}(x - h)) - \frac{\frac{2}{h^2} + 1 + \frac{h^2}{12}}{1 + \frac{h^2}{12}}\tilde{u}(x)$$

$$= x^2 + \frac{\frac{h^2}{12}}{1 + \frac{h^2}{12}}(2)$$

or

7.10 $$\tilde{L}_h[\tilde{u}] = x^2 + \frac{\frac{h^2}{12}}{1 + \frac{h^2}{12}}\frac{d^2}{dx^2}(x^2).$$

Evidently $\tilde{L}_h[\tilde{u}]$ is a linear discrete operator which is different from $L_h[u]$ but is still consistent with $L[u]$. We also note that in our new difference equation we set $\tilde{L}_h[\tilde{u}]$ equal to x^2 plus another quantity rather than equal to x^2 alone.

Let us now consider a somewhat more direct approach for deriving a modified difference equation of the form (7.10). We consider the differential equation

7.11 $$L[u] = u'' + u' - u = G(x)$$

and seek to express the derivatives of $u(x)$ in terms of $u(x + h)$, $u(x - h)$, $u(x)$, and $L[u]$ and various derivatives of $L[u]$. Thus, for example, we have

7.12 $$\frac{\delta u}{2h} = \frac{u(x + h) - u(x - h)}{2h} = u'(x) + \frac{h^2}{6} u^{(3)}(x) + \frac{h^4}{120} u^{(5)}(x) + \cdots$$

or

7.13 $$u'(x) = \frac{\delta u}{2h} - \frac{h^2}{6} u^{(3)}(x) - \frac{h^4}{120} u^{(5)}(x) + \cdots.$$

By differentiation we have

7.14
$$L'[u] = u^{(3)} + u'' - u', \qquad u^{(3)} = L' - u'' + u' = L' - L + 2u' - u,$$
$$L''[u] = u^{(4)} + u^{(3)} - u'', \qquad u^{(4)} = L'' - u^{(3)} + u''$$
$$= L'' - L' + 2L - 3u' + 2u,$$

and hence

7.15
$$u'(x) = \frac{\delta u}{2h} - \frac{h^2}{6}(L' - L + 2u' - u) + O(h^4)$$
$$= \frac{\delta u}{2h} - \frac{h^2}{6}\left(L' - L + 2\frac{\delta u}{2h} - u\right) + O(h^4).$$

Similarly,

7.16 $$u''(x) = \frac{\delta^2 u}{h^2} - \frac{h^2}{12}\left[L'' - L' + 2L - \frac{3\delta u}{2h} + 2u\right] + O(h^4)$$

where

7.17 $$\delta^2 u = u(x + h) + u(x - h) - 2u(x).$$

Therefore we have

7.18 $$L[u] = u'' + u' - u = \tilde{L}_h[u] - \frac{h^2}{12} L'[u] - \frac{h^2}{12} L''[u] + O(h^4)$$

where

7.19 $$\tilde{L}_h[u] = \frac{\delta^2 u}{h^2} - \frac{h^2}{12} \frac{\delta u}{2h} + \frac{\delta u}{2h} - u$$

$$= \left(\frac{1}{h^2} + \frac{1}{2h} - \frac{h}{24} \right) u(x+h) + \left(\frac{1}{h^2} - \frac{1}{2h} + \frac{h}{24} \right) u(x-h)$$

$$+ u(x) \left(-\frac{2}{h^2} - 1 \right).$$

Suppose now that u satisfies the difference equation

7.20 $$\tilde{L}_h[u] = G + \frac{h^2}{12} (G' + G'')$$

and that $\bar{u}$ satisfies $L[\bar{u}] = G$. Then

7.21 $$\tilde{L}_h[u - \bar{u}] = \tilde{L}_h[u] - \tilde{L}_h[\bar{u}] = G + \frac{h^2}{12} G' + \frac{h^2}{12} G''$$

$$- \left\{ L[\bar{u}] + \frac{h^2}{12} L'[\bar{u}] + \frac{h^2}{12} L''[\bar{u}] + O(h^4) \right\}$$

$$= O(h^4)$$

since

7.22 $$L[\bar{u}] = G, \qquad L'[\bar{u}] = G', \qquad L''[\bar{u}] = G''.$$

From this it follows, as we shall show later under more general conditions, that the overall error $u(x) - \bar{u}(x)$ is $O(h^4)$.

We remark that if we have desired the error to be $O(h^6)$ we would have had to represent $u^{(5)}$ and $u^{(6)}$ in terms of u, u', L, and various derivatives of L. One then would replace u' by $\delta u/2h$ or by (7.15), as appropriate, in order to obtain $O(h^6)$ accuracy.

In order to develop a more systematic procedure for constructing higher accuracy difference equations we introduce the following definition:

A linear discrete operator $L_h[u]$ which is consistent with a linear differential operator $L[u]$ has a *relative degree of approximation p* if

7.23 $$L_h[u] - L[u] = O(h^{p+1})$$

for all analytic functions u such that

7.24 $$L[u] = 0.$$

Examples will be given below. First we prove

7.25 Theorem. Let $L_h[u]$ be a discrete linear operator of the form

7.26
$$L_h[u] = \sum_{k=0}^{m} \alpha_k\, u(x_0 + s_k h)$$

where $s_0 = 0$ and $s_1, s_2, \ldots, s_m$ are given distinct nonzero integers and where $\alpha_0, \alpha_1, \ldots, \alpha_m$ are functions of x and h. A necessary and sufficient condition that $L_h[u]$ be consistent with the operator $L[u]$ given by (4.10) and have relative degree of approximation of p is that there exist functions $\beta_0, \beta_1, \beta_2, \ldots, \beta_t$ of x and h such that

7.27
$$\lim_{h \to 0} \beta_i = 0, \qquad i = 0, 1, 2, \ldots, t$$

and such that for functions $u(x)$ which are analytic at x_0

7.28
$$L_h[u] - L[u] - \sum_{i=0}^{t} \beta_i L^{(i)}[u] = O(h^{p+1}).$$

Proof. The sufficiency of the conditions is obvious. To prove the necessity we use the Taylor's series

7.29
$$u(x_0 + s_k h) = \sum_{i=0}^{\infty} \frac{s_k^i h^i}{i!} u^{(i)}(x_0), \qquad k = 1, 2, \ldots, m$$

and obtain

7.30
$$L_h[u] - L[u] = \sum_{i=0}^{\infty} \gamma_i u^{(i)}(x_0)$$

for some $\gamma_0, \gamma_1, \gamma_2, \ldots$. First of all, we show that for i sufficiently large we have

7.31
$$\gamma_i = O(h^{p+1}).$$

Thus, for example, in the case of the three-point operator (7.1) we have by (7.29) and (4.10)

7.32
$$L_h[u] - L[u] = (\alpha_0 + \alpha_1 + \alpha_2 - F)u + (h(\alpha_1 - \alpha_2) - D)u'$$
$$+ \left(\frac{h^2}{2}(\alpha_1 + \alpha_2) - A\right)u'' + \frac{h^3}{6}(\alpha_1 - \alpha_2)u^{(3)} + \frac{h^4}{24}(\alpha_1 + \alpha_2)u^{(4)} + \cdots$$

But by the consistency of $L_h[u]$ we have

7.33
$$\alpha_1 + \alpha_2 = \frac{2A(x)}{h^2} + o\left(\frac{1}{h^2}\right), \qquad \alpha_1 - \alpha_2 = \frac{D(x)}{h} + o\left(\frac{1}{h}\right),$$

so that for i large enough we have

7.34 $\dfrac{h^i}{i!}(\alpha_1 + \alpha_2) = O(h^{p+1}),$ $\dfrac{h^i}{i!}(\alpha_1 - \alpha_2) = O(h^{p+1}).$

More generally, one can show that for some t we have

7.35 $$L_h[u] - L[u] = \sum_{i=0}^{t} \gamma_i u^{(i)}(x_0) + O(h^{p+1}).$$

Next, by repeated differentiation of the identity

7.36 $$L[u] = Au'' + Du' + Fu$$

we can express u'', $u^{(3)}$, ... in terms of $u, u', L[u], L^{(1)}[u], L^{(2)}[u], \ldots$. Thus we have for some $\beta_0, \beta_1, \ldots, \beta_t$,

7.37 $$L_h[u] - L[u] = \delta_0 u + \delta_1 u' + \sum_{i=0}^{t} \beta_i L^{(i)}[u] + O(h^{p+1}).$$

Let $v = v(x)$ be the function such that

7.38 $$\begin{cases} v(x_0) = 1, \\ v'(x_0) = 0, \\ L[v] = 0. \end{cases}$$

Evidently by (7.37) we have

7.39 $$L_h[v](x_0) - L[v](x_0) = \delta_0.$$

In order that (7.23) hold we must have

7.40 $$\delta_0 = O(h^{p+1}).$$

Similarly by considering $w = w(x)$ such that

7.41 $$\begin{cases} w(x_0) = 0 \\ w'(x_0) = 1 \\ L[w] = 0 \end{cases}$$

we can show that

7.42 $$\delta_1 = O(h^{p+1}),$$

and hence (7.28) holds.

It remains to show that (7.27) holds. For each $i = 0, 1, \ldots$ let $z_i(x)$ be the function such that

7.43
$$
\begin{cases}
z_i(x_0) = 0, \\
z_i'(x_0) = 0, \\
L[z_i] = (x - x_0)^i.
\end{cases}
$$

From (7.28) we have

7.44
$$
L_h[z_i](x_0) - L[z_i](x_0) = (i!)\beta_i + O(h^{p+1}),
$$

so that, by the consistency, (7.27) must hold. This completes the proof of Theorem 7.25.

We remark that if $L_h[u]$ is consistent with $L[u]$ and has a relative degree of approximation p, then

7.45
$$
L_h^*[u] = (1 + \beta_0)^{-1} L_h[u]
$$

is also consistent with $L[u]$ and has a relative degree of approximation of p. Here β_0 is any number satisfying the conditions of Theorem 7.25. Moreover,

7.46
$$
L_h^*[u] - L[u] - \sum_{i=1}^{t} \beta_i^* L^{(i)}[u] = O(h^{p+1})
$$

for all analytic functions $u(x)$. Here

7.47
$$
\beta_i^* = \frac{\beta_i}{1 + \beta_0} \qquad i = 1, 2, \ldots t,
$$

where $\beta_0, \beta_1, \ldots \beta_t$ are any coefficients such that (7.28) and (7.27) hold. Hence, in seeking a consistent operator $L_h[u]$ with a high relative degree of approximation, we may as well assume that $\beta_0 = 0$.

Suppose now that $L_h[u]$ is consistent with $L[u]$ and has a relative degree of approximation of p. Let $\beta_0, \beta_1, \ldots, \beta_t$ be functions such that (7.27) and (7.28) hold. We consider the difference equation

7.48
$$
L_h[u] = G + \sum_{i=0}^{t} \beta_i G^{(i)}.
$$

If $L_h[u]$ is a three-point operator of the form (7.1) and if

7.49
$$
\begin{cases}
\alpha_1 > 0, \\
\alpha_2 > 0, \\
-\alpha_0 \geqq \alpha_1 + \alpha_2,
\end{cases}
$$

then the system corresponding to (7.48) and (4.2) has a unique solution. We now show that in this case the accuracy of the solution of (7.48) is $O(h^{p+1})$.

7.50 Theorem. Let $L_h[u]$ be a three-point linear discrete operator of the form (7.1) such that the conditions (7.49) hold for h sufficiently small and such that $L_h[u]$ is consistent with the operator $L[u]$, which is given by (4.10), and has a (uniform) relative degree of approximation of p. Let $u(x)$ be the solution of (7.48) and (4.2) and let $\bar{u}(x)$ be the solution of (4.10) and (4.2). Then

7.51
$$\max_{R_h} |u(x) - \bar{u}(x)| = O(h^{p+1}).$$

Proof. Let $\hat{w}(x)$ be given by (5.48) where m satisfies (5.51). Then by (5.53) we have

7.52
$$- L[\hat{w}] \geq A e^{-2rm},$$

and, by the consistency of $L_h[u]$,

7.53
$$- L_h[\hat{w}] \geq A e^{-2rm} - \varepsilon(h)$$

where $\varepsilon(h) = o(1)$ as $h \to 0$. Consequently, by Lemma 5.18 and (5.54) we have for h sufficiently small

7.54
$$|u(x) - \bar{u}(x)| \leq \max_{R+S} |\hat{w}(x)| \max_{R_h} \left[\frac{|L_h[e]|}{- L_h[\hat{w}]} \right] \leq \max_{R_h} \left[\frac{|L_h[e]|}{- L_h[\hat{w}]} \right],$$

where $e = e(x) = u(x) - \bar{u}(x)$. But by (7.48) and (7.28) we have

7.55
$$L_h[e] = L_h[u - \bar{u}] = L_h[u] - L_h[\bar{u}]$$

$$= G + \sum_{i=0}^{t} \beta_i G^{(i)} - L_h[\bar{u}]$$

$$= G + \sum_{i=0}^{t} \beta_i G^{(i)} - \left\{ L[\bar{u}] + \sum_{i=0}^{t} \beta_i L^{(i)}[\bar{u}] + O(h^{p+1}) \right\}$$

$$= (G - L[\bar{u}]) + \sum_{i=0}^{t} \beta_i (G - L[\bar{u}])^{(i)} + O(h^{p+1})$$

$$= O(h^{p+1}).$$

Therefore, we have

7.56
$$\max_{R_h} |u(x) - \bar{u}(x)| = O(h^{p+1}),$$

and the theorem follows.

We remark that if $F(x) < 0$ for all x in I, then the conditions (7.49) hold for h sufficiently small provided that $L_h[u]$ is consistent with $L[u]$. However, as noted in Section 10.6, it may happen that in the general case $-\alpha_0 < \alpha_1 + \alpha_2$ for at least some points of R_h no matter how small h may be. We shall show in Section 10.8 that even in this case (7.48) is uniquely solvable for h sufficiently small and Theorem 7.50 holds.

Let us now illustrate a procedure for finding a three-point operator $L_h[u]$ of the form (7.1) which is consistent with $L[u]$ and has a degree of approximation of p, where $p > 1$, relative to $L[u]$ where

7.57
$$L[u] = u'' + Pu' + Qu$$

and where

7.58
$$Q(x) \leqq 0 .$$

(We recall that the simple operator

7.59 $$L_h^{(0)}[u] = \frac{u(x_0 + h) + u(x_0 - h) - 2u(x_0)}{h^2} + P(x_0)\frac{u(x_0 + h) - u(x_0 - h)}{2h}$$

$$+ Q(x_0)u(x_0)$$

has an absolute degree of approximation of one as well as a relative degree of approximation of one.)

Our procedure is to substitute the Taylor's series,

7.60 $$u(x_0 \pm h) = u_0 \pm hu_1 + \frac{h^2}{2}u_2 \pm \frac{h^3}{6}u_3 + \frac{h^4}{24}u_4 + \cdots ,$$

where we let

7.61 $$u_i = u^{(i)}(x_0), \qquad i = 0, 1, 2, \ldots ,$$

in the left member of (7.28) obtaining

7.62 $$L_h[u] - L[u] - \sum_{i=0}^{t} \beta_i L^{(i)}[u] = \sum_{i=0}^{\infty} \delta_i u_i .$$

We seek to choose the α_i and the β_i so that $\beta_0 = 0$ and

7.63 $$\begin{cases} \delta_i = O(h^{p+1}), & i = 0, 1, 2, \ldots \\ \beta_i = o(1), & i = 1, 2, \ldots \end{cases}$$

From (7.1) and (7.60) we have

7.64 $$L_h[u] = \alpha_0 u_0 + \alpha_1 \left(u_0 + h u_1 + \frac{h^2}{2!} u_2 + \frac{h^3}{3!} u_3 + \frac{h^4}{4!} u_4 + \cdots \right)$$

$$+ \alpha_2 \left(u_0 - h u_1 + \frac{h^2}{2!} u_2 - \frac{h^3}{3!} u_3 + \frac{h^4}{4!} u_4 + \cdots \right)$$

$$= u_0(\alpha_0 + \alpha_1 + \alpha_2) + u_1 \left[h(\alpha_1 - \alpha_2) \right] + u_2 \left[\frac{h^2}{2}(\alpha_1 + \alpha_2) \right]$$

$$+ u_3 \left[\frac{h^3}{6}(\alpha_1 - \alpha_2) \right] + u_4 \left[\frac{h^4}{24}(\alpha_1 + \alpha_2) \right] + \cdots .$$

By (7.57) we have

7.65
$$\begin{cases}
L[u] = u_2 + P_0 u_1 + Q_0 u_0 \\[4pt]
L'[u] = u_3 + P_0 u_2 + (P_1 + Q_0) u_1 + Q_1 u_0 \\[4pt]
L^{(2)}[u] = u_4 + P_0 u_3 + (2P_1 + Q_0) u_2 + (P_2 + 2Q_1) u_1 + Q_2 u_0 \\[4pt]
L^{(3)}[u] = u_5 + P_0 u_4 + (3P_1 + Q_0) u_3 + (3P_2 + 3Q_1) u_2 + (P_3 + 3Q_2) u_1 \\[4pt]
\hspace{10cm} + Q_3 u_0,
\end{cases}$$

etc. Therefore, the δ_i of (7.62) are given by

7.66
$$\begin{cases}
\delta_0 = \alpha_0 + \alpha_1 + \alpha_2 - Q_0 - \beta_1 Q_1 - \beta_2 Q_2 - \beta_3 Q_3 - \ldots \\[6pt]
\delta_1 = h(\alpha_1 - \alpha_2) - P_0 - \beta_1(P_1 + Q_0) - \beta_2(P_2 + 2Q_1) - \beta_3(P_3 + 3Q_2) - \ldots \\[6pt]
\delta_2 = \frac{h^2}{2}(\alpha_1 + \alpha_2) - 1 - \beta_1 P_0 - \beta_2(2P_1 + Q_0) - \beta_3(3P_2 + 3Q_1) - \ldots \\[6pt]
\delta_3 = \frac{h^3}{6}(\alpha_1 - \alpha_2) \qquad - \beta_1 \quad - \beta_2 P_0 \qquad - \beta_3(3P_1 + Q_0) - \ldots \\[6pt]
\delta_4 = \frac{h^4}{24}(\alpha_1 + \alpha_2) \qquad\qquad - \beta_2 \qquad\qquad - \beta_3 P_0 - \ldots ,
\end{cases}$$

etc. Here, for convenience, for any function $f(x)$ we let

7.67 $$f_i = f^{(i)}(x_0), \qquad i = 0, 1, \ldots .$$

For the case $t = 2$, we can represent the δ_i by the following table:

	α_0	α_1	α_2	θ_0	θ_1	θ_2	-1	$-\beta_1$	$-\beta_2$
δ_0	1	1	1	1	0	0	Q_0	Q_1	Q_2
δ_1	0	h	$-h$	0	0	h	P_0	$P_1 + Q_0$	$P_2 + 2Q_1$
δ_2	0	$\dfrac{h^2}{2!}$	$\dfrac{h^2}{2!}$	0	$\dfrac{h^2}{2!}$	0	1	P_0	$2P_1 + Q_0$
δ_3	0	$\dfrac{h^3}{3!}$	$-\dfrac{h^3}{3!}$	0	0	$\dfrac{h^3}{3!}$	0	1	P_0
δ_4	0	$\dfrac{h^4}{4!}$	$\dfrac{h^4}{4!}$	0	$\dfrac{h^4}{4!}$	0	0	0	1
δ_5	0	$\dfrac{h^5}{5!}$	$-\dfrac{h^5}{5!}$	0	0	$\dfrac{h^5}{5!}$	0	0	0
δ_6	0	$\dfrac{h^6}{6!}$	$\dfrac{h^6}{6!}$	0	$\dfrac{h^6}{6!}$	0	0	0	0
δ_7	0	$\dfrac{h^7}{7!}$	$-\dfrac{h^7}{7!}$	0	0	$\dfrac{h^7}{7!}$	0	0	0
δ_8	0	$\dfrac{h^8}{8!}$	$\dfrac{h^8}{8!}$	0	$\dfrac{h^8}{8!}$	0	0	0	0

Here, for convenience, we let

7.68 $\qquad \theta_0 = \alpha_0 + \alpha_1 + \alpha_2, \qquad \theta_1 = \alpha_1 + \alpha_2, \qquad \theta_2 = \alpha_1 - \alpha_2.$

We show that if h is sufficiently small, we can make $\delta_0 = \delta_1 = \delta_2 = \delta_3 = \delta_4 = 0$ and $\delta_i = 0(h^4)$, for $i \geqq 5$. From the conditions that $\delta_4 = \delta_3 = 0$ we have

7.69 $\qquad \beta_2 = \dfrac{h^4}{24}\theta_1, \qquad \beta_1 = \dfrac{h^3}{6}\theta_2 - P_0\beta_2 = \dfrac{h^3}{6}\theta_2 - P_0\dfrac{h^4}{24}\theta_1.$

We now substitute in the equations $\delta_0 = \delta_1 = \delta_2 = 0$ obtaining the system

7.70
$$
\begin{vmatrix}
1 & -\dfrac{h^2}{12}(Q_2 - Q_1 P_0) & -\dfrac{h^2}{6}Q_1 \\[2ex]
0 & \dfrac{h^2}{12}[P_0(P_1 + Q_0) - (P_2 + 2Q_1)] & 1 - \dfrac{h^2}{6}(P_1 + Q_0) \\[2ex]
0 & 1 - \dfrac{h^2}{12}(2P_1 + Q_0 - P_0^2) & -\dfrac{h^2}{6}P_0
\end{vmatrix}
\begin{pmatrix} \theta_0 \\[2ex] \dfrac{h^2}{2}\theta_1 \\[2ex] h\theta_2 \end{pmatrix}
=
\begin{pmatrix} Q_0 \\[2ex] P_0 \\[2ex] 1 \end{pmatrix}.
$$

For sufficiently small h the determinant of the above system does not vanish; in fact, when $h = 0$, the value of the determinant is -1. Letting $h = 0$ in the matrix of (7.70) we get

7.71 $$\theta_0 = Q_0, \qquad h\theta_2 = P_0, \qquad \frac{h^2}{2}\theta_1 = 1.$$

By the use of Cramer's rule we can show that in general we have

7.72
$$\begin{cases} \dfrac{h^2}{2}\theta_1 = \dfrac{-1 + O(h^2)}{-1 + O(h^2)} = 1 + O(h^2), \qquad \theta_1 = \dfrac{2}{h^2} + O(1) \\[2mm] h\theta_2 = \dfrac{-P_0 + O(h^2)}{-1 + O(h^2)} = P_0 + O(h^2), \qquad \theta_2 = \dfrac{P_0}{h} + O(h) \\[2mm] \theta_0 = Q_0 + O(h^2). \end{cases}$$

From (7.69) it follows that $\beta_1 = O(h^2)$, $\beta_2 = O(h^2)$, and hence (7.27) holds. Moreover, we can easily show that $\delta_5, \delta_6, \ldots$ are all $O(h^4)$. Hence (7.28) holds with $p = 3$ and $L_h[u]$ is consistent with $L[u]$ and has a relative degree of approximation of 3.

In the special case $P = 0, Q = -1$, we have by (7.70)

7.73
$$\begin{pmatrix} 1 & 0 & 0 \\ 0 & 0 & 1 + \dfrac{h^2}{6} \\ 0 & 1 + \dfrac{h^2}{12} & 0 \end{pmatrix} \begin{pmatrix} \theta_0 \\ \dfrac{h^2}{2}\theta_1 \\ h\theta_2 \end{pmatrix} = \begin{pmatrix} -1 \\ 0 \\ 1 \end{pmatrix}$$

or

7.74 $$\theta_0 = -1, \qquad \theta_2 = 0, \qquad \theta_1 = \frac{2}{h^2}\frac{1}{1 + h^2/12} = \frac{24}{h^2(h^2 + 12)}.$$

Therefore, by (7.69)

7.75 $$\beta_2 = \frac{h^2}{h^2 + 12}, \qquad \beta_1 = 0.$$

Moreover, since $\alpha_1 + \alpha_2 = \theta_1$, $\alpha_1 - \alpha_2 = \theta_2$, $\alpha_0 + \alpha_1 + \alpha_2 = \theta_0$ we have

7.76 $$\alpha_1 = \alpha_2 = \frac{12}{h^2(h^2 + 12)}, \qquad \alpha_0 = -\frac{h^4 + 12h^2 + 24}{h^2(h^2 + 12)}.$$

Hence we get the operator $\tilde{L}_h[u]$ of (7.9).

We now show that in the general case, there is no advantage in using only a single β. Carrying out the process described above for the case $t = 1$, we have, upon choosing $\beta_1 = h^3 \theta_2/6$ so that $\delta_3 = 0$,

7.77
$$
\begin{pmatrix}
1 & 0 & -\dfrac{h^2}{6} Q_1 \\[2ex]
0 & 0 & 1 - \dfrac{h^2}{6}(P_1 + Q_0) \\[2ex]
0 & 1 & -P_0 \dfrac{h^2}{6}
\end{pmatrix}
\begin{pmatrix}
\theta_0 \\[2ex]
\dfrac{h^2}{2}\theta_1 \\[2ex]
h\theta_2
\end{pmatrix}
=
\begin{pmatrix}
Q_0 \\[2ex]
P_0 \\[2ex]
1
\end{pmatrix}.
$$

Evidently $\theta_2 = P_0/h + O(h)$, $\theta_1 = 2h^{-2} + O(1)$ so that

7.78
$$
\delta_4 = \frac{h^4}{4!}\theta_1 = \frac{1}{12}h^2 + O(h^4).
$$

Hence $p = 1$. But if we let $\beta_1 = 0$ we have (7.71), and moreover, $\delta_3, \delta_4, \dots$ are all $O(h^2)$. Thus the relative degree of approximation with $\beta_1 = 0$ is the same as with the best choice of β_1. Hence if we wish to use only a single β we may as well let all β_i vanish.

We note that in order to obtain a relative degree of approximation of 3, we need to evaluate the first two derivatives of P, Q, and G. The same order of accuracy could be achieved by using the method of Taylor's series applied to the differential equation

7.79
$$
L[u] = u'' + Pu' + Qu = G.
$$

One would use the formulas

7.80
$$
u(x_{n+1}) = u(x_n) + hu'(x_n) + \frac{h^2}{2}u''(x_n) + \frac{h^3}{6}u^{(3)}(x_n) + \frac{h^4}{24}u^{(4)}(x_n)
$$
$$
u'(x_{n+1}) = u'(x_n) + hu''(x_n) + \frac{h^2}{2}u^{(3)}(x_n) + \frac{h^3}{6}u^{(4)}(x_n).
$$

One could express u'', $u^{(3)}$, and $u^{(4)}$ in terms of u' and u using the differential equation (7.79) and additional equations obtained by differentiating the differential equation. Clearly, one would not have to consider derivatives of P, Q, and G of order greater than two.

The method of superposition could be used based on the solution obtained with $u'(x_0) = 0$ and $u'(x_0) = 1$, choosing an appropriate linear combination so that the right-hand boundary condition is satisfied. Clearly, the overall error will be of order h^4. Since, as we shall later show, we obtain an overall error of order h^4 when $p = 3$, we thus obtain the same accuracy using β_1 and β_2 as with the Taylor series method. Moreover, the work required by the two procedures is comparable.

We now show that if h is sufficiently small, by the use of $\beta_1, \beta_2, \ldots, \beta_t$ we can achieve a relative degree of approximation of at least $t + 1$ if t is even and t if t is odd. We show that we can choose $\alpha_0, \alpha_1, \alpha_2, \beta_1, \beta_2, \ldots, \beta_t$ so that $\delta_0 = \delta_1 = \ldots = \delta_{t+2} = 0$. By letting $\delta_3 = \delta_4 = \ldots = \delta_{t+2} = 0$ we can express each β_i as a linear combination of the θ_i with coefficients involving h^3 or higher-order powers of h. If we substitute in the equations $\delta_0 = \delta_1 = \delta_2 = 0$, we get a system of the form

7.81
$$\begin{pmatrix} 1 & \varepsilon_{1,2} & \varepsilon_{1,3} \\ 0 & \varepsilon_{2,2} & 1 + \varepsilon_{2,3} \\ 0 & 1 + \varepsilon_{3,2} & \varepsilon_{3,3} \end{pmatrix} \begin{pmatrix} \theta_0 \\ \dfrac{h^2}{2}\theta_1 \\ h\theta_2 \end{pmatrix} = \begin{pmatrix} Q_0 \\ P_0 \\ 1 \end{pmatrix}$$

where each $\varepsilon_{i,j} = O(h^2)$. Consequently, for h sufficiently small, we can solve for $\theta_0, \theta_1, \theta_2$ obtaining

7.82
$$\theta_0 = Q_0 + O(h^2), \qquad \theta_1 = \frac{2}{h^2} + O(1), \qquad \theta_2 = \frac{P_0}{h} + O(h).$$

Clearly, the β_i are $O(h)$, so that (7.27) holds. Moreover, if t is even, then

7.83
$$\begin{cases} \delta_{t+3} = \dfrac{h^{t+3}}{(t+3)!}\theta_2 = O(h^{t+2}) \\[2mm] \delta_{t+4} = \dfrac{h^{t+4}}{(t+4)!}\theta_1 = O(h^{t+2}). \end{cases}$$

Similarly, $\delta_i = O(h^{t+2})$ for $i > t + 4$. Hence the relative degree of approximation is at least $t + 1$. If t is odd, then

7.84
$$\delta_{t+3} = \frac{h^{t+3}}{(t+3)!}\theta_1 = O(h^{t+1}).$$

Similarly, $\delta_i = O(h^{t+1})$ for $i > t + 3$, and hence the relative degree of approximation is at least t.

The actual determination of the α_i and β_i can be carried out in several ways. First, we can solve the first $t + 3$ equations $\delta_0 = \delta_1 = \ldots \delta_{t+2} = 0$ analytically for the α_i and β_i as described above. This procedure is very laborious at best. Second, we can, for each value of x solve the equations $\delta_0 = \delta_1 = \ldots = \delta_{t+2} = 0$ numerically using some procedure such as the Gaussian elimination method. Third, we can develop approximate analytic formulas for the coefficients by using an iterative process which we now describe.

The basis of the iterative process is that we do not need to solve the equations $\delta_0 = \delta_1 = \ldots = \delta_{t+2} = 0$ exactly in order to achieve a relative degree of approximation of p. It is sufficient that each equation be satisfied to within $O(h^{p+1})$. We illustrate the process for the case $P = 0$, $Q = -1$, $G = x^2$, $t = 2$. First, we let $\beta_1 = \beta_1^{(0)} = 0$ and $\beta_2 = \beta_2^{(0)} = 0$ and solve $\delta_0 = \delta_1 = \delta_2 = 0$ for θ_0, θ_1, and θ_2 obtaining

7.85
$$\theta_0^{(0)} = -1, \qquad \theta_2^{(0)} = 0, \qquad \theta_1^{(0)} = \frac{2}{h^2},$$

Next we solve $\delta_3 = \delta_4 = 0$ for β_1 and β_2 letting $\theta_i = \theta_i^{(0)}$, $i = 0, 1, 2$, and obtain

7.86
$$\begin{cases} \beta_2^{(1)} = \dfrac{h^4}{24}\theta_1^{(0)} = \dfrac{h^2}{12} \\[3mm] \beta_1^{(1)} = \dfrac{h^3}{6}\theta_2^{(0)} = 0. \end{cases}$$

We then solve $\delta_0 = \delta_1 = \delta_2 = 0$ for $\theta_0^{(1)}, \theta_1^{(1)}$, and $\theta_2^{(1)}$ letting $\beta_1 = \beta_1^{(1)}, \beta_2 = \beta_2^{(1)}$ and obtain

7.87
$$\begin{cases} \theta_0^{(1)} = -1 \\[3mm] \theta_1^{(1)} = \dfrac{2}{h^2}\left(1 - \dfrac{h^2}{12}\right) \\[4mm] \theta_2^{(1)} = 0. \end{cases}$$

If we were to use the $\theta_i^{(1)}$ and $\beta_i^{(1)}$ we would have $\delta_0 = \delta_1 = \delta_2 = \delta_3 = 0$ and

7.88
$$\delta_4 = \frac{h^4}{4!}\theta_1^{(1)} - \beta_2^{(1)} = \frac{h^4}{24}\left(\frac{2}{h^2}\right)\left(1 - \frac{h^2}{12}\right) - \frac{h^2}{12}$$
$$= \frac{-h^4}{144}.$$

Moreover, $\delta_5, \delta_6, \ldots,$ are $O(h^4)$. Thus the values

7.89
$$\begin{cases} \theta_0 = -1, \qquad \theta_2 = 0, \qquad \theta_1 = \dfrac{2}{h^2}\left(1 - \dfrac{h^2}{12}\right) = \dfrac{12 - h^2}{6h^2}, \\[4mm] \beta_1 = 0, \qquad \beta_2 = \dfrac{h^2}{12}, \end{cases}$$

and the corresponding values of α_i yield a relative degree of approximation of 3.

One could, of course, continue the iterative process. This would not improve the relative degree of approximation but the values of the α_i and β_i thus obtained would converge to the values given by (7.75) and (7.76).

With the values of the θ_i and β_i given by (7.89), the difference equation (7.48) becomes

7.90
$$L_h[u] = \frac{1}{h^2}\left(1 - \frac{h^2}{12}\right)[u(x+h) + u(x-h)] - \left[1 + \frac{2}{h^2}\left(1 - \frac{h^2}{12}\right)\right]u(x)$$

$$= x^2 + \frac{h^2}{12}(2).$$

Solving for the case $u(0) = 1, u(2) = 2$, with $h = 1/2$ we get

7.91 $u(1/2) \doteq 0.66043, \qquad u(1) \doteq 0.56395, \qquad u(3/2) \doteq 0.87741.$

On the other hand, if we use (7.9) we get

7.92 $u(1/2) \doteq 0.66083, \qquad u(1) \doteq 0.56453, \qquad u(3/2) \doteq 0.87792.$

Both results agree closely with the exact solution $\bar{u}(x)$ given in Section 10.6.

The reader should show that for the case $L[u] = u'' + u' - u$ the above procedure yields

7.93
$$\begin{cases} \theta_0^{(0)} = -1, \qquad \theta_1^{(0)} = \frac{2}{h^2}, \qquad \theta_2^{(0)} = \frac{1}{h} \\[2mm] \beta_1^{(1)} = \frac{h^2}{12} = \beta_2^{(1)}, \qquad \theta_0^{(1)} = -1, \qquad \theta_1^{(1)} = \frac{2}{h^2}, \qquad \theta_2^{(1)} = \frac{1}{h}\left(1 - \frac{h^2}{12}\right), \end{cases}$$

and the corresponding values of α_k are

7.94 $\alpha_1^{(1)} = \frac{1}{h^2} + \frac{1}{2h} - \frac{h}{24}, \qquad \alpha_2^{(1)} = \frac{1}{h^2} - \frac{1}{2h} + \frac{h}{24}, \qquad \alpha_0^{(1)} = -1 - \frac{2}{h^2}.$

Thus the result of using this procedure agrees with (7.18)–(7.19).

Let us further illustrate the procedure by the case $m = 4$, $s_1 = -s_2 = 1$, $s_3 = -s_4 = 2$, $s_0 = 0$ and $t = 4$. Letting $\theta_0 = \alpha_0 + \alpha_1 + \alpha_2 + \alpha_3 + \alpha_4$; $\theta_1 = \alpha_1 + \alpha_2$, $\theta_2 = \alpha_1 - \alpha_2$, $\theta_3 = \alpha_3 + \alpha_4$, $\theta_4 = \alpha_3 - \alpha_4$, then the δ_i as defined by (7.62) are

	θ_0	θ_1	θ_2	θ_3	θ_4	-1	$-\beta_1$	$-\beta_2$	$-\beta_3$	$-\beta_4$
δ_0	1	0	0	0	0	Q_0	Q_1	Q_2	Q_3	Q_4
δ_1	0	0	h	0	$2h$	P_0	P_1+Q_0	P_2+2Q_1	P_3+3Q_2	P_4+4Q_3
δ_2	0	$\dfrac{h^2}{2}$	0	$4\dfrac{h^2}{2}$	0	1	P_0	$2P_1+Q_0$	$3P_2+3Q_1$	$4P_3+6Q_2$
δ_3	0	0	$\dfrac{h^3}{6}$	0	$8\dfrac{h^3}{6}$	0	1	P_0	$3P_1+Q_0$	$6P_2+4Q_1$
δ_4	0	$\dfrac{h^4}{24}$	0	$16\dfrac{h^4}{24}$	0	0	0	1	P_0	$4P_1+Q_0$
δ_5	0	0	$\dfrac{h^5}{120}$	0	$32\dfrac{h^5}{120}$	0	0	0	1	P_0
δ_6	0	$\dfrac{h^6}{720}$	0	$64\dfrac{h^6}{720}$	0	0	0	0	0	1
δ_7	0	0	$\dfrac{h^7}{5040}$	0	$128\dfrac{h^7}{5040}$	0	0	0	0	0
δ_8	0	$\dfrac{h^8}{40{,}320}$	0	$256\dfrac{h^8}{40320}$	0	0	0	0	0	0

Letting $\beta_1 = \beta_2 = \beta_3 = \beta_4 = 0$ we solve $\delta_0 = \delta_1 = \delta_2 = \delta_3 = \delta_4 = 0$ for the $\theta_i^{(0)}$ obtaining

7.95 $\theta_0^{(0)} = Q_0, \qquad \theta_1^{(0)} = \dfrac{8}{3h^2}, \qquad \theta_2^{(0)} = \dfrac{8P_0}{6h}, \qquad \theta_3^{(0)} = -\dfrac{1}{6h^2}, \qquad \theta_4^{(0)} = -\dfrac{P_0}{6h}.$

This corresponds to the five-point operator (6.36). We then let $\beta_1^{(1)} = \beta_2^{(1)} = 0$ and choose $\beta_3^{(1)}$ and $\beta_4^{(1)}$ so that $\delta_5 = \delta_6 = 0$, thus obtaining

7.96 $$\beta_4^{(1)} = -\frac{h^4}{90}, \qquad \beta_3^{(1)} = -\frac{h^4 P_0}{45}.$$

We then solve $\delta_0 = \delta_1 = \delta_2 = \delta_3 = \delta_4 = 0$ for $\theta_i^{(1)}$ using the above values of $\beta_i^{(1)}$ and with $\beta_1^{(1)} = \beta_2^{(1)} = 0$. One can readily verify that

7.97 $\theta_1^{(1)} = O(h^{-2}), \qquad \theta_3^{(1)} = O(h^{-2}), \qquad \theta_2^{(1)} = O(h^{-1}), \qquad \theta_4^{(1)} = O(h^{-1}),$

and all $\delta_i = O(h^6)$. Therefore $p = 5$.

As an alternative, we may obtain $p = 5$ by using only β_1 and β_2. This, of course, is simpler since one needs to differentiate the differential equation only twice instead of four times as with the other scheme. Here we solve $\delta_0 = \delta_1 = \delta_2 = \delta_5 = \delta_6 = 0$ for $\theta_i^{(0)}$, letting $\beta_1^{(0)} = \beta_2^{(0)} = 0$. We then choose $\beta_1^{(1)}$ and $\beta_2^{(1)}$ so that $\delta_3 = \delta_4 = 0$. Next, we recompute $\theta_i^{(1)}$ so that $\delta_0 = \delta_1 = \delta_2 = \delta_5 = \delta_6 = 0$. Continuing, we compute $\beta_1^{(2)}$ and $\beta_2^{(2)}$ so that $\delta_3 = \delta_4 = 0$ and $\theta_i^{(2)}$ so that $\delta_0 = \delta_1 = \delta_2 = \delta_5 = \delta_6 = 0$.

Thus we have

7.98
$$\theta_0^{(0)} = Q_0, \qquad \theta_1^{(0)} = \frac{64}{30h^2}, \qquad \theta_2^{(0)} = \frac{32}{30}\frac{P_0}{h},$$

$$\theta_3^{(0)} = -\frac{1}{30h^2}, \qquad \theta_4^{(0)} = -\frac{P_0}{30h}$$

and

7.99
$$\beta_1^{(1)} = \frac{P_0 h^2}{15}, \qquad \beta_2^{(1)} = \frac{h^2}{15}.$$

Therefore,

7.100
$$\begin{cases} \theta_0^{(1)} = Q_0 + \dfrac{h^2}{15} P_0 Q_1 + \dfrac{h^2}{15} Q_2 \\[2mm] \theta_1^{(1)} = -64\theta_3^{(1)} \\[2mm] \theta_2^{(1)} = -32\theta_4^{(1)} \\[2mm] \theta_4^{(1)} = -\dfrac{1}{30h}\left\{ P_0 + (P_1 + Q_0)\dfrac{h^2 P_0}{15} + (P_2 + 2Q_1)\dfrac{h^2}{15} \right\} \\[2mm] \theta_3^{(1)} = -\dfrac{1}{30h^2}\left\{ 1 + \dfrac{h^2 P_0^2}{15} + \dfrac{h^2}{15}(2P_1 + Q_0) \right\}. \end{cases}$$

We continue this process, obtaining $\beta_i^{(2)}$ and $\theta_i^{(2)}$. One can easily verify that with the values thus obtained we get $\delta_i = O(h^6)$ for all i and hence $p = 5$.

EXERCISES 10.7

1. Develop a difference equation analogous to (7.20) which leads to $O(h^6)$ accuracy for the differential equation (7.11).

2. Verify that the five-point discrete operator determined by $\theta_i^{(1)}$ and $\beta_i^{(1)}$ given by (7.99) and (7.100) is consistent with $L[u]$, given by (7.57), and has a relative degree of approximation of five. Consider the case where four β_i are used and also the case where two β_i are used.

3. Determine a three-point discrete operator which is consistent with $L[u] = u'' - u$ and has a relative degree of approximation of five. Compare with the exact operator.

4. Find a discrete operator $L_h[u]$ of the form (7.1) which is consistent with

$$L[u] = u'' - \frac{1}{x}u'$$

and has a relative degree of approximation of three. What is the absolute degree of approximation? What difference equations would be used to represent the differential equations $L[u] = 3x$ and $L[u] = x^3$.

5. Consider the two-point boundary value problem

$$L[u] = u'' - 2u' - 3u = 1 + x^2, \qquad 0 < x < 2$$

$$u(0) = 1, \qquad u(2) = 2.$$

Find a linear discrete operator of the form

$$L_h[u] = \alpha_0 u(x) + \alpha_1 u(x + h) + \alpha_2 u(x - h)$$

which is consistent with $L[u]$ and has a relative degree of approximation of three. Construct an appropriate difference equation and solve it for $h = 1$ and $h = \frac{1}{2}$. Verify that the error is $O(h^4)$.

6. Find the absolute degree of approximation of the discrete operator given by (7.1) to the differential operator defined by (7.57) where the coefficients α_i are determined by (7.70) and (7.68).

7. Continue the iterative process described in the text for the case $P = 0, Q = -1, t = 2$, and show that the values of the α_i and β_i converge to those given by (7.75) and (7.76).

8. Verify (7.93) and (7.94).

9. If $L[u] = u'' + u' - u$, find $v(x), w(x)$, and $z_k(x), k = 0, 1, 2$ defined by (7.38), (7.41), and (7.43), respectively. Let $x_0 = 0$.

10. Verify (7.35) for the case of the five-point operator

$$L_h[u] = \alpha_0 u(x) + \alpha_1 u(x + h) + \alpha_3 u(x + 2h) + \alpha_2 u(x - h) + \alpha_4 u(x - 2h).$$

Assume that the relative degree of approximation of $L_h[u]$ to $L[u]$ is p.

10.8 EXACT OPERATORS

In this section we show the existence of a three-point operator

8.1 $$\hat{L}_h[u] = \alpha_0 u(x) + \alpha_1 u(x + h) + \alpha_2 u(x - h)$$

which is consistent with the operator

8.2 $$L[u] = u'' + Pu' + Qu$$

and which is an exact representation of $L[u]$ in the sense that $\hat{L}_h[u] = 0$ for any solution of

8.3 $$L[u] = 0.$$

The existence of such an operator is a consequence of the fact that there exist two linearly independent functions $\phi(x)$ and $\psi(x)$ which satisfy (8.3) and are such that if u is any solution of (8.3) then

8.4 $$u = c_1\phi(x) + c_2\psi(x)$$

for some constants c_1 and c_2. Two functions $\phi(x)$ and $\psi(x)$ are *linearly independent* in $a \leq x \leq b$ if there do not exist constants k_1 and k_2 such that

8.5
$$k_1\phi(x) + k_2\psi(x) \equiv 0$$

unless $k_1 = k_2 = 0$.

Two such linearly independent functions can be constructed by solving the two initial-value problems

8.6
$$\begin{cases} L[\hat{\phi}] = 0, & \hat{\phi}(a) = 1, & \hat{\phi}'(a) = 0 \\ L[\hat{\psi}] = 0, & \hat{\psi}(a) = 0, & \hat{\psi}'(a) = 1. \end{cases}$$

Clearly, any solution u of (8.3) is given by

8.7
$$u(x) = u(a)\hat{\phi}(x) + u'(a)\hat{\psi}(x).$$

If $\phi(x)$ and $\psi(x)$ are any two linearly independent functions satisfying (8.3), then we can write any solution $u(x)$ of (8.3) in the form (8.4) provided we can find $\hat{\phi}(x)$ and $\hat{\psi}(x)$ in terms of $\phi(x)$ and $\psi(x)$. But we can solve the system

8.8
$$\begin{cases} \phi(x) = \phi(a)\hat{\phi}(x) + \phi'(a)\hat{\psi}(x) \\ \psi(x) = \psi(a)\hat{\phi}(x) + \psi'(a)\hat{\psi}(x) \end{cases}$$

for $\hat{\phi}(x)$ and $\hat{\psi}(x)$ in terms of $\phi(x)$ and $\psi(x)$ provided that the *Wronskian* of $\phi(x)$ and $\psi(x)$, namely

8.9
$$W(\phi, \psi) = \phi(x)\psi'(x) - \phi'(x)\psi(x)$$

does not vanish at a. But it is easy to show (see, for instance, Ford [1933], Chapter VII) that if $W(\phi, \psi)$ vanishes at a then $W \equiv 0$ in $a \leq x \leq b$. Moreover, if $W \equiv 0$ then $\phi(x)$ and $\psi(x)$ are not linearly independent. For, in the first place, if $\phi(a)$ and $\psi(a)$ both vanish, then $\phi(x)$ and $\psi(x)$ are clearly not linearly independent since each is proportional to its derivative at a. Let us assume that $\phi(a) \neq 0$. By continuity, for some $\delta > 0$, $\phi(x) \neq 0$ for all $x \in [a, a + \delta]$. Moreover, in the interval $[a, a + \delta]$ we have

8.10
$$\frac{d}{dx}\left[\frac{\psi(x)}{\phi(x)}\right] = -\frac{W(\phi, \psi)}{\phi(x)^2} = 0.$$

Hence $\psi(x) = k\phi(x)$ for some constant k in $[a, a + \delta]$. Since the function

8.11
$$\theta(x) = \psi(x) - k\phi(x)$$

satisfies (8.3) and also the conditions $\theta(a) = \theta'(a) = 0$ it follows that $\theta(x) \equiv 0$ and $\psi(x) \equiv k\phi(x)$ in $[a, b]$. Hence $\phi(x)$ and $\psi(x)$ are not linearly independent.

Construction of Exact Difference Operator

Given linearly independent functions $\phi(x)$ and $\psi(x)$ satisfying (8.3), we can make $\hat{L}_h[u]$ vanish whenever $L[u]$ vanishes simply by requiring that

8.12
$$\hat{L}_h[u] = c_1\hat{L}_h[\phi] + c_2\hat{L}_h[\psi] = 0$$

for all c_1 and c_2, i.e., that

8.13
$$\hat{L}_h[\phi] = \hat{L}_h[\psi] = 0.$$

But these conditions imply

8.14
$$\begin{cases} \alpha_0\phi(x) + \alpha_1\phi(x + h) + \alpha_2\phi(x - h) = 0 \\ \alpha_0\psi(x) + \alpha_1\psi(x + h) + \alpha_2\psi(x - h) = 0. \end{cases}$$

We seek to show that

8.15
$$\phi(x + h)\psi(x - h) - \psi(x + h)\phi(x - h) \neq 0.$$

We note that both $\phi(x)$ and $\psi(x)$ cannot vanish for the same value of x, otherwise $W(\phi, \psi)$ would vanish for all x. Hence, we can assume $\psi(x + h) \neq 0$. If $\phi(x + h)\psi(x - h) - \psi(x + h)\phi(x - h) = 0$, then

8.16
$$\frac{\phi(x + h)}{\psi(x + h)}\psi(x - h) - \phi(x - h) = 0.$$

If $\psi(x - h) = 0$, then $\phi(x - h) = 0$, which is impossible. Therefore, we have

8.17
$$\frac{\phi(x + h)}{\psi(x + h)} = \frac{\phi(x - h)}{\psi(x - h)},$$

and hence, by the mean-value theorem, for some ξ in the range $x - h < \xi < x + h$ we have

8.18
$$\left[\frac{\phi(x)}{\psi(x)}\right]'_{x=\xi} = 0$$

or

8.19
$$\frac{\phi'(\xi)\psi(\xi) - \phi(\xi)\psi'(\xi)}{\psi(\xi)^2} = 0.$$

Hence the Wronskian vanishes at ξ and we have a contradiction. Therefore,

$\phi(x + h)\psi(x - h) - \phi(x - h)\psi(x + h) \neq 0$, and we can solve (8.14) for α_1 and α_2 in terms of α_0 obtaining

8.20
$$\begin{cases} \alpha_1 = \dfrac{\alpha_0[\psi(x)\phi(x - h) - \phi(x)\psi(x - h)]}{\phi(x + h)\psi(x - h) - \phi(x - h)\psi(x + h)} \\[3mm] \alpha_2 = \dfrac{\alpha_0[\phi(x)\psi(x + h) - \psi(x)\phi(x + h)]}{\phi(x + h)\psi(x - h) - \phi(x - h)\psi(x + h)}. \end{cases}$$

We now require that $\hat{L}_h[u]$ be consistent with $L[u]$. Evidently,

8.21 $\quad \hat{L}_h[u] - L[u] = \alpha_0 u_0 + \alpha_1 \left(u_0 + h u_1 + \dfrac{h^2}{2!} u_2 + \cdots \right)$

$$+ \alpha_2 \left(u_0 - h u_1 + \frac{h^2}{2!} u_2 + \cdots \right) - (u_2 + P u_1 + Q u_0)$$

$$= u_0(\alpha_0 + \alpha_1 + \alpha_2 - Q) + u_1(h(\alpha_1 - \alpha_2) - P)$$

$$+ u_2 \left(\frac{h^2}{2!}(\alpha_1 + \alpha_2) - 1 \right) + \cdots$$

Here, as before, we let $u_0 = u$, $u_1 = u'$, $u_2 = u''$, etc. For consistency, we must have

8.22
$$\frac{h^2}{2!}(\alpha_1 + \alpha_2) = 1 + o(1).$$

If we require that $(h^2/2)(\alpha_1 + \alpha_2) = 1$, then by (8.20) α_0 is given by

8.23 $\quad \alpha_0 = \dfrac{2}{h^2} \dfrac{\phi(x + h)\psi(x - h) - \phi(x - h)\psi(x + h)}{\psi(x)[\phi(x - h) - \phi(x + h)] + \phi(x)[\psi(x + h) - \psi(x - h)]}.$

We remark that

8.24 $\quad \psi(x)\phi(x - h) - \phi(x)\psi(x - h) + \phi(x)\psi(x + h) - \psi(x)\phi(x + h)$

$$= \psi(x)[\phi(x - h) - \phi(x + h)] + \phi(x)[\psi(x + h) - \psi(x - h)]$$

$$= 2hW(\phi, \psi) + O(h^3), \quad .$$

which does not vanish for h sufficiently small since the Wronskian does not vanish. Hence for sufficiently small h we can determine α_0.

We now seek to show that for h small enough $\hat{L}_h[u]$ is consistent with $L[u]$. Indeed we have

8.25
$$\begin{cases} \phi(x+h)\psi(x-h) - \psi(x+h)\phi(x-h) = -2hW(\phi,\psi) + O(h^2), \\ \phi(x)\psi(x-h) - \psi(x)\phi(x-h) = -hW(\phi,\psi) + O(h^2), \\ \phi(x+h)\psi(x) - \psi(x+h)\phi(x) = -hW(\phi,\psi) + O(h^2), \end{cases}$$

so that by (8.20)

8.26
$$\alpha_1 = (-\tfrac{1}{2} + O(h))\alpha_0, \qquad \alpha_2 = (-\tfrac{1}{2} + O(h))\alpha_0.$$

Hence $\alpha_1 + \alpha_2 = (-1 + O(h))\alpha_0$ and we have, since $\alpha_1 + \alpha_2 = 2/h^2$,

8.27
$$\begin{cases} \alpha_0 = -\dfrac{2}{h^2}(1 + O(h)) \\[2mm] \alpha_1 = \dfrac{1}{h^2}(1 + O(h)) \\[2mm] \alpha_2 = \dfrac{1}{h^2}(1 + O(h)). \end{cases}$$

Using (8.20) and (8.23) we can show that

8.28
$$\alpha_0 + \alpha_1 + \alpha_2 = \left[\frac{h^3(\phi'\psi'' - \phi''\psi') + O(h^4)}{2h(\phi'\psi - \phi\psi') + O(h^2)}\right]\alpha_0$$
$$= -\frac{h^2}{2}\left[\frac{\phi'\psi'' - \phi''\psi'}{W(\phi,\psi)} + O(h)\right]\alpha_0.$$

But since $L[\phi] = L[\psi] = 0$ we have

8.29
$$\phi'' + P\phi' + Q\phi = 0, \qquad \psi'' + P\psi' + Q\psi = 0$$

and

8.30
$$\psi'\phi'' - \phi'\psi'' + QW(\phi,\psi) = 0,$$
$$Q = \frac{\phi'\psi'' - \phi''\psi'}{W(\phi,\psi)}.$$

Since $\alpha_0 = -2h^{-2} + O(h^{-1})$ we have

8.31
$$\alpha_0 + \alpha_1 + \alpha_2 = Q + O(h)$$

and $\delta_0 = O(h)$.

Next, let us consider $\delta_1 = h(\alpha_1 - \alpha_2) - P$. Again, a straightforward calculation yields

8.32
$$\alpha_1 - \alpha_2 = \left[\frac{h^2(\phi''\psi - \psi''\phi) + O(h^3)}{-2hW(\phi, \psi) + O(h^2)}\right]\alpha_0.$$

But from (8.29) we have

8.33
$$\phi''\psi - \psi''\phi - PW(\phi, \psi) = 0.$$

Since $\alpha_0 = -2h^{-2} + O(h^{-1})$ it follows that

8.34
$$h(\alpha_1 - \alpha_2) = P + O(h),$$

and $\delta_1 = O(h)$. It is easy to verify that $\delta_i = O(h)$ for $i > 2$; hence, since $\delta_2 = 0$, by (8.20) and (8.23), $\hat{L}_h[u]$ is consistent with $L[u]$.

Construction of Exact Difference Equation

It should be noted that while we can construct an "exact" discrete operator $\hat{L}_h[u]$ given the basic solutions $\phi(x)$ and $\psi(x)$ of $L[u] = 0$, in order to construct an exact difference equation we need to find a particular solution $\chi(x)$ of the non-homogeneous equation (7.79). Normally, we would use the difference equation (7.48), but we do not have any β_i. One method of finding a particular solution is to use the method of variation of parameters, i.e., to find $c_1(x)$ and $c_2(x)$ such that

8.35
$$\chi(x) = c_1(x)\phi(x) + c_2(x)\psi(x)$$

satisfies (7.79). One can verify that (7.79) is satisfied if

8.36
$$c_1'(x) = \frac{-G\psi}{W(\phi, \psi)}, \qquad c_2'(x) = \frac{G\phi}{W(\phi, \psi)},$$

since then

8.37
$$\chi'(x) = c_1(x)\phi'(x) + c_2(x)\psi'(x)$$

and

8.38
$$\chi''(x) = c_1(x)\phi''(x) + c_2(x)\psi''(x) + G.$$

Assuming that one can find $\chi(x)$, then the "exact" difference equation is

8.39
$$\hat{L}_h[u] = \hat{L}_h[\chi].$$

To prove the above, we seek to show that if

8.40
$$u(x) = c_1\phi(x) + c_2\psi(x) + \chi(x),$$

then u satisfies (8.39). But

8.41
$$\hat{L}_h[u] = \hat{L}_h[c_1\phi(x) + c_2\psi(x)] + \hat{L}_h[\chi]$$
$$= \hat{L}_h[\chi]$$

since $\hat{L}_h[w]$ vanishes if $L[w] = 0$.

As an example, let us consider the differential equation

8.42
$$u'' - u = e^x.$$

Here we have

8.43
$$\phi(x) = e^x, \qquad \psi(x) = e^{-x}.$$

Evidently,

8.44
$$W(\phi, \psi) = \phi\psi' - \psi\phi' = -2,$$

and we can easily compute a particular solution χ by (8.35) and (8.36) since

8.45
$$\begin{cases} c_1' = \tfrac{1}{2}, & c_1 = \tfrac{1}{2}x \\ c_2' = -\tfrac{1}{2}e^{2x}, & c_2 = -\tfrac{1}{4}e^{2x} \end{cases}$$

so that

8.46
$$\chi(x) = \tfrac{1}{2}xe^x - \tfrac{1}{4}e^x.$$

However, since e^x is a solution of the homogeneous equation $u'' - u = 0$, we can discard the second term of $\chi(x)$ and can use as our particular solution

8.47
$$\chi(x) = \tfrac{1}{2}xe^x.$$

Thus, the general solution of (8.42) is

8.48
$$u = c_1e^x + c_2e^{-x} + \tfrac{1}{2}xe^x.$$

The exact weights are, by (8.20),

8.49
$$\begin{cases} \alpha_1 = \alpha_2 = \left[\dfrac{e^xe^{-x+h} - e^{-x}e^{x-h}}{e^{x+h}e^{-x+h} - e^{x-h}e^{-x-h}} \right](-\alpha_0) \\ \\ = \dfrac{e^h - e^{-h}}{e^{2h} - e^{-2h}}(-\alpha_0) = \dfrac{-\alpha_0}{e^h + e^{-h}} = \dfrac{-\alpha_0}{2\cosh h}. \end{cases}$$

Moreover, by (8.23) we have

8.50
$$\begin{cases} \alpha_0 = -\dfrac{2}{h^2}\cosh h \\[2ex] \alpha_1 = \alpha_2 = \dfrac{1}{h^2}. \end{cases}$$

Hence, an exact difference equation is

8.51 $\hat{L}_h[u] = -\dfrac{2}{h^2}(\cosh h)u(x) + \dfrac{1}{h^2}[u(x+h) + u(x-h)]$

$$= \hat{L}_h[\tfrac{1}{2}xe^x] = -\dfrac{2}{h^2}(\cosh h)(\tfrac{1}{2}xe^x) + \dfrac{1}{h^2}[\tfrac{1}{2}(x+h)e^{x+h} + \tfrac{1}{2}(x-h)e^{x-h}].$$

As an alternative to (8.51), we could let $v = u - \tfrac{1}{2}xe^x$ and solve the homogeneous equation

8.52 $$\hat{L}_h[v] = 0,$$

where v satisfies the boundary conditions

8.53
$$\begin{cases} v(a) = u(a) - \tfrac{1}{2}ae^a \\ v(b) = u(b) - \tfrac{1}{2}be^b. \end{cases}$$

We can relate the linear discrete operator $\hat{L}_h[u]$ to the operator $\tilde{L}_h[u]$ of degree of approximation three given by (7.9) as follows. In (7.9) we have

8.54 $$-\dfrac{\alpha_0}{\alpha_1} = -\dfrac{\alpha_0}{\alpha_2} = 2 + h^2 + \dfrac{h^4}{12} = 2\left(1 + \dfrac{h^2}{2} + \dfrac{h^4}{24}\right),$$

while for (8.51) we have

8.55 $$-\dfrac{\alpha_0}{\alpha_1} = -\dfrac{\alpha_0}{\alpha_2} = 2\cosh h = 2\left(1 + \dfrac{h^2}{2} + \dfrac{h^4}{24} + \cdots\right).$$

For the ordinary 3-point operator $L_h[u]$ of degree of approximation one we have

8.56 $$-\dfrac{\alpha_0}{\alpha_1} = -\dfrac{\alpha_0}{\alpha_2} = 2\left(1 + \dfrac{h^2}{2}\right).$$

Thus, it appears that as the relative degree of approximation increases the values of $-\alpha_0/\alpha_1$ converge to the value of $-\alpha_0/\alpha_1$ for the exact operator.

In the case of the differential equation

8.57
$$u'' - u = x^2,$$

a particular solution is $\chi = -x^2 - 2$, and the exact difference equation (8.41) becomes

8.58
$$\hat{L}_h[u] = \hat{L}_h[-x^2 - 2].$$

where $\hat{L}_h[u]$ is given by (8.51). If the boundary conditions are $u(0) = 1$, $u(2) = 2$, then we let $v = u - \chi = u + x^2 + 2$ and solve the problem given by

8.59
$$\begin{cases} \hat{L}_h[v] = 0 \\ v(0) = 3, \qquad v(2) = 8. \end{cases}$$

Thus we have, with $v_i = v(ih)$, $i = 0, 1, 2, 3, 4$, and $h = \frac{1}{2}$,

8.60
$$\begin{cases} v_0 + v_2 - 2(\cosh \tfrac{1}{2})v_1 = 0 \\ v_1 + v_3 - 2(\cosh \tfrac{1}{2})v_2 = 0 \\ v_2 + v_4 - 2(\cosh \tfrac{1}{2})v_3 = 0 \end{cases}$$

or

8.61
$$\begin{cases} 2(\cosh \tfrac{1}{2})v_1 - v_2 = 3 \\ 2(\cosh \tfrac{1}{2})v_2 - v_1 - v_3 = 0 \\ 2(\cosh \tfrac{1}{2})v_3 - v_2 = 8. \end{cases}$$

Solving we obtain

8.62
$$\begin{cases} v_2 = \dfrac{11}{4 \cosh^2 \tfrac{1}{2} - 2} \doteq 3.56430, \quad u_2 \doteq v_2 + \chi_2 = 3.56430 - 3 \doteq 0.56430 \\[3mm] v_1 = \dfrac{v_2 + 3}{2 \cosh \tfrac{1}{2}} \doteq 2.91067, \qquad u_1 \doteq v_1 + \chi_1 = 2.91067 - 2.25 \doteq 0.66067 \\[3mm] v_3 = \dfrac{v_2 + 8}{2 \cosh \tfrac{1}{2}} \doteq 5.12772, \qquad u_3 \doteq v_3 + \chi_3 = 5.12772 - 4.25 \doteq 0.87772. \end{cases}$$

Thus the values of u_1, u_2, u_3 agree with the exact solution, $\bar{u}(x)$, given at the end of Section 10.6.

In the event that linearly independent solutions $\phi(x)$ and $\psi(x)$ of $L[u] = 0$ are not available, or if a particular integral χ is not available, one can sometimes use the following procedure to find the α_i and β_i. We start with the identity

8.63 $\dfrac{\delta u}{2h} = \dfrac{u(x + h) - u(x - h)}{2h} = u' + \dfrac{h^2}{6} u^{(3)} + \dfrac{h^4}{120} u^{(5)} + \cdots .$

From the relation $L[u] = u'' + Pu' + Qu$ and derivatives of that relation, we can express $u'', u''', \ldots$ in terms of u, u', L and derivatives of L. Thus

8.64 $u^{(k)} = a_k u + b_k u' + \displaystyle\sum_{j=0}^{k-2} c_j L^{(j)}[u].$

Substituting in (8.63) we get

8.65 $\dfrac{\delta u}{2h} = u' + Au + Bu' + \displaystyle\sum_{j=0}^{\infty} d_j L^{(j)}[u].$

Solving for u' we have the exact relation

8.66 $u' = \dfrac{1}{1 + B}\left\{\dfrac{\delta u}{2h} - Au - \displaystyle\sum_{j=0}^{\infty} d_j L^{(j)}[u].\right\}$

Similarly, we have

8.67 $\dfrac{\delta^2 u}{h^2} = u'' + \dfrac{h^2}{12} u^{(4)} + \dfrac{h^4}{360} u^{(6)} + \cdots$

$= u'' + Cu + Du' + \displaystyle\sum_{j=0}^{\infty} e_j L^{(j)}[u].$

Using (8.66) we get

8.68 $u'' = \dfrac{\delta^2 u}{h^2} + E\dfrac{\delta u}{2h} + Fu + \displaystyle\sum_{j=0}^{\infty} f_j L^{(j)}[u].$

Letting

8.69 $\hat{L}_h[u] = \dfrac{\delta^2 u}{h^2} + E\dfrac{\delta u}{2h} + Fu + P\left(\dfrac{\dfrac{\delta u}{2h} - Au}{1 + B}\right) + Qu$

we have the exact relation

8.70 $\hat{L}_h[u] - L[u] = \displaystyle\sum_{j=0}^{\infty} \beta_j L^{(j)}[u]$

for suitable $\beta_0, \beta_1, \ldots .$

For example, for the case $L[u] = u'' - u$ we have

8.71
$$\begin{cases} u''' = u' + L' \\ u^{(4)} = u'' + L'' = u + L + L'' \\ u^{(5)} = u' + L' + L''' \\ u^{(6)} = u + L + L'' + L^{(4)} \end{cases}$$

etc. Moreover,

8.72
$$\begin{cases} \dfrac{\delta^2 u}{h^2} = u'' + 2\left(\dfrac{h^2}{4!} u^{(4)} + \dfrac{h^4}{6!} u^{(6)} + \cdots\right) \\[2mm] \qquad = u'' + 2\left(\dfrac{h^2}{4!}(u + L + L'') + \dfrac{h^4}{6!}(u + L + L'' + L^{(4)}) + \cdots\right) \\[2mm] \qquad = u'' + \dfrac{2}{h^2} u\left(\dfrac{h^4}{4!} + \dfrac{h^6}{6!} + \cdots\right) \\[2mm] \qquad\quad + 2L\left(\dfrac{h^2}{4!} + \dfrac{h^4}{6!} + \cdots\right) + 2L''\left(\dfrac{h^2}{4!} + \dfrac{h^4}{6!} + \cdots\right) + 2L^{(4)}\left(\dfrac{h^4}{6!} + \cdots\right) + \cdots \\[2mm] \qquad = u'' + \dfrac{2}{h^2} u\left(\cosh h - 1 - \dfrac{h^2}{2}\right) + \dfrac{2}{h^2} L\left(\cosh h - 1 - \dfrac{h^2}{2}\right) \\[2mm] \qquad\quad + \dfrac{2}{h^2} L''\left(\cosh h - 1 - \dfrac{h^2}{2}\right) + \dfrac{2}{h^2} L^{(4)}\left(\cosh h - 1 - \dfrac{h^2}{2} - \dfrac{h^4}{4!}\right) + \cdots. \end{cases}$$

Therefore

8.73
$$\begin{cases} u'' - u = \dfrac{\delta^2 u}{h^2} - \dfrac{2}{h^2} u\left(\cosh h - 1 - \dfrac{h^2}{2}\right) - \dfrac{2}{h^2} L\left(\cosh h - 1 - \dfrac{h^2}{2}\right) \\[2mm] \qquad\quad - \dfrac{2}{h^2} L''\left(\cosh h - 1 - \dfrac{h^2}{2}\right) - \dfrac{2}{h^2} L^{(4)}\left(\cosh h - 1 - \dfrac{h^2}{2} - \dfrac{h^4}{4!}\right) + \cdots - u \\[2mm] \qquad = \hat{L}_h[u] - \dfrac{2}{h^2} L\left(\cosh h - 1 - \dfrac{h^2}{2}\right) - \dfrac{2}{h^2} L''\left(\cosh h - 1 - \dfrac{h^2}{2}\right) \\[2mm] \qquad\quad - \dfrac{2}{h^2} L^{(4)}\left(\cosh h - 1 - \dfrac{h^2}{2} - \dfrac{h^4}{4!}\right) + \cdots \end{cases}$$

where $\hat{L}_h[u]$ is given by (8.51). Thus

8.74
$$\hat{L}_h[u] - L[u] = \beta_0 L + \beta_2 L'' + \beta_4 L^{(4)} + \cdots$$

where

8.75
$$\begin{cases} \beta_0 = \dfrac{2}{h^2}\left(\cosh h - 1 - \dfrac{h^2}{2}\right) \\[3mm] \beta_2 = \dfrac{2}{h^2}\left(\cosh h - 1 - \dfrac{h^2}{2}\right) \\[3mm] \beta_4 = \dfrac{2}{h^2}\left(\cosh h - 1 - \dfrac{h^2}{2} - \dfrac{h^4}{4!}\right). \end{cases}$$

etc. We remark that we could, if we choose, replace $\hat{L}_h[u]$ by $(1 + \beta_0)^{-1}\hat{L}_h[u] = L_h^*[u]$ and we would have

8.76
$$\hat{L}_h^*[u] - L[u] = \beta_2^* L''[u] + \beta_4^* L^{(4)}(u) + \cdots$$

where $\beta_i^* \to 0$ as $h \to 0$.

In the example involving the equation $u'' - u = x^2$ we obtain the difference equation

8.77
$$\hat{L}_h[u] = G + \beta_0 G + \beta_2 G'' = x^2 + \frac{2}{h^2}\left(\cosh h - 1 - \frac{h^2}{2}\right)(x^2 + 2)$$

$$= -2 + \frac{2}{h^2}(\cosh h - 1)(x^2 + 2)$$

This agrees with the result obtained above using the particular solution $\chi = -(x^2 + 2)$ since

8.78
$$-\hat{L}_h[x^2 + 2] = -2 + \frac{2}{h^2}(\cosh h - 1)(x^2 + 2).$$

Minimum Principle for Three-point Operators

We now show that if $L_h[u]$ is any three-point operator of the form (6.20) which is consistent with the operator $L[u]$ given by (5.1), then an extension of Lemma 5.10 holds for h sufficiently small. Thus we have

8.79 Lemma. Let $L_h[u]$ be a three-point operator of the form (6.20) which is consistent with $L[u]$, and let

8.80
$$A(x) > 0, \qquad F(x) \leq 0$$

in $R + S$. If h is sufficiently small, then if $u(x) \geq 0$ on S_h and if $-L_h[u] \geq 0$ on R_h then $u \geq 0$ on $R_h + S_h$.

Proof. By the consistency of $L_h[u]$ we have, by (6.27)

8.81
$$\begin{cases} \alpha_0 = -\dfrac{2\tilde{A}(x)}{h^2} + \tilde{F}(x) \\[2mm] \alpha_1 = \dfrac{\tilde{A}(x)}{h^2} + \dfrac{\tilde{D}(x)}{2h} \\[2mm] \alpha_2 = \dfrac{\tilde{A}(x)}{h^2} - \dfrac{\tilde{D}(x)}{2h}, \end{cases}$$

where

8.82
$$\begin{cases} \tilde{A}(x) = A(x) + \varepsilon_2(h) \\[1mm] \tilde{D}(x) = D(x) + \varepsilon_1(h) \\[1mm] \tilde{F}(x) = F(x) + \varepsilon_0(h), \end{cases}$$

and where

8.83
$$\varepsilon_i(h) = o(1), \qquad i = 0, 1, 2,$$

as $h \to 0$.

By (6.20) we have, since $-L_h[u] \geq 0$,

8.84
$$v(x) \leq \left| \frac{1 - \dfrac{h\tilde{D}}{2\tilde{A}}}{1 + \dfrac{h\tilde{D}}{2\tilde{A}}} \right| v(x - h) - \frac{h(\tilde{F}/\tilde{A})}{\left(1 + \dfrac{h\tilde{D}}{2\tilde{A}}\right)} u(x),$$

where we let

8.85
$$v(x) = \frac{u(x + h) - u(x)}{h}.$$

Suppose now that $u(x) < 0$ for some $x \in R_h$. Let

8.86
$$u(\bar{x}) = -M = \min_{R_h} u(x).$$

Let x^* be the smallest $x \in R_h + S_h$ such that $x^* > \bar{x}$ and $u(x^*) \geq 0$. Since $u(b) \geq 0$ we must have $x^* \leq b$. If we let

8.87
$$\gamma = \max\left(0, \max_{R+S} \frac{\tilde{F}(x)}{\tilde{A}(x)}\right),$$

8.88
$$\beta = \tfrac{1}{2} \max_{R+S} \frac{|\tilde{D}(x)|}{\tilde{A}(x)},$$

then we have

8.89
$$
\begin{cases}
v(\bar{x} - h) \leqq 0 \\[2mm]
v(\bar{x}) \qquad \leqq \dfrac{h\gamma M}{1 - h\beta} \\[4mm]
v(\bar{x} + h) \leqq \left(\dfrac{1 + h\beta}{1 - h\beta}\right)\left(\dfrac{h\gamma M}{1 - h\beta}\right) + \dfrac{h\gamma M}{1 - h\beta},
\end{cases}
$$

and in general

8.90
$$
v(\bar{x} + kh) \leqq \left[\left(\frac{1 + h\beta}{1 - h\beta}\right)^{k+1} - 1\right]\left(\frac{\gamma M}{2\beta}\right).
$$

If we choose h so small that

8.91
$$
h\beta \leqq \tfrac{1}{2},
$$

then we have by 9-(5.23)

8.92
$$
\frac{1 + h\beta}{1 - h\beta} \leqq e^{4h\beta}
$$

and

8.93
$$
v(\bar{x} + kh) \leqq \frac{\gamma M}{2\beta}\left[e^{4h\beta(k+1)} - 1\right].
$$

Therefore, if we let $k = (x^* - \bar{x})/h$ we have

8.94
$$
v(x^*) \leqq \frac{\gamma M}{2\beta}\left[e^{4\beta(x^* - \bar{x} + h)} - 1\right].
$$

It is easy to show that for $z > 0, \delta > 0$ we have

8.95
$$
\frac{e^{\delta z} - 1}{\delta} \leqq z e^{\delta z}.
$$

Hence we have for $\bar{x} \leqq x \leqq x^*$,

8.96
$$
\begin{aligned}
v(x) &\leqq 2\gamma M(x^* - \bar{x} + h)e^{4\beta(x^* - \bar{x} + h)} \\
&\leqq 2\gamma M(b - a)e^{4\beta(b-a)}.
\end{aligned}
$$

Moreover, by (8.85),

8.97
$$
\begin{aligned}
u(x^*) &\leqq -M + 2\gamma M(b - a)^2 e^{4\beta(b-a)} \\
&= -M[1 - 2\gamma(b - a)^2 e^{4\beta(b-a)}].
\end{aligned}
$$

Since $F(x) \leq 0$, we have by (8.87) and (8.82)

8.98 $$\lim_{h \to 0} \gamma = 0.$$

Hence, for h sufficiently small,

8.99 $$u(x^*) < 0,$$

which contradicts the assertion that $u(x^*) \geq 0$. The lemma now follows.

The reader should show that from Lemma 8.79 it follows that there exists a unique solution of the difference equation

8.100 $$L_h[u] = \hat{G}(x)$$

for any function $\hat{G}(x)$ and for h sufficiently small. He should also verify that the following generalizations of Lemmas 5.10, 5.15, and 5.18 are valid.

8.101 Lemma. Let $L_h[u]$ be a three-point operator of the form (6.20) which is consistent with $L[u]$ and let (8.80) hold in $R + S$. If

8.102 $$- L_h[u] \geq o(1)$$

in R_h and if $u \geq 0$ on S_h then $u \geq 0$ in R_h for sufficiently small h.

8.103 Lemma. Let $L_h[u]$ satisfy the hypotheses of Lemma 8.101. If $|u| \leq v$ on S_h and if

8.104 $$|L_h[u]| \leq - L_h[v] + o(1)$$

in R_h then $|u| \leq v$ for sufficiently small h.

8.105 Lemma. Let $L_h[u]$ satisfy the hypotheses of Lemma 8.101, and let $w(x)$ be any function such that $- L_h[w] > 0$ in R_h and $w \geq 0$ in S_h. Then (5.19) holds for any function $e(x)$ for sufficiently small h.

EXERCISES 10.8

1. Verify that the Wronskian of two solutions $\phi(x)$ and $\psi(x)$ of (8.3) either vanishes identically or never vanishes.

2. Let $\phi(x)$ and $\psi(x)$ be any two linearly independent solutions of (8.3) for x in $[a, b]$. Show that any solution $u(x)$ of (8.3) can be written in the form

$$u(x) = \frac{u(x_0)\psi'(x_0) - u'(x_0)\psi(x_0)}{W(\phi(x_0), \psi(x_0))} \phi(x)$$

$$+ \frac{u'(x_0)\phi(x_0) - u(x_0)\phi'(x_0)}{W(\phi(x_0), \psi(x_0))} \psi(x)$$

for any x in $[a, b]$. Here W is the Wronskian defined by (8.9).

3. Consider the problem of solving the differential equation

$$L[u] = u'' - \frac{1}{x} u' = 3x$$

in the interval $[1, 2]$ with $u(1)$ and $u(2)$ given. Find a discrete operator $L_h[u]$ of the form

$$L_h[u] = \alpha_0 u(x) + \alpha_1 u(x + h) + \alpha_2 u(x - h)$$

which is consistent with $L[u]$ and such that $L_h[u] = 0$ for all solutions of $L[u] = 0$. Find a function $\Gamma(x)$ such that if u is a solution of $L[u] = 3x$, then u also satisfies $L_h[u] = \Gamma$. Also, find the absolute degree of approximation of $L_h[u]$ to $L[u]$.

Hint: Show that the exact solution of $L[u] = 3x$ is $u = c_1 x^2 + c_2 + x^3$ where c_1 and c_2 are arbitrary constants.

4. For the operator $L[u]$ given in Exercise 5, Section 10.7, find a discrete operator $\hat{L}_h[u]$ which is consistent with $L[u]$ and which is exact in the sense that $\hat{L}_h[u] = 0$ if $L[u] = 0$. What is the absolute degree of approximation of $\hat{L}_h[u]$ to $L[u]$?

5. In the preceding problem find and solve a difference equation with $h = \frac{1}{2}$ which yields exact results.

6. Suppose we have a consistent representation of the self-adjoint operator

$$L[u] = \frac{d}{dx}(Au') + Fu.$$

Do we necessarily have a discrete maximum principle?

7. Consider the discrete operator

$$L_h[u] = -2(1 - h^3)u(x) + u(x + h) + u(x - h).$$

Show that the difference equation

$$L_h[u] = 0$$

with the conditions $u(0) = 1, u(1) = 2$ is uniquely solvable for h sufficiently small. (Assume that h^{-1} is an integer.)

8. Prove Lemmas 8.101, 8.103, and 8.105.

9. Verify that the solution of $w_{k+1} = \alpha w_k + \delta$, where $\alpha \neq 1$, is $w_k = w_0 \alpha^k + \delta(\alpha - 1)^{-1}(\alpha^k - 1)$.

10. Verify that if $c_1(x)$ and $c_2(x)$ satisfy (8.36), then $\chi(x)$, as defined by (8.35), is a particular solution of (7.79).

11. Verify (8.90).

12. Verify (8.95).

13. Verify (8.97).

14. Show that if $L[u] = u'' + Pu' + Qu = 0$ and if $u(a) = u'(a) = 0$, then $u(x) \equiv 0$ in $I = [a, b]$. Here we assume that $P(x)$ and $Q(x)$ are analytic in I.

15. Let $L_h[u]$ be a discrete linear operator of the form (5.11) which is consistent with (5.1) where $A(x) > 0$ and $F(x) \leq 0$ in $[a, b]$. Prove that if $L_h[u] = 0$ in R_h then for h sufficiently small we have

$$\min(u(a), u(b)) \leq u(x) \leq \max(u(a), u(b)).$$

10.9 EIGENVALUE PROBLEMS

Let us now consider the problem of finding a number λ so that the two-point boundary value problem involving the differential equation

9.1 $$L[u] = Au'' + Du' + (F + \lambda H)u = 0, \qquad a < x < b$$

and the boundary conditions

9.2 $$u(a) = u(b) = 0$$

has a nontrivial solution. Here we assume that $A(x) > 0$, $F(x) \leq 0$, and $H(x) > 0$ for x in $[a, b]$. The number λ is said to be an *eigenvalue* and the corresponding function $u(x)$ is said to be an *eigenfunction*.

As an example, consider the problem defined by

9.3 $$\begin{cases} L[u] = u'' + \lambda u = 0, & 0 < x < 1 \\ u(0) = u(1) = 0. \end{cases}$$

Evidently, for each positive integer n,

9.4 $$\lambda = \lambda_n = n^2 \pi^2$$

is an eigenvalue corresponding to the eigenfunction

9.5 $$u(x) = u_n(x) = \sin n\pi x.$$

Method of Superposition

We describe briefly two alternative methods for solving eigenvalue problems. The first method is referred to as the *method of superposition*. In the special case (9.1)–(9.2) the method consists of constructing, for an assumed value of λ, a function $u(x, \lambda)$ satisfying (9.1) and the initial conditions

9.6 $$u(a, \lambda) = 0, \qquad u'(a, \lambda) = 1$$

using any of the methods of Chapter 8. If we define the function $\phi(\lambda)$ by

9.7 $$\phi(\lambda) = u(b, \lambda),$$

then λ is an eigenvalue if and only if the equation

9.8 $$\phi(\lambda) = 0$$

is satisfied. Since $\phi(\lambda)$ can be evaluated for any value of λ we can use any of the methods of Chapter 4 to solve (9.8) which only involve the evaluation of ϕ. Since we are normally interested in real roots of (9.8), the secant method or the method of false position, based on reasonable starting values, would be appropriate in most cases*.

Let us apply the procedure to the example (9.3) where we use the modified Euler method with $h = \frac{1}{3}$. We reduce the second-order equation to the system

9.9
$$\begin{cases} u' = v \\ v' = -\lambda u \end{cases}$$

with the initial condition

9.10
$$u(0) = 0, \qquad v(0) = 1.$$

Letting $u_i = u(ih)$, $v_i = v(ih)$ we have

9.11
$$\begin{cases} u_{i+1} = u_i + \dfrac{h}{2}(v_i + v_{i+1}) \\[2mm] v_{i+1} = v_i - \dfrac{\lambda h}{2}(u_i + u_{i+1}) \end{cases}$$

for $i = 0, 1, 2$, where $u_0 = 0$, $v_0 = 1$. One could determine $\phi(\lambda) = u_3$ for various values of λ and then solve (9.8) using an appropriate method of Chapter 4. However, in this particularly simple case we can find $\phi(\lambda)$ explicitly. Indeed one can easily show that

9.12
$$\phi(\lambda) = u_3 = \frac{\left(3 - \dfrac{h^2\lambda}{4}\right)\left(1 - \dfrac{3h^2\lambda}{4}\right)}{\left(1 + \dfrac{h^2\lambda}{4}\right)^3} h.$$

Hence $\phi(\lambda)$ vanishes when

9.13
$$\begin{cases} \lambda_1 = \dfrac{4}{3h^2} = 12 \\[3mm] \lambda_2 = \dfrac{12}{h^2} = 108. \end{cases}$$

Since the first two eigenvalues of the differential equation are $\pi^2 \doteq 9.86$ and $4\pi^2 = 39.44$, the first eigenvalue is reasonably accurate, but the second is very poor.

To study the method more carefully let us obtain an analytic formula for the eigenvalues obtained for general h. If we assume that

---

* In the work described by Conte and Young [1957], Muller's method was used.

9.14
$$u_k = c\mu^k, \qquad v_k = d\mu^k,$$

we get, upon substitution in (9.11),

9.15
$$\begin{cases} c(\mu - 1) = \dfrac{h}{2} d(1 + \mu) \\[2ex] d(\mu - 1) = -\dfrac{h\lambda}{2} c(1 + \mu). \end{cases}$$

This leads to the conditions

9.16
$$\mu = \frac{1 \pm \dfrac{h}{2}\sqrt{\lambda}\, i}{1 \mp \dfrac{h}{2}\sqrt{\lambda}\, i} \qquad \lambda = -\left(\frac{2}{h}\right)^2 \left(\frac{\mu - 1}{\mu + 1}\right)^2.$$

Thus the general solution of (9.11) is

9.17
$$\begin{cases} u_k = c_1 \mu_1^k + c_2 \mu_2^k \\[2ex] v_k = c_1 \left\{ (\mu_1 - 1)\dfrac{1}{h} + (\mu_1 + 1)\dfrac{\lambda h}{4} \right\} \mu_1^k \\[2ex] \qquad + c_2 \left\{ (\mu_2 - 1)\dfrac{1}{h} + (\mu_2 + 1)\dfrac{\lambda h}{4} \right\} \mu_2^k, \end{cases}$$

where

9.18
$$\mu_1 = \frac{1 + \dfrac{h}{2}\sqrt{\lambda}\, i}{1 - \dfrac{h}{2}\sqrt{\lambda}\, i}, \qquad \mu_2 = \frac{1 - \dfrac{h}{2}\sqrt{\lambda}\, i}{1 + \dfrac{h}{2}\sqrt{\lambda}\, i}.$$

If we require that $u_0 = u_N = 0$, where $N = h^{-1}$, we have $c_1 = -c_2$ and

9.19
$$\mu_1^{2N} = 1, \qquad \mu_1 = e^{i\pi s/N}, \qquad s = 1, 2, \ldots, N - 1.$$

Therefore

9.20
$$\lambda_s = \left(\frac{2}{h}\right)^2 \tan^2 \frac{\pi s}{2N}, \qquad s = 1, 2, \ldots, N - 1.$$

In the case $h = N^{-1} = 1/3$ we get

9.21
$$\lambda_1 = \frac{4}{h^2}(\tfrac{1}{3}) = 12, \qquad \lambda_2 = \frac{4}{h^2}(3) = 108$$

as before. In general, for given s we have

9.22
$$\lambda_s = \left(\frac{2}{h}\right)^2 \tan^2 \frac{\pi s h}{2} \sim \pi^2 s^2 + O(h^2)$$

as $h \to 0$. Thus for small sh there is good agreement between the exact eigenvalue and the values which would be produced by the modified Euler method if the calculations were carried out exactly.

The method of superposition is applicable to the more general class of problems defined by

9.23
$$L[u] = l_0(\lambda, x)u^{(n)} + l_1(\lambda, x)u^{(n-1)} + \cdots + l_n(\lambda, x)u,$$
$$a < x < b$$

with the boundary conditions

9.24
$$U_i[u] = \sum_{j=1}^{n} [a_{i,n-j}(\lambda)u^{(n-j)}(a) + b_{i,n-j}(\lambda)u^{(n-j)}(b)] = 0, \qquad i = 1, 2, \ldots, n,$$

where the $a_{i,k}(\lambda)$ and $b_{i,k}(\lambda)$ are given functions of λ. (See, for instance, Conte and Young [1957].) For a given value of λ one seeks a fundamental set of solutions $u_j(x)$ such that $L[u_j] = 0$ and such that

9.25
$$u_j^{(k-1)}(a) = \begin{cases} 0, & j \neq k \\ 1, & j = k. \end{cases}$$

The most general solution of (9.23) has the form

9.26
$$u(x) = \sum_{j=1}^{n} c_j u_j(x).$$

An attempt to choose the c_j so that the conditions (9.24) are satisfied leads to the requirement that there exist $c_1, c_2, \ldots, c_n$ not all of which vanish such that

9.27
$$\sum_{j=1}^{n} c_j U_i[u_j] = 0, \qquad i = 1, 2, \ldots, n.$$

But this leads to the requirement that the determinant of the matrix

9.28
$$\begin{vmatrix} U_1[u_1] & U_1[u_2] & \cdots & U_1[u_n] \\ U_2[u_1] & U_2[u_2] & \cdots & U_2[u_n] \\ \cdot & \cdot & \cdots & \cdot \\ U_n[u_1] & U_n[u_2] & \cdots & U_n[u_n] \end{vmatrix}$$

vanish. Since, as shown in Chapter 12, the determinant, $\phi(\lambda)$, can be conveniently evaluated for any λ, we can find real roots of $\phi(\lambda) = 0$ by using an appropriate method of Chapter 4.

Method of Finite Differences

With the method of finite differences we choose a mesh size $h = (b - a)/N$, for some positive integer N, and construct a discrete operator $L_h[u]$ which is consistent with $L[u]$. We then seek a function $u(x, \lambda)$ which does not vanish identically on R_h such that

9.29
$$L_h[u] = 0$$

on R_h and such that (9.2) is satisfied.

Let us consider discrete operators of the form

9.30
$$L_h[u] = \alpha_0 u(x) + \alpha_1 u(x + h) + \alpha_2 u(x - h).$$

In the example (9.3) with $h = 1/3$ we get, using the standard three-point representation of $L[u]$,

9.31
$$\begin{cases} \dfrac{u_2 - 2u_1 + u_0}{h^2} = -\lambda u_1, \\[2mm] \dfrac{u_3 - 2u_2 + u_1}{h^2} = -\lambda u_2, \end{cases}$$

or

9.32
$$\begin{cases} (2 - \lambda h^2)u_1 - u_2 = 0, \\ -u_1 + (2 - \lambda h^2)u_2 = 0. \end{cases}$$

Unless $u_1 = u_2 = 0$ we must have $(2 - \lambda h^2)^2 - 1 = 0$ or

9.33
$$\lambda_1 = \frac{1}{h^2} = 9, \qquad \lambda_2 = \frac{3}{h^2} = 27.$$

These values are in reasonable agreement with the first two eigenvalues of the differential equation, namely, $\pi^2 \doteq 9.86$ and $4\pi^2 \doteq 39.44$. Closer agreement could be achieved by using a smaller value of h. If one were to do so for this particular problem one would have

9.34
$$u_{k+1} - 2u_k + u_{k-1} = -\lambda h^2 u_k, \qquad k = 1, 2, \ldots, N - 1,$$

where $N = h^{-1}$ and $u_0 = u_N = 0$. Letting

9.35
$$u_k = u_{n,k} = \sin \frac{n\pi k}{N}$$

for $1 \leq n \leq N - 1$ we obtain

9.36
$$\lambda = \lambda_n = \frac{4}{h^2} \sin^2 \frac{n\pi h}{2} = n^2\pi^2 + O(h^2).$$

One can achieve higher accuracy without increasing the number of terms in the operator $L_h[u]$ by using the methods of Section 10.7. We seek to determine a discrete operator $L_h[u]$ of the form (7.1) and $\beta_1, \beta_2, \ldots$ with the properties specified in Section 10.7 such that

9.37
$$L_h[u] - L[u] = \sum_{i=1}^{\infty} \beta_i L^{(i)}[u] + O(h^{p+1})$$

where p is as large as possible. The resulting coefficients α_i and also the β_i will in general be functions of λ, but, as we shall see, this does not cause any serious complication. The reader should verify that in the example (9.3) if we let $\beta_i = 0$ for $i > 2$ then the coefficients of $u, u', u'', u^{(3)}$, and $u^{(4)}$ in the expansion of

9.38
$$L_h[u] - L[u] - \beta_1 L^{(1)}[u] - \beta_2 L^{(2)}[u]$$

vanish if we let

9.39
$$\begin{cases} \alpha_0 = \lambda - \dfrac{2}{h^2\left(1 - \dfrac{h^2\lambda}{12}\right)} \\[20pt] \alpha_1 = \alpha_2 = \dfrac{1}{h^2\left(1 - \dfrac{h^2\lambda}{12}\right)} \\[20pt] \beta_1 = 0 \\[10pt] \beta_2 = \dfrac{h^2}{12\left(1 - \dfrac{h^2\lambda}{12}\right)}. \end{cases}$$

Moreover, the coefficients of $u^{(5)}, u^{(6)}, \ldots$ are $O(h^4)$. Thus the relative degree of approximation of $L_h[u]$ to $L[u]$ is three.

We now seek to solve, instead of (9.34), the equation

9.40
$$\frac{1}{h^2\left(1 - \dfrac{h^2\lambda}{12}\right)} u_{k+1} - \frac{2}{h^2\left(1 - \dfrac{h^2\lambda}{12}\right)} u_k + \frac{1}{h^2\left(1 - \dfrac{h^2\lambda}{12}\right)} u_{k-1} = -\lambda u_k,$$

$$k = 1, 2, \ldots, N - 1,$$

or equivalently

9.41
$$u_{k+1} - 2u_k + u_{k-1} = -\lambda h^2 \left(1 - \frac{h^2\lambda}{12}\right) u_k.$$

Evidently we get

9.42
$$\lambda = \lambda_n = \frac{4}{h^2} \sin^2 \frac{n\pi h}{2} \left(1 - \frac{h^2\lambda}{12}\right) = n^2\pi^2 + O(h^4).$$

Thus a much greater accuracy has been achieved. As in the case of the two-point boundary value problem, we can achieve arbitrarily high accuracy using a three-point formula by the introduction of more β_i.

In the general case we seek values of λ such that the determinant of a tri-diagonal matrix vanishes. Thus we are led to consider the problem of finding λ such that the matrix appearing in (4.22) vanishes, where the A_i, B_i and C_i are functions of λ. The determinant, $\phi(\lambda)$, can easily be evaluated as follows. Let Δ_k be the determinant of the matrix of size k. Then we have

9.43
$$\begin{cases} \Delta_1 & = B_1 \\ \Delta_2 & = B_1B_2 - A_2C_1 \\ \Delta_{k+1} & = B_{k+1}\Delta_k - A_{k+1}C_k\Delta_{k-1}, \qquad k = 2, 3, \dots . \end{cases}$$

As before, any of the methods of Chapter 4 for finding real roots of the non-linear equation $\phi(\lambda) = 0$ and which involve only functional evaluation can be used.

EXERCISES 10.9

1. For the problem

$$u'' = -\lambda u, \qquad 0 < x < 1$$
$$u(0) = u(1) = 0$$

compute $u(\lambda, x)$ by the modified Euler method with $h = 1/3$ for $\lambda = 11$ and $\lambda = 13$. Then use the method of false position to find λ_1.

2. Find the smallest eigenvalue for the problem

$$L[u] = (xu')' - \frac{1}{1+x}u + \lambda u = 0, \qquad 0 < x < 1$$
$$u(0) = 0, \qquad u(1) = 1$$

using the method of finite differences with $h = 1/3$

a) using the ordinary three-point operator,
b) using a three-point discrete operator with a relative degree of approximation of three.

3. Describe how you would solve the following eigenvalue problem by finite difference methods:

$$\frac{d}{dx}[x^2u'] - (1 + x)u = -\lambda(1 + x^2)u, \qquad 0 < x < 1$$

$$u(0) + 2u'(0) = 0$$

$$3u(1) + u'(1) = 0.$$

4. Find the smallest two eigenvalues for

$$L[u] = u'' + \left(\frac{\lambda}{1 + x} - x^2\right)u = 0, \qquad 0 < x < 1$$

$$u(0) = u(1) = 0$$

using the method of superposition with the Runge-Kutta method and also by the method of finite differences using a three-point discrete operator $L_h[u]$ which is consistent with $L[u]$ and which has a relative degree of approximation of three.

5. Show that one can determine an "exact" operator $L_h[u]$ for the problem given by (9.3) corresponding to the case $h = 1/3$, and that the smallest two eigenvalues can thus be obtained exactly.

6. Verify (9.12).

7. Determine a discrete operator $B_h[u] = \alpha_0 u(a) + \alpha_1 u(a + h)$ such that $B_h[u]$ is consistent with $B[u] = 2au(a) + a^2u'(a)$ and such that for all functions u satisfying

$$L[u] = u'' + (x^2 + \lambda)u = 0$$

we have

$$B_h[u] - B[u] = O(h^2).$$

8. Apply the scheme defined by (9.43) to evaluate the determinant of

$$\begin{vmatrix} 4 & -1 & 0 & 0 \\ -1 & 3 & -1 & 0 \\ 0 & -1 & 2 & -1 \\ 0 & 0 & -1 & 1 \end{vmatrix}.$$

Check the result by direct expansion of the determinant.

9. Determine the α_i in (9.30) so that for the example given by (9.3) the relative degree of approximation of $L_h[u]$ to $L[u]$ is five. Compute the smallest value of λ and compare with the exact value.

SUPPLEMENTARY DISCUSSION

There are many excellent references on methods for solving two-point boundary problems to which the reader should refer. See, for instance, Keller [1968], Ciarlet, Schultz, and Varga [1967], Lees [1966], Henrici [1962], Collatz [1960], Fox [1957], and many others.

An important class of methods not considered in the text involves the use of splines (see Section 6.8). These methods are discussed by Ciarlet, Schultz, and Varga [1966 and subsequent papers].

Section 10.4. The error analysis is based on that given by Gerschgorin [1930] for boundary values with two independent variables involving elliptic partial differential equations.

Section 10.7. The methods described are related to the difference correction method of Fox [1947], the "mehrstellenverfahren" method of Collatz [1960], and the work of Young and Dauwalder [1965].

Section 10.9. For further information on methods for solving the two-point eigenvalue problem, the reader is referred to Birkhoff, deBoor, Swartz, and Wendroff [1966], to Varga [1970], and to Keller [1968].

CHAPTER 11

VECTORS, MATRICES, AND NORMS

11.1 INTRODUCTION

Even though the reader is assumed to have had an exposure to at least an elementary course in linear algebra, we shall review the basic material from linear algebra before introducing a few advanced topics which will be used in subsequent chapters. In most cases either the proofs of the theorems stated are included, or references to the proofs are given. Often, the proofs are left as exercises for the reader. Ordinarily the proof of a theorem is given when not only the theorem itself but the method of proving it exhibit a useful concept in linear algebra. However, if a proof is too lengthy it is omitted in order to save space.

Linear algebra is essentially a study of linear transformations over abstract vector spaces. However, two particular vector spaces, $\mathbf{R}^n$ and $\mathbf{C}^n$ (defined below), are our primary concern in this book and so, for the most part, we shall restrict our review to appropriate properties of linear transformations over $\mathbf{R}^n$ and $\mathbf{C}^n$.

It is assumed that the reader is familiar with the properties of both the real numbers and the complex numbers and so those properties will not be reviewed in any detail. However, we shall mention a few properties of the complex numbers in order to make our notation clear.

11.2 VECTOR SPACES

In order to define a vector space we need first to define a *field*.

2.1 Definition. A field is a set of elements $\mathbf{F} = \{a, b, c, \ldots\}$ along with two binary compositions $+$ and $\cdot$ which satisfy the following postulates:

   i) $a, b \in \mathbf{F} \Rightarrow (a + b) = (b + a) \in \mathbf{F}$.
   ii) $a, b, c \in \mathbf{F} \Rightarrow (a + b) + c = a + (b + c)$.
   iii) $\exists$ unique $z \in \mathbf{F}$ such that $a + z = a$ for all $a \in \mathbf{F}$.
   iv) $a \in \mathbf{F} \Rightarrow \exists$ unique $\underline{a} \in \mathbf{F}$ such that $a + \underline{a} = z$.
   v) $a, b \in \mathbf{F} \Rightarrow (a \cdot b) = (b \cdot a) \in \mathbf{F}$.
   vi) $a, b, c \in \mathbf{F} \Rightarrow (a \cdot b) \cdot c = a \cdot (b \cdot c)$.
   vii) $\exists$ unique $e \in \mathbf{F}$ such that $a \cdot e = a$ for all $a \in \mathbf{F}$.
viii) $a \in \mathbf{F}, a \neq z \Rightarrow \exists$ unique $a^{-1} \in \mathbf{F}$ such that $a \cdot a^{-1} = e$.
   ix) $a, b, c \in \mathbf{F} \Rightarrow a \cdot (b + c) = a \cdot b + a \cdot c$.

The unique element z described in (iii) is called the *additive identity* of $\mathbf{F}$ and the unique element e described in (vii) is called the *multiplicative identity* of **F**. The element $\underline{a}$ described in (iv) is called the *additive inverse* of a in **F**, and the element a^{-1} described in (viii) is called the *multiplicative inverse* of a in **F**.

2.2 EXAMPLE. Let **Q** denote the set of rational numbers and let the binary compositions $+$ and $\cdot$ denote the ordinary addition and multiplication operations of elementary algebra. In this case $z = 0, e = 1, \underline{a} = -a$, and $a^{-1} = 1/a$ for nonzero a. The reader can easily verify that **Q** is a field.

2.3 EXAMPLE. Let **R** denote the set of real numbers and let the binary compositions $+$ and $\cdot$ denote the ordinary addition and multiplication operations of elementary algebra. Again, $z = 0, e = 1, \underline{a} = -a$, and $a^{-1} = 1/a$ for nonzero a. The reader can easily verify that **R** is a field.

2.4 EXAMPLE. Let **C** denote the set of complex numbers which we choose to write as ordered pairs of real numbers, for notational convenience. Thus we consider

$$\mathbf{C} = \{(a, b) : a, b \in \mathbf{R}\}$$

with the binary compositions $\oplus$, called complex addition, and $\circ$, called complex multiplication, defined by

$$(a, b) \oplus (c, d) = (a + c, b + d)$$

and

$$(a, b) \circ (c, d) = (a \cdot c + \underline{b \cdot d}, b \cdot c + a \cdot d),$$

respectively. The symbols $+$ and $\cdot$ on the right are the binary compositions of **R** in (2.3) and $\underline{b \cdot d} = -(b \cdot d)$.

In this example $z = (0, 0), e = (1, 0)$, the additive inverse of (a, b) is $(-a, -b)$, and the multiplicative inverse of $(a, b) \neq (0, 0)$ is

$$\left(\frac{a}{a^2 + b^2}, \quad \frac{-b}{a^2 + b^2} \right).$$

The reader can easily verify that **C** is a field.

Since

2.5
$$(a, 0) \oplus (0, 1) \circ (b, 0) = (a, 0) \oplus (0, b)$$
$$= (a, b)$$

this suggests another way of writing the elements of **C**. If we adopt the notational convention* that $(r, 0) = r \in \mathbf{R}$ and $(0, 1) = i$, where $i^2 = -1$, then we sometimes

---

* With this understanding, then, we note that $\mathbf{R} \subset \mathbf{C}$.

write (a, b) in the form* $a + ib$. The symbol i is usually referred to as the *imaginary unit* and the symbols a and b as the *real* and the *imaginary* parts, respectively, of the complex number (a, b). The symbol "$+$" in $a + ib$ is actually the binary composition $\oplus$ of $\mathbf{C}$ rather than the $+$ of $\mathbf{R}$.

Another form (the polar coordinate form of $a + ib$) is

2.6 $$re^{i\theta} = r(\cos \theta + i \sin \theta)$$

where† $r = (a^2 + b^2)^{\frac{1}{2}}$ and $\theta = \tan^{-1}(b/a)$. Sometimes we prefer a single symbol, for example w, in which case

2.7 $$w = (a, b)$$
$$= a + ib$$
$$= re^{i\theta}.$$

2.8 EXAMPLE.   Let $\mathbf{B} = \{0, 1\}$. If the binary compositions $+$ and $\cdot$ are defined by the tables

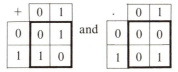

+	0	1
0	0	1
1	1	0

and

·	0	1
0	0	0
1	0	1

then it can be verified that $\mathbf{B}$ is a field.

We shall refer to the elements of a field as *scalars*. At this point we are in a position to define an (abstract) vector space over a field.

2.9 Definition. A vector space $\mathbf{V}$ over a field $\mathbf{F}$ is a set of objects‡ $\{x, y, z, \ldots\}$ called *vectors* such that:

a) For, $x, y \in \mathbf{V}$ there is a binary composition $\boxplus$, for which $x \boxplus y \in \mathbf{V}$, called vector addition, that satisfies the following postulates:

    i) $x, y \in \mathbf{V} \Rightarrow x \boxplus y = y \boxplus x$.
    ii) $x, y, z \in \mathbf{V} \Rightarrow (x \boxplus y) \boxplus z = x \boxplus (y \boxplus z)$.
    iii) $\exists$ unique $\phi \in \mathbf{V}$ such that $x \boxplus \phi = x$ for all $x \in \mathbf{V}$.
    iv) $x \in \mathbf{V} \Rightarrow \exists$ unique $\underline{x} \in \mathbf{V}$ such that $x \boxplus \underline{x} = \phi$.

b) For each scalar $a \in \mathbf{F}$ and each vector $x \in \mathbf{V}$ a binary composition $ax \in \mathbf{V}$, called scalar multiplication, is defined and it satisfies the following postulates:

    i) $a \in \mathbf{F}$ and $x, y \in \mathbf{V} \Rightarrow a(x \boxplus y) = ax \boxplus ay$.
    ii) $a, b \in \mathbf{F}$ and $x \in \mathbf{V} \Rightarrow (a + b)x = ax \boxplus bx$.

---

* See A—(5.1).

† See A—(5.4).

‡ The symbol z used here represents a vector and *not* the additive identity element of $\mathbf{F}$.

iii) $a, b \in \mathbf{F}$ and $x \in \mathbf{V} \Rightarrow (a \cdot b)x = a(bx)$.

iv) $x \in \mathbf{V} \Rightarrow ex = x$, where e is the multiplicative identity of $\mathbf{F}$.

The unique vector ϕ described in (iii) of part (a) is called the *additive identity vector* of $\mathbf{V}$. The vector $\underline{x}$ described in (iv) of part (a) is the *additive inverse vector* of x in $\mathbf{V}$.

Notice that we have been careful to distinguish between the compositions $+$ and $\cdot$ of $\mathbf{F}$ and the compositions $\boxplus$ and ax (indicating addition and scalar multiplication) in $\mathbf{V}$.

It is left as an exercise for the reader to verify that the following example is an (abstract) vector space over $\mathbf{Q}$.

2.10 EXAMPLE. Consider the set

$$\mathbf{V} = \{a + b \cdot \sqrt{2} + c \cdot \sqrt{3} : a, b, c \in \mathbf{Q}\}.$$

If $x, y \in \mathbf{V}$, where

$$x = a_1 + b_1 \cdot \sqrt{2} + c_1 \cdot \sqrt{3}$$
$$y = a_2 + b_2 \cdot \sqrt{2} + c_2 \cdot \sqrt{3},$$

we define vector addition by the expression

$$x \boxplus y = (a_1 + a_2) + (b_1 + b_2) \cdot \sqrt{2} + (c_1 + c_2) \cdot \sqrt{3}.$$

If $k \in \mathbf{Q}$ we define scalar multiplication by the expression

$$kx = (k \cdot a_1) + (k \cdot b_1) \cdot \sqrt{2} + (k \cdot c_1) \cdot \sqrt{3}.$$

The additive identity vector is obtained by selecting $a = 0, b = 0$, and $c = 0$. If

$$u = a + b \cdot \sqrt{2} + c \cdot \sqrt{3}$$

then the additive inverse of u is

$$\underline{u} = (-a) + (-b) \cdot \sqrt{2} + (-c) \cdot \sqrt{3}.$$

EXERCISES 11.2

1. Verify that $\mathbf{Q}$ in (2.2) is a field.

2. Verify that $\mathbf{R}$ in (2.3) is a field.

3. Verify that $\mathbf{C}$ in (2.4) is a field.

4. If complex numbers are written using the notation $w = a + ib$, show how we write

   i) the additive identity;

   ii) the multiplicative identity;

iii) the additive inverse;

iv) the multiplicative inverse.

5. In problem 4 use the notation $w = re^{i\theta}$.

6. In problem 4 use the notation $w = r(\cos\theta + i\sin\theta)$.

7. Verify that **B** in (2.8) is a field.

8. Verify that the set **V** in (2.10) is a vector space over **Q**.

11.3 TWO SPECIAL VECTOR SPACES R^n AND C^n

As we stated earlier we shall not be concerned primarily with abstract vector spaces in general. Instead, we wish to focus our attention on the two vector spaces described in this section. First we need some preliminary definitions.

3.1 Definition. An n-tuple x (also called an n-dimensional coordinate vector) is an *ordered* set of n *components* $x_1, x_2, \ldots, x_n$, which we choose to arrange in a column

$$x = \begin{bmatrix} x_1 \\ x_2 \\ \cdot \\ \cdot \\ \cdot \\ x_n \end{bmatrix}$$

3.2 Definition. For each positive integer n, $\mathbf{R}^n$ will denote the set of all n-tuples whose components $x_i \in \mathbf{R}$. We call $\mathbf{R}^n$ the *n-dimensional real coordinate space*. (The motivation for this terminology is (3.9) below.)

Before we go to our next definition let us recall (2.7) where we used a single symbol w to represent a complex number (a, b). Suppose we let $u = (c, d)$. Then the binary compositions of example (2.4) may be written

3.3
$$w \oplus u = (a, b) \oplus (c, d)$$
$$= (a + c, b + d)$$

and

3.4
$$w \circ u = (a, b) \circ (c, d)$$
$$= (a \cdot c + \underline{b \cdot d}, b \cdot c + a \cdot d)$$

In discussing n-tuples in (3.1) we said nothing about the components other than that we wished to arrange them in a column. In (3.2) we chose them to be real numbers. In (3.5) we choose them to be complex numbers and for simplicity we choose the single letter representation.

3.5 Definition. For each positive integer n, $\mathbf{C}^n$ will denote the set of all n-tuples whose components $x_i \in \mathbf{C}$. We call $\mathbf{C}^n$ the *n-dimensional complex coordinate space*.

If x and y are the n-tuples

$$x = \begin{bmatrix} x_1 \\ x_2 \\ \vdots \\ x_n \end{bmatrix}, \qquad y = \begin{bmatrix} y_1 \\ y_2 \\ \vdots \\ y_n \end{bmatrix}$$

then, in both $\mathbf{R}^n$ and $\mathbf{C}^n$, the equation $x = y$ is equivalent to the system of equations

3.6 $$x_i = y_i$$

for $i = 1, 2, \ldots, n$. This is sometimes called *componentwise equality*.

Since we have implied that $\mathbf{R}^n$ and $\mathbf{C}^n$ are vector spaces we need to define two binary compositions, *addition* and *scalar multiplication*, for the n-tuples in $\mathbf{R}^n$ and $\mathbf{C}^n$. (See (3.9) below.)

3.7 Definition. Let x, y, z be the n-tuples

$$x = \begin{bmatrix} x_1 \\ x_2 \\ \vdots \\ x_n \end{bmatrix}, \qquad y = \begin{bmatrix} y_1 \\ y_2 \\ \vdots \\ y_n \end{bmatrix}, \qquad z = \begin{bmatrix} z_1 \\ z_2 \\ \vdots \\ z_n \end{bmatrix}.$$

i) If $x, y, z \in \mathbf{R}^n$ then *addition* is defined as follows:

$$x \boxplus y = z$$

if and only if

$$x_i + y_i = z_i$$

for $i = 1, 2, \ldots, n$.

ii) If $x, y, z \in \mathbf{C}^n$ then *addition* is defined as follows:

$$x \boxplus y = z$$

if and only if

$$x_i \oplus y_i = z_i$$

for $i = 1, 2, \ldots, n$.

Notice that the symbols $+$, $\oplus$, and $\boxplus$ denote addition of real numbers, addition of complex numbers, and addition of n-tuples, respectively. We should emphasize that addition of n-tuples is *componentwise addition*. Next we define scalar multiplication

3.8 Definition. If $k \in \mathbf{R}$ and $x \in \mathbf{R}^n$ then

$$kx = \begin{bmatrix} k \cdot x_1 \\ k \cdot x_2 \\ \cdot \\ \cdot \\ \cdot \\ k \cdot x_n \end{bmatrix}.$$

If $k \in \mathbf{C}$ and $x \in \mathbf{C}^n$ then

$$kx = \begin{bmatrix} k \circ x_1 \\ k \circ x_2 \\ \cdot \\ \cdot \\ \cdot \\ k \circ x_n \end{bmatrix}.$$

Notice that $k \cdot x_i$, $k \circ x_i$, and kx denote multiplication of real numbers, multiplication of complex numbers, and scalar multiplication, respectively. We emphasize that scalar multiplication is *componentwise multiplication*.

We have now set the stage for the basic result to which we have been alluding.

3.9. Theorem. $\mathbf{R}^n$ and $\mathbf{C}^n$ form vector spaces over $\mathbf{R}$ and $\mathbf{C}$, respectively, if vector addition and scalar multiplication are defined by (3.7) and (3.8).

Proof. To verify postulate (i), part (a), of (2.9) for $\mathbf{R}^n$ let $u, v, x, y \in \mathbf{R}^n$, where

$$u = x \boxplus y$$
$$v = y \boxplus x.$$

Then, from (3.6) and (3.7) and the properties of $\mathbf{R}$, we have

$$u_i = x_i + y_i$$
$$= y_i + x_i$$
$$= v_i.$$

Hence $u = v$ which means $x \boxplus y = y \boxplus x$ in $\mathbf{R}^n$.

To verify this same postulate for $\mathbf{C}^n$ let $u, v, x, y \in \mathbf{C}^n$, where u and v are defined as before. Then, from (3.6) and (3.7) and the properties of $\mathbf{C}$, we have

$$u_i = x_i \oplus y_i$$
$$= y_i \oplus x_i$$
$$= v_i.$$

Hence $u = v$ and so $x \boxplus y = y \boxplus x$ in $\mathbf{C}$.

To verify postulate (i), part (b), of (2.9) for $\mathbf{R}^n$ let $a \in \mathbf{R}$ and $r, s, t, u, v, x, y \in \mathbf{R}^n$, where

$$r = ax$$
$$s = ay$$
$$t = x \boxplus y$$
$$u = at$$
$$v = r \boxplus s.$$

Then, from (3.6), (3.7), (3.8) and the properties of $\mathbf{R}$, we have

$$u_i = a \cdot t_i$$
$$= a \cdot (x_i + y_i)$$
$$= a \cdot x_i + a \cdot y_i$$
$$= r_i + s_i$$
$$= v_i.$$

Hence $u = v$ and so $a(x \boxplus y) = ax \boxplus ay$ in $\mathbf{R}^n$.

To verify this same postulate for $\mathbf{C}^n$ let $a \in \mathbf{C}$ and $r, s, t, u, v, x, y \in \mathbf{C}^n$ where $r, s, t, u,$ and v are defined as before. Then from (3.6), (3.7), (3.8) and the properties of $\mathbf{C}$, we have

$$u_i = a \circ t_i$$
$$= a \circ (x_i \oplus y_i)$$
$$= a \circ x_i \oplus a \circ y_i$$
$$= r_i \oplus s_i$$
$$= v_i.$$

Hence $u = v$ and so $a(x \boxplus y) = ax \boxplus ay$ in $\mathbf{C}^n$.

The reader can verify the remaining postulates of (2.9), including the fact that

3.10
$$\phi = \begin{bmatrix} 0 \\ 0 \\ \vdots \\ 0 \end{bmatrix}$$

in $\mathbf{R}^n$ and

3.11
$$\phi = \begin{bmatrix} (0,0) \\ (0,0) \\ \cdot \\ \cdot \\ \cdot \\ (0,0) \end{bmatrix}$$

in $\mathbf{C}^n$. For obvious reasons we call ϕ the *zero vector*.

Likewise, the reader will be able to show that, for $x \in \mathbf{R}^n$,

3.12
$$\underline{x} = \begin{bmatrix} -x_1 \\ -x_2 \\ \cdot \\ \cdot \\ -x_n \end{bmatrix}$$

and, for

3.13
$$x = \begin{bmatrix} (a_1, b_1) \\ (a_2, b_2) \\ \cdot \\ \cdot \\ (a_n, b_n) \end{bmatrix}$$

in $\mathbf{C}^n$,

3.14
$$\underline{x} = \begin{bmatrix} (-a_1, -b_1) \\ (-a_2, -b_2) \\ \cdot \\ \cdot \\ (-a_n, -b_n) \end{bmatrix}.$$

3.15 *Remark.* Up to this point we have been careful to differentiate among real addition, complex addition, and vector addition by using the symbols $+$, $\oplus$, and $\boxplus$. We also have differentiated among real multiplication, complex multiplication, and scalar multiplication by using the symbols $\cdot$, $\circ$, and the notation ax (for scalar multiplication when $a \in \mathbf{F}$ and $x \in \mathbf{V}$). It is often not necessary to make these distinctions, and usually from this point on we shall use the symbol "$+$" for all types of addition and the notation ax to mean the product of a and x, regardless of the type of multiplication involved.

EXERCISE 11.3

Complete the proof of Theorem 3.9.

11.4 THE REAL COORDINATE SPACE $\mathbf{R}^n$

In the last section we defined $\mathbf{R}^n$ to be

4.1 $$\mathbf{R}^n = \{n\text{-tuples } x : x_i \in \mathbf{R}\}.$$

Theorem 3.9 states that $\mathbf{R}^n$ is a vector space. Hence the n-tuples may be called *vectors* and, for reasons discussed below, they are also called *n-dimensional co-ordinate vectors*.

The motivation for this last terminology is the fact that there is a one-to-one correspondence between the vectors $x \in \mathbf{R}^n$ and the *points* in n-dimensional Euclidean space* with *coordinates* $x_1, x_2, \ldots, x_n$. For example, in 3-dimensional Euclidean space we say the point P has coordinates x_1, x_2, x_3. (See Fig. 4.2).

4.2 Figure

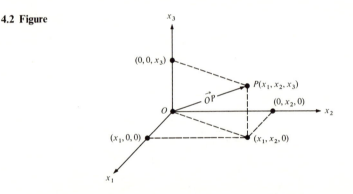

For each such point there is a (unique) directed line segment from the origin O (with coordinates $0, 0, 0$) to the point P, denoted by $\overrightarrow{OP}$, and conversely. Because of the one-to-one correspondence between P and $\overrightarrow{OP}$, and between P (with coordinates x_1, x_2, x_3) and the 3-tuple (3-dimensional coordinate vector)

4.3 $$x = \begin{bmatrix} x_1 \\ x_2 \\ x_3 \end{bmatrix},$$

we *identify* x with both P and $\overrightarrow{OP}$. This is made clear in the next three paragraphs.

By defining addition and scalar multiplication of directed line segments appropriately we can show that the set of directed line segments of the form $\overrightarrow{OP}$ constitutes an abstract vector space. (We leave it as an exercise for the reader

---

* For the formal definition of a Euclidean space see (15.11).

to define addition and scalar multiplication "appropriately.") It can also be shown that, under this one-to-one correspondence, if

$$x \to \overrightarrow{OP}$$

$$y \to \overrightarrow{OQ},$$

then $x + y = w$ and $bx = v$ imply $\overrightarrow{OP} + \overrightarrow{OQ} = \overrightarrow{OW}$ and $b\overrightarrow{OP} = \overrightarrow{OV}$, where

$$w \to \overrightarrow{OW}$$

$$v \to \overrightarrow{OV}.$$

We call such a correspondence a *vector space homomorphism* and we say $\mathbf{R}^3$ is homomorphic with the set of directed line segments of the form $\overrightarrow{OP}$. Actually, in this particular correspondence it turns out that we have a special homomorphism, called a *vector space isomorphism*, which means that every directed line segment of the form $\overrightarrow{OP}$ is the *unique* image of a vector in $\mathbf{R}^3$ and vice versa. When two vector spaces are isomorphic they have identical properties relative to vector addition and scalar multiplication and we say they have the same *structure*. In this case we can identify the elements of one with the elements of the other.

Our statements about the directed line segments of the form $\overrightarrow{OP}$ apply also to the set of points $P(x_1, x_2, x_3)$, and so we often use the terms point, vector, and directed line segment from the origin, interchangeably.

What we have just said relative to $\mathbf{R}^3$ applies equally well to $\mathbf{R}^n$ because we adopt the geometrical language of $\mathbf{R}^3$ when we generalize to n-dimensional Euclidean space. We speak of the "point" P (with coordinates $x_1, x_2, \ldots, x_n$) and the "directed line segment" $\overrightarrow{OP}$ from the "origin" (with coordinates $0, 0, \ldots, 0$) to P. These are both identified with the n-dimensional coordinate vector

4.4
$$x = \begin{bmatrix} x_1 \\ x_2 \\ \cdot \\ \cdot \\ \cdot \\ x_n \end{bmatrix},$$

because of the vector space isomorphisms that exist among the three sets of objects: points P, directed line segments $\overrightarrow{OP}$, and vectors x. Thus, in n-dimensions as in 3-dimensions, we often use the terms point, vector, and directed line segment from the origin, interchangeably.

4.5 *Remark.* In the trivial case, $n = 1$, we have a vector space isomorphism between the elements of $\mathbf{R}^1$ and the elements of $\mathbf{R}$ so that if $x = [x_1]$ we often identify the vector x with its scalar component x_1.

In (2.1) we used the symbol $\underline{a}$ for the additive inverse of a in $\mathbf{F}$ and in (2.9) we used the symbol $\underline{x}$ for the additive inverse vector of x in $\mathbf{V}$. From example (2.3) and from (3.12) and (3.14), we see that in $\mathbf{R}, \mathbf{C}, \mathbf{R}^n$, and $\mathbf{C}^n$ the symbols $-a$ and $-x$ are meaningful and preferred.

In $\mathbf{R}^n$ the vector $-x$ has a geometrical interpretation motivated by our identification of x, P, and $\overrightarrow{OP}$ above. We can illustrate this, for $n = 2$, quite easily. Consider the point $P(x_1, x_2)$ in the Euclidean plane and the directed line segment $\overrightarrow{OP}$ (the vector x).

4.6 Figure

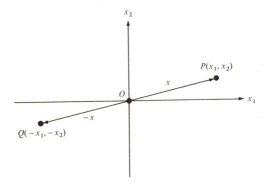

Then $\overrightarrow{OQ}$ in Fig. 4.6 represents $-x$. It is merely a reflection of x through the origin.

In $\mathbf{R}$ we define

4.7
$$a - b = a + (-b)$$

and use the minus sign (the symbol "$-$") on the left to denote the binary operation of *subtraction*. We do a similar thing in $\mathbf{R}^n$.

4.8 Definition. Given $x, y \in \mathbf{R}^n$ we define their *difference* $x - y$ by

$$x - y = x + (-y).$$

The reader should recall (3.15) and note that we are no longer using the symbol $\boxplus$ for vector addition since the meaning of the symbol "$+$" causes no confusion here. Thus we can write.

4.9
$$\begin{bmatrix} x_1 \\ x_2 \\ \vdots \\ x_n \end{bmatrix} - \begin{bmatrix} y_1 \\ y_2 \\ \vdots \\ y_n \end{bmatrix} = \begin{bmatrix} x_1 \\ x_2 \\ \vdots \\ x_n \end{bmatrix} + \begin{bmatrix} -y_1 \\ -y_2 \\ \vdots \\ -y_n \end{bmatrix} = \begin{bmatrix} x_1 - y_1 \\ x_2 - y_2 \\ \vdots \\ x_n - y_n \end{bmatrix},$$

and the subtraction of vectors in $\mathbf{R}^n$ is *componentwise subtraction*.

4.10 Theorem. $x \in \mathbf{R}^n \Rightarrow -x = (-1)x$.

Proof. Let $u = -x$ and $v = (-1)x$. Then, from (3.8), (3.12), and the properties of $\mathbf{R}$,

$$v_i = (-1)x_i$$
$$= -x_i$$
$$= u_i,$$

for $i = 1, 2, \ldots, n$. Hence, $u = v$ and the theorem follows.

4.11 Theorem. $x \in \mathbf{R}^n \Rightarrow 0x = \phi$.

Proof. Let $u = 0x$ and recall (3.10). Then, from (3.8) and the properties of $\mathbf{R}$,

$$u_i = 0 \cdot x_i$$
$$= 0,$$

for $i = 1, 2, \ldots, n$. Hence $u = \phi$ and the theorem is proved.

4.12 Theorem. $a \in \mathbf{R} \Rightarrow a\phi = \phi$.

Proof. Let $u = a\phi$ and recall (3.10). Then

$$u_i = a \cdot 0$$
$$= 0,$$

for $i = 1, 2, \ldots, n$. Hence $u = \phi$ and the theorem is proved.

4.13 *Remark.* Theorems such as 4.11 and 4.12 are trivial to prove for the vector space $\mathbf{R}^n$ (and even for $\mathbf{C}^n$). However, these two theorems are tedious to prove for an arbitrary (abstract) vector space $\mathbf{V}$.

We can now discuss a geometrical interpretation of scalar multiplication; that is, the meaning of the vector ax, where $a \in \mathbf{R}$. From (3.8) we see that if x is identified with the point P, whose coordinates are $x_1, x_2, \ldots, x_n$, then ax must be identified with the point A, whose coordinates are $ax_1, ax_2, \ldots, ax_n$.

If $a = 0$ then, from (4.11), A is the origin. If $a \neq 0$ then A is $|a|$ times as far*

---

* Distance or length are to be interpreted as Euclidean distance or length. Thus the length of $\overrightarrow{OP}$ is $\sqrt{x_1^2 + x_2^2 + \cdots + x_n^2}$ and the length of $\overrightarrow{OA}$ is $\sqrt{a^2(x_1^2 + x_2^2 + \cdots + x_n^2)}$. See, for example, (15.13).

from the origin as P. Thus, if $|a| > 1$, A is farther from the origin than P and the line segment $\overrightarrow{OA}$ is longer than $\overrightarrow{OP}$. If $|a| \leqq 1$, $\overrightarrow{OA}$ has length less than or equal to the length of $\overrightarrow{OP}$.

If $a < 0$ then, from (3.8), (4.10), and our discussion relative to Fig. 4.6, P and A are on opposite sides of the origin but lie on a line through the origin.

4.14 *Remark.* Throughout the remainder of this review we shall concentrate much of our discussion on results relative to $\mathbf{R}^n$, for the sake of simplicity. However, most of the definitions and theorems have their counterparts in an abstract vector space $\mathbf{V}$, and more specifically in $\mathbf{C}^n$. The reader should keep this fact in mind since space does not permit a complete treatment.

EXERCISES 11.4

1. Define addition and scalar multiplication of directed line segments in such a way that the set of directed line segments of the form $\overrightarrow{OP}$ constitutes an abstract vector space.

2. Define addition and scalar multiplication of points in n-dimensional Euclidean space with coordinates $(x_1, x_2, \ldots, x_n)$ in such a way that the set of points constitutes an abstract vector space.

3. Show that the set of directed line segments $\overrightarrow{OP}$ in 3-dimensional Euclidean space is isomorphic with the set of points P with coordinates (x_1, x_2, x_3).

4. Show that the set of points in 3-dimensional Euclidean space is isomorphic with $\mathbf{R}^3$.

5. Do problem 3 for n-dimensional Euclidean space.

6. Do problem 4 for n-dimensional Euclidean space.

7. Prove Theorem 4.10 for a general vector space $\mathbf{V}$ over a field $\mathbf{F}$.

8. Prove Theorem 4.11 for a general vector space $\mathbf{V}$ over a field $\mathbf{F}$.

9. Prove Theorem 4.12 for a general vector space $\mathbf{V}$ over a field $\mathbf{F}$.

11.5 SUBSPACES OF $\mathbf{R}^n$

Suppose $\mathbf{S} \subseteq \mathbf{R}^n$. In this case we call $\mathbf{S}$ a *subset* of $\mathbf{R}^n$. If $\mathbf{S} \subset \mathbf{R}^n$ we call it a *proper subset*. For example, $\mathbf{R}^n$ is a subset, but not a proper subset, of itself. Sometimes $\mathbf{S} \subseteq \mathbf{R}^n$ is also a vector space, in which case we introduce the following terminology:

5.1 Definition. If $\mathbf{S}$ is a nonempty subset of $\mathbf{R}^n$ and if $\mathbf{S}$ is also a vector space over the field $\mathbf{R}$ (with the same rules for vector addition and scalar multiplication as $\mathbf{R}^n$) then $\mathbf{S}$ is called a *subspace* of $\mathbf{R}^n$.

Trivially, $\mathbf{R}^n$ is a subspace of itself. Also, since every subspace must contain ϕ, by definition, it turns out that the subset $\mathbf{Z}$ consisting of ϕ alone, can be shown to be a subspace. This is sometimes called the *zero* or *null* subspace. The following theorem tells us, in general, when a subset of $\mathbf{R}^n$ is a subspace of $\mathbf{R}^n$.

5.2 Theorem. Let $\mathbf{S}$ be a nonempty subset of $\mathbf{R}^n$. Then $\mathbf{S}$ is a subspace of $\mathbf{R}^n$ if and only if

$$a, b \in \mathbf{R} \text{ and } x, y \in \mathbf{S} \Rightarrow ax + by \in \mathbf{S}.$$

Proof. See, for example, Cullen [1966], p. 45.

5.3. EXAMPLE. $\mathbf{S} = \{x \in \mathbf{R}^3 : x_3 = 0\}$.

That $\mathbf{S}$ is a subspace of $\mathbf{R}^3$ is easily verified since, if $a, b \in \mathbf{R}$ and $x, y \in \mathbf{R}^3$, then

$$a \begin{bmatrix} x_1 \\ x_2 \\ 0 \end{bmatrix} + b \begin{bmatrix} y_1 \\ y_2 \\ 0 \end{bmatrix} = \begin{bmatrix} ax_1 + by_1 \\ ax_2 + by_2 \\ 0 \end{bmatrix}$$

and the vector on the right is an element of $\mathbf{S}$. Geometrically, $\mathbf{S}$ is represented by a plane in a 3-dimensional Euclidean space.

It turns out that the intersection of two subspaces is itself a subspace. In fact, this statement can be generalized to an arbitrary number of subspaces.

5.4 Theorem. If $\mathbf{S}_1, \mathbf{S}_2, \ldots, \mathbf{S}_k$ are all subspaces of $\mathbf{R}^n$, then so is their intersection

$$\mathbf{W} = \mathbf{S}_1 \cap \mathbf{S}_2 \cap \ldots \cap \mathbf{S}_k$$

Proof. See, for example, Cullen [1966], p. 45.

5.5. EXAMPLE. Let $\mathbf{S}_1 = \{x \in \mathbf{R}^3 : x_3 = 0\}$ and $\mathbf{S}_2 = \{x \in \mathbf{R}^3 : x_2 = 0\}$. By definition, $x \in \mathbf{S}_1 \cap \mathbf{S}_2$ if and only if $x \in \mathbf{S}_1$ *and* $x \in \mathbf{S}_2$. Thus

$$\mathbf{S}_1 \cap \mathbf{S}_2 = \{x \in \mathbf{R}^3 : x_2, x_3 = 0\}.$$

Since $\mathbf{S}_1$ and $\mathbf{S}_2$ are represented, geometrically, by planes in $\mathbf{R}^3$ it follows that $\mathbf{S}_1 \cap \mathbf{S}_2$ is represented by the line of intersection of the two planes.

EXERCISES 11.5

1. Prove that the set $\{\phi\}$ is a subspace of $\mathbf{R}^n$.

2. Prove Theorem 5.2.

3. Prove that the intersection of two subspaces is a subspace.

4. Prove Theorem 5.4.

11.6 LINEAR COMBINATIONS AND LINEAR DEPENDENCE

Suppose we have a set of vectors $x, y, \ldots, z \in \mathbf{R}^n$. This notation is deficient because it does not give us a clue as to how many vectors there are in the set. Consequently, let us change the notation so that the symbols representing the vectors have indices. For example, we shall introduce superscripts and write $x^{(1)}, x^{(2)}, \ldots, x^{(k)} \in \mathbf{R}^n$. In this case $x^{(i)}$ can be written

$$6.1 \qquad x^{(i)} = \begin{bmatrix} x_1^{(i)} \\ x_2^{(i)} \\ \vdots \\ x_n^{(i)} \end{bmatrix}.$$

In (5.2) an expression of the form $a_1 x^{(1)} + a_2 x^{(2)}$ was used, where $a_1, a_2 \in \mathbf{R}$ and $x^{(1)}, x^{(2)} \in \mathbf{R}^n$. We frequently encounter a generalization of this expression

$$6.2 \qquad x = a_1 x^{(1)} + a_2 x^{(2)} + \ldots + a_t x^{(t)}.$$

It is useful to have a name for the vector x defined in this way.

6.3 Definition. Let $a_1, a_2, \ldots, a_t \in \mathbf{R}$ and $x^{(1)}, x^{(2)}, \ldots, x^{(t)} \in \mathbf{R}^n$. The vector x in (6.2) is called a *linear combination* of the vectors $x^{(1)}, x^{(2)}, \ldots, x^{(t)}$. The scalars $a_1, a_2, \ldots, a_t$ are called the *coefficients* of the linear combination: If $a_i = 0$, for $i = 1, 2, \ldots, t$, the linear combination is said to be *trivial*; otherwise it is called *non-trivial*.

Consider the equation

$$6.4 \qquad \begin{bmatrix} x_1 \\ x_2 \\ x_3 \\ \vdots \\ x_n \end{bmatrix} = x_1 \begin{bmatrix} 1 \\ 0 \\ 0 \\ \vdots \\ 0 \end{bmatrix} + x_2 \begin{bmatrix} 0 \\ 1 \\ 0 \\ \vdots \\ 0 \end{bmatrix} + \ldots + x_n \begin{bmatrix} 0 \\ 0 \\ \vdots \\ 0 \\ 1 \end{bmatrix}$$

which may be written

$$6.5 \qquad x = x_1 e^{(1)} + x_2 e^{(2)} + \ldots + x_n e^{(n)},$$

where $e^{(i)}$ denotes the vector whose components* $e_j^{(i)} = \delta_{ij}, j = 1, 2, \ldots, n$. We call $e^{(1)}, e^{(2)}, \ldots, e^{(n)}$ *unit vectors* in $\mathbf{R}^n$. An obvious restatement of (6.5) is the following:

6.6 Theorem. $x \in \mathbf{R}^n \Rightarrow x$ is a linear combination of the unit vectors

$$\{ e^{(1)}, e^{(2)}, \ldots, e^{(n)} \}.$$

Observe that the components of x are the coefficients of the special linear combination (6.4). Also note that if $x = \phi$ the linear combination is trivial; otherwise, it is non-trivial.

---

* The symbol δ_{ij} is the Kronecker delta. It takes the value "one" when $i = j$ and the value "zero" when $i \neq j$.

Now let us return to the general linear combination (6.2) and introduce the following important concept:

6.7 Definition. The (distinct) vectors $x^{(1)}, x^{(2)}, \ldots, x^{(t)} \in \mathbf{R}^n$ are *linearly dependent* if and only if $\exists$ a non-trivial linear combination of them equal to ϕ, that is, if and only if

$$a_1 x^{(1)} + a_2 x^{(2)} + \ldots + a_t x^{(t)} = \phi$$

where $a_i \neq 0$ for at least one i. On the other hand, if the only linear combination of these vectors which equals ϕ is the trivial one, then they are *linearly independent*.

Note that *the vector ϕ is (obviously) linearly dependent* because the linear combination $a\phi$ for $a \neq 0$ is non-trivial and yet $a\phi = \phi$. This suggests the more general result:

6.8 Theorem. Any set of vectors in $\mathbf{R}^n$ containing ϕ is a linearly dependent set.

Proof. Consider the set $\{\phi, x^{(1)}, x^{(2)}, \ldots, x^{(k)}\}$. Then

$$a\phi + 0 \sum_{i=1}^{k} x^{(i)} = \phi.$$

If $a \neq 0$ this exhibits a non-trivial linear combination which yields the zero vector, and so the theorem is proved.

Another observation is that *a single non-zero vector x is linearly independent* because, in this case, $ax = \phi$ if and only if $a = 0$. Thus, a linearly dependent set of non-zero vectors must contain at least two elements. The next theorem is concerned with the dependence of non-zero vectors.

6.9 Theorem. The non-zero vectors $x^{(1)}, x^{(2)}, \ldots x^{(t)} \in \mathbf{R}^n$ are *linearly dependent* if and only if one of the vectors $x^{(k)}$ is a linear combination of the remaining ones.

Proof. See, for example, Tropper [1969], p. 12.

What this says is that, in the case of linear dependence, there are at least two non-zero coefficients in

6.10 $$\sum_{i=1}^{t} a_i x^{(i)} = \phi$$

and, if a_k is one of these, then we can write

6.11 $$x^{(k)} = \left[\frac{a_1}{a_k}\right] x^{(1)} + \ldots + \left[\frac{a_{k-1}}{a_k}\right] x^{(k-1)} + \left[\frac{a_{k+1}}{a_k}\right] x^{(k+1)} + \ldots + \left[\frac{a_t}{a_k}\right] x^{(t)}.$$

The expression (6.11) suggests a concept which is related to (6.7) but is formally distinct.

6.12 Definition. $y \in \mathbf{R}^n$ is *linearly dependent on the vectors* $y^{(1)}, y^{(2)}, \dots, y^{(s)} \in \mathbf{R}^n$ if and only if y can be expressed as a linear combination of $y^{(1)}, y^{(2)}, \dots y^{(s)}$.

Returning to (6.9), then, we can say that $\{x^{(1)}, x^{(2)}, \dots, x^{(t)}\}$ is a linearly dependent set if and only if one of the vectors is linearly dependent on the remaining vectors in the set. This suggests that we may think of a dependent* set of vectors in $\mathbf{R}^n$ as a *redundant* set and if we delete the redundant vectors we obtain an independent* set.

EXERCISE 11.6

Prove Theorem 6.9.

11.7 SPANNING SETS AND BASES

Example 5.3 describes a subspace $\mathbf{S}$ (a plane) in $\mathbf{R}^3$. Consider an arbitrary vector x in that subspace. From theorem 6.6, x can be written as the linear combination

7.1
$$x = x_1 e^{(1)} + x_2 e^{(2)} + x_3 e^{(3)}$$
$$= x_1 e^{(1)} + x_2 e^{(2)},$$

since $x_3 = 0$ for all vectors in $\mathbf{S}$. In other words, *any vector in $\mathbf{S}$ can be expressed as a linear combination of the two unit vectors $e^{(1)}$ and $e^{(2)}$.* We express this by saying that the subspace $\mathbf{S}$ (or in geometrical language, the plane $\mathbf{S}$) is *spanned by* (generated by) the two vectors $e^{(1)}$ and $e^{(2)}$. We call $\{e^{(1)}, e^{(2)}\}$ a *spanning set* for $\mathbf{S}$.

If x is arbitrary (and not necessarily in $\mathbf{S}$) then we cannot assume, in (7.1), that one of the coefficients vanishes. Thus, a spanning set for $\mathbf{R}^3$ consists of all three of the unit vectors $\{e^{(1)}, e^{(2)}, e^{(3)}\}$. In order to define these terms more precisely let us go to $\mathbf{R}^n$.

Let $\mathbf{B}$ be the vectors $y^{(1)}, y^{(2)}, \dots, y^{(k)} \in \mathbf{R}^n$. Let $\mathbf{P}$ be the set of linear combinations of these vectors

7.2
$$\mathbf{P} = \left\{ y = \sum_{i=1}^{k} b_i y^{(i)} : b_i \in \mathbf{R} \right\}.$$

7.3 Theorem. $\mathbf{P}$ is a subspace of $\mathbf{R}^n$.

Proof. $\mathbf{P} \subseteq \mathbf{R}^n$, since $y \in \mathbf{P}$ is a linear combination of vectors from $\mathbf{R}^n$. Now pick any $d, b \in \mathbf{R}$ and any $u, v \in \mathbf{P}$, where

$$u = \sum_{i=1}^{k} f_i y^{(i)}, \text{ and } v = \sum_{i=1}^{k} g_i y^{(i)}.$$

---

* We often drop the word *linearly* from the phrases "linearly dependent" and "linearly independent."

Then

$$du + bv = d \sum_{i=1}^{k} f_i y^{(i)} + b \sum_{i=1}^{k} g_i y^{(i)}$$

$$= \sum_{i=1}^{k} (df_i + bg_i) y^{(i)}$$

$$= \sum_{i=1}^{k} h_i y^{(i)}.$$

Thus $du + bv \in \mathbf{P}$ and the theorem follows from (5.2).

7.4 Definition. The set of vectors $\{y^{(1)}, y^{(2)}, \ldots, y^{(k)}\}$ in (7.2) is called a *spanning set* for the subspace $\mathbf{P}$ and we say that $\mathbf{P}$ is *spanned by* $\{y^{(1)}, y^{(2)}, \ldots, y^{(k)}\}$.

Obviously, $\mathbf{R}^n$ itself has a spanning set consisting of the unit vectors $\{e^{(1)}, e^{(2)}, \ldots, e^{(n)}\}$, from (6.6).

There is a question as to whether or not spanning sets are linearly independent. The answer is negative, in general, but if they are linearly independent, they are of special interest.

7.5 Definition. If a spanning set of a vector space is linearly independent, then we call it a *basis* for the vector space.

In view of the remarks at the end of the last section this means that *a set of basis vectors is a spanning set with the redundant vectors deleted.* A basis, then, is an optimum spanning set in the sense that it minimizes the number of vectors in the spanning set. See (7.8).

We mentioned above that $\mathbf{R}^n$ has the spanning set $\{e^{(1)}, e^{(2)}, \ldots, e^{(n)}\}$. It turns out that these n unit vectors form a basis for $\mathbf{R}^n$ because of the following result:

7.6 Theorem. The unit vectors $e^{(1)}, e^{(2)}, \ldots, e^{(n)} \in \mathbf{R}^n$ are linearly independent.

Proof. Let

$$\phi = \sum_{i=1}^{n} a_i e^{(i)}.$$

This means that

$$\begin{bmatrix} 0 \\ 0 \\ \vdots \\ 0 \end{bmatrix} = \begin{bmatrix} a_1 \\ a_2 \\ \vdots \\ a_n \end{bmatrix}$$

but this can be true if and only if $a_i = 0$ for all i.

7.7 Definition. A vector space is *finite-dimensional* if it has a finite basis, that is, if it has a basis consisting of a finite number of vectors.

From (7.7), (7.6) and (6.6) then, $\mathbf{R}^n$ is a finite-dimensional vector space. The next theorem is needed in order to clarify the concept of dimension.

7.8 Theorem. All bases of a finite-dimensional vector space contain the same number of elements.

Proof. See, for example, Tropper [1969],p. 13.

Since $\{e^{(1)}, e^{(2)}, \ldots, e^{(n)}\}$ forms a basis for $\mathbf{R}^n$ we have the following:

7.9 Corollary. Any basis of $\mathbf{R}^n$ consists of exactly n vectors.

This leads us to the very important definition of the dimension of a finite-dimensional vector space.

7.10 Definition. The *dimension* of a finite-dimensional vector space is the number of elements in a basis. We use the notation dim $\mathbf{V}$ for this concept.

An obvious result follows from (7.9) and (7.10).

7.11 Theorem. $\mathbf{R}^n$ is an n-dimensional vector space, that is, dim $\mathbf{R}^n = n$.

An important theorem relative to the uniqueness of the representation of an arbitrary vector in a given basis follows:

7.12 Theorem. If the set $\{x^{(1)}, x^{(2)}, \ldots, x^{(n)}\}$ forms a basis for $\mathbf{R}^n$, then every vector $x \in \mathbf{R}^n$ is uniquely expressible as a linear combination of $x^{(1)}, x^{(2)}, \ldots, x^{(n)}$.

Proof. Let x have two representations

$$x = \sum_{i=1}^{n} a_i x^{(i)},$$

and

$$x = \sum_{i=1}^{n} b_i x^{(i)},$$

in terms of the basis. Then

$$\phi = \sum_{i=1}^{n} a_i x^{(i)} - \sum_{i=1}^{n} b_i x^{(i)}$$

$$= \sum_{i=1}^{n} (a_i - b_i) x^{(i)}.$$

However, from (7.5), the vectors $x^{(1)}, x^{(2)}, \ldots, x^{(n)}$ are independent and, from (6.7) this implies

$$a_i - b_i = 0 \qquad (i = 1, 2, \ldots, n).$$

Thus, the representations are identical.

The question arises as to the existence of a basis. We have already seen that $\mathbf{R}^n$ has a basis $\{e^{(1)}, e^{(2)}, \ldots, e^{(n)}\}$. What about the existence of a basis for $\mathbf{S} \subset \mathbf{R}^n$ where $\mathbf{S}$ is a proper subspace of $\mathbf{R}^n$? If $\mathbf{S}$ is the (trivial) zero subspace $\mathbf{Z}$, then it does not have a basis because the only possible basis would be ϕ and we recall from (6.8) that this is linearly dependent. However, we do have a basis when $\mathbf{S} \neq \mathbf{Z}$.

7.13 Theorem. Every non-zero subspace of $\mathbf{R}^n$ has a basis.

Proof. See, for example, Perlis [1952], p. 32.

7.14 *Remark.* We have not shown explicitly that $\mathbf{R}^n$ has bases other than the set $\{e^{(1)}, e^{(2)}, \ldots, e^{(n)}\}$. However, other bases are easy to construct. The set

$$\{e^{(1)}, e^{(2)}, \ldots, e^{(n-1)}, e\},$$

where

$$e = \begin{bmatrix} 1 \\ 1 \\ \cdot \\ \cdot \\ \cdot \\ 1 \end{bmatrix},$$

constitutes an example. This illustrates our next theorem.

The proofs of the next three theorems are left as exercises for the reader.

7.15 Theorem. Any set of n linearly independent vectors in $\mathbf{R}^n$ is a basis for $\mathbf{R}^n$.

7.16 Theorem. Any set of $n + 1$ vectors in $\mathbf{R}^n$ is linearly dependent.

7.17 Theorem. If $\mathbf{S}$ is a subspace of $\mathbf{R}^n$ then dim $\mathbf{S} \leq n$. Moreover, if dim $\mathbf{S} = n$, then $\mathbf{S} = \mathbf{R}^n$.

EXERCISES 11.7

1. Prove Theorem 7.8.

2. Prove Theorem 7.13.

3. Prove that the set of vectors $\{e^{(1)}, e^{(2)}, \ldots, e^{(n-1)}, e\}$ in (7.14) is a basis for $\mathbf{R}^n$.

4. Prove Theorem 7.15.

5. Prove Theorem 7.16.

6. Prove Theorem 7.17.

11.8 ORDERED BASES AND COORDINATES

We prefer to work with finite sets rather than infinite sets. Hence in $\mathbf{R}^n$ we select a set of n basis vectors and use (7.12) to note that any vector $x \in \mathbf{R}^n$ is uniquely expressible as a linear combination of these n basis vectors. This allows us to focus our attention on a finite set of basis vectors while working with the infinite set $\mathbf{R}^n$.

We pointed out earlier that in (6.4) and (6.5) the components of x are the coefficients of the linear combination. However, this statement needs some clarification. First of all, the statement is true only if we use the unit vectors $\{e^{(1)}, e^{(2)}, \ldots, e^{(n)}\}$ as our basis. Also, it is implied that we are using the basis vectors in the order of their ascending superscripts. This suggests the need for the following terminology.

8.1 Definition. An *ordered basis* for a vector space is an ordered set of vectors which forms a basis for the vector space.

Suppose we consider the ordered basis $\{w^{(1)}, w^{(2)}, \ldots, w^{(n)}\}$ for $\mathbf{R}^n$. From (7.12) we know that every vector $y \in \mathbf{R}^n$ can be expressed as a unique linear combination of the vectors in the ordered basis. Thus

8.2
$$y = \sum_{i=1}^{n} c_i w^{(i)},$$

and we can associate the vector y with the unique ordered n-tuple

8.3
$$y \rightarrow \begin{bmatrix} c_1 \\ c_2 \\ \cdot \\ \cdot \\ \cdot \\ c_n \end{bmatrix}.$$

8.4 Definition. The ordered n-tuple in (8.3) associated with the vector y is called the *coordinate vector* of y with respect to the ordered basis $\{w^{(1)}, w^{(2)}, \ldots, w^{(n)}\}$.

We sometimes say this another way: $c_1, c_2, \ldots, c_n$ are the *coordinates* of y with respect to the ordered basis $\{w^{(1)}, w^{(2)}, \ldots, w^{(n)}\}$. Some authors call them the *components* of y with respect to the ordered basis.

Until now we have written $y \in \mathbf{R}^n$ in the form

8.5
$$y = \begin{bmatrix} y_1 \\ y_2 \\ \cdot \\ \cdot \\ \cdot \\ y_n \end{bmatrix}$$

with no further explanation.* Obviously, we have meant, without saying so, that $y_1, y_2, \ldots, y_n$ are the components of y with respect to the ordered basis

$$\{e^{(1)}, e^{(2)}, \ldots, e^{(n)}\}.$$

However, since $\mathbf{R}^n$ has many ordered bases other than the set of unit vectors, we must be more explicit in our statements.

We shall refer to the ordered basis $\{e^{(1)}, e^{(2)}, \ldots, e^{(n)}\}$ as the *natural basis*, or the *standard basis* for $\mathbf{R}^n$, and in equations such as (8.5) the components exhibited are the *natural components*.

Now consider the association (8.3) to be a mapping of the vectors $y \in \mathbf{R}^n$ onto the ordered n-tuples, which we called coordinate vectors in (8.4). We call this the *coordinate mapping* with respect to the ordered basis $\{w^{(1)}, w^{(2)}, \ldots, w^{(n)}\}$.

The reader can verify that if $x, y \in \mathbf{R}^n$, then the coordinate vector of $x + y$ with respect to the ordered basis $\{w^{(1)}, w^{(2)}, \ldots, w^{(n)}\}$ is the sum of the coordinate vectors of x and y with respect to that basis. Likewise, if $k \in \mathbf{R}$ and $y \in \mathbf{R}^n$, the coordinate vector of ky with respect to the ordered basis $\{w^{(1)}, w^{(2)}, \ldots, w^{(n)}\}$ is k times the coordinate vector of y with respect to that basis.

This one-to-one correspondence between the vectors in $\mathbf{R}^n$ and their respective coordinate vectors (with respect to the ordered basis) under addition and scalar multiplication is an example of what we have called† a *vector space isomorphism*.

With this background, we are in a position to restate Theorem 7.12 as follows:

8.6 Theorem. If $\mathbf{B}$ is the ordered basis $\{w^{(1)}, w^{(2)}, \ldots, w^{(n)}\}$ for $\mathbf{R}^n$ then there is a coordinate mapping $\mathscr{F}_\mathbf{B} : \mathbf{R}^n \to \mathbf{R}^n$ such that $y \in \mathbf{R}^n$ implies

$$y \to \begin{bmatrix} c_1 \\ c_2 \\ \vdots \\ c_n \end{bmatrix}$$

if and only if

$$y = c_1 w^{(1)} + c_2 w^{(2)} + \ldots + c_n w^{(n)}.$$

8.7 EXAMPLE. Let $y \in \mathbf{R}^2$ be the vector $\begin{bmatrix} -3 \\ 4 \end{bmatrix}$. Obviously,

$$y = -3 \begin{bmatrix} 1 \\ 0 \end{bmatrix} + 4 \begin{bmatrix} 0 \\ 1 \end{bmatrix}.$$

---

* See (3.1) and (3.7), for example.
† Recall the remarks between (4.3) and (4.4).

However, we also have

$$y = -3\begin{bmatrix} 1 \\ 2 \end{bmatrix} + 10\begin{bmatrix} 0 \\ 1 \end{bmatrix}.$$

Thus we call -3 and 4 the natural components of y, that is, they are the components of y with respect to $e^{(1)}$ and $e^{(2)}$. In like manner we call -3 and 10 the components of y with respect to the ordered basis $\begin{bmatrix} 1 \\ 2 \end{bmatrix}, \begin{bmatrix} 0 \\ 1 \end{bmatrix}$. In other words $\begin{bmatrix} -3 \\ 4 \end{bmatrix}$ is the coordinate vector of y with respect to the natural basis and $\begin{bmatrix} -3 \\ 10 \end{bmatrix}$ is the coordinate vector of y with respect to the ordered basis $\begin{bmatrix} 1 \\ 2 \end{bmatrix}, \begin{bmatrix} 0 \\ 1 \end{bmatrix}$.

EXERCISES 11.8

1. Prove that if $x, y \in \mathbf{R}^n$, then the coordinate vector of $x + y$ with respect to the ordered basis $\{w^{(1)}, w^{(2)}, \ldots, w^{(n)}\}$ is the sum of the coordinate vectors of x and y with respect to that basis.

2. Prove that if $k \in \mathbf{R}$ and $y \in \mathbf{R}^n$, then the coordinate vector of ky with respect to the ordered basis $\{w^{(1)}, w^{(2)}, \ldots, w^{(n)}\}$ is k times the coordinate vector of y with respect to that basis.

3. If -1 and 1 are the natural coordinates of $y \in \mathbf{R}^2$ what are the coordinates of y with respect to the ordered basis $\left\{ \begin{bmatrix} 1 \\ 2 \end{bmatrix}, \begin{bmatrix} 2 \\ 1 \end{bmatrix} \right\}$? Draw a sketch to illustrate this geometrically.

11.9 LINEAR TRANSFORMATIONS OR LINEAR MAPPINGS

In (8.6) we mentioned a coordinate mapping $\mathscr{F}_{\mathbf{B}} : \mathbf{R}^n \to \mathbf{R}^n$. This mapping is a member of a more general class of mappings which associate, with a vector x in some vector space, a unique vector y either in the same or in another vector space. We shall restrict our discussion to mappings, or transformations, which are called linear (defined below).

For example, consider $\mathbf{R}^3$ and the subspace

9.1 $$S = \{y \in \mathbf{R}^3 : y_3 = 0\}.$$

For each vector $x \in \mathbf{R}^3$ there is a unique vector $y \in S$ defined as follows:

9.2 $$x = \begin{bmatrix} x_1 \\ x_2 \\ x_3 \end{bmatrix} \Rightarrow y = \begin{bmatrix} x_1 \\ x_2 \\ 0 \end{bmatrix}.$$

In Fig. 4.2 this can be interpreted geometrically as the *projection* of the point P onto the plane spanned by $e^{(1)}$ and $e^{(2)}$.

It should be pointed out that the number of vectors x mapped onto a single vector y, in this example, is infinite. Thus, this is a many-to-one transformation as opposed to the one-to-one coordinate mapping in (8.6).

It is easily verified that, under this projection,

9.3
$$\begin{bmatrix} u_1 \\ u_2 \\ u_3 \end{bmatrix} + \begin{bmatrix} v_1 \\ v_2 \\ v_3 \end{bmatrix} \rightarrow \begin{bmatrix} u_1 \\ u_2 \\ 0 \end{bmatrix} + \begin{bmatrix} v_1 \\ v_2 \\ 0 \end{bmatrix},$$

and, for $k \in \mathbf{R}$,

9.4
$$k \begin{bmatrix} u_1 \\ u_2 \\ u_3 \end{bmatrix} \rightarrow k \begin{bmatrix} u_1 \\ u_2 \\ 0 \end{bmatrix}.$$

These two properties make this projection a *linear mapping*, or *linear transformation*, according to the formal definition which follows.

9.5 Definition. Let $\mathbf{U}$ and $\mathbf{V}$ be two vector spaces over the same field $\mathbf{F}$. Let $\mathscr{M}$ be the mapping $\mathscr{M} : \mathbf{U} \rightarrow \mathbf{V}$ such that for every $u \in \mathbf{U}$ there exists a unique vector $\mathscr{M} u = v \in \mathbf{V}$. Furthermore, let $\mathscr{M}$ be such that for $u, u^{(1)}, u^{(2)} \in \mathbf{U}$ and $k \in \mathbf{F}$

$$\mathscr{M}(u^{(1)} + u^{(2)}) = \mathscr{M} u^{(1)} + \mathscr{M} u^{(2)}$$

and

$$\mathscr{M}(ku) = k(\mathscr{M} u).$$

Then $\mathscr{M}$ is a *linear transformation*, or *linear mapping*, of $\mathbf{U}$ into $\mathbf{V}$.

The motivation for the name "linear" mapping is the fact that linear combinations are preserved under linear mappings in the following sense.

9.6 Theorem. Consider the linear mapping $\mathscr{M} : \mathbf{U} \rightarrow \mathbf{V}$. Then for $c_1, c_2, \ldots, c_k \in \mathbf{F}$ and $u^{(1)}, u^{(2)}, \ldots, u^{(k)} \in \mathbf{U}$

$$\mathscr{M}[c_1 u^{(1)} + c_2 u^{(2)} + \ldots + c_k u^{(k)}] = c_1 \mathscr{M} u^{(1)} + c_2 \mathscr{M} u^{(2)} + \ldots + c_k \mathscr{M} u^{(k)}.$$

Proof. The proof is left as an exercise.

Let us review some of the terminology associated with linear mappings. We adopt the notation $\mathscr{L}(\mathbf{U}, \mathbf{V})$ for the set of all linear mappings of the vector space $\mathbf{U}$ into the vector space $\mathbf{V}$. Thus, $\mathscr{M} \in \mathscr{L}(\mathbf{U}, \mathbf{V})$. If $\mathscr{M} u = v$ we say that v is the *image* of u under the mapping $\mathscr{M}$.

In the special case where $\mathbf{V} = \mathbf{U}$ we call the mapping *a linear mapping on* $\mathbf{U}$. A mapping is *one-to-one* if distinct elements in $\mathbf{U}$ correspond to distinct images in

V. $\mathcal{M}$ is said to be *onto* if every element in **V** is the image of some element in **U**; in other words, if **V** is the entire *image space** of the mapping.

It should be pointed out that, under linear mappings, linear combinations are preserved, but *dimension may not be preserved.* The example at the beginning of this section exhibits a case where $\mathbf{R}^3$ is mapped onto the plane **S**, that is, a 3-dimensional vector space is mapped onto a 2-dimensional vector space.

9.7 Definition. If a linear mapping $\mathcal{M}$ is both one-to-one and onto it is called *non-singular.*

If $\mathcal{M} \in \mathcal{L}(\mathbf{U}, \mathbf{V})$ is non-singular then $u^{(1)}, u^{(2)} \in \mathbf{U}$, with $u^{(1)} \neq u^{(2)}$, imply that the images $\mathcal{M} u^{(1)} = v^{(1)}$ and $\mathcal{M} u^{(2)} = v^{(2)}$ are distinct in **V**. A mapping can be made in the reverse direction, in this case, so that $v^{(1)} \to u^{(1)}$ and $v^{(2)} \to u^{(2)}$. In other words, for each $v \in \mathbf{V}$ there exists a unique vector $u \in \mathbf{U}$ for which v is its image under $\mathcal{M}$ (we have written this $\mathcal{M} u = v$). We write the *inverse mapping*, that is, the mapping from **V** to **U** as

9.8 $$\mathcal{M}^{-1}v = u.$$

9.10 Theorem. $\mathcal{M}^{-1}$ is a linear mapping of **V** onto **U**, that is, $\mathcal{M}^{-1} \in \mathcal{L}(\mathbf{V}, \mathbf{U})$.

Proof. The proof is left as an exercise.

An interesting question is what happens if we map **U** onto **V** with the non-singular mapping $\mathcal{M}$ and then map **V** onto **U** with $\mathcal{M}^{-1}$. Obviously, $u \to v$ and then $v \to u$ so that each vector $u \in \mathbf{U}$ is unchanged as a result of the two mappings. In other words,

9.11 $$\mathcal{M}^{-1}(\mathcal{M} u) = u.$$

If we introduce the notational convention that, for the linear mappings $\mathcal{N}$ and $\mathcal{M}$

9.12 $$\mathcal{N}(\mathcal{M} u) = (\mathcal{N}\mathcal{M})u,$$

then we have the following:

9.13 Definition. $\mathcal{M}^{-1}\mathcal{M}$ is the *identity* mapping on the space **U**. Likewise, $\mathcal{M}\mathcal{M}^{-1}$ is the identity mapping on **V**.

Notice that (9.12) introduces the concept of the *product of two mappings†* and this can be generalized to any number of factors. According to our convention it is understood that the mappings in the product $\mathcal{N}\mathcal{M}$ are to be applied in the order from right to left.

---

* The image space is the set $\{\mathcal{M} u : u \in \mathbf{U}\}$. See (9.16).

† Some authors use the phrase *composition of two mappings* to mean the same thing. This concept appears later in Theorem 13.4.

An obvious question is whether or not the *sum of two mappings* has meaning. To show that the question has an affirmative answer consider $\mathscr{M}, \mathscr{N}, \mathscr{S} \in \mathscr{L}(\mathbf{U}, \mathbf{V})$. Now suppose $\mathscr{M}u = v^{(1)}$, $\mathscr{N}u = v^{(2)}$ and $\mathscr{S}u = v^{(3)}$. Then we say $\mathscr{S} = \mathscr{M} + \mathscr{N}$ if and only if $v^{(3)} = v^{(1)} + v^{(2)}$, for any choice of $u \in \mathbf{U}$.

9.14 Figure

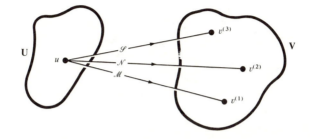

Similarly, we can give meaning to $k\mathscr{M}$, where $k \in \mathbf{F}$. Let $\mathscr{M}u = v^{(1)}$ and $\mathscr{P}u = v^{(2)}$. Then we say that $\mathscr{P} = k\mathscr{M}$ if and only if $v^{(2)} = kv^{(1)}$, for any choice of $u \in \mathbf{U}$. Thus we have *scalar multiplication* for elements of $\mathscr{L}(\mathbf{U}, \mathbf{V})$.

With addition and scalar multiplication defined for linear mappings it is not surprising that we have the following result.

9.15 Theorem. The set of linear mappings $\mathscr{L}(\mathbf{U}, \mathbf{V})$ is an (abstract) vector space under addition and scalar multiplication defined by

i) $\mathscr{M}, \mathscr{N} \in \mathscr{L}(\mathbf{U}, \mathbf{V})$ $\Rightarrow (\mathscr{M} + \mathscr{N})u = \mathscr{M}u + \mathscr{N}u,$

ii) $k \in \mathbf{F}$ and $\mathscr{M} \in \mathscr{L}(\mathbf{U}, \mathbf{V}) \Rightarrow$ $(k\mathscr{M})u = k(\mathscr{M}u),$

for every $u \in \mathbf{U}$.

Proof. The proof is left as an exercise.

Another interesting question is what happens to a subspace of $\mathbf{U}$ under a linear mapping. This is answered as follows.

9.16 Theorem. Let $\mathscr{N} \in \mathscr{L}(\mathbf{U}, \mathbf{V})$. If $\mathbf{W}$ is a subspace of $\mathbf{U}$ then the image space $\mathscr{N}\mathbf{W} = \{\mathscr{N}w : w \in \mathbf{W}\}$ is a subspace of $\mathbf{V}$. In particular, $\mathscr{N}\mathbf{U}$ is a subspace of $\mathbf{V}$.

Proof. The proof is left as an exercise.

Finally, we have a result relative to the behavior of spanning sets under linear mappings.

9.17 Theorem. If $\mathscr{M} \in \mathscr{L}(\mathbf{U}, \mathbf{V})$ and if $\{x^{(1)}, x^{(2)}, \ldots, x^{(n)}\}$ is a spanning set for $\mathbf{U}$, then $\{\mathscr{M}x^{(1)}, \mathscr{M}x^{(2)}, \ldots, \mathscr{M}x^{(n)}\}$ is a spanning set for the image space $\mathscr{M}\mathbf{U}$. In addition, dim $\mathscr{M}\mathbf{U} \leq$ dim $\mathbf{U}$.

Proof. See, for example, Krause [1970], p. 119.

EXERCISES 11.9

1. Show that the mapping defined by (9.2) is a linear mapping of $\mathbf{R}^3$ into $\mathbf{S}$ given by (9.1).

2. Prove Theorem 9.6.   5. Prove Theorem 9.16.

3. Prove Theorem 9.10.   6. Prove Theorem 9.17.

4. Prove Theorem 9.15.

11.10 THE RANK OF A LINEAR MAPPING

If $\mathscr{M} \in \mathscr{L}(\mathbf{U}, \mathbf{V})$ we have already mentioned in (9.14) that the image space $\mathscr{M}\,\mathbf{U}$ is a subspace of $\mathbf{V}$. This subspace is also called the *range* of the mapping $\mathscr{M}$. Another important subspace is defined as follows:

10.1 Definition. $\mathbf{K} = \{u \in \mathbf{U} : \mathscr{M}u = \phi\}$ is called the *kernel* of the mapping $\mathscr{M}$.

Notice that the range of the mapping is a subspace of $\mathbf{V}$ whereas the kernel of the mapping is a subspace of $\mathbf{U}$.

10.2 Definition. The dimension of the range of $\mathscr{M}$ is called the *rank* of $\mathscr{M}$, and the dimension of the kernel of $\mathscr{M}$ is called the *nullity* of $\mathscr{M}$. In symbols

$$\text{rank}\,\mathscr{M} = \dim \mathscr{M}\,\mathbf{U}$$

$$\text{nullity}\,\mathscr{M} = \dim \mathbf{K}.$$

These quantities are related by the following equation*, when the dimensions are finite.

10.3 Theorem. $\dim \mathbf{U} - \dim \mathbf{K} = \dim \mathscr{M}\,\mathbf{U}$.

Proof. See, for example, Krause [1970], p. 125.

If we call $\mathbf{U}$ the *domain* of the mapping then (10.3) states that the dimension of the domain of the mapping equals the sum of the dimensions of the kernel and the range. Figure 10.4 is helpful in this regard.

10.4 Figure

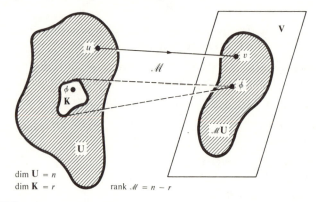

dim $\mathbf{U} = n$
dim $\mathbf{K} = r$ rank $\mathscr{M} = n - r$

---

* This is sometimes called *the rank theorem*.

10.5 EXAMPLE. In (9.2) $U = V = R^3$. The mapping is the projection of R^3 onto the plane S. Thus R^3 is the domain, $\mathscr{M}R^3 = S$ is the range, and the kernel is the set

$$K = \{x \in R^3 : x_1 = x_2 = 0\}.$$

Note that dim $R^3 = 3$, dim $S = 2$, and dim $K = 1$. Hence (10.3) is satisfied.

There is a class of mappings $\mathscr{M} \in \mathscr{L}(U, V)$ for which rank $\mathscr{M} = \dim U = \dim V$. These are particularly important. In this situation $K = \{\phi\}$, and it turns out that these are the nonsingular mappings of (9.7).

10.6 Theorem. If $\mathscr{M} \in \mathscr{L}(U, V)$, then the following statements are equivalent.

i) $\mathscr{M}$ is non-singular.
ii) $\mathscr{M}$ has an inverse, $\mathscr{M}^{-1}$.
iii) rank $\mathscr{M} = \dim U = \dim V$.
iv) If $\{x^{(1)}, x^{(2)}, \ldots, x^{(n)}\}$ is a basis for U, then $\{\mathscr{M}x^{(1)}, \mathscr{M}x^{(2)}, \ldots, \mathscr{M}x^{(n)}\}$ is a basis for V.

Proof. See, for example, Krause [1970], pp. 130–133.

As a consequence of (10.3) we have the following important result.*

10.7 Theorem. The linear mapping of a finite-dimensional vector space into another vector space of the same dimension is *onto* if and only if it is one-to-one.

Proof. The proof is left as an exercise for the reader.

EXERCISES 11.10

1. Show that the kernel K of a linear mapping $\mathscr{M} \in \mathscr{L}(U, V)$ is a subspace of U.
2. Prove Theorem 10.3.
3. Show that, for any linear mapping $\mathscr{M} \in \mathscr{L}(U, V)$ with kernel K, $\phi \in K$.
4. Prove Theorem 10.6.
5. Prove Theorem 10.7.
6. If $K = \{x \in R^3 : x_1 = x_2 = 0\}$ is the kernel of $\mathscr{M} \in \mathscr{L}(R^3, R^3)$, show that dim $K = 1$.

11.11 THE MATRIX OF A LINEAR MAPPING

Consider the vector spaces U and V over a field F, where dim $U = q$ and dim $V = p$. Let $\mathscr{M} \in \mathscr{L}(U, V)$. For U we choose the ordered basis $\{x^{(1)}, x^{(2)}, \ldots, x^{(q)}\}$ and for V we choose the ordered basis $\{y^{(1)}, y^{(2)}, \ldots, y^{(p)}\}$.

Since the basis vectors of U are mapped into V under $\mathscr{M}$, that is, since

11.1 $$\mathscr{M}x^{(j)} = w^{(j)} \in V \qquad (j = 1, 2, \ldots, q),$$

and since each vector in V can be expressed as a linear combination of the basis vectors $y^{(1)}, y^{(2)}, \ldots, y^{(p)}$, we have

---

* For an application, see Theorem 12—(5.6).

$$
\textbf{11.2} \quad
\begin{cases}
w^{(1)} = m_{11}y^{(1)} + m_{21}y^{(2)} + \ldots + m_{p1}y^{(p)} \\
w^{(2)} = m_{12}y^{(1)} + m_{22}y^{(2)} + \ldots + m_{p2}y^{(p)} \\
\cdots\cdots\cdots\cdots\cdots\cdots\cdots\cdots\cdots\cdots\cdots\cdots \\
w^{(j)} = m_{1j}y^{(1)} + m_{2j}y^{(2)} + \ldots + m_{pj}y^{(p)} \\
\cdots\cdots\cdots\cdots\cdots\cdots\cdots\cdots\cdots\cdots\cdots\cdots \\
w^{(q)} = m_{1q}y^{(1)} + m_{2q}y^{(2)} + \ldots + m_{pq}y^{(p)}
\end{cases}
$$

where the coefficients in the linear combinations are elements of **F**.

We direct the reader's attention to (8.4), where the coefficients in the linear combination (8.2) were written as the (column) coordinate vector (8.3). With this in mind we shall write the q sets of coefficients in (11.2) as an array consisting of q columns. Each column, obviously, contains p components. Hence, we have the rectangular array

$$
\textbf{11.3} \qquad M =
\begin{bmatrix}
m_{11} & m_{12} & \ldots & m_{1j} & \ldots & m_{1q} \\
m_{21} & m_{22} & \ldots & m_{2j} & \ldots & m_{2q} \\
\cdots & \cdots & \cdots & \cdots & \cdots & \cdots \\
m_{p1} & m_{p2} & \ldots & m_{pj} & \ldots & m_{pq}
\end{bmatrix} .
$$

We speak of the scalars $m_{ij} \in \mathbf{F}$, $i = 1, 2, \ldots, p, j = 1, 2, \ldots, q$, as the *elements* of the array M, and we call M a *matrix* of *order* p by q (usually written $p \times q$). This terminology is motivated by the fact that the array has p rows and q columns. When we write the arbitrary element m_{ij}, the first index indicates that the element is from row i and the second index indicates that the element is from column j.

The vector $w^{(j)}$ in (11.2) is the image of the basis vector $x^{(j)}$, of the basis $\{x^{(1)}, x^{(2)}, \ldots, x^{(j)}, \ldots, x^{(q)}\}$, under the linear mapping $\mathcal{M}$. The scalars m_{1j}, $m_{2j}, \ldots, m_{pj}$ are the components of the coordinate vector associated with $w^{(j)}$ as defined in (8.4). We sometimes say that the scalars $m_{1j}, m_{2j}, \ldots, m_{pj}$ are the *coordinates* of $w^{(j)}$ with respect to the ordered basis $\{y^{(1)}, y^{(2)}, \ldots, y^{(p)}\}$. With this understanding, then, the matrix M of (11.3) is the matrix of coordinates of $\{w^{(1)}, w^{(2)}, \ldots, w^{(q)}\}$ with respect to the ordered basis $\{y^{(1)}, y^{(2)}, \ldots y^{(p)}\}$.

Since the elements of M depend on both the ordered basis $\{x^{(1)}, x^{(2)}, \ldots, x^{(q)}\}$ for **U** *and* on the ordered basis $\{y^{(1)}, y^{(2)}, \ldots, y^{(p)}\}$ for **V** we have the following definition.

11.4 Definition. The matrix M of (11.3) is called *the matrix of the linear mapping* $\mathcal{M}$ relative to the ordered bases $\{x^{(1)}, x^{(2)}, \ldots, x^{(q)}\}$ and $\{y^{(1)}, y^{(2)}, \ldots, y^{(p)}\}$.

11.5 *Remark*. We have used the concept of the linear transformation, or linear mapping, to motivate our definition of a matrix. This is not essential, and many authors treat matrices as abstract objects independent of their association with linear mappings. In other words, they begin their discussion with an equation such as (11.3) and define matrices simply as rectangular arrays of elements from a field **F**, usually taken to be **R** or **C**. They often introduce the algebra of matrices without any reference to the theory of linear mappings.

We shall not follow that procedure here, however, and we shall usually keep in mind that each matrix can be associated with a linear mapping and vice versa. Whenever it is convenient we shall adopt the notational convention that M is the matrix associated with the mapping $\mathscr{M}$ and $m_{ij} \in$ **F** is the element of M in row i and column j. In fact we often write (11.3) in the form

11.6 $M = (m_{ij})$

when the order of the matrix is understood.

EXERCISES 11.11

1. Consider the linear mapping $\mathscr{M} \in \mathscr{L}(\mathbf{R}^2, \mathbf{R}^2)$ which takes $e^{(1)} \rightarrow \begin{bmatrix} \cos\theta \\ \sin\theta \end{bmatrix}$ and $e^{(2)} \rightarrow \begin{bmatrix} -\sin\theta \\ -\cos\theta \end{bmatrix}$.

What is the mapping usually called? What is the matrix of the linear mapping?

2. Let $\mathscr{M} \in \mathscr{L}(\mathbf{R}^3, \mathbf{R}^2)$, where $\left\{ \begin{bmatrix} 1 \\ 2 \\ 3 \end{bmatrix}, \begin{bmatrix} 2 \\ 3 \\ 1 \end{bmatrix}, \begin{bmatrix} 3 \\ 1 \\ 2 \end{bmatrix} \right\}$ is an ordered basis for $\mathbf{R}^3$ and

$\left\{ \begin{bmatrix} 1 \\ 2 \end{bmatrix}, \begin{bmatrix} 3 \\ -1 \end{bmatrix} \right\}$ is an ordered basis for $\mathbf{R}^2$.

a) What is the matrix of $\mathscr{M}$ with respect to these two ordered bases if the images of the basis vectors for $\mathbf{R}^3$ are $\begin{bmatrix} 1 \\ 2 \end{bmatrix}, \begin{bmatrix} 5 \\ 3 \end{bmatrix}$, and $\begin{bmatrix} -3 \\ 1 \end{bmatrix}$, respectively?

b) What is the image of $\begin{bmatrix} 1 \\ -2 \\ 1 \end{bmatrix}$ under $\mathscr{M}$?

11.12 MATRICES OF THE SAME ORDER

Let $\mathbf{F}^{pq}$ denote the class of all matrices of order $p \times q$ with elements from the field **F**. Then $A \in \mathbf{R}^{pq}$ implies that $A = (a_{ij})$ is such that $a_{ij} \in$ **R** for all i and j. In this case we call A a *real* matrix, or a matrix over the real field. Similarly, if $b_{ij} \in$ **C** we call $B = (b_{ij})$ a *complex* matrix, or a matrix over the complex field.

We deal almost exclusively with real and complex matrices in this book. Consequently, unless a field is specified it will be assumed that the matrices are from either $\mathbf{R}^{pq}$ or $\mathbf{C}^{pq}$.

Let $M = (m_{ij})$ and $N = (n_{ij})$ be matrices of the same order. Then $M = N$ if and only if $m_{ij} = n_{ij}$ for all i and j. Notice that we cannot speak of equality unless the matrices are of the same order.

When we can define it, equality is *elementwise* equality. Next we define a binary composition called *matrix addition*.

12.1 Definition. Let $M = (m_{ij})$, $N = (n_{ij})$ and $P = (p_{ij})$ be matrices of the same order. Then $M + N = P$ if and only if

$$m_{ij} + n_{ij} = p_{ij},$$

for all i and j.

This is *elementwise* addition since the (i, j)-element in P is the sum of the (i, j)-elements of M and N. It is not defined unless the matrices are of the same order.

12.2 EXAMPLE. We can write

$$\begin{bmatrix} 2 & -1 \\ 0 & -2 \end{bmatrix} + \begin{bmatrix} 1 & 3 \\ -1 & 2 \end{bmatrix} = \begin{bmatrix} 3 & 2 \\ -1 & 0 \end{bmatrix},$$

but

$$\begin{bmatrix} 2 & 1 \\ 1 & 4 \\ 0 & -1 \end{bmatrix} + \begin{bmatrix} 1 & 2 & 0 \\ 3 & -1 & 2 \end{bmatrix}$$

is undefined.

In order to motivate definition 12.1 for matrix addition let us consider the linear mappings corresponding to the matrices we are adding. In other words, let $\mathscr{M}, \mathscr{N} \in \mathscr{L}(\mathbf{U}, \mathbf{V})$ and choose $\{x^{(1)}, x^{(2)}, \ldots, x^{(q)}\}$ and $\{y^{(1)}, y^{(2)}, \ldots, y^{(p)}\}$ as ordered bases for $\mathbf{U}$ and $\mathbf{V}$, respectively. From (11.1) and (11.2) we have

12.3
$$\begin{cases} \mathscr{M} x^{(1)} = m_{11} y^{(1)} + m_{21} y^{(2)} + \ldots + m_{p1} y^{(p)} \\ \mathscr{M} x^{(2)} = m_{12} y^{(1)} + m_{22} y^{(2)} + \ldots + m_{p2} y^{(p)} \\ \ldots\ldots\ldots\ldots\ldots\ldots\ldots\ldots\ldots\ldots\ldots\ldots\ldots\ldots\ldots\ldots\ldots\ldots \\ \mathscr{M} x^{(g)} = m_{1q} y^{(1)} + m_{2q} y^{(2)} + \ldots + m_{pq} y^{(g)}. \end{cases}$$

Likewise we have

12.4
$$\begin{cases} \mathscr{N} x^{(1)} = n_{11} y^{(1)} + n_{21} y^{(2)} + \ldots + n_{p1} y^{(p)} \\ \mathscr{N} x^{(2)} = n_{12} y^{(1)} + n_{22} y^{(2)} + \ldots + n_{p2} y^{(p)} \\ \ldots\ldots\ldots\ldots\ldots\ldots\ldots\ldots\ldots\ldots\ldots\ldots\ldots\ldots\ldots\ldots\ldots\ldots \\ \mathscr{N} x^{(q)} = n_{1q} y^{(1)} + n_{2q} y^{(2)} + \ldots + n_{pq} y^{(p)}. \end{cases}$$

From (9.15) we have $(\mathcal{M} + \mathcal{N})u = \mathcal{M}u + \mathcal{N}u$. Hence

12.5
$$
\begin{cases}
(\mathcal{M} + \mathcal{N})x^{(1)} = (m_{11} + n_{11})y^{(1)} + (m_{21} + n_{21})y^{(2)} + \ldots + (m_{p1} + n_{p1})y^{(p)} \\
(\mathcal{M} + \mathcal{N})x^{(2)} = (m_{12} + n_{12})y^{(1)} + (m_{22} + n_{22})y^{(2)} + \ldots + (m_{p2} + n_{p2})y^{(p)} \\
\cdots\cdots\cdots\cdots\cdots\cdots\cdots\cdots\cdots\cdots\cdots\cdots\cdots\cdots\cdots\cdots\cdots\cdots \\
(\mathcal{M} + \mathcal{N})x^{(q)} = (m_{1q} + n_{1q})y^{(1)} + (m_{2q} + n_{2q})y^{(2)} + \ldots + (m_{pq} + n_{pq})y^{(p)}.
\end{cases}
$$

This last set of equations exhibits the elements of the matrix of the linear mapping $\mathcal{M} + \mathcal{N}$. Since the elements are obtained by adding corresponding elements of the matrices of $\mathcal{M}$ and $\mathcal{N}$ it is obvious why matrix addition is defined by (12.1).

What we have done in (12.1) through (12.5) leads us to the following result.

12.6 Theorem. The matrix of the mapping $\mathcal{M} + \mathcal{N}$ is the sum of the matrices of the individual mappings $\mathcal{M}$ and $\mathcal{N}$.

Next we define the multiplication of a matrix by a scalar. Again, we are motivated by (ii) in (9.15) where a linear mapping is multiplied by a scalar.

12.7 Definition. If $k \in \mathbf{F}$ and $M \in \mathbf{F}^{pq}$, then

$$kM = (km_{ij}).$$

Thus the (i,j)-element of the matrix kM is $k \cdot m_{ij}$ and so we say that the multiplication of a matrix by a scalar is *elementwise* multiplication.

By defining the multiplication of a matrix by a scalar as in (12.7) we are led to the following result.

12.8 Theorem. The matrix of the mapping $k\mathcal{M}$ is k times the matrix of the mapping $\mathcal{M}$.

We shall leave it as an exercise for the reader to prove the following important result. It is the matrix analog* of (9.15) for $\mathscr{L}(\mathbf{U}, \mathbf{V})$.

12.9 Theorem. The matrices $\mathbf{F}^{pq}$ form an abstract vector space over $\mathbf{F}$, if vector addition is the elementwise addition of (12.1) and scalar multiplication is the elementwise multiplication of (12.7).

Under this interpretation, it is not surprising to find that the additive identity is

12.10
$$
\phi = \begin{bmatrix}
z & z & \ldots & z \\
z & z & \ldots & z \\
\multicolumn{4}{c}{\cdots\cdots\cdots\cdots} \\
z & z & \ldots & z
\end{bmatrix},
$$

---

* One approach to this proof would be to prove that $\mathscr{L}(\mathbf{U}, \mathbf{V})$ and their matrix representations, $\mathbf{F}^{pq}$, are isomorphic. See, for example, Krause [1970], p. 158.

where z is the additive identity element of $\mathbf{F}$. In $\mathbf{R}^{pq}$, of course, this would be the *zero matrix** of order $p \times q$

12.11
$$\phi = \begin{bmatrix} 0 & 0 & \cdots & 0 \\ 0 & 0 & \cdots & 0 \\ \cdots\cdots\cdots\cdots\cdots \\ 0 & 0 & \cdots & 0 \end{bmatrix}.$$

The additive inverse matrix of $A = (a_{ij})$ is the matrix $(\underline{a}_{ij})$ where $\underline{a}_{ij}$ is the additive inverse of a_{ij} in $\mathbf{F}$. For $\mathbf{R}^{pq}$ and $\mathbf{C}^{pq}$, of course, this is written $(-a_{ij}) = -A$ and so

12.12
$$A + (-A) = \phi.$$

In a manner analogous to (4.8) we introduce the binary operation of *subtraction*. As we stated earlier, when we fail to specify a field it will be assumed that we are discussing either $\mathbf{R}$ or $\mathbf{C}$ or both.

12.13 Definition. Let A and B be matrices of the same order. We define their *difference $A - B$* by

$$A - B = A + (-B).$$

Again, notice that this binary operation is *elementwise* and is defined only for matrices of the same order.

12.14 EXAMPLE.

$$\begin{bmatrix} 2 & 1 \\ 7 & -3 \end{bmatrix} - \begin{bmatrix} 1 & -1 \\ 2 & 4 \end{bmatrix} = \begin{bmatrix} 2 & 1 \\ 7 & -3 \end{bmatrix} + \begin{bmatrix} -1 & 1 \\ -2 & -4 \end{bmatrix}$$

$$= \begin{bmatrix} 1 & 2 \\ 5 & -7 \end{bmatrix}.$$

The reader can easily prove the basic results:

12.15 Theorem. Let $A \in \mathbf{F}^{pq}$:

i) If e is the multiplicative identity in $\mathbf{F}$, then

$$eA = A.$$

---

* We use the symbol ϕ to mean both the zero vector and the zero matrix. However, there should be no confusion as to which we mean in a given instance. In general, of course, it means the additive identity element.

ii) If $\underline{e}$ is the additive inverse of e, then

$$\underline{e}A = (\underline{a}_{ij}).$$

iii) If z is the additive identity in $\mathbf{F}$, then

$$zA = \phi.$$

iv) If $k \in \mathbf{F}$ then

$$k\phi = \phi.$$

The following example illustrates these results when we work in $\mathbf{R}$.

12.16 EXAMPLE. If $A \in \mathbf{R}^{pq}$ and $k \in \mathbf{R}$, then

i) $\qquad\qquad\qquad\qquad\qquad (1)A = \mathrm{A}$

ii) $\qquad\qquad\qquad\qquad\qquad (-1)A = -A$

iii) $\qquad\qquad\qquad\qquad\qquad 0A = \phi$

iv) $\qquad\qquad\qquad\qquad\qquad k\phi = \phi.$

We shall need the concept of *unit matrix* in what follows. Notice the analogy with the notion of unit vector in (6.4) and (6.5).

12.17 Definition. $E^{ij} \in \mathbf{F}^{pq}$ is called a unit matrix if the (i, j)-element is e and all other elements are z.

Thus, for example,*

12.18 $\qquad\qquad E^{pq} = \begin{bmatrix} z & \cdots & z & z & z \\ \multicolumn{5}{c}{\cdots\cdots\cdots\cdots\cdots\cdots\cdots} \\ z & \cdots & z & z & z \\ z & \cdots & z & z & e \end{bmatrix}.$

In $\mathbf{R}^{pq}$ this becomes

12.19 $\qquad\qquad E^{pq} = \begin{bmatrix} 0 & \cdots & 0 & 0 & 0 \\ \multicolumn{5}{c}{\cdots\cdots\cdots\cdots\cdots\cdots\cdots} \\ 0 & \cdots & 0 & 0 & 0 \\ 0 & \cdots & 0 & 0 & 1 \end{bmatrix}.$

---

* See (13.15).

Now (12.9) states that the matrices $\mathbf{F}^{pq}$ form an abstract vector space over $\mathbf{F}$. The dimension of this vector space is of interest. Since dim $\mathbf{F}^{pq}$ is the number of vectors in a basis and since the set of unit matrices $\{E^{11}, \ldots, E^{ij}, \ldots, E^{pq}\}$ can be shown to form a basis, we have the result:

12.20 Theorem. Dim $\mathbf{F}^{pq} = pq$.

Proof. Let $A \in \mathbf{F}^{pq}$. Since

$$A = \sum_{j=1}^{q} \sum_{i=1}^{p} a_{ij} E^{ij}$$

and since the unit matrices are linearly independent (the proof of this statement is left as an exercise) the unit matrices form a basis. There are pq unit matrices in $\{E^{11}, \ldots, E^{ij}, \ldots, E^{pq}\}$ and this proves the theorem.

Thus $\mathbf{F}^{pq}$ is a pq-dimensional vector space and a matrix $A \in \mathbf{F}^{pq}$ is sometimes called a pq-dimensional vector.

There are two special cases that deserve mention. The first is the case $p = n, q = 1$, and we often call the matrix

12.21
$$C = \begin{bmatrix} c_{11} \\ c_{21} \\ \vdots \\ c_{n1} \end{bmatrix}$$

an n-dimensional *column vector*. The second case of interest is where $p = 1, q = m$, and we often call the matrix

12.22
$$R = [r_{11}\, r_{12} \ldots r_{1m}]$$

an m-dimensional *row vector*.

EXERCISES 11.12

1. Prove Theorem 12.8.
2. Prove Theorem 12.9.
3. Prove Theorem 12.15.
4. Show that the unit vectors $\{E^{11}, \ldots, E^{ij}, \ldots, E^{pq}\}$ are linearly independent.

11.13 THE ELEMENTARY ALGEBRA OF MATRICES

We have already defined matrix addition in (12.1) and the multiplication of a matrix by a scalar in (12.7). Subtraction was defined in (12.13), and Theorem 12.9 stated that the matrices $\mathbf{F}^{pq}$ form an abstract vector space over $\mathbf{F}$.

In asking the reader to prove (12.9), part of his task was to verify the following:

13.1 Properties. Let $a, b \in \mathbf{F}$ and $A, B, C \in \mathbf{F}^{pq}$. Then

i) $$A + B = B + A$$

ii) $$(A + B) + C = A + (B + C)$$

iii) $$A + \phi = A$$

iv) $$(a + b)A = aA + bA$$

v) $$a(A + B) = aA + aB$$

vi) $$a(bA) = (ab)A.$$

Theorem 12.15 and Example 12.16 listed additional properties involving multiplication by special scalars and the special matrix ϕ. Unit matrices were defined in (12.17). Finally, since $\mathbf{F}^{pq}$ is a pq-dimensional vector space, we introduced the terms column vector and row vector, respectively, for the matrices C and R of (12.21) and (12.22). Thus far, then, linear mappings and the fact that $\mathbf{F}^{pq}$ is an abstract vector space have motivated our discussion of the algebraic properties of matrices.

From *addition* and *subtraction* of matrices (along with multiplication by a scalar) let us now turn our attention to *matrix multiplication*. Again we provide the motivation for our definition by looking at linear mappings on vector spaces.

Let $\mathbf{U}$, $\mathbf{V}$, and $\mathbf{H}$ be vector spaces over $\mathbf{F}$, of dimension q, p, and r, respectively. Let $\mathscr{M} \in \mathscr{L}(\mathbf{U}, \mathbf{V})$ and $\mathscr{N} \in \mathscr{L}(\mathbf{V}, \mathbf{H})$. In other words, for every $u \in \mathbf{U}$ there exists a unique $v \in \mathbf{V}$ such that $\mathscr{M}u = v$. Also there exists a unique $h \in \mathbf{H}$ such that $\mathscr{N}v = h$.

13.2 Figure

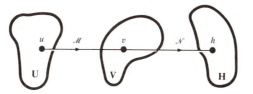

In (9.12) we introduced the concept of the *product* of two mappings,* which means essentially that two mappings are carried out in succession. In terms of (9.12) and Figure (13.2), we can write

13.3 $$(\mathscr{N}\mathscr{M})u = \mathscr{N}(\mathscr{M}u)$$

$$= \mathscr{N}v$$

$$= h.$$

---

* The term *composition* of two mappings is also used.

It turns out that if $\mathcal{M}$ and $\mathcal{N}$ are linear mappings, then so is $\mathcal{N}\mathcal{M}$.

13.4 Theorem. $\mathcal{M} \in \mathcal{L}(\mathbf{U}, \mathbf{V})$ and $\mathcal{N} \in \mathcal{L}(\mathbf{V}, \mathbf{H}) \Rightarrow \mathcal{N}\mathcal{M} \in \mathcal{L}(\mathbf{U}, \mathbf{H})$.

Proof. See, for example, Tropper [1969], p. 26.

Let $\{x^{(1)}, x^{(2)}, \ldots, x^{(q)}\}$ be an ordered basis for $\mathbf{U}$, let $\{y^{(1)}, y^{(2)}, \ldots, y^{(p)}\}$ be an ordered basis for $\mathbf{V}$, and let $\{z^{(1)}, z^{(2)}, \ldots, z^{(r)}\}$ be an ordered basis for $\mathbf{H}$. We shall use the method of section 11 to find the matrices of the mappings $\mathcal{N}$ and $\mathcal{N}\mathcal{M}$.

We recall that (11.1) and (11.2) have already given us M in (11.3). Now we determine N. Let the basis $\{y^{(1)}, y^{(2)}, \ldots, y^{(p)}\}$ have the image set $\{t^{(1)}, t^{(2)}, \ldots, t^{(p)}\}$ in $\mathbf{H}$. Thus

13.5 $\mathcal{N} y^{(j)} = t^{(j)} \in \mathbf{H} \qquad (j = 1, 2, \ldots, p),$

and since each vector in $\mathbf{H}$ can be expressed as a linear combination of

$$z^{(1)}, z^{(2)}, \ldots, z^{(r)},$$

we have

13.6
$$\begin{cases} t^{(1)} = n_{11}z^{(1)} + n_{21}z^{(2)} + \ldots + n_{i1}z^{(i)} + \ldots + n_{r1}z^{(r)} \\[6pt] t^{(2)} = n_{12}z^{(1)} + n_{22}z^{(2)} + \ldots + n_{i2}z^{(i)} + \ldots + n_{r2}z^{(r)} \\[6pt] \cdots\cdots\cdots\cdots\cdots\cdots\cdots\cdots\cdots\cdots\cdots\cdots\cdots\cdots \\[6pt] t^{(j)} = n_{1j}z^{(1)} + n_{2j}z^{(2)} + \ldots + n_{ij}z^{(i)} + \ldots + n_{rj}z^{(r)} \\[6pt] \cdots\cdots\cdots\cdots\cdots\cdots\cdots\cdots\cdots\cdots\cdots\cdots\cdots\cdots \\[6pt] t^{(p)} = n_{1p}z^{(1)} + n_{2p}z^{(2)} + \ldots + n_{ip}z^{(i)} + \ldots + n_{rp}z^{(r)} \end{cases}$$

where the coefficients in the linear combinations are elements of $\mathbf{F}$. Just as in (11.3), we define N to be the $r \times p$ matrix

13.7 $N = \begin{bmatrix} n_{11} & n_{12} & \cdots & n_{1j} & \cdots & n_{1p} \\[4pt] n_{21} & n_{22} & \cdots & n_{2j} & \cdots & n_{2p} \\[4pt] \cdots\cdots & \cdots & \cdots & \cdots & \cdots & \cdots \\[4pt] n_{i1} & n_{i2} & \cdots & n_{ij} & \cdots & n_{ip} \\[4pt] \cdots\cdots & \cdots & \cdots & \cdots & \cdots & \cdots \\[4pt] n_{r1} & n_{r2} & \cdots & n_{rj} & \cdots & n_{rp} \end{bmatrix}.$

Finally, we determine the matrix corresponding to the mapping $\mathcal{N}\mathcal{M}$. In order to simplify the algebra we make use of the following observation. In (13.6) the coefficients in the linear combination representing $t^{(j)}$ are the elements of column j of the matrix N. A similar statement applies to the coefficients of $w^{(j)}$ in (11.2) and the elements of column j of the matrix M.

Consequently, instead of writing out the complete system analogous to (11.2) and (13.6) we shall write only the jth equation, that is, an equation analogous to the linear combinations for $w^{(j)}$ and $t^{(j)}$ in (11.2) and (13.6). This will give us the elements of column j in the matrix representing $\mathcal{N}\mathcal{M}$. Thus

13.8
$$[\mathcal{N}\mathcal{M}]x^{(j)} = \mathcal{N}[\mathcal{M}x^{(j)}]$$
$$= \mathcal{N}\,w^{(j)}$$
$$= \mathcal{N}\sum_{k=1}^{p} m_{kj}y^{(k)}$$
$$= \sum_{k=1}^{p} m_{kj}\mathcal{N}\,y^{(k)}$$
$$= \sum_{k=1}^{p} m_{kj}t^{(k)}$$
$$= \sum_{k=1}^{p} m_{kj}\left[\sum_{i=1}^{r} n_{ik}z^{(i)}\right]$$
$$= \sum_{i=1}^{r}\left[\sum_{k=1}^{p} n_{ik}m_{kj}\right]z^{(i)}.$$

In other words, the jth equation expresses the image of the basis vector $x^{(j)} \in \mathbf{U}$, under the mapping $\mathcal{N}\mathcal{M}$, as a vector in $\mathbf{H}$. This image in $\mathbf{H}$, however, is not expressed specifically, but is expressed as a linear combination of the basis vectors $z^{(1)}, z^{(2)}, \ldots, z^{(r)}$. Thus

13.8a
$$[\mathcal{N}\mathcal{M}]x^{(j)} = \left[\sum_{k=1}^{p} n_{1k}m_{kj}\right]z^{(1)} + \left[\sum_{k=1}^{p} n_{2k}m_{kj}\right]z^{(2)}$$
$$+ \ldots + \left[\sum_{k=1}^{p} n_{ik}m_{kj}\right]z^{(i)}$$
$$+ \ldots + \left[\sum_{k=1}^{p} n_{rk}m_{kj}\right]z^{(r)},$$

and, as before, the coefficients of the linear combination on the right are the elements of column j in the matrix representing $\mathcal{N}\mathcal{M}$.

Suppose we let $B = (b_{ij})$ be the matrix of the linear mapping $\mathcal{N}\mathcal{M}$. Obviously, the element b_{ij} completely describes the matrix and, from (13.8a),

13.9
$$b_{ij} = \sum_{k=1}^{p} n_{ik} m_{kj}$$

$$= n_{i1} m_{1j} + n_{i2} m_{2j} + \ldots + n_{ip} m_{pj}.$$

Notice that b_{ij} is completely defined in terms of elements of row i of N and column j of M. This provides us with the motivation for using the symbols NM for the matrix of the linear mapping $\mathcal{N}\mathcal{M}$ and we write

13.10
$$B = NM.$$

Since we call $\mathcal{N}\mathcal{M}$ the product of linear mappings we also call NM the *product* of the matrices N and M. This leads us to the definition of matrix multiplication:

13.11 Definition. If the $r \times p$ matrix N is described by (13.7), and the $p \times q$ matrix M is described by (11.3), we define the *product* NM to be the $r \times q$ matrix (b_{ij}) where b_{ij} is defined by (13.9).

What we have done in (13.8) through (13.11) leads us to the following result.

13.12 Theorem. The matrix of the mapping $\mathcal{N}\mathcal{M}$ is the product of the matrices of the individual mappings $\mathcal{N}$ and $\mathcal{M}$ as (13.10) indicates.

We should point out that matrix multiplication is defined *only if* the numbers of *columns* in the matrix N is the same as the number of *rows* in the matrix M. Such matrices are called a *conformable* pair of matrices, and we say that the product NM is conformable. What this means in terms of linear mappings is that $\mathcal{M}$ maps **U** into **V** and then $\mathcal{N}$ maps *the same* vector space **V** into **H**.

13.13 EXAMPLE.

$$\begin{bmatrix} 2 & 1 & 3 \\ 4 & -1 & 2 \end{bmatrix} \begin{bmatrix} -3 & 1 \\ 2 & 5 \\ 0 & 2 \end{bmatrix} = \begin{bmatrix} -4 & 13 \\ -14 & 3 \end{bmatrix}.$$

In this case a 2×3 matrix times a 3×2 matrix yields a 2×2 matrix. In general, from (13.11), an $r \times p$ matrix times a $p \times q$ matrix yields an $r \times q$ matrix.

If the matrix product NM is conformable, it does not necessarily follow that MN is conformable. For this to be true we would have to require $r = q$ in the previous sentence. A particularly interesting class of matrices with this property is the class of *square* matrices $\mathbf{F}^{pp}$ since they have the same number of rows as columns. Thus $A, B \in \mathbf{F}^{pp} \Rightarrow$ both AB and BA are conformable.

13.14 *Remark.* One of the first comments that must be made relative to matrix multiplication is that, in general, $AB \neq BA$ even if both AB and BA are conformable. To illustrate this, we merely reverse the order of multiplication in (13.13) and obtain

$$\begin{bmatrix} -3 & 1 \\ 2 & 5 \\ 0 & 2 \end{bmatrix} \begin{bmatrix} 2 & 1 & 3 \\ 4 & -1 & 2 \end{bmatrix} = \begin{bmatrix} -2 & -4 & -7 \\ 24 & -3 & 16 \\ 8 & -2 & 5 \end{bmatrix}.$$

Notice that in one case the result is a 2×2 matrix whereas the other case yields a 3×3 matrix.

13.15 Remark. We have used the symbols z and e to represent the additive and multiplicative identities, respectively, in an arbitrary field **F**. In **R** these symbols become 0 and 1, and in **C** they become (0, 0) and (1, 0) when we use the notation (a, b), and 0 and 1 when we use the single symbol representation of a complex number.

From this point on we shall always use the symbols 0 and 1, regardless of the field, because there will never be any ambiguity in t' eir meaning.

We have mentioned the class of square matrices $\mathbf{F}^{pp}$. This class contains matrices of the form $D = (d_{jj})$, where

13.16
$$D = \begin{bmatrix} d_{11} & 0 & 0 & \ldots & 0 & 0 \\ 0 & d_{22} & 0 & \ldots & 0 & 0 \\ 0 & 0 & d_{33} & \ldots & 0 & 0 \\ \hdotsfor{6} \\ 0 & 0 & 0 & \ldots & 0 & d_{pp} \end{bmatrix}$$

and the only non-zero elements are the scalars on the *main diagonal*.

If $d_{11} = d_{22} = \ldots = d_{pp} = s$, then the diagonal matrix $D = (s)$ is called a *scalar matrix*. This designation is due to the fact that a scalar matrix can be written as the product of a scalar and a special diagonal matrix. Thus $D = (s)$ can be written

13.17
$$\begin{bmatrix} s & 0 & 0 & \ldots & 0 & 0 \\ 0 & s & 0 & \ldots & 0 & 0 \\ 0 & 0 & s & \ldots & 0 & 0 \\ \hdotsfor{6} \\ 0 & 0 & 0 & \ldots & 0 & s \end{bmatrix} = s \begin{bmatrix} 1 & 0 & 0 & \ldots & 0 & 0 \\ 0 & 1 & 0 & \ldots & 0 & 0 \\ 0 & 0 & 1 & \ldots & 0 & 0 \\ \hdotsfor{6} \\ 0 & 0 & 0 & \ldots & 0 & 1 \end{bmatrix}.$$

The special diagonal matrix

13.18
$$I_p = \begin{bmatrix} 1 & 0 & 0 & \dots & 0 & 0 \\ 0 & 1 & 0 & \dots & 0 & 0 \\ 0 & 0 & 1 & \dots & 0 & 0 \\ \hdotsfor{6} \\ 0 & 0 & 0 & \dots & 0 & 1 \end{bmatrix}$$

is called the $p \times p$ *identity matrix* because of the properties it exhibits under matrix multiplication. See, for example, (13.20) and (13.21). We write the symbol I without the subscript whenever p is understood.

In (13.14) we mentioned that $AB \neq BA$, in general. We are now in a position to demonstrate a class of matrices for which multiplication is commutative.*

13.19 Theorem. If $D = (s)$ is a scalar matrix in $\mathbf{F}^{pp}$, then D commutes with all other matrices in $\mathbf{F}^{pp}$. In other words, $A \in \mathbf{F}^{pp} \Rightarrow AD = DA$.

Proof. The proof of this theorem and the proof of the next theorem are left as exercises for the reader.

13.20 Theorem. Let $A \in \mathbf{F}^{rp}$ and $B \in \mathbf{F}^{pq}$, then $AI_p = A$ and $I_p B = B$.

As a special case of both (13.19) and (13.20) we have, for any $A \in \mathbf{F}^{pp}$,

13.21
$$AI_p = I_p A.$$

As we stated earlier, I plays the role of *multiplicative identity* among matrices.† In (13.20) we must use the terms *left identity* and *right identity* because of the conformability requirement for matrix multiplication. For example, if $M \in \mathbf{F}^{pq}$, then I_p is the left identity whereas I_q is the right identity. Thus $I_p M = M = M I_q$. In (13.21), of course, the left identity and the right identity are the same since the matrices are square.

13.22 *Remark.* In (13.14) we point out that, in general, $AB \neq BA$. A second property of matrix multiplication which must be emphasized is that it is possible for

$$AB = \phi$$

with both A and B non-zero matrices. For example,

$$\begin{bmatrix} a_{11} & 0 \\ a_{21} & 0 \end{bmatrix} \begin{bmatrix} 0 & 0 \\ b_{21} & b_{22} \end{bmatrix} = \begin{bmatrix} 0 & 0 \\ 0 & 0 \end{bmatrix},$$

---

* See also, (17.15) and (17.25).

† If we use the Kronecker delta, then we can write $I = (\delta_{ij})$. See, for example, (6.5).

regardless of the values of a_{11}, a_{21}, b_{21} and b_{22}. We describe this property by saying that A and B are "divisors of zero".

Since there is no general commutative law of multiplication, the (most obvious) next question to ask is whether or not the associative and distributive laws hold. The answer to this question is included in the next result.

13.23 Theorem. Whenever the indicated multiplications are defined, that is, whenever the conformability requirements are met, the following results hold (a is a scalar).

i) $$A(BC) = (AB)C$$

ii) $$A(B + C) = AB + AC$$

iii) $$(B + C)A = BA + CA$$

iv) $$a(AB) = (aA)B$$
$$= A(aB).$$

Proof. The proof is left as an exercise.

Having discussed addition, subtraction and multiplication, briefly, we now turn to *the matrix analog of division**. We shall restrict ourselves to *square* matrices for this discussion. The reader can easily prove the following result:

13.24 Theorem. In F^{pp}, the multiplicative identity I_p in (13.21) is unique.

In (2.1) we defined the multiplicative inverse of $a \in F$ ($a \neq 0$) to be the unique element a^{-1} with the property that

13.25 $$a^{-1}a = aa^{-1}$$
$$= 1.$$

This allows us to express division as a product, since

13.26 $$\frac{b}{a} = a^{-1}b.$$

The matrix analog of division involves an approach similar to this. Under a condition analogous to $a \neq 0$, we seek for $A \in F^{pp}$, a unique matrix $A^{-1} \in F^{pp}$ with the property that

13.27 $$A^{-1}A = AA^{-1}$$
$$= I_p.$$

---

* Division is not defined for matrices. However, see (13.28).

13.28 Definition. When it exists, A^{-1} is called the *multiplicative inverse* of A.

In the next section we introduce the concept of the *determinant* of a square matrix. See Definition 14.1. We use the symbols det A to denote this concept and at this point we merely state that, for each $A \in \mathbf{F}^{pp}$, det $A \in \mathbf{F}$. With the reader's indulgence we shall use the concept before we define it.

13.29 Definition. A matrix $A \in \mathbf{F}^{pp}$ is called *singular* if det $A = 0$. If det $A \neq 0$ it is called *non-singular*.

We now state the fundamental theorems (Faddeev and Faddeeva [1963], pp. 9–10) relative to the existence and uniqueness of A^{-1}.

13.30 Theorem. If $A \in \mathbf{F}^{pp}$, then A^{-1} exists if and only if A is non-singular, that is, if and only if det $A \neq 0$.

13.31 Theorem. When A^{-1} exists it is unique.

Notice that for $a \in \mathbf{F}$ the multiplicative inverse exists if and only if $a \neq 0$. For $A \in \mathbf{F}^{pp}$ the analogous condition is det $A \neq 0$. We shall need this condition in (14.8).

In the next section we approach the multiplicative inverse of A (or simply, the inverse of A) from a constructive point of view. This is useful since (13.27) and (13.28) give no indication of how to obtain A^{-1} when it exists.

EXERCISES 11.13

1. Prove Theorem 13.4.
2. Prove Theorem 13.19.
3. Prove Theorem 13.20.
4. Prove Theorem 13.23.
5. Give a numerical example in which $AB = \phi$, with $A \neq \phi$ and $B \neq \phi$, where
   a) A and B are elements of $\mathbf{R}^{33}$.
   b) A and B are elements of $\mathbf{C}^{33}$.
6. Prove Theorem 13.24.
7. Prove Theorem 13.30.
8. Prove Theorem 13.31.

11.14 THE MULTIPLICATIVE INVERSE

In this section we give a formal procedure for constructing the inverse of a non-singular matrix $A \in \mathbf{F}^{pp}$ which shows how each element of A^{-1} is a function of the elements a_{ij} of A. However, this procedure is of theoretical interest only, because the algorithm is completely impractical from a computational point of view, even on the fastest electronic digital computers.

First we define the determinant of a square matrix and we use an inductive definition.

14.1 Definition.

i) If $A \in \mathbf{F}^{11}$ then $\det A = a_{11}$.

ii) If $A \in \mathbf{F}^{pp}$ with $p > 1$ we denote by A_{ij} the matrix in $\mathbf{F}^{p-1,p-1}$ obtained from A by deleting row i and column j, that is

$$A_{ij} = \begin{bmatrix} a_{11} & a_{12} & \cdots & a_{1,j-1} & a_{1,j+1} & \cdots & a_{1p} \\ \cdots\cdots\cdots\cdots\cdots\cdots\cdots\cdots\cdots\cdots\cdots\cdots\cdots\cdots\cdots\cdots \\ a_{i-1,1} & a_{i-1,2} & \cdots & a_{i-1,j-1} & a_{i-1,j+1} & \cdots & a_{i-1,p} \\ a_{i+1,1} & a_{i+1,2} & \cdots & a_{i+1,j-1} & a_{i+1,j+1} & \cdots & a_{i+1,p} \\ \cdots\cdots\cdots\cdots\cdots\cdots\cdots\cdots\cdots\cdots\cdots\cdots\cdots\cdots\cdots\cdots \\ a_{p1} & a_{p2} & \cdots & a_{p,j-1} & a_{p,j+1} & \cdots & a_{pp} \end{bmatrix}.$$

Assuming that the determinant of a $(p-1) \times (p-1)$ matrix has already been defined we define

$$\det A = \sum_{j=1}^{p} a_{1j} c_{1j}$$

where

$$c_{ij} = (-1)^{i+j} \det A_{ij}$$

is called the *cofactor* of a_{ij}.

14.2 EXAMPLE. Compute $\det A$ if $A = \begin{bmatrix} a_{11} & a_{12} \\ a_{21} & a_{22} \end{bmatrix}$.

$$\begin{aligned} \det A &= a_{11} c_{11} + a_{12} c_{12} \\ &= a_{11}(-1)^2 \det A_{11} + a_{12}(-1)^3 \det A_{12} \\ &= a_{11} a_{22} - a_{12} a_{21}. \end{aligned}$$

This procedure is called *expanding det A by means of cofactors of the first row* since we multiply each element of the first row by its cofactor and form the sum. As a matter of fact, we get the same result if we expand $\det A$ by the cofactors of any row or any column. See Hohn [1958], p. 34.

14.3 Theorem.   Using the notation of (14.1)

$$\det A = \sum_{i=1}^{p} a_{ij} c_{ij}$$

for any j.   Likewise,

$$\det A = \sum_{j=1}^{p} a_{ij} c_{ij}$$

for any i.

14.4 EXAMPLE.   Compute $\det A$ in Example 14.2 by means of cofactors of the last column.

$$\begin{aligned} \det A &= a_{12} c_{12} + a_{22} c_{22} \\ &= a_{12}(-1)^3 \det A_{12} + a_{22}(-1)^4 \det A_{22} \\ &= -a_{12} a_{21} + a_{22} a_{11}. \end{aligned}$$

This answer agrees with (14.2).

Next we introduce, for each matrix $A \in \mathbf{F}^{pq}$, a matrix A^T called the *transpose* of A.

14.5 Definition.   If $A \in \mathbf{F}^{pq}$, then $A^T \in \mathbf{F}^{qp}$ is the matrix obtained from A by interchanging its rows and columns.   If $A = (a_{ij})$ we sometimes write $A^T = (a_{ji})$.

14.6 EXAMPLE.

$$A = \begin{bmatrix} 1 & 0 \\ 2 & -1 \\ 0 & 4 \end{bmatrix} \Rightarrow A^T = \begin{bmatrix} 1 & 2 & 0 \\ 0 & -1 & 4 \end{bmatrix}.$$

Now we introduce, for each matrix $A \in \mathbf{F}^{pp}$, a matrix A^{adj} called the *adjoint* of A.

14.7 Definition.   Let $A \in \mathbf{F}^{pp}$.   Form the matrix of cofactors, that is, construct the matrix where a_{ij} is replaced by c_{ij}.   The transpose of the matrix of cofactors is called the adjoint of A.   In other words

$$A^{adj} = \begin{bmatrix} c_{11} & c_{21} & \cdots & c_{p1} \\ c_{12} & c_{22} & \cdots & c_{p2} \\ \cdots\cdots\cdots\cdots\cdots\cdots \\ c_{1p} & c_{2p} & \cdots & c_{pp} \end{bmatrix}.$$

This definition leads us to the result we want.

14.8 Theorem. If $A \in \mathbf{F}^{pp}$ and $\det A \neq 0$, then

$$A^{-1} = \frac{1}{\det A} A^{adj}.$$

Proof. See, for example, Perlis [1952], p. 79.

14.9 Theorem. $A A^{adj} = A^{adj} A = (\det A)I$.

Proof. If A is non-singular, the theorem follows as a corollary to (14.8). The case when A is singular is left as an exercise for the reader.

It is obvious, from (14.8) why it is necessary and sufficient, for the existence of A^{-1}, that $\det A \neq 0$. As we stated earlier, (14.8) gives us a procedure, of theoretical interest, for constructing A^{-1} from the elements of A. However, it is not computationally feasible for arbitrary matrices of order greater than two. Computational methods for higher-order matrices, for use with high-speed digital computers, are discussed in the next chapter.

In Section 9 we discussed the linear mapping $\mathcal{M} \in \mathcal{L}(\mathbf{U}, \mathbf{V})$. A non-singular mapping was defined in (9.7) and we introduced $\mathcal{M}^{-1}$ as the inverse mapping of a non-singular mapping $\mathcal{M}$. In (9.13) $\mathcal{M}^{-1}\mathcal{M}$ and $\mathcal{M}\mathcal{M}^{-1}$ were defined as identity mappings in $\mathbf{U}$ and $\mathbf{V}$, respectively. Consequently, it is not surprising to find that the next two theorems are true.

14.10 Theorem. If $M \in \mathbf{F}^{pp}$ is the matrix of the non-singular linear mapping $\mathcal{M}$, then M^{-1} is the matrix of the inverse mapping $\mathcal{M}^{-1}$.

Proof. The proof is left as an exercise.

14.11 Theorem. If $M \in \mathbf{F}^{pp}$ is the matrix of the non-singular mapping $\mathcal{M} \in \mathcal{L}(\mathbf{U}, \mathbf{V})$ then $I = M^{-1}M$ is the matrix of the identity mapping

$$\mathcal{M}^{-1}\mathcal{M} \in \mathcal{L}(\mathbf{U}, \mathbf{U})$$

and $I = MM^{-1}$ is the matrix of the identity mapping $\mathcal{M}\mathcal{M}^{-1} \in \mathcal{L}(\mathbf{V}, \mathbf{V})$.

Proof. The proof is left as an exercise.

In fact, there is an isomorphism [Krause, p. 158] between $\mathcal{L}(\mathbf{U}, \mathbf{V})$ and $\mathbf{F}^{pq}$ which emphasizes the fact that there is a one-to-one correspondence between most of the theory of linear mappings and the theory of matrices.

It is a lengthy process to emphasize each point of correspondence between the theory of linear mappings and the theory of matrices, and so we shall not do so in each case. Our objective here is to *review* the basic properties of $\mathbf{R}^{pq}$ and $\mathbf{C}^{pq}$ so that we can use these results in subsequent chapters.

EXERCISES 11.14

1. Prove Theorem 14.3.

2. Prove that $\displaystyle\sum_{j=1}^{p} a_{ij}c_{kj} = 0$ $(i \neq k)$ and $\displaystyle\sum_{i=1}^{p} a_{ij}c_{ik} = 0$ $(j \neq k)$.

3. Prove Theorem 14.8.

4. Prove Theorem 14.10.

5. Prove Theorem 14.11.

11.15 THE INNER PRODUCT IN $\mathbf{R}^n$

From (3.9) and (7.9) we know that $\mathbf{R}^n$ is an n-dimensional (real) vector space. Let us examine, for the moment, $\mathbf{R}^3$, and recall Fig. 4.2, where we interpreted the 3-dimensional vector

15.1
$$x = \begin{bmatrix} x_1 \\ x_2 \\ x_3 \end{bmatrix}.$$

as the directed line segment $\overrightarrow{OP}$. We now introduce the concept of *dot product* in $\mathbf{R}^3$.

15.2 Definition. If $x, y \in \mathbf{R}^3$, then

$$(x, y) = x_1y_1 + x_2y_2 + x_3y_3$$

is called the *dot product* of the vectors x and y.

The dot product $(x, y) \in \mathbf{R}$ has a geometric interpretation. Let x and y correspond to $\overrightarrow{OP}$ and $\overrightarrow{OQ}$, respectively. Then we can compute (x, y) as follows.

15.3 Theorem. If $x, y \in \mathbf{R}^3$, then (x, y) in (15.2) can also be expressed as

$$(x, y) = \sqrt{x_1^2 + x_2^2 + x_3^2} \cdot \sqrt{y_1^2} + \sqrt{y_2^2 + y_3^2} \cos \theta$$

where θ is the angle between the directed line segments $\overrightarrow{OP}$ and $\overrightarrow{OQ}$.

Proof. The proof is left as an exercise.

This theorem establishes the criterion for perpendicularity of x and y.

15.4 Corollary. If $x, y \in \mathbf{R}^3$, then $x \perp y$ if and only if $(x, y) = 0$.

Proof. The proof is left as an exercise.

15.5 Corollary. If $x \in \mathbf{R}^3$

$$(x, x) = x_1^2 + x_2^2 + x_3^2.$$

Perpendicularity is a geometric concept. The corresponding algebraic concept is *orthogonality*, and we usually use this term when we generalize the dot product to $\mathbf{R}^n$. When $n > 3$, most authors call the following an *inner product*.*

15.6 Definition. If $x, y \in \mathbf{R}^n$, then

$$(x, y) = x_1 y_1 + x_2 y_2 + \ldots + x_n y_n$$

is called the *inner product* of the vectors x and y.

The following definition is obviously motivated by what we did in $\mathbf{R}^3$.

15.7 Definition. If $x, y \in \mathbf{R}^n$ then x and y are orthogonal if and only if $(x, y) = 0$.

From this definition, a trivial observation is that ϕ is orthogonal to every vector in the vector space.

For completeness we include the definition of an inner product for any real vector space, not necessarily $\mathbf{R}^n$.

15.8 Definition. Let $\mathbf{V}$ be any vector space over $\mathbf{R}$. An inner product on $\mathbf{V}$ is a function that assigns to each ordered pair of vectors $x, y \in \mathbf{V}$ a real number (x, y) satisfying

   i) $x \neq \phi$ $\Rightarrow (x, x) > 0.$

   ii) $x, y \in \mathbf{V}$ $\Rightarrow (x, y) = (y, x).$

   iii) $x, y, z \in \mathbf{V}$ $\Rightarrow (x + y, z) = (x, z) + (y, z).$

   iv) $k \in \mathbf{R}$ and $x, y \in \mathbf{V} \Rightarrow (kx, y) = k(x, y).$

The following additional properties of an inner product can be proved by using only the properties in Definition 15.8.

15.9 Theorem.

   i) $x = \phi$ $\Leftrightarrow (x, x) = 0.$

   ii) $k \in \mathbf{R}$ and $x, y \in \mathbf{V} \Rightarrow (x, ky) = k(x, y).$

   iii) $y \in \mathbf{V} \Rightarrow (\phi, y) = 0.$

Proof. The proof is left as an exercise.

15.10 Theorem. The inner product defined for $\mathbf{R}^n$ in (15.6) is an inner product in the sense of (15.8).

Proof. The proof is left as an exercise.

Now we come to the main point of this section.

15.11 Definition. A real vector space $\mathbf{V}$ together with an inner product defined by (15.8) is called a *Euclidean space* (or *inner product space*).

---

* There are inner products other than the dot product. See, for example, Krause [1970], p. 320.

15.12 Theorem. $\mathbf{R}^n$, with the inner product (15.6), is a Euclidean space.

Proof. The proof is left as an exercise.

15.13 *Remark.* If we think about the geometrical meaning of (15.5) in $\mathbf{R}^3$, we see that one way to define the *length of a vector* in $\mathbf{R}^n$ is to use the positive square root of the inner product

$$\sqrt{(x, x)} = \sqrt{x_1^2 + x_2^2 + \ldots + x_n^2}.$$

In (12.21) and (12.22) we exhibited matrices consisting of a single column and a single row, respectively. We also mentioned that, since $\mathbf{F}^{pq}$ is a vector space, it is customary to call these special matrices column vectors and row vectors, when it is useful to do so.

Sometimes it is convenient (especially in $\mathbf{R}^n$ and $\mathbf{C}^n$) to reverse this procedure and treat vectors as matrices consisting either of a single row or a single column.

Suppose we are working with matrices whose elements are from $\mathbf{R}$. Let us form the product of a $1 \times n$ and an $n \times 1$ matrix.

15.14
$$[x_{11} \ x_{12} \ \ldots \ x_{1n}] \begin{bmatrix} y_{11} \\ y_{21} \\ \cdot \\ \cdot \\ \cdot \\ y_{n1} \end{bmatrix} = [z_{11}]$$

where

15.15
$$z_{11} = x_{11}y_{11} + x_{12}y_{21} + \ldots + x_{1n}y_{n1}.$$

If we compare (15.15) with (15.6) we see that they are identical in form. Consequently, it is not surprising to see many authors use the terms row (column) vector and row (column) matrix interchangeably. Thus, we often write

15.16
$$x = \begin{bmatrix} x_1 \\ x_2 \\ \cdot \\ \cdot \\ \cdot \\ x_n \end{bmatrix} \Rightarrow x^T = [x_1 \ x_2 \ldots x_n].$$

With this notation in mind, in $\mathbf{R}^n$ the inner product (15.6) can be written

15.17
$$(x, y) = y^T x.$$

It should be pointed out that (15.14) produces a 1×1 matrix (vector) and (15.6) produces a scalar (in $\mathbf{R}$). However, in (4.5) we emphasized the one-to-one correspondence that exists here, and very seldom do we need to distinguish between a 1×1 matrix (vector) and its scalar element (component).

In order to handle this identification problem, some authors*, when identifying special matrices with row or column vectors, write (15.17) as

15.17a $$(x, y) = \det y^T x.$$

This is consistent with our definition of the determinant of a 1×1 matrix in (14.1)

15.18 *Remark.* In defining matrix multiplication in (13.11) it was pointed out that the (i, j)-element of NM was b_{ij} in (13.9). With (15.14) and (15.15) before us we can rewrite (13.9) using the notation

$$[n_{i1} \ n_{i2} \ \ldots \ n_{ip}] \begin{bmatrix} m_{1j} \\ m_{2j} \\ \cdot \\ \cdot \\ m_{pj} \end{bmatrix} = b_{ij}$$

EXERCISES 11.15

1. Prove Theorem 15.3.

2. Prove Corollary 15.4.

3. Prove Theorem 15.9.

4. Prove Theorem 15.10.

5. Prove Theorem 15.12.

11.16 THE INNER PRODUCT IN Cⁿ

There is a change necessary in (15.6), (15.8) and most other equations in section 15 if we work in **C** rather than **R**. First of all, let us review the following properties of complex numbers (we are using the single letter notation rather than either of the other (2.7) representations).

A bar over a symbol means the complex conjugate. Thus, if $c = u + iv$, then $\bar{c} = u - iv$. Obviously, $c\bar{c} = u^2 + v^2$, the square of the modulus of c. Hence, $c\bar{c} = |c|^2$. We recall also, that if $v = 0$, then $c \in \mathbf{R}$ which means that the real field is a subset of **C**.

A bar over the symbol representing a vector or a matrix means that each component or element of the corresponding vector or matrix has been replaced by its complex conjugate.

---

* See, for example, Bronson [1969], p. 224.

16.1 Properties.

i) $0 \neq c \in \mathbf{C}$ $\quad\Rightarrow 0 < c\bar{c} \in \mathbf{R}$

ii) $r \in \mathbf{R}$ $\quad\Leftrightarrow r = \bar{r}$

iii) $A \in \mathbf{R}^{pq}$ $\quad\Leftrightarrow A = \bar{A}$

iv) $c \in \mathbf{C}$ $\quad\Rightarrow (c + \bar{c}) \in \mathbf{R}$

v) $A \in \mathbf{C}^{pq}$ $\quad\Rightarrow (A + \bar{A}) \in \mathbf{R}^{pq}$

vi) $c, d \in \mathbf{C}$ $\quad\Rightarrow \overline{cd} = \bar{c}\,\bar{d}$

vii) A, B
conformable $\Rightarrow \overline{AB} = \bar{A}\,\bar{B}$

viii) $c \in \mathbf{C}$ $\quad\Rightarrow \bar{\bar{c}} = c$

ix) $A \in \mathbf{C}^{pq}$ $\quad\Rightarrow \bar{\bar{A}} = A$

x) $c, d \in \mathbf{C}$ $\quad\Rightarrow \overline{c + d} = \bar{c} + \bar{d}$

xi) $A, B \in \mathbf{C}^{pq}$ $\quad\Rightarrow \overline{A + B} = \bar{A} + \bar{B}.$

We now introduce the concept of an inner product in a complex vector space. Notice how this includes (15.8) as a special case.

16.2 Definition.
Let $\mathbf{V}$ be a vector space over $\mathbf{C}$. An inner product on $\mathbf{V}$ is a function that assigns to each ordered pair of vectors $x, y \in \mathbf{V}$ a complex number* (x, y) satisfying

i) $x \neq \phi$ $\quad\Rightarrow 0 < (x, x) \in \mathbf{R}.$

ii) $x, y \in \mathbf{V}$ $\quad\Rightarrow (x, y) = \overline{(y, x)}.$

iii) $x, y, z \in \mathbf{V}$ $\quad\Rightarrow (x + y, z) = (x, z) + (y, z).$

iv) $c \in \mathbf{C}$ and $x, y \in \mathbf{V} \Rightarrow (cx, y) = c(x, y).$

The following additional properties of an inner product can be proved by using only the properties in Definition (16.2).

16.3 Theorem.

i) $x = \phi$ $\quad\Leftrightarrow (x, x) = 0.$

ii) $c \in \mathbf{C}$ and $x, y \in \mathbf{V} \Rightarrow (x, cy) = \bar{c}(x, y).$

iii) $y \in \mathbf{V}$ $\quad\Rightarrow (\phi, y) = 0.$

Proof. The proof is left as an exercise.

We now introduce an inner product in $\mathbf{C}^n$ which includes (15.6) as a special case.

---

* Since $\mathbf{R} \subset \mathbf{C}$ some values of (x, y) may, in fact, be real.

16.4 Theorem. The inner product for $x, y \in \mathbf{C}^n$

$$(x, y) = x_1 \bar{y}_1 + x_2 \bar{y}_2 + \ldots + x_n \bar{y}_n$$

is an inner product in the sense of (16.2).

Proof. The proof is left as an exercise.

16.5 Definition. If $x, y \in \mathbf{C}^n$ then x and y are orthogonal if and only if $(x, y) = 0$.

Again we notice that ϕ is orthogonal to every vector in $\mathbf{C}^n$. Also, we have a definition similar to (15.11).

16.6 Definition. A complex vector space **V** together with an inner product defined by (16.2) is called a *unitary space*.

16.7 Theorem. $\mathbf{C}^n$, with the inner product (16.4), is a unitary space.

Proof. The proof is left as an exercise.

16.8 *Remark.* As we did in (15.13) we observe that one way to define the *length of a vector* in $\mathbf{C}^n$ is to use the positive square root of the inner product

$$\sqrt{(x, x)} = \sqrt{|x_1|^2 + |x_2|^2 + \ldots + |x_n|^2}.$$

As a matter of fact, it is our desire that this length be a real number which motivates (16.2) and (16.4). This would not be real if we used (15.6) rather than (16.4).

Now if we wish to use symbols analogous to those in (15.16) and (15.17) we need the following definition.

16.9 Definition. If $(a_{ij}) = A \in \mathbf{C}^{pq}$, then A^H is the matrix obtained from A by taking the transpose of $\bar{A}$. In other words,

$$A^H = (\bar{a}_{ji})$$

and we call it the *complex conjugate transpose of A* or the *Hermitian conjugate of A*.

Thus, in $\mathbf{C}^n$, (15.16) becomes

16.10
$$x = \begin{bmatrix} x_1 \\ x_2 \\ \cdot \\ \cdot \\ \cdot \\ x_n \end{bmatrix} \Rightarrow x^H = [\bar{x}_1 \ \bar{x}_2 \ldots \bar{x}_n]$$

so that the inner product may be written

16.11
$$(x, y) = y^H x.$$

16.12 *Remark.* In matrix multiplication the (i, j)-element of the product $C = AB$ is

$$c_{ij} = \sum_{k=1}^{p} a_{ik} b_{kj}$$

where we assume that A has p columns and B has p rows. *For real matrices* this can be treated as the inner product of the ith row of A with the jth column of B. However, *for complex matrices* the expression is not the inner product since b_{kj}, rather than $\bar{b}_{kj}$ occurs on the right. This point sometimes causes confusion and the reader should note the difference. For example, if a is the ith row of A, and b is the jth column of B, then for both real and complex matrices $c_{ij} = a^T b$. If the matrices are real, then $c_{ij} = (b, a)$, but if the matrices are complex, $c_{ij} \neq (b, a)$ because $(b, a) = a^H b$ and not $a^T b$ in the complex case.

EXERCISES 11.16

1. Verify the properties in (16.1).

2. Prove Theorem 16.3.

3. Prove Theorem 16.4.

4. Prove Theorem 16.7.

11.17 SOME SPECIAL KINDS OF MATRICES

We shall discuss matrices in $\mathbf{R}^{pq}$ and $\mathbf{C}^{pq}$, in general, although many of our results refer to matrices with elements from any field.

17.1 Definition. If $A \in \mathbf{C}^{pp}$ and $A = A^H$, we call A a *complex Hermitian matrix.*

Notice that a special case occurs for $B \in \mathbf{R}^{pp}$. In this case $B^H = B^T$ and so we have the simpler result. See (17.36).

17.2 Definition. If $B \in \mathbf{R}^{pp}$ and $B = B^T$, we call B a *real symmetric matrix.*

In the complex Hermitian case

17.3 $(a_{ij}) = (\bar{a}_{ji})$

whereas, in the real symmetric case,

17.4 $(b_{ij}) = (b_{ji})$

17.5 EXAMPLES.

$$A = \begin{bmatrix} 2 & 1 - i \\ 1 + i & 1 \end{bmatrix} \Rightarrow A = A^H.$$

$$B = \begin{bmatrix} 2 & 5 \\ 5 & -3 \end{bmatrix} \Rightarrow B = B^T.$$

17.6 *Remark.* If it were a useful concept we would emphasize the fact that there are also complex symmetric matrices such as

$$S = \begin{bmatrix} 2 & 1 + i \\ 1 + i & 1 \end{bmatrix}.$$

However, this concept is seldom of interest and will not be emphasized.

17.7 Theorem. For complex matrices, and $k \in \mathbf{C}$,

i) $(A^H)^H = A$
ii) $(A + B)^H = A^H + B^H$
iii) $(kA)^H = \bar{k}A^H$
iv) $(AB)^H = B^H A^H$,

where we assume that the matrices are such that the indicated operations are all defined.

Proof. The proof is left as an exercise.

This theorem takes on a simpler form if the matrices are real*.

17.8 Corollary. For real matrices, and $k \in \mathbf{R}$,

i) $(A^T)^T = A$
ii) $(A + B)^T = A^T + B^T$
iii) $(kA)^T = kA^T$
iv) $(AB)^T = B^T A^T$

where we assume that the matrices are such that the indicated operations are all defined.

Proof. The proof is left as an exercise.

17.9 Corollary.

i) $A \in \mathbf{C}^{pp} \Rightarrow AA^H$ is Hermitian.
ii) $A \in \mathbf{R}^{pp} \Rightarrow AA^T$ is symmetric.

Proof. The proof is left as an exercise.

17.10 Theorem. Let $A, B \in \mathbf{F}^{pp}$. Then

$$\det(AB) = (\det A)(\det B).$$

Proof. See, for example, Faddeev and Faddeeva [1963], p. 6.

---

* In fact (17.8) holds even if the field is **C**.

17.11 Corollary. Let $A, B \in \mathbf{F}^{pp}$. Then

$$AB \text{ is non-singular} \Leftrightarrow A \text{ is non-singular and } B \text{ is non-singular.}$$

In other words,

$$\det(AB) \neq 0 \Leftrightarrow \det A \neq 0 \text{ and } \det B \neq 0.$$

Proof. The corollary follows directly from (17.10).

17.12 Theorem.

i) $(A^{-1})^{-1} = A$

ii) $(kA)^{-1} = (1/k)A^{-1}$

iii) $(AB)^{-1} = B^{-1}A^{-1}$

where $0 \neq k \in \mathbf{F}$ and $A, B \in \mathbf{F}^{pp}$ are non-singular.

Proof. We shall prove (iii) and leave (i) and (ii) as exercises for the reader. AB is non-singular since A and B are non-singular. Hence, AB has an inverse and

$$I = (AB)(AB)^{-1}.$$

If we pre-multiply (multiply on the left) by A^{-1} we obtain

$$A^{-1}I = A^{-1}[(AB)(AB)^{-1}]$$
$$A^{-1} = [A^{-1}(AB)](AB)^{-1}$$
$$= [(A^{-1}A)B](AB)^{-1}$$
$$= B(AB)^{-1}.$$

If we pre-multiply by B^{-1} we obtain

$$B^{-1}A^{-1} = B^{-1}[B(AB)^{-1}]$$
$$= [B^{-1}B](AB)^{-1}$$
$$= (AB)^{-1}$$

and this proves (iii).

17.13 Definition. If $A \in \mathbf{F}^{pp}$ then we define $A^0 = I, A^1 = A$, and if $n > 1$ is an integer, we define

$$A^n = AA \ldots A$$

where there are n factors on the right.

With this definition the reader can easily prove the following results:

17.14 Theorem. Let $A \in F^{pp}$. Then if n, m are non-negative integers

  i) $A^m A^n = A^{m+n}$

  ii) $(A^n)^m = A^{mn}$.

17.15 Corollary. If $A \in F^{pp}$ and n, m are non-negative integers, then

$$A^m A^n = A^n A^m.$$

17.16 Theorem. If $A \in C^{pp}$ is non-singular, then

$$(A^{-1})^H = (A^H)^{-1}.$$

Proof. By definition

$$I = A^{-1}A.$$

Also,

$$
\begin{aligned}
I &= I^H \\
&= (A^{-1}A)^H \\
&= A^H(A^{-1})^H.
\end{aligned}
$$

If A is non-singular then A^H is non-singular. (The verification of this fact is left as an exercise.)

If we pre-multiply by $(A^H)^{-1}$ we obtain

$$
\begin{aligned}
(A^H)^{-1}I &= (A^H)^{-1}[A^H(A^{-1})^H] \\
(A^H)^{-1} &= [(A^H)^{-1}A^H](A^{-1})^H \\
&= (A^{-1})^H
\end{aligned}
$$

and the theorem is proved.

As a consequence of (17.16) we introduce the notation A^{-H} for either $(A^{-1})^H$ or $(A^H)^{-1}$. Using this notation we have the simple result:

17.17 Corollary. If $A \in C^{pp}$ is non-singular, then

$$A^{-H}A^H = I.$$

17.18 Corollary. If $A \in R^{pp}$ is non-singular, then

  i) $(A^{-1})^T = (A^T)^{-1} = A^{-T}$

  ii) $A^{-T}A^T = I$.

Proof. The proof is left as an exercise.

17.19 Definition. If $A \in \mathbf{R}^{pp}$, if A is non-singular, and if $A^T = A^{-1}$, then A is said to be an *orthogonal* matrix.

The reader can easily prove the following property.

17.20 Theorem. A is orthogonal $\Leftrightarrow AA^T = I$.

It should be noted that the product AA^T involves forming inner products where the vectors used in the inner products are rows of A. What (17.20) says is that, if $a^{(i)}$ is the ith row of A, then

17.21
$$AA^T = \begin{bmatrix} (a^{(1)}, a^{(1)}) & (a^{(1)}, a^{(2)}) & \ldots & (a^{(1)}, a^{(p)}) \\ (a^{(2)}, a^{(1)}) & (a^{(2)}, a^{(2)}) & \ldots & (a^{(2)}, a^{(p)}) \\ \ldots\ldots\ldots\ldots\ldots\ldots\ldots\ldots\ldots\ldots\ldots\ldots\ldots\ldots\ldots \\ (a^{(p)}, a^{(1)}) & (a^{(p)}, a^{(2)}) & \ldots & (a^{(p)}, a^{(p)}) \end{bmatrix}$$
$$= (\delta_{ij}).$$

In other words, the rows of A are *mutually orthogonal* in the sense of (15.7). In addition, each row of A has *unit length* in the sense of (15.13). Vectors with unit length are called *normalized* vectors.

17.22 Definition. A set of vectors $\{x^{(1)}, x^{(2)}, \ldots, x^{(t)}\}$ which is both normalized and mutually orthogonal is called an *orthonormal* set.

Now since $A^T = A^{-1}$ in (17.19) it follows that $A^T A = AA^T$ and everything we have said about the rows of A also applies to the columns. Thus we have the following result.

17.23 Theorem. If A is orthogonal, then both its rows and its columns form orthonormal sets.

In the complex case we have similar results.

17.24 Definition. If $A \in \mathbf{C}^{pp}$, if A is non-singular, and if $A^H = A^{-1}$, then A is said to be a *unitary* matrix.

The next theorem is a generalization of (17.20).

17.25 Theorem. A is unitary $\Leftrightarrow AA^H = I$.

Notice that an orthogonal matrix can be considered to be a special case of a unitary matrix. We have a result analogous to (17.23) for unitary matrices.

17.26 Theorem. If A is unitary, then both its rows and its columns form orthonormal sets.

Proof. The proof is left as an exercise.

17.27 Theorem. Consider $A, B, I \in \mathbf{C}^{pp}$. Then

  i) I is unitary.

  ii) A is unitary $\Rightarrow A^H$ is unitary.

  iii) A and B are unitary $\Rightarrow AB$ is unitary.

  iv) A is unitary $\Rightarrow |\det A| = 1$.

Proof. See, for example, Faddeev and Faddeeva [1963], pp. 25–28.

17.28 Corollary. Consider $A, B, I \in \mathbf{R}^{pp}$. Then

  i) I is orthogonal.

  ii) A is orthogonal $\Rightarrow A^T$ is orthogonal.

  iii) A and B are orthogonal $\Rightarrow AB$ is orthogonal.

  iv) A is orthogonal $\Rightarrow \det A = \pm 1$.

17.29 Definition. If A is orthogonal and $\det A = 1$, then A is called *properly* orthogonal. If $\det A = -1$ it is called *improperly* orthogonal.

17.30 EXAMPLE. The class of orthogonal matrices contains the so-called *elementary rotation* matrices which differ from the identity matrix in only four elements. They have the form

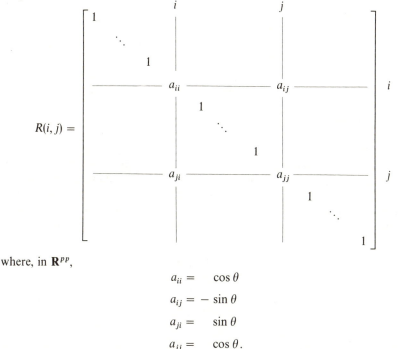

where, in $\mathbf{R}^{pp}$,

$$a_{ii} = \cos \theta$$
$$a_{ij} = -\sin \theta$$
$$a_{ji} = \sin \theta$$
$$a_{jj} = \cos \theta.$$

There is an important class of matrices each of which commutes with its Hermitian conjugate. Notice that these matrices contain real symmetric and complex Hermitian matrices as special cases. They are called *normal matrices*.

17.31 Definition. If $A \in C^{pp}$ and $A^H A = A A^H$, then A is said to be a *normal* matrix.

Another important class of matrices is the class of *permutation matrices*.

17.32 Definition. A matrix $P \in F^{pp}$ whose columns are $e^{(1)}, e^{(2)}, \ldots, e^{(p)}$, but not necessarily in that order, is called a *permutation* matrix.

For example,

17.33
$$P = \begin{bmatrix} 0 & 1 & 0 \\ 0 & 0 & 1 \\ 1 & 0 & 0 \end{bmatrix}$$

is a permutation matrix in $\mathbf{R}^{33}$.

The motivation for the name is the fact that a matrix A, multiplied by P on the left (right), merely has its rows (columns) permuted*.

17.34 EXAMPLE.

$$\begin{bmatrix} 0 & 1 & 0 \\ 0 & 0 & 1 \\ 1 & 0 & 0 \end{bmatrix} \begin{bmatrix} a_{11} & a_{12} & a_{13} \\ a_{21} & a_{22} & a_{23} \\ a_{31} & a_{32} & a_{33} \end{bmatrix} = \begin{bmatrix} a_{21} & a_{22} & a_{23} \\ a_{31} & a_{32} & a_{33} \\ a_{11} & a_{12} & a_{13} \end{bmatrix}.$$

17.35 Theorem. A permutation matrix $P \in C^{pp}$ is unitary, that is, $P^H = P^{-1}$.

Proof. The proof is left as an exercise.

17.36 *Remark.* It is not necessary to give specific results in $\mathbf{R}^{pq}$ when they are merely special cases of results in C^{pq}. However, we often do so for convenience. We could eliminate the terms symmetric and orthogonal, for example, and always use Hermitian and unitary. However, it is convenient many times to know whether the matrices are real or complex, and these superfluous terms are helpful.

EXERCISES 11.17

1. Prove Theorem 17.7.
2. Prove Theorem 17.8.
3. Show that (17.8) holds even if the field is C.

---

* See 12—(11.2).

4. Prove Corollary 17.9.

5. Prove Theorem 17.10.

6. Prove Theorem 17.12 parts (i) and (ii).

7. Prove Theorem 17.14.

8. Prove that if A is non-singular, then A^H is non-singular.

9. Prove Theorem 17.18.

10. Prove Theorem 17.20.

11. Prove Theorem 17.25.

12. Prove Theorem 17.26.

13. Prove Theorem 17.27.

14. Prove Theorem 17.28.

15. Prove Theorem 17.35.

11.18 MATRICES AND COORDINATES

Suppose $\mathscr{M} \in \mathscr{L}(\mathbf{U}, \mathbf{V})$, where $\mathbf{U}$ is a q-dimensional vector space and $\mathbf{V}$ is a p-dimensional vector space. Let $\mathbf{U}$ have the ordered basis $\{x^{(1)}, x^{(2)}, \ldots, x^{(q)}\}$ and $\mathbf{V}$ have the ordered basis $\{y^{(1)}, y^{(2)}, \ldots, y^{(p)}\}$. In (11.4) we defined the matrix M of (11.3) to be the matrix of the linear mapping $\mathscr{M}$ with respect to these two ordered bases. We shall demonstrate how M can be used to compute the coordinates of $v \in \mathbf{V}$, if v is the image of $u \in \mathbf{U}$ under the mapping $\mathscr{M}$.

If u is an arbitrary vector in $\mathbf{U}$ then it can be written as the linear combination

18.1 $$u = u_1 x^{(1)} + u_2 x^{(2)} + \ldots + u_q x^{(q)}.$$

Thus, as in (8.3), we can associate u with its coordinate vector with respect to $\{x^{(1)}, x^{(2)}, \ldots, x^{(q)}\}$,

18.2 $$u \rightarrow \begin{bmatrix} u_1 \\ u_2 \\ \vdots \\ u_q \end{bmatrix}.$$

Next, we form $\mathscr{M}u = v \in \mathbf{V}$ and express v in terms of the basis $\{y^{(1)}, y^{(2)}, \ldots, y^{(p)}\}$, making use of (11.1) and (11.2). Thus

18.3 $$\mathscr{M}u = u_1 \mathscr{M}x^{(1)} + u_2 \mathscr{M}x^{(2)} + \cdots + u_q \mathscr{M}x^{(q)}.$$
$$= u_1 w^{(1)} + u_2 w^{(2)} + \cdots + u_q w^{(q)}$$

$$= u_1[m_{11}y^{(1)} + m_{21}y^{(2)} + \cdots + m_{p1}y^{(p)}]$$
$$+ u_2[m_{12}y^{(1)} + m_{22}y^{(2)} + \cdots + m_{p2}y^{(p)}]$$
$$+ \cdots$$
$$+ u_q[m_{1q}y^{(1)} + m_{2q}y^{(2)} + \cdots + m_{pq}y^{(p)}]$$
$$= (u_1m_{11} + u_2m_{12} + \cdots + u_qm_{1q})y^{(1)}$$
$$+ (u_1m_{21} + u_2m_{22} + \cdots + u_qm_{2q})y^{(2)}$$
$$+ \cdots$$
$$+ (u_1m_{p1} + u_2m_{p2} + \cdots + u_qm_{pq})y^{(p)}$$
$$= v_1y^{(1)} + v_2y^{(2)} + \cdots + v_py^{(p)}.$$
$$= v$$

In (18.2), $u \in \mathbf{U}$ has coordinates relative to the basis $\{x^{(1)}, x^{(2)}, \ldots, x^{(q)}\}$. Likewise $v \in \mathbf{V}$ has coordinates relative to $\{y^{(1)}, y^{(2)}, \ldots, y^{(p)}\}$ and we write

18.4
$$v \rightarrow \begin{bmatrix} v_1 \\ v_2 \\ \vdots \\ v_p \end{bmatrix}.$$

We are interested in computing $v_1, v_2, \ldots, v_p$ if $u_1, u_2, \ldots, u_q$ are given. Obviously, from (18.3)

18.5
$$\begin{cases} v_1 = u_1m_{11} + u_2m_{12} + \cdots + u_qm_{1q} \\ v_2 = u_1m_{21} + u_2m_{22} + \cdots + u_qm_{2q} \\ \cdots\cdots\cdots\cdots\cdots\cdots\cdots\cdots\cdots\cdots\cdots\cdots\cdots\cdots\cdots \\ v_p = u_1m_{p1} + u_2m_{p2} + \cdots + u_pm_{pq} \end{cases}$$

This set of equations can be written in the following form:

18.5a
$$\begin{bmatrix} v_1 \\ v_2 \\ \vdots \\ v_p \end{bmatrix} = \begin{bmatrix} m_{11} & m_{12} & \cdots & m_{1q} \\ m_{21} & m_{22} & \cdots & m_{2q} \\ \cdots\cdots\cdots\cdots\cdots\cdots\cdots \\ m_{p1} & m_{p2} & \cdots & m_{pq} \end{bmatrix} \begin{bmatrix} u_1 \\ u_2 \\ \vdots \\ u_q \end{bmatrix}.$$

Notice that the two coordinate vectors can be treated as $p \times 1$ and $q \times 1$ matrices, and the rules for matrix multiplication and matrix equality make (18.5a) equivalent to (18.5).

We should point out that if we introduce the notation

18.6
$$u^{(x)} = \begin{bmatrix} u_1 \\ u_2 \\ \vdots \\ u_q \end{bmatrix}, \quad v^{(y)} = \begin{bmatrix} v_1 \\ v_2 \\ \vdots \\ v_p \end{bmatrix}$$

then (18.5a) becomes

18.7
$$v^{(y)} = Mu^{(x)},$$

where M is the matrix of the linear transformation $\mathcal{M} \in \mathcal{L}(\mathbf{U}, \mathbf{V})$, relative to the two ordered bases, which we displayed in (11.3). We have proved the following result:

18.8 Theorem. If $u^{(x)}$ is the coordinate vector of $u \in \mathbf{U}$ relative to the ordered basis $\{x^{(1)}, x^{(2)}, \ldots, x^{(q)}\}$, if $v^{(y)}$ is the coordinate vector of the image $v = \mathcal{M}u$, relative to the ordered basis $\{y^{(1)}, y^{(2)}, \ldots, y^{(p)}\}$ and if M is the matrix of $\mathcal{M} \in \mathcal{L}(\mathbf{U}, \mathbf{V})$ relative to the two ordered bases, then (18.7) provides us with an algorithm for computing $v^{(y)}$ from $u^{(x)}$.

Notice that M has dimension $p \times q$, where q and p are the dimensions of $\mathbf{U}$ and $\mathbf{V}$, respectively. The ith column of M is the coordinate vector of the image vector, $\mathcal{M}x^{(i)} \in \mathbf{V}$, relative to the ordered basis $\{y^{(1)}, y^{(2)}, \ldots, y^{(p)}\}$. Obviously, M is completely determined by the choice of bases in $\mathbf{U}$ and $\mathbf{V}$.

For $\mathcal{M}$ to be non-singular (and thus have an inverse $\mathcal{M}^{-1}$) we must have $p = q$ from (10.6). In this case M is a square matrix (non-singular) and M^{-1} is the matrix of the inverse mapping $\mathcal{M}^{-1}$, as we stated in (14.10).

For the most part we shall be using the matrix language of (18.5a) and (18.7) rather than the language of linear mappings from this point on. As a matter of fact, we usually leave off the superscripts and write

$$v = Mu,$$

when the ordered bases for $\mathbf{U}$ and $\mathbf{V}$ are understood. Since we are primarily interested in $\mathbf{R}^n$ and $\mathbf{C}^n$, rather than more general vector spaces, and since we most often use the natural basis $\{e^{(1)}, e^{(2)}, \ldots, e^{(n)}\}$ for an n-dimensional vector space the equation $v = Mu$ very seldom needs further explanation.

Consider, then, the matrix $M \in \mathbf{C}^{pq}$ and the expression

18.9
$$Mu = v.$$

We interpret this as a mapping $\mathbf{C}^q \to \mathbf{C}^p$, and we define several terms associated with the matrix M.

18.10 Definition. Let $M \in \mathbf{C}^{pq}$. The *row space* of M, written $r(M)$ is the subspace of $\mathbf{C}^q$ spanned by the rows of M. The *column space* of M is $r(M^T) \subset \mathbf{C}^p$.

18.11 Definition. The *row rank* of M is the dimension of $r(M)$ and the *column rank* of M is the dimension of $r(M^T)$.

The next theorem states that the row rank of M and the column rank of M are equal and so we introduce the single term, *rank of M*, and write rank M.

18.12 Theorem. $\dim [r(M)] = \dim [r(M^T)] \equiv$ rank M.

Proof. See, for example, Zelinsky [1968], p. 155.

18.13 Corollary. Rank M equals the maximum number of linearly independent rows of M and also the maximum number of linearly independent columns of M.

Proof. The proof is left as an exercise.

18.14 Definition*. The *null space* of M, written $\mathbf{K}(M)$, is the set

$$\mathbf{K}(M) = \{u \in \mathbf{C}^q : Mu = \phi\}.$$

The dimension of $\mathbf{K}(M)$ is called the *nullity* of M, and we write this

$$\text{nullity } M = \dim [\mathbf{K}(M)].$$

18.15 Definition. The *range* of M, written $\mathbf{R}(M)$, is the set

$$\mathbf{R}(M) = \{v \in \mathbf{C}^p : v = Mu \text{ for some } u \in \mathbf{C}^q\}.$$

Now the product Mu is a vector v which is actually a linear combination of the columns of M, where the coefficients in the linear combination are the components of u. Thus, $v \in r(M^T)$ and so the dimension of $\mathbf{R}(M)$ is the column rank of M (or, simply, rank M). Thus, we have a result which is analogous to (10.2).

18.16 Theorem. Rank $M = \dim [\mathbf{R}(M)]$.

We have an equation, analogous to (10.3), which relates the nullity and the rank of M.

18.17 Theorem. Let $M \in \mathbf{C}^{pq}$. Then $q =$ nullity $M +$ rank M or, equivalently,

$$q = \dim [\mathbf{K}(M)] + \dim [\mathbf{R}(M)].$$

Proof. The proof is left as an exercise.

EXERCISES 11.18

1. Prove Theorem 18.12.

2. Prove Corollary 18.13.

3. Prove Theorem 18.17.

---

* See (10.1) and (10.2).

11.19 QUADRATIC, HERMITIAN, AND INNER PRODUCT FORMS

Now that we have introduced the symbols Mu (a matrix times a column vector) we must return to the inner product and add the following property to those listed in (16.2) and (16.3).

19.1 Theorem. Consider the unitary* spaces $\mathbf{C}^m$ and $\mathbf{C}^n$ along with $A \in \mathbf{C}^{mn}$ and $B \in \mathbf{C}^{nm}$. If $x \in \mathbf{C}^m$ and $y \in \mathbf{C}^n$, then

i) $(x, Ay) = (A^H x, y)$

ii) $(Bx, y) = (x, B^H y)$.

Proof. See, for example, Bronson [1969], p.233.

19.2 Corollary. Consider the Euclidean* spaces $\mathbf{R}^m$ and $\mathbf{R}^n$ along with $A \in \mathbf{R}^{mn}$ and $B \in \mathbf{R}^{nm}$. If $x \in \mathbf{R}^m$ and $y \in \mathbf{R}^n$, then

i) $(x, Ay) = (A^T x, y)$

ii) $(Bx, y) = (x, B^T y)$.

The expression (Bx, y) in (19.2) occurs quite frequently, and it is called a *bilinear form* in $x_1, x_2, \ldots, x_m$ and $y_1, y_2, \ldots, y_n$. It can be written

19.3
$$(Bx, y) = y^T Bx$$
$$= \sum_{i=1}^{n} \left(\sum_{j=1}^{m} b_{ij} x_j \right) y_i.$$

A special case occurs when $x = y$. In this case, B is an $n \times n$ matrix and

19.4
$$(Bx, x) = x^T Bx$$
$$= \sum_{i=1}^{n} \left(\sum_{j=1}^{n} b_{ij} x_j \right) x_i.$$

If $n = 2$, for example, we can write

19.5
$$Q = x^T Bx$$
$$= [x_1 x_2] \begin{bmatrix} b_{11} & b_{12} \\ b_{21} & b_{22} \end{bmatrix} \begin{bmatrix} x_1 \\ x_2 \end{bmatrix}$$
$$= b_{11} x_1^2 + b_{22} x_2^2 + b_{12} x_1 x_2 + b_{21} x_2 x_1.$$

Notice that b_{ij} is the coefficient of the term $x_i x_j$. Notice, also, that $x_i x_j = x_j x_i$ and so, in our example,

19.6
$$b_{12} x_1 x_2 + b_{21} x_2 x_1 = (b_{12} + b_{21}) x_1 x_2.$$

---

* See (16.7) and (15.12).

If it should happen that $b_{12} \neq b_{21}$ we can set

19.7
$$\hat{b}_{12} = \hat{b}_{21} = \frac{b_{12} + b_{21}}{2}$$

and rewrite (19.5) in the form

19.8
$$Q = x^T \hat{B} x$$

$$= [x_1 x_2] \begin{bmatrix} b_{11} & \hat{b}_{12} \\ \hat{b}_{21} & b_{22} \end{bmatrix} \begin{bmatrix} x_1 \\ x_2 \end{bmatrix}$$

where $\hat{B}$ is symmetric. In other words, the expression $x^T B x$ with B non-symmetric is equal to the expression $x^T \hat{B} x$ with $\hat{B}$ symmetric. This holds true for arbitrary n.

19.9 Theorem.   If $Q = x^T B x$, with arbitrary $B \in \mathbf{R}^{nn}$, then

$$x^T B x = x^T \hat{B} x,$$

where $\hat{B} = \frac{1}{2}(B^T + B)$ is symmetric.

Proof. The proof is left as an exercise.

Thus, there is no loss in generality in assuming B to be symmetric. This result leads us to the following definition.

19.10 Definition.   If $x \in \mathbf{R}^n$ and $B \in \mathbf{R}^{nn}$ is symmetric, then an expression of the form

$$Q = x^T B x$$

$$= (Bx, x)$$

$$= \sum_{i=1}^{n} \sum_{j=1}^{n} b_{ij} x_j x_i$$

is called a *quadratic form* in $x_1, x_2, \ldots, x_n$. The symmetric matrix B is called the *matrix of the quadratic form.*

When we extend this concept to a unitary space we get the obvious generalization.

19.11 Definition.   If $x \in \mathbf{C}^n$ and $M \in \mathbf{C}^{nn}$ is Hermitian, then an expression of the form

$$H = x^H M x$$

$$= (Mx, x)$$

$$= \sum_{i=1}^{n} \sum_{j=1}^{n} m_{ij} x_j \bar{x}_i$$

is called an *Hermitian form* in $x_1, x_2, \ldots, x_n$. The Hermitian matrix M is called the *matrix of the Hermitian form*.

Notice that we can form expressions such as

19.12
$$(Bx, x) = x^T Bx$$

in Euclidean space and

19.13
$$(Mx, x) = x^H Mx$$

in unitary space with a non-symmetric matrix B and a non-Hermitian matrix M. This leads us to define something called an *inner product form* of which the quadratic form and the Hermitian form are special cases.

19.14 Definition. Let $P \in \mathbf{C}^{nn}$ and $x \in \mathbf{C}^n$. Then we call

$$F = x^H Px$$

$$= (Px, x)$$

$$= \sum_{i=1}^{n} \sum_{j=1}^{n} p_{ij} x_j \bar{x}_i$$

an inner product form in $x_1, x_2, \ldots, x_n$. The matrix P is called the *matrix of the inner product form*.

When P is symmetric and everything is real this reduces to (19.10). When P is Hermitian and everything is complex this reduces to (19.11). Obviously, we have included (19.12) and (19.13) also. See (20.5), (20.9), and (20.10) for related discussions.

EXERCISES 11.19

1. Prove Theorem 19.1.

2. Prove Theorem 19.9.

11.20 POSITIVE DEFINITE AND POSITIVE REAL MATRICES

We assume that the matrix M is Hermitian. Then the Hermitian form

$$x^H Mx = x^H M^H x$$

$$= (x^H Mx)^H$$

$$= \overline{x^H Mx},$$

which implies that $x^H Mx$ is a real number.

20.1 Lemma. The Hermitian form

$$x^H M x = (Mx, x)$$

is real for all $x \in \mathbf{C}^n$.

The converse is also true and we leave its proof as an exercise for the reader. If we combine (20.1) with its converse we have the following result.

20.2 Theorem. If $M \in \mathbf{C}^{nn}$, then $x^H M x = (Mx, x) \in \mathbf{R}$ for all $x \in \mathbf{C}^n$ if and only if M is Hermitian.

This result leads us to a very important concept when $x^H M x = (Mx, x)$ is not only real but positive (nonnegative).

20.3 Definition. If the Hermitian form $(Mx, x) = x^H M x$ is positive (nonnegative) for all non-zero $x \in \mathbf{C}^n$, then we call this a *positive definite (nonnegative definite)* Hermitian form, and we call the Hermitian matrix M a positive definite (nonnegative definite) matrix.

In an obvious way we can introduce the terms *negative definite* and *nonpositive definite*.

We should point out that, as a special case, M could be real in (20.3) even though the space is unitary, in which case the Hermitian matrix M is, more specifically, a real symmetric matrix.

In a further special case everything could be real. In other words, we can consider, in Euclidean space, a quadratic form as defined in (19.10).

20.4 Definition. If the quadratic form $(Bx, x) = x^T B x$ is positive (nonnegative) for all non-zero $x \in \mathbf{R}^n$, then we call this a *positive definite (nonnegative definite)* quadratic form and we call the symmetric matrix B a positive definite (nonnegative definite) matrix.

The question arises as to whether or not a real symmetric matrix which is positive definite (nonnegative definite) according to (20.4) satisfies (20.3) also, since we mentioned earlier that the Hermitian matrix in (20.3) could be real symmetric in a special case. This leads us to the theorem.

20.5 Theorem. If $M \in \mathbf{R}^{nn}$ is symmetric and (Mx, x) is positive (nonnegative) for all non-zero $x \in \mathbf{R}^n$, then (Mx, x) is positive (nonnegative) for all non-zero $x \in \mathbf{C}^n$.

Before proving the theorem we need the following lemma. (See Hohn [1958], p. 257).

20.6 Lemma. If M is real symmetric positive definite, then there exists a nonsingular matrix $B \in \mathbf{R}^{nn}$ such that $M = B^T B$.

With this lemma we can prove (20.5) quite easily.

Proof.
$$(Mx, x) = (B^T Bx, x)$$
$$= (Bx, Bx)$$
$$= (y, y)$$

and this is positive for all non-zero $x \in \mathbf{C}^n$. This takes care of the positive definite part of the proof. We leave the nonnegative definite part as an exercise for the reader.

Now we return to the inner product form of (19.14). Since the inner product form $X^H Mx = (Mx, x)$ is real for all $x \in \mathbf{C}^n$ if and only if M is Hermitian we cannot talk about $x^H Mx$ being positive or nonnegative for non-Hermitian matrices. However, if we restrict x so that $x \in \mathbf{R}^n$, then it *is* possible for $x^T Mx$ to be real for all $x \in \mathbf{R}^n$ even though M is complex non-Hermitian. For example,

20.7 $\quad [x_1 x_2] \begin{bmatrix} 2 & 3-i \\ 2+i & 4 \end{bmatrix} \begin{bmatrix} x_1 \\ x_2 \end{bmatrix} = 2x_1^2 + 4x_2^2 + (3 - i)x_1 x_2 + (2 + i)x_2 x_1$

$$= 2x_1^2 + 4x_2^2 + 5x_1 x_2$$

and this expression is real for all $x \in \mathbf{R}^n$.

In Euclidean space if $B \in \mathbf{R}^{nn}$ is non-symmetric, $x^T Bx$ is always real and we can talk about its sign just as in the complex non-Hermitian example above. Thus, we introduce, a definition due to Wachpress [1966], p. 9.

20.8 Definition. If $M \in \mathbf{C}^{nn}$ and if the inner product form $x^T Mx$ is positive (non-negative) for all non-zero $x \in \mathbf{R}^n$, we call M a *positive real* (*nonnegative real*) matrix.

It should be emphasized that $M \in \mathbf{C}^{nn}$ includes $M \in \mathbf{R}^{nn}$ as a special case. Notice that all positive definite matrices are positive real but the converse is not necessarily true, as Example (20.7) demonstrates. We observe that m_{12} and m_{21} have the property that the sum of their imaginary parts is zero. This leads us to a characterization of positive real matrices.

20.9 Theorem. $M \in \mathbf{C}^{nn}$ is positive real (nonnegative real) if and only if

i) $M^T + M$ is real, and

ii) $M^T + M$ is positive definite (nonnegative definite).

Proof. Let $M = (A + iC)$ where $A, C \in \mathbf{R}^{nn}$. Then, for $x \in \mathbf{R}^n$,
$$x^T Mx = x^T(A + iC)x$$
$$= x^T Ax + ix^T Cx$$
$$= \tfrac{1}{2}[x^T(A^T + A)x + ix^T(C^T + C)x].$$

Now this expression is real if and only if $x^T(C^T + C)x = 0$ for all $x \in \mathbf{R}^n$, and this is true if and only if $C^T + C = \phi$. (The proof is left as an exercise.)

This means, then, that $x^T M x$ is positive (nonnegative) if and only if $A^T + A$ is positive definite (nonnegative definite). Since

$$M^T + M = (A^T + A) + i(C^T + C))$$
$$= (A^T + A)$$

the conclusion follows.

Obviously, it is not an easy task to *test* a matrix for positive definiteness. However, there are ways to *generate* positive definite (nonnegative definite) matrices. Consider an arbitrary matrix $A \in \mathbf{C}^{nn}$. From (17.9), $A A^H$ is Hermitian and the Hermitian form

20.10
$$x^H A A^H x = (A A^H x, x)$$
$$= (A^H x, A^H x)$$
$$= (y, y)$$
$$\geqq 0$$

for all x. Moreover, if equality holds, then $A^H x = 0$. Now if A is nonsingular this implies $x = \phi$. Thus, we have the following theorem.

20.11 Theorem. For any matrix $A \in \mathbf{C}^{nn}$ the matrix $A A^H$ is Hermitian and non-negative definite. If A is nonsingular, then $A A^H$ is positive definite.

EXERCISES 11.20

1. Prove that the diagonal elements of an Hermitian matrix are real.
2. Prove the converse of Theorem 20.1, i.e., prove that if $M \in \mathbf{C}^{nn}$ and if $x^H M x \in \mathbf{R}$ for all $x \in \mathbf{C}^n$, then M is Hermitian.
3. Define the terms negative definite and non-positive definite.
4. Prove Lemma 20.6.
5. Prove the nonnegative part of Theorem 20.5.
6. Prove that $x^T(C^T + C)x = 0$ for all $x \in \mathbf{R}^n$ ($C \in \mathbf{R}^{nn}$) if and only if $C^T + C = \phi$.
7. Show that $A^H x = \phi$ and $\det A \neq 0$ imply $x = \phi$.

11.21 THE ALGEBRAIC EIGENVALUE—EIGENVECTOR PROBLEM

Consider a linear mapping of an n-dimensional vector space into itself. We shall be working in the unitary space $\mathbf{C}^n$, for the most part, but from time to time we shall restrict ourselves to the Euclidean space $\mathbf{R}^n$. Also, we shall use the matrix representation of the linear mapping relative to some ordered basis*, rather than the linear mapping itself.

---

* Recall (18.8).

When $\mathcal{M} \in \mathcal{L}(\mathbf{C}^n, \mathbf{C}^n)$, the matrix M of the linear mapping relative to an ordered basis is a matrix of order $n \times n$. We shall assume an ordered basis for $\mathbf{C}^n$ and work with the coordinate vectors relative to that basis. Therefore, if $x \to y$ under the mapping we write

21.1
$$Mx = y,$$

from Theorem 18.18. This is a computational description of the mapping.

A question arises as to whether or not there are vectors in $\mathbf{C}^n$ which are *invariant* under the mapping.

21.2 Definition. If $\mathbf{U}$ is a vector space over $\mathbf{F}$ and if $\mathcal{M} \in \mathcal{L}(\mathbf{U}, \mathbf{U})$, then a non-zero vector $u \in \mathbf{U}$ is *invariant* under the mapping $\mathcal{M}$ if and only if $u \to \lambda u$, where $\lambda \in \mathbf{F}$.

We should point out that λ is independent of a given matrix representation of $\mathcal{M}$ since each matrix representation depends on a particular choice of an ordered basis for $\mathbf{U}$, and the invariance (or non-invariance) of a vector under the mapping is independent of the choice of basis.

Using this definition, then, the question arises as to whether or not there is at least one non-zero vector $x \in \mathbf{C}^n$ such that $y = \lambda x$ in (21.1), that is, whether or not there is a non-zero vector x such that

21.3
$$Mx = \lambda x.$$

This can be written in the equivalent forms

21.3a
$$(M - \lambda I)x = \phi,$$

and

21.3b
$$\begin{bmatrix} (m_{11}-\lambda) & m_{12} & m_{13} & \cdots & m_{1n} \\ m_{21} & (m_{22}-\lambda) & m_{23} & \cdots & m_{2n} \\ m_{31} & m_{32} & (m_{33}-\lambda) & \cdots & m_{3n} \\ \cdots\cdots\cdots\cdots\cdots\cdots\cdots\cdots\cdots\cdots\cdots\cdots\cdots\cdots \\ m_{n1} & m_{n2} & m_{n3} & \cdots & (m_{nn}-\lambda) \end{bmatrix} \begin{bmatrix} x_1 \\ x_2 \\ x_3 \\ \vdots \\ x_n \end{bmatrix} = \phi.$$

In Chapter 12 we discuss systems of linear algebraic equations and if we ignore, for the moment, the fact that λ is a parameter in (21.3a) and (21.3b), we recognize that (21.3a) and (21.3b) represent a *homogeneous* system of linear algebraic equations according to Definition 12—(5.1). An obvious solution is the zero vector ϕ, but this is seldom of interest and we call $x = \phi$ the *trivial solution*.

Theorem 12—(5.4) states that *non-trivial* solutions exist for such equations if and only if the coefficient matrix is singular; in other words, if and only if the determinant of the coefficient matrix vanishes. What this says, here, is that we can find (non-zero) invariant vectors satisfying (21.3) if and only if

21.4 $$\det (M - \lambda I) = 0.$$

Since $(M - \lambda I)$ contains the parameter λ we can find (non-zero) invariant vectors satisfying (21.3) if and only if we can find values of λ satisfying (21.4). It turns out that $\det (M - \lambda I)$ is a polynomial in λ of degree n (the proof of this fact is left as an exercise for the reader)

21.5 $$\det (M - \lambda I) = a_n + a_{n-1}\lambda + a_{n-2}\lambda^2 + \cdots + a_0\lambda^n$$

and, from Theorem A—(8.7), the polynomial equation (21.4) has exactly* n roots $\lambda_1, \lambda_2, \ldots, \lambda_n$.

21.6 Definition. If $M \in \mathbf{C}^{nn}$, then the polynomial equation

$$\det (M - \lambda I) = 0$$

is called the *characteristic equation* of M. The polynomial $\det (M - \lambda I)$ is called the *characteristic polynomial* of M.

21.7 Definition. If $\det (M - \lambda I) = 0$ is the characteristic equation of $M \in \mathbf{C}^{nn}$, then the n roots $\lambda_1, \lambda_2, \ldots, \lambda_n$ of this polynomial equation are called the *eigenvalues* of M.

Synonyms for the term eigenvalue are *proper value*, *characteristic value*, and (in Britain) *latent root*. These are mentioned because of the lack of uniformity in the use of these terms in the literature.

Notice that, by definition, an eigenvalue $\lambda \in \mathbf{C}$ is a number which causes the matrix $M - \lambda I$ to be a singular matrix. There are exactly n of these numbers although they need not be distinct since the characteristic equation can have multiple roots.

For each (distinct) eigenvalue, then, we have at least one non-trivial solution to (21.3) and this leads us to another definition.

21.8 Definition. If $M \in \mathbf{C}^{nn}$ and if $\lambda_1, \lambda_2, \ldots, \lambda_n$ (not necessarily distinct) are the eigenvalues of M, then any non-trivial solution $x^{(i)}$ to the equation

$$(M - \lambda_i I)x^{(i)} = \phi$$

or, equivalently,

$$Mx^{(i)} = \lambda_i x^{(i)}$$

is called an *eigenvector* corresponding to the eigenvalue λ_i.

---

* Here we include the multiplicity of each root in counting the roots.

The eigenvectors, then, are the invariant vectors we have been seeking. They are the vectors in $\mathbf{C}^n$ which have the property that

21.9 $$x^{(i)} \to \lambda_i x^{(i)}$$

under the mapping represented by the matrix M.

Just as the term eigenvalue is not standard we have the following synonyms for the term eigenvector: *proper vector, characteristic vector,* and *latent vector.*

By definition, ϕ is not an eigenvector. However, nothing in our definition excludes $\lambda = 0$ from being an eigenvalue. Obviously, the eigenvectors corresponding to the eigenvalue $\lambda = 0$ (in those cases where zero is an eigenvalue) are the non-zero vectors in the null space of the matrix M. See (18.14).

We shall list some properties of the eigenvalues and eigenvectors of a matrix $M \in \mathbf{C}^{nn}$. First, however, we introduce the following definitions.

21.10 Definition. The *spectral radius* $S(M)$ of $M \in \mathbf{C}^{nn}$ is $|\lambda_1|$, where λ_1 is the (an) eigenvalue of M with greatest absolute value.

21.11 Definition. The *trace* of a matrix $M \in \mathbf{C}^{nn}$ is the sum of the elements on the main diagonal. We write this

$$tr\, M = \sum_{i=1}^{n} m_{ii}.$$

21.12 Definition. A matrix is called upper (lower) triangular if all elements below (above) the main diagonal are zero.

21.13 Theorem. If $\lambda_1, \lambda_2, \ldots, \lambda_n$ are the eigenvalues of $M \in \mathbf{C}^{nn}$ then

$$tr\, M = \sum_{i=1}^{n} \lambda_i.$$

Proof. See, for example, Gantmacher [1960], p. 87.

21.14 Theorem. $\lambda = 0$ is an eigenvalue of $M \in \mathbf{C}^{nn}$ if and only if M is a singular matrix.

Proof. The proof is left as an exercise.

21.15 Theorem. The eigenvalues of a triangular matrix $M \in \mathbf{C}^{nn}$ (either upper or lower) are the diagonal elements. In other words,

$$\lambda_i = m_{ii} \qquad (i = 1, 2, \ldots, n).$$

Proof. The proof is left as an exercise.

An obvious corollary follows.

21.16 Corollary. The eigenvalues of a diagonal matrix $D \in \mathbf{C}^{nn}$ are

$$d_{11}, d_{22}, \ldots, d_{nn}.$$

21.17 Theorem. If $M \in \mathbf{C}^{nn}$ is non-singular and if $Mx = \lambda x$ for $\lambda \in \mathbf{C}$ and $x \in \mathbf{C}^n$, then

$$M^{-1}x = \frac{1}{\lambda}x.$$

Proof. The proof is left as an exercise.

What this theorem says is that M and M^{-1} have reciprocal eigenvalues but an eigenvector of M corresponding to the eigenvalue λ is an eigenvector of M^{-1} corresponding to the eigenvalue $1/\lambda$.

Suppose $\lambda_1, \lambda_2, \ldots, \lambda_n$ are the eigenvalues (not necessarily distinct) of a matrix $M \in \mathbf{C}^{nn}$. For each eigenvalue there is at least one eigenvector, and so we can write the *n eigenvalue equations*

21.18
$$\begin{cases} Mx^{(1)} = \lambda_1 x^{(1)} \\ Mx^{(2)} = \lambda_2 x^{(2)} \\ \quad\vdots \\ Mx^{(n)} = \lambda_n x^{(n)}. \end{cases}$$

If X is the matrix whose n columns are $x^{(1)}, x^{(2)}, \ldots, x^{(n)}$, then (21.18) can be written as the single matrix equation

21.19
$$MX = X\Lambda,$$

where Λ is the diagonal matrix of eigenvalues of M

21.20
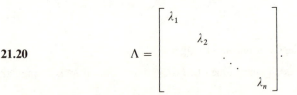
$$\Lambda = \begin{bmatrix} \lambda_1 & & & \\ & \lambda_2 & & \\ & & \ddots & \\ & & & \lambda_n \end{bmatrix}.$$

21.21 Definition. The matrix X in (21.19) whose columns are the vectors $x^{(1)}, x^{(2)}, \ldots, x^{(n)}$ in (21.18) is called the *modal matrix* for M and (21.19) is called the *modal equation* for M.

21.22 *Remark.* When we write the n equations in (21.18) as the single matrix equation (21.19) it should be obvious that the process is reversible. In other words, if we are given the modal equation (21.19) we can write the n individual equations (21.18). However, nothing we have said so far requires the eigenvalues in (21.18) and (21.20) to be distinct, and so we do not require the corresponding eigenvectors to have any special properties when we form the modal matrix. On the other hand, *the vectors might be linearly independent, and this is a most fortunate circumstance.*

21.23 Theorem. If the columns of X in (21.19) form a set of n linearly independent eigenvectors then X is non-singular and we can write the modal equation in the form

$$X^{-1}MX = \Lambda.$$

In order to state a corollary to this theorem we need one definition and one theorem.

21.24 Definition. The matrix $B \in \mathbf{C}^{nn}$ is said to be *similar* to the matrix $A \in \mathbf{C}^{nn}$ if and only if $\exists$ a non-singular matrix $P \in \mathbf{C}^{nn}$ such that $PAP^{-1} = B$. We call this a *similarity transformation* on A.

21.25 Theorem. If $B \in \mathbf{C}^{nn}$ is similar to $A \in \mathbf{C}^{nn}$ then A is similar to B.

Proof. The proof is left as an exercise.

Notice that if we let $Q = P^{-1}$, then the similarity transformation in (21.24) can be written as $Q^{-1}AQ = B$. With (21.24), then, we have an obvious corollary to Theorem 21.23.

21.26 Corollary. If $M \in \mathbf{C}^{nn}$ has a set of n linearly independent eigenvectors then M is similar to a diagonal matrix Λ whose diagonal elements are the eigenvalues of M.

This corollary indicates how important it is for a matrix in $\mathbf{C}^{nn}$ to have n linearly independent eigenvectors. Many matrices, of course, do not have this property, and so it is important to be able to identify those matrices that do.

21.27 Definition. If a matrix $M \in \mathbf{C}^{nn}$ has a set of n linearly independent eigenvectors, then it is called *non-defective*. If it has fewer than n linearly independent eigenvectors, it is called *defective*.

This suggests, then, that it is not *the number of eigenvectors* for a given matrix that is important, because this number is always infinite. (If $Mx = \lambda x$ then $M(kx) = \lambda(kx)$ for $k \in \mathbf{C}$, and so if x is an eigenvector, then so is kx for any $k \neq 0$.) What is important is the *number of linearly independent eigenvectors*.

21.28 Theorem. Eigenvectors of $M \in \mathbf{C}^{nn}$ corresponding to distinct eigenvalues are linearly independent.

Proof. The proof is left as an exercise.

This theorem has two obvious corollaries.

21.29 Corollary. If $M \in \mathbf{C}^{nn}$ has n distinct eigenvalues $\lambda_1, \lambda_2, \ldots, \lambda_n$ then M has a set of n linearly independent eigenvectors and is therefore non-defective.

21.30 Corollary. If $M \in \mathbf{C}^{nn}$ has n distinct eigenvalues $\lambda_1, \lambda_2, \ldots, \lambda_n$, then M is similar to the diagonal matrix

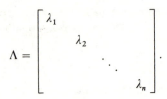

$$\Lambda = \begin{bmatrix} \lambda_1 & & & \\ & \lambda_2 & & \\ & & \ddots & \\ & & & \lambda_n \end{bmatrix}.$$

Matrices which are similar to a diagonal matrix are called *diagonalizable**. Obviously the diagonalizable matrices form an important class because of (21.16) and the next result.

21.31 Theorem. If $A, B \in \mathbb{C}^{nn}$ and there exists a non-singular matrix $P \in \mathbb{C}^{nn}$ such that $PAP^{-1} = B$, then A and B have the same set of eigenvalues.

Proof. Let λ be an eigenvalue of A and let x be an associated eigenvector so that $Ax = \lambda x$. Let $y = Px$, which means $x = P^{-1}y$. If we substitute for x, above, we obtain $AP^{-1}y = \lambda P^{-1}y$, which may be written $PAP^{-1}y = \lambda y$. But $PAP^{-1} = B$ and so $By = \lambda y$. Thus, λ is an eigenvalue of B. In a similar manner it can be shown that if λ is an eigenvalue of B, then it is also an eigenvalue of A.

An obvious corollary follows.

21.32 Corollary. If $A, B \in \mathbb{C}^{nn}$ and $B = PAP^{-1}$ then, if x is an eigenvector of A associated with the eigenvalue λ, $y = Px$ is an eigenvector of B associated with the eigenvalue λ.

Thus, *similar matrices have the same eigenvalues* and their eigenvectors are related by the non-singular linear transformation represented by the non-singular matrix P of the similarity transformation.

It should be pointed out that the converse of (21.31) is not true because $A = \begin{bmatrix} 2 & 0 \\ 0 & 2 \end{bmatrix}$ and $B = \begin{bmatrix} 2 & 1 \\ 0 & 2 \end{bmatrix}$ have the same eigenvalues but no similarity transformation exists† which will transform A into B. Now we return to our earlier statement about the importance of the class of diagonalizable matrices. Because of (21.16) and (21.31) we see that *if a similarity transformation can be found which diagonalizes a matrix, then we have its eigenvalues by inspection.*

Diagonalizable matrices are not easily characterized. However, in doing so we show that normal matrices, defined in (17.31), form an important subset of the diagonalizable matrices.

21.33 Theorem. A matrix $A \in \mathbb{C}^{nn}$ is diagonalizable if and only if $\exists$ a positive definite Hermitian matrix P such that $PAP^{-1} = N$ is a normal matrix.

Proof. See, for example, Mitchell [1953], p. 94.

---

* See (21.33) and (21.36).

† See (21.47), (21.53) and (21.59).

If $A = N$ and $P = I$ in (21.33) we immediately see that the following is true.

21.34 Corollary. If $N \in C^{nn}$ is normal, it is diagonalizable.

Since the normal matrices are diagonalizable, and the real symmetric and complex Hermitian matrices are examples of normal matrices, we have the obvious corollary.

21.35 Corollary. If $S \in R^{nn}$ is symmetric and $H \in C^{nn}$ is Hermitian, then S and H are diagonalizable matrices.

We leave as an exercise the proof of the theorem which relates the terms non-defective and diagonalizable.

21.36 Theorem. A matrix $A \in C^{nn}$ is diagonalizable if and only if it is non-defective.

What about those matrices which are not diagonalizable and hence are defective? From (21.27) they have a set of *fewer than n* linearly independent eigenvectors. Therefore, the eigenvectors of a defective matrix cannot be used as a basis for an n-dimensional vector space.

This creates a problem in applications because there are many important situations where it is desirable to have a basis for C^n which consists entirely of the eigenvectors of some matrix in C^{nn}.

It turns out that the best way to study the *structure* of a matrix $A \in C^{nn}$ (that is, the *multiplicities** of its eigenvalues* and the *number of linearly independent eigenvectors*) is to reduce it by a similarity transformation† to its *Jordan canonical form* (discussed below).

The Jordan canonical form exhibits the basic structure of defective matrices, and in the special case when a Jordan canonical matrix $J \in C^{nn}$ corresponding to a matrix $A \in C^{nn}$ happens to be a diagonal matrix (that is, when A happens to be diagonalizable, or non-defective) the structure is also exhibited. In other words, the Jordan canonical form enables us to tell, by inspection, whether a matrix (similar* to a given Jordan canonical matrix) is defective or non-defective and in either case it exhibits the multiplicities† of its eigenvalues and the number of linearly independent eigenvectors it possesses.

However, before we can discuss the Jordan canonical form we must first introduce the *Smith canonical form*.

21.37 Theorem. Every matrix of the form $A - \lambda I$, with $A \in C^{nn}$, can be transformed into a unique diagonal matrix S of the form

---

* See Theorem 21.49.
† See Definition 21.61.

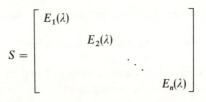

$$S = \begin{bmatrix} E_1(\lambda) & & & \\ & E_2(\lambda) & & \\ & & \ddots & \\ & & & E_n(\lambda) \end{bmatrix}$$

where the diagonal elements are monic polynomials in λ and $E_i(\lambda)$ divides $E_{i+1}(\lambda)$ for $i = 1, 2, \ldots, n - 1$. The transformation has the form $PAQ = S$, where P and Q have elements which are polynomials in λ with coefficients from $\mathbf{C}$, and non-vanishing determinants which do not depend on λ. The matrix S is called the Smith canonical form for $A - \lambda I$.

Proof. See, for example, Gantmacher [1960], pp. 139–144.

21.38 Definition. The polynomials $E_i(\lambda)$ in the Smith canonical form

$$E_1(\lambda) = (\lambda - \alpha_1)^{e_{11}} (\lambda - \alpha_2)^{e_{12}} \cdots (\lambda - \alpha_k)^{e_{1k}}$$
$$E_2(\lambda) = (\lambda - \alpha_1)^{e_{21}} (\lambda - \alpha_2)^{e_{22}} \cdots (\lambda - \alpha_k)^{e_{2k}}$$
$$\vdots$$
$$E_i(\lambda) = (\lambda - \alpha_1)^{e_{i1}} (\lambda - \alpha_2)^{e_{i2}} \cdots (\lambda - \alpha_k)^{e_{ik}}$$
$$\vdots$$
$$E_n(\lambda) = (\lambda - \alpha_1)^{e_{n1}} (\lambda - \alpha_2)^{e_{n2}} \cdots (\lambda - \alpha_k)^{e_{nk}},$$

where $k \leq n$, are called the *invariant factors* of $(A - \lambda I)$.

Note, since $E_i(\lambda)$ divides $E_{i+1}(\lambda)$, the exponents in the invariant factors satisfy the inequalities $e_{ij} \leq e_{i+1,j}$ for $j = 1, 2, \ldots, k$.

21.39 EXAMPLE. If

$$S = \begin{bmatrix} 1 & & & \\ & 1 & & \\ & & (\lambda - \alpha_1) & \\ & & & (\lambda - \alpha_1)(\lambda - \alpha_2)^2 \end{bmatrix}$$

then

$$E_1(\lambda) = 1$$
$$E_2(\lambda) = 1$$
$$E_3(\lambda) = (\lambda - \alpha_1)$$
$$E_4(\lambda) = (\lambda - \alpha_1)(\lambda - \alpha_2)^2.$$

From our example, it is seen that many of the exponents in (21.38) can be (and usually are) zero. To emphasize this, we could have written the four invariant factors in (21.39) as

$$E_1(\lambda) = (\lambda - \alpha_1)^0 (\lambda - \alpha_2)^0$$
$$E_2(\lambda) = (\lambda - \alpha_1)^0 (\lambda - \alpha_2)^0$$
$$E_3(\lambda) = (\lambda - \alpha_1)^1 (\lambda - \alpha_2)^0$$
$$E_4(\lambda) = (\lambda - \alpha_1)^1 (\lambda - \alpha_2)^2 .$$

21.40 Definition. The terms $(\lambda - \alpha_j)^{e_{ij}}$ in (21.38) for which $e_{ij} \neq 0$ are called *the elementary divisors* of A.

Thus, in our example, we have three elementary divisors: $(\lambda - \alpha_1)$, $(\lambda - \alpha_1)$, and $(\lambda - \alpha_2)^2$. The exponents associated with the elementary divisors are very important. For example, it is critical* whether or not any of the exponents is greater than one.

21.41 Definition. If $A \in \mathbf{C}^{nn}$ has an elementary divisor $(\lambda - \alpha_j)^{e_{ij}}$ with $e_{ij} > 1$ we say that A has a *non-linear elementary divisor.*

In our example, then, we have two *linear* elementary divisors $(\lambda - \alpha_1)$ and $(\lambda - \alpha_1)$ as well as one *non-linear* elementary divisor $(\lambda - \alpha_2)^2$. Now we come to the basic theorem which relates similar matrices to the Smith canonical form. It has two obvious (and important) corollaries.

21.42 Theorem. The matrices $A, B \in \mathbf{C}^{nn}$ are similar if and only if $A - \lambda I$ and $B - \lambda I$ have the same Smith canonical form.

Proof. See, for example, Wilkinson [1965], pp. 22–24.

21.43 Corollary. The matrices $A, B \in \mathbf{C}^{nn}$ are similar if and only if $A - \lambda I$ and $B - \lambda I$ have the same invariant factors $E_1(\lambda), E_2(\lambda), \ldots, E_n(\lambda)$.

21.44 Corollary. The matrices $A, B \in \mathbf{C}^{nn}$ are similar if and only if A and B have the same elementary divisors.

The proof of the following theorem is left as an exercise for the reader.

21.45 Theorem. If $A \in \mathbf{C}^{nn}$ is similar to $B \in \mathbf{C}^{nn}$ and B is similar to $C \in \mathbf{C}^{nn}$ then A is similar to C.

Thus, we can group matrices into classes of similar matrices (or *similarity classes.*) Theorem 21.31 states that similar matrices have the same eigenvalues. In other words *all matrices in the same similarity class have the same set of eigenvalues.*

Theorem 21.42 and its first corollary imply, then, that *if $A, B, C, \ldots$ are matrices in the same similarity class, then $A - \lambda I$, $B - \lambda I$, $C - \lambda I$, $\ldots$ all have the same*

---

* See Corollary 21.51.

Smith canonical form, and hence the same set of invariant factors. The second corollary implies then that *if A, B, C, ... are matrices in the same similarity class, then they all have the same set of elementary divisors.*

Once we have the elementary divisors for a matrix $A \in \mathbf{C}^{nn}$ (and hence for all matrices similar to A) we can construct the Jordan canonical matrix corresponding to A. To do this, we associate with each elementary divisor* $(\lambda - \alpha)^e$ an *elementary Jordan block* of the form

21.46

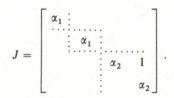

$$
J_\alpha =
\begin{bmatrix}
\alpha & 1 & & & & \\
 & \alpha & 1 & & & \\
 & & \alpha & \ddots & & \\
 & & & \ddots & 1 & \\
 & & & & \alpha &
\end{bmatrix}
$$

where J_α is of order $e \times e$. If the elementary divisor is linear, that is, if $e = 1$, then J_α contains the single element α.

21.47 Definition. The Jordan canonical form is a *block diagonal matrix* whose diagonal blocks are the elementary Jordan blocks of the form J_α.

. Thus, in Example 21.39, since there are three elementary divisors there are three elementary Jordan blocks, and we have

21.48

$$
J =
\begin{bmatrix}
\alpha_1 & \vdots & & \\
\cdots & \vdots & & \\
 & \alpha_1 & \vdots & \\
 & \cdots & & \\
 & \vdots & \alpha_2 & 1 \\
 & \vdots & & \alpha_2
\end{bmatrix}.
$$

Since there are two *linear* elementary divisors $(\lambda - \alpha_1)$ and $(\lambda - \alpha_1)$ and one *non-linear* elementary divisor $(\lambda - \alpha_2)^2$ the three elementary Jordan blocks on the diagonal of J are of orders 1×1, 1×1, and 2×2.

Notice that J is an upper triangular matrix and, by (21.15), the diagonal elements are its eigenvalues. In our example, α_1 is an eigenvalue of multiplicity two and so is α_2. However, we shall notice a difference between the two cases when we examine their respective eigenvectors. See (21.5), part (iv).

We can now state the basic theorem of this entire discussion.

---

* The exponent e is defined in (21.38) and (21.40). See also (21.64).

21.49 Theorem. If $A \in \mathbf{C}^{nn}$ and J is its Jordan canonical matrix, then $J \in \mathbf{C}^{nn}$ and A is similar to J.

Proof. See, for example, Faddeev and Faddeeva [1963], pp. 70–71 and Gantmacher [1960], pp. 151–152.

It should be obvious, by the method of construction of J, that J is *unique* except for possible permutations of the elementary Jordan blocks. The next result follows directly from (21.49).

21.50 Corollary. If $(\lambda - \lambda_1)^{e_1}, (\lambda - \lambda_2)^{e_2}, \ldots, (\lambda - \lambda_t)^{e_t}$ are the elementary divisors* of $A \in \mathbf{C}^{nn}$, where $\lambda_1, \lambda_2, \ldots, \lambda_t$ need not be distinct, then

$$n = e_1 + e_2 + \cdots + e_t.$$

In this corollary we are changing the notation from that which was used in (21.38). In those equations $\alpha_1, \alpha_2, \ldots, \alpha_k$ were *distinct* eigenvalues. However, since an eigenvalue can appear in more than one elementary divisor the notation of (21.50) is more useful if we want to *count* elementary divisors.

Notice that $e_1 = e_2 = \cdots = e_t = 1 \Rightarrow t = n$. In other words, if all elementary divisors are *linear*, the Jordan canonical form is a diagonal matrix. In this case we have the following result.

21.51 Corollary. If $A \in \mathbf{C}^{nn}$ has only linear elementary divisors then A is diagonalizable (and non-defective). If A has at least one non-linear elementary division, then A is not diagonalizable (and is defective).

Proof. The proof is left as an exercise.

In view of our earlier remarks, Theorem 21.49 can be interpreted to mean that all matrices $A, B, C, \ldots$ in the same similarity class have the same Jordan canonical matrix. In other words each similarity class contains a matrix in Jordan canonical form which can be said to *represent* the similarity class in the sense that it reveals the structure of all the matrices in the similarity class.

Now we proceed to investigate how the structure is revealed by examining the Jordan canonical form. *Since the eigenvalues are exhibited as the diagonal elements we can obtain the multiplicity†* of each eigenvalue by inspection. We need the next theorem to partially understand the picture with regard to the number of linearly independent eigenvectors.

21.52 Theorem. If $A \in \mathbf{C}^{nn}$ has $t \leq n$ elementary divisors (and hence t elementary Jordan blocks in its Jordan canonical form) then A has a set of t linearly independent eigenvectors.

Proof. See, for example, Faddeev and Faddeeva [1963], p. 70.

---

* See (21.38) and (21.40).

† See Definition 21.61.

What this theorem says is that the number of linearly independent eigenvectors corresponds to the number of elementary Jordan blocks in its Jordan canonical form. What it does not say is how many linearly independent eigenvectors are associated with each eigenvalue. The next theorem discusses this question.

21.53 Theorem.   If α is an eigenvalue of $M \in \mathbf{C}^{nn}$ which appears in r elementary divisors of M (or in r elementary Jordan blocks of the Jordan canonical form), then there are r linearly independent eigenvectors of M associated with the eigenvalue α.

Proof. See, for example, Hodge and Pedoe [1947], p. 336.

21.54 Corollary.   The r linearly independent eigenvectors corresponding to the eigenvalue α in Theorem 21.53 form the basis of an r-dimensional subspace. Each vector in this subspace is an eigenvector of M corresponding to the eigenvalue α.

The subspace of eigenvectors corresponding to an eigenvalue α is an example of a more general subspace called an *invariant subspace* of the linear mapping $\mathscr{M} \in \mathscr{L}(\mathbf{C}^n, \mathbf{C}^n)$.

21.55 Definition.   An invariant subspace of the linear mapping $\mathscr{M} \in \mathscr{L}(\mathbf{C}^n, \mathbf{C}^n)$ is any subspace $\mathbf{S}$ such that $\mathscr{M}\mathbf{S} \subseteq \mathbf{S}$.

Now we examine a concept which is related to Theorem 21.53 in that it is concerned with whether or not an eigenvalue appears in more than one elementary divisor. We shall use the notation of (21.50) where $\lambda_1, \lambda_2, \ldots, \lambda_t$ need not be distinct.

21.56 Definition.   Let $(\lambda - \lambda_1)^{e_1}, (\lambda - \lambda_2)^{e_2}, \ldots, (\lambda - \lambda_t)^{e_t}$ be the elementary divisors of $A \in \mathbf{C}^{nn}$. If $\lambda_1, \lambda_2, \ldots, \lambda_t$ are distinct, then A is called *non-derogatory*. On the other hand, if any two of them are equal, then A is called *derogatory*.

Since each elementary divisor corresponds to an elementary Jordan block of the form (21.46) we have an equivalent statement of this concept.

21.57 Theorem.*   A matrix is derogatory if and only if the same eigenvalue appears in more than one elementary Jordan block of its Jordan canonical matrix.

Proof. The proof is left as an exercise.

This last theorem, together with (21.51), give us the following corollary.

21.58 Corollary.   A matrix $M \in \mathbf{C}^{nn}$ has at least one multiple eigenvalue if and only if it is defective or derogatory, or both.

It turns out that there are four possibilities, then, for a matrix $M \in \mathbf{C}^{nn}$ and we can give an example which illustrates each case.

21.59 EXAMPLES

   i) $M^{(1)}$ is *non-defective* and *non-derogatory*.

---

* See (21.53).

Let $M^{(1)}$ have the elementary divisors $(\lambda - \alpha_1), (\lambda - \alpha_2), (\lambda - \alpha_3),$ and $(\lambda - \alpha_4),$ with distinct eigenvalues $\alpha_1, \alpha_2, \alpha_3,$ and α_4. Then

$$
J^{(1)} = \begin{bmatrix} \alpha_1 & & & \\ & \alpha_2 & & \\ & & \alpha_3 & \\ & & & \alpha_4 \end{bmatrix}.
$$

If we solve the equations

$$(J^{(1)} - \alpha_1 I)x^{(1)} = \phi$$

$$(J^{(1)} - \alpha_2 I)x^{(2)} = \phi$$

$$(J^{(1)} - \alpha_3 I)x^{(3)} = \phi$$

$$(J^{(1)} - \alpha_4 I)x^{(4)} = \phi$$

we find that

$$
x^{(1)} = \begin{bmatrix} x_1 \\ 0 \\ 0 \\ 0 \end{bmatrix}, \quad x^{(2)} = \begin{bmatrix} 0 \\ x_2 \\ 0 \\ 0 \end{bmatrix}, \quad x^{(3)} = \begin{bmatrix} 0 \\ 0 \\ x_3 \\ 0 \end{bmatrix}, \text{ and } x^{(4)} = \begin{bmatrix} 0 \\ 0 \\ 0 \\ x_4 \end{bmatrix}
$$

constitute a linearly independent set of eigenvectors corresponding to $\alpha_1, \alpha_2,$ $\alpha_3,$ and α_4. See (21.28).

ii) $M^{(2)}$ is defective but non-derogatory.

Let $M^{(2)}$ have the elementary divisors $(\lambda - \alpha_1), (\lambda - \alpha_2),$ and $(\lambda - \alpha_3)^2,$ with distinct eigenvalues $\alpha_1, \alpha_2,$ and α_3. Then

$$
J^{(2)} = \begin{bmatrix} \alpha_1 & & & \\ & \alpha_2 & & \\ & & \alpha_3 & 1 \\ & & & \alpha_3 \end{bmatrix}.
$$

If we solve the equations

$$(J^{(2)} - \alpha_1 I)x^{(1)} = \phi$$

$$(J^{(2)} - \alpha_2 I)x^{(2)} = \phi$$

$$(J^{(2)} - \alpha_3 I)x^{(3)} = \phi$$

we find that

$$x^{(1)} = \begin{bmatrix} x_1 \\ 0 \\ 0 \\ 0 \end{bmatrix}, \quad x^{(2)} = \begin{bmatrix} 0 \\ x_2 \\ 0 \\ 0 \end{bmatrix}, \text{ and } x^{(3)} = \begin{bmatrix} 0 \\ 0 \\ x_3 \\ 0 \end{bmatrix}$$

constitute a linearly independent set of eigenvectors corresponding to α_1, α_2, and α_3. See (21.28). Notice that α_3 appears in a non-linear elementary divisor and even though α_3 is an eigenvalue of multiplicity two there is only one linearly independent eigenvector corresponding to α_3. This is in agreement with theorems 21.52 and 21.53. Notice, however, how this case differs from the next case, where α_3 also has multiplicity two.

iii) $M^{(3)}$ is non-defective but derogatory.

Let $M^{(3)}$ have the elementary divisors $(\lambda - \alpha_1), (\lambda - \alpha_2), (\lambda - \alpha_3)$ and $(\lambda - \alpha_3)$, with distinct eigenvalues α_1, α_2, and α_3. Then

$$J^{(3)} = \begin{bmatrix} \alpha_1 & \vdots & & \\ \cdots & \cdots & \cdots & \cdots & \\ & \vdots & \alpha_2 & \vdots & \\ & & \cdots & \cdots & \cdots & \\ & & & \vdots & \alpha_3 & \vdots \\ & & & & \cdots & \cdots & \cdots \\ & & & & & \vdots & \alpha_3 \end{bmatrix}.$$

If we solve the equations

$$(J^{(3)} - \alpha_1 I)x^{(1)} = \phi$$

$$(J^{(3)} - \alpha_2 I)x^{(2)} = \phi$$

$$(J^{(3)} - \alpha_3 I)x^{(3)} = \phi,$$

we find that

$$x^{(1)} = \begin{bmatrix} x_1 \\ 0 \\ 0 \\ 0 \end{bmatrix}, \quad x^{(2)} = \begin{bmatrix} 0 \\ x_2 \\ 0 \\ 0 \end{bmatrix}$$

are eigenvectors corresponding to α_1 and α_2, respectively. The situation with regard to α_3 is very interesting. Any vector of the form

$$x^{(3)} = \begin{bmatrix} 0 \\ 0 \\ x_3 \\ x_4 \end{bmatrix}$$

is an eigenvector corresponding to α_3. However,

$$x^{(3)} = x_3 \begin{bmatrix} 0 \\ 0 \\ 1 \\ 0 \end{bmatrix} + x_4 \begin{bmatrix} 0 \\ 0 \\ 0 \\ 1 \end{bmatrix}$$

which means that $x^{(3)}$ is any vector in the two-dimensional subspace (a plane) spanned by the unit vectors $e^{(3)}$ and $e^{(4)}$. It is no coincidence that α_3 has multiplicity *two* and $x^{(3)}$ is any vector in a certain *two-dimensional* subspace. See Theorem 21.53 and Corollary 21.54.

iv) $M^{(4)}$ *is both defective and derogatory.*
Let $M^{(4)}$ have the elementary divisors $(\lambda - \alpha_1), (\lambda - \alpha_1)$, and $(\lambda - \alpha_2)^2$ with distinct eigenvalues α_1 and α_2, each of multiplicity two. This is example 21.39 and $J^{(4)}$ is exhibited in (21.48). If we solve the equations

$$(J^{(4)} - \alpha_1 I)x^{(1)} = \phi$$
$$(J^{(4)} - \alpha_2 I)x^{(2)} = \phi$$

we find that

$$x^{(1)} = \begin{bmatrix} x_1 \\ x_2 \\ 0 \\ 0 \end{bmatrix}, \text{ and } x^{(2)} = \begin{bmatrix} 0 \\ 0 \\ x_3 \\ 0 \end{bmatrix}$$

are eigenvectors corresponding to α_1 and α_2, respectively. However,

$$x^{(1)} = x_1 \begin{bmatrix} 1 \\ 0 \\ 0 \\ 0 \end{bmatrix} + x_2 \begin{bmatrix} 0 \\ 1 \\ 0 \\ 0 \end{bmatrix}$$

and so $x^{(1)}$ can be any vector in the two-dimensional subspace spanned by the unit vectors $e^{(1)}$ and $e^{(2)}$. Thus, corresponding to α_1 (of multiplicity two) we have a two-dimensional subspace of eigenvectors. Corresponding to α_2 (also of multiplicity two), on the other hand, we have $x^{(2)} = x_3 e^{(3)}$ which is a one-dimensional subspace spanned by the unit vector $e^{(3)}$.

In these examples we see the difference between the case where a multiple eigenvalue comes from a derogatory matrix and the case where a multiple eigenvalue comes from a defective matrix. The reader should examine examples such as

21.60
$$\begin{bmatrix} \alpha_1 & 1 & \vdots & & \\ & \alpha_1 & \vdots & & \\ \hdotsfor{5} \\ & & \vdots & \alpha_1 & 1 \\ & & \vdots & & \alpha_1 \end{bmatrix} \text{ and } \begin{bmatrix} \alpha_1 & \vdots & & & \\ \hdotsfor{5} \\ & \vdots & \alpha_1 & & \\ \hdotsfor{5} \\ & \vdots & & \alpha_1 & 1 \\ & \vdots & & & \alpha_1 \end{bmatrix}$$

to be sure he understands the four cases we have described.

Both of these matrices have a single eigenvalue α_1 and in each case the multiplicity is four. However, the first matrix listed has a two-dimensional subspace of eigenvectors, mentioned in (21.55), whereas the second matrix has a three-dimensional subspace of eigenvectors.

This leads us to introduce the descriptive adjectives *algebraic* and *geometric* when referring to the multiplicity of an eigenvalue.

21.61 Definition. The *geometric multiplicity* of an eigenvalue is the dimension of the subspace of eigenvectors associated with the eigenvalue. The *algebraic multiplicity* of an eigenvalue is the number of times it appears on the diagonal of the Jordan canonical form.

In the two examples just mentioned both matrices have α_1 as an eigenvalue with algebraic multiplicity four, whereas their geometric multiplicities are two and three, respectively.

We observe that it is not necessarily "bad" for a matrix to have multiple eigenvalues if the algebraic multiplicities and the geometric multiplicities are the same. This occurs when a matrix is diagonalizable (non-defective). Among the four examples in (21.59) both $M^{(1)}$ and $M^{(3)}$ have this property and each has a set of n linearly independent eigenvectors which can be used as a basis for the vector space.

From this point of view, then, $M^{(2)}$ and $M^{(4)}$ could be called "bad" matrices since each matrix has the feature that, for at least one eigenvalue, the algebraic multiplicity exceeds the geometric multiplicity. Thus, in each case, there is not a full set of linearly independent eigenvectors with which to span the vector space.

However, the situation is not as bad as it might seem because of the following definition and theorem.

21.62 Definition. A vector $y^{(m)} \in \mathbf{C}^n$ is a *principal vector (generalized eigenvector)* of degree m corresponding to the matrix $M \in \mathbf{C}^{nn}$ and the eigenvalue $\lambda \in \mathbf{C}$, if $(M - \lambda I)^m y^{(m)} = \phi$ but $(M - \lambda I)^{m-1} y^{(m)} \neq \phi$.

Notice that an ordinary eigenvector is a principal vector (generalized eigenvector) of degree $m = 1$. Now we have the following existence theorem.

21.63 Theorem. Every matrix $M \in \mathbf{C}^{nn}$ possesses n linearly independent principal vectors. Principal vectors corresponding to distinct eigenvalues are linearly independent. If λ is an eigenvalue of M of algebraic multiplicity k, then M will have k linearly independent principal vectors corresponding to λ.

Proof. See, for example, Wedderburn [1934], Chapter 3 and Friedman [1956], Chapter 2.

21.64 Remark. If $P^{-1}MP = J$, the Jordan canonical matrix of M, then $MP = PJ$ and the columns of P give us n linearly independent principal vectors of M. *If M is non-defective, then all of its principal vectors are eigenvectors.* On the other hand, if M is defective we can assume the existence of at least one non-linear elementary divisor $(\lambda - \alpha)^e$ with its corresponding elementary Jordan block* J_α of order $e \times e$. If J_α appears in columns $t, t + 1, t + 2, \ldots, t + e - 1$ of J, then columns $t, t + 1, t + 2, \ldots, t + e - 1$ of P are linearly independent principal vectors of M corresponding to $\lambda = \alpha$. Furthermore, column t of P is a principal vector of degree† $m = 1$ (an eigenvector) and subsequent columns of P are principal vectors of degrees $m = 2, 3, \ldots, e$, respectively.

We close this section on eigenvalues and eigenvectors by listing some basic results, most of which are related to Hermitian matrices. We emphasize the fact that results stated for Hermitian matrices also apply to real symmetric matrices.‡ For proofs of Theorems 21.66 through 21.72, see, for example, Bronson [1969], pp. 236–250.

In (17.22) we defined an orthonormal set of vectors. If the set is *complete* and the vectors are eigenvectors of some matrix we introduce the following terminology:

21.65 Definition. A set of vectors $x^{(1)}, x^{(2)}, \ldots, x^{(t)} \in \mathbf{C}^n$ is a complete orthonormal set of eigenvectors for $M \in \mathbf{C}^{nn}$ if and only if

  i) $t = n$.

 ii) $Mx^{(i)} = \lambda_i x^{(i)}$ $(i = 1, 2, \ldots, n)$.

iii) $\{x^{(1)}, x^{(2)}, \ldots, x^{(n)}\}$ is an orthonormal set.

---

* Recall (21.46).

† See (21.6).

‡ See (17.37).

21.66 Theorem. If $M \in \mathbf{C}^{nn}$ is Hermitian, then

i) The eigenvalues of M are real.

ii) Eigenvectors of M corresponding to distinct eigenvalues are orthogonal.

iii) M possesses a complete orthonormal set of eigenvectors.

iv) There exists a similarity transformation $UMU^{-1} = \Lambda$, where $U \in \mathbf{C}^{nn}$ is unitary and $\Lambda \in \mathbf{R}^{nn}$ is diagonal.

21.67 Definition. A similarity transformation with a unitary matrix is called a *unitary similarity transformation.*

21.68 Corollary. If $M \in \mathbf{R}^{nn}$ is symmetric, then (i), (ii), and (iii) of Theorem 21.66 apply as they are worded.

The last part of Theorem 21.66 can be simplified if we first state the following result.

21.69 Theorem. If $M \in \mathbf{R}^{nn}$ is symmetric, then the eigenvectors of M can always be chosen to be real.

Hence, part (iv) of Theorem 21.66 can be simplified.

21.70 Corollary. If $M \in \mathbf{R}^{nn}$ is symmetric, then there exists a similarity transformation $QMQ^{-1} = \Lambda$ where $Q \in \mathbf{R}^{nn}$ is orthogonal and $\Lambda \in \mathbf{R}^{nn}$ is diagonal.

21.71 Definition. A similarity transformation with an orthogonal matrix is called an *orthogonal similarity transformation.*

If, in addition to being Hermitian, we also have positive (nonnegative) definiteness we have the following result.

21.72 Theorem. $M \in \mathbf{C}^{nn}$ is positive definite (nonnegative definite), if and only if M is Hermitian and all eigenvalues of M are positive (nonnegative).

This theorem includes the real symmetric positive definite matrices as special cases.

In our discussion of the Jordan canonical form we indicated in (21.49) that every matrix can be transformed into a triangular matrix (of special type), by a similarity transformation, which exhibits the basic structure of the matrix. A related theorem states that every matrix can be transformed by a unitary similarity transformation into a triangular matrix (not of special type) which exhibits the eigenvalues on the main diagonal.

21.73 Theorem (Schur). Let $M \in \mathbf{C}^{nn}$. Then $\exists$ a unitary matrix U such that

$$UMU^H = T$$

where T is triangular and $t_{11}, t_{22}, \dots, t_{nn}$ are the eigenvalues of M.

Proof. See, for example, Bellman [1960], p. 195.

21.74 Theorem. If $A, B \in \mathbf{C}^{nn}$, then AB and BA have the same set of eigenvalues.

Proof. The proof is left as an exercise.

21.75 Theorem. If $M \in \mathbf{C}^{nn}$ and if $\lambda_1, \lambda_2, \ldots, \lambda_n$ are its eigenvalues, then

$$\det M = \prod_{i=1}^{n} \lambda_i .$$

Proof. See, for example, Bronson [1969], p. 96.

In (20.12) we stated that for arbitrary $A \in \mathbf{C}^{nn}$ the matrix $A^H A$ is Hermitian and nonnegative definite. In (21.66) and (21.72) we see that the eigenvalues of $A^H A$ are real and nonnegative. This leads us to the following definition.

21.76 Definition. If μ is an eigenvalue of $A^H A$ for $A \in \mathbf{C}^{nn}$ then $\sqrt{\mu}$ is called a *singular value* of A.

The singular values of a matrix are useful in many areas. For example, in (27.21) the largest singular value is used as one of the basic matrix norms. Obviously, because of (21.74), the eigenvalues of AA^H are the same as the eigenvalues of $A^H A$. Hence, either matrix can be used in the definition above.

We now point out that, given a positive definite matrix $A \in \mathbf{C}^{nn}$, we can find a "square root" of A.

21.77 Theorem. If the Hermitian matrix $A \in \mathbf{C}^{nn}$ is positive definite (nonnegative definite), then $\exists$ a unique positive definite (nonnegative definite) matrix B such that $B^2 = A$. If, more specifically, $A \in \mathbf{R}^{nn}$, then $B \in \mathbf{R}^{nn}$. We usually write $B = A^{\frac{1}{2}}$

Proof. See, for example, Young [1971a], pp. 22–23.

EXERCISES 11.21

1. If $M \in \mathbf{C}^{nn}, \lambda \in \mathbf{C}$, show that $\det(M - \lambda I)$ is a polynomial of degree n in λ. (Use induction on n.)

2. Prove Theorem 21.13.

3. Prove Theorem 21.14.

4. Prove Theorem 21.15.

5. Prove Theorem 21.17.

6. Prove Theorem 21.23.

7. Prove Theorem 21.25.

8. Prove Theorem 21.28.

9. Complete the proof of Theorem 21.31.

10. Prove Theorem 21.33.

11. Prove Theorem 21.36.

12. Prove Theorem 21.37.

13. Prove Theorem 21.42.

14. Prove Theorem 21.45.

15. Prove Theorem 21.49.

16. Prove Corollary 21.50.

17. Prove Corollary 21.51.

18. Prove Theorem 21.52.

19. Prove Theorem 21.53.

20. Prove Corollary 21.54.

21. Prove Theorem 21.57.

22. Prove Corollary 21.58.

23. Prove Theorem 21.63.

24. Prove Theorem 21.66.

25. Prove Theorem 21.69.

26. Prove Corollary 21.70.

27. Prove Theorem 21.72.

28. Prove Theorem 21.73.

29. Prove Theorem 21.74.

30. Prove Theorem 21.75.

31. Prove Theorem 21.77.

11.22 SEQUENCES AND INFINITE SERIES OF VECTORS

Consider the *sequence* of vectors from $\mathbf{C}^n$

22.1 $$\{x^{(i)}\} = x^{(1)}, x^{(2)}, \ldots, x^{(i)}, \ldots$$

Suppose the n sequences, formed by the respective components of the vectors in the above sequence, converge* to finite limits in $\mathbf{C}$. In other words, assume that, for $k = 1, 2, \ldots, n$,

22.2 $$x_k^{(i)} \to x_k \qquad (i \to \infty).$$

Then we can define *convergence of the sequence* of vectors as follows:

22.3 Definition. The sequence of vectors $\{x^{(i)}\}$ converges to the vector x as i tends to ∞, or in symbols,

$$x^{(i)} \to x \qquad (i \to \infty)$$

if and only if (22.2) holds for all k.

---

* See Theorem A—(2.9) and comments following A—(5.9).

We sometimes call this *componentwise convergence* and we often write

22.4
$$\lim_{i \to \infty} x^{(i)} = x.$$

Next, consider the *infinite series* of vectors, from $\mathbf{C}^n$,

22.5
$$\sum_{i=1}^{\infty} x^{(i)} = x^{(1)} + x^{(2)} + \cdots + x^{(i)} + \cdots$$

along with the sequence of partial sums

22.6
$$\begin{cases} s^{(1)} = x^{(1)} \\ s^{(2)} = x^{(1)} + x^{(2)} \\ \quad \vdots \\ s^{(i)} = x^{(1)} + x^{(2)} + \cdots + x^{(i)} \\ \quad \vdots \end{cases}$$

Then we can define *convergence** *of the infinite series* as follows:

22.7 Definition. The infinite series of vectors $\sum_{i=1}^{\infty} x^{(i)}$ converges to the vector s if and only if the sequence of partial sums $\{s^{(i)}\}$ converges to s, that is, if and only if $s^{(i)} \to s$ as $i \to \infty$.

We usually indicate this convergence by writing

22.8
$$\sum_{i=1}^{\infty} x^{(i)} = s.$$

11.23 SEQUENCES AND INFINITE SERIES OF MATRICES

Consider the *sequence* of matrices from $\mathbf{C}^{mn}$

23.1
$$\{A^{(k)}\} = A^{(1)}, A^{(2)}, \cdots, A^{(k)}, \cdots .$$

Assume that the mn sequences, formed by the respective elements of the matrices in the above sequence, converge to finite limits in $\mathbf{C}$. In other words, assume that, for all i and j,

23.2
$$a_{ij}^{(k)} \to a_{ij} \qquad (k \to \infty).$$

---

* See Theorem A—(2.14) and comments following A—(5.9).

Then we can define *convergence of the sequence* of matrices as follows:

23.3 Definition. The sequence of matrices $\{A^{(k)}\}$ converges to the matrix $A \in \mathbf{C}^{mn}$ as k tends to ∞, or in symbols,

$$A^{(k)} \to A \qquad (k \to \infty)$$

if and only if (23.2) holds for all i and j.

We sometimes call this elementwise convergence and we often write

23.4
$$\operatorname*{Lim}_{k \to \infty} A^{(k)} = A.$$

Next, consider the *infinite series* of matrices, from $\mathbf{C}^{mn}$,

23.5
$$\sum_{k=1}^{\infty} A^{(k)} = A^{(1)} + A^{(2)} + \cdots + A^{(k)} + \cdots$$

along with the sequence of partial sums

23.6
$$\begin{cases} S^{(1)} = A^{(1)} \\ S^{(2)} = A^{(1)} + A^{(2)} \\ \qquad \vdots \\ S^{(k)} = A^{(1)} + A^{(2)} + \cdots + A^{(k)} \\ \qquad \vdots \end{cases}$$

Then we can define *convergence of the infinite series* as follows:

23.7 Definition. The infinite series of matrices $\Sigma_{k=1}^{\infty} A^{(k)}$ converges to the matrix $S \in \mathbf{C}^{mn}$ if and only if the sequence of partial sums $\{S^{(k)}\}$ converges to S, that is, if and only if $S^{(k)} \to S$ as $k \to \infty$.

We usually indicate this convergence by writing

23.8
$$\sum_{k=1}^{\infty} A^{(k)} = S.$$

11.24 VECTOR NORMS

It is often necessary to consider the "magnitude" or length* of a vector in $\mathbf{C}^n$. What we mean is that we wish to associate a single, nonnegative, real number with each vector (which is meaningful in some sense) in much the same way as we associate the modulus $|w|$ with each complex number† w. To do this we use the notion of *vector norm*.

Since there are infinitely many vector norms, and space limitations prevent a complete discussion, we shall merely introduce the concept and then restrict our discussion to a few norms which are widely used. It is assumed throughout that we are working with vectors in $\mathbf{C}^n$.

24.1 Definition. The vector norm $\| \cdot \|_\alpha$ is a nonnegative function on the space $\mathbf{C}^n$ with the following properties:

  a) $\|x\|_\alpha > 0$ if $x \neq \phi$

  b) $\|cx\|_\alpha = |c| \cdot \|x\|_\alpha$ for any $c \in \mathbf{C}$ and any $x \in \mathbf{C}^n$

  c) $\|x + y\|_\alpha \leq \|x\|_\alpha + \|y\|_\alpha$ for all vectors $x, y \in \mathbf{C}^n$.

Property (c) is often called the *triangle inequality*. As a direct consequence of (a) and (b), with $c = 0$, we have, for every vector norm, the trivial result:

24.2 Theorem. $\|x\|_\alpha = 0$ if and only if $x = \phi$.

We shall demonstrate later that there are infinitely many vector norms which satisfy (24.1). These norms can be shown to satisfy the inequality:

24.3 Theorem. For every vector norm and for every $x, y \in \mathbf{C}^n$

$$\left| \, \|x\|_\alpha - \|y\|_\alpha \, \right| \leq \|x - y\|_\alpha.$$

Proof. If we use the triangle inequality we can write

$$\|x\|_\alpha = \|(x - y) + y\|_\alpha$$
$$\leq \|x - y\|_\alpha + \|y\|_\alpha.$$

This allows us to write

$$\|x\|_\alpha - \|y\|_\alpha \leq \|x - y\|_\alpha.$$

Similarly, we obtain

$$\|y\|_\alpha - \|x\|_\alpha \leq \|y - x\|_\alpha.$$

---

* See (16.8).

† See (2.7).

However, property (b) of the definition, with $c = -1$, allows us to equate the right-hand sides of the last two inequalities. Thus,

$$\left| \|x\|_\alpha - \|y\|_\alpha \right| \leq \|x - y\|_\alpha.$$

The reader should note the importance of this form of the triangle inequality. It is used, for example, in proving Theorem 31.5 and Theorem 32.3. Now we shift our attention to the norms of square matrices.

11.25 MATRIX NORMS

Since $A \in \mathbf{C}^{nn}$ can be considered to be an n^2 dimensional vector*, it is not surprising that we assume the same set of properties in defining a *matrix norm* for an $n \times n$ matrix. However, a fourth property is desirable.

25.1 Definition. The matrix norm $\| \cdot \|_\beta$ is a nonnegative function on the space $\mathbf{C}^{nn}$ with the following properties:

a) $\|A\|_\beta > 0$ if $A \neq \phi$

b) $\|cA\|_\beta = |c| \cdot \|A\|_\beta$ for any $c \in \mathbf{C}$ and any $A \in \mathbf{C}^{nn}$

c) $\|A + B\|_\beta \leq \|A\|_\beta + \|B\|_\beta$ for all $A, B \in \mathbf{C}^{nn}$

d) $\|AB\|_\beta \leq \|A\|_\beta \|B\|_\beta$ for all $A, B \in \mathbf{C}^{nn}$.

Property (c) is called the *triangle inequality*, and property (d) is called the *product inequality*. We shall demonstrate later that there are infinitely many norms satisfying (25.1). It is assumed that a matrix $A \in \mathbf{C}^{nn}$ unless we state otherwise.

The following theorems are proved in a manner analogous to the way we proved (24.2) and (24.3) above.

25.2 Theorem. $\|A\|_\beta = 0$ if and only if $A = \phi$.

25.3 Theorem. For every matrix norm and for every $A, B \in \mathbf{C}^{nn}$

$$\left| \|A\|_\beta - \|B\|_\beta \right| \leq \|A - B\|_\beta.$$

Suppose $A \in \mathbf{C}^{nn}$ represents a linear mapping of each vector $x \in \mathbf{C}^n$ into the vector

25.4 $y = Ax$

of $\mathbf{C}^n$. Then for any vector norm $\| \cdot \|_\alpha$ we can define the function

25.5 $$\sup_{x \neq \phi} \frac{\|y\|_\alpha}{\|x\|_\alpha} = \sup_{x \neq \phi} \frac{\|Ax\|_\alpha}{\|x\|_\alpha} = \sup_{\|x\|_\alpha = 1} \|Ax\|_\alpha.$$

---

* See (12.9).

This is a function of A, and it can be shown that it satisfies the four conditions required of a matrix norm (see exercises). Thus we may write*

25.6
$$\|A\|_\sigma = \sup_{x \neq \phi} \frac{\|Ax\|_\alpha}{\|x\|_\alpha},$$

where $\sigma = \sigma(\alpha)$.

Given any vector norm $\|\cdot\|_\alpha$, a matrix norm $\|\cdot\|_\sigma$ is determined in this way. This leads us to the concept of a *subordinate matrix norm*.

25.7 Definition. The matrix norm $\|\cdot\|_\sigma$ defined by (25.6) is said to be *subordinate* to the corresponding vector norm $\|\cdot\|_\alpha$.

One of the main advantages in choosing a matrix norm which is subordinate to a vector norm is that the inequality

25.8
$$\|Ax\|_\alpha \leqq \|A\|_\beta \cdot \|x\|_\alpha$$

is satisfied for all $A \in \mathbf{C}^{nn}$ and $x \in \mathbf{C}^n$. This follows directly from (25.6).

25.9 Definition. A matrix norm $\|A\|_\beta$ and a vector norm $\|x\|_\alpha$ for which the inequality (25.8) holds for all A and x are called *consistent* (some authors use *compatible*).

Thus, a vector norm $\|\cdot\|_\alpha$ and its subordinate matrix norm $\|\cdot\|_\sigma$ are always consistent, and so we can always find at least one matrix norm which is consistent with a given vector norm. One might ask whether or not a vector norm can be found which is consistent with a given matrix norm. It turns out that the answer is in the affirmative. [Collatz, 1966, p. 170].

25.10 Theorem. If $x \in \mathbf{C}^n$ is an arbitrary vector and $B = [x, \phi, \phi, \ldots, \phi]$ then

$$\|x\|_\alpha = \|B\|_\beta$$

defines a vector norm $\|\cdot\|_\alpha$ which is consistent with the matrix norm $\|\cdot\|_\beta$.

Proof. Clearly, this defines a vector norm. (This step is left as an exercise.) Let $A \in \mathbf{C}^{nn}$ and $x \in \mathbf{C}^n$ be arbitrary. Then

$$\|Ax\|_\alpha = \|[Ax, \phi, \phi, \ldots, \phi]\|_\beta$$
$$= \|AB\|_\beta$$
$$\leqq \|A\|_\beta \|B\|_\beta$$
$$= \|A\|_\beta \|x\|_\alpha.$$

Thus, the consistency condition (25.8) is satisfied.

---

* Norms defined by (25.6) are sometimes called *natural norms*. See, for example, John [1967], p. 7.

EXERCISES 11.25

1. Prove Theorem 25.2.

2. Prove Theorem 25.3.

3. Show that the function defined in (25.5) satisfies the four conditions required of a matrix norm.

4. Show that $\| \cdot \|_a$ defined in Theorem 25.10 is a vector norm.

11.26 EXAMPLES OF VECTOR NORMS

It was mentioned earlier that there are infinitely many vector norms. To illustrate this fact consider the l_p-norms (Hölder norms)

26.1
$$\|x\|_p = \begin{cases} \sqrt[p]{\sum_{i=1}^{n} |x_i|^p} & p = 1, 2, 3, \cdots \\[2ex] \max_i |x_i| & p = \infty \end{cases}$$

Among these norms the most widely used are the l_1-norm, the l_2-norm, and the l_∞-norm. Consequently, we shall restrict further discussion to the norms

26.2
$$\|x\|_1 = \sum_{i=1}^{n} |x_i|,$$

26.3
$$\|x\|_2 = \sqrt{\sum_{i=1}^{n} |x_i|^2},$$

and

26.4
$$\|x\|_\infty = \max_i |x_i|.$$

It should be pointed out that $\|x\|_1$ is also called the *sum norm*, $\|x\|_2$ is also called the *Euclidean norm*, and $\|x\|_\infty$ is also called the *maximum norm*. It is easy to derive the following inequalities among these norms:

26.5 Theorem.

i) $\|x\|_\infty \leqq \|x\|_2 \leqq \|x\|_1$

ii) $\|x\|_\infty \leqq \|x\|_1 \leqq n\|x\|_\infty$

iii) $\|x\|_\infty \leqq \|x\|_2 \leqq n^{\frac{1}{2}} \|x\|_\infty$

iv) $n^{-\frac{1}{2}} \|x\|_1 \leqq \|x\|_2 \leqq \|x\|_1$

Proof. The proof is left as an exercise.

EXERCISES 11.26

1. Show that for $p = 1, 2, \infty$ the l_p-norms satisfy the conditions (24.1) for a vector norm. For $p = 2$ use the inequality

$$\left[\sum_{i=1}^{n} x_i y_i \right]^2 \leq \sum_{i=1}^{n} x_i^2 \sum_{i=1}^{n} y_i^2 .$$

2. Prove Theorem 26.5.

11.27 THE CORRESPONDING MATRIX NORMS

The natural question to ask at this point is, what are the matrix norms which are subordinate to the three vector norms $\|x\|_1$, $\|x\|_2$, and $\|x\|_\infty$, respectively? We must refer to (25.6), of course. However, it is simpler to use an equivalent form of (25.6), namely,

27.1
$$\|A\|_\sigma = \sup_{\|x\|_\alpha = 1} \|Ax\|_\alpha .$$

Thus, we may define the subordinate matrix norms as

27.2
$$\|A\|_1 = \sup_{\|x\|_1 = 1} \|Ax\|_1 ,$$

27.3
$$\|A\|_2 = \sup_{\|x\|_2 = 1} \|Ax\|_2 ,$$

and

27.4
$$\|A\|_\infty = \sup_{\|x\|_\infty = 1} \|Ax\|_\infty ,$$

respectively.

Notice that we have chosen the value of σ to equal the corresponding value of α. This makes it easy to identify the matrix norm with the vector norm to which it is subordinate.

In order to see what $\|A\|_\infty$ becomes, let us examine $\|Ax\|_\infty$ more closely. Since

27.5
$$Ax = \begin{bmatrix} \sum_{j=1}^{n} a_{1j}x_j \\ \vdots \\ \sum_{j=1}^{n} a_{ij}x_j \\ \vdots \\ \sum_{j=1}^{n} a_{nj}x_j \end{bmatrix} ,$$

we can use (26.4) to write

27.6
$$\|Ax\|_\infty = \max_i \left| \sum_{j=1}^n a_{ij}x_j \right|.$$

Now, since

27.7
$$\max_j |x_j| = 1,$$

we find that

27.8
$$\left| \sum_{j=1}^n a_{ij}x_j \right| \leqq |a_{i1}| \cdot |x_1| + |a_{i2}| \cdot |x_2| + \cdots + |a_{in}| \cdot |x_n| \leqq \sum_{j=1}^n |a_{ij}|,$$

which says that the magnitude of the ith component of the vector Ax is bounded by the row sum of absolute values of the ith row of A. Let r be a value of i which produces the largest row sum of absolute values. It follows then that

27.9
$$\|Ax\|_\infty = \max_i \left| \sum_{j=1}^n a_{ij}x_j \right|$$

$$\leqq \sum_{j=1}^n |a_{rj}|,$$

and we have an upper bound for $\|Ax\|_\infty$.

If we choose $i = r$ and x_j according to the rule

27.10
$$x_j = \begin{cases} +1 \text{ if } a_{rj} \geqq 0 \\ -1 \text{ if } a_{rj} < 0, \end{cases}$$

then we achieve equality in (27.9). In other words,

27.11
$$\left| \sum_{j+1}^n a_{rj}x_j \right| = \sum_{j=1}^n |a_{rj}|,$$

in this case, and the upper bound on $\|Ax\|_\infty$ is achieved. Since $\|A\|_\infty$ is the supremum of $\|Ax\|_\infty$, for $\|x\|_\infty = 1$, we have the final result

27.12
$$\|A\|_\infty = \max_i \sum_{j=1}^n |a_{ij}|.$$

Thus, $\|A\|_\infty$ may be described as the *maximum row sum of absolute values.* In a similar manner it can be shown that

27.13
$$\|A\|_1 = \max_j \sum_{i=1}^n |a_{ij}|,$$

which says that $\|A\|_1$ is the *maximum column sum of absolute values.*

To see what $\|A\|_2$ becomes we note, from (26.3), that

27.14
$$\|x\|_2 = \sqrt{(x, x)}$$
$$= \sqrt{x^H x}.$$

Hence,

$$\|A\|_2 = \sup_{\|x\|_2 = 1} \|Ax\|_2$$

$$= \sup_{\|x\|_2 = 1} \sqrt{(Ax, Ax)}$$

$$= \sup_{\|x\|_2 = 1} \sqrt{x^H A^H Ax}.$$

The matrix $A^H A$ is Hermitian and nonnegative definite*. Consequently, its eigenvalues μ_i are real and nonnegative, and we may write†

27.15
$$\mu_1 \geqq \mu_2 \geqq \cdots \geqq \mu_n \geqq 0.$$

Also, $A^H A$ has a complete orthonormal set of eigenvectors‡

$$v^{(1)}, v^{(2)}, \ldots, v^{(n)},$$

which form a basis for the vector space $\mathbf{C}^n$. Therefore, any vector x with $\|x\|_2 = 1$, but otherwise arbitrary, can be written

27.16
$$x = \sum_{j=1}^n c_j v^{(j)}$$

where, since $(x, x) = 1$,

27.17
$$\sum_{j=1}^n |c_j|^2 = 1.$$

---

* See (20.12).

† See (21.76), (21.66), and (21.72).

‡ See (21.66).

Thus,

27.18
$$A^H A x = A^H A \left[\sum_{j=1}^{n} c_j v^{(j)} \right]$$

$$= \sum_{j=1}^{n} c_j \mu_j v^{(j)}$$

and, because of the orthonormal property of the eigenvectors,

27.19
$$(Ax, Ax) = x^H (A^H A x)$$

$$= \sum_{j=1}^{n} |c_j|^2 \mu_j$$

$$\leq \mu_1 \sum_{j=1}^{n} |c_j|^2$$

$$= \mu_1 .$$

If x is taken to be $v^{(1)}$, the eigenvector corresponding to the largest eigenvalue, then

27.20
$$(Av^{(1)}, Av^{(1)}) = (v^{(1)}, A^H A v^{(1)})$$

$$= (v^{(1)}, \mu_1 v^{(1)})$$

$$= \mu_1 (v^{(1)}, v^{(1)})$$

$$= \mu_1$$

and equality is achieved. Since $\|A\|_2$ is the supremum of $\sqrt{(Ax, Ax)}$, for $\|x\|_2 = 1$, we have the final result

27.21
$$\|A\|_2 = \sqrt{\mu_1} .$$

Thus $\|A\|_2$ may be described either as the *positive square root of the largest eigenvalue of $A^H A$* or as *the largest singular value* of A*.

It is interesting to note that even though $\|A\|_1$, $\|A\|_2$, and $\|A\|_\tau$ are subordinate to the vector norms $\|x\|_1$, $\|x\|_2$, and $\|x\|_\tau$, respectively, the formulas

---

* See (21.76).

$$\|A\|_1 = \max_j \sum_{i=1}^n |a_{ij}|,$$

$$\|A\|_2 = \sqrt{\mu_1},$$

$$\|A\|_\infty = \max_i \sum_{j=1}^n |a_{ij}|,$$

define these norms without reference to the corresponding vector norms.

Because of the obvious connection with the spectral radius* of $A^H A$ we also call $\|A\|_2$ the *spectral norm* of A. Note that if A is unitary, $A^H A = I$ and $\|A\|_2 = 1$. Note also that $\|I\|_1 = \|I\|_2 = \|I\|_\infty = 1$.

We close this section with the following theorem and corollary.

27.22 Theorem. Let $A \in \mathbf{C}^{nn}$ have eigenvalues $\lambda_1, \lambda_2, \ldots, \lambda_n$ (not necessarily distinct) and let $P_m(t)$ be a given polynomial. Then the eigenvalues of $P_m(A)$ are the numbers $P_m(\lambda_1), P_m(\lambda_2), \ldots, P_m(\lambda_n)$.

Proof. See Faddeev and Faddeeva [1963], p. 14.

As a corollary, we can prove that *the spectral norm of a real symmetric matrix is equal to its spectral radius.*

27.23 Corollary. If $A \in \mathbf{R}^{nn}$ is symmetric, then

$$\|A\|_2 = S(A).$$

Proof. $\|A\|_2 = \sqrt{\mu_1}$, where μ_1 is the largest eigenvalue of $A^T A = A^2$. Since the eigenvalues of A^2 are the squares of the eigenvalues A this means that $\sqrt{\mu_1} = S(A)$.

EXERCISES 11.27

1. Show that

$$\sup_{\|x\|_\alpha = 1} \|Ax\|_\alpha = \sup_{x \neq \phi} \frac{\|Ax\|_\alpha}{\|x\|_\alpha}.$$

2. Derive the result 27.13.

3. Prove that if A is unitary, $\|A\|_2 = 1$.

4. Prove Theorem 27.22.

11.28 THE EUCLIDEAN MATRIX NORM

Another matrix norm which is often used is the *Euclidean matrix norm* defined by

28.1
$$\|A\|_E = \sqrt{\sum_{j=1}^n \sum_{i=1}^n |a_{ij}|^2}.$$

* See (21.10) and (27.23).

It is also called the *Schur norm* and the *Frobenius norm*. It can be verified that $\|A\|_E$ is a matrix norm and that it is consistent with $\|x\|_2$, the Euclidean vector norm. However, $\|A\|_E$ is not the norm which is subordinate to $\|x\|_2$ (it is $\|A\|_2$). As a matter of fact, we have the result:

28.2 Theorem. $\|A\|_E$ *is not subordinate to any vector norm.*

Proof (by contradiction). Assume that there exists a vector norm $\|x\|_\alpha$ such that

$$\|A\|_E = \sup_{\|x\|_\alpha = 1} \|Ax\|_\alpha.$$

Let $A = I$. Then

$$\sup_{\|x\|_\alpha = 1} \|Ix\|_\alpha = 1.$$

However,

$$\|I\|_E = \sqrt{n},$$

and this is a contradiction. Thus the original assumption is false.

In some ways it is unfortunate that the Euclidean matrix norm is not the norm which is subordinate to the Euclidean vector norm. Moreover, the property described in Theorem 28.2 makes $\|A\|_E$ unsatisfactory for many purposes. Consequently, in most *theoretical* work, $\|A\|_2$, rather than $\|A\|_E$, is used in conjunction with $\|x\|_2$ since $\|A\|_2$ is the norm subordinate to $\|x\|_2$.

On the other hand, in *numerical* work and error analysis $\|A\|_E$ is sometimes preferred to $\|A\|_2$. This is true because

i) $\|A\|_E$ is easy to compute, whereas $\|A\|_2$ is very difficult to compute.

ii) The Euclidean norm of $A = (a_{ij})$ is the same as the Euclidean norm of $|A| = (|a_{ij}|)$.

Both $\|A\|_E$ and $\|A\|_2$ are consistent with $\|x\|_2$ and sometimes $\|A\|_E$ is not a bad approximation to $\|A\|_2$. When this is the case it is very practical to use the Euclidean matrix norm with $\|x\|_2$. In general, though, the following inequality can be derived

28.3 $$\|A\|_2 \leqq \|A\|_E \leqq \sqrt{n}\,\|A\|_2,$$

where both limits are attainable.

The two advantages mentioned above for $\|A\|_E$ also apply to $\|A\|_1$ and $\|A\|_\infty$. However, as far as A and $|A|$ are concerned*, the best we can say about the spectral norm is that

28.4 $$\|A\|_2 \leqq \|\,|A|\,\|_2.$$

---

* The matrix $|A|$ is defined above in (ii).

If we combine (28.3), (28.4), and (ii) above, we have the following result.

28.5 Theorem.

$$\|A\|_2 \leqq \| \, |A| \, \|_2 \leqq \| \, |A| \, \|_E = \|A\|_E \leqq \sqrt{n} \, \|A\|_2 .$$

EXERCISES 11.28

1. Show that $\| \cdot \|_E$ in (28.1) is a matrix norm and show also that it is consistent with $\|x\|_2$, the Euclidean vector norm.
2. Derive (28.3).   Give an example for which $\|A\|_2 = \|A\|_E$ and one for which $\|A\|_E = \sqrt{n} \, \|A\|_2$.
3. Derive (28.4).   Give an example for which equality holds and one for which strict inequality holds.

11.29 SOME USEFUL THEOREMS

The following theorem gives a bound on the eigenvalues of A in terms of arbitrary matrix norms:

29.1 Theorem.   Let λ be any eigenvalue of $A \in \mathbf{C}^{nn}$.   Then $|\lambda| \leqq \|A\|_\beta$ for every matrix norm.   In other words, the spectral radius* of A is bounded by the norms of A.

Proof.   Let λ be any eigenvalue of A and let x be a corresponding eigenvector. Then

$$\lambda x = Ax .$$

Now choose an arbitrary matrix norm $\| \cdot \|_\beta$ and with it a consistent† vector norm $\| \cdot \|_\alpha$.   Using these norms we may write

$$|\lambda| \cdot \|x\|_\alpha \leqq \|A\|_\beta \|x\|_\alpha$$

from which we get the result

$$|\lambda| \leqq \|A\|_\beta .$$

Since this inequality holds for every eigenvalue, the spectral radius of A is bounded by every norm of A.

It can be shown [John, 1967, pp 7–8] that the equality sign does not always hold in (29.1).   On the other hand, we can show that a *natural norm‡ exists which is arbitrarily close to the spectral radius.*

---

* See (21.10).

† See (25.10)

‡ See (25.6).

29.2 Theorem. Let $A \in \mathbf{C}^{nn}$ with spectral radius $S(A)$. Then for every $\varepsilon > 0 \; \exists$ a natural norm $\| \cdot \|$ such that $\|A\| \leq S(A) + \varepsilon$.

Proof. See, for example, John [1967], p. 8.

The following theorems involve unitary* matrices in $\mathbf{C}^{nn}$.

29.3 Theorem. Let $A^H A = I$. Then for arbitrary $x \in \mathbf{C}^{nn}$

$$\|Ax\|_2 = \|x\|_2 .$$

Proof.

$$
\begin{aligned}
\|Ax\|_2^2 &= (Ax, Ax) \\
&= (x, A^H Ax) \\
&= (x, x) \\
&= \|x\|_2^2 .
\end{aligned}
$$

Note that $\|x\|_2$ represents the length of x. Hence, we see from this theorem that length is preserved under a unitary transformation.

29.4 Theorem. Let $A^H A = I$. Then, for arbitrary $B \in \mathbf{C}^{nn}$,

i) $\|AB\|_2 = \|B\|_2$,

ii) $\|ABA^H\|_2 = \|B\|_2$,

iii) $\|AB\|_E = \|B\|_E$.

Proof of (i):

$$
\begin{aligned}
\|AB\|_2 &= \sup_{\|x\|_2 = 1} \|ABx\|_2 \\
&= \sup_{\|x\|_2 = 1} \sqrt{(ABx, ABx)} \\
&= \sup_{\|x\|_2 = 1} \sqrt{(x, B^H A^H ABx)} \\
&= \sup_{\|x\|_2 = 1} \sqrt{(x, B^H Bx)} \\
&= \sup_{\|x\|_2 = 1} \sqrt{(Bx, Bx)} \\
&= \|B\|_2 .
\end{aligned}
$$

The proofs of (ii) and (iii) are left as exercises for the reader.

---

* See (17.24) and (17.25).

EXERCISES 11.29

1. Give an example for which strict inequality holds in (29.1).

2. Prove Theorem 29.2.

3. Prove parts (ii) and (iii) of Theorem 29.4.

11.30 A GENERALIZATION

In Section 11.25 we interpreted $y = Ax$ as a linear mapping of each vector x of **U** into a vector y of **V**. We then introduced the matrix norm

30.1
$$\|A\|_\alpha = \sup_{x \neq \phi} \frac{\|y\|_\alpha}{\|x\|_\alpha}$$
$$= \sup_{x \neq \phi} \frac{\|Ax\|_\alpha}{\|x\|_\alpha}$$

for an arbitrary vector norm $\|\cdot\|_\alpha$. Notice that $\|\cdot\|_\alpha$ is the vector norm for both **U** and **V**.

Suppose, on the other hand, we choose the vector norm $\|\cdot\|_r$ for **U** and the vector norm $\|\cdot\|_s$ for **V**. This leads us to a more general matrix norm defined by

30.2
$$\|A\|_{r,s} = \sup_{x \neq \phi} \frac{\|y\|_s}{\|x\|_r}$$
$$= \sup_{x \neq \phi} \frac{\|Ax\|_s}{\|x\|_r}$$

which is equivalent to

30.3
$$\|A\|_{r,s} = \sup_{\|x\|_r = 1} \|Ax\|_s .$$

Obviously, when $r = s$, we have the subordinate matrix norm as a special case.

In this book we shall be mostly concerned with the matrix norms $\|A\|_1$, $\|A\|_2$, $\|A\|_\infty$, and $\|A\|_E$, previously discussed.

EXERCISE 11.30

Show that

$$\sup_{\|x\|_r = 1} \|Ax\|_s = \sup_{x \neq \phi} \frac{\|Ax\|_s}{\|x\|_r} .$$

11.31 CRITERIA FOR THE CONVERGENCE OF SEQUENCES TO THE NULL VECTOR AND TO THE NULL MATRIX

In Section 22 and Section 23 we defined the convergence of sequences for vectors and for matrices, respectively. On the other hand, we did not establish criteria for convergence. However, now that we have discussed norms we can establish such criteria. In this section we discuss convergence to the null vector and to the null matrix, and in the next section we discuss convergence in general.

31.1 Lemma. If the sequence $\{x^{(i)}\}$ converges to the null vector, or in symbols, if $x^{(i)} \to \phi$ as $i \to \infty$, then $\|x^{(i)}\|_\alpha \to 0$ for every vector norm $\|\cdot\|_\alpha$. We should point out that this is often referred to as *convergence with respect to the α-norm or as α-norm convergence*.

Proof. We can express $x^{(i)}$ as a linear combination of the unit vectors $e^{(k)}$ as follows:

31.2
$$x^{(i)} = \begin{bmatrix} x_1^{(i)} \\ \cdot \\ \cdot \\ \cdot \\ x_k^{(i)} \\ \cdot \\ \cdot \\ \cdot \\ x_n^{(i)} \end{bmatrix} = \sum_{k=1}^{n} x_k^{(i)} e^{(k)}.$$

This allows us to write, for every vector norm,

31.3
$$\|x^{(i)}\|_\alpha = \| \sum_{k=1}^{n} x_k^{(i)} e^{(k)} \|_\alpha$$

$$\leq \sum_{k=1}^{n} |x_k^{(i)}| \cdot \|e^{(k)}\|_\alpha$$

$$\leq q \max_k |x_k^{(i)}|$$

where q is the constant (for a given norm)

31.4
$$q = \sum_{k=1}^{n} \|e^{(k)}\|_\alpha.$$

Now suppose $x^{(i)} \to \phi$ as $i \to \infty$. This means, for every k, that $x_k^{(i)} \to 0$. Consequently, $\max_{(k)} |x_k^{(i)}| \to 0$ as $i \to \infty$. Thus, from (31.3), we see that $\|x^{(i)}\|_\alpha \to 0$ as $i \to \infty$.

31.5 Theorem (*continuity*). Let $\{x^{(i)}\}$ be any sequence of vectors such that $x^{(i)} \to \phi$ as $i \to \infty$. Then, for every vector norm $\|\cdot\|_\alpha$

$$\|y + x^{(i)}\|_\alpha \to \|y\|_\alpha \qquad (i \to \infty).$$

Proof. From Theorem 24.3 we have

31.6
$$\big|\ \|y + x^{(i)}\|_\alpha - \|y\|_\alpha\ \big| \leq \|y + x^{(i)} - y\|_\alpha$$
$$= \|x^{(i)}\|_\alpha$$

which means we can write

31.7
$$- \|x^{(i)}\|_\alpha \leq \|y + x^{(i)}\|_\alpha - \|y\|_\alpha \leq \|x^{(i)}\|_\alpha.$$

Now assume $x^{(i)} \to \phi$ as $i \to \infty$. From (31.1), this implies $\|x^{(i)}\|_\alpha \to 0$ as $i \to \infty$, and so $\|y + x^{(i)}\|_\alpha - \|y\|_\alpha \to 0$. But this means

31.8
$$\|y + x^{(i)}\|_\alpha \to \|y\|_\alpha \qquad (i \to \infty)$$

and continuity is proved.

31.9 Theorem (*comparability*)*. Let $\|\cdot\|_\alpha$ be any vector norm defined on the n-dimensional vector space $\mathbf{C}^n$. Then there exist positive constants p and q, independent of x, such that

$$p \max_k |x_k| \leq \|x\|_\alpha \leq q \max_k |x_k|$$

for all x. In other words, for all x,

$$p\|x\|_\infty \leq \|x\|_\alpha \leq q\|x\|_\infty.$$

Proof. We already have q from (31.3) and (31.4). Thus it remains to find p. See, for example, Franklin [1968], p. 167.

31.10 Lemma. If $\|x^{(i)}\|_\alpha \to 0$ as $i \to \infty$ for at least one vector norm, then $x^{(i)} \to \phi$.

Proof. Assume $\|x^{(i)}\|_\alpha \to 0$ as $i \to \infty$ for some $\|\cdot\|_\alpha$. Then, from (31.9)

31.11
$$\max_k |x_k^{(i)}| \leq p^{-1}\|x^{(i)}\|_\alpha,$$

where $p > 0$ is a constant. This inequality, coupled with the hypothesis, implies

31.12
$$\max_k |x_k^{(i)}| \to 0 \qquad (i \to \infty).$$

Thus $x_k^{(i)} \to 0$ for $k = 1, 2, \ldots, n$ which means $x^{(i)} \to \phi$.

If we combine (31.10) and (31.1) we immediately have the result:

See, for example, (ii) and (iii) of (26.5).

31.13 Theorem. $x^{(i)} \to \phi$ as $i \to \infty$ if and only if $\|x^{(i)}\|_\alpha \to 0$ for every vector norm.

What this theorem says is that *componentwise convergence** *is equivalent to any vector-norm convergence.* The following theorems are proved in a manner analogous to the way we proved (31.1), (31.5), (31.9), and (31.10), respectively.

31.14 Lemma. If the sequence $\{A^{(k)}\}$ converges to the null matrix, or in symbols, if $A^{(k)} \to \phi$ as $k \to \infty$, then $\|A^{(k)}\|_\beta \to 0$ for every matrix norm $\|\cdot\|_\beta$.

We should point out that this is often referred to as *convergence with respect to the β-norm or as β-norm convergence.*

31.15 Theorem (*continuity*). Let $\{A^{(k)}\}$ be any sequence of matrices such that $A^{(k)} \to \phi$ as $k \to \infty$. Then, for every matrix norm $\|\cdot\|_\beta$,

$$\|B + A^{(k)}\|_\beta \to \|B\|_\beta \qquad (k \to \infty).$$

31.16 Theorem (*comparability*). Let $\|\cdot\|_\beta$ be any matrix norm defined over $\mathbf{C}^{nn}$. Then there exist positive constants ρ and σ independent of A, such that

$$\rho \max_{i,j} |a_{ij}| \leqq \|A\|_\beta \leqq \sigma \max_{i,j} |a_{ij}|$$

for all $A \in \mathbf{C}^{nn}$.

31.17 Lemma. If $\|A^{(k)}\|_\beta \to 0$, as $k \to \infty$, for at least one matrix norm, then $A^{(k)} \to \phi$.

Again, (31.14) and (31.17) immediately give us the result:

31.18 Theorem. $A^{(k)} \to \phi$ as $k \to \infty$ if and only if $\|A^{(k)}\|_\beta \to 0$ for every matrix norm.

What this theorem says is that *elementwise convergence†* *is equivalent to any matrix-norm convergence.*

EXERCISES 11.31

1. Complete the proof of Theorem 31.9.

2. Prove Lemma 31.14.

3. Prove Theorem 31.15.

4. Prove Theorem 31.16.

5. Prove Lemma 31.17.

11.32 CRITERIA FOR CONVERGENCE, IN GENERAL

In the previous sections we were primarily interested in the convergence of

---

* See (22.3).

† See (23.3).

sequences to the null vector and the null matrix. We now turn to convergence, in general.

32.1 Lemma. $x^{(i)} \to x$, as $i \to \infty$, if and only if $(x^{(i)} - x) \to \phi$.

Proof. From (22.3), $x^{(i)} \to x$, if and only if $x_k^{(i)} \to x_k$ for $k = 1, 2, \ldots, n$. This is true if and only if $(x_k^{(i)} - x_k) \to 0$, for $k = 1, 2, \ldots, n$. But this is true if and only if $(x^{(i)} - x) \to \phi$.

32.2 Theorem. $x^{(i)} \to x$, as $i \to \infty$, if and only if $\|x^{(i)} - x\|_\alpha \to 0$ for every vector norm $\| \cdot \|_\alpha$.

Proof. From (32.1) $x^{(i)} \to x$ if and only if $(x^{(i)} - x) \to \phi$. But, from (31.7), this is true if and only if $\|x^{(i)} - x\|_\alpha \to 0$ for every vector norm $\| \cdot \|_\alpha$.

32.3 Corollary. If $x^{(i)} \to x$, as $i \to \infty$, then $\|x^{(i)}\|_\alpha \to \|x\|_\alpha$ for every vector norm.

Proof. Assume $x^{(i)} \to x$, as $i \to \infty$. Then, from Theorem 32.2, $\|x^{(i)} - x\|_\alpha \to 0$ for every vector norm. However, from Theorem 24.3

$$\left| \|x^{(i)}\|_\alpha - \|x\|_\alpha \right| \leq \|x^{(i)} - x\|_\alpha,$$

for every vector norm, and so $| \|x^{(i)}\|_\alpha - \|x\|_\alpha| \to 0$. But this means $\|x^{(i)}\|_\alpha \to \|x\|_\alpha$.

The "converse" of this corollary, $\|x^{(i)}\|_\alpha \to \|x\|_\alpha$ implies $x^{(i)} \to x$, is not necessarily true because we can easily find examples for which $\|x^{(i)}\|_\alpha \to \|x\|_\alpha$ but $x^{(i)} \to y$, where $\|y\|_\alpha = \|x\|_\alpha$ with $x \neq y$.

In a similar manner we can prove the following results for matrices:

32.4 Lemma. $A^{(k)} \to A$, as $k \to \infty$, if and only if $(A^{(k)} - A) \to \phi$.

32.5 Theorem. $A^{(k)} \to A$, as $k \to \infty$ if and only if $\|A^{(k)} - A\|_\beta \to 0$ for every matrix norm $\| \cdot \|_\beta$.

32.6 Corollary. If $A^{(k)} \to A$, as $i \to \infty$, then $\|A^{(k)}\|_\beta \to \|A\|_\beta$ for every matrix norm.

Again, we do not necessarily have a converse to the corollary because we can find examples for which $\|A^{(k)}\|_\beta \to \|A\|_\beta$ but $A^{(k)} \to B$, where $\|A\|_\beta = \|B\|_\beta$ with $A \neq B$.

32.7 Theorem. If $\|A\|_\beta < 1$ for at least one matrix norm, then $A^m \to \phi$ as $m \to \infty$.

Proof. Assume $\|A\|_\beta < 1$ for some matrix norm $\| \cdot \|_\beta$. Then

$$\|A^m\|_\beta = \|AA^{m-1}\|_\beta$$

$$\leq \|A\|_\beta \cdot \|A^{m-1}\|_\beta$$

$$\leq \|A\|_\beta^2 \cdot \|A^{m-2}\|_\beta$$

$$\leq \cdots$$

$$\leq \|A\|_\beta^m.$$

Since $\|A\|_\beta < 1$, both $\|A\|_\beta^m$ and $\|A^m\|_\beta \to 0$ as m $\to \infty$. Thus $A^m \to \phi$, from (31.17).

32.8 Theorem. $A^m \to \phi$, as $m \to \infty$, if and only if $S(A) < 1$.

Proof. (i) Assume $S(A) < 1$. Now consider the Jordan canonical form J produced by the similarity transformation

32.9 $$J = SAS^{-1}.$$

Then

$$A^m = (S^{-1}JS)^m$$

32.10 $$= S^{-1}J^mS,$$

and so $A^m \to \phi$, as $m \to \infty$, if $J^m \to \phi$. To show that $J^m \to \phi$ we examine

$$J = \begin{bmatrix} J_1 & & & & \\ & \ddots & & & \\ & & J_i & & \\ & & & \ddots & \\ & & & & J_t \end{bmatrix} \text{ and } J^m = \begin{bmatrix} J_1^m & & & & \\ & \ddots & & & \\ & & J_i^m & & \\ & & & \ddots & \\ & & & & J_t^m \end{bmatrix}$$

where the elementary Jordan blocks $J_i\,(i = 1, 2, \ldots, t)$ have the form

32.11 $$J_i = \begin{bmatrix} \lambda_i & 1 & & & \\ & \lambda_i & 1 & & \\ & & \lambda_i & \ddots & \\ & & & \ddots & 1 \\ & & & & \lambda_i \end{bmatrix}$$

with $|\lambda_i| < 1$. Hence, if J_i is of order e and $m \geq e$,

32.12 $$J_i^m = \begin{bmatrix} \lambda_i^m & c_1\lambda_i^{m-1} & c_2\lambda_i^{m-2} & \cdots & c_{e-1}\lambda_i^{m-e+1} \\ & \lambda_i^m & c_1\lambda_i^{m-1} & \cdots & c_{e-2}\lambda_i^{m-e+2} \\ & & \lambda_i^m & \cdots & c_{e-3}\lambda_i^{m-e+3} \\ & & & \cdots & \cdots \\ & & & \cdots & c_1\lambda_i^{m-1} \\ & & & & \lambda_i^m \end{bmatrix}$$

where the constants $c_i = \dbinom{m}{i}$.

Obviously each J_i^m vanishes as $m \to \infty$. Consequently, $A^m \to \phi$.

ii) Assume $A^m \to \phi$ as $m \to \infty$. Then since

32.13 $$J^m = SA^mS^{-1},$$

$J^m \to \phi$ and so each elementary Jordan block tends to a zero submatrix. From (32.12), then, $|\lambda_i| < 1$ for all i. In other words, $S(A) < 1$.

32.14 Theorem. The infinite series

$$\sum_{m=0}^{\infty} A^m = I + A + A^2 + \cdots + A^m + \cdots$$

converges if and only if $A^m \to \phi$ as $m \to \infty$. In fact, when it does converge,

$$\sum_{m=0}^{\infty} A^m = (I - A)^{-1}.$$

Proof. (i) Assume $A^m \to \phi$ as $m \to \infty$. Then, from Theorem 32.8, $S(A) < 1$, and $\lambda = 1$ is not an eigenvalue of A. This means that $\det(I - A) \neq 0$ and so $(I - A)^{-1}$ exists. Consider the identity

32.15 $$(I - A)(I + A + A^2 + \cdots + A^m) \equiv I - A^{m+1}.$$

Hence,

32.16 $$I + A + A^2 + \cdots + A^m \equiv (I - A)^{-1} - (I - A)^{-1}A^{m+1}.$$

The second term on the right tends to ϕ and so

32.17 $$I + A + A^2 + \cdots + A^m \to (I - A)^{-1}.$$

ii) Assume that

32.18 $$S^{(m)} = I + A + A^2 + \cdots + A^m \to (I - A)^{-1}.$$

This means that, for all i and j,

$$S_{ij}^{(m)} \to (I - A)_{ij}^{-1} \qquad (m \to \infty),$$

and we have n^2 convergent infinite series of real or complex numbers. A necessary condition for convergence of such series is that the mth element in the sum tends to zero. Thus, for all i and j, we have $(A^m)_{ij} \to 0$ as $m \to \infty$. This implies $A^m \to \phi$.

32.19 Corollary. $\sum_{m=0}^{\infty} A^m = (I - A)^{-1}$ if and only if

$$S(A) < 1.$$

Proof. The result is a direct consequence of (32.8) and (32.14).

32.20 Corollary. If there exists at least one matrix norm $\| \cdot \|_\beta$ for which $\|A\|_\beta < 1$ then

$$\sum_{m=0}^{\infty} A^m = (I - A)^{-1}.$$

Proof. Assume $\|A\|_\beta < 1$ for some matrix norm $\| \cdot \|_\beta$. Then, from (32.7), $A^m \to \phi$ as $m \to \infty$. Therefore, from (32.14), the result follows.

Matrices do exist for which some of the norms are greater than 1, and yet

32.21
$$\sum_{m=0}^{\infty} A^m = (I - A)^{-1}.$$

Obviously, however, $S(A) < 1$ in each case.

32.22 EXAMPLE. The matrix

$$A = \begin{bmatrix} 0 & 2 \\ 0 & 0 \end{bmatrix}$$

is such that $\|A\|_\infty = \|A\|_2 = \|A\|_1 = \|A\|_E = 2$ and yet $A^m = 0$ for $m \geq 2$. Thus

$$(I - A)^{-1} = I + A$$

$$= \begin{bmatrix} 1 & 2 \\ 0 & 1 \end{bmatrix}.$$

In this case $S(A) = 0$.

EXERCISES 11.32

1. Prove Lemma 32.4.
2. Prove Theorem 32.5.
3. Prove Corollary 32.6.
4. Give an example which disproves the "converse" of (32.3).
5. Give an example which disproves the "converse" of (32.6).
6. Verify equation (32.12).
7. Prove the following: If $G \in \mathbf{C}^{nn}$ and $u^{(i)} \in \mathbf{C}^n$, then $u^{(i)} \to u$, as $i \to \infty$, implies $Gu^{(i)} \to Gu$.

CHAPTER 12

THE SOLUTION OF SYSTEMS OF LINEAR ALGEBRAIC EQUATIONS BY DIRECT METHODS

12.1 INTRODUCTION

In this chapter we are concerned with methods for solving the matrix equation

1.1
$$AX = B,$$

where $A \in \mathbf{C}^{mn}$ and $B \in \mathbf{C}^{mt}$. Any matrix $X \in \mathbf{C}^{nt}$ satisfying (1.1) is called a *solution* of the matrix equation.

If $A \in \mathbf{C}^{nn}$ and $B = I_n$, then $X = A^{-1}$, and so *the computation of a matrix inverse is a special case of the general problem.* We have another special case when $t = 1$, in which case we usually call the matrices (consisting of a single column) vectors* and we write

1.2a
$$Ax = b$$

to emphasize this fact. Here, $A \in \mathbf{C}^{mn}, b \in \mathbf{C}^m$, and any solution is a vector $x \in \mathbf{C}^n$.
This can be written

1.2b
$$\begin{bmatrix} a_{11} & a_{12} & \cdots & a_{1n} \\ a_{21} & a_{22} & \cdots & a_{2n} \\ \cdots\cdots\cdots\cdots\cdots\cdots\cdots \\ a_{m1} & a_{m2} & \cdots & a_{mn} \end{bmatrix} \begin{bmatrix} x_1 \\ x_2 \\ \cdot \\ \cdot \\ x_n \end{bmatrix} = \begin{bmatrix} b_1 \\ b_2 \\ \cdot \\ \cdot \\ b_m \end{bmatrix}$$

or, as *a system of linear algebraic equations,*

1.2c
$$\sum_{j=1}^{n} a_{ij}x_j = b_i \qquad (i = 1, 2, \ldots, m).$$

---

* See 11-(12.21).

If we write this in expanded form we have

1.2d
$$\begin{cases} a_{11}x_1 + a_{12}x_2 + \cdots + a_{1n}x_n = b_1 \\ a_{21}x_1 + a_{22}x_2 + \cdots + a_{2n}x_n = b_2 \\ \cdots\cdots\cdots\cdots\cdots\cdots\cdots\cdots\cdots\cdots\cdots\cdots\cdots \\ a_{m1}x_1 + a_{m2}x_2 + \cdots + a_{mn}x_n = b_m. \end{cases}$$

Observe that there are m equations and n unknowns (the n components of x).

We should point out that solving (1.1) is equivalent to solving t systems of the form (1.2) since

1.3
$$\begin{cases} Ax^{(1)} = b^{(1)} \\ Ax^{(2)} = b^{(2)} \\ \quad\vdots \\ Ax^{(t)} = b^{(t)} \end{cases}$$

can be written as (1.1) if X has the columns $x^{(1)}, x^{(2)}, \ldots, x^{(t)}$ and B has the columns $b^{(1)}, b^{(2)}, \ldots, b^{(t)}$. Consequently, we shall concentrate our discussion on (1.2), for simplicity. We keep in mind, however, the more general problem of solving (1.3) or, what is equivalent, solving (1.1).

1.4 Definition. A system of linear algebraic equations is called *consistent* if it has at least one solution. It is called *inconsistent* if it has no solution.

EXERCISE 12.1

Prove that the solution $x^{(i)}$ to the system of equations*

$$Ax^{(i)} = e^{(i)},$$

is the ith column of A^{-1}. Assume $\det A \neq 0$.

12.2 SOME EXAMPLES AND THEIR GEOMETRIC INTERPRETATION

Consider the three matrix equations, with $A \in \mathbf{R}^{22}$ and $b \in \mathbf{R}^2$,

2.1
$$\begin{bmatrix} 1 & 1 \\ 1 & -1 \end{bmatrix} \begin{bmatrix} x_1 \\ x_2 \end{bmatrix} = \begin{bmatrix} 5 \\ 1 \end{bmatrix}.$$

2.2
$$\begin{bmatrix} 1 & 1 \\ 1 & 1 \end{bmatrix} \begin{bmatrix} x_1 \\ x_2 \end{bmatrix} = \begin{bmatrix} 5 \\ 0 \end{bmatrix}$$

---

* Recall 11-(6.5).

and

2.3
$$\begin{bmatrix} 1 & 1 \\ 1 & 1 \end{bmatrix} \begin{bmatrix} x_1 \\ x_2 \end{bmatrix} = \begin{bmatrix} 5 \\ 5 \end{bmatrix}.$$

We observe that (2.1) can be written as the two linear algebraic equations in two unknowns

$$\begin{cases} x_1 + x_2 = 5 \\ x_1 - x_2 = 1 \end{cases}$$

whose graphs are plotted in Fig. 2.4.

2.4 Figure

The point* of intersection (in this example it is unique) of the two lines is the geometric representation of the solution. In other words, (2.1) is *consistent* and

2.5
$$x = \begin{bmatrix} 3 \\ 2 \end{bmatrix}$$

is the unique solution.

The matrix equation (2.2) can be written as the two linear algebraic equations

$$\begin{cases} x_1 + x_2 = 5 \\ x_1 + x_2 = 0 \end{cases}$$

---

* See Section 11.4.

which are *inconsistent* by inspection. This is demonstrated geometrically by the fact that the two straight lines in Fig. 2.4 corresponding to these equations have no point of intersection.

In (2.3) we notice that the two equations represented are identical. Hence, the straight lines corresponding to the two equations coincide (they intersect at an infinite number of points) and so (2.3) is *consistent*.

We have exhibited three examples of systems of two equations in two unknowns which have (i) a unique solution, (ii) no solution, and (iii) an infinity of solutions. We shall see below that an arbitrary system (1.2) also yields either a unique solution, no solution, or an infinity of solutions. There are no other possibilities*.

Observe that (1.2a) can be interpreted as a linear mapping of $\mathbf{R}^n$ into $\mathbf{R}^m$, where the ordered bases in the two vector spaces consist of the unit vectors. From this point of view, then, (2.1), (2.2), and (2.3) can be interpreted as linear mappings of $\mathbf{R}^2$ into $\mathbf{R}^2$. In (2.1), then,

2.6
$$\begin{bmatrix} 3 \\ 2 \end{bmatrix} \rightarrow \begin{bmatrix} 5 \\ 1 \end{bmatrix}$$

under the mapping whose matrix is

2.7
$$A = \begin{bmatrix} 1 & 1 \\ 1 & -1 \end{bmatrix}.$$

In both (2.2) and (2.3) the matrix of the mapping is

2.8
$$A = \begin{bmatrix} 1 & 1 \\ 1 & 1 \end{bmatrix}.$$

We can interpret (2.2) and (2.3) as follows. There is no vector whose image is the vector $\begin{bmatrix} 5 \\ 0 \end{bmatrix}$ but all vectors

2.9
$$x = \begin{bmatrix} x_1 \\ x_2 \end{bmatrix}$$

with the property that $x_1 + x_2 = 5$ are mapped onto $\begin{bmatrix} 5 \\ 5 \end{bmatrix}$. Clearly, there exists an infinity of such vectors.

---

* See (3.4).

Next, consider the matrix equation

2.10
$$\begin{bmatrix} 1 & -1 & -1 \\ 1 & 1 & -1 \end{bmatrix} \begin{bmatrix} x_1 \\ x_2 \\ x_3 \end{bmatrix} = \begin{bmatrix} 0 \\ 2 \end{bmatrix},$$

where $A \in \mathbf{R}^{23}$ and $b \in \mathbf{R}^2$. We seek $x \in \mathbf{R}^3$ which has as its image the vector $\begin{bmatrix} 0 \\ 2 \end{bmatrix} \in \mathbf{R}^2$.

Observe that (2.10) can be written as the pair of linear algebraic equations in three unknowns

2.11
$$\begin{cases} x_1 - x_2 - x_3 = 0 \\ x_1 + x_2 - x_3 = 2 \end{cases}$$

and these may be interpreted geometrically as two planes in $\mathbf{R}^3$ whose inter-section is a straight line. See Fig. 2.12. Since *any point* on the line of intersection of the planes represents a solution to (2.11) we say that we have a *single infinity* of solutions. (If the solutions to some system of equations correspond to the points of a plane we say that we have a *double infinity* of solutions.)

2.12 Figure

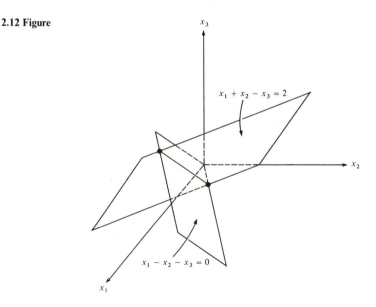

Suppose we write (2.11) in the form

2.13
$$\begin{cases} x_1 - x_2 = x_3 \\ x_1 + x_2 = x_3 + 2 \end{cases}$$

and treat x_3 as a parameter. It is easily verified that

2.14
$$\begin{cases} x_1 = x_3 + 1 \\ x_2 = 1 \end{cases}$$

constitute a solution. Since we can assign *any value* to x_3 in order to compute x_1 and x_2, we say that we have a *one-parameter family* of solutions with x_3 as the parameter. This is another way of saying that we have a single infinity of solutions.

We must point out that the choice of the parameter is restricted to x_3 and x_1 in this example. To see this, suppose we select x_2 as the parameter. This means writing (2.11) in the form

2.15
$$\begin{cases} x_1 - x_3 = x_2 \\ x_1 - x_3 = 2 - x_2 \end{cases}$$

and we cannot arbitrarily assign values to the "parameter" x_2 because, unless $x_2 = 1$ (which is the constant value obtained above) these last equations are inconsistent, by inspection.

12.3 THE EXISTENCE AND UNIQUENESS OF SOLUTIONS

The examples of the last section make it clear that we need a criterion for the existence of solutions of $Ax = b$, and when solutions exist, a criterion for uniqueness.

In order to present these criteria we need to introduce the definition below. The matrix A in (1.2a) is called the *coefficient matrix* of the system of linear algebraic equations described by (1.2a), (1.2b), (1.2c), and (1.2d). The vector b is called the *vector of constant terms* of the system. The matrix obtained by adjoining the vector of constant terms to the coefficient matrix is of special importance.

3.1 Definition. The matrix of order $m \times (n + 1)$

$$[A, b] = \begin{bmatrix} a_{11} & a_{12} & \cdots & a_{1n} & b_1 \\ a_{21} & a_{22} & \cdots & a_{2n} & b_2 \\ \hdotsfor{5} \\ a_{m1} & a_{m2} & \cdots & a_{mn} & b_m \end{bmatrix}$$

is called the *augmented matrix* of (1.2).

If we use the concept of the rank* of a matrix, along with this definition, we can describe completely the uniqueness and the existence properties of any system $Ax = b$.

3.2 Theorem. If $A \in \mathbf{C}^{mn}$ and $b \in \mathbf{C}^m$, then the system of equations $Ax = b$ is consistent if and only if rank A = rank $[A, b]$.

Proof. See, for example, Krause [1970], p. 203.

3.3 Theorem. Let $A \in \mathbf{C}^{mn}$ and $b \in \mathbf{C}^m$. If rank A = rank $[A, b] = r$ and, in addition,

  i) if $r = n$, then $Ax = b$ has a unique solution.

 ii) if $r < n$, then $Ax = b$ has an $(n - r)$-parameter family of solutions.

Proof. See, for example, Uspensky [1948], pp. 244–246.

These two theorems take care of all possibilities since the case $r > n$ cannot exist. (The verification of this fact is left as an exercise for the reader.) Thus, we have the result:

3.4 Corollary. If $A \in \mathbf{C}^{mn}$ and $b \in \mathbf{C}^m$, then $Ax = b$ has either no solution, a unique solution, or an infinity of solutions. There are no other possibilities.

If we have an infinity of solutions, then the difference between any two solutions belongs to the null space† of A.

3.5 Theorem. If x and y are solutions of (1.2) then $(x - y) \in \mathbf{K}(A)$.

Proof. The proof is left as an exercise.

3.6 Corollary. Let x be a solution of (1.2). Then y is a solution of (1.2) if and only if $y = x + w$, where $w \in \mathbf{K}(A)$.

Proof. i) Let $y = x + w$ where $Ax = b$ and $Aw = \phi$. Then

$$Ay = A(x + w)$$
$$= Ax + Aw$$
$$= b.$$

Thus, y is a solution of (1.2).

ii) Let $Ay = b$. Then, since $Ax = b, (y - x) \in \mathbf{K}(A)$. However,

$$y = x + (y - x)$$
$$= x + w$$

where $w \in \mathbf{K}(A)$, and this proves the corollary.

---

* See Definition 11-(18.12).

† See 11-(18.14).

We can also relate the uniqueness of a solution of (1.2) to the null space of A.

3.7 Corollary. A solution of (1.2) is unique if and only if $\mathbf{K}(A) = \{\phi\}$.

Proof. The proof is left as an exercise.

EXERCISES 12.3

1. Prove Theorem 3.2.

2. Prove Theorem 3.3.

3. Show that Theorem 3.2 and Theorem 3.3 take care of all possibilities, that is, verify that the case $r > n$ cannot exist.

4. Prove Theorem 3.5.

5. Prove Corollary 3.7.

12.4 THE CASE WHEN THE SOLUTION IS UNIQUE

We shall primarily be interested in solving a system $Ax = b$, with $A \in \mathbf{C}^{mn}$ and $b \in \mathbf{C}^m$, which has a unique solution, that is, a system of equations for which

4.1
$$\text{rank } A = \text{rank } [A, b] = n.$$

4.2 Theorem. Let $A \in \mathbf{C}^{mn}$. Then rank $A = r$, where r is the largest integer such that A has at least one $r \times r$ submatrix S_r with det $S_r \neq 0$.

Proof. See, for example, Franklin [1968], p. 47.

Since

4.3
$$\text{rank } A \leq \min (m, n),$$

we cannot have a unique solution when $m < n$. If $m > n$ we can only have a solution if at least $m - n$ of the equations are redundant (why?) in which case we can discard the redundant equations and reduce the system to the case where $m = n$. Consequently, let us focus our attention, during the remainder of this section, on *the case where the number of equations equals the number of unknowns,* i.e., where $m = n$.

4.4 Theorem. If $A \in \mathbf{C}^{nn}$, then rank $A = n$ if and only if det $A \neq 0$, that is, if and only if A is non-singular.

Proof. i) Suppose det $A \neq 0$, that is, suppose A is non-singular. Then, by Theorem 4.2, rank $A = n$.

ii) Suppose rank $A = n$. Then, by Theorem 4.2, there exists at least one $n \times n$ submatrix S_n with det $S_n \neq 0$. But A is the only $n \times n$ submatrix. Hence det $A \neq 0$, that is A is non-singular.

4.5 Corollary. If $A \in \mathbf{C}^{nn}$ and $b \in \mathbf{C}^n$, then $Ax = b$ has a unique solution if and only if det $A \neq 0$, that is, if and only if A is non-singular.

Proof. From (4.4), det $A \neq 0$ if and only if rank $A = n$. But from (3.3) this means $Ax = b$ has a unique solution. Hence, the solution is unique if and only if det $A \neq 0$.

The unique solution, when it exists, may be expressed formally in two rather simple forms. See (4.7) and (4.8) below.

Assume that $Ax = b$ has a unique solution so that

4.6 $$D = \det A \neq 0.$$

Then we have the formal result

4.7 Theorem. (Cramer's Rule). Let $Ax = b$, with $A \in \mathbf{C}^{nn}$ and $b \in \mathbf{C}^n$, have a *unique* solution x. If we form the determinants

$$D_j = \det \begin{bmatrix} a_{11} & a_{12} & \cdots & a_{1,j-1} & b_1 & a_{1,j+1} & \cdots & a_{1n} \\ a_{21} & a_{22} & \cdots & a_{2,j-1} & b_2 & a_{2,j+1} & \cdots & a_{2n} \\ \multicolumn{8}{c}{\dotfill} \\ a_{n1} & a_{n2} & \cdots & a_{n,j-1} & b_n & a_{n,j+1} & \cdots & a_{nn} \end{bmatrix}$$

for $j = 1, 2, \ldots, n$, where the jth column of A has been replaced by the vector b, then

$$x_1 = \frac{D_1}{D}, \ldots, x_j = \frac{D_j}{D}, \ldots, x_n = \frac{D_n}{D}$$

are the components of x.

Proof. See, for example, Uspensky [1948], pp. 231–232.

Another way to express the unique solution, when it exists, uses the matrix inverse. If we pre-multiply $Ax = b$ by A^{-1}, we obtain

4.8 $$x = A^{-1}b.$$

The reader can show that (4.7) and (4.8) are algebraically equivalent by using the formal definition of A^{-1} in 11-(14.8).

EXERCISES 12.4

1. Prove Theorem 4.2.

2. If $A \in \mathbf{C}^{mn}$ with $m > n$, prove that $Ax = b$ is consistent only if at least $m - n$ of the equations are redundant (linear combinations of the remaining equations).

3. Show that (4.7) and (4.8) are algebraically equivalent.

4. Solve the system of equations (2.1) by using Cramer's Rule.

5. Solve the system of equations (13.1) by using Cramer's Rule.

6. Solve the system of equations (17.3) by using Cramer's Rule.

12.5 HOMOGENEOUS SYSTEMS

In this section we discuss the interesting special case in which $A \in \mathbf{C}^{nn}$ but $b = \phi$. These equations are of theoretical interest and are given a special designation.

5.1 Definition. A system of linear algebraic equations of the form

$$
\begin{bmatrix}
a_{11} & a_{12} & \cdots & a_{1n} \\
a_{21} & a_{22} & \cdots & a_{2n} \\
\multicolumn{4}{c}{\dotfill} \\
a_{n1} & a_{n2} & \cdots & a_{nn}
\end{bmatrix}
\begin{bmatrix}
x_1 \\
x_2 \\
\vdots \\
x_n
\end{bmatrix}
=
\begin{bmatrix}
0 \\
0 \\
\vdots \\
0
\end{bmatrix}
$$

is called a *homogeneous* system.

When we write this in matrix language, $Ax = \phi$, and interpret it as a linear mapping, then any solution, x, turns out to be an element of the null space $\mathbf{K}(A)$. Thus, when we solve a homogeneous system of linear algebraic equations we are finding the null space of the corresponding linear mapping.*

Obviously, $x = \phi$ always solves $Ax = \phi$. This is in agreement with the fact that the null space of a linear mapping always contains the zero vector. This immediately enables us to state the following result.

5.2 Theorem. Every homogeneous system of linear algebraic equations, $Ax = \phi$, with $A \in \mathbf{C}^{nn}$, is consistent.

Of course the solution $x = \phi$ is seldom of interest and so we call it the *trivial* solution. It is the *nontrivial* solutions that are of interest (when they exist).

5.3 Lemma. If $A \in \mathbf{C}^{nn}$ and $Ax = \phi$ has a unique solution, it is necessarily the trivial solution.

Proof. If the unique solution were not the trivial solution, $x = \phi$, this would contradict the obvious fact that $x = \phi$ is always a solution of $Ax = \phi$.

This leads us to an important result concerning the existence of nontrivial solutions (nonzero elements of the null space).

5.4 Theorem. Nontrivial solutions exist for a homogeneous system $Ax = \phi$ if and only if $\det A = 0$, that is to say, if and only if A is singular.

---

* Recall Fig. 10.4 of Chapter 11.

Proof. From Lemma 5.3 a nontrivial solution cannot be unique. From Corollary 4.5, then, nontrivial solutions exist if and only if $\det A = 0$.

5.5 EXAMPLE. Let $A \in \mathbf{R}^{33}$ and $b = \phi$. Consider the equations

$$\begin{cases} x_1 + x_2 + 2x_3 = 0 \\ x_1 - x_2 - x_3 = 0 \\ 2x_1 + x_3 = 0. \end{cases}$$

The coefficient matrix is singular, that is,

$$\det \begin{bmatrix} 1 & 1 & 2 \\ 1 & -1 & -1 \\ 2 & 0 & 1 \end{bmatrix} = 0,$$

and so nontrivial solutions exist.

Since at least one 2 x 2 determinant does not vanish, for example,

$$\det \begin{bmatrix} 1 & 1 \\ 1 & -1 \end{bmatrix} \neq 0,$$

we know that

$$\text{rank } A = 2.$$

Thus, in (3.3),

$$n - r = 1,$$

which means a one-parameter family of solutions exists. One of the equations is redundant (the last equation is the sum of the first two) and may be deleted from the system. We observe that if we delete the last equation and treat x_3 as a parameter, we obtain from

$$\begin{cases} x_1 + x_2 = -2x_3 \\ x_1 - x_2 = x_3 \end{cases}$$

the one-parameter family of solutions

$$\begin{cases} x_1 = -[\tfrac{1}{2}]x_3 \\ x_2 = -[\tfrac{3}{2}]x_3. \end{cases}$$

Theorem 11-(10.7) states that a linear mapping $\mathcal{M} \in \mathcal{L}(\mathbf{V}, \mathbf{W})$, where $\dim \mathbf{V} = \dim \mathbf{W}$ (finite), is onto if and only if it is one-to-one. As a consequence of this theorem we have the result:

5.6 Theorem. If $A \in \mathbf{C}^{nn}$ and $b \in \mathbf{C}^{n}$, then $Ax = b$ has a unique solution for all $b \neq \phi$ if and only if the homogeneous system $Ax = \phi$ has no solution other than the trivial solution.

Proof. The proof is left as an exercise.

EXERCISE 12.5

Prove Theorem 5.6.

12.6 PRACTICAL METHODS FOR SOLVING LINEAR EQUATIONS

We are interested at this point in discussing practical computational schemes for solving large systems of linear algebraic equations, and we shall restrict ourselves to systems where $m = n$ and where a unique solution exists.

The solution (4.7) provided by Cramer's rule is almost impossible to obtain in practice because it involves the evaluation of $n + 1$ determinants of order n. The evaluation of each of these determinants, in general, involves $p_n n!$ multiplications, if expanded in terms of minors [Kuntz, 1957, pp. 216–217], where

6.1
$$p_n = \sum_{j=2}^{n} \frac{1}{(j - 1)!}$$

and where

6.2
$$\lim_{n \to \infty} p_n = e - 1.$$

About the same number of additions is also required. Thus, the solution (4.7) involves approximately $2p_n(n + 1)!$ arithmetic operations.

This number is excessive even with the high computational speeds of modern computers. For example, if $n = 20$, the number of arithmetic operations would be approximately 16×10^{19}. A computer, capable of performing 2 million operations per second, running continuously, would spend more than 2 million years on this problem.

Even if more sophisticated methods, such as the method of Chio [Kuntz, 1957, p. 217], are used for evaluating determinants, the fantastic reduction in the number of operations is not sufficient to make the method competitive with methods discussed in the next few sections. The number of operations is not merely important from the standpoint of computation time. It is important also from the standpoint of accuracy of computation due to the possibility of roundoff-error accumulation when the computation is carried out by machine.

The simplest of all methods, from the computational point of view, for solving a general system of linear algebraic equations is the elimination method usually attributed to Gauss. An extremely elementary treatment of this method is presented in most high school algebra courses.

In order to understand the Gaussian elimination method, consider the system of equations

6.3

$$\begin{cases} a_{11}^{(1)}x_1 + a_{12}^{(1)}x_2 + a_{13}^{(1)}x_3 + \cdots + a_{1n}^{(1)}x_n = b_1^{(1)} \\ a_{21}^{(1)}x_1 + a_{22}^{(1)}x_2 + a_{23}^{(1)}x_3 + \cdots + a_{2n}^{(1)}x_n = b_2^{(1)} \\ a_{31}^{(1)}x_1 + a_{32}^{(1)}x_2 + a_{33}^{(1)}x_3 + \cdots + a_{3n}^{(1)}x_n = b_3^{(1)} \\ \cdots\cdots\cdots\cdots\cdots\cdots\cdots\cdots\cdots\cdots\cdots\cdots\cdots\cdots \\ a_{n1}^{(1)}x_1 + a_{n2}^{(1)}x_2 + a_{n3}^{(1)}x_3 + \cdots + a_{nn}^{(1)}x_n = b_n^{(1)} \end{cases}$$

which may be written

6.3a

$$\begin{bmatrix} a_{11}^{(1)} & a_{12}^{(1)} & a_{13}^{(1)} & \cdots & a_{1n}^{(1)} \\ a_{21}^{(1)} & a_{22}^{(1)} & a_{23}^{(1)} & \cdots & a_{2n}^{(1)} \\ a_{31}^{(1)} & a_{32}^{(1)} & a_{33}^{(1)} & \cdots & a_{3n}^{(1)} \\ \cdots\cdots\cdots\cdots\cdots\cdots\cdots \\ a_{n1}^{(1)} & a_{n2}^{(1)} & a_{n3}^{(1)} & \cdots & a_{nn}^{(1)} \end{bmatrix} \begin{bmatrix} x_1 \\ x_2 \\ x_3 \\ \vdots \\ x_n \end{bmatrix} = \begin{bmatrix} b_1^{(1)} \\ b_2^{(1)} \\ b_3^{(1)} \\ \vdots \\ b_n^{(1)} \end{bmatrix}$$

or, more simply,

6.3b

$$A^{(1)}x = b^{(1)}.$$

We assume that $\det A^{(1)} \neq 0$ so that a unique solution exists. We also assume that $b^{(1)} \neq \phi$.

The first step in the procedure is to replace (6.3) by an *equivalent* system of equations (that is, a system with the same solution) which is *simpler* than (6.3) in some sense. The system

6.4

$$\begin{cases} a_{11}^{(1)}x_1 + a_{12}^{(1)}x_2 + a_{13}^{(1)}x_3 + \cdots + a_{1n}^{(1)}x_n = b_1^{(1)} \\ a_{22}^{(2)}x_2 + a_{23}^{(2)}x_3 + \cdots + a_{2n}^{(2)}x_n = b_2^{(2)} \\ a_{32}^{(2)}x_2 + a_{33}^{(2)}x_3 + \cdots + a_{3n}^{(2)}x_n = b_3^{(2)} \\ \cdots\cdots\cdots\cdots\cdots\cdots\cdots\cdots\cdots\cdots\cdots\cdots \\ a_{n2}^{(2)}x_2 + a_{n3}^{(2)}x_3 + \cdots + a_{nn}^{(2)}x_n = b_n^{(2)} \end{cases}$$

is *simpler* than (6.3) in the sense that the variable x_1 has been eliminated from all except the first equation. This may be written

6.4a

$$\left[\begin{array}{c|cccc} a_{11}^{(1)} & a_{12}^{(1)} & a_{13}^{(1)} & \cdots & a_{1n}^{(1)} \\ \hline & a_{22}^{(2)} & a_{23}^{(2)} & \cdots & a_{2n}^{(2)} \\ & a_{32}^{(2)} & a_{33}^{(2)} & \cdots & a_{3n}^{(2)} \\ \bigcirc & \multicolumn{4}{c}{\cdots\cdots\cdots\cdots\cdots} \\ & a_{n2}^{(2)} & a_{n3}^{(2)} & \cdots & a_{nn}^{(2)} \end{array}\right] \left[\begin{array}{c} x_1 \\ x_2 \\ x_3 \\ \vdots \\ x_n \end{array}\right] = \left[\begin{array}{c} b_1^{(1)} \\ b_2^{(2)} \\ b_3^{(2)} \\ \vdots \\ b_n^{(2)} \end{array}\right]$$

or, more simply,

6.4b
$$A^{(2)}x = b^{(2)}.$$

The elimination procedure which produces $A^{(2)}$ from $A^{(1)}$ is described in Section 7.

The second step in the procedure is to replace (6.4) by an equivalent system of equations which is *simpler* than (6.4). The system

6.5
$$\begin{cases} a_{11}^{(1)}x_1 + a_{12}^{(1)}x_2 + a_{13}^{(1)}x_3 + \cdots + a_{1n}^{(1)}x_n = b_1^{(1)} \\ \qquad\quad a_{22}^{(2)}x_2 + a_{23}^{(2)}x_3 + \cdots + a_{2n}^{(2)}x_n = b_2^{(2)} \\ \qquad\qquad\qquad\quad a_{33}^{(3)}x_3 + \cdots + a_{3n}^{(3)}x_n = b_3^{(3)} \\ \qquad\qquad\qquad\quad \cdots\cdots\cdots\cdots\cdots\cdots\cdots\cdots \\ \qquad\qquad\qquad\quad a_{n3}^{(3)}x_3 + \cdots + a_{nn}^{(3)}x_n = b_n^{(3)} \end{cases}$$

is *simpler* than (6.4) in the sense that the variable x_2 has been eliminated from all except the first two equations. This may be written

6.5a

$$\left[\begin{array}{cc|cccc} a_{11}^{(1)} & a_{12}^{(1)} & a_{13}^{(1)} & \cdots & a_{1n}^{(1)} \\ 0 & a_{22}^{(2)} & a_{23}^{(2)} & \cdots & a_{2n}^{(2)} \\ \hline & & a_{33}^{(3)} & \cdots & a_{3n}^{(3)} \\ & \bigcirc & \multicolumn{3}{c}{\cdots\cdots\cdots\cdots} \\ & & a_{n3}^{(3)} & \cdots & a_{nn}^{(3)} \end{array}\right] \left[\begin{array}{c} x_1 \\ x_2 \\ x_3 \\ \vdots \\ x_n \end{array}\right] = \left[\begin{array}{c} b_1^{(1)} \\ b_2^{(2)} \\ b_3^{(3)} \\ \vdots \\ b_n^{(3)} \end{array}\right]$$

or more simply,

6.5b
$$A^{(3)}x = b^{(3)}.$$

After $r - 1$ such simplifications we have

6.6
$$
\begin{cases}
a_{11}^{(1)}x_1 + a_{12}^{(1)}x_2 + \cdots + a_{1,r-1}^{(1)}x_{r-1} + a_{1r}^{(1)}x_r + \cdots + a_{1n}^{(1)}x_n = b_1^{(1)} \\
\qquad\quad a_{22}^{(2)}x_2 + \cdots + a_{2,r-1}^{(2)}x_{r-1} + a_{2r}^{(2)}x_r + \cdots + a_{2n}^{(2)}x_n = b_2^{(2)} \\
\qquad\qquad\qquad \cdots\cdots\cdots\cdots\cdots\cdots\cdots\cdots\cdots\cdots\cdots\cdots \\
\qquad\qquad\qquad\qquad a_{r-1,r-1}^{(r-1)}x_{r-1} + a_{r-1,r}^{(r-1)}x_r + \cdots + a_{r-1,n}^{(r-1)}x_n = b_{r-1}^{(r-1)} \\
\qquad\qquad\qquad\qquad\qquad\qquad\qquad\quad a_{rr}^{(r)}x_r + \cdots + a_{rn}^{(r)}x_n = b_r^{(r)} \\
\qquad\qquad\qquad\qquad\qquad\qquad\qquad\quad \cdots\cdots\cdots\cdots\cdots\cdots\cdots\cdots \\
\qquad\qquad\qquad\qquad\qquad\qquad\qquad\quad a_{nr}^{(r)}x_r + \cdots + a_{nn}^{(r)}x_n = b_n^{(r)}
\end{cases}
$$

which may be written

6.6a
$$
\begin{bmatrix}
a_{11}^{(1)} & a_{12}^{(1)} \cdots a_{1,r-1}^{(1)} & a_{1r}^{(1)} & \cdots a_{1n}^{(1)} \\
 & a_{22}^{(2)} \cdots a_{2,r-1}^{(2)} & a_{2r}^{(2)} & \cdots a_{2n}^{(2)} \\
 & \cdots\cdots\cdots & \cdots\cdots & \cdots \\
O & a_{(r-1,r-1)}^{(r-1)} & a_{r-1,r}^{(r-1)} & \cdots a_{r-1,n}^{(r-1)} \\
\hline
 & & a_{rr}^{(r)} & \cdots a_{rn}^{(r)} \\
 O & & \cdots\cdots\cdots \\
 & & a_{nr}^{(r)} & \cdots a_{nn}^{(r)}
\end{bmatrix}
\begin{bmatrix}
x_1 \\ x_2 \\ \vdots \\ x_{r-1} \\ x_r \\ \vdots \\ x_n
\end{bmatrix}
=
\begin{bmatrix}
b_1^{(1)} \\ b_2^{(2)} \\ \vdots \\ b_{r-1}^{(r-1)} \\ b_r^{(r)} \\ \vdots \\ b_n^{(r)}
\end{bmatrix}
$$

or, more simply,

6.6b
$$
A^{(r)}x = b^{(r)}.
$$

Finally, after $n - 1$ such simplifications we have the *triangular* system of equations

6.7
$$
\begin{cases}
a_{11}^{(1)}x_1 + a_{12}^{(1)}x_2 + \cdots + a_{1,n-1}^{(1)}x_{n-1} + a_{1n}^{(1)}x_n = b_1^{(1)} \\
\qquad\quad a_{22}^{(2)}x_2 + \cdots + a_{2,n-1}^{(2)}x_{n-1} + a_{2n}^{(2)}x_n = b_2^{(2)} \\
\qquad\qquad\qquad \cdots\cdots\cdots\cdots\cdots\cdots\cdots\cdots\cdots\cdots\cdots \\
\qquad\qquad\qquad\qquad a_{n-1,n-1}^{(n-1)}x_{n-1} + a_{n-1,n}^{(n-1)}x_n = b_{n-1}^{(n-1)} \\
\qquad\qquad\qquad\qquad\qquad\qquad\qquad a_{nn}^{(n)}x_n = b_n^{(n)}
\end{cases}
$$

which may be written

6.7a

$$\begin{bmatrix} a_{11}^{(1)} & a_{12}^{(1)} & \cdots & a_{1,n-1}^{(1)} & a_{1n}^{(1)} \\ & a_{22}^{(2)} & \cdots & a_{2,n-1}^{(2)} & a_{2n}^{(2)} \\ & & \cdots\cdots\cdots\cdots\cdots & & \\ \text{\Large O} & & & a_{n-1,n-1}^{(n-1)} & a_{n-1,n}^{(n-1)} \\ & & & & a_{nn}^{(n)} \end{bmatrix} \begin{bmatrix} x_1 \\ x_2 \\ \vdots \\ x_{n-1} \\ x_n \end{bmatrix} = \begin{bmatrix} b_1^{(1)} \\ b_2^{(2)} \\ \vdots \\ b_{n-1}^{(n-1)} \\ b_n^{(n)} \end{bmatrix}$$

or, more simply,

6.7b $$A^{(n)}x = b^{(n)},$$

where $A^{(n)}$ is an *upper triangular* matrix.

It should be pointed out that a (finite) sequence of systems of equations has been described, and it is indicated in (7.13) and (8.12) that these systems can be formed in such a way that each system of equations in the sequence is equivalent to the system (6.3), which means that each system in the sequence has the same solution as (6.3).

Now the triangular system (6.7) is the *simplest* system in the sequence in the sense that it is the easiest system to solve. The well-known *back-substitution method* is obviously applicable here.

The first step in the back-substitution method is to solve the last equation for x_n. The computed value of x_n is then substituted into the next-to-last equation to give a simple equation in x_{n-1}. As soon as this equation is solved for x_{n-1}, the computed values of x_n and x_{n-1} are substituted into the second-from-last equation to give a simple equation in x_{n-2}. This equation is solved for x_{n-2}, and the procedure is continued until all the unknowns have been computed.

What has been described is a plan to replace a general system of linear algebraic equations by an equivalent triangular system which is extremely simple to solve by the method of back-substitution. The procedure for finding the triangular system has not yet been spelled out, but essentially it involves computing the sequence of matrices

$$A^{(1)}, A^{(2)}, \ldots, A^{(r)}, \ldots, A^{(n)}$$

(where $A^{(n)}$ is an upper triangular matrix) and the sequence of vectors

$$b^{(1)}, b^{(2)}, \ldots, b^{(r)}, \ldots, b^{(n)},$$

by a process which is known as Gaussian elimination (see the next few sections for details).

In matrix language, then, the system $A^{(1)}x = b^{(1)}$ is solved by going to an equivalent triangular system $A^{(n)}x = b^{(n)}$ and by solving the triangular system by the back-substitution method.

It turns out that the amount of work involved in the solution of the triangular system by the back-substitution method is trivial when compared with the amount of work involved in the elimination steps which produce the triangular system.

12.7 THE ELIMINATION PROCEDURE—STEP ONE

The first equation in (6.3) can be used to eliminate x_1 from each of the remaining $n - 1$ equations in the system if $a_{11}^{(1)} \neq 0$. In this case we call the first equation the *pivotal equation* and $a_{11}^{(1)}$ the *pivotal element*, or simply the *pivot*. (What to do if the candidate for pivot vanishes is discussed in Section 12.10.)

To eliminate x_1 from the second equation in (6.3), add m_{21} times the pivotal equation to the second equation, where

7.1
$$m_{21} = - \frac{a_{21}^{(1)}}{a_{11}^{(1)}}.$$

This gives us the equation in $x_2, x_3, \ldots, x_n$

7.2
$$[a_{22}^{(1)} + m_{21}a_{12}^{(1)}]x_2 + \cdots + [a_{2n}^{(1)} + m_{21}a_{1n}^{(1)}]x_n = [b_2^{(1)} + m_{21}b_1^{(1)}],$$

which may be written

7.3
$$a_{22}^{(2)}x_2 + \cdots + a_{2n}^{(2)}x_n = b_2^{(2)}$$

if we set

7.4
$$a_{2j}^{(2)} = a_{2j}^{(1)} + m_{21}a_{1j}^{(1)}, \qquad (j \geq 2)$$

and

7.5
$$b_2^{(2)} = b_2^{(1)} + m_{21}b_1^{(1)}.$$

Notice that (7.3) is merely the second equation in (6.4).

In general, to eliminate x_1 from the ith equation in (6.3), where $1 < i \leq n$, add m_{i1} times the pivotal equation to the ith equation, where

7.6
$$m_{i1} = - \frac{a_{i1}^{(1)}}{a_{11}^{(1)}}, \qquad (i \geq 2).$$

This gives us

7.7
$$[a_{i2}^{(1)} + m_{i1}a_{12}^{(1)}]x_2 + \cdots + [a_{in}^{(1)} + m_{i1}a_{1n}^{(1)}]x_n = [b_i^{(1)} + m_{i1}b_1^{(1)}]$$

which may be written

7.8 $$a_{i2}^{(2)}x_2 + \cdots + a_{in}^{(2)}x_n = b_i^{(2)}, \quad (i \geqq 2)$$

if we set

7.9 $$a_{ij}^{(2)} = a_{ij}^{(1)} + m_{i1}a_{1j}^{(1)}, \quad (i,j \geqq 2)$$

and

7.10 $$b_i^{(2)} = b_i^{(1)} + m_{i1}b_1^{(1)}, \quad (i \geqq 2).$$

If all but the pivotal equation in (6.3) are replaced by the $n - 1$ equations (7.8), we get exactly the system (6.4).

It is easily verified by direct multiplication that if $M^{(1)}$ is the matrix

7.11 $$M^{(1)} = \begin{bmatrix} 1 & & \bigcirc \\ m_{21} & & \\ m_{31} & & I_{n-1} \\ \vdots & & \\ m_{n1} & & \end{bmatrix}$$

then the first elimination step can be carried out *formally* by multiplying (6.3a) by the matrix $M^{(1)}$. Thus

7.12 $$M^{(1)}A^{(1)}x = M^{(1)}b^{(1)}$$

becomes

$$A^{(2)}x = b^{(2)},$$

which we previously labeled (6.4b).

7.13 Theorem. *The system $A^{(1)}x = b^{(1)}$ is equivalent to the system $A^{(2)}x = b^{(2)}$.*

Proof. We have assumed that (6.3) has a unique solution and so $\det A^{(1)} \neq 0$. Now $\det M^{(1)} = 1$ and

7.14 $$A^{(2)} = M^{(1)}A^{(1)}.$$

If we use 11-(17.10), we obtain

7.15 $$\det A^{(2)} = [\det M^{(1)}][\det A^{(1)}]$$
$$= \det A^{(1)}.$$

Hence, $A^{(2)}$ is nonsingular and so (6.4) has a unique solution. But (7.12) demonstrates that the solution to (6.3) satisfies (6.4). Therefore, the unique solutions to (6.3) and (6.4) are identical.

12.8 THE rth ELIMINATION STEP

The first elimination step is typical. For instance, after $r - 1$ eliminations if $a_{rr}^{(r)} \neq 0$, then this element becomes the pivot and

8.1
$$a_{rr}^{(r)}x_r + \cdots + a_{rn}^{(r)}x_n = b_r^{(r)}$$

becomes the pivotal equation.

To eliminate x_r from any one of the $n - r$ remaining equations in (6.6), add m_{ir} times the pivotal equation to the ith equation, where $r < i \leq n$, and where

8.2
$$m_{ir} = -\frac{a_{ir}^{(r)}}{a_{rr}^{(r)}}.$$

This gives us

8.3
$$[a_{i,r+1}^{(r)} + m_{ir}a_{r,r+1}^{(r)}]x_{r+1} + \cdots + [a_{in}^{(r)} + m_{ir}a_{rn}^{(r)}]x_n = [b_i^{(r)} + m_{ir}b_r^{(r)}],$$

which may be written

8.4
$$a_{i,r+1}^{(r+1)}x_{r+1} + \cdots + a_{in}^{(r+1)}x_n = b_i^{(r+1)}, \qquad (i > r)$$

if we set

8.5
$$a_{ij}^{(r+1)} = a_{ij}^{(r)} + m_{ir}a_{rj}^{(r)} \qquad (i,j > r),$$

and

8.6
$$b_i^{(r+1)} = b_i^{(r)} + m_{ir}b_r^{(r)} \qquad (i > r).$$

If the last $n - r$ equations in (6.6) are replaced by (8.4), for $r < i \leq n$, then we get a system which can be written

8.7
$$A^{(r+1)}x = b^{(r+1)}.$$

It is easily verified by direct multiplication that if

8.8
$$M^{(r)} = \begin{bmatrix} I_{r-1} & & & & & \\ & 1 & & & & \\ \mathbf{O} & m_{r+1,r} & 1 & & & \\ & m_{r+2,r} & 0 & 1 & & \\ & \cdots\cdots\cdots\cdots\cdots\cdots & & & \\ & m_{nr} & 0 & 0 & \cdots & 1 \end{bmatrix}$$

then the rth elimination step can be carried out formally by multiplying (6.6a) by $M^{(r)}$. Thus

8.9 $$M^{(r)} A^{(r)} x = M^{(r)} b^{(r)}$$

becomes

$$A^{(r+1)} x = b^{(r+1)}$$

which we previously labeled (8.7).

With this background it is easy to see that the entire triangularization process can be described formally by multiplying (6.3b) by

8.10 $$M = M^{(n-1)} \cdots M^{(2)} M^{(1)}.$$

Thus

8.11 $$M A^{(1)} x = M b^{(1)}$$

becomes

$$A^{(n)} x = b^{(n)},$$

which we previously labeled (6.7b).

It is left to the reader to verify the equivalence of (6.3) and (6.7), as is stated in the following result.

8.12 Theorem. *The system $A^{(n)} x = b^{(n)}$ is equivalent to the system $A^{(1)} x = b^{(1)}$.*

The proof is similar to the proof of (7.13).

EXERCISES 12.8

1. Verify, by direct multiplications, that $A^{(2)} = M^{(1)} A^{(1)}$, where $M^{(1)}$ is given by (7.11).
2. Verify, by direct multiplication, that $A^{(r+1)} = M^{(r)} A^{(r)}$, where $M^{(r)}$ is given by (8.8).
3. Prove Theorem 8.12.

12.9 DETERMINANT EVALUATION

A by-product of Gaussian elimination is a method for determinant evaluation. Since the triangular matrix $A^{(n)}$ is given by

9.1 $$A^{(n)} = M A^{(1)},$$

we may write

9.2 $$\det A^{(n)} = [\det M][\det A^{(1)}].$$

From (8.10) we may write

9.3
$$\det M = \prod_{r=1}^{n-1} \det M^{(r)},$$

and since $M^{(r)}$, for $1 \leqq r \leqq n - 1$, is *unit triangular**, this means $\det M = 1$. Hence

9.4
$$\det A^{(1)} = \det A^{(n)}$$
$$= a_{11}^{(1)} a_{22}^{(2)} \cdots a_{nn}^{(n)},$$

since $A^{(n)}$ is a triangular matrix. This says that the determinant of a matrix is the product of the pivotal elements† in the Gaussian elimination process.

12.10 FAILURE OF UNMODIFIED GAUSSIAN ELIMINATION

In our preceding discussion we have ignored the possibility that, at some stage in the elimination, the candidate for pivot might be zero. We now discuss this case.

Suppose the first zero pivotal candidate is encountered at the rth stage. In (6.6) the last $(n - r + 1)$ equations are

10.1
$$\begin{cases} a_{rr}^{(r)}x_r + \cdots + a_{rn}^{(r)}x_n = b_r^{(r)} \\ \dots\dots\dots\dots\dots\dots\dots\dots\dots \\ a_{ir}^{(r)}x_r + \cdots + a_{in}^{(r)}x_n = b_i^{(r)} \\ \dots\dots\dots\dots\dots\dots\dots\dots \\ a_{nr}^{(r)}x_r + \cdots + a_{nn}^{(r)}x_n = b_n^{(r)} \end{cases}$$

and we are assuming that

10.2
$$a_{rr}^{(r)} = 0.$$

Since $a_{rr}^{(r)}$ appears in the denominator of (8.2), this means that it cannot serve as pivot, and so the first equation in (10.1) cannot serve as the pivotal equation. This is not serious, however. All we need to do is to inspect the leading coefficients of the remaining equations until we find one, for example $a_{ir}^{(r)}$, which is not zero. Then we interchange the rth and ith equations and use the new rth equation as the pivotal equation.

This certainly does not change the mathematical problem, because the formal solution of a system of linear algebraic equations is independent of the order of the equations in the system. (The order may be important from a computational point of view, however, as the previous paragraph shows.)

---

* This means a triangular matrix T with $t_{ii} = 1$ for all i.

† In this discussion we are assuming that all of the pivotal candidates are nonzero.

One might ask whether or not we can always find a nonzero $a_{ir}^{(r)}$ among the leading coefficients in (10.1). The next theorem gives us the answer in the affirmative.

10.3 Theorem. *If $A^{(1)}$ is nonsingular, then there exists a nonzero $a_{ir}^{(r)}$ among the leading coefficients in* (10.1).

Proof. We can use an argument similar to that used in section 9 to show that $\det A^{(r)} = \det A^{(1)}$. Thus $\det A^{(r)} \neq 0$. But from (6.6a) we see that

$$\det A^{(r)} = [a_{11}^{(1)} a_{22}^{(2)} \cdots a_{r-1,r-1}^{(r-1)}]\det A_{n-r+1}^{(r)}$$

where

10.4
$$A_{n-r+1}^{(r)} = \begin{bmatrix} a_{rr}^{(r)} \cdots a_{rn}^{(r)} \\ \cdots\cdots\cdots \\ a_{ir}^{(r)} \cdots a_{in}^{(r)} \\ \cdots\cdots\cdots \\ a_{nr}^{(r)} \cdots a_{nn}^{(r)} \end{bmatrix}.$$

Consequently, $\det A_{n-r+1}^{(r)} \neq 0$, which means that $A_{n-r+1}^{(r)}$ cannot have all zeros in its first column. Thus we can always find a pivot in the first column of $A_{n-r+1}^{(r)}$, and so the elimination procedure can always be continued to completion.

EXERCISE 12.10
Show that $\det A^{(r)} = \det A^{(1)}$ in the proof of Theorem 10.3.

12.11 MODIFIED GAUSSIAN ELIMINATION
This last discussion suggests that we must always consider the possibility that the pivotal candidate might be zero. Consequently, we should modify the elimination procedure so that an exchange of the rth and ith equations is provided for at each elimination step with the case of no interchange described as an exchange of the rth equation with itself. The notation $\mathscr{A}^{(r)}$ and $\beta^{(r)}$ instead of $A^{(r)}$ and $b^{(r)}$ will be used with the modified elimination procedure.

In matrix language, the exchange of the rth and ith equations corresponds to an interchange in the rth and ith rows of $\mathscr{A}^{(r)}$ and the rth and ith components of $\beta^{(r)}$. This is equivalent to multiplying

11.1
$$\mathscr{A}^{(r)}x = \beta^{(r)}$$

by the matrix $(r < i)$

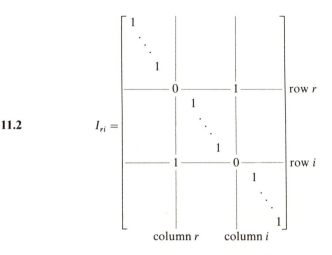

11.2

where I_{ri} can be described as the identity matrix with the rth and ith rows interchanged. Such matrices are called *permutation matrices**, and it is easily verified that

$$I_{ri}^2 = I.$$

The generic rth step of the elimination then becomes, formally, a multiplication of (11.1) by I_{ri} followed by a multiplication of the resulting equation by $M^{(r)}$. Thus, analogous to (8.9), we can write

11.3 $$[M^{(r)}I_{ri}]\mathscr{A}^{(r)}x = [M^{(r)}I_{ri}]\beta^{(r)},$$

which becomes

11.4 $$\mathscr{A}^{(r+1)}x = \beta^{(r+1)}.$$

In case no interchange of rows is called for, $i = r$ and so

11.5 $$I_{ri} = I_{rr} = I.$$

The entire triangularization process now can be described formally by multiplying

11.6 $$\mathscr{A}^{(1)}x = \beta^{(1)}$$

by

11.7 $$\mathscr{M} = [M^{(n-1)}I_{n-1,i_{n-1}}] \cdots [M^{(2)}I_{2,i_2}][M^{(1)}I_{1,i_1}].$$

---

* See 11-(17.32).

Thus,

11.8
$$\mathscr{M}\mathscr{A}^{(1)}x = \mathscr{M}\beta^{(1)}$$

becomes the triangular system

11.9
$$\mathscr{A}^{(n)}x = \beta^{(n)}.$$

EXERCISE 12.11

Verify that $I_{ri}^2 = I$ in (11.2).

12.12 THE EQUIVALENCE OF THE TRIANGULAR SYSTEM AND THE ORIGINAL SYSTEMS

We can show the equivalence of the triangular system (11.9) and the original system (11.6).

12.1 Theorem.   *The system $\mathscr{A}^{(1)}x = \beta^{(1)}$ is equivalent to the system $\mathscr{A}^{(n)}x = \beta^{(n)}$.*

Proof.   We have assumed that (11.6) has a unique solution and so $\det \mathscr{A}^{(1)} \neq 0$. Now we must show that a unique solution exists for (11.9) and that the solutions are the same.   To do this we see from (11.8) and (11.9) that

12.2
$$\mathscr{A}^{(n)} = \mathscr{M}\mathscr{A}^{(1)}$$

and so

12.3
$$\det \mathscr{A}^{(n)} = [\det \mathscr{M}][\det \mathscr{A}^{(1)}].$$

Thus, $\det \mathscr{A}^{(n)} \neq 0$ if and only if $\det \mathscr{M} \neq 0$.
     From (11.7) we see that

12.4
$$\det \mathscr{M} = \left[\prod_{r=1}^{n-1} \det M^{(r)}\right]\left[\prod_{r=1}^{n-1} \det I_{r,i_r}\right],$$

and we have already seen (Section 9) that

12.5
$$\prod_{r=1}^{n-1} \det M^{(r)} = 1.$$

All that remains is to show that

12.6
$$\prod_{r=1}^{n-1} \det I_{r,i_r} \neq 0.$$

That this is true follows from the fact that, for $r = 1, 2, \ldots, n - 1$, whenever $r \neq i_r$,

12.7 $$\det I_{r, i_r} = -1,$$

because I_{r, i_r} can be obtained from I by interchanging row r with row i_r. (If $r = i_r$, the determinant has the value 1.)

Thus, (12.4), (12.5), and (12.6) imply that

12.8 $$\det \mathcal{M} \neq 0,$$

and this result along with (12.3) implies that

12.9 $$\det \mathcal{A}^{(n)} \neq 0,$$

which means (11.9) has a unique solution.

The solution to (11.6)

12.10 $$x = [\mathcal{A}^{(1)}]^{-1} \beta^{(1)}$$

satisfies (11.9), or what is more convenient, (11.8) for, by direct substitution,

12.11 $$\mathcal{M} \mathcal{A}^{(1)} \{ [\mathcal{A}^{(1)}]^{-1} \beta^{(1)} \} = \mathcal{M} \beta^{(1)}.$$

This means (11.9) and (11.6) have the same solution and therefore they are equivalent.

12.13 NUMERICAL INSTABILITY

When an algorithm, such as the Gaussian elimination procedure, is applied in practice, there are considerations over and above those already mentioned which must be taken into account*. These additional considerations are necessary due to the fact that we are forced to work with an automatic digital computer. This means, then, that our mathematical universe is a finite set of machine representable numbers, for example, the set mentioned in Section 2.6. Notice that many "simple" numbers such as $\frac{1}{3}$ are excluded from the set of representable numbers just mentioned because they do not have a finite binary representation.

We have already mentioned earlier that this fact, the fact that arithmetic performed by computers is inexact†, and the fact that the representable numbers do not possess the properties of a field due to the rounding properties of computers make this entire discussion interesting. If we could represent all real numbers exactly and if we could do infinite-precision arithmetic, then the linear equation problem could be solved by the modified Gaussian elimination algorithm

---

* See 1-(3.5).

† Recall Section 2.7.

of Section 12.11, followed by back substitution, and we could forget about the problem from this point on.

However, we must consider the effects of the limitations mentioned above. In the mathematical description of Gaussian elimination we were concerned merely with the question of whether or not the pivotal candidate vanished at each step. In practice, it is just as important to ask whether or not the pivotal candidate, $a_{rr}^{(r)}$, is "small" compared with some of the elements $a_{ir}^{(r)}$, $(i > r)$.

If $a_{rr}^{(r)}$ is small, then, from (8.2), we see that the corresponding multipliers m_{ir} will be large, and the equations produced by the elimination step will have coefficients that are much larger than the coefficients from the previous step. Wilkinson [1965, p. 205] points out that this is likely to cause the process to become *numerically unstable**. His example [1965, p. 216] might be appropriate at this point. Consider the system of three equations in three unknowns,

13.1
$$\begin{bmatrix} 0.000003 & 0.213472 & 0.332147 \\ 0.215512 & 0.375623 & 0.476625 \\ 0.173257 & 0.663257 & 0.625675 \end{bmatrix} \begin{bmatrix} x_1 \\ x_2 \\ x_3 \end{bmatrix} = \begin{bmatrix} 0.235262 \\ 0.127653 \\ 0.285321 \end{bmatrix}$$

where we assume that we are working with a 6-decimal digit machine.

If 0.000003 is the first pivot, then the multipliers for the first elimination step are

13.2
$$\begin{cases} m_{21} \doteq -71837.3 \\ m_{31} \doteq -57752.3, \end{cases}$$

and the reduced system of equations becomes

13.3
$$\begin{bmatrix} 0.000003 & 0.213472 & 0.332147 \\ 0 & -15334.9 & -23860.0 \\ 0 & -12327.8 & -19181.7 \end{bmatrix} \begin{bmatrix} x_1 \\ x_2 \\ x_3 \end{bmatrix} \doteq \begin{bmatrix} 0.235262 \\ -16900.5 \\ -13586.6 \end{bmatrix}.$$

For the second elimination step, if -15334.9 is the pivot, the multiplier is

13.4
$$m_{32} \doteq -0.803905,$$

and the triangular system of equations becomes

13.5
$$\begin{bmatrix} 0.000003 & 0.213472 & 0.332147 \\ 0 & -15334.9 & -23860.0 \\ 0 & 0 & -0.500000 \end{bmatrix} \begin{bmatrix} x_1 \\ x_2 \\ x_3 \end{bmatrix} \doteq \begin{bmatrix} 0.235262 \\ -16900.5 \\ -0.20000 \end{bmatrix}.$$

---

* Recall Section 1.4.

Using back substitution, the solution is found to be

13.6
$$\begin{cases} x_3 \doteq \quad 0.400000 \\ x_2 \doteq \quad 0.479723 \\ x_1 \doteq \ -1.33333. \end{cases}$$

To see how poor these answers are, compare (13.6) with (13.14).

These answers can be improved considerably if we resort to *row interchanges* whenever the pivotal candidates are smaller than other coefficients in the same column. In fact, suppose we search the first column of the coefficient matrix in (13.1) and select as pivot *the coefficient with largest magnitude*, 0.215512. If we interchange the first two equations, then we begin with the system

13.7
$$\begin{bmatrix} 0.215512 & 0.375623 & 0.476625 \\ 0.000003 & 0.213472 & 0.332147 \\ 0.173257 & 0.663257 & 0.625675 \end{bmatrix} \begin{bmatrix} x_1 \\ x_2 \\ x_3 \end{bmatrix} = \begin{bmatrix} 0.127653 \\ 0.235262 \\ 0.285321 \end{bmatrix}.$$

This time we obtain multipliers which are less than 1 in magnitude. To be specific, we obtain

13.8
$$\begin{cases} \hat{m}_{21} \doteq \ -0.000014 \\ \hat{m}_{31} \doteq \ -0.803932, \end{cases}$$

and the reduced system of equations becomes

13.9
$$\begin{bmatrix} 0.215512 & 0.375623 & 0.476625 \\ 0 & 0.213467 & 0.332140 \\ 0 & 0.361282 & 0.242501 \end{bmatrix} \begin{bmatrix} x_1 \\ x_2 \\ x_3 \end{bmatrix} \doteq \begin{bmatrix} 0.127653 \\ 0.235260 \\ 0.182697 \end{bmatrix}.$$

Now if the coefficient in the second column which has the largest magnitude (ignoring the first equation) is chosen as the pivot for the second step, we must interchange the last two equations. Thus we begin the second elimination step with the system

13.10
$$\begin{bmatrix} 0.215512 & 0.375623 & 0.476625 \\ 0 & 0.361282 & 0.242501 \\ 0 & 0.213467 & 0.332140 \end{bmatrix} \begin{bmatrix} x_1 \\ x_2 \\ x_3 \end{bmatrix} \doteq \begin{bmatrix} 0.127653 \\ 0.182697 \\ 0.235260 \end{bmatrix}.$$

Again, the multiplier is less than 1 in magnitude,

13.11
$$\hat{m}_{32} \doteq \ -0.590860,$$

and the triangular system of equations becomes

13.12
$$
\begin{bmatrix}
0.215512 & 0.375623 & 0.476625 \\
0 & 0.361282 & 0.242501 \\
0 & 0 & 0.188856
\end{bmatrix}
\begin{bmatrix}
x_1 \\ x_2 \\ x_3
\end{bmatrix}
\doteq
\begin{bmatrix}
0.127653 \\ 0.182697 \\ 0.127312
\end{bmatrix}.
$$

Using back substitution, the solution is found to be

13.13
$$
\begin{cases}
x_3 \doteq & 0.674122 \\
x_2 \doteq & 0.0532050 \\
x_1 \doteq & -0.991291.
\end{cases}
$$

Since the correct answers, to 10 figures, are

13.14
$$
\begin{cases}
x_3 \doteq & 0.6741214694 \\
x_2 \doteq & 0.05320393391 \\
x_1 \doteq & -0.9912894252
\end{cases}
$$

this shows the remarkable improvement brought about by the change in strategy in selecting the pivots.

If the strategy used in selecting pivots is to select the element with largest magnitude in the first column of the matrix $A^{(r)}_{n-r+1}$ in (10.4), then we describe this as *partial pivoting for size* or simply *partial pivoting*. If, on the other hand, the pivot is selected as the element with largest magnitude in the entire matrix $A^{(r)}_{n-r+1}$, then the strategy is called *complete pivoting for size* or simply *complete pivoting*. Whereas partial pivoting merely involves row interchanges, complete pivoting is more complicated in that column interchanges are also involved. The complications involved in column interchanges, plus the time required to search the entire matrix $A^{(r)}_{n-r+1}$ for the element with largest magnitude, cause Wilkinson [1967, p. 81] to state that ". . . it is doubtful whether the use of complete pivoting is justified. . .". Certainly, it makes sense to use partial pivoting most of the time and to resort to complete pivoting only in those rare cases where partial pivoting proves to be inadequate.

Businger [1971] proposes a scheme for monitoring a computation under partial pivoting in order to detect whether or not it is necessary to shift to complete pivoting. His algorithm does not require re-starting the computation but allows the shift in pivotal strategy to take place whenever it becomes evident that partial pivoting is going to prove inadequate.

In this book we discuss the details of partial pivoting but not complete pivoting. This is motivated by Wilkinson's remark (quoted above) and by the fact that inaccurate solutions* obtained by partial pivoting can be refined by the *method of iterative improvement* (see Section 12.21).

EXERCISE 12.13

Write a computer program to solve (1.1) or, equivalently, (1.3) using Gaussian elimination with partial pivoting. Include a provision for inverting a nonsingular matrix, as a special case. Make your program a subroutine.

12.14 THE CAUSE OF THE DIFFICULTY

It is not difficult to demonstrate what causes the fantastic loss in accuracy when the multipliers are allowed to become excessively large relative to the elements in the coefficient matrix. In the example above we see from (13.2) that, when 0.000003 is selected as pivot, $m_{21} = -71837.3$.

In order to compute $a_{21}^{(2)}$, $a_{22}^{(2)}$, and $a_{23}^{(2)}$ we find, using floating-point arithmetic with 6 significant digits, that

$$a_{21}^{(2)} = 0.215512 - (71837.3)(0.000003)$$

$$= 0.215512 - 0.2155119$$

$$\doteq 0.215512 - 0.215512 \quad \cdots\cdots\cdots (6\ \text{Digits})$$

$$\doteq 0.000000$$

$$a_{22}^{(2)} = 0.375623 - (71837.3)(0.213472)$$

$$= 0.375623 - 15335.2521056$$

$$\doteq 0.375623 - 15335.3 \quad \cdots\cdots\cdots (6\ \text{Digits})$$

$$\doteq -15334.9$$

$$a_{23}^{(2)} = 0.476625 - (71837.3)(0.332147)$$

$$= 0.476625 - 23860.5436731$$

$$\doteq 0.476625 - 23860.5 \quad \cdots\cdots\cdots (6\ \text{Digits})$$

$$\doteq -23860.0$$

Notice that even with floating-point arithmetic we must line up decimal points in order to add. Therefore, the coefficients $a_{22}^{(1)} = 0.375623$ and $a_{23}^{(1)} = 0.476625$ in the second row of (13.1) could have been *any numbers* satisfying the inequalities

14.1
$$\begin{cases} 0.350000 < a_{22}^{(1)} < 0.450000 \\ 0.450000 < a_{23}^{(1)} < 0.550000 \end{cases}$$

---

* Here we are assuming that each component of the solution vector contains at least one significant digit of accuracy.

and we would have obtained the same computed values for $a_{22}^{(2)}$ and $a_{23}^{(2)}$.

Therefore, if we use Gaussian elimination without partial pivoting, we find ourselves unable to take full advantage of the six significant digits available in $a_{22}^{(1)}$ and $a_{23}^{(1)}$ (as far as this example is concerned).

14.2 *Remark.* Here is a good example of what happens when an unstable computational algorithm* is applied to a well-conditioned system of linear algebraic equations†.

It is true that Gaussian elimination with partial pivoting is a more stable algorithm than Gaussian elimination without partial pivoting, at least for a class of problems which includes Wilkinson's example (above). On the other hand, there exists a class of problems for which it can be shown [Wilkinson, 1965, p. 220] that partial pivoting is not at all necessary for numerical stability. Problems in this class include those for which the matrix A, in (1.2a), is real, symmetric, and positive definite‡.

EXERCISE 12.14

In remark 14.2 it is stated that partial pivoting is not necessary for numerical stability if the coefficient matrix is real, symmetric, and positive definite. Prove this assertion.

12.15 ILL-CONDITIONED SYSTEMS

The difficulty resulting from the use of large multipliers in Gaussian elimination is something which can be controlled by choosing a proper pivotal strategy, as the example in Section 12.13 demonstrates. As we mentioned in Chapter 1, there is also the possibility of difficulty if the system of equations

15.1 $Ax = b$

is ill-conditioned§. This means that the solution is very sensitive to small perturbations in the elements of the augmented matrix

15.2 $B = [A, b]$.

If a system of linear algebraic equations is ill-conditioned, then a progressive loss of significant digits during the computations associated with the elimination process and a resulting loss of accuracy are more or less inevitable. If we take the point of view that the *computed result is the exact solution of a slightly perturbed problem*¶, then we can replace our concern over rounding errors introduced

---

* Recall Section 1.4.

† We discuss the *condition* of a system of linear algebraic equations in the next three sections.

‡ See 11-(20.3).

§ See 1-(4.7).

¶ See 1-(4.10).

during the computation by a concern over the effect of introducing perturbations in the original data. Consequently, we need to analyze the effect on the solution when we introduce small perturbations in the data.

Suppose the nonsingular matrix A is perturbed by E and the vector $b \neq \phi$ is perturbed by k. These perturbations produce a perturbation h in the solution x. The perturbed system can be written

15.3 $$(A + E)(x + h) = b + k,$$

and it is possible to derive the following bound on the overall relative change in the solution.

15.4 Theorem. If $\|E\|_\beta \|A^{-1}\|_\beta < 1$ and $\|I\|_\beta = 1$, then for any vector norm $\|\cdot\|_\alpha$ consistent with $\|\cdot\|_\beta$

$$\frac{\|h\|_\alpha}{\|x\|_\alpha} \leq K(\beta)\|A\|_\beta\|A^{-1}\|_\beta \left[\frac{\|k\|_\alpha}{\|b\|_\alpha} + \frac{\|E\|_\beta}{\|A\|_\beta}\right]$$

where

$$K(\beta) = [1 - \|E\|_\beta\|A^{-1}\|_\beta]^{-1}.$$

Proof (Noble [1969], p. 434). From (15.3) we obtain

$$(A + E)h = k - Ex.$$

In order to solve for h we must show that $(A + E) = A(I + A^{-1}E)$ is non-singular. Since $\|E\|_\beta\|A^{-1}\|_\beta < 1$ we know that

$$S(A^{-1}E) \leq \|A^{-1}E\|_\beta < 1,$$

and so $I + A^{-1}E$ is non-singular*, which implies $A + E$ is non-singular. Thus

$$h = (I + A^{-1}E)^{-1}A^{-1}(k - Ex)$$

and, for any pair of consistent vector and matrix norms,

$$\|h\|_\alpha = \|(I + A^{-1}E)^{-1}A^{-1}(k - Ex)\|_\alpha$$
$$\leq \|(I + A^{-1}E)^{-1}\|_\beta\|A^{-1}\|_\beta\|k - Ex\|_\alpha.$$

As long as $\|I\|_\beta = 1$ (this is not true for $\|I\|_E$, for example) and $\|A^{-1}E\|_\beta < 1$ we can write*

$$\|(I + A^{-1}E)^{-1}\|_\beta \leq \frac{1}{1 - \|A^{-1}E\|_\beta}.$$

---

* The verification of this step is left as an exercise for the reader.

Hence,

$$\|h\|_\alpha \leq \frac{\|A^{-1}\|_\beta}{1 - \|A^{-1}\|_\beta\|E\|_\beta}[\|k\|_\alpha + \|E\|_\beta\|x\|_\alpha]$$

and so

$$\frac{\|h\|_\alpha}{\|x\|_\alpha} \leq \frac{\|A\|_\beta\|A^{-1}\|_\beta}{1 - \|A^{-1}\|_\beta\|E\|_\beta}\left[\frac{\|k\|_\alpha}{\|A\|_\beta\|x\|_\alpha} + \frac{\|E\|_\beta}{\|A\|_\beta}\right]$$

$$\leq K(\beta)\|A\|_\beta\|A^{-1}\|_\beta\left[\frac{\|k\|_\alpha}{\|b\|_\alpha} + \frac{\|E\|_\beta}{\|A\|_\beta}\right].$$

This completes the proof.

The following corollaries describe two (obvious) special cases of this result.

15.5 Corollary. If $E = \phi$ in (15.4), then

$$\frac{\|h\|_\alpha}{\|x\|_\alpha} \leq \|A\|_\beta\|A^{-1}\|_\beta\frac{\|k\|_\alpha}{\|b\|_\alpha}.$$

15.6 Corollary. If $k = \phi$ in (15.4), then

$$\frac{\|h\|_\alpha}{\|x\|_\alpha} \leq K(\beta)\|A\|_\beta\|A^{-1}\|_\beta\frac{\|E\|_\beta}{\|A\|_\beta}.$$

We observe that the overall relative change in x is bounded by $K(\beta)\|A\|_\beta\|A^{-1}\|_\beta$ multiplied by

i) the overall relative change in A, if b is not perturbed.

ii) the overall relative change in b, if A is not perturbed.

iii) the overall relative change in A plus the overall relative change in b if both perturbations are present.

Of course $K(\beta) = 1$ in (15.5) and, if E is sufficiently small, $K(\beta)$ is close to 1 in (15.4) and (15.6). This shows us the importance of the constant

15.7 $$\kappa_\beta(A) = \|A\|_\beta\|A^{-1}\|_\beta.$$

If $\kappa_\beta(A)$ is small, then a small overall relative change in A (introduced by the perturbation E) or a small overall relative change in b (introduced by the perturbation k) or small overall relative changes in both A and b are guaranteed to produce small overall relative changes in x.

Forsythe and Moler [1967, p. 21] show that equality can be achieved in (15.5) and so *no sharper bound can be given* for arbitrary vectors b and k. On the other hand, it should be pointed out that (15.4), (15.5), and (15.6) give us *upper bounds* on

$\|h\|_\alpha/\|x\|_\alpha$ and these bounds are not always sharp. Consequently, a large value of $\kappa_\beta(A)$ tells us nothing unless we know that the bound is sharp.

EXERCISES 12.15

1. Verify that $I + A^{-1}E$ is nonsingular in the proof of Theorem 15.4.
2. Verify that

$$\|(I + A^{-1}E)^{-1}\|_\beta \leq [1 - \|A^{-1}E\|_\beta]^{-1}$$

in the proof of Theorem 15.4.

12.16 CONDITION NUMBERS

The results of the previous section provide the motivation for the following concept of a condition number for the problem under discussion.

16.1 Definition. If A is nonsingular, then

$$\kappa_\beta(A) = \|A\|_\beta \|A^{-1}\|_\beta$$

is a *condition number* for the linear equations problem

$$Ax = b,$$

where $\|\cdot\|_\beta$ is any matrix norm. If $\|\cdot\|_\beta$ is the spectral norm $\|\cdot\|_2$ then

16.2
$$\kappa_2(A) = \|A\|_2 \|A^{-1}\|_2$$
$$= \sqrt{\frac{\mu_1}{\mu_n}},$$

the ratio of the largest and smallest singular values* of A.

This is sometimes called the *spectral condition number* for the problem† and is one of the most widely used.

Observe that if $\kappa_\beta(A)$ is small, then small perturbations in either A or b can introduce only small perturbations in x and the problem is well-conditioned. On the other hand, if $\kappa_\beta(A)$ is large, then small perturbations in either A or b *might* introduce large perturbations in x. *If this happens* the problem is ill-conditioned. Some authors prefer to talk about the matrix A rather than the linear equations problem, so they use the terminology "A is well-conditioned (or ill-conditioned) with respect to the linear equations problem." Using this terminology, then, we

---

* See 11-(27.21).

† Compare the spectral condition number of A relative to the linear equations problem with the spectral condition number of A relative to the eigenvalue problem. See 14-(5.28).

can say that $\kappa_2(A)$ is the spectral condition number of A (with respect to the linear equations problem). Obviously, from (16.2),

16.3 $$\kappa_2(A) \geqq 1.$$

As an item of interest, it will be pointed out later* that the spectral condition number of a matrix A with respect to its eigenvalue problem is merely the spectral condition number of its matrix of eigenvectors with respect to the linear equations problem.

16.4 Theorem. If A is real and symmetric, then

$$\kappa_2(A) = \left| \frac{\lambda_1}{\lambda_n} \right|,$$

where λ_1 and λ_n are the eigenvalues of A of largest and smallest modulus, respectively.

The proof of this theorem is left to the reader.

EXERCISE 12.16
Prove Theorem 16.4.

12.17 A MISCONCEPTION ABOUT AN ILL-CONDITIONED SYSTEM
It must be mentioned that there is a popular feeling that A is ill-conditioned if and only if det A is small. However, unless A is normalized the situation is not that simple (see, e.g., Forsythe and Moler [1967], p. 22). Examples can be constructed for which det A is extremely small and yet

17.1 $$\frac{\|h\|_2}{\|x\|_2} = \frac{\|k\|_2}{\|b\|_2}.$$

On the other hand, if $\|A\|_2 = 1$ (we can always multiply $Ax = b$ by an appropriate scale factor to achieve this) then

17.2 $$\kappa_2(A) = \|A^{-1}\|_2$$

$$= \frac{1}{\sqrt{\mu_n}}$$

$$= \frac{1}{|\lambda_n|} \qquad \text{(if } A \text{ is symmetric).}$$

---

* See 14-(5.28).

Therefore, a normalized matrix has a large spectral condition number if and only if $\|A^{-1}\|_2$ is large, that is, if and only if μ_n is small (λ_n in the case of a symmetric matrix).

If we hold μ_1 and n fixed, then det $A \to 0$ implies that $\mu_n \to 0$, and this means, of course, that $\kappa_2(A) \to \infty$. Hence, if A is normalized so as to keep μ_1 fixed, then the smallness of det A is related to the ill-condition of A, but if A is not normalized, the situation is not clear. Consequently, we should emphasize that the *largeness of $\kappa_2(A)$ is more important than the smallness of det A as a basis for determining when $Ax = b$ is ill-conditioned.*

Another point to emphasize is that $\kappa_2(A)$ can be very large even if n is small.

17.3 EXAMPLE.   (Forsythe and Moler [1967], p. 24).

$$\begin{bmatrix} 1.00 & 0.99 \\ 0.99 & 0.98 \end{bmatrix} \begin{bmatrix} x_1 \\ x_2 \end{bmatrix} = \begin{bmatrix} 1.99 \\ 1.97 \end{bmatrix}.$$

In this example the eigenvalues of A are

$$\lambda_1 \doteq 1.98005$$

$$\lambda_2 \doteq -0.00005$$

and so

$$\kappa_2(A) \doteq 39,601.$$

This problem is badly ill-conditioned since $\kappa_2(A)$ is quite large. Geometrically this is borne out by the fact that the two straight lines whose intersection we seek almost coincide. Obviously $\kappa_2(A)$ is a more important criterion than the size of n for determining the condition of a system of linear algebraic equations.

EXERCISE 12.17

Construct an example for which det A is extremely small in $A(x + h) = b + k$ and yet $\dfrac{\|h\|_2}{\|x\|_2} = \dfrac{\|k\|_2}{\|b\|_2}$. See (15.5).

12.18 SCALING

In the previous section it was mentioned that we can always multiply $Ax = b$ by an appropriate scale factor so as to have $\|A\|_2 = 1$. However, having $\|A\|_2 = 1$ is not really our objective, in general.

Wilkinson's approach [1961, p. 284] to scaling is to insist that A be *equilibrated* (see definition below) before any attempt is made to solve the system $Ax = b$. He points out [1965, p. 194] that "since we almost invariably decrease the condition number of a matrix by equilibration the case for it would appear very strong."

18.1 Definition. A matrix A for which every row and every column has a length of order unity will be called an equilibrated matrix.

The term is not used too precisely, and, in practice, an equilibrated matrix $A = (a_{ij})$ is one whose rows and columns have been scaled so that $|a_{ij}| \leq 1$ for all i and j and each row and each column has at least one element greater than or equal to $1/\beta$, where β is the base for floating-point computation on the digital computer used. Notice that scaling by integer powers of β minimizes the rounding errors, and this is extremely important if the problem is ill-conditioned.

Unfortunately, the equilibrated form of a matrix is not unique. Forsythe and Moler [1967, p. 45] give an example due to Hamming.

18.2
$$A = \begin{bmatrix} 1 & 1 & 2(10^9) \\ 2 & -1 & 10^9 \\ 1 & 2 & 0 \end{bmatrix}.$$

If the *columns* are scaled by powers of 10, we obtain the equilibrated matrix

18.3
$$A' = \begin{bmatrix} 0.1 & 0.1 & 0.2 \\ 0.2 & -0.1 & 0.1 \\ 0.1 & 0.2 & 0.0 \end{bmatrix}$$

(and no row scaling is necessary). If the *rows* are scaled first, however, we get a different equilibrated matrix

18.4
$$A'' = \begin{bmatrix} 10^{-10} & 10^{-10} & 0.2 \\ 2(10^{-9}) & -10^{-9} & 1.0 \\ 0.1 & 0.2 & 0.0 \end{bmatrix}$$

(and no column scaling is necessary). Notice that A' and A'' are quite different, and the systems of equations corresponding to these two matrices will have different condition numbers and a different choice of pivots for Gaussian elimination.

All of this suggests that we examine the problem of scaling more carefully. We recall that we can scale the rows and columns of a matrix A by pre- and post-multiplying A by diagonal matrices. To see how this is related to solving the system of linear equations $Ax = b$, let D_1 and D_2 be non-singular diagonal matrices and substitute

18.5
$$b = D_1 b'$$

and

18.6
$$x = D_2 x'$$

for b and x, respectively. This gives us

18.7 $$A(D_2 x') = D_1 b',$$

which may be written

18.8 $$(D_1^{-1} A D_2) x' = b',$$

or finally,

18.9 $$A' x' = b',$$

where

18.10 $$A' = D_1^{-1} A D_2.$$

We may consider (18.9) the *scaled equivalent* of $Ax = b$ since we may obtain x from x' through (18.6).

The most logical goal to achieve, in any scaling procedure, would seem to be to improve the condition of the system of equations. Thus, it seems appropriate to ask the following questions (Forsythe and Moler [1967], pp. 42–43).

18.11 Given a matrix A, what choice of nonsingular diagonal matrices D_1 and D_2 will cause $\kappa_\beta(D_1^{-1} A D_2)$ to be minimum?

18.12 If minimizing diagonal matrices D_1 and D_2 exist for (18.11), can they (or good approximations) be determined with a sufficiently fast computer algorithm?

Due to the fact that $\kappa_\beta(M) = \|M\|_\beta \|M^{-1}\|_\beta$, any answer to the first question must be in terms of the particular norm $\|\cdot\|_\beta$ used. It turns out that an answer has been found for certain matrix norms but not for all matrix norms. For example, Bauer [1963] gives a solution for the maximum norm $\|\cdot\|_\infty$. However, before we can state his solution we must introduce certain concepts.

18.13 Definition. A matrix $A \in \mathbf{C}^{nn}$ is called *partly decomposable* if and only if there exist permutation* matrices P and Q such that

$$PAQ = \begin{bmatrix} M & R \\ \phi & N \end{bmatrix},$$

where M and N are square and ϕ is null.

18.14 Definition. If $Q = P^T$ in the previous definition, then A is called *decomposable*.

---

* See Definition 11-17.32.

18.15 Definition.   If A is not decomposable, it is called *indecomposable.*

Since every decomposable matrix is partly decomposable and since some partly decomposable matrices (those for which $Q \neq P^T$) are indecomposable, we need one additional term.

18.16 Definition.   If A is not partly decomposable, it is called *fully indecomposable.*

Thus, indecomposable matrices are either partly decomposable or fully indecomposable. This terminology is used by Marcus and Minc [1964, pp. 122–123]. Other authors use the terms reducible and irreducible but not in a consistent manner. For example, Forsythe and Moler [1967, p. 15] call *partly decomposable* matrices reducible matrices. On the other hand, Forsythe and Wasow [1960, p. 208], Householder [1964, p. 48], and Varga [1962, p. 18] use the term reducible to mean *decomposable* rather than partly decomposable. Our Definition 16-2.4 equates irreducibility with indecomposability.

18.17 *Remark.*   The concept of decomposability is extremely useful. For example, if A is decomposable, the system $Ax = b$ can be transformed into the partitioned system

$$\begin{bmatrix} M & R \\ \phi & N \end{bmatrix} \begin{bmatrix} x' \\ x'' \end{bmatrix} = \begin{bmatrix} b' \\ b'' \end{bmatrix}$$

by an appropriate interchange of rows (equations) and columns (unknowns). This enables us to write the pair of equations

$$\begin{cases} Mx' = b' - Rx'' \\ Nx'' = b''. \end{cases}$$

Consequently, a large system can be decomposed into two smaller systems.

18.18 Definition.   If $A \in \mathbf{R}^{nn}$ and if $a_{ij} \geq 0$, for all i and j, we say that $A = (a_{ij})$ is nonnegative and we write $A \geq 0$. When these are strict inequalities, we use the term positive rather than nonnegative.

18.19 Definition.   If $A = (a_{ij})$, then $|A|$ denotes the nonnegative matrix with elements $|a_{ij}|$.

18.20 Theorem (*Perron-Frobenius*).   If $A \geq 0$ is indecomposable then

i) A has a positive real eigenvalue ρ (the Perron root of A) which is a simple eigenvalue of A.

ii) $|\lambda_i| \leq \rho$ for every eigenvalue λ_i of A.

iii) there is a positive eigenvector $x > 0$ (the Perron vector of A) associated with ρ, and x is unique (to within a constant factor).

Proof. See Varga [1962], pp. 30–31.

18.21 Theorem. If $A \in \mathbf{C}^{nn}$ is nonsingular and fully indecomposable, then $|A| \cdot |A^{-1}|$ is indecomposable.

Proof. See Businger [1968], p. 348.

18.22 Theorem. Under the same hypotheses, $|A^{-1}| \cdot |A|$ is indecomposable.

Proof. The proof is left as an exercise.

Now we can return to Bauer's solution to (18.11) for the maximum norm $\| \cdot \|_{\infty}$. Let A be nonsingular and fully indecomposable and form the two non-negative matrices

18.23
$$\begin{cases} U = |A| \cdot |A^{-1}| \\ V = |A^{-1}| \cdot |A|. \end{cases}$$

From (18.21) and (18.22) both U and V are indecomposable, and so we can apply the Perron-Frobenius Theorem, along with Theorem 11-21.74, to declare the existence of a Perron root, $\rho > 0$, common to both U and V. They will not have common Perron vectors, however. Let $u > 0$ and $v > 0$, with

18.24
$$u = \begin{bmatrix} u_1 \\ u_2 \\ \cdot \\ \cdot \\ u_n \end{bmatrix} \text{ and } v = \begin{bmatrix} v_1 \\ v_2 \\ \cdot \\ \cdot \\ v_n \end{bmatrix},$$

be the respective Perron vectors (scale factors are irrelevant, here) and form the diagonal matrices

18.25
$$D_1 = \begin{bmatrix} u_1 & & & \\ & u_2 & & \\ & & \cdot & \\ & & & u_n \end{bmatrix}, D_2 = \begin{bmatrix} v_1 & & & \\ & v_2 & & \\ & & \cdot & \\ & & & v_n \end{bmatrix}.$$

According to Bauer, these matrices are the scaling matrices for which $\kappa_{\infty}(D_1^{-1}AD_2)$ is a minimum. We even know the minimum value.

18.26 Theorem. If we use the scaling matrices D_1^{-1} and D_2 above, then the minimum condition number obtained is

$$\kappa_{\infty}(D_1^{-1}AD_2) = \rho,$$

where $\rho > 0$ is the common Perron root of U and V in (18.23).

Proof. See Noble [1969], p. 441.

18.27 EXAMPLE.   The matrix

$$A = \begin{bmatrix} a & b \\ c & d \end{bmatrix}$$

is nonsingular and fully indecomposable if $a, b, c, d \in \mathbf{R}^+$ and $ad \neq bc$. Under these assumptions, then, we can write

$$A^{-1} = \frac{1}{\det A} \begin{bmatrix} d & -b \\ -c & a \end{bmatrix}.$$

Hence,

$$U = \frac{1}{|\det A|} \begin{bmatrix} (ad+bc) & 2ab \\ 2cd & (ad+bc) \end{bmatrix}$$

and

$$V = \frac{1}{|\det A|} \begin{bmatrix} (ad+bc) & 2bd \\ 2ac & (ad+bc) \end{bmatrix}.$$

The (common) Perron root of U and V is

$$\rho = \frac{ad + bc + 2\sqrt{abcd}}{|ad - bc|}.$$

The Perron vectors for U and V are

$$u = \begin{bmatrix} \sqrt{abcd} \\ cd \end{bmatrix}, \text{ and } v = \begin{bmatrix} \sqrt{abcd} \\ ac \end{bmatrix}$$

which means

$$D_1 = \begin{bmatrix} \sqrt{abcd} & 0 \\ 0 & cd \end{bmatrix}, \text{ and } D_2 = \begin{bmatrix} \sqrt{abcd} & 0 \\ 0 & ac \end{bmatrix}.$$

Hence,

$$D_1^{-1} A D_2 = \begin{bmatrix} a & \sqrt{\dfrac{abc}{d}} \\ \sqrt{\dfrac{abc}{d}} & a \end{bmatrix}$$

and

$$\kappa_\infty(D_1^{-1}AD_2) = \frac{ad + bc + 2\sqrt{abcd}}{|ad - bc|}.$$

To observe the improvement in the condition number, due to optimum scaling, we need to compare $\kappa_\infty(D_1^{-1}AD_2)$ with

$$\kappa_\infty(A) = \|A\|_\infty\|A^{-1}\|_\infty$$

$$= \frac{\max(a + b, c + d)\max(b + d, a + c)}{|ad - bc|}.$$

Suppose

$$a = 100$$

$$b = 0.01$$

$$c = 99$$

$$d = 0.01.$$

The condition number of A, with respect to the ∞-norm, is

$$\kappa_\infty(A) = \frac{(100.01)(199)}{0.01}$$

$$= 1,990,199.$$

Optimum scaling reduces the condition number to

$$\kappa_\infty(D_1^{-1}AD_2) = \frac{1.99 + 2\sqrt{0.99}}{0.01}$$

$$\doteq 397.$$

Notice that Bauer's solution is not easy to compute since $|A^{-1}|$, as well as the Perron vectors u and v, must be determined. This fact, along with the fact that solutions to (18.11) have not been found for arbitrary norms, lead us to the statement that, *at this time, there is no satisfactory practical solution to the scaling problem* (18.11) *which will work for arbitrary matrices and arbitrary norms.*

A fairly simple scaling procedure has been suggested by McKeeman [1962] with the following comment: "scaling the rows of the matrix A to roughly the same maximum (here, dividing by the largest element) allows the procedure ... to select effective pivotal elements for the Gaussian decomposition of the matrix. ... If the matrix is badly conditioned then the solution is sensitive to perturbations in the input and the scaling division must be done not by the largest element, but rather by the power of the machine number base (2 and 10 for binary and decimal machines, respectively) nearest the largest element, so as to avoid rounding errors." This approach appears to be reasonably effective as a practical procedure, and until further investigation uncovers more effective procedures, this is a reasonable initial approach.

EXERCISES 12.18

1. If $A = (a_{ij})$ is such that $a_{ij} \neq 0$ for all i and j, can A be decomposable? Can A be partly decomposable?

2. Classify the matrix $\begin{bmatrix} 1 & 1 \\ 1 & 0 \end{bmatrix}$ according to the definitions in (18.13), (18.14), (18.15) and (18.16).

3. Prove Theorem 18.21.

4. Prove Theorem 18.22.

5. Prove Theorem 18.26.

6. Use (18.27) to invert the coefficient matrix in Example 17.3. Solve (17.3) and compare the answer with the answer obtained using Cramer's Rule. (See Exercise 6, Section 12.4.)

7. Use the method of (18.27) to scale the equations in Example 17.3. Compute $\kappa_\infty(A)$ and $\kappa_\infty(D_1^{-1}AD_2)$. Solve the scaled equations.

12.19 TRIANGULAR DECOMPOSITION

Suppose we return to unmodified Gaussian elimination (i.e., assume that none of the pivotal candidates vanishes) where $A^{(1)}x = b^{(1)}$ is transformed into the triangular system $MA^{(1)}x = Mb^{(1)}$ or $A^{(n)}x = b^{(n)}$. This enables us to write

19.1 $$A^{(1)} = M^{-1}A^{(n)}.$$

If we use (8.8) and (8.10) we have

19.2 $$M^{-1} = [M^{(1)}]^{-1}[M^{(2)}]^{-1} \cdots [M^{(n-1)}]^{-1}$$

where

19.3 $$[M^{(r)}]^{-1} = \begin{bmatrix} I_{r-1} & & & & & \\ & 1 & & & & \\ & -m_{r+1,r} & 1 & & & \\ O & -m_{r+2,r} & 0 & 1 & & \\ & \cdots\cdots\cdots\cdots\cdots & & & \\ & -m_{nr} & 0 & 0 & \cdots & 1 \end{bmatrix}$$

By direct computation, we have then

19.4

$$M^{-1} = \begin{bmatrix} 1 & & & & & \\ -m_{21} & 1 & & & & \\ -m_{31} & -m_{32} & 1 & & & \\ -m_{41} & -m_{42} & -m_{43} & 1 & & \\ \multicolumn{6}{c}{\dotfill} \\ -m_{n1} & -m_{n2} & -m_{n3} & -m_{n4} & \cdots & 1 \end{bmatrix}$$

and, consequently, M^{-1} is unit lower triangular. We leave the verification of the last two relations to the reader.

Since $A^{(n)}$ is an upper triangular matrix, this means that if none of the pivotal candidates vanishes, in the process of forming $A^{(n)}$ from $A^{(1)}$, then $A^{(1)}$ can be decomposed into the product of a unit lower triangular matrix, M^{-1}, and an upper triangular matrix, $A^{(n)}$. The next theorem gives sufficient conditions, in terms of principal minors, for the existence of such a decomposition.

19.5 Theorem. An $n \times n$ matrix $A = (a_{ij})$ may be written as the product LU, where L is lower triangular and U is upper triangular, if the leading principal minors $\det [a_{11}] \neq 0, \det \begin{bmatrix} a_{11} & a_{12} \\ a_{21} & a_{22} \end{bmatrix} \neq 0, \ldots, \det A \neq 0$. Furthermore, the triangular decomposition is *unique* if the diagonal elements of either L or U are specified.

Proof. We prefer to carry out a constructive proof for the special case in which L is chosen to be unit lower triangular and in which we are interested. Consider the equation $A = LU$, with L chosen to be unit lower triangular, so that

19.6

$$\begin{bmatrix} a_{11} & a_{12} & a_{13} & \cdots & a_{1n} \\ a_{21} & a_{22} & a_{23} & \cdots & a_{2n} \\ a_{31} & a_{32} & a_{33} & \cdots & a_{3n} \\ \multicolumn{5}{c}{\dotfill} \\ a_{n1} & a_{n2} & a_{n3} & \cdots & a_{nn} \end{bmatrix} = \begin{bmatrix} 1 & & & \\ l_{21} & 1 & & \\ l_{31} & l_{32} & 1 & \\ \multicolumn{4}{c}{\dotfill} \\ l_{n1} & l_{n2} & l_{n3} & \cdots & 1 \end{bmatrix} \begin{bmatrix} u_{11} & u_{12} & u_{13} & \cdots & u_{1n} \\ & u_{22} & u_{23} & \cdots & u_{2n} \\ & & u_{33} & \cdots & u_{3n} \\ \multicolumn{5}{c}{\dotfill} \\ & & & & u_{nn} \end{bmatrix}.$$

If we perform the matrix multiplication on the right and equate corresponding elements, we obtain n^2 non-linear equations in n^2 unknowns. For example, if we compute the elements of the *first row*, we obtain the equations

19.7
$$\begin{cases} u_{11} = a_{11} \\ u_{12} = a_{12} \\ \cdots\cdots\cdots \\ u_{1n} = a_{1n} \end{cases}$$

which tell us immediately that the first rows of A and U are identical. Likewise, if we compute the elements of the *first column*, below the main diagonal, we obtain the equations

19.8
$$\begin{cases} l_{21}u_{11} = a_{21} \\ l_{31}u_{11} = a_{31} \\ \cdots\cdots\cdots\cdots \\ l_{n1}u_{11} = a_{n1}. \end{cases}$$

Since $u_{11} = a_{11}$, we observe immediately that if $a_{11} \neq 0$, then these equations give us the elements in the first column of L.

In general, if we compute the elements of the rth *row*, beginning with the element on the main diagonal, we obtain

19.9
$$\begin{cases} \sum_{j=1}^{r-1} l_{rj}u_{jr} + u_{rr} = a_{rr} \\ \\ \sum_{j=1}^{r-1} l_{rj}u_{j,r+1} + u_{r,r+1} = a_{r,r+1} \\ \\ \cdots\cdots\cdots\cdots\cdots\cdots\cdots\cdots\cdots\cdots\cdots \\ \\ \sum_{j=1}^{r-1} l_{rj}u_{jn} + u_{rn} = a_{rn}, \end{cases}$$

and since l_{rj} and u_{jk} are known, for $j \leq r - 1$, we can solve immediately for the unknown elements of U in the rth row

19.10
$$\begin{cases} u_{rr} = a_{rr} - \sum_{j=1}^{r-1} l_{rj}u_{jr} \\ \\ u_{r,r+1} = a_{r,r+1} - \sum_{j=1}^{r-1} l_{rj}u_{j,r+1} \\ \\ \cdots\cdots\cdots\cdots\cdots\cdots\cdots\cdots\cdots\cdots \\ \\ u_{rn} = a_{rn} - \sum_{j=1}^{r-1} l_{rj}u_{jn}. \end{cases}$$

Likewise, if we compute the elements of the rth *column*, below the main diagonal, we obtain

19.11
$$\left\{ \begin{array}{c} \displaystyle\sum_{j=1}^{r-1} l_{r+1,j}u_{jr} + l_{r+1,r}u_{rr} = a_{r+1,r} \\[2ex] \displaystyle\sum_{j=1}^{r-1} l_{r+2,j}u_{jr} + l_{r+2,r}u_{rr} = a_{r+2,r} \\[2ex] \cdots\cdots\cdots\cdots\cdots\cdots\cdots\cdots\cdots\cdots\cdots \\[2ex] \displaystyle\sum_{j=1}^{r-1} l_{nj}u_{jr} + l_{nr}u_{rr} = a_{nr} \end{array} \right.$$

and if $u_{rr} \neq 0$ (guaranteed by the hypothesis that det $A \neq 0$), then these equations give us the elements in the rth column of L

19.12
$$\left\{ \begin{array}{c} \displaystyle l_{r+1,r} = \frac{1}{u_{rr}}\left[a_{r+1,r} - \sum_{j=1}^{r-1} l_{r+1,j}u_{jr} \right] \\[3ex] \displaystyle l_{r+2,r} = \frac{1}{u_{rr}}\left[a_{r+2,r} - \sum_{j=1}^{r-1} l_{r+2,j}u_{jr} \right] \\[3ex] \cdots\cdots\cdots\cdots\cdots\cdots\cdots\cdots\cdots\cdots\cdots \\[3ex] \displaystyle l_{nr} = \frac{1}{u_{rr}}\left[a_{nr} - \sum_{j=1}^{r-1} l_{nj}u_{jr} \right]. \end{array} \right.$$

The case in which U is unit upper triangular is similar. There is a complete (but nonconstructive) proof in Faddeev and Faddeeva [1963], p. 17.

Consider the case $n = 3$ where the conditions of Theorem 19.5 are satisfied. If we write

19.13
$$\begin{bmatrix} a_{11} & a_{12} & a_{13} \\ a_{21} & a_{22} & a_{23} \\ a_{31} & a_{32} & a_{33} \end{bmatrix} = \begin{bmatrix} 1 & & \\ l_{21} & 1 & \\ l_{31} & l_{32} & 1 \end{bmatrix}\begin{bmatrix} u_{11} & u_{12} & u_{13} \\ & u_{22} & u_{23} \\ & & u_{33} \end{bmatrix},$$

we obtain nine nonlinear equations in nine unknowns

19.14
$$
\begin{cases}
u_{11} = a_{11} \\[4pt]
u_{12} = a_{12} \\[4pt]
u_{13} = a_{13} \\[4pt]
l_{21}u_{11} = a_{21} \\[4pt]
l_{31}u_{11} = a_{31} \\[4pt]
l_{21}u_{12} + u_{22} = a_{22} \\[4pt]
l_{21}u_{13} + u_{23} = a_{23} \\[4pt]
l_{31}u_{12} + l_{32}u_{22} = a_{32} \\[4pt]
l_{31}u_{13} + l_{32}u_{23} + u_{33} = a_{33}.
\end{cases}
$$

The first three equations tell us that the first row of U is identical with the first row of A. Now, since we know u_{11}, the fourth and fifth equations give us

19.15
$$
\begin{cases}
l_{21} = \dfrac{a_{21}}{a_{11}} \\[12pt]
l_{31} = \dfrac{a_{31}}{a_{11}},
\end{cases}
$$

since $a_{11} \neq 0$, by hypothesis. From the sixth equation of the system we obtain u_{22}, since everything else in that equation is known. In fact, we now have, from the sixth and seventh equations,

19.16
$$
\begin{cases}
u_{22} = a_{22} - \dfrac{a_{21}a_{12}}{a_{11}} \\[12pt]
u_{23} = a_{23} - \dfrac{a_{21}a_{13}}{a_{11}}.
\end{cases}
$$

All that remains is to find l_{32} and u_{33}, and there are two equations we have not used.

Before continuing, however, let us associate our 3 x 3 matrix A with the system of equations $A^{(1)}x = b^{(1)}$ of (6.3b). In other words, we let $A \equiv A^{(1)}$ and so $a_{ij} \equiv a_{ij}^{(1)}$. Now compare (19.15) with (7.6). This enables us to write

19.17
$$
\begin{cases}
l_{21} = -m_{21} \\[4pt]
l_{31} = -m_{31}.
\end{cases}
$$

Likewise, if we compare (19.16) with (7.9), we observe that

19.18
$$\begin{cases} u_{22} = a_{22}^{(2)} \\ u_{23} = a_{23}^{(2)}. \end{cases}$$

Note that u_{22} also has the form

19.19
$$u_{22} = \frac{1}{a_{11}} \det \begin{bmatrix} a_{11} & a_{12} \\ a_{21} & a_{22} \end{bmatrix}.$$

From the eighth equation in (19.14) we can solve for l_{32} if $u_{22} \neq 0$. However, both $\det \begin{bmatrix} a_{11} & a_{12} \\ a_{21} & a_{22} \end{bmatrix} \neq 0$ and $a_{11} \neq 0$, by hypothesis. Thus, $u_{22} \neq 0$, and if we use (8.2) and (8.5), we obtain

19.20
$$l_{32} = \frac{1}{u_{22}} [a_{32} - l_{31}u_{12}]$$

$$= \frac{1}{a_{22}^{(2)}} [a_{32}^{(1)} + m_{31}a_{12}^{(1)}]$$

$$= \frac{a_{32}^{(2)}}{a_{22}^{(2)}}$$

$$= -m_{32}.$$

From the last equation in (19.14) we obtain, using (7.9) and (8.5),

19.21
$$u_{33} = a_{33} - [l_{31}u_{13} + l_{32}u_{23}]$$

$$= [a_{33}^{(1)} + m_{31}a_{13}^{(1)}] + m_{32}a_{23}^{(2)}$$

$$= a_{33}^{(2)} + m_{32}a_{23}^{(2)}$$

$$= a_{33}^{(3)}.$$

In this example, then, we obtain the unique decomposition

19.22
$$\begin{bmatrix} a_{11}^{(1)} & a_{12}^{(1)} & a_{13}^{(1)} \\ a_{21}^{(1)} & a_{22}^{(1)} & a_{23}^{(1)} \\ a_{31}^{(1)} & a_{32}^{(1)} & a_{33}^{(1)} \end{bmatrix} = \begin{bmatrix} 1 & & \\ -m_{21} & 1 & \\ -m_{31} & -m_{32} & 1 \end{bmatrix} \begin{bmatrix} a_{11}^{(1)} & a_{12}^{(1)} & a_{13}^{(1)} \\ & a_{22}^{(2)} & a_{23}^{(2)} \\ & & a_{33}^{(3)} \end{bmatrix}.$$

19.23 *Remark.* Obviously, because of the uniqueness of the triangular decomposition when L is unit lower triangular, $A^{(1)} = LU$ with* $M^{-1} = L$ and $A^{(n)} = U$.

---

* Recall (19.1).

In other words, Gaussian elimination without pivoting is mathematically equivalent to the unique triangular decomposition LU with L chosen to be unit lower triangular.

Wilkinson [1965, p. 223] goes further and states that not only are the mathematical results equivalent but, if the computations in both processes are carried out on an automatic digital computer in standard floating-point arithmetic, then "even the rounding errors are the same in the two processes" and failure of triangular decomposition occurs in the same circumstances as failure of Gaussian elimination without pivoting.

He makes it clear on that same page, however, that there is a computational advantage in triangular decomposition over elimination if scalar products can be accumulated using $fl_2(\ )$ arithmetic* rather than standard floating-point arithmetic. (See Wilkinson [1965, p. 114 and p. 117] for a treatment of $fl(\ )$ versus $fl_2(\ )$ accumulation of scalar products.) This is illustrated by the fact that in (19.21), for example,

$$u_{33} = [-l_{31}, -l_{32}, a_{33}] \begin{bmatrix} u_{13} \\ u_{22} \\ 1 \end{bmatrix},$$

and the scalar product can be accumulated without intermediate rounding, whereas the equivalent computation, using Gaussian elimination, involves the formation of $a_{33}^{(2)}$ (with subsequent rounding) before completing the computation. (See, also, Fox and Mayers [1968], pp. 86–87.)

19.24 *Remark.* As we stated in (14.2), the class of matrices for which no pivoting is required (and for which the LU decomposition is assured, therefore) includes those matrices which are real, symmetric, and positive definite. It also includes those nonsingular matrices which are *diagonally dominant* (see Wendroff [1966, p. 122], for example), that is, nonsingular matrices $A = (a_{ij})$ for which

$$|a_{ii}| > \sum_{j \ne i} |a_{ij}| \qquad (i = 1, 2, \ldots, n).$$

If A is real, symmetric, and positive definite then $U = L^T$, which means $A = LL^T$. This is called the *Choleski decomposition.* (See Wilkinson [1967, p. 71], for example.)

19.25 *Remark.* Suppose we can decompose A. In this case we are solving the system

$$(LU)x = b$$

_____

* See 2-(7.47) for remarks about $fl_2(\ )$ arithmetic.

or, equivalently, the two triangular systems

$$Ly = b,$$

and

$$Ux = y.$$

Since these last two systems are triangular, the first can be solved for y by forward-substitution and the second can be solved for x by back-substitution (both relatively simple processes compared to the process of triangular decomposition).

If several systems of equations with the same coefficient matrix* are to be solved, then it is possible to compute L and U once and for all and to repeat the solution of $Ly^{(i)} = b^{(i)}$ and $Ux^{(i)} = y^{(i)}$ for each vector $b^{(i)}$. This is useful for computing A^{-1} if we set $t = n$ and $b^{(i)} = e^{(i)}$ in (1.3), for example.

EXERCISES 12.19

1. Verify (19.3).

2. Verify (19.4).

3. Carry out a constructive proof of Theorem 19.5 for the special case in which U is unit upper triangular.

4. Carry out the triangular decomposition (19.13) for the matrix

$$A = \begin{bmatrix} 1 & 1 & 2 \\ 2 & -1 & 1 \\ 1 & 2 & 0 \end{bmatrix}.$$

5. Prove the truth of Wilkinson's assertion that "even the rounding errors are the same in the two processes," which we quoted in Remark 19.23.

6. In Remark 19.24 it is stated that partial pivoting is not necessary for numerical stability if the coefficient matrix is nonsingular and diagonally dominant. Prove this assertion.

12.20 TRIANGULAR DECOMPOSITION WITH PARTIAL PIVOTING

In the last section we discussed the mathematical equivalence of triangular decomposition and Gaussian elimination *without* pivoting. An obvious question is whether or not there is a triangular decomposition equivalent to Gaussian elimination *with* partial pivoting. This is discussed by Forsythe and Moler [1967, p. 36] and others.

Let P denote a permutation matrix†. Then we have the following result for PA (rather than A).

---

* Recall (1.1) and (1.3).

† See 11-(17.32).

20.1 Theorem. For an arbitrary $n \times n$ matrix A there exist a permutation matrix P, a unit lower triangular matrice L, and an upper triangular matrix U such that PA has the triangular decomposition $PA = LU$. However, L and U are not always uniquely determined by P and A.

Proof. See Wendroff [1966, pp. 127–129].

In his proof, Wendroff shows that the effect of partial pivoting during the process of elimination is mathematically the same as carrying out Gaussian elimination without pivoting on some matrix PA obtained from A by permuting its rows. This means, then, that the process is equivalent to the triangular decomposition of PA. Obviously, since $\det P \neq 0$, the systems $Ax = b$ and

20.2
$$PAx = Pb$$

are equivalent, and nothing is lost by decomposing PA rather than A.

It can be shown that the following scheme, described by Wilkinson [1967, p. 70], accomplishes the decomposition of PA without the necessity of knowing P in advance. Just as in Gaussian elimination with pivoting, we examine the first column of A and do a row interchange if a_{11} is not the element in the first column with greatest magnitude. At this point we can use (19.5) and (19.6) to compute* the first row of U and the first column of L. If these elements are overwritten on the corresponding elements of A, we have the augmented array (the reason for the column of zeros is explained below)

20.3
$$\begin{bmatrix} u_{11} & u_{12} & u_{13} & \cdots & u_{1n} & b_1 & 0 \\ l_{21} & a_{22} & a_{23} & \cdots & a_{2n} & b_2 & 0 \\ l_{31} & a_{32} & a_{33} & \cdots & a_{3n} & b_3 & 0 \\ \cdots & & & & & & \cdots \\ l_{n1} & a_{n2} & a_{n3} & \cdots & a_{nn} & b_n & 0 \end{bmatrix}$$

stored in the computer. It should be pointed out that, for notational simplicity, we do not introduce any special notation for the symbols a_{ij} and b_i to indicate a row interchange (if one is necessary). The reader can think of the subscripts as referring to a position in the *current* array rather than to a position in some previous array. However, the symbols a_{ij} and b_i refer to the *original* data (probably with altered row indices) and l_{ij} and u_{ij} refer to computed results.

---

* Obviously, since the first rows of A (after the row interchange, if it is needed) and U are identical, nothing needs to be done to "compute" the elements of the first row of U.

The second step is typical of the rth step in which we compute the quantities

20.4
$$\begin{cases} s_r & = a_{rr} \quad - \displaystyle\sum_{j=1}^{r-1} l_{rj}u_{jr} \\[2ex] s_{r+1} & = a_{r+1,r} - \displaystyle\sum_{j=1}^{r-1} l_{r+1,j}u_{jr} \\[1ex] \cdots\cdots\cdots\cdots\cdots\cdots\cdots\cdots\cdots \\[1ex] s_n & = a_{nr} \quad - \displaystyle\sum_{j=1}^{r-1} l_{nj}u_{jr} \end{cases}$$

and overwrite them on the elements of the last column of the augmented matrix. Thus, if $r = 2$, we obtain

20.5
$$\left[\begin{array}{cccccc|c|c} u_{11} & u_{12} & u_{13} & \cdots & u_{1n} & b_1 & 0 \\ l_{21} & a_{22} & a_{23} & \cdots & a_{2n} & b_2 & s_2 \\ l_{31} & a_{32} & a_{33} & \cdots & a_{3n} & b_3 & s_3 \\ \cdots & \cdots & \cdots & & \cdots & \cdots & \cdots \\ l_{n1} & a_{n2} & a_{n3} & \cdots & a_{nn} & b_n & s_n \end{array} \right].$$

Notice that (in the notation of Gaussian elimination) $s_r, s_{r+1}, \ldots, s_n$ are merely $a_{rr}^{(r)}, a_{r+1,r}^{(r)}, \ldots, a_{nr}^{(r)}$. Hence, the search for the pivot involves finding p such that $|s_p| = \max_{r \le i \le n} |s_i|$. If s_r is not the pivot, then we interchange rows r and p $(r < p)$ of the *entire* augmented matrix so as to place s_p in the location previously occupied by s_r. Again, for notational simplicity, we do not introduce new notation but assume that a row interchange, if one was needed, has been performed, and the notation (in (20.5), for example) represents the array *after* the interchange. At this point, then, we see that (for the rth step)

20.6
$$u_{rr} = s_r$$

and the remaining elements of the rth row of U are computed using

20.7
$$\begin{cases} u_{r,r+1} = a_{r,r+1} - \displaystyle\sum_{j=1}^{r-1} l_{rj}u_{j,r+1} \\[2ex] u_{r,r+2} = a_{r,r+2} - \displaystyle\sum_{j=1}^{r-1} l_{rj}u_{j,r+2} \\[1ex] \cdots\cdots\cdots\cdots\cdots\cdots\cdots\cdots\cdots \\[1ex] u_{rn} = a_{rn} \quad - \displaystyle\sum_{j=1}^{r-1} l_{rj}u_{jn}. \end{cases}$$

Likewise, the elements of the rth column of L are given by

20.8
$$\begin{cases} l_{r+1,r} = \dfrac{s_{r+1}}{u_{rr}} \\[2mm] l_{n+2,r} = \dfrac{s_{r+2}}{u_{rr}} \\[2mm] \cdots\cdots\cdots\cdots \\[2mm] l_{nr} = \dfrac{s_n}{u_{rr}}. \end{cases}$$

These equations, of course, are exactly (19.8) and (19.10), if no row interchanges are necessary.

20.9 *Remark.* If the notation of (11.8) and (11.9) is used to describe Gaussian elimination with partial pivoting, then we can identify the elements of L and U (from $PA = LU$) with those of $\mathcal{M}$ and $\mathcal{A}^{(n)}$, respectively. Again, if the computations are carried out on a computer with which the right sides of (20.4) and (20.7) can be accumulated using $fl_2(\ )$ arithmetic, then the triangular decomposition $PA = LU$ will generate fewer rounding errors than the mathematically equivalent Gaussian elimination with partial pivoting.

20.10 *Remark.* We mentioned above that we are solving a system $PAx = Pb$ rather than $Ax = b$. In this section we have discussed the decomposition of PA. It should be obvious (although we have not said so, specifically) that the scheme we have described for decomposing PA also gives us Pb, since the correct permutations on the components of b are performed by row interchanges on the augmented matrix.

Another observation needs to be made. When the rows of the augmented matrix are interchanged, we interchange rows in the partially computed matrix L. That this is proper stems from the fact that the elements l_{ij} are related to the multipliers in Gaussian elimination, and in (11.7) we see that each step has associated with it a possible row interchange which affects all previous steps.

EXERCISES 12.20

1. Prove Theorem 20.1.

2. Carry out triangular decomposition with partial pivoting for the matrix

$$A = \begin{bmatrix} 1 & 1 & 2 \\ 2 & -1 & 1 \\ 1 & 2 & 0 \end{bmatrix}.$$

12.21 ITERATIVE IMPROVEMENT OF AN APPROXIMATE SOLUTION

The purpose of this section is to discuss an algorithm for improving the accuracy of an approximation to the solution to a system $Ax = b$, once it is obtained (by whatever means). This is particularly useful if the problem is ill-conditioned* and rounding errors have badly contaminated the computed solution.

To begin the first stage in the iteration we compute the residual vector

21.1
$$r^{(1)} = b - Ax^{(1)},$$

where $x^{(1)}$ is an initial approximation to the solution. Hopefully, $r^{(1)}$ has components which are relatively "small," in the sense that $x^{(1)}$ has some correct digits in each component.

The next part of the first iteration stage is motivated by the fact that, if the error in $x^{(1)}$ is

21.2
$$e^{(1)} = x - x^{(1)},$$

then

21.3
$$\begin{aligned} Ae^{(1)} &= Ax - Ax^{(1)} \\ &= b - Ax^{(1)} \\ &= r^{(1)}, \end{aligned}$$

which means that $e^{(1)}$ is the solution of the system of equations $Ae^{(1)} = r^{(1)}$. Note that $x = x^{(1)} + e^{(1)}$. However, we cannot solve (21.3) and obtain $e^{(1)}$ (even if we compute $r^{(1)}$ exactly) because, if we could, we could also solve $Ax = b$ for x. (It is our inability to solve $Ax = b$ for x which makes this entire study necessary.)

If we denote by $\varepsilon^{(1)}$ the approximation to $e^{(1)}$ which we do compute, then it is reasonable to conjecture that

21.4
$$x^{(2)} = x^{(1)} + \varepsilon^{(1)}$$

is an improved approximation to x.

In general then, if we have an approximation $x^{(i)}$ and we wish to compute a better approximation $x^{(i+1)}$, we proceed as follows. First, we compute the residual

21.5
$$r^{(i)} = b - Ax^{(i)}$$

as accurately as possible. Next, we use this approximation to $r^{(i)}$ as the right hand side of a system of equations whose coefficient matrix is A. We denote the solution of this system by $\varepsilon^{(i)}$ and compute $x^{(i+1)}$ by forming

21.6
$$x^{(i+1)} = x^{(i)} + \varepsilon^{(i)}.$$

---

* See 1-(4.7).

21.7 *Remark.* It is imperative that the residual vector $r^{(i)}$ be computed accurately. This means using higher precision than is used in most of the other parts of the computation. Wilkinson [1967, p. 86] states that if the vector $Ax^{(i)}$ cannot be computed by accumulating scalar products in $fl_2(\ )$ arithmetic,* then it must be computed in true double-precision arithmetic. He warns that the components of the vector $Ax^{(i)}$ must *not* be rounded to single length *before* they are subtracted from the components of b but that the rounding must take place *after* the subtraction.

21.8 *Remark.* Notice that the iterative improvement algorithm involves (at each stage) the solution of a system of equations whose coefficient matrix is A. If the initial approximation to the solution, $x^{(1)}$, was obtained by triangular decomposition of A, and if L and U of that decomposition are preserved, then very little additional work is involved in iterative improvement. We merely use the procedure described in (19.23) when we solve for $\varepsilon^{(i)}$.

EXERCISES 12.21

1. Write a computer program to solve $Ax = b$ using triangular decomposition with partial pivoting. Let part two of your program carry out iterative improvement on the solution obtained in part one.

2. Construct an example to demonstrate that a "small" residual $\|Ax - b\|_\alpha$ does not necessarily mean that x is a "good" solution to $Ax = b$. In other words, find an example for which $\|Ax - b\|_\alpha \le \|Ay - b\|_\alpha$ and yet y is a "better" solution, in some sense, than x. (See Wilkinson [1965] p. 249.)

12.22 CONVERGENCE OF ITERATIVE IMPROVEMENT

Suppose we look at a simplified version of the iterative improvement algorithm where the computations indicated in both (21.5) and (21.6) are assumed to be exact. (This is not too bad an assumption if double-length arithmetic is used.) Then

22.1
$$\begin{cases} r^{(i)} = b - Ax^{(i)} \\ (A + E^{(i)})\varepsilon^{(i)} = r^{(i)} \\ x^{(i+1)} = x^{(i)} + \varepsilon^{(i)} \end{cases}$$

describes the ith iteration stage, where the middle step is assumed to be the source of errors which are introduced.

What we do, when we write the middle equation, is to assume that $\varepsilon^{(i)} = e^{(i)} + p^{(i)}$ is the exact solution of a perturbed system which can be written

22.2
$$\begin{aligned} r^{(i)} &= (A + E^{(i)})(e^{(i)} + p^{(i)}) \\ &= (A + E^{(i)})\varepsilon^{(i)} \\ &= A(I + F^{(i)})\varepsilon^{(i)}, \end{aligned}$$

---

* See 2-(7.47) for a comment about $fl_2(\ )$ arithmetic.

if we define

22.3 $$F^{(i)} = A^{-1}E^{(i)}.$$

The following theorem enables us to prove the convergence of the simplified version of the algorithm.

22.4 Theorem. Let $\{x^{(1)}, x^{(2)}, \ldots, x^{(i)}, \ldots\}$ be the sequence of vectors generated by the algorithm (22.1). Let $x = A^{-1}b$ be the true solution of $Ax = b$. Assume that, for some matrix norm $\| \cdot \|_\beta$,

$$\|F^{(i)}\|_\beta \leqq \delta < \tfrac{1}{2}$$

for all i. Then, for any vector norm $\| \cdot \|_\alpha$ consistent with $\| \cdot \|_\beta$ we can write

$$\|x^{(i)} - x\|_\alpha \leqq \left(\frac{\delta}{1 - \delta}\right)^i \|x\|_\alpha.$$

Proof. See Forsythe and Moler [1967], pp. 109–111.
As a corollary, then, $x^{(i)} \to x$ as $i \to \infty$, since $\delta < \tfrac{1}{2}$ means $\delta/(1 - \delta) < 1$. (Recall Theorems (31.10), (31.13), and (32.2) of Chapter 11.)

22.5 *Remark.* If we do not assume that $r^{(i)}$ and $x^{(i+1)}$ are computed exactly, in (22.1), but are still computed with high accuracy, then the convergence proof above is more complicated. See, for example, Wilkinson [1963, pp. 124–125]. Moreover, it cannot be shown that $x^{(i)} \to x$ as $i \to \infty$. However, it can be shown that $x^{(i)}$ does converge to a rounded value of x.

Notice that a sufficient condition for convergence is that $\|F^{(i)}\|_\beta < \tfrac{1}{2}$ for all i. Since

$$\|F^{(i)}\|_\beta = \|A^{-1}E^{(i)}\|_\beta$$
$$\leqq \|A^{-1}\|_\beta \|E^{(i)}\|_\beta,$$

it is sufficient that $\|A^{-1}\|_\beta \|E^{(i)}\|_\beta < \tfrac{1}{2}$. If we assume that A has been normalized so that $\|A\|_\beta = 1$, then the condition number $\kappa_\beta(A)$ of (16.1) reduces to $\|A^{-1}\|_\beta$, and we can write the sufficiency condition as

$$\|E^{(i)}\|_\beta < \frac{1}{2\kappa_\beta(A)}$$

for all i. In other words, the size of the condition number $\kappa_\beta(A)$ is critical, and if it is too large we may not be able to guarantee convergence.

EXERCISE 12.22

Prove Theorem 22.4.

SUMMARY DISCUSSION

We have presented basically an expository treatment of Gaussian elimination with partial pivoting and indicated that if scalar products can be accumulated accurately using $fl_2(\ )$ arithmetic*, then the equivalent triangular decomposition should be used (instead of elimination), in order to improve the computational accuracy. (Without $fl_2(\ )$ accumulation of scalar products the accuracy will be the same in either case.) We have also described iterative improvement, a procedure for refining a solution obtained by the method above (or by any method, for that matter).

There are variations of the elimination algorithm associated with such names as Jordan,† Doolittle, Crout, and Banachiewicz (see Fox [1965], for example), but we have not included a discussion of them here. The reader who is interested in writing a computer program to solve $Ax = b$ along the lines we have suggested should read the excellent treatise of Wilkinson [1967] before he begins his work.

One final word of caution is in order. For a general matrix A (with no special properties such as diagonal dominance, positive definiteness, or sparseness), pivoting is a reasonable strategy *only* if A is equilibrated, in some sense (see Section 12.18), even though there is no foolproof algorithm for equilibration known at this time.

---

* See 2-(7.47).

† See Section 13.10 for a description of Gauss-Jordan elimination.

CHAPTER 13

SOLVING SYSTEMS OF
LINEAR ALGEBRAIC EQUATIONS
USING RESIDUE ARITHMETIC

13.1 INTRODUCTION

In Chapter 12 we consider the problem of solving a system of linear algebraic equations *approximately* and, as a consequence, we must consider the *numerical stability* of the algorithm used, as well as the *condition* of the system of equations being solved. It is evident that much care is required if we are going to compute answers for which reasonable error bounds can be guaranteed. Consequently, it is interesting to consider the possibility that some of the difficulties usually associated with the numerical solution of a system of linear algebraic equations can be avoided *completely* if residue arithmetic (rather than fixed-radix arithmetic) is used. This is our purpose in this chapter.

In residue arithmetic we work only with integers and so the computations are performed *exactly*. Consequently, there are *no rounding errors* and so ill-conditioned problems can be solved very easily. The fact that A and b, in

1.1
$$Ax = b$$

must be integral is no serious restriction due to the fact that if the elements of A and the components of b are rational numbers, they can be converted to integers simply by scaling. Note that, in a fixed-word-length computer, the only numbers stored are rational numbers.

We shall survey the elementary material relative to residue arithmetic and then apply the theoretical results to the Gauss-Jordan algorithm for solving systems of linear equations. Many of the theorems are found in Szabó and Tanaka [1967], and if the proofs of these theorems are not included, or left as exercises for the reader, they will be starred.

Theoretical Background

13.2 SOME BASIC THEOREMS

The following material can be found in any introductory book on number theory, and so the proofs of the theorems will be left to the reader.

2.1 Definition. If a and b are integers and if an integer $m \neq 0$ divides $a - b$, we write

$$a \equiv b \,(\mathrm{mod}\, m)$$

and say *a is congruent to b* modulo *m*. If *m* does not divide $a - b$, we write

$$a \not\equiv b \,(\mathrm{mod}\ m)$$

and say that *a is not congruent to b* modulo *m*.

Since *m* divides $a - b$ if and only if $- m$ divides $a - b$, we can assume throughout this discussion that the *modulus m* is a positive integer.

2.2 Theorem. Let a, b, c, d, x, y, and $m > 0$ be integers. Then

a) The following statements are equivalent:

    i) $a \equiv b \,(\mathrm{mod}\ m)$

    ii) $b \equiv a \,(\mathrm{mod}\ m)$

    iii) $a - b \equiv 0 \,(\mathrm{mod}\ m)$.

b) If

    i) $a \equiv b \,(\mathrm{mod}\ m)$

    ii) $b \equiv c \,(\mathrm{mod}\ m)$,

  then

$$a \equiv c \,(\mathrm{mod}\ m).$$

c) If

    i) $a \equiv b \,(\mathrm{mod}\ m)$

    ii) $c \equiv d \,(\mathrm{mod}\ m)$,

  then

$$ax + cy \equiv bx + dy \,(\mathrm{mod}\ m).$$

d) If

    i) $a \equiv b \,(\mathrm{mod}\ m)$

    ii) $c \equiv d \,(\mathrm{mod}\ m)$,

  then

$$ac \equiv bd \,(\mathrm{mod}\ m).$$

e) If $d > 0$ divides *m* and if

$$a \equiv b \,(\mathrm{mod}\ m),$$

  then

$$a \equiv b \,(\mathrm{mod}\ d).$$

For our purposes we shall find it convenient to follow Szabó and Tanaka [1967, Chapter 2] with regard to additional concepts and notation. For example, we shall define the term *residue* as follows.

2.3 Definition.   Given any integer x and any integer $m > 0$, if

$$r \equiv x \,(\text{mod } m)$$

and if $0 \leq r < m$, then we write

$$r = |x|_m$$

and say r is a *residue* of x modulo m.

The question of uniqueness arises, and it is answered by the following theorem.

2.4 Theorem.   Given any integer x and any integer $m > 0$, $|x|_m$ is unique.

Hence we can say, in (2.3), r is *the* residue of x modulo m. However, more than one integer can have the same residue. In fact, we can be quite specific.

2.5 Theorem.   If a, b, and $m > 0$ are integers, then

$$|a|_m = |b|_m$$

if and only if

$$a \equiv b \,(\text{mod } m).$$

Since all the integers congruent to a given integer constitute what is called a *residue class*, then, obviously, all members of a given residue class have the same residue.

Tables 2.6 and 2.7 exhibit the elements of the residue classes for the cases $m = 7$ (a prime) and $m = 8$ (not a prime), respectively. All integers in a vertical column constitute a residue class, and the *residue* for the elements of a given residue class is the positive integer in the residue class which is enclosed in the rectangle.

EXERCISES 13.2

1. Prove Theorem 2.2.
2. Prove Theorem 2.4.
3. Prove Theorem 2.5.

13.3 RESIDUE ARITHMETIC WITH A SINGLE MODULUS

The definitions and theorems in this section† establish the basic rules for doing arithmetic using a single modulus m.

---

† As we mentioned in Section 13.1 the symbol * means that the theorem and its proof can be found in Szabó and Tanaka [1967].

2.6 Table. The Residue Classes Modulo 7.

.	.	.	.	.	.	.
.	.	.	.	.	.	.
.	.	.	.	.	.	.
-70	-69	-68	-67	-66	-65	-64
-63	-62	-61	-60	-59	-58	-57
-56	-55	-54	-53	-52	-51	-50
-49	-48	-47	-46	-45	-44	-43
-42	-41	-40	-39	-38	-37	-36
-35	-34	-33	-32	-31	-30	-29
-28	-27	-26	-25	-24	-23	-22
-21	-20	-19	-18	-17	-16	-15
-14	-13	-12	-11	-10	-9	-8
-7	-6	-5	-4	-3	-2	-1
0	1	2	3	4	5	6
7	8	9	10	11	12	13
14	15	16	17	18	19	20
21	22	23	24	25	26	27
28	29	30	31	32	33	34
35	36	37	38	39	40	41
42	43	44	45	46	47	48
49	50	51	52	53	54	55
56	57	58	59	60	61	62
63	64	65	66	67	68	69
70	71	72	73	74	75	76
.	.	.	.	.	.	.
.	.	.	.	.	.	.
.	.	.	.	.	.	.

2.7 Table. The Residue Classes Modulo 8.

.	.	.	.	.	.	.	.
.	.	.	.	.	.	.	.
.	.	.	.	.	.	.	.
−72	−71	−70	−69	−68	−67	−66	−65
−64	−63	−62	−61	−60	−59	−58	−57
−56	−55	−54	−53	−52	−51	−50	−49
−48	−47	−46	−45	−44	−43	−42	−41
−40	−39	−38	−37	−36	−35	−34	−33
−32	−31	−30	−29	−28	−27	−26	−25
−24	−23	−22	−21	−20	−19	−18	−17
−16	−15	−14	−13	−12	−11	−10	−09
−08	−07	−06	−05	−04	−03	−02	−01
0	1	2	3	4	5	6	7
8	9	10	11	12	13	14	15
16	17	18	19	20	21	22	23
24	25	26	27	28	29	30	31
32	33	34	35	36	37	38	39
40	41	42	43	44	45	46	47
48	49	50	51	52	53	54	55
56	57	58	59	60	61	62	63
64	65	66	67	68	69	70	71
72	73	74	75	76	77	78	79
.	.	.	.	.	.	.	.
.	.	.	.	.	.	.	.
.	.	.	.	.	.	.	.

3.1 Theorem. * Let $a, k,$ and $m > 0$ be integers. Then

a) $|km|_m = 0,$

b) $k|a|_m = |ka|_{km},$

c) $|a|_m = a,$

if and only if $0 \leqq a < m$.

d) $|a \pm km|_m = |a|_m,$

e) $|-a|_m = |m - a|_m,$

f) If m is a prime, then

$$|a^m|_m = |a|_m.$$

3.2 Theorem. *If $x, y,$ and $m > 0$ are integers, then

$$|x \pm y|_m = ||x|_m \pm |y|_m|_m = |x \pm |y|_m|_m = ||x|_m \pm y|_m.$$

Thus there are four ways to compute the *sum or difference modulo m* of two integers. This result is easily generalized to contain an arbitrary number of terms.

3.3 Theorem. * If $x, y,$ and $m > 0$ are integers, then

$$|xy|_m = ||x|_m|y|_m|_m = |x|y|_m|_m = ||x|_m y|_m.$$

Thus there are four ways to compute the *product modulo m* of two integers. This result also can be generalized to contain an arbitrary number of factors.

These last two theorems state that there is complete arbitrariness in deciding when to reduce, modulo m, an operand or an intermediate result in a calculation which uses additions, subtractions, and multiplications.

Division, on the other hand, is an arithmetic operation which causes some difficulty, and we approach this subject with care. First, there is a *cancellation law of multiplication for integers modulo m*.

3.4 Theorem. * If $k, a, b,$ and $m > 0$ are integers, and if

i) $|ka|_m = |kb|_m,$

ii) $(k, m) = 1,$

then

$$|a|_m = |b|_m.$$

Next we introduce the concept of multiplicative inverse modulo m.

3.5 Definition. * If $a, b,$ and $m > 1$ are integers, and if

i) $0 < b < m,$

ii) $|ab|_m = |ba|_m = 1,$

then we write

$$b = a^{-1}(m)$$

and say b is a *multiplicative inverse of a modulo m.*

Notice that regardless of the sign of a, the multiplicative inverse of a modulo m is defined to be a positive integer in the range

3.6 $$0 < a^{-1}(m) < m.$$

There are questions of *existence* and *uniqueness* of the multiplicative inverse modulo m, and these questions are settled by the following theorem.

3.7 Theorem.* If a and $m > 0$ are integers, then $a^{-1}(m)$ exists if and only if

    i) $|a|_m \neq 0$

    ii) $(a, m) = 1$.

If $a^{-1}(m)$ exists, it is unique.

However, despite the uniqueness of the multiplicative inverse, when it exists, it is possible for more than one integer to have the same inverse. For example, since

$$|(-16)(3)|_7 = 1$$

and

$$|(12)(3)|_7 = 1,$$

both 12 and -16 have 3 as their multiplicative inverse modulo 7.

In Table 2.6 observe that 12 and -16 belong to the same residue class, the residue class whose elements have 5 as their residue. It is easily verified that all integers in this residue class have 3 as their multiplicative inverse modulo 7. This suggests the general result that all members of a residue class have the same inverse.

3.8 Theorem. If

    i) $a \equiv b \pmod{m}$

    ii) $a^{-1}(m)$ exists,

then $b^{-1}(m)$ exists and

$$a^{-1}(m) = b^{-1}(m).$$

Proof. From (ii) we can write

$$1 = |a^{-1}(m)a|_m$$
$$= |a^{-1}(m)|a|_m|_m.$$

From (i) and Theorem 2.5 this becomes

$$1 = |a^{-1}(m)|b|_m|_m$$
$$= |a^{-1}(m)b|_m.$$

Thus $b^{-1}(m)$ exists and, as a matter of fact,

$$b^{-1}(m) = a^{-1}(m).$$

This completes the proof.

Again, in Table 2.6, observe that not only do all integers congruent to 5 modulo 7 have 3 as their multiplicative inverse modulo 7, but all integers congruent to 3 modulo 7 have 5 as their multiplicative inverse modulo 7. This suggests the general result:

3.9 Theorem.* If $a, b,$ and $m > 1$ are integers, then

$$a^{-1}(m) = |b|_m,$$

if and only if

$$b^{-1}(m) = |a|_m.$$

The picture is clear, then, when the multiplicative inverse modulo m exists. When $m = 7$ (a prime) the multiplicative inverse exists for each integer except those which belong to the residue class containing zero. (For the elements of this residue class, condition (i) of Theorem 3.7 fails to hold.) When $m = 8$ (not a prime), condition (ii) of Theorem 3.7 fails to hold for the integers belonging to the residue classes containing $x = 2, x = 4,$ and $x = 6$. Thus all integers congruent to 0, 2, 4, and 6 fail to have multiplicative inverses modulo 8. In order to avoid this kind of problem, *we usually choose our moduli from the set of primes.*

This choice is even more important when we realize that, if m is not a prime, no convenient explicit expression for $a^{-1}(m)$ has been discovered whereas if m is a prime, one has been found.

3.10 Theorem. If a and $m > 1$ are integers, and if

i) m is a prime,

ii) $a^{-1}(m)$ exists,

then

$$a^{-1}(m) = |a^{m-2}|_m.$$

Proof. The proof is based on Fermat's theorem. See, for example, Szabó and Tanaka [1967], p. 24.

Thus we have a procedure for computing $a^{-1}(m)$. It may be useful, at times, to use the equivalent result:

3.11 Corollary. If a is an integer and m is a prime, then, when $a^{-1}(m)$ exists,

$$a^{-1}(m) = \left| \left(|a|_m \right)^{m-2} \right|_m.$$

Proof. The proof uses Theorem 3.10 and Theorem 3.3.

Now we can perform division to a limited extent. For example, if the quotient of two integers is again an integer, we can compute this quotient.

3.12 Theorem.* If a, b, and $m > 1$ are integers, and if

    i) $a|b$,

    ii) $a^{-1}(m)$ exists,

then

$$\left| \frac{b}{a} \right|_m = |ba^{-1}(m)|_m.$$

If, in addition, m is a prime, then

$$\left| \frac{b}{a} \right|_m = |ba^{m-2}|_m.$$

In general, we should like to find a solution for an equation of the form $ax = b$, where a and b are given. In residue arithmetic we work with integers, and so the corresponding problem is to find an integer $\bar{x}$ such that†

3.13 $$|a\bar{x}|_m = |b|_m.$$

The next two theorems discuss the existence and uniqueness of $\bar{x}$.

3.14 Theorem.* If a, b, and $m > 1$ are integers, and if $a^{-1}(m)$ exists, then (3.13) has a unique solution, $|\bar{x}|_m$, given by

$$|\bar{x}|_m = |ba^{-1}(m)|_m.$$

If, in addition, m is a prime, then

$$|\bar{x}|_m = |ba^{m-2}|_m.$$

Since $|\bar{x}|_m$ is unique but $\bar{x}$ is not, we might suspect that any integer in the same residue class with $\bar{x}$ would satisfy (3.13). This is indeed the case.

---

† In this chapter we use a bar over a symbol to emphasize that the symbol represents an integer.

3.15 Theorem. If $\bar{x}$ satisfies (3.13), and if

$$\bar{y} \equiv \bar{x} \pmod{m},$$

then $\bar{y}$ satisfies (3.13).

Proof. Since $\bar{x}$ satisfies (3.13), we may write

$$|b|_m = |a\bar{x}|_m$$
$$= |a|\bar{x}|_m|_m.$$

From our hypothesis and Theorem 2.5, this becomes

$$|b|_m = |a|\bar{y}|_m|_m$$
$$= |a\bar{y}|_m.$$

Thus, $\bar{y}$ satisfies (3.13) and the proof is complete.

3.16 EXAMPLE. (a divides b)

$$a = 4 \qquad\qquad |\bar{x}|_7 = |24 \cdot 4^5|_7$$
$$b = 24 \qquad\qquad\qquad\quad = |3 \cdot 1024|_7$$
$$m = 7 \qquad\qquad\qquad\quad = |3 \cdot 2|_7$$
$$= 6.$$

3.17 EXAMPLE. (a does not divide b)

$$a = 10 \qquad\qquad |\bar{x}|_7 = |15 \cdot 10^5|_7$$
$$b = 15 \qquad\qquad\qquad\quad = |1 \cdot 3^5|_7$$
$$m = 7 \qquad\qquad\qquad\quad = |243|_7$$
$$= 5.$$

The reader can easily demonstrate the validity of Theorem 3.15 for these examples.

EXERCISES 13.3

1. Prove Theorem 3.1.
2. Prove Theorem 3.2.
3. Prove Theorem 3.3.
4. Prove Theorem 3.4.
5. Prove Theorem 3.7.
6. Prove that all elements of the residue class with 5 as the residue modulo 7 have 3 as their multiplicative inverse modulo 7.

7. Prove Theorem 3.9.

8. If $m = 12$, the integers of which residue classes fail to have multiplicative inverses? Which part of Theorem 3.7 fails to hold in each case?

9. Prove Theorem 3.10.

10. Prove Theorem 3.12.

11. Prove Theorem 3.14.

13.4 RESIDUE NUMBER SYSTEMS

Our previous discussion has been confined to the case in which a single modulus $m > 0$ was used. We should like to broaden the discussion so as to consider more than one modulus. By a *residue number system* we usually mean a system in which more than one modulus is used.

In a fixed-radix number system, the number system is completely specified by stating the radix or base. Likewise, a residue number system is completely specified by stating its base. However, in this case, the *base* is an s-tuple of moduli $m_1, m_2, \ldots, m_s$. As we indicated earlier, there is no loss in generality if we select all the moduli to be positive.

4.1 Definition. A *residue representation* of an integer x, for the base $m_1, m_2, \ldots, m_s$ is the s-tuple

$$x \sim \{|x|_{m_1}, |x|_{m_2}, \ldots, |x|_{m_s}\}.$$

4.2 Theorem (uniqueness). For a given base, the residue representation of each integer is unique.

Proof. The uniqueness is a direct consequence of the uniqueness of $|x|_{m_i}$ for all i.

On the other hand, even though $x \neq y$, it is possible to have

4.3 $$|x|_{m_i} = |y|_{m_i} \qquad (i = 1, 2, \ldots, s),$$

in which case residue representations of x and y are identical. Thus the correspondence between the integers and their representations, for a given base, is not one-to-one. This is made clear by the following result.

4.4 Theorem.* Two integers x and y have the same residue representation for the base $m_1, m_2, \ldots, m_s$ if and only if

$$x \equiv y \,(\mathrm{mod}\, L),$$

where L is the least common multiple of the moduli.

In other words, all members of a given residue class modulo L are mapped onto the same residue representation. An obvious corollary follows.

4.5 Corollary. The residue representations of the integers w in the range

$$0 \leqq w \leqq L - 1$$

are distinct.

EXERCISES 13.4
1. Prove Theorem 4.4.
2. Illustrate Theorem 4.4 with several numerical examples.

13.5 RESIDUE ARITHMETIC WITH MORE THAN ONE MODULUS
First we consider addition and subtraction. These operations are described as follows.

5.1 Theorem.* Let $m_1, m_2, \ldots, m_s$ be the base for a residue number system, where $(m_i, m_j) = 1$ for $i \neq j$, and let

$$M = m_1 m_2 \ldots m_s.$$

Then the residue representation of $|x \pm y|_M$ is given by

$$
\begin{aligned}
x &\sim \{|x|_{m_1}, |x|_{m_2}, \ldots, |x|_{m_s}\} \\
y &\sim \{|y|_{m_1}, |y|_{m_2}, \ldots, |y|_{m_s}\} \\
\hline
|x \pm y|_M &\sim \{|z_1|_{m_1}, |z_2|_{m_2}, \ldots, |z_s|_{m_s}\}
\end{aligned}
$$

where

$$z_i = |x|_{m_i} \pm |y|_{m_i} \qquad (i = 1, 2, \ldots, s).$$

This result gives us a clue as to why it is desirable to have more than one modulus. Notice that we obtain the residue representation of $|x \pm y|_M$. Obviously, we want M to be large. Also, we want our moduli to be primes because, if this is so, then $L = M$ in Corollary 4.5 and we have a one-to-one correspondence between the integers w, in the range

5.2 $$0 \leqq w \leqq M - 1,$$

and their residue representations. Without this one-to-one correspondence it is difficult to relate $\{|z_1|_{m_1}, |z_2|_{m_2}, \ldots, |z_s|_{m_s}\}$ to the proper integer.

In a similar manner we can describe multiplication for residue arithmetic with more than one modulus.

5.3 Theorem.* Let $\{m_1, m_2, \ldots, m_s\}$ be the base for a residue number system where $(m_i, m_j) = 1$ for $i \neq j$, and let

$$M = m_1 m_2 \ldots m_s.$$

Then the residue representation of $|xy|_M$ is given by

$$x \sim \{|x|_{m_1}, \; |x|_{m_2}, \; \ldots, \; |x|_{m_s}\}$$
$$\frac{y \sim \{|y|_{m_1}, \; |y|_{m_2}, \; \ldots, \; |y|_{m_s}\}}{|xy|_M \sim \{|w_1|_{m_1}, |w_2|_{m_2}, \ldots, |w_s|_{m_s}\}}$$

where

$$w_i = |x|_{m_i} \cdot |y|_{m_i} \qquad (i = 1, 2, \ldots, s).$$

This takes care of three of the four basic arithmetic operations. Now we consider division. As before, we approach the subject with care. Since our results above were integers modulo M, we need to discuss the multiplicative inverse modulo M. To be more specific, since (3.5) defines the multiplicative inverse, we need a residue representation for the multiplicative inverse modulo M.

From (3.5) we know that, when it exists, $x^{-1}(M)$ is an integer z which satisfies

5.4 $$0 < z < M,$$

and

5.5 $$|xz|_M = |zx|_M = 1.$$

Our question, then, is whether or not there is a residue representation for $x^{-1}(M)$. The answer follows.

5.6 Theorem. Let $m_1, m_2, \ldots, m_s$ be the base for a residue number system, where $(m_i, m_j) = 1$ for $i \neq j$, and let

$$M = m_1 m_2 \ldots m_s.$$

If x is an integer such that $x^{-1}(M)$ exists, then $x^{-1}(M)$ has a residue representation in this system if and only if $x^{-1}(m_i)$ exists for all i. When the representation exists, it is unique and

$$x^{-1}(M) \sim \{x^{-1}(m_1), \; x^{-1}(m_2), \ldots, x^{-1}(m_s)\}.$$

Proof. (i) Assume $x^{-1}(m_i)$ exists for all i. Then form the s-tuple whose ith component is $x^{-1}(m_i)$. From the Chinese Remainder Theorem,† we can find a unique integer y satisfying (5.4) for which

5.7 $$y \sim \{x^{-1}(m_1), \; x^{-1}(m_2), \ldots, x^{-1}(m_s)\}.$$

---

† See (15.1).

By direct multiplication, using (5.3), we have

$$
\begin{array}{rl}
x & \sim \{|x|_{m_1} \quad , |x|_{m_2} \quad , \ldots, |x|_{m_s}\} \\
y & \sim \{x^{-1}(m_1), x^{-1}(m_2), \ldots, x^{-1}(m_s)\} \\
\hline
|xy|_M & \sim \{ \quad 1 \quad , \quad 1 \quad , \ldots, \quad 1 \quad \}
\end{array}
$$

Since $xy = yx$ this means that y satisfies (5.5). From the uniqueness of the multiplicative inverse this means

5.8 $$y = x^{-1}(M).$$

Thus the residue representation exists and, from (4.2), it is unique.

(ii) We shall use a proof by contradiction. Assume that, for $i = p, x^{-1}(m_p)$ does not exist and yet there is a residue representation for $x^{-1}(M)$,

5.9 $$x^{-1}(M) \sim \{r_1, \ldots, r_{p-1}, r_p, r_{p+1}, \ldots, r_s\}.$$

Then, by (5.3) and (5.5), the pth component of $|xx^{-1}(M)|_M$ is

5.10 $$|w_p|_{m_p} = ||x|_{m_p} r_p|_{m_p} = 1.$$

But this implies that

5.11 $$r_p = x^{-1}(m_p),$$

and contradicts our hypothesis that $x^{-1}(m_p)$ doesn't exist. Thus, our assumption is false, which implies that if $x^{-1}(m_p)$ does not exist, then no residue representation for $x^{-1}(M)$ exists in this system. Thus, $x^{-1}(M)$ has a residue representation in this system if and only if $x^{-1}(m_i)$ exists for all i.

Despite the uniqueness of the residue representation of $x^{-1}(M)$, when it exists, it is possible for x and y to be unequal and yet have the same residue representations for $x^{-1}(M)$ and $y^{-1}(M)$. The next theorem describes the general result.

5.12 Theorem. If $x \equiv y \pmod{L}$ and if $x^{-1}(M)$ has a residue representation in a residue system with base $m_1, m_2, \ldots, m_s$, then $y^{-1}(M)$ also has a residue representation in this system, and they are identical. In other words,

$$\{x^{-1}(m_1), x^{-1}(m_2), \ldots, x^{-1}(m_s)\} = \{y^{-1}(m_1), y^{-1}(m_2), \ldots, y^{-1}(m_s)\}.$$

Proof. If $x \equiv y \pmod{L}$, then $x \equiv y \pmod{m_i}$ for all i, from part (e) of (2.2). Since a residue representation for $x^{-1}(M)$ exists, it follows that $x^{-1}(m_i)$ exists for all i, from (5.6). Therefore $y^{-1}(m_i)$ exists for all i, from (3.8), and, in addition,

$$x^{-1}(m_i) = y^{-1}(m_i)$$

for all i. Hence

$$\{x^{-1}(m_1), x^{-1}(m_2), \ldots, x^{-1}(m_s)\} = \{y^{-1}(m_1), y^{-1}(m_2), \ldots, y^{-1}(m_s)\}.$$

We have already mentioned the advantages of choosing our moduli from the set of primes. If we make this choice, then we have an explicit formula for obtaining the residue representation of the multiplicative inverse.

5.13 Theorem. Let $m_1, m_2, \ldots, m_s$ (all primes) be the base of a residue number system where

$$M = m_1 m_2 \ldots m_s.$$

If x is an integer for which $x^{-1}(M)$ has a residue representation, then

$$x^{-1}(M) \sim \{|x^{m_1-2}|_{m_1}, \quad |x^{m_2-2}|_{m_2}, \ldots, |x^{m_s-2}|_{m_s}\}.$$

Proof. The theorem is a direct consequence of (5.6) and (3.10).

Now we are prepared to discuss division to a limited extent. We have a theorem analogous to (3.12).

5.14 Theorem.* If a and b are integers, if $m_1, m_2, \ldots, m_s$ is the base of a residue number system with

$$M = m_1 m_2 \ldots m_s,$$

and if

i) $a^{-1}(M)$ exists

ii) $a|b$

then

$$\left|\frac{b}{a}\right|_M \sim \{|ba^{-1}(m_1)|_{m_1}, \quad |ba^{-1}(m_2)|_{m_2}, \ldots, |ba^{-1}(m_s)|_{m_s}\}.$$

If, in addition, the moduli are all primes, then

$$\left|\frac{b}{a}\right|_M \sim \{|ba^{m_1-2}|_{m_1}, \quad |ba^{m_2-2}|_{m_2}, \ldots, |ba^{m_s-2}|_{m_s}\}.$$

Earlier, in our discussion of division, using a single modulus m, we turned from the case in which $a|b$ to the case in which we could not necessarily assume that $a|b$. This led us to (3.13) and (3.14). We know, from (3.14), that

5.15 $$|a\bar{x}|_M = |b|_M$$

has the unique solution, when $a^{-1}(M)$ exists, given by

5.16
$$|\bar{x}|_M = |ba^{-1}(M)|_M.$$

We immediately question whether or not $|\bar{x}|_M$ has a residue representation in a residue number system with base $m_1, m_2, \ldots, m_s$ and M defined in the usual way. The answer is contained in the following theorem.

5.17 Theorem. If a and b are integers in (5.15) and if $a^{-1}(M)$ exists and has a residue representation in the system under consideration, then $|\bar{x}|_M$ has the residue representation

$$|\bar{x}|_M \sim \{|ba^{-1}(m_1)|_{m_1},\ |ba^{-1}(m_2)_{m_2}, \ldots, |ba^{-1}(m_s)|_{m_s}\}.$$

If, in addition, all the moduli are primes, then

$$|\bar{x}|_M \sim \{|ba^{m_1-2}|_{m_1},\ |ba^{m_2-2}|_{m_2}, \ldots, |ba^{m_s-2}|_{m_s}\}.$$

Proof. From (5.16) and (5.3)

$$
\begin{aligned}
b &\sim \{|b|_{m_1}\ \ ,|b|_{m_2}\ \ \ ,\ldots,|b|_{m_s}\} \\
a^{-1}(M) &\sim \{a^{-1}(m_1), a^{-1}(m_2), \ldots, a^{-1}(m_s)\} \\
\hline
|ba^{-1}(M)|_M &\sim \{|ba^{-1}(m_1)|_{m_1}, |ba^{-1}(m_2)|_{m_2}, \ldots, |ba^{-1}(m_s)|_{m_s}\}.
\end{aligned}
$$

If the moduli are all primes, then the second part of the theorem follows immediately from (5.13).

EXERCISES 13.5

1. Prove Theorem 5.1.
2. Consider a residue number system with base $2, 3, 5, 7$. What are the residue representations of 62, 12, 37, and 25? Find the residue representations of $(37 + 25)$ and $(37 - 25)$ using Theorem 5.1. What happens if you form a sum which exceeds 210?
3. Prove Theorem 5.3.
4. Consider a residue number system with base $2, 3, 5, 7$. What are the residue representations of 15, 6, 13, 7, 90, and 91? Find the residue representations of $(15)(6)$ and $(13)(7)$ using Theorem 5.3. What happens if you form a product which exceeds 210?
5. Prove Theorem 5.14.

13.6 BASIC THEOREMS FOR MATRICES
The following material is a brief survey of some of the definitions and theorems for integral matrices. Most of these results are analogous to results for integers.

6.1 Definition. If $A = (a_{ij})$ and $B = (b_{ij})$ are $p \times q$ integral matrices and $m > 0$ is an integer, and if

$$a_{ij} \equiv b_{ij} \,(\text{mod } m)$$

for all i and j, then we write

$$A \equiv B \,(\text{mod } m)$$

and say A *is congruent to* B modulo m. If

$$a_{ij} \not\equiv b_{ij} \,(\text{mod } m)$$

for all i and j, then we write

$$A \not\equiv B \,(\text{mod } m)$$

and say A *is not congruent to* B modulo m.

As before, there is no loss in generality in assuming $m > 0$. The reader can easily prove the next theorem by referring to Theorem 2.2.

6.2 Theorem. Let A, B, C, and D be $p \times q$ integral matrices. Also, let x, y, d, and $m > 0$ be integers. Then

a) The following statements are equivalent:

    i) $A \equiv B \,(\text{mod } m)$

    ii) $B \equiv A \,(\text{mod } m)$

    iii) $A - B \equiv \phi \,(\text{mod } m)$,

where ϕ is the null matrix.

b) If

    i) $A \equiv B \,(\text{mod } m)$

    ii) $B \equiv C \,(\text{mod } m)$,

then

$$A \equiv C \,(\text{mod } m).$$

c) If

    i) $A \equiv B \,(\text{mod } m)$

    ii) $C \equiv D \,(\text{mod } m)$,

then

$$xA + yC \equiv xB + yD \,(\text{mod } m).$$

d) If A and C and also B and D are conformable, and if

   i) $A \equiv B \,(\mathrm{mod}\ m)$

   ii) $C \equiv D \,(\mathrm{mod}\ m),$

then

$$AC \equiv BD \,(\mathrm{mod}\ m).$$

e) If $d > 0$ divides m and if

$$A \equiv B \,(\mathrm{mod}\ m),$$

then

$$A \equiv B \,(\mathrm{mod}\ d).$$

Next we define a *residue matrix* modulo m. This definition is analogous to Definition 2.3.

6.3 Definition. Let $X = (x_{ij})$ be a $p \times q$ integral matrix and $m > 0$ be an integer. If $R = (r_{ij})$ is the matrix with elements defined by

$$r_{ij} = |x_{ij}|_m,$$

for all i, and j, then we write

$$R = |X|_m,$$

and say R is a *residue of X* modulo m.

The question of uniqueness arises, and it is answered by the following theorem.

6.4 Theorem. Given any $p \times q$ integral matrix X and any integer $m > 0$, $|X|_m$ is unique.

Proof. The uniqueness is a direct consequence of the uniqueness of $|x_{ij}|_m$.

Hence we can say, in (6.3), R is *the* residue of X modulo m. However, it is possible for $X \neq Y$ and yet $|X|_m = |Y|_m$, as the following theorem states.

6.5 Theorem. If X and Y are integral $p \times q$ matrices and $m > 0$ is an integer, then

$$|X|_m = |Y|_m$$

if and only if

$$X \equiv Y \,(\mathrm{mod}\ m).$$

Proof. $|X|_m = |Y|_m$ if and only if $|x_{ij}|_m = |y_{ij}|_m$ for all i and j. From Theorem 2.5 this is true if and only if $x_{ij} \equiv y_{ij} \,(\mathrm{mod}\ m)$. The theorem follows from this congruence and Definition 6.1.

We can write $|X|_m$ in several ways. For example, the following expressions are equivalent ways of writing the residue of X modulo m:

6.6 $$|X|_m = |(x_{ij})|_m = (|x_{ij}|_m).$$

EXERCISES 13.6

1. Prove Theorem 6.2.

2. If $A = \begin{bmatrix} 12 & -16 \\ -4 & 13 \end{bmatrix}$, what is $|A|_5$?

13.7 RESIDUE ARITHMETIC FOR MATRICES—SINGLE MODULUS

The definitions and theorems in this section establish the basic rules for doing matrix arithmetic using a single modulus m.

7.1 Theorem. If A and B are $p \times q$ integral matrices and if k and $m > 0$ are integers, then

a) $|mA|_m = \phi$ (the null matrix).

b) $k|A|_m = |kA|_{km}$.

c) $|A|_m = A$

   if and only if $0 \leq a_{ij} < m$ for all i and j.

d) $|A \pm mB|_m = |A|_m$.

e) $| - A|_m \quad = |mE - A|_m$.

Proof. This theorem is an immediate consequence of Theorem 3.1.

7.2 Theorem. If A and B are $p \times q$ integral matrices and if $m > 0$ is an integer, then

$$|A \pm B|_m = ||A|_m \pm |B|_m|_m = |A \pm |B|_m|_m = ||A|_m \pm B|_m.$$

Proof. Consider the (i, j) term in each of the four expressions given. They are, respectively,

$$|a_{ij} \pm b_{ij}|_m, \quad ||a_{ij}|_m \pm |b_{ij}|_m|_m, \quad |a_{ij} \pm |b_{ij}|_m|_m, \quad ||a_{ij}|_m \pm b_{ij}|_m$$

and these are equal, by Theorem 3.2.

   This result is easily generalized to contain an arbitrary number of terms.

7.3 Theorem. If A and B are conformable integral matrices and if $m > 0$ is an integer, then

$$|AB|_m = ||A|_m|B|_m|_m = |A|B|_m|_m = ||A|_mB|_m.$$

Proof. To prove the first part of the theorem we proceed as follows:

$$|AB|_m = \left| \left(\sum_{k=1}^{n} a_{ik}b_{kj} \right) \right|_m$$

$$= \left(\left| \sum_{k=1}^{n} |a_{ik}|_m |b_{kj}|_m \right|_m \right)$$

$$= \left| \left(\sum_{k=1}^{n} |a_{ik}|_m |b_{kj}|_m \right) \right|_m$$

$$= ||A|_m |B|_m|_m.$$

The other parts are proved in a similar manner.

This result, also, is easily generalized to contain an arbitrary number of factors. These last two theorems state that there is complete arbitrariness in deciding when to reduce, modulo m, an operand or an intermediate result in a matrix calculation involving additions, subtractions, and multiplications.

In a manner analogous to Theorem 3.4 we can prove the *cancellation law for scalar multiplication*:

7.4 Theorem. If A and B are $p \times q$ integral matrices and k and $m > 0$ are integers, and if

i) $|kA|_m = |kB|_m$
ii) $(k, m) = 1$

then

$$|A|_m = |B|_m.$$

Proof. From (i) we may write $|ka_{ij}|_m = |kb_{ij}|_m$ for all i and j. This equation, along with (ii), allows us to deduce $|a_{ij}|_m = |b_{ij}|_m$ from Theorem 3.4. Hence, the conclusion follows.

Up until now our comments have dealt with rectangular matrices. However, now we wish to discuss matrix inversion and so we must restrict our discussion to square matrices.

7.5 Definition. If A and C are $n \times n$ integral matrices and $m > 1$ is an integer, and if

i) $|AC|_m = I = |CA|_m$
ii) $|C|_m = C,$

then we write

$$C = A^{-1}(m)$$

and call C a *multiplicative inverse of A modulo m.*

The question of uniqueness of the multiplication inverse is settled by the following theorem.

7.6 Theorem. If A is an $n \times n$ integral matrix and if $A^{-1}(m)$ exists, then it is unique.

Proof. Assume that C and D are any multiplicative inverses of A modulo m. Then we may write

$$\begin{aligned}
C &= |C|_m \\
&= |CI|_m \\
&= |C|AD|_m|_m \\
&= ||CA|_m D|_m \\
&= |ID|_m \\
&= D.
\end{aligned}$$

Thus $C = D$ and the multiplicative inverse modulo m is unique.

We still have the question of existence, however, but before this question can be settled we must introduce two definitions and one theorem.·

7.7 Definition. An $n \times n$ integral matrix A is said to be *nonsingular modulo m* if and only if both

   i) $|\det A|_m \neq 0$
  ii) $(\det A, m) = 1.$

Otherwise, A is called *singular modulo m.*

7.8 Theorem. If A is an $n \times n$ integral matrix, then

$$|\det A|_m = |\det|A|_m|_m.$$

Proof.

$$|\det A|_m = |a_{11}A_{11} + a_{12}A_{12} + \ldots + a_{1n}A_{1n}|_m$$

where A_{ij} is the cofactor of the element a_{ij}. Then

$$\begin{aligned}
|\det A|_m &= ||a_{11}|_m|A_{11}|_m + |a_{12}|_m|A_{12}|_m + \cdots + |a_{1n}|_m|A_{1n}|_m|_m \\
&= |\det |A|_m|_m.
\end{aligned}$$

7.9 Definition. If A is an $n \times n$ integral matrix, then the *adjoint matrix modulo* m, $|A^{adj}|_m$, is defined to be

$$|A^{adj}|_m = |(A_{ji})|_m$$

where A_{ij} is the *cofactor* of a_{ij}. (Here (A_{ij}) is the transpose of the matrix of cofactors.)

Now we are ready to discuss the question of the existence of the multiplicative inverse modulo m.

7.10 Theorem. $A^{-1}(m)$ exists if and only if A is nonsingular modulo m, that is, if and only if

   i) $|d|_m \neq 0$
   ii) $(d, m) = 1,$

where $d = \det A$. In this case

$$A^{-1}(m) = |d^{-1}(m)|A^{adj}|_m|_m.$$

Proof. (a) Assume that $C = A^{-1}(m)$ exists. Then by Definition 7.5

$$|AC|_m = |CA|_m = I.$$

Thus

$$
\begin{aligned}
1 &= \det |AC|_m \\
&= |\det|AC|_m|_m \\
&= |\det(AC)|_m \\
&= |d \cdot \det C|_m \\
&= ||d|_m|\det C|_m|_m,
\end{aligned}
$$

and so $|d|_m \neq 0$.

We note, also, that we may write

$$1 = |d|\det C|_m|_m$$

and, by Definition 3.5, this implies

$$|\det C|_m = d^{-1}(m).$$

Consequently, $(d, m) = 1$, by Theorem 3.7.

(b) Assume $|d|_m \neq 0$ and $(d, m) = 1$. Hence, by Theorem 3.7, $d^{-1}(m)$ exists and is unique. Now consider the matrix

$$C = |d^{-1}(m)|A^{adj}|_m|_m.$$

Since

$$dI = AA^{adj}$$

then

$$|dI|_m = |AA^{adj}|_m$$
$$= |A|A^{adj}|_m|_m.$$

If we multiply both sides of this equation by $d^{-1}(m)$, then we obtain

$$d^{-1}(m)|dI|_m = d^{-1}(m)|A|A^{adj}|_m|_m.$$

Therefore,

$$|d^{-1}(m)dI|_m = |d^{-1}(m)|A|A^{adj}|_m|_m|_m.$$

The left side reduces immediately and we have

$$I = |A\ d^{-1}(m)|A^{adj}|_m|_m$$
$$= |A|d^{-1}(m)|A^{adj}|_m|_m|_m$$
$$= |AC|_m.$$

If we begin with the equation

$$dI = A^{adj}A$$

and go through an analogous sequence of steps, then we arrive at the result

$$I = |CA|_m.$$

Thus $C = A^{-1}(m)$ and the proof is complete.

We have discussed the uniqueness and the existence of $A^{-1}(m)$. However, there is one more point that must be made: even though $A^{-1}(m)$ is unique when it exists, it is possible for more than one matrix to have the same inverse. This is specified in the next theorem.

7.11 Theorem. If A and B are $n \times n$ integral matrices and if

  i) $A \equiv B$
  ii) $A^{-1}(m)$ exists,

then $B^{-1}(m)$ exists and

$$A^{-1}(m) = B^{-1}(m).$$

Proof. From (ii) we may write

$$I = |A^{-1}(m) A|_m$$
$$= |A^{-1}(m)|A|_m|_m.$$

From (i) and (6.5), this becomes

$$= |A^{-1}(m)|B|_m|_m$$
$$= |A^{-1}(m)B|_m.$$

Likewise, using a similar sequence of steps, we obtain

$$I = |AA^{-1}(m)|_m$$
$$= |BA^{-1}(m)|_m.$$

Hence $B^{-1}(m)$ exists and

$$B^{-1}(m) = A^{-1}(m).$$

Thus all matrices in the same residue class modulo m have the same multiplicative inverse modulo m, when the inverse exists. This result is analogous to (3.8) for integers.

EXERCISES 13.7

1. Prove Theorem 7.1.

2. Complete the proof of Theorem 7.3.

3. a) If $A = \begin{bmatrix} 4 & 13 & 27 \\ 3 & 15 & -21 \\ -3 & 2 & 9 \end{bmatrix}$ what is $|A^{adj}|_7$?

   b) Compute $|\det A|_7$.

   c) Compute $A^{-1}(7)$ and verify the correctness of your answer by multiplication.

4. Construct a matrix which has the same inverse modulo 7 as the matrix of the previous problem.

13.8 A RESIDUE SYSTEM OF EQUATIONS

Consider the system of linear algebraic equations $Ax = b$. Even if A and b are required to be integral there is no guarantee that x, the solution vector, will be integral. On the other hand, when we write the *residue system of equations*

8.1 $$|A\bar{x}|_m = |b|_m,$$

with A and b integral, we seek an integral vector $\bar{x}$ which satisfies (8.1). In general, then, x and $\bar{x}$ are different.

Thus, at first glance, it would appear that $\bar{x}$ would not aid us in finding x. However, it turns out that this is not the case. We can use residue arithmetic in solving (8.1) and this will lead us to a solution of $Ax = b$, where A and b are integral.

The following theorem gives us an expression for $|\bar{x}|_m$. Note the analogy with (3.14).

8.2 Theorem. If

    i) A is an $n \times n$ integral matrix

    ii) A is nonsingular modulo m

    iii) b is an integral vector

    iv) $|A\bar{x}|_m = |b|_m$,

then

$$|\bar{x}|_m = |A^{-1}(m)|b|_m|_m.$$

Proof. Let

$$|b|_m = |A\bar{x}|_m.$$

Since A is nonsingular, we can multiply by $A^{-1}(m)$. Thus

$$A^{-1}(m)|b|_m = A^{-1}(m)|A\bar{x}|_m$$

and so

$$|A^{-1}(m)|b|_m|_m = |A^{-1}(m)|A\bar{x}|_m|_m$$
$$= ||A^{-1}(m)A|_m\bar{x}|_m$$
$$= |\bar{x}|_m.$$

The next theorem describes the situation that exists if A is replaced by a matrix in the same residue class modulo m, and b is replaced by a vector in the same residue class modulo m.

8.3 Theorem. If A and C are $n \times n$ integral matrices and b and d are integral vectors, and if

    i) $|A\bar{x}|_m = |b|_m$

    ii) $|C\bar{y}|_m = |d|_m$

    iii) $|b|_m = |d|_m$

    iv) $|A|_m = |C|_m$

    v) A and C are nonsingular modulo m,

then

$$|\bar{x}|_m = |\bar{y}|_m.$$

Proof. From Theorem 8.2 and (i) we are able to write

$$|\bar{x}|_m = \left| A^{-1}(m)|b|_m \right|_m.$$

From (iii), (iv), (v), and Theorem 7.11, this becomes

$$|\bar{x}|_m = \left| A^{-1}(m)|d|_m \right|_m.$$

From (ii) and Theorem 8.2 we have, finally,

$$|\bar{x}|_m = |\bar{y}|_m.$$

EXERCISE 13.8

If $A = \begin{bmatrix} 4 & 13 & 27 \\ 5 & 15 & -21 \\ -3 & 2 & 9 \end{bmatrix}$ and $b = \begin{bmatrix} 9 \\ -1 \\ 8 \end{bmatrix}$, solve the residue system of equations $|A\bar{x}|_7 = |b|_7$.

Use the result obtained in Exercise 3, Section 7.

The Computational Procedure

13.9 INTRODUCTION

First, we shall review the computational procedure for solving

9.1 $Ax = b,$

using ordinary (fixed-radix) arithmetic and then we shall describe the analogous procedure, using residue arithmetic, for solving

9.2 $|A\bar{x}|_m = |b|_m.$

Finally, we shall show how this will lead us to an exact solution of (9.1).

It is more appropriate for our purposes here to solve (9.2) using the analog of the Gauss-Jordan elimination rather than Gaussian elimination. Consequently, we shall briefly discuss the Jordan variation of Gaussian elimination in order to develop an analogous procedure for (9.2).

13.10 GAUSS-JORDAN ELIMINATION

Assume that A is an $n \times n$ nonsingular matrix in (9.1) and the vector b is not the null vector. We seek a nonsingular $n \times n$ matrix J for which

10.1 $JAx = Jb,$

with

$$JA = I.$$

Thus,

10.2
$$J = A^{-1},$$

and

10.3
$$x = Jb$$

is the solution.

The reduction of A to the identity matrix consists of n major steps which correspond to n matrix row operations. These are similar to the n steps in ordinary Gaussian elimination which reduce A to an upper triangular matrix. However, Jordan's variation eliminates the elements above the diagonal as it eliminates the elements below the diagonal and also scales the diagonal elements to unity. Thus, there is no need for back substitution.

Let the original set of equations be written

10.4
$$A^{(1)}x = b^{(1)}.$$

Each step of the reduction leads to a new set of equations which is equivalent to (10.4) and which is simpler than the previous set in that the new set has a column of the identity matrix (where there was not one before) in its matrix of coefficients. Thus the second set of equations

10.5
$$J_1 A^{(1)}x = J_1 b^{(1)}$$

may be written

10.6
$$A^{(2)}x = b^{(2)},$$

where

10.7
$$A^{(2)} = \begin{bmatrix} 1 & * \ldots * \\ \hline 0 & \\ \vdots & * \\ 0 & \end{bmatrix}.$$

The $(i + 1)$st equivalent set is

10.8
$$(J_i \ldots J_2 J_1)A^{(1)}x = (J_i \ldots J_2 J_1)b^{(1)}$$

which may be written

10.9
$$A^{(i+1)}x = b^{(i+1)},$$

with

10.10
$$A^{(i+1)} = \left[\begin{array}{c|c} I_i & * \\ \hline \phi & * \end{array}\right].$$

Finally,

10.11
$$(J_n \ldots J_2 J_1)A^{(1)}x = (J_n \ldots J_2 J_1)b^{(1)}$$

may be written

10.12
$$J A^{(1)}x = J b^{(1)},$$

if we define

10.13
$$J = J_n \ldots J_2 J_1.$$

We have already mentioned that our objective is to have $JA^{(1)} = I$ in order that

10.14
$$X = (J_n \ldots J_2 J_1)b^{(1)}.$$

To achieve this result, we must choose J_i as follows:

10.15
$$J_1 = \left[\begin{array}{c|c} \mu_{11} & 0 \ldots 0 \\ \hline \mu_{21} & \\ \vdots & I_{n-1} \\ \mu_{n1} & \end{array}\right],$$

10.16
$$J_i = \left[\begin{array}{c|c|c} & \mu_{1i} & \\ I_{i-1} & \vdots & O \\ & \mu_{i-1,i} & \\ \hline 0 \ldots 0 & \mu_{ii} & 0 \ldots 0 \\ \hline & \mu_{i+1,i} & \\ O & \vdots & I_{n-i} \\ & \mu_{ni} & \end{array}\right], \qquad (i = 2, 3, \ldots, n-1)$$

10.17
$$J_n = \left[\begin{array}{c|c} I_{n-1} & \begin{matrix} \mu_{1n} \\ \vdots \\ \vdots \\ \mu_{n-1,n} \end{matrix} \\ \hline 0 \dots 0 & \mu_{nn} \end{array}\right],$$

where

10.18
$$\mu_{ii} = \frac{1}{a_{ii}^{(i)}},$$

if the pivotal candidate does not vanish, that is, if

10.19
$$a_{ii}^{(i)} \neq 0.$$

For $t \neq i$, in this case,

$$\mu_{ti} = -\frac{a_{ti}^{(i)}}{a_{ii}^{(i)}}.$$

If (10.19) is not satisfied, that is, if the pivotal candidate does vanish, we must do row interchanges before we can carry out that step of the elimination.

EXERCISE 13.10

Solve

$$\begin{bmatrix} 2 & 2 & -1 \\ -3 & 0 & 2 \\ 4 & -5 & -1 \end{bmatrix} \begin{bmatrix} x_1 \\ x_2 \\ x_3 \end{bmatrix} = \begin{bmatrix} 5 \\ -5 \\ 0 \end{bmatrix}$$

by Gauss-Jordan elimination.

13.11 THE ANALOG OF GAUSS-JORDAN ELIMINATION FOR RESIDUE ARITHMETIC USING A SINGLE MODULUS

Assume that A is an $n \times n$ integral matrix, nonsingular modulo m, and b is an integral vector (not null). In this case we seek an $n \times n$ integral matrix $\bar{J}$, nonsingular modulo m, with

11.1
$$|\bar{J}|_m = \bar{J},$$

such that

11.2
$$|\bar{J} A \bar{x}|_m = |\bar{J} b|_m,$$

and

11.3
$$|\bar{J}A|_m = I.$$

Thus,

11.4
$$\bar{J} = A^{-1}(m),$$

and so

11.5
$$|\bar{x}|_m = |\bar{J}b|_m.$$

Let the residue system (9.2) be rewritten using the notation

11.6
$$|\mathscr{A}^{(1)}\bar{x}|_m = \beta^{(1)}$$

where

11.7
$$\mathscr{A}^{(1)} = |A|_m,$$

and

11.8
$$\beta^{(1)} = |b|_m.$$

The reduction of $\mathscr{A}^{(1)}$ to the identity matrix is accomplished by n major steps which correspond to n row operations (using residue matrices) to be explained below. Each step of the reduction leads to a new set of equations which is equivalent to (11.6) and which is simpler than the previous set in that the new set has a column of the identity matrix (where there was not one before) in its matrix of coefficients. Thus the second set of equations is obtained by writing

11.9
$$|\bar{J}_1|\mathscr{A}^{(1)}\bar{x}|_m|_m = |\bar{J}_1\beta^{(1)}|_m,$$

which may be written

11.10
$$||\bar{J}_1\mathscr{A}^{(1)}|_m\bar{x}|_m = \beta^{(2)},$$

or

11.11
$$|\mathscr{A}^{(2)}\bar{x}|_m = \beta^{(2)},$$

if we define

11.12
$$\mathscr{A}^{(2)} = |\bar{J}_1\mathscr{A}^{(1)}|_m,$$

and

11.13
$$\beta^{(2)} = |\bar{J}_1\beta^{(1)}|_m.$$

The ith step produces

11.14
$$\big|\,|\bar{J}_i|\mathscr{A}^{(i)}\bar{x}\big|_m\big|_m = |\bar{J}_i\beta^{(i)}|_m,$$

which may be written

11.15
$$\big|\,|\bar{J}_i\mathscr{A}^{(i)}|_m\bar{x}\big|_m = \beta^{(i+1)},$$

or

11.16
$$|\mathscr{A}^{(i+1)}\bar{x}|_m = \beta^{(i+1)},$$

if we define

11.17
$$\mathscr{A}^{(i+1)} = |\bar{J}_i\mathscr{A}^{(i)}|_m,$$

and

11.18
$$\beta^{(i+1)} = |\bar{J}_i\beta^{(i)}|_m.$$

Here, as in (10.10),

11.19
$$\mathscr{A}^{(i+1)} = \left[\begin{array}{c|c} I_i & * \\ \hline \phi & * \end{array}\right].$$

Finally,

11.20
$$\big|\,|\bar{J}_n|\,\mathscr{A}^{(n)}\bar{x}\big|_m\big|_m = |\bar{J}_n\beta^{(n)}|_m,$$

which may be written

11.21
$$\big|\,|\bar{J}_n\mathscr{A}^{(n)}|_m\bar{x}\big|_m = \beta^{(n+1)},$$

or

11.22
$$|\mathscr{A}^{(n+1)}\bar{x}|_m = \beta^{(n+1)},$$

if we define

11.23
$$\mathscr{A}^{(n+1)} = |\bar{J}_n\mathscr{A}^{(n)}|_m,$$

and

11.24
$$\beta^{(n+1)} = |\bar{J}_n\beta^{(n)}|_m.$$

If we achieve our objective so that, in (11.22),

11.25
$$\mathscr{A}^{(n+1)} = I,$$

then

11.26
$$|\bar{x}|_m = \beta^{(n+1)}$$
$$= |\bar{J}_n \ldots |\bar{J}_2|\bar{J}_1 \beta^{(1)}|_m|_m \ldots|_m$$
$$= ||\bar{J}_n \ldots \bar{J}_2 \bar{J}_1|_m \beta^{(1)}|_m$$
$$= |\bar{J}\beta^{(1)}|_m$$
$$= |A^{-1}(m)b|_m.$$

In order to achieve this objective, we must have

11.27
$$\bar{J}_1 = \begin{bmatrix} \bar{\mu}_{11} & 0 \ldots 0 \\ \hline \bar{\mu}_{21} & \\ \vdots & I_{n-1} \\ \bar{\mu}_{n1} & \end{bmatrix},$$

11.28
$$\bar{J}_i = \begin{bmatrix} & \bar{\mu}_{1i} & \\ I_{i-1} & \vdots & O \\ & \bar{\mu}_{i-1,i} & \\ \hline 0\ldots0 & \bar{\mu}_{ii} & 0\ldots0 \\ \hline & \bar{\mu}_{i+1,i} & \\ O & \vdots & I_{n-1} \\ & \bar{\mu}_{ni} & \end{bmatrix}, \qquad (i = 2, 3, \ldots, n-1)$$

and

11.29
$$\bar{J}_n = \begin{bmatrix} & \bar{\mu}_{1n} \\ I_{n-1} & \vdots \\ & \bar{\mu}_{n-1,n} \\ \hline 0\ldots0 & \bar{\mu}_{nn} \end{bmatrix},$$

where, if the pivotal candidate $a_{ii}^{(i)}$ (we shall call it p_i) has a multiplicative inverse modulo m,

11.30
$$\bar{\mu}_{ii} = p_i^{-1}(m),$$

and, for $t \neq i$,

11.31 $$\bar{\mu}_{ti} = |-p_i^{-1}(m)a_{ti}^{(i)}|_m.$$

If the pivotal candidate does not have a multiplicative inverse modulo m, for some i, then we do a row interchange in order to produce a pivotal candidate which does have a multiplicative inverse modulo m. The only effect of the row interchange is a change in the sign of the determinant of A.

13.12 OBTAINING x FROM $|\bar{x}|_m$.

In the previous section we described a procedure for obtaining $|\bar{x}|_m$. However, we have already indicated that $|\bar{x}|_m$ will not be the same as x, the solution of (9.1). Fortunately, we can obtain x if we are willing to do the additional work described below.

Let

12.1 $$d = \det A$$

and

12.2 $$y = A^{adj}b.$$

Then we shall need to compute $|d|_m$ and $|y|_m$ as we are computing $|\bar{x}|_m$. From $|d|_m$ and $|y|_m$ we can compute d and y, respectively (see below). Once we have d and y we can find x immediately, since

12.3 $$x = A^{-1}b$$
$$= \frac{1}{d} A^{adj}b$$
$$= \frac{1}{d} y.$$

Since x is obtained from y by dividing the components of y by d, it should be pointed out that *only at this point do we leave residue arithmetic, and so only at this point do we introduce rounding errors*. Actually, if the division is merely indicated, but never carried out, then there will be no errors introduced at all.

We indicated above that we must compute $|d|_m$ and $|y|_m$ in addition to computing $|\bar{x}|_m$. The next two theorems tell us how this can be done.

12.4 Theorem. If $a_{11}^{(1)}, a_{22}^{(2)}, \ldots, a_{nn}^{(n)}$ are the pivots in the Gauss-Jordan elimination, then

$$|d|_m = |a_{11}^{(1)} a_{22}^{(2)} \ldots a_{nn}^{(n)}|_m.$$

Proof.

$$|d|_m = |\det A|_m$$
$$= |\det|A|_m|_m$$
$$= |\det \mathscr{A}^{(1)}|_m$$
$$= |a_{11}^{(1)} a_{22}^{(2)} \ldots a_{nn}^{(n)}|_m.$$

12.5 Theorem. We can compute $|y|_m$ using the equation

$$|y|_m = ||d|_m|\bar{x}|_m|_m.$$

Proof.

$$|\bar{x}|_m = |A^{-1}(m)|b|_m|_m$$
$$= |d^{-1}(m)|A^{adj}|_m b|_m$$
$$= |d^{-1}(m)A^{adj}b|_m$$
$$= |d^{-1}(m)y|_m.$$

Hence,

$$|d|_m|\bar{x}|_m = |d|_m|d^{-1}(m)y|_m$$

and so

$$||d|_m|\bar{x}|_m|_m = ||d|_m|d^{-1}(m)y|_m|_m$$
$$= ||dd^{-1}(m)|_m y|_m$$
$$= |y|_m.$$

Newman [1967] shows that, with proper choice of m, we can find d and y from $|d|_m$ and $|y|_m$, respectively. The next two theorems cover this point. Before we state them, however, we should recall that we have already imposed one condition on m in order to guarantee the existence of $d^{-1}(m)$, namely,

12.6 $(m, d) = 1.$

12.7 Theorem. [Newman, 1967, pp. 171–172]. If the modulus m is chosen so that

  i) $m > 2|d|,$

and if d' is formed from $|d|_m$ so as to satisfy

  ii) $|d'|_m = |d|_m$
  iii) $|d'| < m/2$

then

$$d' = d.$$

Proof. From (ii) there exists an integer k such that

$$mk = d - d'.$$

Thus

$$|mk| = |d - d'|$$

and so

$$m|k| \leqq |d| + |d'|$$

$$< \frac{m}{2} + \frac{m}{2},$$

from (i) and (iii). Thus

$$m|k| < m.$$

But k is an integer and so this last statement implies that it can only be zero. Thus

$$d = d'.$$

12.8 Theorem. [Newman, 1967, p. 172]. If, in addition to (12.6) the modulus m is chosen so that

i) $\quad m > 2 \max_i |y_i|,$

and if y' is formed from $|y|_m$ so as to satisfy

ii) $\quad |y'|_m = |y|_m,$

iii) $\max_i |y'_i| < \frac{m}{2},$

then

$$y' = y.$$

Proof. From (ii), (12.5), and (12.6), we know that

$$m|(y'_i - y_i) \qquad (i = 1, 2, \ldots, n),$$

and so integers k_i exist, for all i, such that

$$mk_i = y'_i - y_i.$$

Thus, for all i,

$$|mk_i| = |y_i' - y_i|$$

and so

$$m|k_i| \leq |y_i'| + |y_i|$$

$$< \frac{m}{2} + \frac{m}{2},$$

from (i) and (iii). Thus, for all i,

$$m|k_i| < m.$$

But k_i is an integer, for each value of i, and so each k_i must equal zero. Thus, y' and y have identical components and so

$$y' = y.$$

Notice that we now use (ii) and (iii) of Theorem 12.7 and (ii) and (iii) of Theorem 12.8, respectively, to compute d and y from $|d|_m$ and $|y|_m$ whenever m satisfies both (12.6) and

12.9 $$m > 2 \max (|d|, \ \max_i |y_i|).$$

13.13 AN EXAMPLE

The following example demonstrates the algorithm we have described in the last two sections. Let $Ax = b$ be the system

13.1 $$\begin{bmatrix} 12 & 3 \\ -3 & -1 \end{bmatrix} \begin{bmatrix} x_1 \\ x_2 \end{bmatrix} = \begin{bmatrix} -1 \\ -2 \end{bmatrix}.$$

Because of the simplicity of the example it is possible to compute d and y in order to select a modulus m which satisfies (12.6) and (12.9). This is not possible in the general case. Moreover, if we knew d and y, we would have the solution from (12.3) and there would be no reason to continue.

We easily observe that

13.2 $$d = -3,$$

and

13.3 $$y = A^{adj}b$$

$$= \begin{bmatrix} -1 & -3 \\ 3 & 12 \end{bmatrix} \begin{bmatrix} -1 \\ -2 \end{bmatrix}$$

$$= \begin{bmatrix} 7 \\ -27 \end{bmatrix}.$$

Since

13.4 $$\max_i |y_i| = 27,$$

we choose m so that

13.5 $$m > 2 \max (3, 27).$$

The smallest prime satisfying (13.5) is

13.6 $$m = 59,$$

so we shall use 59 as our modulus.

First, we wish to solve

13.7 $$|A\bar{x}|_{59} = |b|_{59},$$

where

13.8 $$|A|_{59} = \begin{bmatrix} 12 & 3 \\ 56 & 58 \end{bmatrix}$$

and

13.9 $$|b|_{59} = \begin{bmatrix} 58 \\ 57 \end{bmatrix}.$$

We begin with the augmented matrix

13.10 $$[\mathscr{A}^{(1)}, \beta^{(1)}] = \begin{bmatrix} 12 & 3 & 58 \\ 56 & 58 & 57 \end{bmatrix}.$$

We multiply the first row by $12^{-1}(59) = 5$, and reduce the results modulo 59. This gives us

$$\begin{bmatrix} 1 & 15 & 54 \\ 56 & 58 & 57 \end{bmatrix}.$$

Next we subtract 56 times the first row from the second row and reduce the results modulo 59. This gives us

13.11 $$[\mathscr{A}^{(2)}, \beta^{(2)}] = \begin{bmatrix} 1 & 15 & 54 \\ 0 & 44 & 42 \end{bmatrix}.$$

Now we multiply row two by $44^{-1}(59) = 55$ and reduce the results modulo 59. This gives us

$$\begin{bmatrix} 1 & 15 & 54 \\ 0 & 1 & 9 \end{bmatrix}.$$

Finally, we subtract 15 times row two from row one and reduce the results modulo 59. This gives us

13.12
$$[\mathscr{A}^{(3)}, \beta^{(3)}] = \begin{bmatrix} 1 & 0 & 37 \\ 0 & 1 & 9 \end{bmatrix},$$

and so

13.13
$$|\bar{x}|_{59} = \begin{bmatrix} 37 \\ 9 \end{bmatrix}.$$

To compute $|d|_{59}$ we use Theorem 12.4 and so

13.14
$$|d|_{59} = |a_{11}^{(1)} a_{22}^{(2)}|_{59}$$
$$= |12 \cdot 44|_{59}$$
$$= 56.$$

To compute $|y|_{59}$ we use Theorem 12.5 and so

13.15
$$|y|_{59} = \left| 56 \cdot \begin{bmatrix} 37 \\ 9 \end{bmatrix} \right|_{59}$$
$$= \begin{bmatrix} 7 \\ 32 \end{bmatrix}.$$

Now we use Theorem 12.7 and choose d' so that $|d'|_{59} = 56$ and $|d'| \leq 29$. Hence

13.16
$$d' = -3.$$

Notice that this agrees with (13.2) as we should expect.

Next we use Theorem 12.8 and choose y' so that both

$$|y'|_{59} = \begin{bmatrix} 7 \\ 32 \end{bmatrix},$$

and $\max_{i} |y_i'| \leqq 29$. Hence

13.17
$$y' = \begin{bmatrix} 7 \\ -27 \end{bmatrix},$$

which, of course, agrees with (12.3).

Therefore,

13.18
$$x = \frac{1}{d'} y'$$

$$= -\tfrac{1}{3} \begin{bmatrix} 7 \\ -27 \end{bmatrix}$$

$$= \begin{bmatrix} -\tfrac{7}{3} \\ 9 \end{bmatrix}.$$

As a check we observe that

13.19
$$\begin{bmatrix} 12 & 3 \\ -3 & -1 \end{bmatrix} \begin{bmatrix} -\tfrac{7}{3} \\ 9 \end{bmatrix} = \begin{bmatrix} -1 \\ -2 \end{bmatrix}.$$

EXERCISE 13.13

Solve

$$\begin{bmatrix} 2 & 2 & -1 \\ -3 & 0 & 2 \\ 4 & -5 & -1 \end{bmatrix} \begin{bmatrix} x_1 \\ x_2 \\ x_3 \end{bmatrix} = \begin{bmatrix} 5 \\ -5 \\ 0 \end{bmatrix}$$

using the method of Section 13.13.

The Computational Procedure Using More Than One Modulus

13.14 INTRODUCTION

In the last section we gave a simple example which illustrates the basic procedure for solving $Ax = b$ using a single modulus m. Actually, it is more practical in a general situation to select a set of moduli $m_1, m_2, \ldots, m_s$, with

14.1
$$M = m_1 m_2 \ldots m_s,$$

because, as we shall see in subsequent sections, this enables us to obtain results modulo M by doing most of the arithmetic modulo m_i, for $i = 1, 2, \ldots, s$.

To be more specific, we select a set of moduli $m_1, m_2, \ldots, m_s$, with†

14.2
$$(m_i, m_j) = 1,$$

for $i \neq j$. M is defined by (14.1). We shall assume that there are sufficient moduli (of sufficient size) so that M satisfies

14.3
$$(d, M) = 1,$$

and

14.4
$$M > 2 \max (|d|, \max_i |y_i|),$$

where d and y are defined by (12.1) and (12.2), respectively.

We solve the residue systems

14.5
$$|A\bar{x}|_{m_i} = |b|_{m_i},$$

for $i = 1, 2, \ldots, s$, by the single-modulus procedure described in the previous sections, and obtain the residue representations

14.6
$$d \sim \{|d|_{m_1}, |d|_{m_2}, \ldots, |d|_{m_s}\}$$

and

14.7
$$y \sim \{|y|_{m_1}, |y|_{m_2}, \ldots, |y|_{m_s}\}.$$

From these two s-tuples we can determine $|d|_M$ and $|y|_M$ (see (15.1), for example), and this means that if M is large enough (i.e., if (14.4) is satisfied), we then can determine d and y and, ultimately, $x = A^{-1}b$. Algorithms using this general procedure are described in Section 13.16 and in Howell and Gregory [1970].

13.15 THE CHINESE REMAINDER THEOREM

There are various algorithms for obtaining $|d|_M$ and $|y|_M$ from (14.6) and (14.7), respectively. Perhaps the best known (but not the fastest) procedure makes use of a classic theorem from the theory of numbers called the *Chinese Remainder Theorem*.

15.1 Theorem. * Let $m_1, m_2, \ldots, m_s$ be the base for a residue number system where $(m_i, m_j) = 1$ for $i \neq j$, and let

$$M = m_1 m_2 \ldots m_s.$$

---

† For the necessity of (14.2), see (15.1).

Also, let

$$\hat{m}_j = \frac{M}{m_j}.$$

Now, if q has the residue representation

$$q \sim \{r_1, r_2, \ldots, r_s\},$$

where

$$r_i = |q|_{m_i} \qquad i = 1, 2, \ldots, s,$$

then

$$|q|_M = \left| \sum_{j=1}^{s} \hat{m}_j |r_j \hat{m}_j^{-1}(m_j)|_{m_j} \right|_M$$

$$= |\hat{m}_1 |r_1 \hat{m}_1^{-1}(m_1)|_{m_1} + \cdots + \hat{m}_s |r_s \hat{m}_s^{-1}(m_s)|_{m_s}|_M.$$

15.2 PROBLEM. If $m_1 = 13$, $m_2 = 11$, $m_3 = 7$, and $m_4 = 9$, find $|q|_M$ from the residue representation

$$q \sim \{4, 2, 4, 7\}.$$

Solution. First we compute

$m_1 = 13$		$\hat{m}_1 = 693$
$m_2 = 11$		$\hat{m}_2 = 819$
$m_3 = 7$	$M = 9009$	$\hat{m}_3 = 1287$
$m_4 = 9$		$\hat{m}_4 = 1001.$

From (2.5) and (3.8) we find it useful to compute

$$|\hat{m}_1|_{m_1} = |693|_{13} = 4$$
$$|\hat{m}_2|_{m_2} = |819|_{11} = 5$$
$$|\hat{m}_3|_{m_3} = |1287|_7 = 6$$
$$|\hat{m}_4|_{m_4} = |1001|_9 = 2$$

in order to find

$$\hat{m}_1^{-1}(m_1) = 4^{-1}(13) = 10$$
$$\hat{m}_2^{-1}(m_2) = 5^{-1}(11) = 9$$

$$\hat{m}_3^{-1}(m_3) = 6^{-1}(7) = 6$$

$$\hat{m}_4^{-1}(m_4) = 2^{-1}(9) = 5.$$

Thus

$$|q|_{9009} = \left| 693|(4)(10)|_{13} + 819|(2)(9)|_{11} + 1287|(4)(6)|_7 + 1001|(7)(5)|_9 \right|_{9009}$$

$$= |693(1) + 819(7) + 1287(3) + 1001(8)|_{9009}$$

$$= |18295|_{9009}$$

$$= 277.$$

EXERCISE 13.15

Let $2, 3, 5, 7$ be the base for a residue number system. If q has the residue representation $\{1, 2, 3, 2\}$, find $|q|_{210}$ using the Chinese Remainder Theorem.

13.16 ALGORITHM I

Given a set of moduli $m_1, m_2, \ldots, m_s$, we indicated in Section 13.14 how to obtain residue representations for $d = \det A$ and $y = A^{adj}b$. In Section 13.15 we indicated how to use the Chinese Remainder Theorem to compute $|d|_M$ and $|y|_M$ from these residue representations. Back in Section 13.12 we indicated how to obtain d and y from $|d|_M$ and $|y|_M$, respectively. Since, from (12.3),

16.1 $$x = \frac{1}{d} y,$$

this completely outlines a computational procedure for solving $Ax = b$. This procedure will be called Algorithm I throughout the remainder of this chapter.†

A discussion of methods for selecting the moduli $m_1, m_2, \ldots, m_s$ is postponed until Section 13.17.

To illustrate Algorithm I we return to the example in Section 13.13, where

16.2 $$A = \begin{bmatrix} 12 & 3 \\ -3 & -1 \end{bmatrix},$$

and

16.3 $$b = \begin{bmatrix} -1 \\ -2 \end{bmatrix}.$$

We shall choose two moduli, $m_1 = 7$ and $m_2 = 11$. Hence $M = 77$. Notice that we are safe because M is larger than the single modulus used in Section 13.13.

---

† An algorithm called Algorithm II is described in Howell and Gregory [1970].

First we solve the residue system

16.4 $$|A\bar{x}|_7 = |b|_7,$$

for $|\bar{x}|_7$, $|d|_7$, and $|y|_7$. Since

16.5 $$|A|_7 = \begin{bmatrix} 5 & 3 \\ 4 & 6 \end{bmatrix}$$

and

16.6 $$|b|_7 = \begin{bmatrix} 6 \\ 5 \end{bmatrix},$$

we begin with the augmented matrix

16.7 $$[\mathscr{A}^{(1)}, \beta^{(1)}] = \begin{bmatrix} 5 & 3 & 6 \\ 4 & 6 & 5 \end{bmatrix}.$$

This matrix can be reduced to

16.8 $$[\mathscr{A}^{(3)}, \beta^{(3)}] = \begin{bmatrix} 1 & 0 & 0 \\ 0 & 1 & 2 \end{bmatrix},$$

using the single-modulus procedure which was illustrated in detail in Section 13.13. The two pivots are 5 and 5. Thus

16.9 $$|\bar{x}|_7 = \begin{bmatrix} 0 \\ 2 \end{bmatrix},$$

16.10 $$|d|_7 = |(5)(5)|_7$$
$$= 4,$$

and

16.11 $$|y|_7 = ||d|_7|\bar{x}|_7|_7$$
$$= \begin{bmatrix} 0 \\ 1 \end{bmatrix}.$$

Next, we solve the residue system

16.12 $$|A\bar{x}|_{11} = |b|_{11},$$

for $|\bar{x}|_{11}$, $|d|_{11}$, and $|y|_{11}$. Since

16.13
$$|A|_{11} = \begin{bmatrix} 1 & 3 \\ 8 & 10 \end{bmatrix},$$

and

16.14
$$|b|_{11} = \begin{bmatrix} 10 \\ 9 \end{bmatrix},$$

we begin with the augmented matrix

16.15
$$[\mathscr{A}^{(1)}, \beta^{(1)}] = \begin{bmatrix} 1 & 3 & 10 \\ 8 & 10 & 9 \end{bmatrix}.$$

This matrix can be reduced to

16.16
$$[\mathscr{A}^{(3)}, \beta^{(3)}] = \begin{bmatrix} 1 & 0 & 5 \\ 0 & 1 & 9 \end{bmatrix}$$

where the two pivots are 1 and 8. Thus,

16.17
$$|\bar{x}|_{11} = \begin{bmatrix} 5 \\ 9 \end{bmatrix},$$

16.18
$$|d|_{11} = |(1)(8)|_{11}$$
$$= 8,$$

and

16.19
$$|y|_{11} = ||d|_{11}|\bar{x}|_{11}|_{11}$$
$$= \begin{bmatrix} 7 \\ 6 \end{bmatrix}.$$

We now have the residue representations

16.20
$$d \sim \{4, 8\}$$

and

16.21
$$y \sim \left\{ \begin{bmatrix} 0 \\ 1 \end{bmatrix}, \begin{bmatrix} 7 \\ 6 \end{bmatrix} \right\}$$

or, what is more appropriate from this point on,

16.21a $$y_1 \sim \{0, 7\}$$

and

16.21b $$y_2 \sim \{1, 6\}.$$

To use the Chinese Remainder Theorem, we first compute

$$\begin{array}{ccc} m_1 = & 7 & \hat{m}_1 = 11 \\ & M = 77 & \\ m_2 = & 11 & \hat{m}_2 = & 7, \end{array}$$

which means

$$|\hat{m}_1|_{m_1} = |11|_7 = 4$$
$$|\hat{m}_2|_{m_2} = |7|_{11} = 7,$$

and so

$$\hat{m}_1^{-1}(m_1) = 4^{-1}(7) = 2$$
$$\hat{m}_2^{-1}(m_2) = 7^{-1}(11) = 8.$$

Therefore,

$$\begin{aligned} |d|_{77} &= \big| |11|(4)(2)|_7 + 7|(8)(8)|_{11} \big|_{77} \\ &= |11(1) + 7(9)|_{77} \\ &= 74, \\ |y_1|_{77} &= \big| |11|(0)(2)|_7 + 7|(7)(8)|_{11} \big|_{77} \\ &= |7(1)|_{77} \\ &= 7, \end{aligned}$$

and

$$\begin{aligned} |y_2|_{77} &= \big| |11|(1)(2)|_7 + 7|(6)(8)|_{11} \big|_{77} \\ &= |11(2) + 7(4)|_{77} \\ &= 50. \end{aligned}$$

We now use Theorems 12.7 and 12.8 to compute d and y, respectively. We seek numbers less than $\frac{77}{2}$ in magnitude which are congruent, modulo 77, to 74,

7, and 50, respectively. These are easily seen to be -3, 7, and -27. Thus,

16.22
$$d = -3,$$

and

16.23
$$y = \begin{bmatrix} 7 \\ -27 \end{bmatrix},$$

which, of course, agree with (13.16) and (13.17).

From (16.1) we obtain the solution

16.24
$$x = \begin{bmatrix} -\frac{7}{3} \\ 9 \end{bmatrix},$$

which obviously agrees with (13.18).

Notice that in this section and in Section 13.13 we have solved the same set of equations. Here we used the moduli $m_1 = 7$ and $m_2 = 11$ and most of the arithmetic was performed either modulo 7 or modulo 11. Only when we used the Chinese Remainder Theorem, in computing $|d|_{77}$ and $|y|_{77}$, did we do any arithmetic modulo 77. In Section 13.13, on the other hand all the arithmetic was done modulo 59.

EXERCISE 13.16

Solve

$$\begin{bmatrix} 2 & 2 & -1 \\ -3 & 0 & 2 \\ 4 & -5 & -1 \end{bmatrix} \begin{bmatrix} x_1 \\ x_2 \\ x_3 \end{bmatrix} = \begin{bmatrix} 5 \\ -5 \\ 0 \end{bmatrix}$$

using the method of Section 13.16 with two moduli.

Selection of the Moduli

13.17 INTRODUCTION

We recall from (12.9) that if we choose m so that

17.1
$$m > 2 \max (|d|, \ \max |y_i|),$$

then solving the residue system

$$|A\bar{x}|_m = |b|_m$$

for $|d|_m$ and $|y|_m$ enables us to find d and y, respectively. In other words, by choosing m large enough, we can guarantee that $|d|$ and $\max_i |y_i|$ are in the interval $\left(-\dfrac{m}{2}, \dfrac{m}{2}\right)$, and therefore guarantee that we can find a solution vector x by the method described in this chapter. An ideal computer program for this algorithm would first determine a permissible value for m, that is, a lower bound for m, then choose a set of moduli so that their product is greater than or equal to the permissible value for m.

We shall consider a method, described by Newman [1967], for finding a lower bound for m, and then describe how the set of moduli are chosen.

13.18 CALCULATION OF A PERMISSIBLE VALUE FOR m

For any vector z or matrix A define

18.1
$$M(z) = \max_i |z_i|$$

and

18.2
$$M(A) = \max_{i,j} |a_{ij}|.$$

From (17.1) we see that a lower bound for m can be found by first computing a maximum value for $|d|$ and $M(y)$, and then choosing the lower bound for m as twice the larger of the upper bound for $|d|$ and the upper bound for $M(y)$.

Considering first an upper bound for $|d|$, we note that by Hadamard's inequality

18.3
$$|d|^2 \leq \sum_{j=1}^{n} |a_{1j}|^2 \sum_{j=1}^{n} |a_{2j}|^2 \cdots \sum_{j=1}^{n} |a_{nj}|^2.$$

Since

18.4
$$|a_{ij}| \leq M(A), \quad \begin{cases} 1 \leq i \leq n \\ 1 \leq j \leq n, \end{cases}$$

then clearly

18.5
$$|a_{ij}|^2 \leq M(A)^2$$

and

18.6
$$\sum_{j=1}^{n} |a_{ij}|^2 \leq n M(A)^2$$

for $1 \leq i \leq n$. Since A is real, then (18.6) may be written

18.7
$$\sum_{j=1}^{n} a_{ij}^2 \leq nM(A)^2.$$

Then from (18.3) and (18.7) we have

18.8
$$|d|^2 \leq \sum_{j=1}^{n} a_{1j}^2 \sum_{j=1}^{n} a_{2j}^2 \cdots \sum_{j=1}^{n} a_{nj}^2$$

$$\leq (nM(A)^2)^n$$

$$= n^n M(A)^{2n}$$

and so

18.9
$$|d| \leq n^{n/2} M(A)^n,$$

and we have an upper bound for $|d|$.

Considering next an upper bound for $M(y)$, we recall that

18.10
$$M(y) = \max_{j} |y_j|$$

and

$$y = A^{adj} b.$$

Then

18.11
$$M(y) = \max_{j} \left| \sum_{i=1}^{n} A_{ij} b_i \right|,$$

where A_{ij} is the cofactor of the element a_{ij} of the matrix A. Hence

18.12
$$M(y) \leq \max_{j} \left| \sum_{i=1}^{n} A_{ij} \right| \cdot M(b)$$

$$\leq n \cdot \max_{i,j} |A_{ij}| \cdot M(b).$$

From (18.9), we obtain

18.13
$$|A_{ij}| \leq (n-1)^{(n-1)/2} M(P_{ij})^{n-1}$$

where P_{ij} is the $n - 1$ by $n - 1$ submatrix of A formed by deleting row i and column j and

18.14
$$|\det P_{ij}| = |A_{ij}|.$$

Since

18.15
$$M(P_{ij}) \leqq M(A),$$

then, for all i and j,

18.16
$$|A_{ij}| \leqq (n - 1)^{(n-1)/2} M(A)^{n-1}.$$

Hence, from (18.12) and (18.16)

18.17
$$M(y) \leqq n(n - 1)^{(n-1)/2} M(A)^{n-1} M(b),$$

and we have an upper bound for $M(y)$. Thus, if we want

$$m > 2 \max (|d|, \qquad M(y)),$$

then by (18.9) and (18.17), we can choose

18.18 $\qquad m > 2 \max \left(n^{n/2} M(A)^n, \qquad n(n - 1)^{(n-1)/2} M(A)^{n-1} M(b) \right)$

where $(m, d) = 1$.

This latter choice for m is unnecessarily conservative in most cases. It would indeed become time consuming in a computer program to scan A and b to find $M(A)$ and $M(b)$, and to compute upper bounds for $|d|$ and $M(y)$ before applying the theory for each problem to be solved.

A practical alternative to the procedure outlined above would be to operate with a sufficiently large predetermined set of moduli, $m_1, m_2, \ldots, m_r$, and to choose $m = m_1 m_2 \ldots m_s$ $(s \leqq r)$ such that

18.19
$$m \geqq 2 \prod_{i=1}^{n} \left(\sum_{k=1}^{n} a_{ik}^2 \right)^{1/2} \sum_{j=1}^{n} |b_j|.$$

One advantage of (18.19) is that the bound is easily computed.

We can show that the above lower bound is indeed a good lower bound by showing that it satisfies (17.1). From (18.3)

18.20
$$|d|^2 = \prod_{i=1}^{n} \left(\sum_{k=1}^{n} a_{ik}^2 \right).$$

Then

18.21
$$|d| \leqq \left(\prod_{i=1}^{n} \left(\sum_{k=1}^{n} a_{ik}^2 \right) \right)^{1/2}$$

$$= \prod_{i=1}^{n} \left(\sum_{k=1}^{n} a_{ik}^2 \right)^{1/2},$$

and so

18.22
$$2|d| \leqq 2 \prod_{i=1}^{n} \left(\sum_{k=1}^{n} a_{ik}^2 \right)^{1/2}$$

$$< 2 \prod_{i=1}^{n} \left(\sum_{k=1}^{n} a_{ik}^2 \right)^{1/2} \sum_{j=1}^{n} |b_j|$$

$$\leqq m.$$

Hence, if we choose m by (18.19), then

18.23
$$m > 2|d|.$$

From (18.3), we see that

18.24
$$|A_{jr}| \leqq \prod_{\substack{i=1 \\ i \neq j}}^{n} \left(\sum_{\substack{k=1 \\ k \neq r}}^{n} a_{ik}^2 \right)^{1/2}.$$

Also, from (18.11)

18.25
$$M(y) = \max_r \left| \sum_{j=1}^{n} A_{jr} b_j \right|$$

$$\leqq \max_r \left| \sum_{j=1}^{n} |A_{jr}| |b_j| \right|.$$

Then, from (18.24)

18.26
$$M(y) \leqq \max_r \left| \left(\sum_{j=1}^{n} \prod_{\substack{i=1 \\ i \neq j}}^{n} \left(\sum_{\substack{k=1 \\ k \neq r}}^{n} a_{ik}^2 \right)^{1/2} \right) |b_j| \right|$$

$$< \max_r \left| \sum_{j=1}^{n} \left(\prod_{i=1}^{n} \left(\sum_{\substack{k=1 \\ k \neq r}}^{n} a_{ik}^2 \right)^{1/2} \right) |b_j| \right|$$

$$= \max_r \left| \left(\prod_{i=1}^{n} \left(\sum_{\substack{k=1 \\ k \neq r}}^{n} a_{ik}^2 \right)^{1/2} \right) \cdot \sum_{j=1}^{n} |b_j| \right|$$

$$= \max_r \left| \prod_{i=1}^{n} \left(\sum_{\substack{k=1 \\ k \neq r}}^{n} a_{ik}^2 \right)^{1/2} \right| \cdot \sum_{j=1}^{n} |b_j|$$

$$< \max_r \left| \prod_{i=1}^{n} \left(\sum_{k=1}^{n} a_{ik}^2 \right)^{1/2} \right| \cdot \sum_{j=1}^{n} |b_j|$$

$$= \prod_{i=1}^{n} \left(\sum_{k=1}^{n} a_{ik}^2 \right)^{1/2} \sum_{j=1}^{n} |b_j|.$$

Then

18.27
$$2M(y) < 2 \prod_{i=1}^{n} \left(\sum_{k=1}^{n} a_{ik}^2 \right)^{1/2} \sum_{j=1}^{n} |b_j|$$

$$\leqq m,$$

and (17.1) is satisfied.

Borosh and Fraenkel [1966] suggest a different method for choosing enough moduli so that m satisfies (17.1). They propose solving the residue system

18.28
$$|A\bar{x}|_{m_k} = |b|_{m_k},$$

and then combining results, by the Chinese Remainder Theorem, with previously combined results (in other words, results combined modulo $m_1 m_2 \ldots m_{k-1}$). The procedure is continued until the solutions produced in the k-th and $(k+1)$-st steps are the same. When this is the case, a substitution check is made to determine whether still another modulus is required.

EXERCISE 13.18
Verify (18.19).

13.19 SELECTION OF THE m_i AS PRIMES
In practice, the moduli are chosen as large prime numbers. This choice increases the probability that

19.1
$$(d, m_i) = 1$$

and that

19.2
$$|d|_{m_i} \neq 0.$$

We recall that if these two conditions are satisfied, then A is nonsingular modulo m_i, and the residue system

$$|A\bar{x}|_{m_i} = |b|_{m_i}$$

can be solved for $|d|_{m_i}$ and $|y|_{m_i}$. If the two conditions are not satisfied, then we simply select another prime for a modulus. Furthermore, by choosing the moduli as prime numbers, we guarantee that

$$(m_i, m_j) = 1$$

for $i \neq j$, and hence, by (4.4), there is a unique integer in the interval $(0, m - 1)$ with a given residue representation.

If a computer for which a program is being written is capable of multiplying or adding together any pair of integers a and b, such that

19.3 $$|a| < K$$

and

19.4 $$|b| < K$$

where K is a given constant for that computer, then Newman [1967] suggests choosing as moduli the s largest primes less than K. Thus, the sum or product of any two numbers, which normally fills a computer word, will, when reduced modulo m, be less than K. This prevents large numbers from accumulating during computations and hence prevents the possibility of overflow.

EXERCISE 13.19

Write a computer program to solve a system of linear algebraic equations with integral coefficients. Use the algorithm described in this chapter.

13.20 NUMERICAL RESULTS

A computer program for solving linear algebraic equations, using the procedure of this chapter, has been written for the CDC 6600 computer of The University of Texas at Austin. The following results were obtained from systems of equations where coefficient matrices are known to be ill-conditioned.

20.1 EXAMPLE.

$$A = \begin{bmatrix} 5 & 7 & 6 & 5 \\ 7 & 10 & 8 & 7 \\ 6 & 8 & 10 & 9 \\ 5 & 7 & 9 & 10 \end{bmatrix}, \quad b = \begin{bmatrix} 23 \\ 32 \\ 33 \\ 31 \end{bmatrix}.$$

	Machine Results	Exact Solution
	$1.0000000E + 00$	
	$1.0000000E + 00$	
	$1.0000000E + 00$	
	$1.0000000E + 00$	

$$x = \begin{bmatrix} 1 \\ 1 \\ 1 \\ 1 \end{bmatrix}.$$

20.2 EXAMPLE.

$$A = \begin{bmatrix}
10 & 9 & 8 & 7 & 6 & 5 & 4 & 3 & 2 & 1 \\
9 & 9 & 8 & 7 & 6 & 5 & 4 & 3 & 2 & 1 \\
8 & 8 & 8 & 7 & 6 & 5 & 4 & 3 & 2 & 1 \\
7 & 7 & 7 & 7 & 6 & 5 & 4 & 3 & 2 & 1 \\
6 & 6 & 6 & 6 & 6 & 5 & 4 & 3 & 2 & 1 \\
5 & 5 & 5 & 5 & 5 & 5 & 4 & 3 & 2 & 1 \\
4 & 4 & 4 & 4 & 4 & 4 & 4 & 3 & 2 & 1 \\
3 & 3 & 3 & 3 & 3 & 3 & 3 & 3 & 2 & 1 \\
2 & 2 & 2 & 2 & 2 & 2 & 2 & 2 & 2 & 1 \\
1 & 1 & 1 & 1 & 1 & 1 & 1 & 1 & 1 & 1
\end{bmatrix}, \quad b = \begin{bmatrix} 1 \\ 2 \\ -5 \\ 9 \\ 15 \\ 1 \\ 6 \\ 14 \\ 3 \\ 1 \end{bmatrix}.$$

	Machine Results	Exact Solution
	$-1.0000000E + 00$	
	$8.0000000E + 00$	
	$-2.1000000E + 01$	
	$8.0000000E + 00$	
	$2.0000000E + 01$	
	$-1.9000000E + 01$	
	$-3.0000000E + 00$	
	$1.9000000E + 01$	
	$-9.0000000E + 00$	
	$-1.0000000E + 00$	

$$x = \begin{bmatrix} -1 \\ 8 \\ -21 \\ 8 \\ 20 \\ -19 \\ -3 \\ 19 \\ -9 \\ -1 \end{bmatrix}.$$

20.3 EXAMPLE.

$$A = \begin{bmatrix} 2 & -1 & 0 & 0 & 0 & 0 & 0 & 0 \\ -1 & 2 & -1 & 0 & 0 & 0 & 0 & 0 \\ 0 & -1 & 2 & -1 & 0 & 0 & 0 & 0 \\ 0 & 0 & -1 & 2 & -1 & 0 & 0 & 0 \\ 0 & 0 & 0 & -1 & 2 & -1 & 0 & 0 \\ 0 & 0 & 0 & 0 & -1 & 2 & -1 & 0 \\ 0 & 0 & 0 & 0 & 0 & -1 & 2 & -1 \\ 0 & 0 & 0 & 0 & 0 & 0 & -1 & 2 \end{bmatrix}, \quad b = \begin{bmatrix} -1 \\ 1 \\ -1 \\ 1 \\ -1 \\ 1 \\ -1 \\ 1 \end{bmatrix}.$$

Machine Results	Exact Solution
$-4.4444444E - 01$	
$1.1111111E - 01$	
$-3.3333333E - 01$	
$2.2222222E - 01$	
$-2.2222222E - 01$	
$3.3333333E - 01$	
$-1.1111111E - 01$	
$4.4444444E - 01$	

$$x = \begin{bmatrix} -4/9 \\ 1/9 \\ -1/3 \\ 2/9 \\ -2/9 \\ 1/3 \\ -1/9 \\ 4/9 \end{bmatrix}.$$

SUPPLEMENTARY DISCUSSION

In the supplementary discussion at the end of Chapter 2 we mentioned the work of Takahasi and Ishibashi [1961], Lindamood [1964], Borosh and Fraenkel [1966], Newman [1967], Howell and Gregory [1970], and McClellan [1971]. We should also mention Howell and Gregory [1969b] which contains a condensed version of the material in this chapter. An improved version of the algorithm (described in this chapter) has been published by Howell [1971] and is available on magnetic tape. For information about how to obtain copies of the tape write to the Editor, Algorithms Section, *Communications of the Association for Computing Machinery.*

CHAPTER 14

THE ALGEBRAIC EIGENVALUE—EIGENVECTOR PROBLEM

14.1 INTRODUCTION

In Section 11.21 we discussed the mathematical formulation of the algebraic eigenvalue-eigenvector problem.* In the present chapter we discuss the mathematical and the computational aspects† of algorithms for solving the problem.

From a mathematical point of view we might assume that a good algorithm for finding the eigenvalues of a matrix $A \in \mathbf{C}^{nn}$ would be to obtain the characteristic polynomial

1.1
$$\det(A - \lambda I) = (-1)^n[\lambda^n - c_{n-1}\lambda^{n-1} - \cdots - c_1\lambda - c_0]$$

and then to compute the zeros of this polynomial. However, Wilkinson [1963, Chapter 2, section 7] discusses the *condition* of a polynomial with respect to the computation of its zeros and he exhibits an ill-conditioned ** polynomial

1.2
$$P_{20}(x) = (x - 1)(x - 2)(x - 3)\cdots(x - 20)$$

whose zeros $x_1 = 1, x_2 = 2, \ldots, x_{20} = 20$, are quite different from the zeros of the slightly perturbed polynomial $Q_{20}(x)$, which is defined by

1.3
$$Q_{20}(x) = P_{20}(x) - 2^{-23}x^{19}.$$

In fact, $Q_{20}(x)$ has only ten real zeros (the other ten consist of five complex conjugate pairs, four of which lie in the complex plane between 1.6 and 2.9 units away from the real axis).

Since the coefficients $c_0, c_1, \ldots, c_{n-1}$ in (1.1) must be computed from the elements of A, it is quite possible for rounding errors during the computation to introduce perturbations in these coefficients and (1.2) and (1.3) demonstrate that if the polynomial is ill-conditioned, this can be disastrous! Consequently, *we never compute the coefficients of the characteristic polynomial of a matrix in order to compute its eigenvalues.* As a matter of fact, we sometimes try to find the zeros of a polynomial by computing the eigenvalues of its associated *companion matrix.*‡

---

* See 11-(21.3).
† Recall Section 1.3.
** Recall 1-(4.7).
‡ See Section 5.6.

The property of matrices most widely used in constructing algorithms for computing the eigenvalues is the property that *all matrices in the same similarity class have the same set of eigenvalues**. This fact enables us to take an arbitrary matrix and to transform it (using a similarity transformation) into a matrix of simpler form, in some sense. What we usually mean by "simpler form" is a form with more off-diagonal elements equal to zero.

In Section 2 we discuss computational algorithms which are especially designed for solving the eigenvalue-eigenvector problem for general (not necessarily sparse) Hermitian matrices. Methods for arbitrary (in particular, non-Hermitian) matrices are discussed in Section 3.

This is a natural way to classify matrices for computational purposes because the problem of computing eigenvalues and eigenvectors for Hermitian matrices is much simpler than the problem for non-Hermitian matrices. There are several reasons for this. To be specific, every Hermitian matrix $A \in \mathbf{C}^{nn}$ is diagonalizable† under a unitary similarity transformation, that is, for every A there exists a unitary matrix R such that

1.4 $R^H A R = \Lambda$

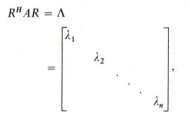

$$= \begin{bmatrix} \lambda_1 & & & \\ & \lambda_2 & & \\ & & \cdot & \\ & & & \cdot \\ & & & & \lambda_n \end{bmatrix},$$

where the eigenvalues $\lambda_1, \lambda_2, \ldots, \lambda_n$ are real, and A has a complete, orthonormal set of eigenvectors (the columns of R). Furthermore, when $A = A^H$ the matrix is well-conditioned** with respect to the eigenvalue problem which means that small perturbations in the matrix elements produce only small perturbations in the eigenvalues.†

None of these statements applies, in general, to a non-Hermitian matrix, which means we have to worry about whether it is defective or not (we may have to compute generalized eigenvectors if it is defective) and whether or not the eigenvalues are real (we sometimes have to compute in complex arithmetic). Also, we may have difficulty because of the accumulation of rounding errors, since the problem may be ill-conditioned.

Localization Theorems

It might be appropriate at this point to state two theorems due to Gerschgorin

---

* See 11-(21.31).
† See 11-(21.66).
** See 1-(4.7) as well as (5.14) and (5.31).
‡ See Section 5.

[1931] which give us a rough idea as to where (in the complex plane) the eigenvalues are located.

1.5 Theorem. Every eigenvalue of $A \in \mathbf{C}^{nn}$ lies in at least one of the circular discs (called Gerschgorin discs) with centers at a_{ii} and radii

$$\rho_i = \sum_{\substack{j=1 \\ j \neq i}}^{n} |a_{ij}|,$$

respectively, for $i = 1, 2, \ldots, n$.

Proof. Consider any eigenvalue λ of A. Then, for some $x \neq \phi$ we can write $Ax = \lambda x$ which implies

$$\sum_{j=1}^{n} a_{ij}x_j = \lambda x_i$$

for $i = 1, 2, \ldots, n$. If we normalize x so that $\max_i |x_i| = 1$ and if $i = r$ yields the component with maximum modulus, then

$$\lambda - a_{rr} = a_{r1}x_1 + \cdots + a_{r,r-1}x_{r-1} + a_{r,r+1}x_{r+1} + \cdots + a_{rn}x_n$$

which implies

$$|\lambda - a_{rr}| \leqq \sum_{\substack{j=1 \\ j \neq r}}^{n} |a_{rj}|.$$

In other words, $|\lambda - a_{rr}| \leqq \rho_r$ and λ lies in the Gerschgorin disc with centre at a_{rr}.

The second theorem gives more specific information about the location of the eigenvalues.

1.6 Theorem. If t of the Gerschgorin discs of the previous theorem form a connected domain which is isolated from the other discs, then exactly t of the eigenvalues of A lie in this connected domain.

Proof. See Wilkinson [1965], p. 71.

1.7 Corollary. If any Gerschgorin disc is isolated, then it contains exactly one eigenvalue of A.

1.8 Corollary If, in row i of $A \in \mathbf{C}^{nn}$, the only non-zero element is a_{ii}, then a_{ii} is an eigenvalue of A.

EXERCISES 14.1

1. Prove Theorem 1.6.

2. Prove Corollary 1.7.

3. Prove Corollary 1.8.
4. Construct the Gerschgorin discs for:

$$A = \begin{bmatrix} -5 & 0 & 1 \\ 1 & 1 & -2 \\ -3 & 0 & 8 \end{bmatrix}$$

$$B = \begin{bmatrix} 4 & 1 & -1 \\ -2 & 0 & 1 \\ 1 & 0 & -4 \end{bmatrix}$$

$$C = \begin{bmatrix} 3 & 1 & 0 & 2i \\ 1 & 3 & -2i & 0 \\ 0 & 2i & 1 & 1 \\ -2i & 0 & 1 & 1 \end{bmatrix}.$$

14.2 HERMITIAN MATRICES

We mentioned in the introduction that this section would deal with Hermitian matrices. As a matter of fact we shall restrict ourselves to real symmetric matrices, for simplicity, because the results are easily extended to complex Hermitian matrices. Moreover, in this case, the eigenvectors can always be chosen to be real* and so *all* computations can be done in real arithmetic.

2.1 *Remark.* If we do not wish to use complex arithmetic for complex Hermitian matrices (even though the eigenvalues are real, the eigenvectors are usually complex) we can write the eigenvalue equation

$$Ax = \lambda x,$$

for $A^H = A$ and $A \in \mathbf{C}^{nn}$, in the form

$$(C + iD)(u + iv) = \lambda(u + iv)$$

where $C \in \mathbf{R}^{nn}$ is symmetric and $D \in \mathbf{R}^{nn}$ is *skew symmetric*, that is, $D^T = -D$. If we equate the real and imaginary parts of the equation above, we obtain the real symmetric problem

$$\begin{bmatrix} C & -D \\ D & C \end{bmatrix} \begin{bmatrix} u \\ v \end{bmatrix} = \lambda \begin{bmatrix} u \\ v \end{bmatrix}$$

---

*See 11-(21.69).

where the matrix belongs to $\mathbf{R}^{2n,2n}$. It is easily verified that if A has the eigenvalues $\lambda_1, \lambda_2, \ldots, \lambda_n$, then the new problem has $2n$ eigenvalues $\lambda_1, \lambda_1, \lambda_2, \lambda_2, \ldots, \lambda_n, \lambda_n$.

The Jacobi Algorithm

This algorithm was suggested by Jacobi in 1846 (see Greenstadt [1960], for example). To motivate the algorithm, consider the quadratic form* in two real variables

2.2
$$Q_2 = a_{11}x_1^2 + a_{12}x_1x_2 + a_{21}x_2x_1 + a_{22}x_2^2$$

$$= [x_1x_2]\begin{bmatrix} a_{11} & a_{12} \\ a_{21} & a_{22} \end{bmatrix}\begin{bmatrix} x_1 \\ x_2 \end{bmatrix}$$

$$= x^T A x.$$

If we set $Q_2 = k$ (where $k \in \mathbf{R}$) we have the equation of a conic section (for example an ellipse) as in Fig. 2.3.

A basic problem in analytic geometry is to "rotate the axes" through the angle θ so as to make the w_1-axis and the w_2-axis coincide with the major and minor axes of the conic section. This has the effect of removing the cross-product terms w_1w_2 and w_2w_1 when the quadratic form is written

2.3 Figure

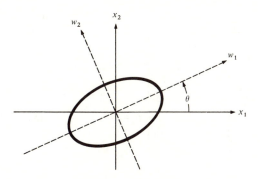

in terms of the new variables. We recall that the plane rotation

2.4
$$\begin{bmatrix} x_1 \\ x_2 \end{bmatrix} = \begin{bmatrix} \cos\theta & -\sin\theta \\ \sin\theta & \cos\theta \end{bmatrix}\begin{bmatrix} w_1 \\ w_2 \end{bmatrix}$$

or

2.4a
$$x = Rw$$

---

* See 11-(19.10).

accomplishes this. Thus, if $A = A^T$, then

2.5
$$Q_2 = x^T A x$$
$$= w^T (R^T A R) w$$
$$= w^T D w$$
$$= [w_1 w_2] \begin{bmatrix} d_{11} & 0 \\ 0 & d_{22} \end{bmatrix} \begin{bmatrix} w_1 \\ w_2 \end{bmatrix}$$
$$= d_{11} w_1^2 + d_{22} w_2^2.$$

Since $R^T = R^{-1}$, the transformation

2.6
$$R^T A R = D$$
$$= \begin{bmatrix} d_{11} & 0 \\ 0 & d_{22} \end{bmatrix}$$

is an orthogonal similarity transformation.

Now let us introduce a *plane (or elementary) rotation* matrix*, $R(p, q)$, which can be considered to be a generalization of the matrix in (2.4) in the sense that an orthogonal similarity transformation using

2.7 $R(p, q) =$

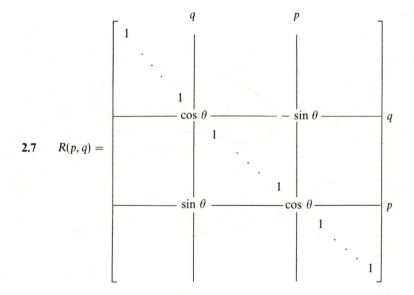

---

*See 11-(17.30).

where $q < p$ and where

2.8
$$\begin{cases} r_{qq} = r_{pp} = \cos\theta \\ -r_{qp} = r_{pq} = \sin\theta \\ r_{ii} = 1 \qquad\qquad i \neq p, i \neq q, \\ r_{ij} = 0 \qquad\qquad \text{otherwise,} \end{cases}$$

will annihilate the cross product terms, $x_p x_q$ and $x_q x_p$, in the quadratic form

2.9
$$Q_n = \sum_{i=1}^{n} \sum_{j=1}^{n} a_{ij} x_i x_j$$

$$= [x_1 x_2 \ldots x_n] \begin{bmatrix} a_{11} & a_{12} & \cdots & a_{1n} \\ a_{21} & a_{22} & \cdots & a_{2n} \\ \cdots\cdots\cdots\cdots\cdots\cdots\cdots \\ a_{n1} & a_{n2} & \cdots & a_{nn} \end{bmatrix} \begin{bmatrix} x_1 \\ x_2 \\ \vdots \\ x_n \end{bmatrix}$$

$$= x^T A x.$$

What this means, as far as the symmetric matrix A is concerned, is that the orthogonal similarity transformation

2.10
$$A' = R^T(p, q)\, A\, R(p, q)$$

annihilates both a_{pq} and a_{qp}, the coefficients of $x_p x_q$ and $x_q x_p$, respectively. For this reason we call (2.10) a *plane rotation in the* (p, q)-*plane*.

2.11 Theorem. If A is symmetric in (2.10), then A' is also symmetric.

Proof. The proof of this theorem follows directly from (2.16), (2.17), and (2.19).

The relationship between the symmetric matrix $A \in \mathbf{R}^{nn}$ and the quadratic form Q_n, and the effect of the plane rotation (2.10), provide the motivation for the Jacobi algorithm. From time to time we drop the symbols p and q in $R(p, q)$ and introduce a single subscript m to indicate the mth step in a sequence. Using this notation, then, suppose we form the sequence of symmetric matrices

2.12
$$\begin{cases} A_1 = A \\ A_2 = R_1^T A_1 R_1 \\ A_3 = R_2^T A_2 R_2 \\ \vdots \\ A_{m+1} = R_m^T A_m R_m \\ \vdots \end{cases}$$

where each orthogonal similarity transformation is designed to annihilate a symmetric pair of off-diagonal elements. We need the following theorem and definition before continuing.

2.13 Theorem. The plane rotation (2.10), which annihilates the symmetric pair a_{pq} and a_{qp}, reduces the sum of the squares of the off-diagonal elements of A by exactly $2a_{pq}^2$.

Proof. The proof is a direct consequence of (2.16) and (2.17).

2.14 Definition. The element a_{pq} of (2.13) is called the *pivot* for the plane rotation in the (p, q)-plane.

Goldstine, Murray, and von Neumann [1959, p. 72] prove that if the pivotal candidate in each transformation $R_m^T A_m R_m$ has magnitude greater than the average magnitude of the off-diagonal elements of A_m, then

2.15
$$A_m \to \Lambda = \begin{bmatrix} \lambda_1 & & & & \\ & \lambda_2 & & & \\ & & \cdot & & \\ & & & \cdot & \\ & & & & \lambda_n \end{bmatrix}$$

as $m \to \infty$. Since the symmetric matrices $A_1, A_2, \ldots, A_m, \ldots,$ are *similar* they all have the same set of eigenvalues, and these are shown to be $\lambda_1, \lambda_2, \ldots, \lambda_n$.

As a practical computational algorithm, the sequence (2.12) should be continued until A_m has the sum of the squares of its off-diagonal elements less than or equal to some small $\varepsilon > 0$ because, at this stage, the diagonal elements of A_m are good approximations to the eigenvalues as a consequence of the Gerschgorin theorems (and their corollaries) in Section 1.

It was demonstrated experimentally by Gregory [1953] and proved by Henrici [1958], that the convergence indicated in (2.15) takes place even if the pivots are not selected on the basis of their magnitude but are selected in "typewriter" fashion, that is, in the order $(2, 1), (3, 1), (3, 2), (4, 1), (4, 2), (4, 3), \ldots, (n, 1), (n, 2), \ldots, (n, n - 1)$, $(2, 1), \ldots,$ provided that $|\theta| \leq \pi/4$. This version of the algorithm is called the *special cyclic Jacobi algorithm.** It has the advantage that no special search is necessary when selecting the pivots.

One further refinement is suggested by Pope and Tompkins [1957]. Because of (2.13) they suggest that using a small pivot would be wasteful and so they introduce a constant, called a *threshold*, during each cyclic *sweep* through the off-diagonal elements and a pivotal candidate is ignored unless its magnitude is greater than or equal to the threshold. The threshold is lowered during each sweep, of course. (See Greenstadt [1960] and Wilkinson [1965, p. 277], for details.)

---

* An ordinary cyclic Jacobi algorithm is one in which the off-diagonal elements are chosen cyclically but not necessarily in typewriter fashion.

We conclude this discussion of Jacobi algorithms by describing the details of the plane rotation displayed in (2.10). It can be verified by direct multiplication that the only elements of $A = (a_{ij})$ which are changed when we compute $A' = (a'_{ij})$ are those in rows p and q and those in columns p and q. To be specific, the only transformed elements are

2.16
$$\begin{cases} a'_{qi} = a_{qi} \cos \theta + a_{pi} \sin \theta \\ a'_{pi} = -a_{qi} \sin \theta + a_{pi} \cos \theta \end{cases} \qquad i \neq p, q$$

2.17
$$\begin{cases} a'_{iq} = a_{iq} \cos \theta + a_{ip} \sin \theta \\ a'_{ip} = -a_{iq} \sin \theta + a_{ip} \cos \theta \end{cases} \qquad i \neq p, q$$

2.18
$$\begin{cases} a'_{qq} = a_{qq} \cos^2 \theta + 2a_{qp} \sin \theta \cos \theta + a_{pp} \sin^2 \theta \\ a'_{pp} = a_{qq} \sin^2 \theta - 2a_{pq} \sin \theta \cos \theta + a_{pp} \cos^2 \theta \end{cases}$$

and

2.19
$$\begin{cases} a'_{pq} = a_{pq} \cos 2\theta + \frac{1}{2}(a_{pp} - a_{qq}) \sin 2\theta \\ a'_{qp} = a_{qp} \cos 2\theta + \frac{1}{2}(a_{pp} - a_{qq}) \sin 2\theta. \end{cases}$$

Since the purpose of the plane rotation in the (p, q)-plane is to annihilate a_{pq} and a_{qp} we set their transforms $a'_{pq} = a'_{qp} = 0$ and obtain the following trigonometric equation in θ

2.20
$$\tan 2\theta = \frac{2a_{pq}}{a_{qq} - a_{pp}}, \qquad q < p.$$

If $a_{qq} = a_{pp}$ we use the angle of rotation $\theta = \pi/4$.

Actually, we do not need to solve (2.20) for θ, itself, because (2.16), (2.17), and (2.18) only contain $\sin \theta$ and $\cos \theta$. Hence, we use (2.20) to compute $\sin \theta$ and $\cos \theta$. If we let*

2.21
$$\begin{cases} \beta = |a_{qq} - a_{pp}| \\ \alpha = 2a_{pq} \, \text{sgn}(a_{qq} - a_{pp}), \end{cases}$$

then

2.22
$$\tan 2\theta = \frac{\alpha}{\beta}.$$

_____

* The symbol sgn x means $\dfrac{x}{|x|}$ and is either $+1$ or -1.

To get $\cos \theta$, we note that

2.23 $$\sec^2 2\theta = 1 + \frac{\alpha^2}{\beta^2},$$

and so

2.24 $$\cos^2 2\theta = \frac{\beta^2}{\alpha^2 + \beta^2}.$$

Since we want $|\theta| \leq \pi/4$, for convergence of the algorithm,* we choose $\cos 2\theta$ to be positive. Hence, we choose the positive square root and write

2.25 $$\cos 2\theta = \frac{\beta}{\sqrt{\alpha^2 + \beta^2}}.$$

It is easy to verify that, with this choice,

2.26 $$\sin 2\theta = \frac{\alpha}{\sqrt{\alpha^2 + \beta^2}}.$$

From the identity, $\cos 2\theta = 2\cos^2\theta - 1$, we obtain

2.27 $$\cos \theta = \left[\frac{1}{2}\left(1 + \frac{\beta}{\sqrt{\alpha^2 + \beta^2}} \right) \right]^{\frac{1}{2}},$$

and again we choose the positive square root. From the identity

$$\sin 2\theta = 2 \sin \theta \cos \theta$$

we obtain

2.28 $$\sin \theta = \frac{\alpha}{2 \cos \theta \sqrt{\alpha^2 + \beta^2}},$$

and we are through with the derivation.

In summary, then, if we call a_{pq} our *pivot*, we can carry out a plane rotation in the (p, q)-plane

  i) by computing α and β, using (2.21),

 ii) by computing $\cos \theta$, using (2.27),

iii) by computing $\sin \theta$, using (2.28),

---

* See the second paragraph following (2.15).

and

iv) by computing A' from A, using (2.16), (2.17), and (2.18).

Instead of using (2.19) we complete step (iv) by setting $a'_{pq} = a'_{qp} = 0$ directly, thus avoiding any residue due to rounding errors. We should, however, use (2.19) as a computational check, along with a test of the size of $|\sin^2 \theta + \cos^2 \theta - 1|$.

We have been careful to choose our computational equations in a form in keeping with considerations of numerical stability. (See 1-(4.8) and Section 1.2.) However, it is important that we compute the two square roots in (2.27) accurately so that $|\sin^2\theta + \cos^2\theta - 1|$ is very small. Otherwise, we cannot safely set $a'_{pq} = a'_{qp} = 0$.

2.29 *Remark.* We can write (2.12) in the form

$$A_{m+1} = (R_1 R_2 R_3 \ldots R_m)^T A_1 (R_1 R_2 R_3 \ldots R_m)$$
$$= R^T A_1 R,$$

where the product of the first m plane rotation matrices is

$$R = R_1 R_2 R_3 \ldots R_m,$$

and if m is sufficiently large, then Goldstine, Murray, and von Neumann [1959] show that R is close to the matrix whose columns are the eigenvectors of A. In other words

$$R^T A R \doteq \Lambda$$

and so

$$A R \doteq R \Lambda$$

is an approximation to the modal* equation.

Now R is orthogonal, since it is the product of orthogonal matrices,† and so the columns of R are approximations to a complete orthogonal set of eigenvectors of A. The ith column of R approximates the eigenvector corresponding to the ith diagonal element of Λ and so the approximations to the eigenvalues and eigenvectors appear in the same order.

The Givens Algorithm

We did not state the fact, explicitly, but it should be obvious that the elements in the (p, q) and (q, p) positions, annihilated by a plane rotation in the (p,q)-plane, do

---

* See 11-(21.21).

† See 11-(17.28).

not necessarily remain zero during subsequent transformations of the Jacobi algorithm. This led Givens [1954] to propose an algorithm (using plane rotations) which preserves the zeros in the off-diagonal positions, once they are created.

He begins by choosing a_{32} as the pivot, but he selects θ to annihilate a'_{31} and a'_{13} rather than to annihilate a'_{32} and a'_{23}, as we do in Jacobi's algorithm. By choosing the pivots in positions $(3, 2), (4, 2), \ldots, (n, 2)$ and angles so as to annihilate the symmetric pairs of elements in positions $(3, 1)$ and $(1, 3)$, $(4, 1)$ and $(1, 4), \ldots,$ $(n, 1)$ and $(1, n)$, respectively, he transforms a symmetric matrix A into a symmetric matrix of the form

$$
\left[
\begin{array}{c|ccccccc}
x & x & 0 & 0 & 0 & \cdots & 0 \\
\hline
x & x & x & x & x & \cdots & x \\
0 & x & x & x & x & \cdots & x \\
0 & x & x & x & x & \cdots & x \\
0 & x & x & x & x & \cdots & x \\
\cdot\,\cdot & \multicolumn{6}{c}{\dotfill} \\
0 & x & x & x & x & \cdots & x
\end{array}
\right],
$$

which has $n - 2$ zeros in both the first row and the first column (in the off-tri-diagonal positions).

By selecting his next pivots from the third column in the positions $(4, 3)$, $(5, 3), \ldots, (n, 3)$ and his angles so as to annihilate the symmetric pairs of elements in positions $(4, 2)$ and $(2, 4)$, $(5, 2)$ and $(2, 5), \ldots, (n, 2)$ and $(2, n)$, respectively, he produces a matrix of the form

$$
\left[
\begin{array}{cc|ccccc}
x & x & 0 & 0 & 0 & \cdots & 0 \\
x & x & x & 0 & 0 & \cdots & 0 \\
\hline
0 & x & x & x & x & \cdots & x \\
0 & 0 & x & x & x & \cdots & x \\
0 & 0 & x & x & x & \cdots & x \\
\cdot\,\cdot & \cdot\,\cdot & \multicolumn{5}{c}{\dotfill} \\
0 & 0 & x & x & x & \cdots & x
\end{array}
\right]
$$

which has $n - 3$ zeros in both the second row and the second column (again in the off-tridiagonal positions). He continues this process of selecting a_{pq} as pivot and θ to annihilate $a'_{p,q-1}$ and $a'_{q-1,p}$ until, after exactly $(n-1)(n-2)/2$ plane rotations, he produces a symmetric tridiagonal matrix.

Whereas the process of *diagonalizing a matrix* (using, for example, the Jacobi algorithm) produces a matrix whose eigenvalues are exhibited on the main diagonal we are not so fortunate when we *tridiagonalize a matrix* (using, for example, the Givens algorithm). However, a tridiagonalization procedure has the advantage that it requires only a finite number of similarity transformations and the symmetric tridiagonal matrix which it produces is a matrix which can be handled quite easily. Thus, the tridiagonalization procedure is only *the first step* in Givens' algorithm.

2.30 *Remark.* Due to the fact that Householder [1958a] discovered a tridiagonalization procedure which requires essentially half as much computation as Givens' tridiagonalization procedure we shall substitute Householder's procedure for the tridiagonalization step. Ortega [1967] calls the resulting combination the Givens-Householder method.

The Givens-Householder Algorithm

First, we describe the tridiagonalization procedure proposed by Householder [1958a]. This procedure is strongly endorsed by Wilkinson [1960, 1962, and 1968] and Ortega [1967], and the reader is urged to read their papers before attempting to write a computer program for this algorithm.

Instead of using $n-k-1$ plane rotations to create $n-k-1$ zeros in the kth row and $n-k-1$ zeros in the kth column of A (in the off-tridiagonal positions) as Givens does, Householder uses a *single* orthogonal similarity transformation to to the same job. Since tridiagonalization requires the creation of zeros in the off-tridiagonal positions of $n-2$ rows and columns, tridiagonalization can be completed using exactly $n-2$ Householder transformations. These transformations are more complicated than the plane rotations proposed by Givens, but the complete tridiagonalization requires only about half as much computation, as we mentioned in (2.30).

The Householder transformation uses matrices $P \in \mathbf{R}^{nn}$ of the form

2.31 $$P = I - 2ww^T$$

where $w \in \mathbf{R}^n$ is such that

2.32 $$w^T w = w_1^2 + w_2^2 + \cdots + w_n^2$$
$$= 1.$$

2.33 Theorem. If P is defined by (2.31) and (2.32), then

  i) $P = P^T$

  ii) $P = P^{-1}$.

Proof. i) By definition,

$$P = \begin{bmatrix} (1 - 2w_1^2) & -2w_1w_2 & -2w_1w_3 & \cdots & -2w_1w_n \\ -2w_2w_1 & (1 - 2w_2^2) & -2w_2w_3 & \cdots & -2w_2w_n \\ -2w_3w_1 & -2w_3w_2 & (1 - 2w_3^2) & \cdots & -2w_3w_n \\ \multicolumn{5}{c}{\dotfill} \\ -2w_nw_1 & -2w_nw_2 & -2w_nw_3 & \cdots & (1 - 2w_n^2) \end{bmatrix}$$

and $P = P^T$ by inspection.

  ii) $P^T P = P^2$

$$= (I - 2ww^T)(I - 2ww^T)$$
$$= I - 4ww^T + 4w(w^Tw)w^T$$
$$= I - 4ww^T + 4ww^T$$
$$= I.$$

Hence, $P = P^T = P^{-1}$ and P is both symmetric and orthogonal.

    Now consider the $n - 2$ Householder transformations, beginning with a symmetric matrix $A \in \mathbf{R}^{nn}$,

2.34
$$\begin{cases} A_1 = A \\ A_2 = P_2 A_1 P_2 \\ A_3 = P_3 A_2 P_3 \\ \quad\vdots \\ A_{n-1} = P_{n-1} A_{n-2} P_{n-1}, \end{cases}$$

where we define P_r, for $r = 2, 3, \ldots, n - 1$, to be the matrix

2.35
$$P_r = I - 2w^{(r)}w^{(r)^T},$$

with

2.36
$$w^{(r)} = \begin{bmatrix} 0 \\ \vdots \\ 0 \\ w_r \\ w_{r+1} \\ \vdots \\ w_n \end{bmatrix},$$

and

2.37
$$w_r^2 + w_{r+1}^2 + \cdots + w_n^2 = 1.$$

The object, of course, is to transform the real symmetric matrix A_1 into a real symmetric tridiagonal matrix A_{n-1} by the $n-2$ orthogonal similarity transformations shown.

To compute A_r in (2.34), we must first compute components $w_r, w_{r+1}, \ldots, w_n$ of the vector $w^{(r)}$ so that $P_r A_{r-1} P_r$ annihilates the $n-r$ off-tridiagonal elements in row $r-1$ and column $r-1$ of A_{r-1}. In order to motivate the derivation of formulas for computing $w_r, w_{r+1}, \ldots, w_n$, let us examine, in detail, the case $n=4$ and $r=2$. In other words, we wish to annihilate the symmetric pairs $a_{13} = a_{31}$ and $a_{14} = a_{41}$ in

2.38
$$A_1 = \begin{bmatrix} a_{11} & a_{12} & a_{13} & a_{14} \\ a_{21} & a_{22} & a_{23} & a_{24} \\ a_{31} & a_{32} & a_{33} & a_{34} \\ a_{41} & a_{42} & a_{43} & a_{44} \end{bmatrix}$$

by the transformation $A_2 = P_2 A_1 P_2$, where

2.39
$$w^{(2)} = \begin{bmatrix} 0 \\ w_2 \\ w_3 \\ w_4 \end{bmatrix}, \qquad w_2^2 + w_3^2 + w_4^2 = 1,$$

and, consequently,

2.40
$$P_2 = \begin{bmatrix} 1 & 0 & 0 & 0 \\ 0 & (1 - 2w_2^2) & - 2w_2w_3 & - 2w_2w_4 \\ 0 & - 2w_3w_2 & (1 - 2w_3^2) & - 2w_3w_4 \\ 0 & - 2w_4w_2 & - 2w_4w_3 & (1 - 2w_4^2) \end{bmatrix}.$$

Since A_2 is symmetric, we need to exhibit only its first column

2.41
$$A_2 = \begin{bmatrix} a_{11} & a_{12}' & a_{13}' & a_{14}' \\ a_{21}(1-2w_2^2) + a_{31}(-2w_2w_3) + a_{41}(-2w_2w_4) & a_{22}' & a_{23}' & a_{24}' \\ a_{21}(-2w_3w_2) + a_{31}(1-2w_3^2) + a_{41}(-2w_3w_4) & a_{32}' & a_{33}' & a_{34}' \\ a_{21}(-2w_4w_2) + a_{31}(-2w_4w_3) + a_{41}(1-2w_4^2) & a_{42}' & a_{43}' & a_{44}' \end{bmatrix}.$$

For simplicity of notation we use a prime to differentiate an element of A_2 from the corresponding element of A_1. If we let

2.42
$$p = w_2a_{21} + w_3a_{31} + w_4a_{41},$$

then we can write the elements of the first column of A_2 in the simpler form

2.43
$$\begin{cases} a_{11}' = a_{11} \\ a_{21}' = a_{21} - 2w_2p \\ a_{31}' = a_{31} - 2w_3p \\ a_{41}' = a_{41} - 2w_4p. \end{cases}$$

Let s_k^2 be the sum of the squares of the elements below the main diagonal in column k. Then for A_2 we have in the first column

2.44
$$\begin{aligned} s_1^2 &= (a_{21}')^2 + (a_{31}')^2 + (a_{41}')^2 \\ &= (a_{21} - 2w_2p)^2 + (a_{31} - 2w_3p)^2 + (a_{41} - 2w_4p)^2 \\ &= a_{21}^2 + a_{31}^2 + a_{41}^2 - 4p(a_{21}w_2 + a_{31}w_3 + a_{41}w_4) + 4p^2(w_2^2 + w_3^2 + w_4^2) \\ &= a_{21}^2 + a_{31}^2 + a_{41}^2 - 4p^2 + 4p^2 \\ &= a_{21}^2 + a_{31}^2 + a_{41}^2. \end{aligned}$$

Thus, s_1^2 is the same for both A_1 and A_2. As a matter of fact, there is a general result.

2.45 Theorem. The sum of the squares of the elements below the main diagonal, in column $r - 1$ of A_{r-1}, is invariant under the Householder transformation $A_r = P_r A_{r-1} P_r$. Due to symmetry, the same statement applies to the sum of the squares of the elements to the right of the main diagonal in row $r - 1$.

Proof. The proof is left as an exercise for the reader.

Now we return to (2.43) and (2.44) and set $a'_{31} = a'_{41} = 0$. This gives us the three equations in w_2, w_3, and w_4

2.46
$$\begin{cases} a_{21} - 2w_2 p = \pm s_1 \\ a_{31} - 2w_3 p = 0 \\ a_{41} - 2w_4 p = 0, \end{cases}$$

where s_1 represents the positive square root

2.47
$$s_1 = \sqrt{a_{21}^2 + a_{31}^2 + a_{41}^2}.$$

If we multiply the three equations in (2.46) by w_2, w_3, and w_4, respectively, and add, we obtain

2.48
$$w_2 a_{21} + w_3 a_{31} + w_4 a_{41} - 2p(w_2^2 + w_3^2 + w_4^2) = \pm w_2 s_1$$

and, from (2.39) and (2.42), this reduces to

2.48a
$$p = \mp w_2 s_1.$$

Thus, (2.46) can be written

2.49
$$\begin{cases} a_{21} \pm 2w_2^2 s_1 = \pm s_1 \\ a_{31} \pm 2w_2 w_3 s_1 = 0 \\ a_{41} \pm 2w_2 w_4 s_1 = 0, \end{cases}$$

and we have three non-linear equations to solve for the three unknowns w_2, w_3, and w_4.

These equations yield the results

2.50
$$\begin{cases} w_2^2 = \dfrac{s_1 \mp a_{21}}{2s_1} \\ w_3 = \mp \dfrac{a_{31}}{2w_2 s_1} \\ w_4 = \mp \dfrac{a_{41}}{2w_2 s_1} \end{cases}$$

where a decision now has to be made about which sign to use. To avoid cancellation in the numerator of the first equation we choose the sign to agree with* sgn a_{21}. Thus

2.51
$$\begin{cases} w_2^2 = \dfrac{a_{21}(\text{sgn }a_{21}) + s_1}{2s_1} \\[2ex] w_3 = a_{31}\left(\dfrac{\text{sgn }a_{21}}{2w_2 s_1}\right) \\[2ex] w_4 = a_{41}\left(\dfrac{\text{sgn }a_{21}}{2w_2 s_1}\right). \end{cases}$$

It turns out that we do not need to take the square root of w_2^2 and compute w_2 explicitly since each term in (2.40) is quadratic in w_2. If we divide the first equation by w_2 and write

2.52
$$w_2 = \frac{a_{21}(\text{sgn }a_{21}) + s_1}{2w_2 s_1},$$

then $w^{(2)}$ has the particularly simple form

2.53
$$w^{(2)} = \left(\frac{\text{sgn }a_{21}}{2w_2 s_1}\right)\begin{bmatrix} 0 \\ a_{21} + s_1\,(\text{sgn }a_{21}) \\ a_{31} \\ a_{41} \end{bmatrix}$$
$$= \beta_2 v^{(2)},$$

where

2.54
$$\beta_2 = \frac{\text{sgn }a_{21}}{2w_2 s_1}$$

and

2.55
$$v^{(2)} = \begin{bmatrix} 0 \\ a_{21} + s_1\,(\text{sgn }a_{21}) \\ a_{31} \\ a_{41} \end{bmatrix}.$$

---

*The symbol sgn x means $\dfrac{x}{|x|}$ and is either $+1$ or -1.

Notice how simple it is to compute $v^{(2)}$.

With (2.53), (2.54), and (2.55) we can write P_2 in terms of $v^{(2)}$ rather than $w^{(2)}$ and this allows us to avoid computing β_2 (we compute $2\beta_2^2$ instead).

2.56
$$\begin{aligned} P_2 &= I - 2w^{(2)}w^{(2)T} \\ &= I - 2\beta_2^2 v^{(2)}v^{(2)T} \\ &= I - \alpha_2 v^{(2)}v^{(2)T} \end{aligned}$$

where $\alpha_2 = 2\beta_2^2$ and furthermore,

2.57
$$\alpha_2 = \frac{1}{s_1^2 + s_1|a_{21}|}.$$

By using $v^{(2)}$, rather than $w^{(2)}$, we avoid computing a second square root during each transformation.

In general, we wish to carry out the Householder transformation $P_r A_{r-1} P_r$ so as to annihilate the $n - r$ off-tri-diagonal elements in row $r - 1$ and column $r - 1$ of A_{r-1}, where P_r is defined in (2.35). This leads us to write the general result

2.58 Theorem. The transformation $P_r A_{r-1} P_r$, with P_r defined by the equations

$$P_r = I - \alpha_r v^{(r)} v^{(r)T}$$

$$\alpha_r = \frac{1}{s_{r-1}^2 + s_{r-1}|a_{r,r-1}|},$$

$$v^{(r)} = \begin{bmatrix} 0 \\ \vdots \\ 0 \\ \hat{a}_{r,r-1} \\ a_{r+1,r-1} \\ a_{r+2,r-1} \\ \vdots \\ a_{n,r-1} \end{bmatrix}$$

where

$$\hat{a}_{r,r-1} = a_{r,r-1} + s_{r-1}\,(\text{sgn}\ a_{r,r-1})$$

and where

$$s_{r-1} = \sqrt{a_{r,r-1}^2 + a_{r+1,r-1}^2 + \cdots + a_{n,r-1}^2},$$

will annihilate the $n - r$ off-tridiagonal elements in row $r - 1$ and column $r - 1$ of A_{r-1}.

Proof. The proof is left as an exercise for the reader.

This theorem is the basis for a computationally stable procedure for carrying out the $n - 2$ Householder transformations in (2.34). We should point out, however, that we do not have to form P_r, explicitly, in order to carry out the transformation $P_r A_{r-1} P_r$. To see this, we simplify the notation by dropping the index r and by writing

2.59
$$A' = PAP$$

$$= (I - \alpha vv^T)A(I - \alpha vv^T)$$

$$= A - \alpha vv^T A - \alpha Avv^T + \alpha^2 v(v^T Av)v^T$$

$$= A - \alpha vv^T A - \alpha Avv^T + \alpha^2 (v^T Av)vv^T$$

$$= A - (vq^T + qv^T),$$

where

2.60
$$\begin{cases} q = u - \mu v \\ u = \alpha Av \\ \mu = \dfrac{\alpha}{2} v^T u. \end{cases}$$

Thus, once v is known, we use (2.60) to compute the vector q, and then we use (2.59) to carry out the transformation. There is a good opportunity here to achieve high accuracy by accumulating scalar products in $fl_2(\ )$ arithmetic.

2.61 *Remark.* Following the transformation $P_r A_{r-1} P_r$, the transformed elements in column $r - 1$ (with similar results for row $r - 1$) are

$$\begin{cases} a'_{r-1,r-1} = a_{r-1,r-1} \\ a'_{r,r-1} \quad\; = s_{r-1} \\ a'_{r+1,r-1} = 0 \\ \quad\ldots\ldots\ldots\ldots \\ a'_{n,r-1} \quad\; = 0. \end{cases}$$

The second equation above is a direct consequence of Theorem 2.45. These equations can be used as a check on the accuracy of the computation, and, as a matter of fact, they should replace the computed values of these particular elements.

Symmetric Tridiagonal Matrices

Now we turn our attention to the Givens procedure for finding the eigenvalues of the symmetric tridiagonal matrix T, produced above. We shall write its characteristic polynomial using the following notation, for simplicity:

2.62 $P_n(\lambda) = \det(T - \lambda I)$

$$= \det \begin{bmatrix} (a_1 - \lambda) & b_1 & & & \\ b_1 & (a_2 - \lambda) & b_2 & & \\ & & \dotsb & & \\ & & b_{n-2} & (a_{n-1} - \lambda) & b_{n-1} \\ & & & b_{n-1} & (a_n - \lambda) \end{bmatrix}.$$

We assume that $b_i \neq 0$ for $i = 1, 2, \ldots, n - 1$, because whenever $b_i = 0$, for some i, the matrix is block tridiagonal and we can work with each block separately.

Despite the fact that we mentioned in Section 1 that it is not advisable to compute the coefficients of the characteristic polynomial (as a means of finding eigenvalues) we sometimes find it acceptable to evaluate the characteristic polynomial directly (without computing its coefficients). This we do as a basic step in the algorithm which follows.

Givens [1954] proposes the following recurrence procedure for expressing the characteristic polynomial of T. Let $P_i(\lambda)$ be the determinant of the leading principal submatrix (of order i) of $T - \lambda I$. Then, for $i = 1, 2, 3$ we obtain

2.63
$$\begin{cases} P_1(\lambda) = a_1 - \lambda \\ P_2(\lambda) = (a_2 - \lambda)P_1(\lambda) - b_1^2 \\ P_3(\lambda) = (a_3 - \lambda)P_2(\lambda) - b_2^2 P_1(\lambda). \end{cases}$$

This suggests a general recurrence formula.

2.64 Theorem. If T is a real symmetric tri-diagonal matrix and if $T - \lambda I$ is given by (2.62), then, for $i = 1, 2, 3, \ldots, n$,

$$P_i(\lambda) = (a_i - \lambda)P_{i-1}(\lambda) - b_{i-1}^2 P_{i-2}(\lambda)$$

where we define $P_{-1}(\lambda) \equiv 0$, $P_0(\lambda) \equiv 1$, and $b_0 = 0$.

Proof. Let

$$P_i(\lambda) = \det \begin{bmatrix} (a_1 - \lambda) & b_1 & & & \\ b_1 & (a_2 - \lambda) & b_2 & & \\ & & \cdots\cdots\cdots\cdots\cdots & & \\ & & b_{i-2} & (a_{i-1} - \lambda) & b_{i-1} \\ & & & b_{i-1} & (a_i - \lambda) \end{bmatrix}$$

and expand the determinant in terms of the elements of the last row. This gives us, for $i \geq 3$,

$$P_i(\lambda) = (a_i - \lambda)P_{i-1}(\lambda) - b_{i-1}^2 P_{i-2}(\lambda).$$

By artificially introducing $P_{-1}(\lambda) \equiv 0$, $P_0(\lambda) \equiv 1$, and $b_0 = 0$, we can write

$$P_1(\lambda) = (a_1 - \lambda)P_0(\lambda) - b_0^2 P_{-1}(\lambda)$$

$$P_2(\lambda) = (a_2 - \lambda)P_1(\lambda) - b_1^2 P_0(\lambda)$$

and the formula holds, formally, for $i = 1, 2, \ldots, n$.

Clearly, the polynomial $P_i(\lambda)$ is the characteristic polynomial of the leading principal submatrix of T, of order i. Also, since each principal submatrix of the symmetric matrix T is also symmetric, then each principal submatrix has real eigenvalues. In other words, the roots of each polynomial equation, $P_i(\lambda) = 0$, are real. Since $P_n(\lambda) = \det(T - \lambda I)$ is the characteristic polynomial of T, *our goal is to find the roots of $P_n(\lambda) = 0$.* (We shall do this without computing the polynomial coefficients.) First, we need the following result.

2.65 Theorem. (Givens). Let T be the real symmetric tri-diagonal matrix

$$T = \begin{bmatrix} a_1 & b_1 & & & \\ b_1 & a_2 & b_2 & & \\ & & \cdots\cdots\cdots\cdots & & \\ & & b_{n-2} & a_{n-1} & b_{n-1} \\ & & & b_{n-1} & a_n \end{bmatrix}$$

with $b_1, b_2, \ldots, b_{n-1}$ different from zero. Then the roots of each equation $P_i(\lambda) = 0$ are distinct and are separated by the roots of $P_{i-1}(\lambda) = 0$.

Proof. Since $b_k \neq 0$ for $k = 1, 2, \ldots, n - 1$, no two successive equations $P_{i-1}(\lambda) = 0$ and $P_i(\lambda) = 0$ can have a common root because, if they did, then $P_{i-2}(\lambda) = 0$, $P_{i-3}(\lambda) = 0, \ldots, P_1(\lambda) = 0$ would also have the same root. However, a_1 is the only root of $P_1(\lambda) = 0$ and a_1 is not a root of $P_2(\lambda) = 0$ since $b_1 \neq 0$ by the hypothesis.

The root separation property can be proved by mathematical induction. It is easily verified that a_1, the root of $P_1(\lambda) = 0$, lies between the two distinct roots of $P_2(\lambda) = 0$. Assume that the roots of $P_{i-2}(\lambda) = 0$ and $P_{i-1}(\lambda) = 0$ are distinct and that the roots of the former separate the roots of the latter. Let $r_1 < r_2 < \ldots < r_{i-1}$ be the roots of $P_{i-1}(\lambda) = 0$. Then, from the recurrence formula in (2.64), for $k = 1, 2, \ldots, i - 1$,

$$P_i(r_k) = -b_{i-1}^2 P_{i-2}(r_k)$$

which implies $P_i(r_k)$ and $P_{i-2}(r_k)$ have opposite signs. However, by the induction hypothesis, $P_{i-2}(\lambda)$ changes sign between r_k and $r_{k+1}, k = 1, 2, \ldots, i - 2$, which implies $P_i(\lambda)$ also changes sign. In other words, $P_i(\lambda) = 0$ has a root between each pair of adjacent roots of $P_{i-1}(\lambda) = 0$. Since

$$P_i(\lambda) \to \begin{cases} \infty & \lambda \to -\infty \\ (-1)^i \infty & \lambda \to \infty \end{cases}$$

for $i = 1, 2, \ldots, n$, it follows that $P_i(\lambda) = 0$ has a root to the right of r_{i-1} and a root to the left of r_1. If the roots of $P_i(\lambda) = 0$ are $s_1, s_2, \ldots, s_i$, we have shown that

$$s_1 < r_1 < s_2 < r_2 < \ldots < s_{i-1} < r_{i-1} < s_i.$$

Hence the roots of $P_i(\lambda) = 0$ are distinct and are separated by the roots of

$$P_{i-1}(\lambda) = 0.$$

We observe that the sequence of polynomials in (2.64) constitutes a *Sturm sequence** in the interval $(-\infty, \infty)$ according to the following:

2.66 Definition. (Gantmacher [1960] vol. II, p. 175.) Consider the sequence of real polynomials $p_0(x), p_1(x), p_2(x), \ldots, p_m(x)$ with the following two properties with respect to an open interval (a, b), where a may be $-\infty$ and b may be $+\infty$.

i) For every value $x_0 \in (a, b)$ if $p_k(x_0) = 0$, then $p_{k-1}(x_0)p_{k+1}(x_0) < 0$ which means that $p_{k-1}(x_0)$ and $p_{k+1}(x_0)$ have opposite signs.

ii) $p_0(x) \neq 0$ for any $x \in (a, b)$.

Such a sequence of polynomials is called *a Sturm sequence* in the interval (a, b).

---

*We mentioned Sturm sequences in Section 5.5 of Volume I.

We illustrate the root separation property, for $n = 3$, in the figure below. Let a_1 be the root of $P_1(\lambda) = 0$, μ_1 and μ_2 be the roots of $P_2(\lambda) = 0$, and λ_1, λ_2, and λ_3 be the roots of $P_3(\lambda) = 0$.

2.67 Figure

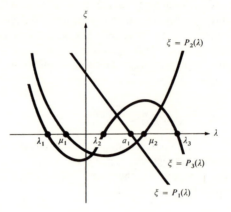

It is unimportant whether $\lambda_2 < a_1$ or $a_1 < \lambda_2$ in this diagram as long as they both lie in the interval (μ_1, μ_2).

2.68 Definition. Let $N(c)$ be the number of sign agreements in the sequence, $1, P_1(c), P_2(c), \ldots, P_n(c)$. If $P_i(c) = 0$, for some i, we take its sign to be the sign of $P_{i-1}(c)$.

For example, in Fig. 2.67 we observe the following sign patterns when c lies in the intervals indicated:

2.69 Table

	$(-\infty, \lambda_1]$	$(\lambda_1, \mu_1]$	$(\mu_1, \lambda_2]$	$(\lambda_2, a_1]$	$(a_1, \mu_2]$	$(\mu_2, \lambda_3]$	(λ_3, ∞)
1	+	+	+	+	+	+	+
$P_1(c)$	+	+	+	+	−	−	−
$P_2(c)$	+	+	−	−	−	+	+
$P_3(c)$	+	−	−	+	+	+	−
$N(c)$	3	2	2	1	1	1	0

From Fig. 2.67 we observe that the number $N(c)$ is equal to the number of roots of $P_3(\lambda) = 0$ which are greater than or equal to c. This is true, in general.

2.70 Theorem. $N(c)$ is the number of roots of $P_n(\lambda) = 0$ which are greater than or equal to c.

Proof. The proof is left as an exercise for the reader.

With this background we are ready to use Theorem 2.70 to find approximations to the distinct roots of $P_n(\lambda) = 0$. Since* $S(T) \leqq \|T\|_\beta$ for every matrix norm it is obviously true for the infinity norm. Hence, all of the eigenvalues of T (roots of $P_n(\lambda) = 0$) lie in the closed interval $[-\|T\|_\infty, \|T\|_\infty]$.

We can bisect this interval and use (2.70) to find out how many eigenvalues of T lie to the right of the midpoint. This enables us to locate the half interval which contains the particular eigenvalue we seek (for example, the largest or possibly the smallest). Since the eigenvalues are distinct we can continue bisecting the intervals, always keeping the interval containing the eigenvalue we seek, until we obtain an interval (c, d) containing *only* the eigenvalue we seek. At this point, the process of bisection can be continued until we have the desired accuracy for the eigenvalue we have isolated. Obviously, after m bisections we are down to an interval of size $2^{-m}[-\|T\|_\infty, \|T\|_\infty]$.

Another possibility is to use the Gerschgorin Theorems,† which tell us that the eigenvalues lie in the closed intervals on the real axis

2.71
$$
\begin{cases}
[a_1 - |b_1|, & a_1 + |b_1|] \\
[a_i - |b_{i-1}| - |b_i|, & a_i + |b_{i-1}| + |b_i|] \qquad i = 2, 3, \ldots, n-1 \\
[a_n - |b_{n-1}|, & a_n + |b_{n-1}|].
\end{cases}
$$

2.72 *Remark.* In (2.29) we mentioned how to find the eigenvectors using the Jacobi algorithm. Finding the eigenvectors, when the Givens-Householder algorithm is used, is not quite as simple, and we postpone the discussion until Section 4.

2.73 *Remark.* Obviously, the Givens-Householder algorithm finds the eigenvalues one at a time and if, as sometimes is the case, we only want *one* eigenvalue, then this algorithm is quite appropriate. Even if we want *all* the eigenvalues it requires less computation time than the cyclic-threshold-Jacobi algorithm, and for this reason it is widely used for arbitrary, real, symmetric matrices. The Jacobi algorithm, on the other hand, is well suited for finding all the eigenvalues simultaneously, plus the complete orthonormal set of eigenvectors (if n is not so large that computation time becomes a critical factor).

2.74 *Remark.* The Givens-Householder algorithm consists of two independent steps; the tridiagonalization step followed by the computation of the eigenvalues of the tridiagonal matrix. Since there are several algorithms (we have mentioned only two) for carrying out the first step, it is not surprising to find that additional algorithms have been proposed for carrying out the second step. We shall mention one that seems especially promising. Stewart [1970a, 1970b] has written a program

---

* See 11-(29.1).

† See (1.5), (1.6), and (1.7).

which uses Householder's algorithm for the tridiagonalization step and the QR algorithm* with origin shifts for the second step, and the reader should look at this algorithm along with the ones we have mentioned.

EXERCISES 14.2

1. Write the quadratic form

$$Q_2 = x^2 + 4xy - y^2$$

   in matrix language.

2. Carry out a plane rotation in the xy-plane to reduce Q_2 to the form

$$Q_2 = a\hat{x}^2 + b\hat{y}^2.$$

3. Prove that if A_1 is symmetric in (2.12) then so are $A_2, A_3, \ldots, A_{m+1}, \ldots$.

4. Prove Theorem 2.13.

5. Verify (2.16), (2.17), (2.18), and (2.19).

6. Derive (2.20).

7. Write a computer program to compute the eigenvalues and eigenvectors of a real symmetric matrix using the threshold-cyclic-Jacobi algorithm. Include in your program an option for computing the eigenvalues but not the eigenvectors.

8. Write a computer program to "tridiagonalize" a real symmetric matrix using the Givens algorithm. (This program should be a subroutine.)

9. Prove Theorem 2.45.

10. Prove Theorem 2.58.

11. Write a computer program to "tridiagonalize" a real symmetric matrix using the Householder algorithm. (This program should be a subroutine.)

12. Verify (2.60).

13. Prove Theorem 2.70.

14. Write a computer program to compute the eigenvalues of a real symmetric, tridiagonal matrix using the Sturm sequence property and the method of bisection, as proposed by Givens. (This program should be a subroutine.)

15. Write a computer program to compute the eigenvalues of a real symmetric matrix using the subroutines of Exercise 8 and Exercise 14. (This program should be a subroutine.)

16. Repeat Exercise 15 by replacing the subroutine in Exercise 8 with the subroutine in Exercise 11.

14.3 NON-HERMITIAN MATRICES

We deal with arbitrary complex matrices in this section and we assume that, in general, they are non-Hermitian. (Of course any method for solving the

---

* See Section 3 for a discussion of the QR algorithm.

eigenproblem for arbitrary matrices will work for Hermitian matrices, although not necessarily as efficiently as the methods of the previous section.)

The Power Method

This algorithm (see, for example, Wilkinson [1954], p. 537, and [1965], p. 570) yields the dominant eigenvalue of $A \in \mathbf{C}^{nn}$, along with its corresponding eigenvector, by means of the simple iterative procedure which follows. (Notice that A is not completely arbitrary, however.) Assume that A has linear elementary divisors,* which means that it is non-defective and has a set of n linearly independent eigenvectors,† (In other words, A is diagonalizable.) Also, assume that

3.1 $$|\lambda_1| > |\lambda_j|, \qquad j = 2, 3, \ldots, n$$

so that λ_1 is the dominant eigenvalue. Let $\{v^{(1)}, v^{(2)}, \ldots, v^{(n)}\}$, normalized so that the maximum component of each vector is unity, be the set of linearly independent eigenvectors. If we use this set as a basis, then an arbitrary vector $y \in \mathbf{C}^n$ can be written as a linear combination of these vectors, that is,

3.2 $$y = c_1 v^{(1)} + c_2 v^{(2)} + \cdots + c_n v^{(n)}.$$

From a theoretical standpoint we need to require that y have a component in the direction of $v^{(1)}$, (that is, that $c_1 \neq 0$) but y can be completely arbitrary, otherwise. (However, from the discussion below, y should not be the zero vector.)

Now we let $y = y^{(0)}$ and we form the sequence of vectors $y^{(0)}, y^{(1)}, \ldots, y^{(k)}, \ldots$ defined as follows:

3.3
$$
\left\{
\begin{aligned}
y^{(1)} &= A y^{(0)} \\
&= c_1 \lambda_1 v^{(1)} + c_2 \lambda_2 v^{(2)} + \cdots + c_n \lambda_n v^{(n)} \\
y^{(2)} &= A y^{(1)} \\
&= A^2 y^{(0)} \\
&= c_1 \lambda_1^2 v^{(1)} + c_2 \lambda_2^2 v^{(2)} + \cdots + c_n \lambda_n^2 v^{(n)} \\
&\vdots \\
y^{(k)} &= A y^{(k-1)} \\
&= A^k y^{(0)} \\
&= c_1 \lambda_1^k v^{(1)} + c_2 \lambda_2^k v^{(2)} + \cdots + c_n \lambda_n^k v^{(n)} \\
&\vdots
\end{aligned}
\right.
$$

---

* See (3.16).
† See 11-(21.52) and 11-(21.58).

We can write

3.4
$$A^k y^{(0)} = \lambda_1^k \left[c_1 v^{(1)} + c_2 \left(\frac{\lambda_2}{\lambda_1} \right)^k v^{(2)} + \cdots + c_n \left(\frac{\lambda_n}{\lambda_1} \right)^k v^{(n)} \right]$$

or, more simply,

3.4a
$$y^{(k)} = \lambda_1^k [c_1 v^{(1)} + \varepsilon^{(k)}]$$

where, as $k \to \infty$, $\varepsilon^{(k)} \to \phi$ because of (3.1). If we assume that k is sufficiently large, so that $\varepsilon^{(k)}$ can be ignored, then

3.5
$$\begin{cases} y^{(k)} \doteq \lambda_1^k c_1 v^{(1)} \\ y^{(k+1)} \doteq \lambda_1^{k+1} c_1 v^{(1)} \end{cases}$$

which means that we can write

3.6
$$y^{(k+1)} \doteq \lambda_1 y^{(k)}.$$

We cannot divide vectors but we can divide their respective components. Hence

3.7
$$\lambda_1 \doteq \frac{y_i^{(k+1)}}{y_i^{(k)}}, \qquad i = 1, 2, \ldots, n.$$

In other words, we form the sequence of vectors $y^{(0)}, A y^{(0)}, A^2 y^{(0)}, \ldots, A^k y^{(0)}$, $A^{k+1} y^{(0)}$, until the ratios of the respective components approach the same constant value and this constant value is an approximation to the dominant eigenvalue λ_1. The vector $y^{(k+1)}$ is an unnormalized approximation to the corresponding eigenvector. The rate of convergence, of course, depends on how rapidly the vector $\varepsilon^{(k)} \to \phi$ which really means how rapidly the expressions $c_j(\lambda_j/\lambda_1)^k$ in (3.4) approach zero.

Theoretically, if it should happen that our choice of y is such that $c_1 = 0$ in (3.2), but $|\lambda_2| > |\lambda_j|$ for $j \geqq 3$, then our iteration converges to λ_2 and a multiple of $v^{(2)}$. However, in practice, we can probably count on rounding errors to aid us here. It is most likely that rounding errors will introduce a small component in the direction of $v^{(1)}$, and the effect of this component will be to eventually turn the convergence to λ_1 rather than to λ_2.

It is good computational procedure to normalize the vectors at each stage. One procedure is to make the largest component equal to unity. We can describe this modified iteration by writing

3.8
$$
\begin{cases}
y^{(1)} = Ay^{(0)} & z^{(1)} = \dfrac{1}{m_1}\, y^{(1)} \\[2.5ex]
y^{(2)} = Az^{(1)} & z^{(2)} = \dfrac{1}{m_2}\, y^{(2)} \\[2.5ex]
\quad\vdots & \quad\vdots \\[1.5ex]
y^{(k)} = Az^{(k-1)} & z^{(k)} = \dfrac{1}{m_k}\, y^{(k)} \\[2.5ex]
\quad\vdots & \quad\vdots
\end{cases}
$$

where m_k is the component of $y^{(k)}$ of maximum modulus. This means that when $z^{(k)} \doteq v^{(1)}$ we can assume that

3.9
$$
y^{(k+1)} \doteq Av^{(1)}
$$
$$
= \lambda_1 v^{(1)}
$$

so that the element of $y^{(k+1)}$ of maximum modulus satisfies

3.10
$$
m_{k+1} \doteq \lambda_1.
$$

The advantage here, of course, is that the sequence of normalizing factors $m_1, m_2, m_3, \ldots$, converges to λ_1, and it is easier to monitor this sequence than it is to test the ratios (3.7).

Power Method with a Shift of Origin

The following lemma relates the eigenvalues and eigenvectors of A and $A - qI$, where $A \in \mathbf{C}^{nn}$ and $q \in \mathbf{C}$.

3.11 Lemma. $A - qI$ and A have the same set of eigenvectors and for each eigenvalue λ_i of A we have, for $A - qI$, the eigenvalue $\lambda_i - q$.

Proof. Let $Av^{(i)} = \lambda_i v^{(i)}$. Then we can write

$$
(A - qI)v^{(i)} = Av^{(i)} - qIv^{(i)}
$$
$$
= (\lambda_i - q)v^{(i)}.
$$

Subtracting the quantity q from the diagonal elements of A, then, has the effect of subtracting q from its eigenvalues. Another interpretation is the geometrical one in which the *origin has been shifted* by the amount q in the complex plane containing the eigenvalues.

Suppose we make the simplifying assumptions that $A \in \mathbf{R}^{nn}$, that A has linear elementary divisors,* that all the eigenvalues are real, and that

3.12
$$|\lambda_1| > |\lambda_2| \geq |\lambda_3| \geq \cdots \geq |\lambda_{n-1}| > |\lambda_n|.$$

Then subtracting $q \in \mathbf{R}$ from the diagonal elements of A has the effect of shifting the origin q units to the right along with the real axis.

Regardless of the value of q we choose, the dominant eigenvalue of $A - qI$ will always be either $\lambda_1 - q$ or $\lambda_n - q$. It turns out that if we choose $q = \frac{1}{2}[\lambda_2 + \lambda_n]$ we have the maximum rate of convergence to $\lambda_1 - q$ when we use $A - qI$ (instead of A) as our iteration matrix. Likewise, $q = \frac{1}{2}[\lambda_1 + \lambda_{n-1}]$ is optimum for convergence to $\lambda_n - q$.

To show the former, let $y^{(0)} = \sum_{i=1}^{n} c_i v^{(i)}$ and write the equation analogous to (3.4),

3.13
$$(A - qI)^k y^{(0)} = (\lambda_1 - q)^k \left[c_1 v^{(1)} + c_2 \left(\frac{\lambda_2 - q}{\lambda_1 - q} \right)^k v^{(2)} + \cdots + c_n \left(\frac{\lambda_n - q}{\lambda_1 - q} \right)^k v^{(n)} \right].$$

The rate of convergence is determined by how rapidly

3.14
$$\left(\frac{\lambda_2 - q}{\lambda_1 - q} \right)^k = \left(\frac{\lambda_2 - \lambda_n}{2\lambda_1 - \lambda_2 - \lambda_n} \right)^k$$

tends to zero. In the latter case the rate of convergence is determined by how rapidly

3.15
$$\left(\frac{\lambda_{n-1} - q}{\lambda_n - q} \right)^k = \left(\frac{\lambda_1 - \lambda_{n-1}}{\lambda_1 + \lambda_{n-1} - 2\lambda_n} \right)^k$$

tends to zero. Notice that with an origin shift we can find λ_n as well as λ_1.

3.16 *Remark.* We should point out that we have covered the previous material in this section primarily for the purpose of providing background material for the inverse power method. We have not been complete and we have made unnecessary assumptions for simplicity. For example, (3.1) is sufficient for convergence of the algorithm and it is not necessary that the matrix have all non-linear elementary divisors.

We discuss the inverse power method next. It has definite advantages over the power method as the reader will see. We have mentioned both methods in Chapter 5.

The Inverse Power Method
The inverse power method has the advantage that it provides for computing an

---

* See (3.16).

approximation to *any* eigenvalue, not merely λ_1 and λ_n. (See Wilkinson [1965], p. 619.) Before we begin a discussion of the method, however, let us recall* that if λ is an eigenvalue of a non-singular matrix A, corresponding to an eigenvector v, then λ^{-1} is an eigenvalue of A^{-1} corresponding to the same eigenvector v.

Again, for simplicity, assume that $A \in \mathbf{R}^{nn}$, that A has linear elementary divisors†, and that the eigenvalues are all real. As before, choose any non-zero vector $y^{(0)} \in \mathbf{R}^n$ and express it as a linear combination of the eigenvectors $v^{(1)}$, $v^{(2)}, \dots, v^{(n)}$. Now use A^{-1} instead of A in (3.3). This gives us

3.17
$$\begin{cases} y^{(1)} = A^{-1}y^{(0)} \\ y^{(2)} = A^{-1}y^{(1)} \\ \quad \vdots \\ y^{(k)} = A^{-1}y^{(k-1)} \\ \quad \vdots \end{cases}$$

and we can find an approximation to the dominant eigenvalue of A^{-1} (the eigenvalue of A with smallest modulus).

If we introduce an origin shift and write

3.18
$$\begin{cases} y^{(1)} = (A - qI)^{-1}y^{(0)} \\ y^{(2)} = (A - qI)^{-1}y^{(1)} \\ \quad \vdots \\ y^{(k)} = (A - qI)^{-1}y^{(k-1)} \\ \quad \vdots \end{cases}$$

then the equation analogous to (3.13) becomes‡

3.19
$$\begin{aligned} y^{(k)} &= (A - qI)^{-1}y^{(k-1)} \\ &= (A - qI)^{-k}y^{(0)} \\ &= \frac{c_1}{(\lambda_1 - q)^k}v^{(1)} + \frac{c_2}{(\lambda_2 - q)^k}v^{(2)} + \cdots + \frac{c_n}{(\lambda_n - q)^k}v^{(n)} \\ &= \frac{1}{(\lambda_i - q)^k}\left[c_1\left(\frac{\lambda_i - q}{\lambda_1 - q}\right)^k v^{(1)} + \cdots + c_i v^{(i)} + \cdots + c_n\left(\frac{\lambda_i - q}{\lambda_n - q}\right)^k v^{(n)}\right] \\ &= \frac{c_i}{(\lambda_i - q)^k}[v^{(i)} + \varepsilon^{(k)}], \end{aligned}$$

---

* See 11-(21.17).
† This is not a necessary condition for convergence.
‡ $(A - qI)^{-k}$ means $[(A - qI)^{-1}]^k$.

where λ_i is the eigenvalue closest to q. As $k \to \infty$, $\varepsilon^{(k)} \to \phi$. Hence

3.20
$$
\begin{cases}
y^{(k)} \doteq \dfrac{c_i}{(\lambda_i - q)^k}\, v^{(i)} \\[2ex]
y^{(k+1)} \doteq \dfrac{c_i}{(\lambda_i - q)^{k+1}}\, v^{(i)},
\end{cases}
$$

which means that we can write

3.21
$$
y^{(k+1)} \doteq \left(\frac{1}{\lambda_i - q} \right) y^{(k)}.
$$

Hence,

3.22
$$
\frac{1}{\lambda_i - q} \doteq \frac{y_j^{(k+1)}}{y_j^{(k)}}, \qquad j = 1, 2, \ldots, n.
$$

Since q can be located arbitrarily, we can, therefore (by carefully selecting q), find an approximation to *any eigenvalue of A*.

We point out that (3.18) can also be written in the form

3.23
$$
\begin{cases}
(A - qI)y^{(1)} = y^{(0)} \\
(A - qI)y^{(2)} = y^{(1)} \\
\qquad\quad \vdots \\
(A - qI)y^{(k)} = y^{(k-1)} \\
\qquad\quad \vdots
\end{cases}
$$

and we find $y^{(t)}$ by solving the system of linear algebraic equations

3.24
$$
(A - qI)y^{(t)} = y^{(t-1)}, \qquad t = 1, 2, \ldots, k.
$$

In practice, of course, it is advisable to normalize the vectors at each stage of the iteration as we did in (3.8).

3.25 *Remark.* Wilkinson [1965, Chapter 9] has a rather extensive treatment of both the power method (direct iteration) and the inverse power method (inverse iteration). The reader should also read Ostrowski [1958–1959]. We have applied these methods to the polynomial root-finding problem in Chapter 5. In the next section the inverse power method is mentioned in connection with finding eigenvectors. Wilkinson [1965, p. 622] states "I have found inverse iteration to be by far the most powerful and accurate of methods I have used for computing eigenvectors."

The QR Algorithm

Rutishauser [1958] proposed an algorithm for solving the eigenvalue problem by means of what he called LR transformations described as follows: Given an arbitrary matrix $A = A_1$, form the sequence $A_1, A_2, A_3, \ldots$ first by carrying out a triangular decomposition* on A_i

3.26 $$A_i = L_i R_i$$

(L_i is unit lower triangular and R_i is upper triangular) and then by multiplying these factors in reverse order, to form

3.27 $$A_{i+1} = R_i L_i.$$

Since L_i is non-singular, we can write

3.27a $$A_{i+1} = L_i^{-1} A_i L_i,$$

which shows that A_{i+1} and A_i are related by a similarity transformation. Rutishauser shows that under certain conditions the sequence $A_1, A_2, \ldots, A_k, \ldots$ converges to an upper triangular matrix which displays the eigenvalues of A on the main diagonal.

There are difficulties associated with the LR algorithm (see Wilkinson [1965], p. 498 and p. 538), and this fact led Francis [1961–1962] and Kublanovskaya [1961] to propose that L be replaced by a unitary† matrix Q. This replacement gives us the QR algorithm which can be described by the iteration

3.28 $$\begin{cases} A_i = Q_i R_i \\ A_{i+1} = Q_i^{-1} A_i Q_i \qquad i = 1, 2, 3, \ldots \\ \qquad = R_i Q_i. \end{cases}$$

The following theorem gives us the conditions under which a unique QR decomposition exists.

3.29 Theorem. If A is nonsingular, then there exists a decomposition

$$A = QR$$

for which Q is unitary and R is upper triangular. Furthermore, if the diagonal elements $r_{ii} \in \mathbf{R}^+$, the decomposition is unique.

Proof. (by construction). We seek a unitary matrix Q and an upper triangular

---

* See 12-(19.3). Rutishauser uses the notation LR (left, right) rather than LU (lower, upper) for a triangular decomposition.
† See (5.35).

matrix R, with $r_{ii} > 0$, such that $A = QR$. Let $a^{(k)}$, $q^{(k)}$, and $r^{(k)}$ be the kth columns of A, Q, and R, respectively.

For $k = 1$ we have

$$a^{(1)} = Qr^{(1)}$$
$$= r_{11}q^{(1)},$$

since $r^{(1)} = r_{11}e^{(1)}$. In other words, once we select r_{11}, we can determine $q^{(1)}$ uniquely by writing

$$q^{(1)} = r_{11}^{-1} a^{(1)}.$$

However, we want Q to be unitary, which means we want

$$1 = \|q^{(1)}\|_2$$
$$= \|r_{11}^{-1}a^{(1)}\|_2$$
$$= |r_{11}^{-1}| \cdot \|a^{(1)}\|_2,$$

and so $|r_{11}| = \|a^{(1)}\|_2$. For $r_{11} \in \mathbf{R}^+$, then, we must choose $r_{11} = \|a^{(1)}\|_2$. Notice that $\|a^{(1)}\|_2 \neq 0$ since A is non-singular.

For $k = 2$ we have

$$a^{(2)} = Qr^{(2)}$$
$$= r_{12}q^{(1)} + r_{22}q^{(2)}.$$

Since Q is to be unitary

$$[q^{(1)}]^H a^{(2)} = r_{12}[q^{(1)}]^H q^{(1)} + r_{22}[q^{(1)}]^H q^{(2)}$$
$$= r_{12}$$

and r_{12} is determined. Now

$$\|a^{(2)} - r_{12}q^{(1)}\|_2 = \|r_{22}q^{(2)}\|_2$$
$$= |r_{22}| \cdot \|q^{(2)}\|_2$$
$$= |r_{22}|,$$

and we want $r_{22} \in \mathbf{R}^+$. Thus, we must choose $r_{22} = \|a^{(2)} - r_{12}q^{(1)}\|_2$. The non-singularity of A guarantees that $r_{22} \neq 0$ since $a^{(2)} - r_{12}q^{(1)}$ is a non-trivial linear combination of $a^{(1)}$ and $a^{(2)}$.

For $k = s$ we have

$$a^{(s)} = Qr^{(s)}$$
$$= r_{1s}q^{(1)} + r_{2s}q^{(2)} + \cdots + r_{ss}q^{(s)}.$$

In other words, once $r_{1s}, r_{2s}, \ldots, r_{ss}$ are known, we can determine $q^{(s)}$ uniquely as

$$q^{(s)} = r_{ss}^{-1} [a^{(s)} - r_{1s}q^{(1)} - r_{2s}q^{(2)} - \cdots - r_{s-1,s}q^{(s-1)}].$$

We use the fact that Q is unitary to determine r_{is}, for $i = 1, 2, \ldots, s - 1$, by writing

$$[q^{(i)}]^H a^{(s)} = r_{1s}[q^{(i)}]^H q^{(1)} + r_{2s}[q^{(i)}]^H q^{(2)} + \cdots + r_{ss}[q^{(i)}]^H q^{(s)}$$
$$= r_{is}.$$

All that remains is to find r_{ss}. Now

$$\|r_{ss}q^{(s)}\|_2 = \|a^{(s)} - r_{1s}q^{(1)} - r_{2s}q^{(2)} - \cdots - r_{s-1,s}q^{(s-1)}\|_2$$
$$= \|p^{(s)}\|_2,$$

and $r_{ss} \in \mathbf{R}^+$ implies that we must choose $r_{ss} = \|p^{(s)}\|_2$.

The construction can break down only if $p^{(s)} = \phi$, but this vector is a non-trivial linear combination of $a^{(1)}, a^{(2)}, \ldots, a^{(s)}$, and these are linearly independent because A is nonsingular. This completes the constructive proof.

We should point out that the construction used in this proof is not an efficient computational procedure and it is not recommended for practical consideration.*

The QR algorithm is not quite as simple in its practical implementation as (3.28) would have us believe, however. First of all, both Francis [1961, p. 270] and Wilkinson [1965, p. 524] state that the QR algorithm is practical *only* when applied to a matrix A in (upper) Hessenberg form (almost triangular form)

3.30
$$A = \begin{bmatrix} a_{11} & a_{12} & a_{13} & a_{14} & \cdots & a_{1n} \\ a_{21} & a_{22} & a_{23} & a_{24} & \cdots & a_{2n} \\ & a_{32} & a_{33} & a_{34} & \cdots & a_{3n} \\ & & a_{43} & a_{44} & \cdots & a_{4n} \\ & & & \cdots\cdots\cdots\cdots \\ & & & & a_{n,n-1} & a_{nn} \end{bmatrix}.$$

The basic consideration here is the fact that the number of multiplications and additions involved in one QR transformation is proportional to n^3 for a full matrix whereas it is only n^2 for a Hessenberg matrix. Thus, they recommend a two-step procedure in which the matrix is first condensed to upper Hessenberg form and then the QR algorithm is applied to the upper Hessenberg matrix.

This is analogous to the two-step procedure described earlier for Hermitian matrices in which the first step is to condense the matrix to tri-diagonal form.

---

* See (3.71).

The reader should recall (2.72) where we mentioned Stewart's program which uses Householder's algorithm for the tri-diagonalization step and the QR algorithm for the second step.

Another recommendation (see the treatment by Parlett [1967], for example) for the practical implementation of the QR algorithm is the use of *origin shifts*. The reader is strongly advised to examine Parlett's paper before attempting to implement this algorithm on a computer.

Since any matrix can be transformed by similarity transformations to Hessenberg form, the two-step procedure has much to recommend it. Obviously we should expect the following to be true.

3.31 Theorem. The Hessenberg form of a matrix is preserved under a QR transformation.

Proof. The proof is left as an exercise.

Reducing a Matrix to Hessenberg Form

Wilkinson [1965, pp. 347–355] discusses various methods for reducing a matrix to Hessenberg form. He shows that Givens' method, Householder's method, and a method using *elementary similarity transformations* require approximately $10n^3/3$, $5n^3/3$, and $5n^3/6$ multiplications, respectively. The elementary similarity transformations, therefore, use only half the multiplications used by Householder's method and one-fourth the number used by Givens' method. Wilkinson [1965, p. 353] calls the matrices *stabilized elementary matrices* because the transformations appear to be stable in most instances. However, he issues a word of caution [1965, 364–65] about potential danger from (highly improbable) pivotal growth and states that "... anyone of a cautious disposition might well prefer to use Householder's method."

The situation is similar to the one that exists in Gaussian elimination (see the paragraph following 12-(13.14) relative to partial pivoting versus complete pivoting) and, although Businger [1969] has exhibited a class of matrices which cannot be stably reduced to Hessenberg form by elementary similarity transformations, the method is widely used just as Gaussian elimination with partial (rather than complete) pivoting is widely used. In fact (when the improbable pivotal growth does not exist) if scalar products are accumulated using $\mathrm{fl}_2(\ )$ arithmetic,* the accuracy of transformations using stabilized elementary matrices usually is slightly superior to that obtained using Householder transformations. With this background, we discuss elementary similarity transformations at this point.

Two types of elementary matrices are used—permutation† matrices, and matrices almost like $M^{(r)}$ in 12-(8.8). They are used together in a manner very much

---

* See 2-(7.47) for remarks about $\mathrm{fl}_2(\ )$ arithmetic.
† See 12-(11.2).

like the matrices are used in 12-(11.3) to carry out Gaussian elimination with partial pivoting. The difference, of course, is that (in this case) we must use similarity transformations so as to preserve eigenvalues.

We can illustrate the process for $n = 4$ and

3.32
$$A_1 = \begin{bmatrix} a_{11} & a_{12} & a_{13} & a_{14} \\ a_{21} & a_{22} & a_{23} & a_{24} \\ a_{31} & a_{32} & a_{33} & a_{34} \\ a_{41} & a_{42} & a_{43} & a_{44} \end{bmatrix}.$$

We shall assume that row and column interchanges are required at each stage of the reduction. Thus, suppose $|a_{41}| > |a_{i1}|$ for $i = 2$ and 3. In this case we interchange rows 2 and 4 and columns 2 and 4 by the similarity transformation*

3.33 $A_1' = I_{24} A_1 I_{24}$

$$= \begin{bmatrix} 1 & 0 & 0 & 0 \\ 0 & 0 & 0 & 1 \\ 0 & 0 & 1 & 0 \\ 0 & 1 & 0 & 0 \end{bmatrix} \begin{bmatrix} a_{11} & a_{12} & a_{13} & a_{14} \\ a_{21} & a_{22} & a_{23} & a_{24} \\ a_{31} & a_{32} & a_{33} & a_{34} \\ a_{41} & a_{42} & a_{43} & a_{44} \end{bmatrix} \begin{bmatrix} 1 & 0 & 0 & 0 \\ 0 & 0 & 0 & 1 \\ 0 & 0 & 1 & 0 \\ 0 & 1 & 0 & 0 \end{bmatrix}$$

$$= \begin{bmatrix} a_{11} & a_{14} & a_{13} & a_{12} \\ a_{41} & a_{44} & a_{43} & a_{42} \\ a_{31} & a_{34} & a_{33} & a_{32} \\ a_{21} & a_{24} & a_{23} & a_{22} \end{bmatrix}$$

$$= \begin{bmatrix} a_{11}' & a_{12}' & a_{13}' & a_{14}' \\ a_{21}' & a_{22}' & a_{23}' & a_{24}' \\ a_{31}' & a_{32}' & a_{33}' & a_{34}' \\ a_{41}' & a_{42}' & a_{43}' & a_{44}' \end{bmatrix}$$

Next we annihilate a_{31}' and a_{41}' by the similarity transformation

---

*Recall the result, $I_{pq}^2 = I$.

3.34 $A_2 = M_2 A_1' M_2^{-1}$

$$
= \begin{bmatrix} 1 & & & \\ 0 & 1 & & \\ 0 & m_{32} & 1 & \\ 0 & m_{42} & 0 & 1 \end{bmatrix}
\begin{bmatrix} a_{11}' & a_{12}' & a_{13}' & a_{14}' \\ a_{21}' & a_{22}' & a_{23}' & a_{24}' \\ a_{31}' & a_{32}' & a_{33}' & a_{34}' \\ a_{41}' & a_{42}' & a_{43}' & a_{44}' \end{bmatrix}
\begin{bmatrix} 1 & & & \\ 0 & 1 & & \\ 0 & -m_{32} & 1 & \\ 0 & -m_{42} & 0 & 1 \end{bmatrix}
$$

$$
= \begin{bmatrix} a_{11}'' & a_{12}'' & a_{13}'' & a_{14}'' \\ a_{21}'' & a_{22}'' & a_{23}'' & a_{24}'' \\ 0 & a_{32}'' & a_{33}'' & a_{34}'' \\ 0 & a_{42}'' & a_{43}'' & a_{44}'' \end{bmatrix}
$$

$$
= \begin{bmatrix} b_{11} & b_{12} & b_{13} & b_{14} \\ b_{21} & b_{22} & b_{23} & b_{24} \\ 0 & b_{32} & b_{33} & b_{34} \\ 0 & b_{42} & b_{43} & b_{44} \end{bmatrix},
$$

where

3.35
$$
\begin{cases} m_{32} = -\dfrac{a_{31}'}{a_{21}'} \\[2mm] m_{42} = -\dfrac{a_{41}'}{a_{21}'}. \end{cases}
$$

Suppose $|b_{42}| > |b_{32}|$. Then we interchange the last two rows and the last two columns by the similarity transformation

3.36 $A_2' = I_{34} A_2 I_{34}.$

Finally, we reduce A_2' to Hessenberg form by the similarity transformation

3.37 $A_3 = M_3 A_2' M_3^{-1}$

$$
= \begin{bmatrix} 1 & & & \\ 0 & 1 & & \\ 0 & 0 & 1 & \\ 0 & 0 & m_{43} & 1 \end{bmatrix}
\begin{bmatrix} b_{11}' & b_{12}' & b_{13}' & b_{14}' \\ b_{21}' & b_{22}' & b_{23}' & b_{24}' \\ 0 & b_{32}' & b_{33}' & b_{34}' \\ 0 & b_{42}' & b_{43}' & b_{44}' \end{bmatrix}
\begin{bmatrix} 1 & & & \\ 0 & 1 & & \\ 0 & 0 & 1 & \\ 0 & 0 & -m_{43} & 1 \end{bmatrix}
$$

$$= \begin{bmatrix} b''_{11} & b''_{12} & b''_{13} & b''_{14} \\ b''_{21} & b''_{22} & b''_{23} & b''_{24} \\ 0 & b''_{32} & b''_{33} & b''_{34} \\ 0 & 0 & b''_{43} & b''_{44} \end{bmatrix}$$

$$= \begin{bmatrix} h_{11} & h_{12} & h_{13} & h_{14} \\ h_{21} & h_{22} & h_{23} & h_{24} \\ 0 & h_{32} & h_{33} & h_{34} \\ 0 & 0 & h_{43} & h_{44} \end{bmatrix},$$

where

3.38
$$m_{43} = - \frac{b'_{42}}{b'_{32}}.$$

To carry out the reduction, in general, let

3.39
$$A_1 = \begin{bmatrix} a_{11} & a_{12} & a_{13} & \cdots & a_{1n} \\ a_{21} & a_{22} & a_{23} & \cdots & a_{2n} \\ a_{31} & a_{32} & a_{33} & \cdots & a_{3n} \\ \multicolumn{5}{c}{\dotfill} \\ a_{n1} & a_{n2} & a_{n3} & \cdots & a_{nn} \end{bmatrix}$$

and examine the magnitudes of the elements $a_{21}, a_{31}, \ldots, a_{i_1 1}, \ldots, a_{n1}$. Let $a_{i_1 1}$ be the element* with largest magnitude (the pivot). This means we must interchange† row 2 with row i_1 and column 2 with column i_1 by the similarity transformation

3.40
$$A'_1 = I_{2i_1} A_1 I_{2i_1}.$$

Next, we annihilate $a'_{31}, a'_{41}, \ldots, a'_{n1}$ with the transformation

3.41
$$A_2 = M_2 A'_1 M_2^{-1},$$

---

* If several elements have the same (largest) magnitude then any one of them can serve as the pivot.
† If $i_1 = 2$, then there is no interchange since $I_{pp} = I$.

where

3.42
$$M_2 = \begin{bmatrix} 1 & & & & & & \\ 0 & 1 & & & & & \\ 0 & m_{32} & 1 & & & & \\ 0 & m_{42} & 0 & 1 & & & \\ & & \cdots\cdots\cdots\cdots\cdots & & & \\ 0 & m_{n2} & 0 & 0 & \cdots & 1 \end{bmatrix}$$

and

3.43
$$m_{t2} = -\frac{a'_{t1}}{a'_{21}}, \quad (t = 3, 4, \ldots, n).$$

This completes the first step.

We begin the kth step ($k \leq n - 2$) with the matrix

3.44
$$A_k = \begin{bmatrix} f_{11} & f_{12} & \cdots & f_{1,k-2} & f_{1,k-1} & f_{1k} & \cdots & f_{1n} \\ f_{21} & f_{22} & \cdots & f_{2,k-2} & f_{2,k-1} & f_{2k} & \cdots & f_{2n} \\ 0 & f_{32} & \cdots & f_{3,k-2} & f_{3,k-1} & f_{3k} & \cdots & f_{3n} \\ & & \cdots\cdots\cdots\cdots\cdots\cdots & & & & \\ 0 & 0 & \cdots & 0 & f_{k,k-1} & f_{kk} & \cdots & f_{kn} \\ 0 & 0 & \cdots & 0 & 0 & f_{k+1,k} & \cdots & f_{k+1,n} \\ 0 & 0 & \cdots & 0 & 0 & f_{k+2,k} & \cdots & f_{k+2,n} \\ & & \cdots\cdots\cdots\cdots\cdots\cdots & & & & \\ 0 & 0 & \cdots & 0 & 0 & f_{nk} & \cdots & f_{nn} \end{bmatrix}.$$

We select the element* with greatest magnitude (the pivot) among the elements $f_{k+1,k}, f_{k+2,k}, \ldots, f_{nk}$ and denote it by $f_{i_k,k}$. This means that we interchange† row i_k with row $k + 1$ and column i_k with column $k + 1$ with the similarity transformation

3.45
$$A'_k = I_{k+1,i_k} A_k I_{k+1,i_k}.$$

---

*If several elements have the same (greatest) magnitude, then any one of them can serve as the pivot.

† If $i_k = k + 1$, then there is no interchange.

Next, we annihilate $f'_{k+2,k}, f'_{k+3,k}, \ldots, f'_{nk}$ with the transformation

3.46
$$A_{k+1} = M_{k+1} A'_k M_{k+1}^{-1}$$

where

3.47
$$M_{k+1} = \begin{bmatrix} 1 & & & & & & & \\ & \cdot & & & & & & \\ & & \cdot & & & & & \\ & & & \cdot & & & & \\ & & & & 1 & & & \\ & & & & m_{k+2,k+1} & & & \\ & & & & m_{k+3,k+1} & \cdot & & \\ & & & & \vdots & & \cdot & \\ & & & & m_{n,k+1} & & & \cdot & 1 \end{bmatrix}$$

and

3.48
$$m_{t,k+1} = -\frac{f'_{tk}}{f'_{k+1,k}}, \qquad (t = k+2, k+3, \ldots, n).$$

In summary, then,

3.49
$$A_{k+1} = [M_{k+1} I_{k+1,i_k}] A_k [M_{k+1} I_{k+1,i_k}]^{-1} \qquad (k = 1, 2, \ldots, n-2)$$

where the upper Hessenberg matrix A_{n-1} is similar to the given matrix A_1.

3.50 *Remark.* One or more of the sub-diagonal elements of an upper Hessenberg matrix $H = (h_{ij})$ may be zero. By partitioning with respect to these elements, H may be written in block upper-triangular form where each diagonal block is a Hessenberg matrix with nonzero sub-diagonal elements. For example, we might have

$$H = \begin{bmatrix} H_{11} & H_{12} & H_{13} \\ & H_{22} & H_{23} \\ & & H_{33} \end{bmatrix}$$

where H_{11}, H_{22}, and H_{33} are Hessenberg matrices with nonzero sub-diagonal elements.

It can be shown that the QR transformation acts independently on each diagonal block H_{ii} so that if $H_{ii} = Q_i R_i$ and $H = QR$, then Q is the direct sum of the Q_i and the diagonal blocks of RQ are just $R_i Q_i$. For this reason we only need to consider *unreduced* Hessenberg matrices with nonzero sub-diagonal elements.

The Case where H is Singular

One of the advantages of applying the QR algorithm to an unreduced, *singular*, Hessenberg matrix H_1 is that after *one step* of the algorithm a zero eigenvalue is revealed. (See Parlett [1966], p. 612.) Following one application of (3.28) to H_1 we obtain

3.51
$$H_2 = \begin{bmatrix} h_{11}^{(2)} & h_{12}^{(2)} & \cdots & h_{1,n-2}^{(2)} & h_{1,n-1}^{(2)} & h_{1n}^{(2)} \\ h_{21}^{(2)} & h_{22}^{(2)} & \cdots & h_{2,n-2}^{(2)} & h_{2,n-1}^{(2)} & h_{2n}^{(2)} \\ & h_{32}^{(2)} & \cdots & h_{3,n-2}^{(2)} & h_{3,n-1}^{(2)} & h_{3n}^{(2)} \\ & & \cdots\cdots\cdots\cdots\cdots\cdots\cdots\cdots\cdots\cdots \\ & & & h_{n-1,n-2}^{(2)} & h_{n-1,n-1}^{(2)} & h_{n-1,n}^{(2)} \\ & & & & 0 & 0 \end{bmatrix}$$

and H_2 may be *deflated* by deleting the last row and column so as to continue with an unreduced Hessenberg matrix of order $(n - 1) \times (n - 1)$. If the deflated matrix is singular the process is repeated until all the zero eigenvalues have been removed.

Convergence Theorems

Parlett [1967, p. 117] and [1968] discusses the convergence of the QR algorithm, that is, the convergence of the sequence $\{A_i\}$ described in (3.28). When the sequence converges (see the definition below), it converges either to a matrix in upper triangular form (which displays the eigenvalues of A_1 on the main diagonal) or to a matrix in block triangular form* with all diagonal blocks being either 1×1 or 2×2 submatrices.

3.52 Definition. The QR algorithm *converges* for an $n \times n$ Hessenberg matrix $H_1 = (h_{ij}^{(1)})$ if the sequence $\{H_i\}$ generated by (3.28) satisfies $h_{j+1,j}^{(i)} h_{j,j-1}^{(i)} \to 0$ as $i \to \infty$, for $j = 2, 3, \ldots, n - 1$.

Notice that *each* sub-diagonal element is not required to vanish (in which case convergence is to an upper triangular matrix) but merely that *at least one element out of each adjacent pair* of sub-diagonal elements must vanish. We cannot predict to which of the two forms above $\{H_i\}$ converges (when it does converge) without knowing something about H_1 in advance. (For example, is it real with pairs of complex conjugate eigenvalues?) We now state necessary and sufficient conditions for convergence.

---

* See (3.55), for example.

3.53 Theorem. The QR algorithm applied to an unreduced Hessenberg matrix H converges if, and only if, among each set of eigenvalues (of H) with equal modulus, there are at most two of even multiplicity and two of odd multiplicity.

Proof. See Parlett [1968].

3.54 EXAMPLE. Let H be an unreduced, complex, Hessenberg matrix with four distinct eigenvalues of equal modulus $|\lambda_1| = |\lambda_2| = |\lambda_3| = |\lambda_4|$ with multiplicities 4, 3, 2, and 1, respectively. Since λ_1 and λ_3 have even multiplicity and λ_2 and λ_4 have odd multiplicity, the theorem guarantees that the QR algorithm will converge when it is applied to H.

It is easy to show that an unreduced Hessenberg matrix is nonderogatory* which implies that if it does have multiple eigenvalues, it must be defective† (and have nonlinear elementary divisors**). In Example 3.54, then, H has the elementary divisors $(\lambda - \lambda_1)^4$, $(\lambda - \lambda_2)^3$, $(\lambda - \lambda_3)^2$, and $(\lambda - \lambda_4)$ and so the geometric multiplicity‡ of each eigenvalue is 1. Consequently, it is the algebraic multiplicity we are referring to in Theorem (3.53).

Notice that only *distinct eigenvalues of equal modulus* can prevent the convergence of the QR algorithm and then only when more than two have odd algebraic multiplicities or more than two have even algebraic multiplicities. Thus, if H has fewer than 3 distinct eigenvalues of equal modulus (this is always true if the eigenvalues are real), the QR algorithm *must converge*. Likewise, if H has more than 4 distinct eigenvalues of equal modulus, the QR algorithm *cannot converge*.

In the case of a real matrix H, for example, with several pairs of complex conjugate eigenvalues (but no more than 4 of equal modulus) the sequence $\{H_i\}$ produced by the QR algorithm tends to a matrix in upper block-triangular form with several 2×2 blocks on the main diagonal. To illustrate this let $n = 6$ with $\bar{\lambda}_1 = \lambda_2$ and $\bar{\lambda}_4 = \lambda_5$. Then the sequence tends to a matrix of the form

3.55
$$W = \begin{bmatrix} x & x & x & x & x & x \\ x & x & x & x & x & x \\ & & \lambda_3 & x & x & x \\ & & & x & x & x \\ & & & x & x & x \\ & & & & & \lambda_6 \end{bmatrix}$$

with the upper left-hand 2×2 diagonal block having eigenvalues λ_1 and $\bar{\lambda}_1$ and the other 2×2 diagonal block having eigenvalues λ_4 and $\bar{\lambda}_4$.

---

* See 11-(21.56).
† See 11-(21.27).
** See 11-(21.51).
‡ See 11-(21.61).

In practice we are often interested in what Wilkinson [1965a, p. 83] calls *essential convergence*, that is, the convergence of the sequence $\{A_i\}$ to upper triangular form (even though certain elements above the main diagonal may not converge). A basic result follows.

3.56 Theorem   If A_1 in (3.28) has eigenvalues satisfying $|\lambda_1| > \cdots > |\lambda_n| > 0$, then the sequence $\{A_i\}$, produced by the QR algorithm, *converges essentially* to an upper triangular matrix (which displays the eigenvalues on the main diagonal).

Proof.   See Francis [1961].

Notice that the eigenvalues here are required not only to be distinct but, in addition, they are required to be of different modulus. On the other hand, if some of the eigenvalues are of equal modulus but *all of the eigenvalues with the same modulus are equal*, then the conditions of Theorem (3.53) are not violated and convergence is achieved.

QR with Origin Shifts

We mentioned earlier that (3.28) cannot be implemented, as it stands, for computational reasons. We have described the reduction to Hessenberg form which will enable us to carry out each step of the QR algorithm with order n^2, rather than n^3, multiplications and additions. However, for some matrices this is not sufficient to insure rapid convergence of the algorithm and so shifts of origin are necessary. (See Francis [1961, p. 271] and Parlett [1967], p. 118.)

Another reason for introducing origin shifts is to get around the problem of convergence when distinct eigenvalues have the same modulus. Such eigenvalues lie on a circle in the complex plane (with center at the origin) and shifting the origin appropriately can change the fact that their moduli are equal.

It should be pointed out that when the algorithm converges (see, for example, Naiser [1967], p. 40) it is observed that the convergence of the sub-diagonal element

$h_{i,i-1}$ depends on the ratio of the eigenvalues $\left|\dfrac{\lambda_i}{\lambda_{i-1}}\right|$, and unless this ratio is small

the *rate* of convergence can be very unsatisfactory.

Consider the case of an unreduced Hessenberg matrix H whose eigenvalues have distinct moduli. For this matrix there exists a similarity transformation such that

3.57
$$UHU^{-1} = \Lambda$$

$$= \begin{bmatrix} \lambda_1 & & & \\ & \lambda_2 & & \\ & & \ddots & \\ & & & \lambda_n \end{bmatrix}.$$

If we form $H - qI$, that is, if we shift the origin by the amount q, then

3.58 $U(H - qI)U^{-1} = \Lambda - qI$

$$= \begin{bmatrix} (\lambda_1 - q) & & & \\ & (\lambda_2 - q) & & \\ & & \ddots & \\ & & & (\lambda_n - q) \end{bmatrix}$$

and we have subtracted q from each eigenvalue. This agrees with (3.11).

If we apply the QR algorithm to $H - qI$, then the convergence of the sub-diagonal element $h_{n,n-1}$ to zero depends on the ratio of the eigenvalues $\left| \dfrac{\lambda_n - q}{\lambda_{n-1} - q} \right|$ and, if we can find a way to choose q close to λ_n, we can accelerate the convergence. Observe that as $h_{n,n-1} \to 0$, the corner element $h_{nn} \to \lambda_n - q$. When $h_{n,n-1}$ is sufficiently small it can be assumed that we have a reduced matrix* and, at this point, we can record λ_n, deflate, and continue. (See (3.51) and associated remarks.)

The difficulty, of course, is that we do not know λ_n and so we cannot choose q intelligently. However, this discussion provides the motivation for the following scheme in which we use a different value of q at each step of the iteration.

Let H_1 be an unreduced Hessenberg matrix and form the following sequence:

3.59 $\begin{cases} H_j - q_j I = Q_j R_j \\ \quad H_{j+1} = R_j Q_j + q_j I \end{cases}$ $(j = 1, 2, 3, \ldots)$.

This is sometimes called QR *with origin shifts and restoring* since the shift is "added back" (or restored) in the second equation above.

To see that H_{j+1} is similar to H_j, observe that

3.60 $R_j = Q_j^H (H_j - q_j I)$

and so

3.61 $H_{j+1} = Q_j^H (H_j - q_j I)Q_j + q_j I$

$\qquad\qquad\qquad\quad = Q_j^H H_j Q_j - q_j I + q_j I$

$\qquad\qquad\qquad\quad = Q_j^H H_j Q_j.$

Notice that, by restoring the shifts, the matrices in the sequence $\{H_j\}$ all have the same eigenvalues $\lambda_1, \lambda_2, \ldots, \lambda_n$ as H_1. Consequently, if these eigenvalues all have different moduli we are expecting $h_{n,n-1}^{(j)}$ to approach zero and $h_{nn}^{(j)}$ to approach

---

*Recall (3.50).

λ_n. It seems logical, therefore, for rapid convergence, to take $q_j = h_{nn}^{(j)}$ as soon as $h_{n,n-1}^{(j)}$ shows signs of convergence.

If the eigenvalues do not have different moduli (for example, a real matrix for which at least one pair of complex conjugate eigenvalues λ_t and $\bar{\lambda}_t$ exists) we may find $h_{n-1,n-2}^{(j)}$ (rather than $h_{n,n-1}^{(j)}$) approaching zero. In this case our sequence tends to a reduced Hessenberg matrix of the form*

$$\textbf{3.62} \qquad H = \left[\begin{array}{c|cc} H_{11} & & H_{12} \\ \hline & h_{n-1,n-1} & h_{n-1,n} \\ \bigcirc & h_{n,n-1} & h_{nn} \end{array} \right],$$

and λ_{n-1} and λ_n are the eigenvalues of the 2×2 block in the lower right-hand corner.

Thus, if $h_{n,n-1}^{(j)}$ becomes effectively zero, we can record λ_n and deflate by deleting the *last* row and column before continuing. On the other hand, if $h_{n-1,n-2}^{(j)}$ becomes effectively zero, we can compute λ_{n-1} and λ_n from the 2×2 submatrix in the lower right-hand corner and deflate by deleting the *last two* rows and columns.

We mentioned a procedure for selecting q_j if $h_{n,n-1}^{(j)} \to 0$. How do we select the shift if $h_{n-1,n-2}^{(j)} \to 0$, instead? One procedure is as follows: Compute the eigenvalues μ_{1j} and μ_{2j} of the 2×2 submatrix

$$\textbf{3.63} \qquad S_j = \left[\begin{array}{cc} h_{n-1,n-1}^{(j)} & h_{n-1,n}^{(j)} \\ h_{n,n-1}^{(j)} & h_{nn}^{(j)} \end{array} \right],$$

where $|\mu_{1j}| \geqq |\mu_{2j}|$. There are several cases to be considered, depending on whether μ_{1j} and μ_{2j} are real or complex. (See Francis [1962, p. 339] and Naiser [1967, p. 45], for example.) An interesting case occurs when μ_{1j} and μ_{2j} are complex conjugates, and one possibility, in this case, is to shift once with μ_{1j} and once with μ_{2j}.

The suggestion to carry out two shifts, using μ_{1j} and μ_{2j}, is the motivation for the introduction of the double QR algorithm. Consequently, we interrupt our discussion of the selection of origin shifts until we have introduced the double QR step.

The Double QR Algorithm

This algorithm is particularly advantageous when H_1 is real, and not all of its

---

* Recall (3.55).

eigenvalues are real, because it provides us with a procedure for doing all of the computations in real arithmetic. Initially, then, we assume H_1 to be real.

Consider any two steps in (3.59) (of which the first two are typical). Thus, we can write

3.64
$$\begin{cases} H_1 - q_{11}I = Q_1 R_1 \\ \quad\quad H_2 = R_1 Q_1 = q_{11}I \\ H_2 - q_{21}I = Q_2 R_2 \\ \quad\quad H_3 = R_2 Q_2 + q_{21}I. \end{cases}$$

From (3.61) we have

3.65
$$\begin{aligned} H_3 &= Q_2^H H_2 Q_2 \\ &= Q_2^H (Q_1^H H_1 Q_1) Q_2 \\ &= (Q_1 Q_2)^H H_1 (Q_1 Q_2). \end{aligned}$$

An interesting result is the fact that

3.66
$$\begin{aligned} Q_1 Q_2 R_2 R_1 &= Q_1 (H_2 - q_{21}I) R_1 \\ &= Q_1 H_2 R_1 - q_{21} Q_1 R_1 \\ &= Q_1 (R_1 Q_1 + q_{11}I) R_1 - q_{21} Q_1 R_1 \\ &= Q_1 R_1 Q_1 R_1 + q_{11} Q_1 R_1 - q_{21} Q_1 R_1 \\ &= (Q_1 R_1 + q_{11}I) Q_1 R_1 - q_{21} Q_1 R_1 \\ &= H_1 Q_1 R_1 - q_{21} Q_1 R_1 \\ &= (H_1 - q_{21}I)(H_1 - q_{11}I) \end{aligned}$$

because, if H_1 is real and q_{11} and q_{21} are chosen to satisfy

3.67
$$q_{21} = \bar{q}_{11},$$

then the matrix on the right, in (3.66), is real and this implies $Q_1 Q_2$ and $R_2 R_1$ are real.* Thus, from (3.65), H_3 is real.

Since the double QR step which computes H_{2t+1} from H_{2t-1} is exactly like the first double QR step we do not need to repeat the details. We leave that as an exercise for the reader.

Notice that if q_{11} and q_{21} are complex (but (3.67) is satisfied), then H_2 will be complex but H_3 will be real. In general, then, the sequence $H_1, H_3, H_5, \ldots$ is real.

---

*See the proof of (3.29).

Fortunately, we can avoid computing $H_2, H_4, \ldots$ by using the following device. Instead of finding the decompositions $Q_1 R_1$ and $Q_2 R_2$ in (3.64), in order to have $Q_1 Q_2$ for (3.65), we form $Q_1 Q_2$ directly. This can be done by forming the product $(H_1 - \bar{q}_{11} I)(H_1 - q_{11} I)$ and by carrying out the decomposition

3.68
$$(H_1 - \bar{q}_{11} I)(H_1 - q_{11} I) = \hat{Q}_1 \hat{R}_1.$$

Because of the uniqueness* of this decomposition, when the diagonal elements of $\hat{R}_1$ are real and positive, we have

3.69
$$\hat{Q}_1 = Q_1 Q_2,$$

and we can by-pass H_2 and go to H_3 directly. The apparent price we pay for this convenience is the computation of the matrix product

$$(H_{2t-1} - q_{2,2t-1} I)(H_{2t-1} - q_{1,2t-1} I)$$

at step t of the double QR algorithm.

However, it turns out that if $H_1, H_3, \ldots, H_{2t-1}$ are Hessenberg matrices,† there is a way to avoid paying this price. The details of this simplification are lengthy and we shall not include them here. The reader is referred to Parlett [1967, pp. 120–121].

Now we return to (3.63) and to $\mu_{1,2t-1}$ and $\mu_{2,2t-1}$, the eigenvalues of S_{2t-1}. There seems to be no known optimal choice for the shifts $q_{1,2t-1}$ and $q_{2,2t-1}$, and possibly the simplest proposal is to let

3.70
$$\begin{cases} q_{1,2t-1} = \mu_{1,2t-1} \\ q_{2,2t-1} = \mu_{2,2t-1} \end{cases}$$

at step t. A more complicated scheme is proposed by Parlett [1967, p. 123], but he concludes by saying, "There is no clear cut optimum and convergence rates do not seem sensitive to the choice."

3.71 Remark. Following the proof of (3.29) we state that the procedure used in the constructive proof is not recommended for practical consideration. This is due to the fact that, in practice, we use the double QR algorithm and only on Hessenberg matrices.

EXERCISES 14.3

1. Write a computer program to compute the eigenvalues of an arbitrary real matrix using the inverse power method with origin shifts. (Normalize the vectors at each stage of the iteration.)

---

* See (3.30).
† Recall (3.31).

2. Prove that the number of multiplications and additions involved in one QR transformation is proportional to n^3 for a full matrix whereas it is only n^2 for a Hessenberg matrix.

3. Prove Theorem 3.31.

4. Write a computer program to reduce an arbitrary real matrix to upper Hessenberg form using elementary similarity transformations. (This program should be a subroutine.)

5. Prove the statement in the last paragraph of Remark 3.50.

6. Prove that one step of the QR algorithm applied to an unreduced, singular, Hessenberg matrix reveals a zero eigenvalue.

7. What is the mathematical justification for deflating the matrix in (3.51)?

8. Prove Theorem 3.53.

9. Prove that an unreduced Hessenberg matrix is nonderogatory.

10. Prove Theorem 3.56.

11. Prove that (3.67) implies that the matrix on the right in (3.66) is real (when H_1 is real) and that this fact implies $Q_1 Q_2$ and $R_2 R_1$ are both real.

12. In this section we have given the details of the first step of the double QR algorithm which produces H_3 from H_1 (assumed to be real). Write out the details of the step which produces H_{2t+1} from H_{2t-1}.

13. Show that if q_{11} and q_{21} are complex but satisfy (3.67) and if H_1 is real, then H_2 will be complex but H_3 will be real.

14. Following (3.69) we point out that if $H_1, H_3, H_5, \ldots, H_{2t-1}$ are Hessenberg matrices we can avoid "paying the price" which enables us to by-pass H_{2t} and go to H_{2t+1} directly, during step t of the double QR algorithm. Prove this assertion.

15. Write a computer program to compute the eigenvalues of an arbitrary real matrix using the double QR algorithm. Use the subroutine in Exercise 4 to reduce the matrix to Hessenberg form. (This program should be a subroutine.)

14.4 COMPUTING THE EIGENVECTORS

In our discussions of algorithms for computing the eigenvalues we have not mentioned the corresponding procedures for computing the eigenvectors (except for the Jacobi algorithm for real symmetric matrices). We are now ready to do so, however. Again, we discuss the Hermitian case before we go to the non-Hermitian case.

Hermitian Matrices

We shall restrict the discussion to real symmetric matrices just as we did in the previous section. We assume that we have computed one or more eigenvalues of a symmetric matrix $A \in \mathbf{R}^{nn}$ by first reducing it to a symmetric tri-diagonal matrix T. Let

4.1
$$\begin{cases} A = PTP^T \\ PP^T = I. \end{cases}$$

Now if λ is an eigenvalue of T with corresponding eigenvector y, then $Ty = \lambda y$ and so

4.2
$$\lambda(Py) = PTy$$
$$= PTP^T(Py)$$
$$= A(Py),$$

which shows that λ is an eigenvalue of A with corresponding eigenvector

4.3
$$x = Py.$$

In other words, we can compute the eigenvector x (for A) by first computing the eigenvector y (for T) and then using (4.3). Since T is tri-diagonal, the computation of y is not too difficult. For reasons of computational stability Wilkinson [1963, p. 142] recommends* the use of the inverse power method† for computing y. (See, also, Ortega [1967], pp. 98–99.)

If μ is an approximation to λ (but $\mu \neq \lambda$) we pick an arbitrary normalized‡ vector $y^{(0)} \in \mathbf{R}^n$ and form the sequence $\{z^{(k)}\}$ as follows:

4.4
$$\begin{cases} (T - \mu I)y^{(1)} = y^{(0)} \qquad z^{(1)} = \dfrac{1}{m_1} y^{(1)} \\[2mm] (T - \mu I)y^{(2)} = z^{(1)} \qquad z^{(2)} = \dfrac{1}{m_2} y^{(2)} \\[2mm] \qquad \vdots \qquad\qquad\qquad\qquad \vdots \\[2mm] (T - \mu I)y^{(k)} = z^{(k-1)} \qquad z^{(k)} = \dfrac{1}{m_k} y^{(k)} \\[2mm] \qquad \vdots \qquad\qquad\qquad\qquad \vdots \end{cases}$$

Here, the normalizing factor, m_k, is the component of $y^{(k)}$ of maximum modulus, and so

4.5
$$|m_k| = \max_i |y_i^{(k)}|.$$

With mild assumptions on μ and $y^{(0)}$, $z^{(k)}$ converges to y (normalized).

---

* Recall (3.25).
† Recall (3.18) and (3.23).
‡ That is, $\| y^{(0)} \|_\infty = 1$.

Since T is symmetric it cannot be defective* and so there exists a set of n linearly independent eigenvectors of T (including the eigenvector y) which forms a basis for $\mathbf{R}^n$. Hence we can write an equation similar to (3.19). It is easily verified that the *rate* of convergence of (4.4) depends on the size of the ratio

4.6
$$\max_{\lambda_i \neq \lambda} \left| \frac{\lambda - \mu}{\lambda_i - \mu} \right|,$$

and convergence itself depends on the fact that $y^{(0)}$ has a component in the direction of y. Thus, the two "mild assumptions" on μ and $y^{(0)}$ are that

4.7
$$|\lambda - \mu| < |\lambda_i - \mu|, \quad \lambda_i \neq \lambda,$$

and that $y^{(0)}$ have a component in the direction of y.

Notice that λ can be a multiple eigenvalue (in which case T is derogatory†) but we must require that μ be closer to λ than to any other eigenvalue. If λ is an eigenvalue of multiplicity t, there exists a t-dimensional invariant subspace of eigenvectors corresponding to λ. To have the process yield t linearly independent eigenvectors we must either use t different starting vectors $y^{(0)}$, or use t slightly different values of the approximation μ. Otherwise, the convergence will be to the same vector each time. Once a set of t linearly independent vectors has been obtained, if we want them to be orthonormal we must apply the Schmidt orthogonalization procedure at this point.

One way to (hopefully) insure that $y^{(0)}$ has a component in the direction of y is to choose its components as random numbers in the interval $[-1, 1]$. However, we cannot *guarantee* that this will always give us a suitable starting vector, and any computer program must provide for the rare case when this procedure fails.

Wilkinson [1963, pp. 142–143] shows that, unless $|\lambda - \mu|$ is extremely close to $|\lambda_i - \mu|$ for some i, or $y^{(0)}$ has a pathologically small component in the direction of the eigenvector y, the sequence (4.4) usually converges after only two steps.

The sequence (4.4) involves the solution of a system of linear algebraic equations at each step, and since the coefficient matrix is tri-diagonal, either Gaussian elimination with partial pivoting, or the corresponding triangular decomposition, will be extremely simple to carry out.

4.8 *Remark.* If the tri-diagonal matrix T is produced by a sequence of Householder transformations then, in (4.3),

$$x = (P_2 P_3 \ldots P_{n-1})y$$
$$= [I - \alpha_2 v^{(2)} v^{(2)T}] \cdots [I - \alpha_{n-1} v^{(n-1)} v^{(n-1)T}]y$$

---

* See (21.27), (21.35) and (21.36) of Chapter 11.
† See 11-(21.56).

and these multiplications can be carried out quite simply, if we observe that

$$(I - \alpha vv^T)y = y - \alpha(v^T y)v.$$

The simplicity of the computation is aided by the fact that, for $v^{(r)}$, the first $r - 1$ components are zero. Notice that we never form P_r explicitly and this agrees with our comment before (2.59).

Non-Hermitian Matrices

In this case the condensed form is an upper Hessenberg matrix H and, if $A \in \mathbf{C}^{nn}$ is the original matrix,

4.9
$$\begin{cases} A = UHU^H \\ UU^H = I. \end{cases}$$

Just as in (4.2), if λ is an eigenvalue of H with corresponding eigenvector y, then $Hy = \lambda y$ and so

4.10
$$\begin{aligned} \lambda(Uy) &= UHy \\ &= UHU^H(Uy) \\ &= A(Uy), \end{aligned}$$

which shows that λ is an eigenvalue of A with corresponding eigenvector

4.11
$$x = Uy.$$

As before, we use the inverse power method for finding y. Let σ be an approximation to λ (but $\sigma \neq \lambda$) and pick an arbitrary normalized vector $y^{(0)} \in \mathbf{C}^n$. Form the sequence $\{z^{(k)}\}$, as in (4.4),

4.12
$$\begin{cases} (H - \sigma I)y^{(1)} = y^{(0)} \qquad z^{(1)} = \dfrac{1}{m_1} y^{(1)} \\[2ex] (H - \sigma I)y^{(2)} = z^{(1)} \qquad z^{(2)} = \dfrac{1}{m_2} y^{(2)} \\[1ex] \qquad \vdots \qquad\qquad\qquad\quad \vdots \\[1ex] (H - \sigma I)y^{(k)} = z^{(k-1)} \qquad z^{(k)} = \dfrac{1}{m_k} y^{(k)} \\[1ex] \qquad \vdots \qquad\qquad\qquad\quad \vdots \end{cases}$$

Again, the normalizing factor, m_k, is the component of $y^{(k)}$ of maximum modulus.

Also, as in (4.7), if

4.13 $|\lambda - \sigma| < |\lambda_i - \sigma|, \quad \lambda_i \neq \lambda,$

and if $y^{(0)}$ has a component in the direction of y, then $z^{(k)}$ tends to y. Solving the linear equations in (4.12) is not too difficult, since H is an upper Hessenberg matrix.

4.14 *Remark.* If H has multiple eigenvalues there are two cases to consider. First, if λ is an eigenvalue whose algebraic multiplicity* and geometric multiplicity are both equal to t (this is the only possibility if H is Hermitian) then there exists a t-dimensional subspace of eigenvectors corresponding to λ, and we can treat this case the way we treated the multiple eigenvalue case for Hermitian matrices.

If λ is a multiple eigenvalue associated with at least one non-linear elementary divisor then its algebraic multiplicity t is greater than its geometric multiplicity $s < t$ and the matrix is defective. In this case there exist only s linearly independent eigenvectors associated with λ, and these can be found as above.

There are $t - s$ additional vectors (called generalized eigenvectors or principal vectors† which exist and which (together with the s linearly independent eigenvectors already mentioned) provide us with a basis for the t-dimensional invariant subspace associated with λ. Actually, the term principal vector includes the term eigenvector as a special case so that there are t linearly independent principal vectors in all.

Computing those principal vectors which are not eigenvectors is a very complicated computational problem and the reader is referred to recent work by Varah [1967] for a discussion of this problem. (See also Ruhe [1970], Poole [1970], and Wilkinson [1965], pp. 182–187.)

EXERCISES 14.4

1. Assume that you have a subroutine (Exercise 16 of Section 14.2) using the Householder-Givens algorithm for computing the eigenvalues of a real symmetric matrix. Write a complete program which uses this subroutine to compute the eigenvalues and then computes the eigenvectors using the inverse power method.

2. Work out the mathematical details of the claim in Remark 4.8 that we never have to form P_r explicitly.

14.5 ILL-CONDITIONED EIGENPROBLEMS

In Section 12.15 we discussed ill-conditioned systems of linear algebraic equations, and this discussion led to the definition of a class of condition numbers for the problem.‡ In this section we do something similar for the matrix eigenvalue-eigenvector problem.

---

* See 11-(21.61).
† See 11-(21.62).
‡ See 12-(16.1).

Two examples are worth mentioning before we begin the theoretical discussion. The first exhibits the sensitivity of eigenvalues, and the second exhibits the sensitivity of eigenvectors, to perturbations in matrix elements.

5.1 EXAMPLE (Forsythe). Let $A \in \mathbf{R}^{nn}$ and $A + E$ be the upper Hessenberg matrices (A is also lower triangular)

$$A = \begin{bmatrix} a & & & & & \\ 1 & a & & & & \\ & 1 & a & & & \\ & & 1 & a & & \\ & & & \cdots & & \\ & & & & 1 & a \end{bmatrix}, \qquad A + E = \begin{bmatrix} a & & & & & \varepsilon \\ 1 & a & & & & \\ & 1 & a & & & \\ & & 1 & a & & \\ & & & \cdots & & \\ & & & & 1 & a \end{bmatrix},$$

which differ by the single element ε in the $(1, n)$ position. The characteristic equation for A is

$$(a - \lambda)^n = 0$$

and so $\lambda = a$ is an eigenvalue of algebraic multiplicity n. On the other hand

$$(a - \lambda)^n + \varepsilon(-1)^{n+1} = 0$$

is the characteristic equation for $A + E$. Thus, the eigenvalues are

$$\lambda_r = a + \omega^r \varepsilon^{1/n} \qquad (r = 0, 1, 2, \ldots, n - 1)$$

where ω is any primitive nth root of unity. In other words, $A + E$ has n distinct complex eigenvalues lying on a circle with center at a and radius $\varepsilon^{1/n}$.

Suppose $n = 10$ and $\varepsilon = 10^{-10}$, for example. In this case, the radius of the circle is 10^{-1} which means that the change in the eigenvalues is 10^9 times as great as the perturbation introduced into a single matrix element. If $n = 100$ and $\varepsilon = 10^{-100}$, on the other hand, the radius of the circle is still 10^{-1} but *the change in the eigenvalues is 10^{99} times as great as the perturbation.*

5.2 EXAMPLE (Givens). Let $A \in \mathbf{R}^{22}$ be the matrix

$$A = \begin{bmatrix} 1 + \varepsilon \cos \dfrac{2}{\varepsilon} & -\varepsilon \sin \dfrac{2}{\varepsilon} \\[3mm] -\varepsilon \sin \dfrac{2}{\varepsilon} & 1 - \varepsilon \cos \dfrac{2}{\varepsilon} \end{bmatrix}.$$

A is real and symmetric with eigenvalues

$$\begin{cases} \lambda_1 = 1 - \varepsilon \\ \lambda_2 = 1 + \varepsilon \end{cases}$$

and eigenvectors

$$x^{(1)} = \begin{bmatrix} \sin\dfrac{1}{\varepsilon} \\[2mm] \cos\dfrac{1}{\varepsilon} \end{bmatrix}, \qquad x^{(2)} = \begin{bmatrix} -\cos\dfrac{1}{\varepsilon} \\[2mm] \sin\dfrac{1}{\varepsilon} \end{bmatrix}.$$

Observe that small changes in ε make small changes in λ_1 and λ_2 but as $\varepsilon \to 0$ the eigenvectors do not tend to a limit. Thus, *small changes in the elements of A can produce large changes in the eigenvectors.* In this example, the eigenvectors are not even continuous functions of the matrix elements.

Condition Numbers

For the class of diagonalizable* matrices (matrices with linear elementary divisors) we can, following Wilkinson, introduce a set of n condition numbers as follows. Consider a matrix $A \in \mathbf{C}^{nn}$ with linear elementary divisors. There exists a nonsingular matrix X for which

5.3
$$X^{-1}AX = \begin{bmatrix} \lambda_1 & & & \\ & \lambda_2 & & \\ & & \cdot & \\ & & & \lambda_n \end{bmatrix} = \Lambda.$$

The matrix X is not unique because, for any nonsingular diagonal matrix D,

$$(XD)^{-1}A(XD) = D^{-1}(X^{-1}AX)D$$

5.4
$$= \begin{bmatrix} \lambda_1 & & & \\ & \lambda_2 & & \\ & & \cdot\cdot & \\ & & & \lambda_n \end{bmatrix}.$$

If A has a multiple eigenvalue (since A is diagonalizable, this implies it is derogatory), there is an even wider choice for X.

---

* See (21.33) and (21.36) of Chapter 11.

Since the columns of X are eigenvectors* of A we have a set of n linearly independent *right* eigenvectors $x^{(1)}, x^{(2)}, \ldots, x^{(n)}$. We also have a set of n linearly independent *left* eigenvectors $y^{(1)}, y^{(2)}, \ldots, y^{(n)}$, which are the rows of X^{-1}. To verify this fact, multiply (5.3) on the right by X^{-1} and obtain

5.5
$$X^{-1}A = \Lambda X^{-1},$$

which is equivalent to

5.6
$$y^{(j)T}A = \lambda_j y^{(j)T}, \qquad j = 1, 2, \ldots, n.$$

The left eigenvectors of A are merely the right eigenvectors of A^T since the transpose of (5.6) is

5.7
$$A^T y^{(j)} = \lambda_j y^{(j)}, \qquad j = 1, 2, \ldots, n.$$

5.8 Theorem. If $x^{(1)}, x^{(2)}, \ldots, x^{(n)}$ and $y^{(1)}, y^{(2)}, \ldots, y^{(n)}$ are right and left eigenvectors, respectively, of $A \in C^{nn}$, then

$$y^{(j)T}x^{(i)} = 0, \qquad \lambda_i \neq \lambda_j.$$

Proof. Multiply (5.6) on the right by $x^{(i)}$ and $Ax^{(i)} = \lambda_i x^{(i)}$ on the left by $y^{(j)T}$. If the resulting equations are subtracted we obtain

$$0 = (\lambda_i - \lambda_j)y^{(j)T}x^{(i)}.$$

The result follows, when $\lambda_i \neq \lambda_j$.

If the vectors are normalized so that

5.9
$$\|y^{(i)}\|_2 = \|x^{(i)}\|_2 = 1,$$

then Wilkinson [1963, p. 137] introduces the symbols

5.10
$$s_i = y^{(i)T}x^{(i)}, \qquad i = 1, 2, \ldots, n,$$

where, for all i,

5.11
$$|s_i| = |y^{(i)T}x^{(i)}|$$
$$\leq \|y^{(i)}\|_2 \|x^{(i)}\|_2$$
$$= 1.$$

He shows [1963, p. 138, and 1965, Chapter 2] that the sensitivity of λ_i to small perturbations in the matrix elements is dependent primarily on the number $|s_i|^{-1}$.

---

*Recall 11-(21.18).

Thus, the n numbers $|s_1|^{-1}, |s_2|^{-1}, \ldots, |s_n|^{-1}$ provide us with a set of n *condition numbers* for the individual eigenvalues.

If the vectors are real, then we have the geometric interpretation

5.12 $s_i = \cos \theta_i$,

where θ_i is the angle between $x^{(i)}$ and $y^{(i)}$. In this case, the condition number for λ_i improves as $\theta_i \to 0$ and worsens as $x^{(i)}$ and $y^{(i)}$ approach orthogonality. We should point out that if the vectors are complex, then the expression in (5.8) is not an inner product. The true inner product* is $(x, y) = y^H x$, in this case.

5.13 Remark. Theoretically, we would need the n^3 numbers $\partial \lambda_i / \partial a_{jk}$ to completely understand the sensitivity of the eigenvalues to perturbations in the individual matrix elements. However, these numbers are not easy to determine, and so we settle for $|s_i|^{-1}, i = 1, 2, \ldots, n$, instead.

5.14 Theorem. If A is Hermitian, then

$$|s_i|^{-1} = 1, \qquad i = 1, 2, \ldots, n.$$

Proof. If $A = A^H$, then there exists† a unitary similarity transformation, $U^H A U = \Lambda$, which diagonalizes A. In this case, then, if $Au^{(i)} = \lambda_i u^{(i)}$, then $A^T \bar{u}^{(i)} = \lambda_i \bar{u}^{(i)}$ or, equivalently, $u^{(i)H} A = \lambda_i u^{(i)H}$. Hence,

$$s_i = \bar{u}^{(i)T} u^{(i)}$$
$$= u^{(i)H} u^{(i)}$$
$$= 1$$

and so $|s_i|^{-1} = 1$, the smallest possible value.

What this means, if A is real and symmetric, is that $\theta_i = 0$ in (5.12). In general, then, *every Hermitian matrix is perfectly conditioned with respect to its eigenvalue problem.* However, Example 5.2 demonstrates that no such claim can be made with respect to the eigenvector problem.

Suppose A is diagonalizable but not necessarily Hermitian. If λ_i, for some i, is a multiple eigenvalue (with linear elementary divisors this implies that the matrix is derogatory), then there is a subspace of right eigenvectors and a subspace of left eigenvectors from which $x^{(i)}$ and $y^{(i)}$ can be chosen to form the condition number $|s_i|^{-1}$. The dimension of each subspace is equal to the (algebraic) multiplicity of λ_i, and some particular choice of $x^{(i)}$ and $y^{(i)}$ must be prescribed.

If the matrix has non-linear elementary divisors the situation is more

---

* Recall 11-(16.2) and 11-(16.11).
† Recall Theorem 11-21.66.

complicated, of course, and the reader is referred to Wilkinson [1965, Chapter 2] for a discussion of the various cases.

In Section 12.16 we use a single constant $\kappa_\beta(A) = \|A\|_\beta \|A^{-1}\|_\beta$ to measure the sensitivity of the solution of the linear equations problem to perturbations in the data. We do have the freedom to choose our favorite matrix norm, however, and $\kappa_2(A)$, the spectral condition number, is one of the most widely used.

We might ask whether or not there exists a single constant, in lieu of the n constant $|s_i|^{-1}$, which measures the overall sensitivity of the eigenvalues of A to small perturbations in the elements of A. Such a measure does exist, but it certainly cannot provide us with information about the sensitivity of individual eigenvalues. For example, the upper Hessenberg matrix (see Frank [1958], p. 385)

$$
5.15 \qquad H_n = \begin{bmatrix}
n & n-1 & n-2 & \cdots & 3 & 2 & 1 \\
n-1 & n-1 & n-2 & \cdots & 3 & 2 & 1 \\
 & n-2 & n-2 & \cdots & 3 & 2 & 1 \\
 & & & \cdots\cdots\cdots\cdots\cdots\cdots\cdots \\
 & & & & 2 & 2 & 1 \\
 & & & & & 1 & 1
\end{bmatrix}
$$

is such that the sensitivity of its smallest eigenvalues is considerably greater than the sensitivity of its largest eigenvalues. Despite this fact we find it useful, in some cases, to have a single condition number rather than n condition numbers.

If A has linear elementary divisors, then (5.3) gives us $H^{-1}AH = \Lambda$. Now consider the perturbed matrix $A + E$ and let μ be one of its eigenvalues. Thus,

$$
5.16 \qquad (A + E)y = \mu y
$$

for some $y \neq \phi$. This also can be written

$$
5.17 \qquad (\mu I - A)y = Ey.
$$

There are two possibilities: μ is either an eigenvalue of A or it is not. Suppose μ is not an eigenvalue of A. Then, from (5.17), we may write

$$
5.18 \qquad H^{-1}(\mu I - A)H(H^{-1}y) = H^{-1}EH(H^{-1}y),
$$

and if we simplify this equation we obtain

$$
5.19 \qquad (\mu I - \Lambda)w = (H^{-1}EH)w,
$$

where w has the obvious definition. Since μ is not an eigenvalue of A, the matrix

5.20

$$\mu I - \Lambda = \begin{bmatrix} (\mu - \lambda_1) & & & & \\ & (\mu - \lambda_2) & & & \\ & & \cdot & & \\ & & & \cdot & \\ & & & & (\mu - \lambda_n) \end{bmatrix}.$$

is nonsingular and we may write

5.21

$$w = (\mu I - \Lambda)^{-1}(H^{-1}EH)w.$$

Thus, for any pair of consistent vector and matrix norms

5.22

$$\|w\|_\alpha \leq \|(\mu I - \Lambda)^{-1}\|_\beta \|H^{-1}EH\|_\beta \|w\|_\alpha$$

and, since $w \neq \phi$, we may write

5.23

$$1 \leq \|(\mu I - \Lambda)^{-1}\|_\beta \|H^{-1}EH\|_\beta,$$

or equivalently,

5.24

$$\|(\mu I - \Lambda)^{-1}\|_\beta^{-1} \leq \|H^{-1}EH\|_\beta.$$

Since

5.25

$$\|(\mu I - \Lambda)^{-1}\|_2 = \max_i |\mu - \lambda_i|^{-1},$$

we have

5.26

$$\|(\mu I - \Lambda)^{-1}\|_2^{-1} = \min_i |\mu - \lambda_i|,$$

and so (5.24) gives us the important result

5.27

$$\min_i |\mu - \lambda_i| \leq \|H^{-1}\|_2 \|H\|_2 \|E\|_2.$$

We have assumed that μ is not an eigenvalue of A. On the other hand, if $\mu = \lambda_i$ for some i, then $\min_i |\mu - \lambda_i| = 0$ and so (5.27) holds in any case.

It is easily verified that (5.25) is true, also, for the matrix norms $\| \cdot \|_1$, $\| \cdot \|_\infty$, and $\| \cdot \|_E$ which implies that (5.27) is true for these additional norms. The relation (5.27) is due to Bauer and Fike [1960] and may be interpreted to mean that every eigenvalue of the perturbed matrix $A + E$ lies in at least one of the circular discs with center λ_i and radius $\|H^{-1}\|_2 \|H\|_2 \|E\|_2$.

Notice that the radius of each disc is bounded by $\|E\|_2$ multiplied by the quantity $\|H\|_2 \|H^{-1}\|_2$ which means that $\|H\|_2 \|H^{-1}\|_2$ determines the overall

sensitivity of the eigenvalues to the perturbation matrix E. Since H, in the transformation $H^{-1}AH = \Lambda$, is not unique we define the *spectral condition number**
of A with respect to the eigenvalue problem to be

5.28
$$\kappa_H(A) = \min_H \|H\|_2 \|H^{-1}\|_2.$$

Since

5.29
$$1 = \|HH^{-1}\|_\beta$$
$$\leq \|H\|_\beta \|H^{-1}\|_\beta$$

the condition number can never be less than one. In other words, one is the best possible condition number.

If A is Hermitian, then H can be taken to be unitary, in which case

5.30
$$\|H\|_2 = \|H^{-1}\|_2 = 1.$$

Thus $\kappa_H(A) = 1$ and so

5.31
$$\min_i |\mu - \lambda_i| \leq \|E\|_2.$$

Notice that small perturbations in A imply that $\|E\|_2$ is small, and this implies that $\min_i|\mu - \lambda_i|$ is small. Since $\kappa_H(A) = 1$ we can state that the determination of the eigenvalues of every Hermitian matrix is a perfectly conditioned problem. This agrees with our interpretation of Theorem 5.14. The reader is warned again, however, that we cannot make any such claim about the condition of the eigenvector problem.

An obvious question at this point is whether or not we can relate $\kappa_H(A)$ to the quantities $|s_i|^{-1}, i = 1, 2, \ldots, n$. The following theorem gives us the answer.

5.32 Theorem. If $A \in \mathbf{C}^{nn}$ has linear elementary divisors, then $|s_1|^{-1}, |s_2|^{-1}, \ldots,$ $|s_n|^{-1}$ and $\kappa_H(A)$ exist and are related by the inequalities

$$\begin{cases} |s_i|^{-1} \leq \kappa_H(A), & i = 1, 2, \ldots, n \\ \kappa_H(A) \leq \sum_{t=1}^{n} |s_t|^{-1}. \end{cases}$$

Proof. See, for example, Wilkinson [1965, pp. 88–89].

Another important question is related to the invariant properties of the condition numbers under unitary similarity transformations.

---

* This name is motivated by the fact that $\|\cdot\|_2$ is called the spectral norm. Compare the spectral condition number of A relative to the eigenvalue problem with the spectral condition number of A relative to the linear equations problem. See 12-(16.2).

5.33 Theorem. If $A = Q^H BQ$ is a unitary similarity transformation and if A is diagonalizable, then A and B have the same spectral condition number.

Proof. Since $H^{-1}AH = \Lambda$ we can write

$$H^{-1}(Q^H BQ)H = \Lambda$$

or, equivalently,

$$(QH)^{-1}B(QH) = \Lambda.$$

Now, from Theorem 11-29.4,

$$\|QH\|_2\|(QH)^{-1}\|_2 = \|H\|_2\|H^{-1}\|_2$$
$$= \kappa_H(A)$$

and this completes the proof.

5.34 Theorem. If $A = Q^H BQ$ is a unitary similarity transformation, and if A is diagonalizable, then A and B have the same condition numbers $|s_i|^{-1}, i = 1, 2, \ldots, n$.

Proof. From the proof of the previous theorem we write

$$(QH)^{-1}B(QH) = \Lambda.$$

If $x^{(i)}$ is a right eigenvector of A then $\hat{x}^{(i)} = Qx^{(i)}$ is a right eigenvector of B. Likewise, if $y^{(i)}$ is the corresponding right eigenvector of A^T (left eigenvector of A), then $\hat{y}^{(i)} = \bar{Q}y^{(i)}$ is the corresponding right eigenvector of B^T (left eigenvector of B). Hence

$$\hat{y}^{(i)T}\hat{x}^{(i)} = y^{(i)T}Q^H Qx^{(i)}$$
$$= y^{(i)T}x^{(i)}$$
$$= s_i$$

and this completes the proof.

5.35 *Remark.* These two theorems demonstrate the fact that the condition of a diagonalizable matrix A, with respect to its eigenvalue problem, is invariant under a unitary similarity transformation. Thus, computational algorithms which use unitary similarity transformations cannot worsen the condition numbers. This is a real advantage if the matrix is ill-conditioned.

On the other hand, these transformations cannot improve the condition numbers either, whereas similarity transformations which do not involve unitary matrices can *either improve or worsen* the condition numbers. Consequently, unless we can guarantee improvement it is safer to use unitary transformations.

5.36 *Remark.* We have restricted the discussion in this section to matrices with linear elementary divisors because these matrices are diagonalizable and have a complete set of linearly independent eigenvectors. This fact simplifies the analysis considerably.

Also, we have avoided the difficult problem of measuring the sensitivity of the eigenvectors to perturbations in the matrix elements. The reader is referred to Wilkinson [1965, Chapter 2] for a treatment of this subject.

5.37 *Remark.* Notice that $\kappa_2(A)$, the spectral condition number of A with respect to the linear equations problem, is expressed in terms of the spectral norms of A and A^{-1} whereas $\kappa_H(A)$, the spectral condition number of A with respect to the eigenvalue problem, is expressed in terms of the spectral norms of H and H^{-1}, where $H^{-1}AH = \Lambda$.

EXERCISES 14.5

1. Verify the mathematical details of Example 5.2.

2. Verify (5.25) for the matrix norms $\| \cdot \|_1, \| \cdot \|_\infty$, and $\| \cdot \|_E$.

3. Prove Theorem 5.32.

SUPPLEMENTARY DISCUSSION

In Section 12.18 we discuss scaling of the linear system $Ax = b$ prior to attempting a numerical solution. The objective is to reduce $\kappa_\beta(A) = \|A\|_\beta \|A^{-1}\|_\beta$, for some norm $\| \cdot \|_\beta$, by scaling the rows and the columns appropriately. (The procedure is sometimes called pre-conditioning the system.) The difficulty of finding an optimum scaling algorithm is clearly indicated.

Osborne [1960] discusses scaling relative to the eigenvalue-eigenvector problem, and Parlett and Reinsch [1969] recommend a modified version, called balancing, before attempting a solution using the QR algorithm. Here, the objective is not to reduce $\kappa_\beta(A)$ but to reduce $\|A\|_\beta$, for some norm $\| \cdot \|_\beta$, by appropriate diagonal similarity transformations.

There are classes of matrices that do not need balancing; for example, normal matrices are already balanced with respect to $\| \cdot \|_2$.

CHAPTER 15

PARTIAL DIFFERENTIAL EQUATIONS: ELLIPTIC BOUNDARY VALUE PROBLEMS

15.1 INTRODUCTION

In this chapter we shall be primarily concerned with linear second-order partial differential equations of the form

1.1 $$L[u] = Au_{xx} + 2Bu_{xy} + Cu_{yy} + Du_x + Eu_y + Fu = G$$

where A, B, C, D, E, F, and G are given functions which are continuous in some region in the (x, y) plane. A typical problem is the following: given a region R, finite or infinite, with a boundary S, find a function $u(x, y)$ which is twice differentiable and satisfies (1.1) in R, which is continuous in $R + S$ and which satisfies prescribed conditions on S. (For example, we might require that $u(x, y) = g(x, y)$ on S, where $g(x, y)$ is a given function.) There is, of course, an analogy with second-order ordinary differential equations. For initial-value problems, $S = \{x_0\}$ and R is the set of all x such that $x > x_0$. For two-point boundary value problems, S is the set of the two end points of an interval $[a, b]$ and R is the open interval (a, b).

Linear second-order partial differential equations may be classified as *elliptic*, *hyperbolic*, or *parabolic* depending on the behavior of the coefficients A, B, and C. Thus the equation is said to be

a) *elliptic* if $B^2 - AC < 0$ in R,

b) *hyperbolic* if $B^2 - AC > 0$ in R,

c) *parabolic* if $B^2 - AC = 0$ in R.

If the "discriminant" $B^2 - AC$ changes sign in R, then the equation is said to be of *mixed type*. Typical examples are

1.2 $$u_{xx} + u_{yy} = 0 \quad \text{(Laplace's equation)—elliptic}$$

1.3 $$u_{xx} - u_{yy} = 0 \quad \text{(vibrating string equation)—hyperbolic}$$

1.4 $$u_{xx} - u_y = 0 \quad \text{(diffusion equation)—parabolic.}$$

The equation $u_{xx} + yu_{yy} = 0$ is of mixed type in the region $0 < x < 1, -1 < y < 1$.

Depending on the classification of the equation, certain types of problems are "well set" in the sense that there exists a unique solution which varies continuously with the boundary data. Let us consider the *generalized Dirichlet problem* involving a bounded connected region R and a continuous function $g(x, y)$ prescribed on S. The function $u(x, y)$ is required to be continuous in $R + S$, to satisfy (1.1) in R and to satisfy

1.5 $u(x, y) = g(x, y)$

on S. If (1.1) is an elliptic equation and if $F \leq 0$ in R, then the generalized Dirichlet problem has a unique solution under fairly general conditions. (See, for instance, Courant and Hilbert [1962], vol. II, Chapter IV.)

As the name implies, the generalized Dirichlet problem is a generalization of the Dirichlet problem, which is a very classical problem in applied mathematics. The *Dirichlet problem* involves Laplace's equation

1.6 $L[u] = u_{xx} + u_{yy} = 0.$

It can be solved analytically in certain special cases. For the circle $x^2 + y^2 = R^2$ one can use the Poisson integral formula

1.7 $$u(x, y) = \frac{R^2 - \rho^2}{2\pi} \int_0^{2\pi} \frac{g(Re^{i\phi})d\phi}{R^2 + \rho^2 - 2R\rho \cos(\theta - \phi)}$$

where $x = \rho \cos \theta, y = \rho \sin \theta$. Analytic solutions can be given for the rectangle, using Fourier series, and for the half plane.* Solutions for certain other regions which can conveniently be transformed, by conformal mapping, into a circle or a half plane can also be obtained.

Generally speaking, however, it is very unusual that an analytic solution of a problem involving (1.1) can be found. Even if solutions of (1.1) are known, it may be very difficult to satisfy the boundary conditions. Thus, for example, in the case of Laplace's equation, we can easily verify that if ϕ and ψ are any twice-differentiable functions of a single argument, then

1.8 $u(x, y) = \phi(x + iy) + \psi(x - iy)$

is a solution. By choosing ϕ and ψ such that $\phi(x + iy) + \psi(x - iy)$ is real for all x and y we have a real solution of (1.6). For instance, if $\phi(z) = z^2$, $\psi(z) = (\bar{z})^2$, where $z = x + iy$ and $\bar{z} = x - iy$, we have

1.9 $u(x, y) = \phi(x + iy) + \psi(x - iy) = 2(x^2 - y^2)$

which satisfies (1.6). One can let ϕ be any polynomial in z with real coefficients and let $\psi(z) = \phi(\bar{z})$. Unfortunately, however, even if there were no differential equation

---

* Since the region is not bounded, we impose the additional requirement that $u(x, y)$ must be bounded in R.

at all, it would not be easy to find a function defined and continuous in $R + S$ which satisfies (1.5). The requirement that the function also satisfy the differential equation makes the determination even more difficult. Thus one is usually forced to use numerical methods. This chapter is devoted to a study of one such method, namely, the method of finite differences.

15.2 FINITE DIFFERENCE METHODS

In the application of finite difference methods one replaces the region R by a finite set of points R_h where $R_h \subseteq R$ and also replaces the boundary S by a set of points, say S_h, which may or may not belong to $R + S$. For each point P of R_h we develop a linear relation involving the value of u at P and the values of u at certain other points of R_h and at certain points of S_h. In the case of the generalized Dirichlet problem the values of u on S_h are determined in terms of the prescribed boundary values.* If there are N points of R_h, one obtains in this way a system of N linear algebraic equations with N unknowns. If the system of linear equations can be solved uniquely, as is frequently the case, then the values of u at points of R_h are accepted as approximate values of the true solution. By the use of various interpolation schemes described in Section 6.14, one can define an approximate solution for all points of R.

We shall be primarily concerned with a square mesh. We define the set Ω_h as follows: Given (x_0, y_0) and a mesh size $h > 0$, we let Ω_h be the set of all points $(x_0 + ih, y_0 + jh), i, j = 0, \pm 1, \pm 2, \dots$. We say that two points (x, y) and (x', y') of Ω_h are *adjacent* if $(x - x')^2 + (y - y')^2 = h^2$. Two adjacent points P and Q of $\Omega_h \cap (R + S)$ are *properly adjacent* if the open segment joining them is in R. Either or both of the points P and Q may be in R or S. We define R_h^* as the set of all *regular points* of $\Omega_h \cap R$, i.e., the set of all points P of $\Omega_h \cap R$ such that the four points of Ω_h adjacent to P belong to $R + S$ and are properly adjacent to P. We let $R_h' = (\Omega_h \cap R) - R_h^*$.

We shall be considering three alternative procedures for setting up discrete analogs of the generalized Dirichlet problem. In Procedure A we let $R_h = R_h^*$ and we let S_h be the set of all points of Ω_h not in R_h which are adjacent to at least one point of R_h. We seek a function u defined on $R_h + S_h$ such that

2.1
$$L_h[u] = G$$

on R_h and such that on S_h we have

2.2
$$u(P) = g(Q).$$

Here $Q = P$ if $P \in S$; otherwise, Q is a point of S which is closest to P. $L_h[u]$ is a

---

* We discuss the case where the outward normal derivative $\partial u/\partial n$ or a linear combination of u and $\partial u/\partial n$ are prescribed on S in Section 15.6.

discrete operator, to be specified later, which represents the differential operator $L[u]$.

In Procedures B and C we let $R_h = \Omega_h \cap R$, and we construct S_h as follows. For each point P of R_h consider the four adjacent points of Ω_h. Let Q be any such point. If the segment PQ contains a point of S, let the closest such point to P belong to S_h. If P is a regular point, then no point other than, possibly, Q can belong to S. We seek a function u defined on $R_h + S_h$ such that

2.3 $$L_h[u] = G$$

on R_h^*, such that

2.4 $$B_h[u] = H$$

on R_h' where $R_h' = R_h - R_h^*$, and

2.5 $$u(P) = g(P)$$

on S_h. Here the equation (2.4) is determined as explained below. In Procedure C we let $H = G$ and $B_h[u]$ is a representation of $L[u]$ for the "irregular" point in question.

Difference Equations for Regular Points

Let us now consider the construction of a discrete representation $L_h[u]$ of the differential operator $L[u]$ given by (1.1) for a (regular) point of R_h^*. We shall assume that the coefficient $B(x, y)$ of the mixed derivative u_{xy} vanishes identically in $R + S$. Later we shall show how, in principle at least, one can make a change of independent variables so that the coefficient of the mixed derivative vanishes. Probably the simplest procedure for constructing $L_h[u]$ is to simply replace the partial derivatives appearing in (1.1) by the usual three-point central difference quotients. Thus we have

2.6
$$\begin{cases} u_x \sim \dfrac{u(x+h, y)-u(x-h, y)}{2h}, & u_y \sim \dfrac{u(x, y+h)-u(x, y-h)}{2h} \\[2mm] u_{xx} \sim \dfrac{u(x+h, y)+u(x-h, y)-2u(x, y)}{h^2}, & u_{yy} \sim \dfrac{u(x, y+h)+u(x, y-h)-2u(x, y)}{h^2}. \end{cases}$$

Substituting in (1.1) we obtain the difference equation

2.7
$$L_h[u] = A\,\frac{u(x+h, y)+u(x-h, y)-2u(x, y)}{h^2} + C\,\frac{u(x, y+h)+u(x, y-h)-2u(x, y)}{h^2}$$
$$+ D\,\frac{u(x+h, y)-u(x-h, y)}{2h} + E\,\frac{u(x, y+h)-u(x, y-h)}{2h} + Fu(x, y) = G$$

or

2.8 $L_h[u] = \alpha_0 u(x, y) + \alpha_1 u(x+h, y) + \alpha_2 u(x, y+h) + \alpha_3 u(x-h, y) + \alpha_4 u(x, y-h) = G$

where

2.9
$$\begin{cases} \alpha_1 = \dfrac{A}{h^2} + \dfrac{D}{2h}, & \alpha_3 = \dfrac{A}{h^2} - \dfrac{D}{2h} \\[2mm] \alpha_2 = \dfrac{C}{h^2} + \dfrac{E}{2h}, & \alpha_4 = \dfrac{C}{h^2} - \dfrac{E}{2h} \\[2mm] \alpha_0 = -(\alpha_1 + \alpha_2 + \alpha_3 + \alpha_4) + F. \end{cases}$$

If we multiply by $-h^2$ we obtain the difference equation

2.10 $a_0 u(x, y) - a_1 u(x+h, y) - a_2 u(x, y+h) - a_3 u(x-h, y) - a_4 u(x, y-h) = t(x, y)$

where

2.11
$$\begin{cases} a_1 = A + \dfrac{h}{2}D, & a_3 = A - \dfrac{h}{2}D \\[2mm] a_2 = C + \dfrac{h}{2}E, & a_4 = C - \dfrac{h}{2}E \\[2mm] a_0 = a_1 + a_2 + a_3 + a_4 - h^2 F \\[2mm] t(x, y) = -h^2 G(x, y). \end{cases}$$

As an example, let us consider the differential equation

2.12
$$u_{xx} + \frac{k}{y} u_y + u_{yy} - \gamma u = G(x, y)$$

where k and γ are constants and $G(x, y)$ is a given function. Evidently, we have $A = C = 1, D = 0, E = ky^{-1}, F = -\gamma$, and

2.13
$$\begin{cases} \alpha_1 = \alpha_3 = \dfrac{1}{h^2}, & \alpha_2 = \dfrac{1}{h^2} + \dfrac{k}{2hy}, & \alpha_4 = \dfrac{1}{h^2} - \dfrac{k}{2hy} \\[2mm] \alpha_0 = -\dfrac{4}{h^2} - \gamma. \end{cases}$$

Moreover,

2.14
$$\begin{cases} a_1 = a_3 = 1, & a_2 = 1 + \dfrac{hk}{2y}, & a_4 = 1 - \dfrac{hk}{2y} \\[2mm] a_0 = 4 + \gamma h^2, & t(x, y) = -h^2 G(x, y) \end{cases}$$

Suppose, now, that $k = 0$ and we wish to solve the Dirichlet problem for the region $0 \le x \le 1$, $0 \le y \le 1$, where the values on the boundary are given, by finite difference methods with $h = 1/3$.

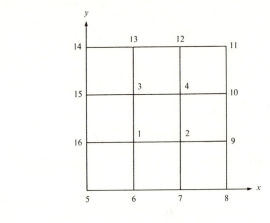

2.15 Figure

With the points numbered as indicated in Figure 2.15 we have

2.16
$$
\begin{cases}
\dfrac{u_2 + u_3 + u_{16} + u_6 - 4u_1}{h^2} - \gamma u_1 = G_1 \\[2mm]
\dfrac{u_9 + u_4 + u_1 + u_7 - 4u_2}{h^2} - \gamma u_2 = G_2 \\[2mm]
\dfrac{u_4 + u_{13} + u_{15} + u_1 - 4u_3}{h^2} - \gamma u_3 = G_3 \\[2mm]
\dfrac{u_{10} + u_{12} + u_3 + u_2 - 4u_4}{h^2} - \gamma u_4 = G_4.
\end{cases}
$$

Here the values of $u_6, u_7, u_9, u_{10}, u_{12}, u_{13}, u_{15}$, and u_{16} are given.

Corresponding to (2.10) we have the alternative form

2.17
$$
\begin{cases}
(4 + \gamma h^2)u_1 - u_2 - u_3 - u_{16} - u_6 = -h^2 G_1 \\
(4 + \gamma h^2)u_2 - u_9 - u_4 - u_1 - u_7 = -h^2 G_2 \\
(4 + \gamma h^2)u_3 - u_4 - u_{13} - u_{15} - u_1 = -h^2 G_3 \\
(4 + \gamma h^2)u_4 - u_{10} - u_{12} - u_3 - u_2 = -h^2 G_4.
\end{cases}
$$

If we transfer to the right-hand side of the equation the known values, we obtain the following system of equations for the unknown values u_1, u_2, u_3, and u_4

2.18
$$
\begin{cases}
(4+\gamma h^2)u_1 & -u_2 & -u_3 & & = -h^2 G_1 + u_{16} + u_6 \\
-u_1 & +(4+\gamma h^2)u_2 & & -u_4 & = -h^2 G_2 + u_9 + u_7 \\
-u_1 & & +(4+\gamma h^2)u_3 & -u_4 & = -h^2 G_3 + u_{13} + u_{15} \\
& -u_2 & -u_3 & +(4+\gamma h^2)u_4 & = -h^2 G_4 + u_{10} + u_{12}.
\end{cases}
$$

This system can be written in the matrix form

2.19
$$
\begin{pmatrix}
4+\gamma h^2 & -1 & -1 & 0 \\
-1 & 4+\gamma h^2 & 0 & -1 \\
-1 & 0 & 4+\gamma h^2 & -1 \\
0 & -1 & -1 & 4+\gamma h^2
\end{pmatrix}
\begin{pmatrix}
u_1 \\ u_2 \\ u_3 \\ u_4
\end{pmatrix}
=
\begin{pmatrix}
-h^2 G_1 + u_{16} + u_6 \\
-h^2 G_2 + u_9 + u_7 \\
-h^2 G_3 + u_{13} + u_{15} \\
-h^2 G_4 + u_{10} + u_{12}
\end{pmatrix}
$$

or

2.20
$$
Au = b.
$$

Suppose now that $\gamma = 0$ and $G(x, y) \equiv 0$ and that the boundary values are 1000 for $y = 1$ and 0 elsewhere on S. Then the system becomes

2.21
$$
\begin{pmatrix}
4 & -1 & -1 & 0 \\
-1 & 4 & 0 & -1 \\
-1 & 0 & 4 & -1 \\
0 & -1 & -1 & 4
\end{pmatrix}
\begin{pmatrix}
u_1 \\ u_2 \\ u_3 \\ u_4
\end{pmatrix}
=
\begin{pmatrix}
0 \\ 0 \\ 1000 \\ 1000
\end{pmatrix}
$$

whose solution is

2.22
$$
u_1 = u_2 = 125, \qquad u_3 = u_4 = 375.
$$

Self-adjoint and Essentially Self-adjoint Equations

In the case where the equation (1.1) is *self-adjoint* we can derive a difference equation such that the matrix A occurring in (2.20) is symmetric. Assuming as before that $B(x, y) \equiv 0$, we say that (1.1) is self-adjoint if

2.23
$$
D = A_x, \qquad E = C_y.
$$

In this case we can write (1.1) in the form

2.24
$$
L[u] = (Au_x)_x + (Cu_y)_y + Fu = G.
$$

We remark that even if (1.1) is not self-adjoint, it may be possible to obtain a self-adjoint equation by multiplying by an "integrating factor" $\mu(x, y)$. Evidently, the equation

2.25 $$\mu L[u] = \mu G$$

is self-adjoint if

2.26 $$\mu D = (\mu A)_x, \qquad \mu E = (\mu C)_y.$$

If such a function exists, the following conditions must hold

2.27 $$\frac{\partial}{\partial y}\left(\frac{D - A_x}{A}\right) = \frac{\partial}{\partial x}\left(\frac{E - C_y}{C}\right)$$

since $(\log \mu)_{xy} = (\log \mu)_{yx}$. If (2.27) holds, we say that the equation (1.1) is *essentially self-adjoint*. If (1.1) is essentially self-adjoint, we can determine a function $\mu(x, y)$ such that $\mu(x_0, y_0) = 1$ for some (x_0, y_0) and such that (2.26) holds. From the condition $\mu D = (\mu A)_x$ we have

2.28 $$\log \mu(x, y) = \log \mu(x_0, y) + \int_{x_0}^{x}\left(\frac{D - A_x}{A}\right) dx.$$

Differentiating with respect to y and using (2.27) we get

2.29 $$\frac{\mu_y}{\mu} = \frac{\partial}{\partial y}\log \mu(x_0, y) + \int_{x_0}^{x}\frac{\partial}{\partial y}\left(\frac{D - A_x}{A}\right) dx$$

$$= \frac{\partial}{\partial y}\log \mu(x_0, y) + \int_{x_0}^{x}\frac{\partial}{\partial x}\left(\frac{E - C_y}{C}\right) dx$$

$$= \frac{\partial}{\partial y}\log \mu(x_0, y) + \left(\frac{E - C_y}{C}\right)\Bigg|_{\substack{x=x \\ y=y}} - \left(\frac{E - C_y}{C}\right)\Bigg|_{\substack{x=x_0 \\ y=y}}$$

Evidently $\mu_y/\mu = (E - C_y)/C$ provided

2.30 $$\frac{\partial}{\partial y}\log \mu(x_0, y) = \frac{E(x_0, y) - C_y(x_0, y)}{C(x_0, y)}$$

and

2.31 $$\log \mu(x_0, y) = \int_{y_0}^{y}\left(\frac{E(x_0, y) - C_y(x_0, y)}{C(x_0, y)}\right) dy.$$

Therefore, we have

2.32 $$\log \mu(x, y) = \int_{y_0}^{y}\left(\frac{E(x_0, y) - C_y(x_0, y)}{C(x_0, y)}\right) dy + \int_{x_0}^{x}\left(\frac{D(x, y) - A_x(x, y)}{A(x, y)}\right) dx.$$

As an example, consider the differential equation

2.33
$$u_{xx} + \frac{k}{y}u_y + u_{yy} = 0.$$

Evidently, since $A = C = 1$, $D = 0$, $E = ky^{-1}$ we have

2.34
$$\frac{\partial}{\partial y}\left(\frac{D - A_x}{A}\right) = 0 = \frac{\partial}{\partial x}\left(\frac{E - C_y}{C}\right)$$

and the equation is essentially self-adjoint. From (2.32) we have, letting $x_0 = y_0 = 1$

2.35
$$\log \mu(x, y) = \int_1^y ky^{-1}dy = k \log y$$

so that

2.36
$$\mu(x, y) = y^k.$$

If we multiply (2.33) by y^k we get the self-adjoint equation

2.37
$$(y^k u_x)_x + (y^k u_y)_y = 0.$$

Given the self-adjoint equation (2.24) we derive a symmetric difference equation by replacing $(Au_x)_x$ by the difference quotient

2.38
$$(Au_x)_x \sim \frac{1}{h^2}\left\{A\left(x + \frac{h}{2}, y\right)[u(x + h, y) - u(x, y)]\right.$$
$$\left. - A\left(x - \frac{h}{2}, y\right)[u(x, y) - u(x - h, y)]\right\}$$

and replacing $(Cu_y)_y$ by a similar expression. Substituting in (2.24) we obtain a difference equation of the form (2.8) or (2.10) where

2.39
$$\begin{cases} \alpha_1 = \frac{1}{h^2}A\left(x + \frac{h}{2}, y\right), & \alpha_3 = \frac{1}{h^2}A\left(x - \frac{h}{2}, y\right) \\ \alpha_2 = \frac{1}{h^2}C\left(x, y + \frac{h}{2}\right), & \alpha_4 = \frac{1}{h^2}C\left(x, y - \frac{h}{2}\right) \\ \alpha_0 = -(\alpha_1 + \alpha_2 + \alpha_3 + \alpha_4) + F \end{cases}$$

and

2.40
$$\begin{cases} a_1 = A\left(x + \frac{h}{2}, y\right), \qquad a_3 = A\left(x - \frac{h}{2}, y\right) \\[2mm] a_2 = C\left(x, y + \frac{h}{2}\right), \qquad a_4 = C\left(x, y - \frac{h}{2}\right) \\[2mm] a_0 = a_1 + a_2 + a_3 + a_4 - h^2 F \\[2mm] t(x, y) = -h^2 G. \end{cases}$$

Procedures for Handling Irregular Boundaries

So far, we have considered finite difference representations of (1.1) or (2.24) for regular points. In Procedure A, we have only regular points of R_h. In Procedures B and C, however, we have both regular points (of R_h^*) and irregular points (of R_h'). In Procedure B, which is based on a scheme suggested by Collatz [1933], we use linear interpolation for a point of R_h'. If $P \in R_h'$ we know that on at least one of the segments joining P to an adjacent point there is a point of S. The closest such point, say Q, belongs to S_h. We then choose the point T as follows. We consider the segment from P to an adjacent point of Ω_h in the opposite direction from PQ, and let T be the adjacent mesh point in that direction if that point is properly adjacent to P. Otherwise, we let T be the closest point of S_h on that segment. We determine $u(P)$ by the linear interpolation formula

2.41
$$u(P) = \frac{h_2}{h_1 + h_2} u(T) + \frac{h_1}{h_1 + h_2} u(Q)$$

where h_1 and h_2 are the lengths of the segments PT and PQ, respectively. If T belongs to S_h, then $u(P)$ is completely determined. Otherwise, we obtain a relation involving $u(T)$ and $u(P)$. The totality of all equations for points of R_h^* together with the linear interpolation relations for points of R_h' gives us a set of linear equations.

In Procedure C, we use a difference equation derived from the differential equation for points of R_h' as well as for points of R_h. Let us consider the configuration in Fig. 2.44 where $h_i = s_i h$ and $0 \leqq s_i \leqq 1$, $i = 1, 2, 3, 4$.

Let us develop formulas for $u_x(x_0, y_0)$ and $u_{xx}(x_0, y_0)$ in terms of $u(x_0, y_0)$, $u(x_0 + h_1, y_0)$ and $u(x_0 - h_3, y_0)$. Evidently we have

2.42
$$\begin{cases} u_1 = u(x_0 + h_1, y_0) = u + h_1 u_x + \frac{h_1^2}{2} u_{xx} + \frac{h_1^3}{6} u_{xxx} + \frac{h_1^4}{24} u_{xxxx} + \cdots \\[3mm] u_3 = u(x_0 - h_3, y_0) = u - h_3 u_x + \frac{h_3^2}{2} u_{xx} - \frac{h_3^3}{6} u_{xxx} + \frac{h_3^4}{24} u_{xxxx} + \cdots. \end{cases}$$

Eliminating u_{xx} we get

2.43 $$u_x = \frac{h_3^2 u_1 - h_1^2 u_3 - (h_3^2 - h_1^2)u_0}{h_1 h_3 (h_1 + h_3)} - \frac{h_1 h_3}{6} u_{xxx} + \cdots$$

where $u = u(x_0, y_0)$, $u_x = u_x(x_0, y_0)$, etc. Similarly, eliminating u_x, we get

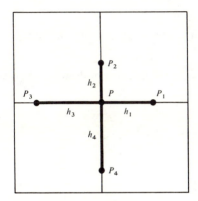

2.44 Fig. Treatment of an Irregular Mesh Point.

2.45 $$u_{xx} = \frac{h_3 u_1 + h_1 u_3 - (h_1 + h_3)u_0}{\frac{1}{2}h_1 h_3 (h_1 + h_3)} - \frac{(h_1 - h_3)}{3} u_{xxx} - \frac{h_1^2 - h_1 h_3 + h_3^2}{12} u_{xxxx} + \cdots$$

Substituting the first terms of the above expressions and similar expressions for u_y and u_{yy} in (1.1) we get the difference equation

2.46 $\alpha_0 u(x, y) + \alpha_1 u(x + h_1, y) + \alpha_2 u(x, y + h_2) + \alpha_3 u(x - h_3, y) + \alpha_4 u(x, y - h_4) = G$

where

2.47
$$\begin{cases} \alpha_1 = \frac{1}{h^2}\left[\frac{2A}{s_1(s_1 + s_3)} + \frac{hs_3 D}{s_1(s_1 + s_3)}\right], \quad \alpha_3 = \frac{1}{h^2}\left[\frac{2A}{s_3(s_1 + s_3)} - \frac{hs_1 D}{s_3(s_1 + s_3)}\right] \\[2mm] \alpha_2 = \frac{1}{h^2}\left[\frac{2C}{s_2(s_2 + s_4)} + \frac{hs_4 E}{s_2(s_2 + s_4)}\right], \quad \alpha_4 = \frac{1}{h^2}\left[\frac{2C}{s_4(s_2 + s_4)} - \frac{hs_2 E}{s_4(s_2 + s_4)}\right] \\[2mm] \alpha_0 = -(\alpha_1 + \alpha_2 + \alpha_3 + \alpha_4) + F. \end{cases}$$

For the self-adjoint equation (2.24) we have

2.48
$$(Au_x)_x \sim \frac{2}{h_1 + h_3}\left\{ A\left(x + \frac{h_1}{2}, y\right)\left[\frac{u(x + h_1, y) - u(x, y)}{h_1}\right]\right.$$
$$\left. - A\left(x - \frac{h_3}{2}, y\right)\left[\frac{u(x, y) - u(x - h_3, y)}{h_3}\right]\right\}.$$

Substituting this and a similar expression for $(Cu_y)_y$ we get (2.46) with

2.49
$$\begin{cases} \alpha_1 = \dfrac{2}{h^2} \dfrac{A\left(x + \dfrac{h_1}{2}, y\right)}{s_1(s_1 + s_3)}, \qquad \alpha_3 = \dfrac{2}{h^2} \dfrac{A\left(x - \dfrac{h_3}{2}, y\right)}{s_3(s_1 + s_3)} \\[3mm] \alpha_2 = \dfrac{2}{h^2} \dfrac{C\left(x, y + \dfrac{h_2}{2}\right)}{s_2(s_2 + s_4)}, \qquad \alpha_4 = \dfrac{2}{h^2} \dfrac{C\left(x, y - \dfrac{h_4}{2}\right)}{s_4(s_2 + s_4)} \\[3mm] \alpha_0 = -(\alpha_1 + \alpha_2 + \alpha_3 + \alpha_4) + F. \end{cases}$$

As an example, consider the Dirichlet problem for the quarter circle of radius unity shown below.

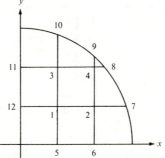

Suppose we wish to solve the discrete Dirichlet problem using Procedure C with $h = 1/3$. The first step is to compute the s_i for the irregular points 2, 3, and 4. Thus we need to compute x_7, x_8, y_9, and y_{10}. Evidently, we have

2.50
$$\begin{cases} x_7 = \sqrt{1 - (1/3)^2} = \dfrac{2\sqrt{2}}{3} = y_{10} \\[3mm] x_8 = \sqrt{1 - (2/3)^2} = \dfrac{\sqrt{5}}{3} = y_9. \end{cases}$$

Thus we have

Point	s_1	s_2	s_3	s_4
1	1	1	1	1
2	$2\sqrt{2} - 2 = .82843$	1	1	1
3	1	$2\sqrt{2} - 2 = .82843$	1	1
4	$\sqrt{5} - 2 = .23607$	$\sqrt{5} - 2 = .23607$	1	1

The corresponding values of the a_i are

Point	a_0	a_1	a_2	a_3	a_4
1	4	1	1	1	1
2	4.41421	1.32038	1.00000	1.09383	1.00000
3	4.41421	1.00000	1.32038	1.00000	1.09383
4	16.94427	6.85410	6.85410	1.61803	1.61803

Here the a_i are $h^2\alpha_i$, $i = 1, 2, 3, 4$ and $a_0 = -h^2\alpha_0$. Thus we are led to the linear system

2.51

$$
\begin{pmatrix}
4 & -1 & -1 & 0 \\
-1.09383 & 4.41421 & 0 & -1.00000 \\
-1.09383 & 0 & 4.41421 & -1.00000 \\
0 & -1.61803 & -1.61803 & 16.94427
\end{pmatrix}
\begin{pmatrix} u_1 \\ u_2 \\ u_3 \\ u_4 \end{pmatrix}
=
\begin{pmatrix}
u_5 + u_{12} \\
1.32038u_7 + u_6 \\
1.32038u_{10} + u_{11} \\
6.85410u_8 + 6.85410u_9
\end{pmatrix}.
$$

Elimination of the Mixed Derivative

Let us now consider the case where the coefficient $B(x, y)$ of (1.1) does not vanish identically. We consider the use of new independent variables ξ and η defined by

2.52 $$\xi = \xi(x, y), \qquad \eta = \eta(x, y),$$

where $\xi(x, y)$ and $\eta(x, y)$ are well-behaved functions such that the *Jacobian*

2.53 $$J(\xi, \eta; x, y) = \xi_x\eta_y - \eta_x\xi_y$$

does not vanish in $R + S$. Given ξ and η one can determine x and y from (2.52) and we let

2.54
$$
\begin{cases}
x = x(\xi, \eta), \qquad y = y(\xi, \eta) \\
v(\xi, \eta) = u(x(\xi, \eta), y(\xi, \eta)).
\end{cases}
$$

Evidently $u(x, y) = v(\xi(x, y), \eta(x, y))$, and

2.55
$$
\begin{cases}
u_x = v_\xi\xi_x + v_\eta\eta_x, \qquad u_y = v_\xi\xi_y + v_\eta\eta_y \\
u_{xx} = v_{\xi\xi}\xi_x^2 + 2v_{\xi\eta}\xi_x\eta_x + v_{\eta\eta}\eta_x^2 + v_\xi\xi_{xx} + v_\eta\eta_{xx} \\
u_{xy} = v_{\xi\xi}\xi_x\xi_y + v_{\xi\eta}(\xi_x\eta_y + \xi_y\eta_x) + v_{\eta\eta}\eta_x\eta_y + v_\xi\xi_{xy} + v_\eta\eta_{xy} \\
u_{yy} = v_{\xi\xi}\xi_y^2 + 2v_{\xi\eta}\xi_y\eta_y + v_{\eta\eta}\eta_y^2 + v_\xi\xi_{yy} + v_\eta\eta_{yy}.
\end{cases}
$$

Substituting in (1.1) we have

2.56 $$\alpha v_{\xi\xi} + 2\beta v_{\xi\eta} + \gamma v_{\eta\eta} + \delta v_\xi + \varepsilon v_\eta + Fv = G$$

where

2.57
$$\begin{cases} \alpha = A\xi_x^2 + 2B\xi_x\xi_y + C\xi_y^2 \\ \beta = A\xi_x\eta_x + B(\xi_x\eta_y + \xi_y\eta_x) + C\xi_y\eta_y \\ \gamma = A\eta_x^2 + 2B\eta_x\eta_y + C\eta_y^2 \\ \delta = A\xi_{xx} + 2B\xi_{xy} + C\xi_{yy} + D\xi_x + E\xi_y \\ \varepsilon = A\eta_{xx} + 2B\eta_{xy} + C\eta_{yy} + D\eta_x + E\eta_y. \end{cases}$$

By direct calculation we can verify that

2.58
$$\begin{aligned} \beta^2 - \alpha\gamma &= (B^2 - AC)(\xi_x\eta_y - \eta_x\xi_y)^2 \\ &= (B^2 - AC)J(\xi, \eta; x, y)^2. \end{aligned}$$

Hence if the Jacobian does not vanish, the new equation (2.56) will have the same classification (elliptic, hyperbolic, or parabolic) as the original equation.

Evidently, we can make β vanish by imposing the conditions

2.59
$$\begin{cases} A\xi_x\eta_x + C\xi_y\eta_y = 0 \\ \xi_x\eta_y + \xi_y\eta_x = 0. \end{cases}$$

But these conditions reduce to

2.60 $$\frac{\xi_y}{\xi_x} = \sqrt{\frac{A}{C}}, \qquad \frac{\eta_y}{\eta_x} = -\sqrt{\frac{A}{C}}.$$

Let us consider the differential equation

2.61 $$\xi_y = q(x, y)\xi_x.$$

Suppose that we have determined a function $\psi(x, y)$ such that for any constant c, the function $y(x)$ defined implicitly by

2.62 $$\psi(x, y(x)) = c$$

satisfies the differential equation

2.63 $$\frac{dy}{dx} = -\frac{1}{q(x, y)}.$$

Then we have

2.64
$$\psi_x + \psi_y y' = 0$$

or

2.65
$$\psi_x - \frac{1}{q(x, y)} \psi_y = 0.$$

Hence $\psi(x, y)$ is a solution of (2.61). We can therefore determine $\xi(x, y)$ by solving the ordinary differential equation

2.66
$$\frac{dy}{dx} = -\sqrt{\frac{C(x, y)}{A(x, y)}}$$

and writing the solution in the form

2.67
$$\psi(x, y) = c.$$

We then let $\xi(x, y) = \psi(x, y)$. Similarly, to find $\eta(x, y)$ we solve the ordinary differential equation

2.68
$$\frac{dy}{dx} = \sqrt{\frac{C(x, y)}{A(x, y)}}.$$

As an example, consider the elliptic equation

2.69
$$u_{xx} - \tfrac{1}{2}u_{xy} + u_{yy} = 0.$$

We solve the ordinary differential equation

2.70
$$\frac{dy}{dx} = -1$$

obtaining the general solution

2.71
$$y + x = c.$$

Thus we let

2.72
$$\xi = x + y.$$

Similarly, we get

2.73
$$\eta = y - x.$$

Thus, letting $\xi = x + y$, $\eta = y - x$ or equivalently,

2.74
$$x = \frac{\xi - \eta}{2}, \qquad y = \frac{\xi + \eta}{2}$$

we obtain the equation

2.75
$$3v_{\xi\xi} + 5v_{\eta\eta} = 0$$

where the mixed derivative is absent.

If the ordinary differential equation (2.66) cannot be conveniently solved in closed form, we can use the following numerical procedure. Choose any point (x_0, y_0) in $R + S$. We construct $\xi(x, y,)$ and $\eta(x, y)$ as follows. We let

2.76
$$\xi(x, y_0) = x, \qquad \eta(x_0, y) = y.$$

Given a point (x, y) we integrate the differential equation

2.77
$$\frac{dx}{dy} = -\sqrt{\frac{C(x, y)}{A(x, y)}}$$

backwards if necessary until we reach (x^*, y_0). We then let

2.78
$$\xi(x, y) = x^*.$$

Similarly, we integrate the equation

2.79
$$\frac{dy}{dx} = \sqrt{\frac{C(x, y)}{A(x, y)}}$$

until we reach (x_0, y^*) and we let

$$\eta(x, y) = y^*.$$

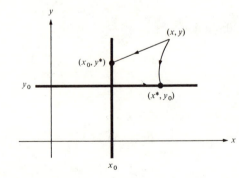

2.80 Figure

The procedure described above will yield ξ and η for given x and y. To get x and y for given ξ and η we proceed as follows. Starting at the point (ξ, y_0) we integrate (2.77). Starting at (x_0, η) we integrate (2.79). The intersection (x, y) of these two curves gives $x(\xi, \eta)$, $y(\xi, \eta)$.

In the example (2.69), let $x_0 = y_0 = 0$. Starting at the point $(\bar{x}, \bar{y})$ we integrate

2.81
$$\frac{dx}{dy} = -1$$

obtaining $x - \bar{x} = -(y - \bar{y})$. Letting $y = 0$ we get

2.82
$$x^* - \bar{x} = \bar{y}$$

and

2.83
$$\xi = x^* = \bar{y} + \bar{x}.$$

Similarly, we get

2.84
$$\eta = \bar{y} - \bar{x}.$$

Suppose now that ξ and η are given. Integrating,

2.85
$$\frac{dx}{dy} = -1$$

starting at $(\xi, 0)$ gives

2.86
$$x - \xi = -y.$$

Integrating

2.87
$$\frac{dy}{dx} = 1$$

starting at $(0, \eta)$ gives

2.88
$$y - \eta = x.$$

The intersection of these two curves is

2.89
$$x = \frac{\xi - \eta}{2}, \qquad y = \frac{\xi + \eta}{2}.$$

EXERCISES 15.2

1. Consider the problem of solving $u_{xx} - (1/y)u_y + u_{yy} = 0, 0 < x < 1, 0 < y < 1$ with $u = 0$ on the boundary of the unit square except for $y = 1$ where $u = 1$. Use the standard five-point difference equation with $h = \frac{1}{3}$. Do the same for the self-adjoint equation $(y^{-1}u_x)_x + (y^{-1}u_y)_y = 0$, using the symmetric difference equation.

2. In the preceding problem determine an approximate value of $u(\frac{1}{4}, \frac{1}{2})$ by bilinear interpolation.

3. Show that the differential equation

$$L[u] = u_{xx} + u_{yy} + 2(x + y)^{-1}(u_x + u_y) = 0$$

is essentially self-adjoint. Determine a suitable integrating factor $\mu(x, y)$ so that the equation

$$\mu(x, y)L[u] = 0$$

is self-adjoint.

4. For the generalized Dirichlet problem for the differential equation of the preceding exercise for the unit square with $h = \frac{1}{3}$, construct the difference equation for $L[u] = 0$ and also the symmetric difference equation for the self-adjoint equation $\mu L[u] = 0$. In each case test whether $HV = VH$, where H and V correspond to $-h^2$ times a sum of terms involving derivatives with respect to x and y, respectively, and $H + V$ corresponds to $-h^2L[u]$, or $-h^2\mu L[u]$, as appropriate.

5. Consider the generalized Dirichlet problem involving the differential equation

$$u_{xx} + \frac{k}{y}u_y + u_{yy} = 0$$

in the region $0 \leq x \leq 1, 1 \leq y \leq 2$. Show that the differential equation is essentially self-adjoint for each value of the constant k. Make the equation self-adjoint and derive a difference equation corresponding to a symmetric matrix. Derive the difference equation for the case $h = \frac{1}{4}$ and show that the matrix is symmetric.

6. Find a function satisfying Laplace's equation in the region $x^2 + y^2 < 1, y > 0$, vanishing for $y = 0, -1 \leq x \leq 1$, and assuming the values $(2 - x)^2y^2$ for $x^2 + y^2 = 1, y > 0$. Use the method of finite differences with $h = \frac{1}{2}$. Use each of the three procedures described in the text for handling points near the boundary. *Optional*: Determine the exact solution of the differential equation at the three interior mesh points.

7. Write down the five-point difference equation for the Dirichlet problem for the region $x^2 + y^2 < 1, x > 0, y > 0$ with boundary values zero except when $x^2 + y^2 = 1, x > 0, y > 0$ where the values are unity. Let $h = \frac{1}{3}$ and use Procedure C for points near the boundary. Express the difference equation in the form $Au = b$ (after multiplying by $-h^2$).

8. Given the differential equation

$$3u_{xx} - \frac{1}{2}u_{xy} + 4u_{yy} = 0,$$

indicate the classification (elliptic, parabolic, or hyperbolic). Perform a suitable change of independent variables to obtain a new equation without the mixed derivative.

15.3 SOLVABILITY OF THE DIFFERENCE EQUATION
In Section 15.2 we have shown how the application of finite difference methods to solve the generalized Dirichlet problem leads to a system of linear algebraic equations of the form

3.1
$$Au = b,$$

where A is an $N \times N$ matrix, b is a known $N \times 1$ matrix, and u is an unknown $N \times 1$ matrix. Here N is the number of points of R_h.

We assume throughout this chapter that the coefficients of the differential equation (1.1) satisfy the following conditions in $R + S$:

3.2
$$\begin{cases} A(x, y) > 0 \\ C(x, y) > 0 \\ B(x, y) \equiv 0 \\ F(x, y) \leqq 0. \end{cases}$$

We also assume that either we are dealing with the self-adjoint differential equation (2.24) or else the mesh size is small enough so that

3.3
$$h < \min\left(\min_{R+S} \frac{2A}{|D|}, \ \min_{R+S} \frac{2C}{|E|}\right).$$

Thus it follows from (2.47) and (2.49) that in the difference equation (2.46) we have

3.4
$$\begin{cases} \alpha_i > 0, \quad i = 1, 2, 3, 4 \\ -\alpha_0 \geqq \alpha_1 + \alpha_2 + \alpha_3 + \alpha_4. \end{cases}$$

We now seek to show that the difference equation has a unique solution.* To do this, as in Section 10.4, it is sufficient to show that the homogeneous system

3.5
$$Au = 0$$

has only the trivial solution $u = 0$. The homogeneous system corresponds to the case of zero boundary values and $G(x, y) \equiv 0$. Suppose that there exists $u \neq 0$ with $Au = 0$. Then either $u(x, y) > 0$ for some point of R_h or else we can work with $-u(x, y)$ which also satisfies the homogeneous system. Let

3.6
$$M = \max_{R_h} u(x, y) = u(\bar{x}, \bar{y}).$$

---

* An alternative proof based on certain properties of the matrix A of (3.1) will be given in Section 16.3.

Evidently, we have, by (2.46)

3.7 $\alpha_0 M = -\alpha_1 u(\bar{x} + h_1, \bar{y}) - \alpha_2 u(\bar{x}, \bar{y} + h_2) - \alpha_3 u(\bar{x} - h_3, \bar{y}) - \alpha_4 u(\bar{x}, \bar{y} - h_4)$

and, by (3.4)

$$0 = (-\alpha_0)M - [\alpha_1 u(\bar{x} + h_1, \bar{y}) + \alpha_2 u(\bar{x}, \bar{y} + h_2) + \alpha_3 u(\bar{x} - h_3, \bar{y}) + \alpha_4 u(\bar{x}, \bar{y} - h_4)]$$
$$\geqq \alpha_1 M + \alpha_2 M + \alpha_3 M + \alpha_4 M - [\alpha_1 u(\bar{x} + h_1, \bar{y}) + \alpha_2 u(\bar{x}, \bar{y} + h_2)$$
$$+ \alpha_3 u(\bar{x} - h_3, \bar{y}) + \alpha_4 u(\bar{x}, \bar{y} - h_4)]$$
$$= \alpha_1(M - u(\bar{x} + h_1, \bar{y})) + \alpha_2(M - u(\bar{x}, \bar{y} + h_2)) + \alpha_3(M - u(\bar{x} - h_3, \bar{y}))$$
$$+ \alpha_4(M - u(\bar{x}, \bar{y} - h_4)).$$

But the right side is strictly positive unless

3.9 $u(\bar{x} + h_1, \bar{y}) = u(\bar{x}, \bar{y} + h_2) = u(\bar{x} - h_3, \bar{y}) = u(\bar{x}, \bar{y} - h_4) = M.$

Similarly, we can show that for all points adjacent to the four points $(\bar{x} + h_1, \bar{y})$, $(\bar{x}, \bar{y} + h_2)$, $(\bar{x} - h_3, \bar{y})$, $(\bar{x}, \bar{y} - h_4)$ the function $u = M$. Continuing, we can show that at some point (x, y) of S_h we have $u(x, y) = M$. This contradiction shows that $u(x, y) \equiv 0$, and therefore the homogeneous system (3.5) has only the trivial solution. Thus the determinant of A does not vanish and hence (3.1) has a unique solution.

If h is very small, the problem of actually solving (3.1) may present serious practical difficulties even though a unique solution is known to exist. Since the matrix A is in general very large but has only a few non-zero elements in any given row, it seems advisable to use some method of solution which does not cause the introduction of new non-zero elements during the computation. Thus if direct methods are to be used, it would seem that some modification of the Gauss elimination method should be developed which preserves the sparseness of A. Rather than attempting to do this, however, we shall consider in Chapter 16 various iterative methods for solving (3.1) which are designed to take advantage of the sparseness of A.

15.4 ACCURACY OF THE DIFFERENCE EQUATION SOLUTION

We now investigate the accuracy of the solution of the difference equation. Our analysis is based on that of Gerschgorin [1930] and we assume that the exact solution of the differential equation has partial derivatives of all orders up to and including the fourth which are bounded in $R + S$. The analysis is very similar to that given in Section 10.5 for the one-dimensional case.

Let us first carry out the analysis for Procedure A. The following lemmas are analogous to the corresponding lemmas in Section 10.5.

4.1 Lemma. Let $L_h[u]$ be a discrete operator of the form (2.8) where $-\alpha_0, \alpha_1, \alpha_2, \alpha_3,$ and α_4 are positive functions such that

4.2
$$-\alpha_0 \geqq \alpha_1 + \alpha_2 + \alpha_3 + \alpha_4.$$

If $u \geqq 0$ on S_h and $-L_h[u] \geqq 0$ on R_h, then $u \geqq 0$ in R_h.

4.3 Lemma. If $|u| \leqq v$ on S_h and $|L_h[u]| \leqq -L_h[v]$ in R_h, then $|u| \leqq v$ in $R_h + S_h$.

4.4 Lemma. Let $w(x, y)$ be any function such that $-L_h[w] > 0$ in R_h and $w \geqq 0$ on S_h. For any function $e(x, y)$ we have

4.5
$$|e(x, y)| \leqq W \max_{R_h} \left[\frac{|L_h[e]|}{-L_h[w]} \right] + \max_{S_h} |e(x)|$$

where

4.6
$$W = \max_{R_h + S_h} |w(x, y)|.$$

4.7 Lemma. Let $\bar{u}(x, y)$ satisfy (1.1), with (3.2), and (1.5). Let $u(x, y)$ satisfy (2.8) in R_h. If $\bar{u}(x, y)$ has partial derivatives of all orders up to including the fourth which are continuous and bounded in $R + S$, then

4.8
$$|L_h[u - \bar{u}]| \leqq \left(\frac{Ah^2}{12} + \frac{Ch^2}{12} \right) M_4 + \left(\frac{|D|h^2}{6} + \frac{|E|h^2}{6} \right) M_3$$

where for $i = 3, 4$ we have

4.9
$$M_i = \max \left\{ \max_{R+S} \left| \frac{\partial^i u}{\partial x^i} \right|, \max_{R+S} \left| \frac{\partial^i u}{\partial y^i} \right| \right\}.$$

4.10 Theorem. Under the hypotheses of Lemma 4.7, for all $(x, y) \in R_h + S_h$ we have

4.11
$$|u(x, y) - \bar{u}(x, y)| \leqq \frac{h^2 r^2}{24} \max_{R+S} \left[\frac{(A + C)M_4 + 2(|D| + |E|)M_3}{A + C - r(|D| + |E|)} \right]$$
$$+ \max_{S_h} |u(x, y) \doteq \bar{u}(x, y)|$$

provided that for all $(x, y) \in R + S$

4.12
$$A + C - r(|D| + |E|) > 0.$$

Moreover, if $D(x, y)$ and $E(x, y)$ are one-signed for all $(x, y) \in R$, then

4.13 $|u(x, y) - \bar{u}(x, y)| \leqq h^2 \left\{ \frac{r^2 M_4}{6} \sqrt{2} + \frac{r M_3}{3} \sqrt{2} \right\} + \max_{S_h} |u(x, y) - \bar{u}(x, y)|.$

Here r is the radius of a circle containing $R + S$.

Proof. If (4.12) holds we apply Lemma 4.4 with the function

4.14 $$w(x, y) = 1 - \frac{(x - x_0)^2 + (y - y_0)^2}{r^2}$$

where the circle $(x - x_0)^2 + (y - y_0)^2 = r^2$ contains $R + S$. Evidently

4.15 $$-L_h[w] = \frac{2}{r^2}\{A + C + D(x - x_0) + E(y - y_0)\} - Fw(x, y)$$

$$\geqq \frac{2}{r^2}(A + C - |D|r - |E|r)$$

since $F \leqq 0$. Moreover, $W \leqq 1$. The result (4.11) now follows from Lemmas 4.4 and 4.7.

Suppose now that both D and E are nonnegative in $R + S$. We let

4.16 $$w(x, y) = \frac{r^2 M_4}{24}\left\{(1 + 2\sqrt{2}) - \frac{2(x - x_0)}{r} - \frac{2(y - y_0)}{r} - \frac{(x - x_0)^2}{r^2} - \frac{(y - y_0)^2}{r^2}\right\}$$

$$+ \frac{rM_3}{6}\left\{\sqrt{2} - \frac{x - x_0}{r} - \frac{y - y_0}{r}\right\}.$$

We easily verify that $w(x, y) \geqq 0$ in $R + S$ by letting $x - x_0 = \rho \cos \theta$, $y - y_0 = \rho \sin \theta$, and observing that w is minimized by letting $\rho = r$, $\theta = \pi/4$. Evidently

4.17 $$-L_h[w] = \frac{r^2 M_4}{24}\left\{\frac{2A}{r^2} + \frac{2C}{r^2} + \frac{2D}{r^2}(r + (x - x_0)) + \frac{2E}{r^2}(r + (y - y_0))\right\}$$

$$+ \frac{rM_3}{6}\left\{\frac{D}{r} + \frac{E}{r}\right\} - Fw(x, y)$$

$$\geqq \frac{M_4}{12}(A + C) + \frac{M_3}{6}(D + E).$$

By Lemmas 4.4 and 4.7 the result (4.13) follows. To handle the case where $D(x, y) \leqq 0$ in $R + S$ we replace $x - x_0$ by $-(x - x_0)$ in (4.16). Similarly, we replace $y - y_0$ by $-(y - y_0)$ if $E(x, y) \leqq 0$ in $R + S$.

Let us now consider the case where $D(x, y)$ and $E(x, y)$ are not necessarily one-signed in $R + S$ and do not necessarily satisfy (4.12). Let us assume that $R + S$ is included in the region

4.18 $$\hat{x} \leqq x \leqq \hat{x} + 2r, \qquad \hat{y} \leqq y \leqq \hat{y} + 2r.$$

We apply Lemma 4.4 with the function

4.19
$$\hat{w}(x, y) = 1 - \frac{e^{m[(x - \hat{x}) + (y - \hat{y})]}}{e^{2rm}}$$

where m is a positive number to be chosen. Since $F(x, y) \leq 0$ we have

4.20
$$-L[\hat{w}] \geq (m^2 A + mD)e^{m[(x - \hat{x}) - 2r]} + (m^2 C + mE)e^{m[(y - \hat{y}) - 2r]}$$
$$\geq [m^2(A + C) - m(|D| + |E|)]e^{-2rm}.$$

We choose m large enough so that

4.21
$$m^2(A + C) - m(|D| + |E|) \geq A + C$$

for all $(x, y) \in R + S$, i.e., so that

4.22
$$m \geq \sigma + \sqrt{1 + \sigma^2}$$

where

4.23
$$\sigma = \max_{R+S} \frac{|D| + |E|}{2(A + C)}.$$

Evidently, we have

4.24
$$-L[\hat{w}] \geq e^{-2rm}(A + C)$$

and

4.25
$$\hat{W} = \max_{R+S} |\hat{w}(x, y)| \leq 1.$$

Moreover, by the direct application of Taylor's theorem we have

4.26
$$-L_h[\hat{w}] \geq -L[\hat{w}] - \frac{(A + C)h^2}{12} N_4 - \frac{(|D| + |E|)}{6} N_3$$

where for $k = 3, 4$ we have

4.27
$$N_k = \max \left\{ \max_{R+S} \left| \frac{\partial^k \hat{w}}{\partial x^k} \right|, \max_{R+S} \left| \frac{\partial^k \hat{w}}{\partial y^k} \right| \right\}.$$

Therefore, by Lemma 4.7 and Lemma 4.4 we have, for h sufficiently small,

4.28 $|u(x, y) - \bar{u}(x, y)| \leq \dfrac{h^2}{12} \max_{R_h} \left[\dfrac{(A + C)M_4 + 2M_3(|D| + |E|)}{e^{-2rm}(A + C) - \dfrac{(A + C)h^2}{12} N_4 - \dfrac{(|D| + |E|)h^2}{6} N_3} \right]$

$$+ \max_{S_h} |u(x, y) - \bar{u}(x, y)|.$$

But since

4.29 $$N_3 \leq m^3, \qquad N_4 \leq m^4$$

we have

4.30 $|u(x, y) - \bar{u}(x, y)| \leq \dfrac{h^2}{12} \dfrac{M_4 + 4\sigma M_3}{e^{-2rm} - \dfrac{h^2}{12}[m^4 + 4m^3\sigma]} + \max_{S_h} |u(x, y) - \bar{u}(x, y)|$

where σ is given by (4.23). Thus if $u(x, y) = \bar{u}(x, y)$ on S_h, then the error varies as the square of the mesh size.

We now give a bound on $|u(x, y) - \bar{u}(x, y)|$ on S_h. By Procedure A, the value of $u(x, y)$ on S_h is taken as the value of $g(x^*, y^*)$ at a point (x^*, y^*) on S which is closest to (x, y). Clearly, the point must be within h of (x, y), for otherwise the four points adjacent to (x, y) would be in R and (x, y) would be in R_h. Evidently, we have

4.31 $$|u(x, y) - \bar{u}(x, y)| = |\bar{u}(x^*, y^*) - \bar{u}(x, y)|.$$

But by Taylor's theorem

4.32 $\bar{u}(x^*, y^*) = \bar{u}(x, y) + (x^* - x)u_x(x + \theta(x^* - x), y + \theta(y^* - y))$

$$+ (y^* - y)u_y(x + \theta(x^* - x), y + \theta(y^* - y))$$

where $0 < \theta < 1$. Hence we have

4.33 $|\bar{u}(x^*, y^*) - \bar{u}(x, y)| \leq M_1\{|x^* - x| + |y^* - y|\}$

$$\leq \sqrt{2}M_1 h$$

since $|x^* - x| + |y^* - y| \leq \sqrt{2}\{(x^* - x)^2 + (y^* - y)^2\}^{1/2}$. Therefore, we have

4.34 $$\max_{S_h} |u(x, y) - \bar{u}(x, y)| \leq \sqrt{2}h M_1.$$

Thus, because of our treatment of the boundary conditions we may introduce an error of order h.

We now study the accuracy of Procedures B and C in terms of the discrete Dirichlet problem. For Procedure B, let R_h^* be the set of regular points of R_h and let $R_h' = R_h - R_h^*$. Also, we let $R_h^{(1)}$ be the set of all points of R_h' which are adjacent to a point of R_h^*. By the first part of Theorem 4.10 we have, for all $(x, y) \in R_h^* + R_h^{(1)}$

4.35
$$|u(x, y) - \bar{u}(x, y)| \leq \frac{h^2 r^2}{24} M_4 + \max_{R_h'} |u(x, y) - \bar{u}(x, y)|.$$

Now, for each point P of R_h' we determine $u(P)$ by the linear interpolation formula (2.41). By the theory of linear interpolation we have

4.36
$$\bar{u}(P) = \frac{h_2}{h_1 + h_2} \bar{u}(T) + \frac{h_1}{h_1 + h_2} \bar{u}(Q) + \varepsilon$$

where

4.37
$$|\varepsilon| \leq \frac{(h_1 + h_2)^2}{8} M_2.$$

Therefore, we have, since $u(Q) = \bar{u}(Q)$

4.38
$$|u(P) - \bar{u}(P)| \leq \frac{h_2}{h_1 + h_2} |u(T) - \bar{u}(T)| + \frac{(h_1 + h_2)^2}{8} M_2.$$

If $T \in S$ then $u(T) = \bar{u}(T)$ and we have

4.39
$$|u(P) - \bar{u}(P)| \leq \frac{h^2}{2} M_2.$$

Otherwise $h_1 = h$ and we have

4.40
$$|u(P) - \bar{u}(P)| \leq \tfrac{1}{2}|u(T) - \bar{u}(T)| + \frac{h^2}{2} M_2.$$

It therefore follows that for all $(x, y) \in R_h'$ we have

4.41
$$|u(x, y) - \bar{u}(x, y)| \leq \tfrac{1}{2} \max_{R_h} |u(x, y) - \bar{u}(x, y)| + \frac{h^2}{2} M_2.$$

Let

4.42
$$\begin{cases} \alpha = \max_{R_h^* + R_h^{(1)}} |u(x, y) - \bar{u}(x, y)| \\ \beta = \max_{R_h'} |u(x, y) - \bar{u}(x, y)|. \end{cases}$$

By (4.35) and (4.41)

4.43
$$\begin{cases} \alpha \leq \dfrac{h^2 r^2}{24} M_4 + \beta \\[4mm] \beta \leq \tfrac{1}{2} \max(\alpha, \beta) + \dfrac{h^2}{2} M_2. \end{cases}$$

If $\alpha \geq \beta$ we have $\beta \leq \tfrac{1}{2}\alpha + \dfrac{h^2}{2} M_2$, and

4.44
$$\alpha \leq \dfrac{h^2 r^2}{24} M_4 + \tfrac{1}{2}\alpha + \dfrac{h^2}{2} M_2$$

so that

4.45
$$\alpha \leq \dfrac{h^2 r^2}{12} M_4 + h^2 M_2.$$

On the other hand, if $\alpha \leq \beta$, then $\beta \leq \tfrac{1}{2}\beta + \dfrac{h^2}{2} M_2$ and $\beta \leq h^2 M_2$ so that

4.46
$$\alpha \leq \dfrac{h^2 r^2}{24} M_4 + h^2 M_2.$$

Therefore we have, since if $\beta \geq \alpha$, then $\beta \leq h^2 M_2$,

4.47
$$\max_{R_h} |u(x, y) - \bar{u}(x, y)| \leq \dfrac{h^2 r^2}{12} M_4 + h^2 M_2.$$

In Procedure C we use (2.46) for points of R_h' where

4.48
$$\begin{cases} \alpha_1 = \dfrac{2}{h^2 s_1(s_1 + s_3)}, \qquad \alpha_3 = \dfrac{2}{h^2 s_3(s_1 + s_3)} \\[4mm] \alpha_2 = \dfrac{2}{h^2 s_2(s_2 + s_4)}, \qquad \alpha_4 = \dfrac{2}{h^2 s_4(s_2 + s_4)} \\[4mm] \alpha_0 = -\dfrac{1}{h^2}\left(\dfrac{2}{s_1 s_3} + \dfrac{2}{s_2 s_4} \right). \end{cases}$$

Solving for $u(x, y)$ we have

4.49 $\quad u(x, y) = \dfrac{s_2 s_4}{s_1 s_3 + s_2 s_4}\left[\dfrac{s_3}{s_1 + s_3} u(x + h_1, y) + \dfrac{s_1}{s_1 + s_3} u(x - h_3, y)\right]$

$$+ \dfrac{s_1 s_3}{s_1 s_3 + s_2 s_4}\left[\dfrac{s_4}{s_2 + s_4} u(x, y + h_2) + \dfrac{s_2}{s_2 + s_4} u(x, y - h_4)\right].$$

The first expression in brackets corresponds to linear interpolation in the points $u(x + h_1, y)$ and $u(x - h_3, y)$ while the second expression corresponds to linear interpolation in the points $u(x, y + h_2)$ and $u(x, y - h_4)$. The overall expression represents linear interpolation in the two interpolated values.

By the properties of linear interpolation we have

4.50 $\begin{cases} \left| \bar{u}(x, y) - \left[\dfrac{s_3}{s_1 + s_3} \bar{u}(x + h_1, y) + \dfrac{s_1}{s_1 + s_3} \bar{u}(x - h_3, y)\right] \right| \leq \dfrac{h^2(s_1 + s_3)^2}{8} M_2 \\[4mm] \left| \bar{u}(x, y) - \left[\dfrac{s_4}{s_2 + s_4} \bar{u}(x, y + h_2) + \dfrac{s_2}{s_2 + s_4} \bar{u}(x, y - h_4)\right] \right| \leq \dfrac{h^2(s_2 + s_4)^2}{8} M_2. \end{cases}$

Therefore,

4.51 $\quad \left| \bar{u}(x, y) - \dfrac{s_2 s_4}{s_1 s_3 + s_2 s_4}\left[\dfrac{s_3}{s_1 + s_3} \bar{u}(x + h_1, y) + \dfrac{s_1}{s_1 + s_3} \bar{u}(x - h_3, y)\right] \right.$

$$\left. - \dfrac{s_1 s_3}{s_1 s_3 + s_2 s_4}\left[\dfrac{s_4}{s_2 + s_4} \bar{u}(x, y + h_2) + \dfrac{s_2}{s_2 + s_4} \bar{u}(x, y - h_4)\right] \right| \leq \dfrac{h^2}{2} M_2$$

since $s_i \leq 1$, $i = 1, 2, 3, 4$. Thus we have

4.52 $\quad |u(x, y) - \bar{u}(x, y)| \leq \gamma_1 |e(x + h_1, y)| + \gamma_2 |e(x, y + h_2)| + \gamma_3 |e(x - h_3, y)|$

$$+ \gamma_4 |e(x, y - h_4)| + \dfrac{h^2}{2} M_2$$

where

4.53 $$e(x, y) = u(x, y) - \bar{u}(x, y)$$

4.54 $\begin{cases} \gamma_1 = \dfrac{s_2 s_4}{s_1 s_3 + s_2 s_4}\left(\dfrac{s_3}{s_1 + s_3}\right), \qquad \gamma_3 = \dfrac{s_2 s_4}{s_1 s_3 + s_2 s_4}\left(\dfrac{s_1}{s_1 + s_3}\right) \\[4mm] \gamma_2 = \dfrac{s_1 s_3}{s_1 s_3 + s_2 s_4}\left(\dfrac{s_4}{s_2 + s_4}\right), \qquad \gamma_4 = \dfrac{s_1 s_3}{s_1 s_3 + s_2 s_4}\left(\dfrac{s_2}{s_2 + s_4}\right). \end{cases}$

We now seek to show that for $(x, y) \in R'_h$

4.55
$$|u(x, y) - \bar{u}(x, y)| \leq \tfrac{3}{4} \max_{R_h} |e(x, y)| + \frac{h^2}{2} M_2.$$

To do this, we first consider the case where one of the points, say $u(x + h_1, y)$, is in S and the other points are not. Hence $s_2 = s_3 = s_4 = 1$. In this case $e(x + h_1, y) = 0$. Since $\gamma_1 + \gamma_2 + \gamma_3 + \gamma_4 = 1$ and since

4.56
$$\gamma_2 + \gamma_4 = \frac{s_1 s_3}{s_1 s_3 + s_2 s_4} \leq \frac{s_2 s_4}{s_1 s_3 + s_2 s_4} = \gamma_1 + \gamma_3$$

we have $\gamma_1 + \gamma_3 \geq \tfrac{1}{2}$. Moreover, since

4.57
$$\frac{s_3}{s_1 + s_3} \geq \frac{s_1}{s_1 + s_3}$$

we have $\gamma_1 \geq \gamma_3$ and $\gamma_1 \geq \tfrac{1}{4}$. Thus $\gamma_2 + \gamma_3 + \gamma_4 \leq \tfrac{3}{4}$. Therefore (4.55) follows from (4.52).

Let us now consider the case where two of the points are on S. There are essentially two different cases. In the first case $(x + h_1, y)$ and $(x - h_3, y)$ are in S and the other two points are not. Clearly $\gamma_1 + \gamma_3 \geq \gamma_2 + \gamma_4$ and $\gamma_2 + \gamma_4 \leq \tfrac{1}{2}$. Thus we have

4.58
$$|u(x, y) - \bar{u}(x, y)| \leq \tfrac{1}{2} \max_{R_h} |e(x, y)| + \frac{h^2}{2} M_2.$$

A similar argument holds if $(x, y + h_2)$ and $(x, y - h_4)$ are in S. If $(x + h_1, y)$ and $(x, y + h_2)$ are in S we have $\gamma_1 \geq \gamma_3$, $\gamma_2 \geq \gamma_4$ and $\gamma_1 + \gamma_2 \geq \gamma_3 + \gamma_4$ so that $\gamma_3 + \gamma_4 \leq \tfrac{1}{2}$. Again, (4.58) holds.

Next, let us consider the case where only one of the points, say $(x + h_1, y)$, is not in S. Evidently

4.59
$$\gamma_1 = \frac{s_2 s_4}{s_1 s_3 + s_2 s_4} \left(\frac{s_3}{s_1 + s_3} \right) \leq \left(\frac{1}{1 + s_3} \right) \left(\frac{s_3}{1 + s_3} \right) \leq \tfrac{1}{4}$$

since $s_2 s_4 (s_1 s_3 + s_2 s_4)^{-1}$ is an increasing function of $s_2 s_4$ and $s_2 s_4 \leq 1$, and since $s_3 (1 + s_3)^{-2}$ is an increasing function of s_3 and $s_3 \leq 1$. Thus we have

4.60
$$|u(x, y) - \bar{u}(x, y)| \leq \tfrac{1}{4} \max_{R_h} |e(x, y)| + \frac{h^2}{2} M_2.$$

Finally, if all four points $(x + h_1, y)$, $(x, y + h_2)$, etc. are on S we have

4.61
$$|u(x, y) - \bar{u}(x, y)| \leq \frac{h^2}{2} M_2.$$

Therefore (4.55) holds in all cases.

We now let

4.62
$$\alpha = \max_{R_h^*} |e(x, y)|, \qquad \beta = \max_{R_h'} |e(x, y)|.$$

Evidently, from (4.55) and the first part of Theorem 4.10 we have

4.63
$$\alpha \leqq \frac{h^2 r^2 M_4}{24} + \beta, \qquad \beta \leqq \tfrac{3}{4} \max(\alpha, \beta) + \frac{h^2}{2} M_2.$$

If $\alpha \geqq \beta$, then $\beta \leqq \tfrac{3}{4}\alpha + \dfrac{h^2 M_2}{2}$, and

4.64
$$\alpha \leqq \frac{h^2 r^2}{24} M_4 + \tfrac{3}{4}\alpha + \frac{h^2 M_2}{2}$$

or

4.65
$$\alpha \leqq \frac{h^2 r^2}{6} M_4 + 2h^2 M_2.$$

On the other hand, if $\beta \geqq \alpha$, then

4.66
$$\beta \leqq \tfrac{3}{4}\beta + \frac{h^2}{2} M_2$$

and

4.67
$$\beta \leqq 2h^2 M_2.$$

Therefore, since $\beta \leqq \alpha$ or else $\beta \leqq 2h^2 M_2$, we have

4.68
$$\max_{R_h} |u(x, y) - \bar{u}(x, y)| \leqq \frac{h^2 r^2}{6} M_4 + 2h^2 M_2.$$

Thus, even though the local accuracy of the discrete representation of the differential equation at points of R_h' is less in Procedures B and C than for regular points, nevertheless the overall accuracy is of the same order as in the case where all points are regular.

We can extend the above analysis of Procedure C to the more general differential equations (1.1) or (2.24) using the methods of Section 10.6. We can easily show that for points of R_h' we have

4.69
$$\tilde{L}_h[u] - L[u] = O(h)$$

where $\tilde{L}_h[u]$ is the discrete representation of $L[u]$ used for points of R_h. Therefore,

4.70 $\tilde{L}_h[e] = O(h).$

We can solve for $e(x, y)$ obtaining

4.71 $e(x, y) = \gamma_1 e(x + h_1, y) + \gamma_2 e(x, y + h_2) + \gamma_3 e(x - h_3, y) + \gamma_4 e(x, y - h_4) + O(h^3).$

It can be shown that the sum of those γ_i associated with points not on S_h is bounded, for h small enough, by a number $\gamma < 1$. We therefore have for all $(x, y) \in R_h'$

4.72 $|e(x, y)| \leq \gamma \max_{R_h} |e(x, y)| + O(h^3).$

Since we also know, by (4.30) that for h sufficiently small

4.73 $\max_{R_h^*} |e(x, y)| \leq Kh^2 + \max_{R_h'} |e(x, y)|$

for some constant K, independent of h, it follows that

4.74 $\max_{R_h} |e(x, y)| = O(h^2).$

EXERCISES 15.4

1. For the example (2.21) involving four interior mesh points, estimate the error by estimating M_4 by means of a suitable difference quotient. Note that $u_{xxxx} = u_{yyyy} = -u_{xxyy}$. Compute the exact solution of the differential equation by Fourier series methods and compare the exact error with the estimated error.

2. Find a function $v(x, y)$ which vanishes on the boundary of the ellipse

$$\frac{(x - x_0)^2}{p^2} + \frac{(y - y_0)^2}{q^2} = 1$$

and is such that

$$L_h(v) = h^{-2}[v(x + h, y) + v(x, y + h) + v(x - h, y) + v(x, y - h) - 4v(x, y)] = -\alpha$$

where α is a positive constant.

3. Consider the solution of the differential equation $u_{xx} - (k/y)u_y + u_{yy} = 0$ in the region $0 < x < 1$, $1 < y < 2$, where u satisfies prescribed boundary values. Assume the usual five-point difference equation is used with $h = \frac{1}{20}$. Find a bound for $|u(x, y) - \bar{u}(x, y)|$, under the hypotheses of Lemma 4.7 and assuming

   a) $k = 1$,
   b) $k = 4$.

   Do the same for the differential equations

   i) $u_{xx} + (y - \frac{3}{2})u_y + u_{yy} = 0,$

ii) $u_{xx} + 4(y - \frac{3}{2})u_y + u_{yy} = 0$.

In each case express the result in terms of M_3 and M_4.

15.5 HIGHER-ORDER METHODS

For the analysis of the effectiveness of various discrete representations of the differential equation (1.1) we use methods similar to those employed in Chapter 10 for two-point boundary value problems involving ordinary differential equations. Given a linear differential operator $L[u]$ we consider approximating $L[u]$ by linear discrete operators $L_h[u]$ of the form

$$5.1 \qquad L_h[u] = \sum_{k=0}^{m} \alpha_k u(x_0 + s_k h, y_0 + t_k h)$$

where m is a given integer, $s_1, s_2, \ldots, s_m$, are distinct nonzero integers, as are $t_1, t_2, \ldots, t_m$ and $s_0 = t_0 = 0$. We can define consistency, absolute degree of approximation, and relative degree of approximation exactly as in Chapter 10. We can also prove

5.2 Theorem. Let $L_h[u]$ be a discrete linear operator of the form (5.1). A necessary and sufficient condition that $L_h[u]$ be consistent with the elliptic operator $L[u]$ given by (1.1) and have a relative degree of approximation of p is that there exist functions $\beta_{i,j}$, $i, j = 0, 1, 2, \ldots, t$, of x, y, and h, such that

$$5.3 \qquad \lim_{h \to 0} \beta_{i,j} = 0, \qquad i, j = 0, 1, 2, \ldots, t$$

and such that for all functions $u(x, y)$ which are analytic at (x_0, y_0) we have

$$5.4 \qquad L_h[u] - L[u] - \sum_{i,j=0}^{t} \beta_{i,j} L^{(i,j)}[u] = O(h^{n+1}).$$

Here we let

$$5.5 \qquad L^{(i,j)}[u] = \frac{\partial^{i+j} L[u]}{\partial x^i \partial y^j}.$$

Proof. The sufficiency of the conditions is obvious. To prove the necessity we seek to express all derivatives of u in terms of the following "basic" derivatives

$$5.6 \qquad \begin{array}{cccccc} u_{0,0} & u_{1,0} & u_{1,1} & u_{1,2} & u_{1,3} & u_{1,4} \cdots \\ & u_{0,1} & u_{0,2} & u_{0,3} & u_{0,4} & u_{0,5} \cdots \end{array}$$

and L and partial derivatives of L. Here $u_{i,j} = \partial^{i+j} u/(\partial x^i \partial y^j)$ for $i, j = 0, 1, 2, \ldots$. Differentiating L we have

5.7
$$\begin{cases} L^{(1,0)}[u] = Au_{3,0} + 2Bu_{2,1} + Cu_{1,2} + \cdots \\ L^{(0,1)}[u] = Au_{2,1} + 2Bu_{1,2} + Cu_{0,3} + \cdots \end{cases}$$

where terms involving derivatives of order two and lower are omitted. Since $A \neq 0$, by assumption, we can solve

5.8
$$L[u] = Au_{2,0} + 2Bu_{1,1} + Cu_{0,2} + Du_{1,0} + Eu_{0,1} + Fu$$

for $u_{2,0}$. We can then express $u_{3,0}$ and $u_{2,1}$ in terms of $u_{0,0}, u_{1,0}, u_{0,1}, u_{1,1}, u_{0,2}, u_{1,2}$, and $u_{0,3}$. By repeating this process we can express any derivative in terms of the basic derivatives (and $L[u]$ and its derivatives). By the use of Taylor's series we have, letting $u_{i,j} = u_{i,j}(x_0, y_0)$,

5.9
$$L_h[u] - L[u] = \sum_{i,j=0}^{\infty} \gamma_{i,j} u_{i,j}.$$

One can show that for any p, if i and j are large enough $\gamma_{i,j} = 0(h^{p+1})$ provided $L_h[u]$ is consistent with $L[u]$. For example, with the five-point operator

5.10
$$\begin{aligned} L_h[u] = \alpha_0 u(x, y) + \alpha_1 u(x + h, y) + \alpha_2 u(x, y + h) \\ + \alpha_3 u(x - h, y) + \alpha_4 u(x, y - h) \end{aligned}$$

and the differential operator

5.11
$$L[u] = Au_{2,0} + Cu_{0,2} + Du_{1,0} + Eu_{0,1} + Fu$$

we have

5.12 $L_h[u] - L[u] = (\alpha_0 + \alpha_1 + \alpha_2 + \alpha_3 + \alpha_4 - F)u$

$$\begin{aligned} &+ (h(\alpha_1 - \alpha_3) - D)u_{1,0} + (h(\alpha_2 - \alpha_4) - E)u_{0,1} \\ &+ \left(\frac{h^2}{2}(\alpha_1 + \alpha_3) - A\right)u_{2,0} + \left(\frac{h^2}{2}(\alpha_2 + \alpha_4) - C\right)u_{0,2} \\ &+ \frac{h^3}{6}(\alpha_1 - \alpha_3)u_{3,0} + \frac{h^3}{6}(\alpha_2 - \alpha_4)u_{0,3} + \frac{h^4}{24}(\alpha_1 + \alpha_3)u_{4,0} \\ &+ \frac{h^4}{24}(\alpha_2 + \alpha_4)u_{0,4} + \cdots \end{aligned}$$

and hence, for consistency we must have

5.13
$$
\begin{cases}
\alpha_0 + \alpha_1 + \alpha_2 + \alpha_3 + \alpha_4 - F = \varepsilon_{0,0} \\[4pt]
h(\alpha_1 - \alpha_3) - D = \varepsilon_{1,0} \\[4pt]
h(\alpha_2 - \alpha_4) - E = \varepsilon_{0,1} \\[4pt]
\dfrac{h^2}{2}(\alpha_1 + \alpha_3) - A = \varepsilon_{2,0} \\[4pt]
\dfrac{h^2}{2}(\alpha_2 + \alpha_4) - C = \varepsilon_{0,2}
\end{cases}
$$

where $\varepsilon_{i,j} = \mathrm{o}(h)$. Therefore

5.14 $\alpha_1 - \alpha_3 = \mathrm{O}(h^{-1})$, $\alpha_2 - \alpha_4 = \mathrm{O}(h^{-1})$, $\alpha_1 + \alpha_3 = \mathrm{O}(h^{-2})$, $\alpha_2 + \alpha_4 = \mathrm{O}(h^{-2})$.

Hence, for any p, the terms beyond a certain point in the expression (5.12) for $L[u] - L[u]$ are certainly $\mathrm{O}(h^{p+1})$.

Upon neglecting all but a finite number of terms of (5.9) and eliminating non-basic derivatives we get

5.15
$$
L_h[u] - L[u] = a_0 u_{0,0} + \sum_{i=1}^{s} (a_i u_{1,i-1} + b_i u_{0,i})
$$
$$
+ \sum_{i,j=0}^{t} \beta_{i,j} L^{(i,j)}[u] + \mathrm{O}(h^{p+1}).
$$

We now seek to show that for any basic derivative $u_{i,j}$ there exists a solution w of $L[u] = 0$ such that every basic derivative of w is zero except $w_{i,j}$ which is unity. But if $L[w] \equiv 0$, then all derivatives of $L[w]$ vanish and we can express all derivatives of w in terms of the basic derivatives. Thus we can determine all derivatives of w and we have

5.16
$$
w(x, y) = \sum_{r,s=0}^{\infty} \frac{(x - x_0)^r (y - y_0)^s}{r! \, s!} w_{r,s}.
$$

In order that $L_h[u] - L[u] = \mathrm{O}(h^{p+1})$ for all solutions of $L[u] = 0$ it is clearly necessary and sufficient that the coefficients a_i and b_i in (5.15) be $\mathrm{O}(h^{p+1})$. If the a_i and b_i are $\mathrm{O}(h^{p+1})$, we have

5.17
$$
L_h[u] - L[u] - \sum_{i,j=0}^{t} \beta_{i,j} L^{(i,j)}[u] = \mathrm{O}(h^{p+1}).
$$

To show that the $\beta_{i,j} \to 0$ as $h \to 0$ we consider the function w such that $L[w] = 1$ and all basic derivatives of w vanish. This can be constructed as in the case of (5.16).

Evidently by consistency we must have $\beta_{0,0} \to 0$ as $h \to 0$. Similarly, if we consider w such that $L[w] = x - x_0$ and all basic derivatives of w vanish, we can show that $\beta_{1,0} \to 0$ as $h \to 0$. Likewise we can show that (5.3) holds for all i, j. This completes the proof of Theorem 5.2.

We remark that if $L_h[u]$ is consistent with $L[u]$ and has a relative degree of approximation p, then $L_h^*[u] = (1 + \beta_{0,0})^{-1} L_h[u]$ is also consistent with $L[u]$ and has a relative degree of approximation p. Here $\beta_{i,j}$, $i, j = 0, 1, \ldots, t$ are any numbers satisfying (5.3) and (5.4). Thus in constructing a consistent operator $L_h[u]$ with a high relative degree of approximation to $L[u]$ we may as well assume that $\beta_{0,0} = 0$. See the remark following 10-(7.44).

Having determined α_i and $\beta_{i,j}$ so that (5.3) and (5.4) hold for some p we use the difference equation

5.18
$$L_h[u] = G + \sum_{i,j=0}^{t} \beta_{i,j} G^{(i,j)}.$$

Since

5.19
$$L_h[\bar{u}] = L[\bar{u}] + \sum_{i,j=0}^{t} \beta_{i,j} L^{(i,j)}[\bar{u}] + O(h^{p+1}),$$

where $\bar{u}$ is any solution of $L[u] = G$, it follows that

5.20
$$L_h[u - \bar{u}] = O(h^{p+1}).$$

With the difference operator (5.1), the notions of adjacency and regular points of R_h must be modified in an obvious way. Let us assume that all points of $R_h = \Omega_h \cap R$ are regular points and that the set of all points of Ω_h adjacent to these points but not in R lie on S. We assume* that $L_h[u]$ is a linear discrete operator of the form (5.1) such that $-\alpha_0, \alpha_1, \ldots, \alpha_m$ are positive and

5.21
$$-\alpha_0 \geqq \alpha_1 + \alpha_2 + \cdots + \alpha_m.$$

Moreover, we assume that $L_h[u]$ is consistent with $L[u]$ and that (5.4) and (5.3) hold. From Lemma 4.4 and (5.20) we can show using the function $\hat{w}(x, y)$ given by (4.19) that for $(x, y) \in R_h$ we have

$$|u(x, y) - \bar{u}(x, y)| = O(h^{p+1}).$$

One could, of course, develop linear discrete operators of high absolute degree of approximation to $L[u]$ using high-order representations of the derivatives involved in $L[u]$. Thus, for example, one could use the nine-point configuration

---

* In some cases if one can prove that $\alpha_i > 0$ for h sufficiently small, for $i = 1, 2, \ldots, m$, then the case where $-\alpha_0 \geqq \alpha_1 + \alpha_2 + \cdots + \alpha_m + o(1)$ can be treated, as in Section 10.8.

$(x, y), (x \pm h, y)(x \pm 2h, y), (x, y \pm h), (x, y \pm 2h)$. This, of course, would mean that special procedures would be needed near the boundary. In this section we shall consider the nine-point configuration $(x + rh, y + sh)$, $r, s = -1, 0, 1$. Here there is a possibility that the regular difference equation can be used for all interior mesh points. Moreover, as we shall see, even though we cannot get a higher absolute degree of approximation than unity, the same as with the usual five-point approximation, we are in most cases able to get a higher relative degree of approximation.

We now seek to determine $\alpha_0, \alpha_1, \alpha_2, \alpha_3, \alpha_4, \alpha_5, \alpha_6, \alpha_7, \alpha_8$ and $\beta_{i,j}$ such that $\beta_{0,0} = 0$ and such that (5.4) holds for some $p > 1$ with

5.22 $L_h[u] = \alpha_0 u(x, y) + \alpha_1 u(x + h, y) + \alpha_2 u(x, y + h) + \alpha_3 u(x - h, y)$

$$+ \alpha_4 u(x, y - h) + \alpha_5 u(x + h, y + h) + \alpha_6 u(x - h, y + h)$$

$$+ \alpha_7 u(x + h, y - h) + \alpha_8 u(x - h, y - h).$$

We introduce the new variables

5.23
$$\begin{cases} \theta_{0,0} = \alpha_0 + \alpha_1 + \alpha_2 + \alpha_3 + \alpha_4 + \alpha_5 + \alpha_6 + \alpha_7 + \alpha_8 \\ \theta_{1,0} = \quad \alpha_1 \quad - \alpha_3 \quad + \alpha_5 - \alpha_6 + \alpha_7 - \alpha_8 \\ \theta_{0,1} = \quad \alpha_2 \quad - \alpha_4 + \alpha_5 + \alpha_6 - \alpha_7 - \alpha_8 \\ \theta_{2,0} = \quad \alpha_1 \quad + \alpha_3 \quad + \alpha_5 + \alpha_6 + \alpha_7 + \alpha_8 \\ \theta_{1,1} = \quad \alpha_5 - \alpha_6 - \alpha_7 + \alpha_8 \\ \theta_{0,2} = \quad \alpha_2 \quad + \alpha_4 + \alpha_5 + \alpha_6 + \alpha_7 + \alpha_8 \\ \theta_{2,1} = \quad \alpha_5 + \alpha_6 - \alpha_7 - \alpha_8 \\ \theta_{1,2} = \quad \alpha_5 - \alpha_6 + \alpha_7 - \alpha_8 \\ \theta_{2,2} = \quad \alpha_5 + \alpha_6 + \alpha_7 + \alpha_8. \end{cases}$$

Once we have determined the $\theta_{i,j}$ we can get the α_k from

5.24
$$\begin{cases} \alpha_0 = \theta_{0,0} - \theta_{2,0} - \theta_{0,2} + \theta_{2,2} \\ 2\alpha_1 = \theta_{1,0} + \theta_{2,0} - \theta_{1,2} - \theta_{2,2} \\ 2\alpha_2 = \theta_{0,1} + \theta_{0,2} - \theta_{2,1} - \theta_{2,2} \\ 2\alpha_3 = -\theta_{1,0} + \theta_{2,0} + \theta_{1,2} - \theta_{2,2} \\ 2\alpha_4 = -\theta_{0,1} + \theta_{0,2} + \theta_{2,1} - \theta_{2,2} \\ 4\alpha_5 = \theta_{1,1} + \theta_{2,1} + \theta_{1,2} + \theta_{2,2} \\ 4\alpha_6 = -\theta_{1,1} + \theta_{2,1} - \theta_{1,2} + \theta_{2,2} \\ 4\alpha_7 = -\theta_{1,1} - \theta_{2,1} + \theta_{1,2} + \theta_{2,2} \\ 4\alpha_8 = \theta_{1,1} - \theta_{2,1} - \theta_{1,2} + \theta_{2,2}. \end{cases}$$

Evidently we have

5.25
$$L_h[u] - L[u] - \sum_{\substack{i,j=0 \\ i+j \geq 1}}^{t} \beta_{i,j} L^{(i,j)}[u] = \sum_{i,j=0}^{\infty} \delta_{i,j} u_{i,j}$$

where the $\delta_{i,j}$ are determined by Table 5.30 (see pages 988–989). Thus, for example,

5.26 $\quad \delta_{0,0} = \theta_{0,0} - F - F_{1,0}\beta_{1,0} - F_{0,1}\beta_{0,1} - F_{2,0}\beta_{2,0} - F_{1,1}\beta_{1,1} - F_{0,2}\beta_{0,2}$

etc. We now show that if $p \geq 2$, then we must have

5.27
$$A = C \quad \text{or} \quad B = 0.$$

Since $L_h[u]$ is consistent with $L[u]$ we must have

5.28
$$\begin{cases} \theta_{0,0} = F + o(1), \quad \theta_{2,0} = \dfrac{2A}{h^2} + o(h^{-2}), \quad \theta_{2,1} = o(h^{-3}) \\[2mm] \theta_{1,0} = \dfrac{D}{h} + o(h^{-1}), \quad \theta_{1,1} = \dfrac{2B}{h^2} + o(h^{-2}), \quad \theta_{1,2} = o(h^{-3}) \\[2mm] \theta_{0,1} = \dfrac{E}{h} + o(h^{-1}), \quad \theta_{0,2} = \dfrac{2C}{h^2} + o(h^{-2}), \quad \theta_{2,2} = o(h^{-4}). \end{cases}$$

We first show that if $i + j \geq 3$ and $p \geq 2$, then $\beta_{i,j} = o(h^2)$. For example, consider $\beta_{4,0}, \beta_{3,1}, \ldots, \beta_{0,4}$. Since $\delta_{6,0} = O(h^3)$ we have $\beta_{4,0} = O(h^3)$. Similarly, $\beta_{3,1} = O(h^3)$, $\beta_{1,3} = O(h^3)$, $\beta_{0,4} = O(h^3)$. Also $\beta_{2,2} = o(h^2)$. Similarly, we show that $\beta_{3,0} = O(h^3)$, $\beta_{0,3} = O(h^3)$, $\beta_{2,1} = o(h^2)$, $\beta_{1,2} = o(h^2)$.

Since $\delta_{4,0} = O(h^3)$, we have

5.29
$$\frac{h^2}{12} A - A\beta_{2,0} = o(h^2)$$

and hence

5.31
$$\beta_{2,0} = \frac{h^2}{12} + o(h^2).$$

Similarly, since $\delta_{0,4} = O(h^3)$ we have

5.32
$$\beta_{0,2} = \frac{h^2}{12} + o(h^2).$$

Since $\delta_{3,1} = O(h^3)$ we have

5.33
$$\frac{h^2 B}{3} + o(h^2) - 2B\beta_{2,0} - A\beta_{1,1} = O(h^3)$$

and

5.34
$$\beta_{1,1} = \frac{Bh^2}{6A} + o(h^2).$$

If now we have $\delta_{1,3} = O(h^3)$ then

5.35
$$\begin{cases} \dfrac{h^4}{6}\theta_{1,1} = C\beta_{1,1} + 2B\beta_{0,2} + O(h^3) \\[2mm] \dfrac{h^2 B}{3} = \dfrac{BCh^2}{6A} + \dfrac{Bh^2}{6} + o(h^2) \\[2mm] \dfrac{h^2}{6}B\left(1 - \dfrac{C}{A}\right) = o(h^2) \end{cases}$$

and hence $A = C$ or $B = 0$.

We now show that if $B = 0$ then we can find $\theta_{i,j}$ and $\beta_{i,j}$ such that $p \geq 3$. (It is also shown in Young and Dauwalder [1965] that this is also true if $A = C$ even if $B \neq 0$.) To do this we use an iterative process similar to that used in Section 10.7. First we let $\beta_{i,j}^{(0)} = 0$ and determine $\theta_{i,j}^{(0)}$ so as to make those $\delta_{i,j}$ vanish which are labeled as "Class A" in Table 5.30. This gives

5.36
$$\begin{cases} \theta_{0,0}^{(0)} = F, \qquad \theta_{2,0}^{(0)} = \dfrac{2A}{h^2}, \qquad \theta_{2,1}^{(0)} = 0 \\[3mm] \theta_{1,0}^{(0)} = \dfrac{D}{h}, \qquad \theta_{1,1}^{(0)} = 0 \qquad \theta_{1,2}^{(0)} = 0 \\[3mm] \theta_{0,1}^{(0)} = \dfrac{E}{h}, \qquad \theta_{0,2}^{(0)} = \dfrac{2C}{h^2}, \qquad \theta_{2,2}^{(0)} = 0. \end{cases}$$

5.30 Table   Coefficients for the Nine-point Configuration

Class		$\theta_{0,0}$	$\theta_{1,0}$	$\theta_{0,1}$	$\theta_{2,0}$	$\theta_{1,1}$	$\theta_{0,2}$	$\theta_{2,1}$	$\theta_{1,2}$	$\theta_{2,2}$	-1	$-\beta_{1,0}$	$-\beta_{0,1}$
A	$\delta_{0,0}$	1									F	$F_{1,0}$	$F_{0,1}$
A	$\delta_{1,0}$		h								D	$F+D_{1,0}$	$D_{0,1}$
A	$\delta_{0,1}$			h							E	$E_{1,0}$	$F+E_{0,1}$
A	$\delta_{2,0}$				$h^2/2$						A	$D+A_{1,0}$	$A_{0,1}$
A	$\delta_{1,1}$					h^2					$2B$	$E+2B_{1,0}$	$D+2B_{0,1}$
A	$\delta_{0,2}$						$h^2/2$				C	$C_{1,0}$	$E+C_{0,1}$
B	$\delta_{3,0}$		$h^3/6$									A	0
A	$\delta_{2,1}$							$h^3/2$				$2B$	A
A	$\delta_{1,2}$								$h^3/2$			C	$2B$
B	$\delta_{0,3}$			$h^3/6$									C
B	$\delta_{4,0}$				$h^4/24$								
B	$\delta_{3,1}$					$h^4/6$							
A	$\delta_{2,2}$									$h^4/4$			
C	$\delta_{1,3}$					$h^4/6$							
B	$\delta_{0,4}$						$h^4/24$						
D	$\delta_{5,0}$		$h^5/120$										
D	$\delta_{4,1}$						$h^5/24$						
D	$\delta_{3,2}$							$h^5/12$					
D	$\delta_{2,3}$							$h^5/12$					
D	$\delta_{1,4}$								$h^5/24$				
D	$\delta_{0,5}$			$h^5/120$									
D	$\delta_{6,0}$				$h^6/720$								
D	$\delta_{5,1}$					$h^6/120$							
D	$\delta_{4,2}$									$h^6/48$			
D	$\delta_{3,3}$					$h^6/36$							
D	$\delta_{2,4}$									$h^6/48$			
D	$\delta_{1,5}$					$h^6/120$							
D	$\delta_{0,6}$						$h^6/720$						

$-\beta_{2,0}$	$-\beta_{1,1}$	$-\beta_{0,2}$	$-\beta_{3,0}$	$-\beta_{2,1}$	$-\beta_{1,2}$	$-\beta_{0,3}$	$-\beta_{4,0}$	$-\beta_{3,1}$	$-\beta_{2,2}$	$-\beta_{1,3}$	$-\beta_{0,4}$
$F_{2,0}$	$F_{1,1}$	$F_{0,2}$									
$2F_{1,0}+D_{2,0}$	$F_{0,1}+D_{1,1}$	$D_{0,2}$									
$E_{2,0}$	$F_{1,0}+E_{1,1}$	$2F_{0,1}+E_{0,2}$									
$F+2D_{1,0}+A_{2,0}$	$D_{0,1}+A_{1,1}$	$A_{0,2}$									
$2E_{1,0}+2B_{2,0}$	$F+D_{1,0}+E_{0,1}+2B_{1,1}$	$2D_{0,1}+2B_{0,2}$									
$C_{2,0}$	$E_{1,0}+C_{1,1}$	$F+2E_{0,1}+C_{0,2}$									
$D+2A_{1,0}$	$A_{0,1}$	0									
$E+4B_{1,0}$	$D+A_{1,0}+2B_{0,1}$	$2A_{0,1}$									
$2C_{1,0}$	$E+C_{0,1}+2B_{1,0}$	$D+4B_{0,1}$									
0	$C_{1,0}$	$E+2C_{0,1}$									
A	0	0									
$2B$	A	0	A	0	0	0					
C	$2B$	A	$2B$	A	0	0					
	C	$2B$	C	$2B$	A	0					
		C		C	$2B$	A	A	0	0	0	0
					C	$2B$	$2B$	A	0	0	0
						C	C	$2B$	A	0	0
								C	$2B$	A	0
									C	$2B$	A
										C	$2B$
											C

Next, we determine $\beta_{1,0}^{(1)}$, $\beta_{0,1}^{(1)}$, $\beta_{2,0}^{(1)}$, $\beta_{1,1}^{(1)}$, $\beta_{0,2}^{(1)}$ such that the $\delta_{i,j}$ labeled "Class B" vanish. This gives

5.37
$$\left\{ \begin{array}{ll} \beta_{1,0}^{(1)} = \dfrac{h^2}{6A}\left(\dfrac{D}{2} - A_{1,0}\right), & \beta_{2,0}^{(1)} = \dfrac{h^2}{12} \\[4mm] \beta_{0,1}^{(1)} = \dfrac{h^2}{6C}\left(\dfrac{E}{2} - C_{0,1}\right), & \beta_{1,1}^{(1)} = 0 \end{array} \right.$$

$$\beta_{0,2}^{(1)} = \dfrac{h^2}{12}.$$

We then determine $\theta_{i,j}^{(1)}$ by requiring again that the Class A equations be satisfied. Thus, for example, we obtain

5.38 $\theta_{0,0}^{(1)} = F + \dfrac{h^2}{12}\left[\dfrac{F_{1,0}}{A}(D - 2A_{1,0}) + \dfrac{F_{0,1}}{C}(E - 2C_{0,1}) + F_{2,0} + F_{0,2}\right]$.

The reader should compute the other values of $\theta_{i,j}^{(1)}$ and then verify that all $\delta_{i,j}$ are $O(h^4)$. Hence $p = 3$.

Let us now consider the possibility of obtaining a higher degree of approximation. It is shown by Young and Dauwalder [1971] that if $p \geq 4$, then one of the following pairs of conditions must hold:

5.39 $A = C$ and $B = 0$

5.40 $A = C$ and $B = A$

5.41 $A = C$ and $B = -A$

5.42 $A = -C$ and $B = 0$.

It is also shown that in the case of primary interest to us, namely the case (5.39), one must have

5.43 $\dfrac{\partial}{\partial y}\left(\dfrac{D}{A}\right) = \dfrac{\partial}{\partial x}\left(\dfrac{E}{A}\right)$

which, by (2.27), implies that $L[u]$ is *essentially self-adjoint*.

If, on the other hand, conditions (5.39) hold and if (5.43) also holds, then one can indeed determine $L_h[u]$ with a relative degree of approximation of five.

In the case of the Laplacian operator $L[u] = u_{xx} + u_{yy}$ if we let

5.44
$$\left\{ \begin{array}{ll} \alpha_0 = -\dfrac{10}{3h^2}, & \alpha_1 = \alpha_2 = \alpha_3 = \alpha_4 = \dfrac{2}{3h^2} \\[4mm] \alpha_5 = \alpha_6 = \alpha_7 = \alpha_8 = \dfrac{1}{6h^2} \end{array} \right.$$

then we have $p = 5$. The corresponding nonzero $\beta_{i,j}$ are

5.45
$$\begin{cases} \beta_{2,0} = \beta_{0,2} = \dfrac{h^2}{12} \\[2mm] \beta_{4,0} = \beta_{0,4} = \dfrac{h^4}{360}, \qquad \beta_{2,2} = \dfrac{h^4}{90}. \end{cases}$$

Thus, to solve the differential equation

5.46
$$L[u] = G$$

we would use the difference equation

5.47
$$L_h[u] = G + \frac{h^2}{12}(G_{2,0} + G_{0,2}) + \frac{h^4}{360}(G_{4,0} + 4G_{2,2} + G_{0,4}).$$

EXERCISES 15.5

1. Find a discrete representation of the Laplacian operator $L[u] = u_{xx} + u_{yy}$ of the form

$$L_h[u] = \alpha_0 u(x, y) + \alpha_1 u(x + h_1, y) + \alpha_2 u(x + h_2, y + k_2) + \alpha_3 u(x + h_3, y) + \alpha_4 u(x + h_2, y + k_4)$$

where

$$h_1 \ne 0, \qquad h_3 \ne 0, \qquad h_1 \ne h_3$$
$$k_2 \ne 0, \qquad k_4 \ne 0, \qquad k_2 \ne k_4$$
$$h_i = s_i h, \qquad k_i = t_i h.$$

The s_i and t_i are independent of h and $0 \le s_i \le 1$, $0 < t_i \le 1$. Choose the α_i so that $L_h[u]$ is consistent with $L[u]$ and such that the relative degree of approximation is as large as possible. Find the absolute degree of approximation.

2. Find a discrete operator based on the nine points $(x + ih, y + jh)$, $i, j = -1, 0, 1$, which is consistent with $L[u] = u_{xx} + (x^2 + y^2)u_{yy}$ and has a relative degree of approximation of 4. What is the absolute degree of approximation? What difference equation would be used to represent the differential equation

$$L[u] = e^x + y^3?$$

3. Consider the problem of solving the differential equation

$$L[u] = u_{xx} - \frac{1}{y}u_y + u_{yy} = 2x^2 + y^3$$

in the region $0 < x < 1$, $1 < y < 2$, with zero boundary values. Using a discrete operator involving the nine points used in Exercise 2 which is consistent with $L[u]$ and has a relative degree of approximation of at least three, construct and solve an accurate difference equation with $h = \frac{1}{3}$.

4. For any consistent discrete operator involving the nine points of Exercise 2 for representing

$L[u]$, given by (1.1), with relative degree of approximation three, show that $\alpha_0 < 0$, $\alpha_i > 0$, $i = 1, 2, \ldots, 8$ if h is sufficiently small, if $A > 0$, $C > 0$, $B = 0$, $F \leq 0$, $5A - C > 0$, and $5C - A > 0$ (Bramble and Hubbard [1963]).

5. Construct a discrete operator involving the points indicated which is consistent with $L[u]$ and which has as high a relative degree of approximation as possible. Find the absolute degree of approximation. What difference equation would you use to solve the differential equation

$$L[u] = G?$$

Consider each of the differential operators

$$L[u] = u_{xx} + u_{yy}$$

$$L[u] = u_{xx} - \frac{1}{y}u_y + u_{yy}$$

$$L[u] = u_{xx} + 2u_{yy}$$

and the following point configurations

(A) (x, y), $(x + h, y)$, $(x, y + h)$, $(x - h, y)$, $(x, y - h)$.

(B) $(x + ih, y)$, $i = -2, -1, 0, 1, 2$; $(x, y + jh)$, $j = -2, -1, 0, 1, 2$.

(C) $(x + ih, y + jh)$, $i, j = -1, 0, 1$.

(D) $(x + h\cos\theta_i, y + h\sin\theta_i)$, $\theta_i = \frac{\pi}{6}i$, $i = 0, 1, 2, 3, 4, 5$.

(E) $(x + h\cos\theta_i, y + h\sin\theta_i)$, $\theta_i = \frac{\pi}{3}i$, $i = 0, 1, 2$.

Leave answers in terms of derivatives of G.

6. Give necessary and sufficient conditions on s_i and t_i, $i = 2, 3, 4$ so that there exists a discrete operator of the form

$$L_h[u] = \alpha_0 u(x, y) + \alpha_1 u(x + h, y) + \sum_{i=2}^{4} \alpha_i u(x + s_i h, y + t_i h)$$

which is consistent with $L[u] = u_{xx} + u_{yy}$.

7. Prove that if $L_h[u]$ is a linear discrete operator of the form (5.1) which is consistent with a differential operator $L[u]$ of the form (1.1) with $B = 0$ then

$$\alpha_i = O(h^{-2}),   i = 0, 1, \ldots, m.$$

15.6 NORMAL DERIVATIVE BOUNDARY CONDITIONS

Let us now consider problems where the desired function u is required to satisfy the condition

6.1
$$\alpha(P)u(P) + \beta(P)\frac{\partial u(P)}{\partial n} = g(P)$$

for each point P of S. If $\beta \equiv 0$ we have the Dirichlet problem; if $\alpha \equiv 0$ we have the Neumann problem; otherwise, we have a mixed problem.

Let us assume that we wish to use a modification of Procedure C described in Section 15.1 to handle points near the boundary. At any point P of S_h for which $\beta \neq 0$, however, the value of $u(P)$ is not known as in the case of the Dirichlet problem. We seek a linear relation involving $u(P)$ and values of u at certain points of $R_h + S_h$.

We first describe a procedure based on that introduced by Fox [1944] (see also Fox [1962]). Given a point P on S_h we simply construct a line through P (see Fig. 6.4) and normal to the boundary S. Let C be the point at which the normal cuts a nearby mesh line. We replace the normal derivative at P by

6.2
$$\frac{\partial u(P)}{\partial n} \sim \frac{u(P) - u(C)}{\overline{PC}}.$$

Here $u(C)$ is found by linear interpolation between $u(A)$ and $u(B)$. Substituting in the condition

6.3
$$\alpha(P)u(P) + \beta(P)\frac{\partial u(P)}{\partial n} = g(P)$$

we get a linear relation involving $u(P)$, $u(A)$, and $u(B)$.

In general, one can expect to obtain less accuracy for a given mesh size for normal derivative problems than for Dirichlet-type problems since the representation of the normal derivative is rather inaccurate. A more accurate but considerably more complicated scheme has been proposed by Viswanathan [1957] (see also Fox [1962]).

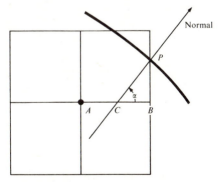

6.4 Fig. Determination of Normal Derivative.

In cases where a point P of S_h lies on a horizontal or a vertical side of S one can, of course, use the crude formula

6.5
$$\frac{\partial u}{\partial n} \sim \frac{u(P) - u(Q)}{\overline{PQ}}$$

where Q is the nearest mesh point to P on the mesh line through P which is normal to S. Alternatively, one can use a more accurate formula taking into account the differential equation at P. Thus in the example of Fig. 6.10 we have

6.6
$$\frac{u(P) - u(Q)}{h} = u_x(P) - \frac{h}{2}u_{xx}(P) + O(h^2)$$

where $h = \overline{PQ}$. If u is assumed to satisfy Poisson's equation $u_{xx} + u_{yy} = f$ at P, then we have

6.7
$$u_{xx}(P) = f(P) - u_{yy}(P)$$

so that if we replace $u_{yy}(P)$ by the usual central difference quotient we have

6.8
$$u_x = \left\{ \frac{u(P) - u(Q)}{h} + \frac{h}{2}\left[f(P) - \frac{u(S) + u(R) - 2u(P)}{h^2} \right] \right\} + O(h^2).$$

If $f \equiv 0$ and if $\partial u/\partial n = 0$ at P, then we get the condition

6.9
$$u(P) \sim \tfrac{1}{4}[2u(Q) + u(S) + u(R)].$$

It should be noted that in the case of the Neumann problem where $\alpha \equiv 0$ the solution may not be unique. Thus let u be any solution of (1.1) subject to the condition

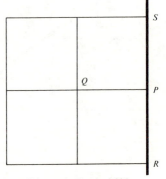

6.10 Fig. Normal Derivative Condition on a Vertical Side.

6.11
$$\frac{\partial u}{\partial n} = g$$

on S where $F \equiv 0$. Then, for any constant $c, v = u + c$ is also a solution of the Neumann problem.

The function g must satisfy a consistency condition if a solution is to exist. For example, for Laplace's equation we must have

6.12
$$\int_S g(x, y)\, ds = 0.$$

This follows from the identity*

6.13
$$0 = \iint_R \left(\frac{\partial^2 u}{\partial x^2} + \frac{\partial^2 u}{\partial y^2} \right) dx\, dy = \int_S \frac{\partial u}{\partial n}\, ds.$$

In the case of the finite difference analog, the matrix of the associated linear system is singular and the solution is not unique. Even if the compatibility condition is satisfied for g, nevertheless, the analogous discrete condition may not be satisfied exactly unless special care is taken. If the discrete condition fails to be satisfied, one may have to be content with a numerical solution where the discrete boundary conditions are not completely satisfied.

It should be noted, however, that if for some part of S the condition $\alpha u + \beta(\partial u/\partial n) = g$ where $\alpha \neq 0$, then we no longer need to be concerned with the consistency condition for g.

EXERCISES 15.6

1. Indicate which of the following problems has a unique solution. In each case u is required to be harmonic in $-1 < x < 1, 0 < y < 1$.

   i) $\partial u/\partial n = x^2$ on S
   ii) $\partial u/\partial n = x$ on S
   iii) $\partial u/\partial n = 0$ on S for $x = -1, 1$
          $u = 1$ on S for $y = 0, 1$.

2. Obtain a representation of $\partial u/\partial n$ at P (see Fig. 6.4) in terms of $u(P)$, $u(A)$, and $u(B)$ using the method described in the text. Show that one obtains the same result by letting

$$\frac{\partial u}{\partial n} = \frac{\partial u}{\partial x} \cos \alpha + \frac{\partial u}{\partial y} \sin \alpha$$

and estimating $\partial u/\partial x$ by $[u(B) - u(A)]/\overline{AB}$ and $\partial u/\partial y$ by $[u(P) - u(B)]/\overline{PB}$.

3. Find an approximate solution of Laplace's equation in the region $0 < x < 1, 0 < y < 1$ where on S, $u = 0$ for $y = 0$ and for $x = 0$, $u(x, 1) = 1$, and $u_x(1, y) = 1$. Let $h = \frac{1}{3}$.

4. Find a function satisfying Laplace's equation in the region $x^2 + y^2 < 1, y > 0$, vanishing for $y = 0, -1 \leq x \leq 1$, and with $\partial u/\partial n = x$ for $y > 0, x^2 + y^2 = 1$. Use the method of finite differences with $h = \frac{1}{2}$. Use the simple scheme described in the text to represent $\partial u/\partial n$.

---

* See, for instance, Widder [1947, p. 195, exercise 15].

15.7 VARIABLE MESH SIZE

Often one wishes to use different mesh sizes in different subregions. It may happen that in most of R the function $u(x, y)$ varies slowly and a coarse mesh would suffice, whereas in one part of R, for example near a corner, the function changes more rapidly and it would be desirable to use a finer mesh. We shall discuss briefly two procedures for changing the mesh size: one involving a variable spacing of mesh lines and the other involving graded nets.

7.1 Fig. Variable Spacing of Mesh Lines.

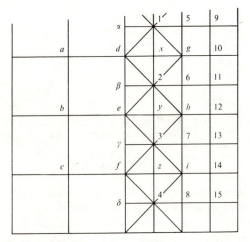

7.2 Fig. The Use of a Graded Net.

An illustration of the procedure of varying the spacing between mesh lines is shown in Fig. 7.1. Here there is no particular complication involved since one can still use central difference formulas as in the case of equal spacing. It should be noted that because of the unequal spacing the degree of approximation of the discrete operators will be one rather than two. This does not in general reduce the overall accuracy but, if the changes in mesh spacing are too abrupt, there may be an adverse effect on the convergence of certain iterative methods. (See Varga [1962, p. 192, Exercise 2.])

With graded nets, one assigns mesh sizes in various subregions of R. Usually the mesh sizes differ by a factor of two. Thus, for example, in Fig. 7.2 we have a case where one mesh size is used in the left-half of the square and a mesh size half as large is used in the right-half. The following procedure can be used to set up the difference equation. (See, for instance, Allen [1954], p. 70.)

The points a, b, c, d, e, f are treated as belonging to the coarse mesh and the appropriate difference equation is used. Points 5, g, 6, h, 7, i, 8 and points to the right are treated as belonging to the fine mesh. The same is true of points x, y, z. For each such point there are four points which are neighbors with respect to the fine mesh. Points 1, 2, 3 and 4 belong to the intermediate mesh. Difference equations for these points are derived based on diagonal neighbors. For example, for point 2 we develop a difference equation based on d, g, e, h and, if needed, 6 and possibly other points. It should be noted that

$$7.3 \qquad \frac{u(x + h, y + h) + u(x + h, y - h) + u(x - h, y + h) + u(x - h, y - h) - 4u(x, y)}{2h^2}$$

$$= u_{xx} + u_{yy} + O(h^2)$$

so that for Poisson's equation the points d, g, e, h would suffice. The points α, β, γ, and δ are discarded. One obtains one linear equation for each mesh point. Moreover, one can verify, at least for Poisson's equation, that the conditions needed to insure the existence of a unique solution of the difference equation are satisfied.

While the use of graded nets can give greater accuracy for a given number of mesh points, nevertheless it has been observed that the numerical solution of the difference equations by iterative methods may be slowed considerably (see, e.g., Young, et al. [1955]). This is because certain properties of the matrix of the system may be lost when one uses graded nets. In particular, one loses symmetry. It is expected that one could modify the choice of the difference equation, perhaps using variational techniques, to obtain a system of equations for which the iterative methods converge rapidly. However, for a problem to be solved on the computer the added complication involved in the use of graded nets make it less attractive than would otherwise be the case. It would seem better to use variable spacing of mesh lines rather than graded nets whenever possible.

EXERCISES 15.7

1. Develop a discrete representation of each of the following differential operators
   a) $L[u] = u_{xx} + 2u_{yy}$

b) $L[u] = u_{xx} + y\,u_y + u_{yy}$

involving the following sets of points:

i) $(x, y), (x + h, y + h), (x + h, y - h), (x - h, y + h), (x - h, y - h)$
ii) same as (i) but add the point $(x + h, y)$.

In each case develop a consistent operator with the highest relative degree of approximation and indicate the absolute degree of approximation.

2. Solve Poisson's equation $u_{xx} + u_{yy} = -2$ in the rectangle $0 < x < 1, 0 < y < \frac{1}{2}$ and zero boundary values, using a mesh size of $\frac{1}{4}$ in the left half of the rectangle and $\frac{1}{8}$ in the right half.

3. Verify that, for the procedure described in Section 15.7 for treating graded nets, the linear system which one obtains with Poisson's equation has properties which guarantee a unique solution of the difference equation.

SUPPLEMENTARY DISCUSSION

We have limited our treatment of methods for solving elliptic partial differential equations to finite difference methods based on the use of Taylor's series. Other methods for deriving difference equations, such as those based on variational methods and integration, can be used, see, e.g., Forsythe and Wasow [1960] and Varga [1962]. The use of finite element methods which are based on variational methods and which allow greater freedom in the choice of mesh points are finding increasing usage. A good discussion of finite element methods is given by Strang [1970].

Section 15.3.
The use of direct methods is becoming more frequent as the size and capacity of computing machines increase. Fox [1962] discusses the use of direct methods for partitioned matrices.

Section 15.4.
It should be noted that the assumption that the solutions should have bounded fourth derivatives in $R + S$ fails in many cases. Among such cases are those involving a corner with an interior angle greater than $180°$, and those where the coefficients or the boundary data are not sufficiently differentiable. In some cases one can subtract out the effects of such "singularities" by approximate analytic methods so that the remaining problem will lead to a well-behaved solution. (See, for example, Gerschgorin [1930].)

Section 15.6.
It is common practice (see, for instance, Allen [1954], Forsythe and Wasow [1960], and Fox [1962]) to replace the boundary S by a polygonal boundary with vertices at mesh points. Some of these vertices may lie outside of the region. For additional discussion on the treatment of normal derivative conditions see Batschelet [1952].

With variable mesh spacing the use of integration methods, as described by Varga [1962, Chapter 6], has many advantages, especially if the coefficients of the differential equation are discontinuous.

CHAPTER 16

ITERATIVE METHODS FOR SOLVING
LARGE LINEAR SYSTEMS

16.1 INTRODUCTION

In this chapter we are concerned with the problem of solving systems of linear algebraic equations of the form

1.1
$$\sum_{j=1}^{N} a_{i,j} u_j = b_i, \qquad i = 1, 2, \ldots, N$$

where the N^2 coefficients $a_{i,j}$ and the N values b_i are given and where $u_1, u_2, \ldots, u_N$ are to be determined. For the case $N = 3$ this system becomes

1.2
$$\begin{cases} a_{1,1}u_1 + a_{1,2}u_2 + a_{1,3}u_3 = b_1 \\ a_{2,1}u_1 + a_{2,2}u_2 + a_{2,3}u_3 = b_2 \\ a_{3,1}u_1 + a_{3,2}u_2 + a_{3,3}u_3 = b_3. \end{cases}$$

One may also write (1.1) as a matrix equation

1.3
$$Au = b$$

where $A = (a_{i,j})$ is an $N \times N$ matrix whose elements are the $a_{i,j}$ and where b is an $N \times 1$ matrix whose elements are the b_i. We seek to determine the $N \times 1$ matrix u whose elements are the u_i. Thus, in the case $N = 3$ we have

1.4
$$\begin{pmatrix} a_{1,1} & a_{1,2} & a_{1,3} \\ a_{2,1} & a_{2,2} & a_{2,3} \\ a_{3,1} & a_{3,2} & a_{3,3} \end{pmatrix} \begin{pmatrix} u_1 \\ u_2 \\ u_3 \end{pmatrix} = \begin{pmatrix} b_1 \\ b_2 \\ b_3 \end{pmatrix}.$$

We shall be primarily concerned here with cases where N is large, say in the range 10^3–10^6 and where A is a "sparse" matrix, i.e., where most of the elements of A vanish. Thus, for example, in Section 15.2 we considered the difference equation

1.5 $\quad a_0 u(x, y) - a_1 u(x + h, y) - a_2 u(x, y + h) - a_3 u(x - h, y) - a_4 u(x, y - h) = t(x, y)$

where

1.6
$$\begin{cases} a_1 = A + \tfrac{1}{2}hD, \qquad a_3 = A - \tfrac{1}{2}hD \\ a_2 = C + \tfrac{1}{2}hE, \qquad a_4 = C - \tfrac{1}{2}hE \\ a_0 = a_1 + a_2 + a_3 + a_4 - h^2F = 2(A + C - \tfrac{1}{2}h^2F) \\ t(x, y) = -h^2G. \end{cases}$$

The diagonal elements $a_{i,i}$ of A correspond to the positive function $a_0(x, y)$, while certain other elements of A correspond to the other functions $a_1(x, y)$, $a_2(x, y)$, $a_3(x, y)$, and $a_4(x, y)$. The b_i correspond to the sum of the function $t(x, y)$ plus terms involving products of values of u at points of S_h adjacent to (x, y) and an appropriate function $a_i(x, y)$.

In the case of the problem given by 15-(2.17) we have, for the case $\gamma = 0$,

1.7
$$\begin{cases} 4u_1 - u_2 - u_3 \quad - u_6 \quad - u_{16} = -h^2G_1 \\ 4u_2 - u_9 - u_4 \quad - u_1 \quad - u_7 \; = -h^2G_2 \\ 4u_3 - u_1 - u_4 \quad - u_{13} - u_{15} = -h^2G_3 \\ 4u_4 - u_{12} - u_{10} - u_3 \quad - u_2 \; = -h^2G_4 \end{cases}$$

or

1.8
$$\begin{pmatrix} 4 & -1 & -1 & 0 \\ -1 & 4 & 0 & -1 \\ -1 & 0 & 4 & -1 \\ 0 & -1 & -1 & 4 \end{pmatrix} \begin{pmatrix} u_1 \\ u_2 \\ u_3 \\ u_4 \end{pmatrix} = \begin{pmatrix} -h^2G_1 + u_6 + u_{16} \\ -h^2G_2 + u_9 + u_7 \\ -h^2G_3 + u_{13} + u_{15} \\ -h^2G_4 + u_{10} + u_{12} \end{pmatrix}.$$

If the number of equations is very large and the matrix is sparse, then the use of iterative methods appears to be much preferable to the use of direct methods such as the Gauss elimination method. With an iterative method one selects an arbitrary initial approximation $u^{(0)}$ to the true solution, $\bar{u}$, of (1.3) and determines a sequence $u^{(1)}, u^{(2)}, \ldots$ according to some algorithm. Under suitable conditions, the sequence $u_i^{(0)}, u_i^{(1)}, \ldots$ converges to $\bar{u}_i$ for all i.

In Section 16.2 we discuss various properties of matrices which frequently hold for problems evolving from elliptic differential equations. Then we show in Section 16.3 that when certain of these properties hold, the system (1.3) has a unique solution. Sections 16.4–16.7 are devoted to a study of the convergence and rates of convergence of various iterative methods for solving (1.3). We shall consider the Jacobi method, the Gauss-Seidel method, the successive overrelaxation method, and the Peaceman-Rachford alternating direction implicit method. Other methods will be discussed briefly in Section 16.8.

16.2 MATRIX PROPERTIES

We shall be concerned with matrices which, in addition to being large and sparse, have other properties which assure us that there exists a unique solution of (1.3) and, moreover, that certain iterative techniques for solving (1.3) can be effectively applied. Some of these properties which will be considered are the following:

2.1 Definition (*Weak diagonal dominance*). A complex matrix $A = (a_{i,j})$ of order N has *weak diagonal dominance* if

2.2
$$|a_{i,i}| \geq \sum_{\substack{j=1 \\ j \neq i}}^{N} |a_{i,j}|$$

for all i and for at least one i

2.3
$$|a_{i,i}| > \sum_{\substack{j=1 \\ j \neq i}}^{N} |a_{i,j}|.$$

2.4 Definition (*Irreducibility*)*. A complex matrix $A = (a_{i,j})$ of order N is *irreducible* if $N = 1$ or if $N > 1$ and given any two nonempty disjoint subsets S and T of W, the set of the first N positive integers, such that $S + T = W$, there exists $i \in S$ and $j \in T$ such that $a_{i,j} \neq 0$.

2.5 Definition (*L-matrix*). A real matrix $A = (a_{i,j})$ of order N is an *L-matrix* if $a_{i,i} > 0, i = 1, 2, \ldots, N$ and $a_{i,j} \leq 0$ if $i \neq j$ for $i,j = 1, 2, \ldots, N$.

Evidently, for h sufficiently small the coefficients a_1, a_2, a_3, and a_4 of (1.6) are positive, and since $F \leq 0$, we have

2.6
$$a_0 \geq a_1 + a_2 + a_3 + a_4.$$

Hence A is an L-matrix which has weak diagonal dominance. (The strict inequality (2.3) holds for any equation corresponding to a point adjacent to the boundary.)

We shall also consider matrices which are *positive definite* (see Definition 11-20.3). From Theorem 11-21.72 we have

2.7 Theorem. A (complex) matrix is positive definite if and only if it is Hermitian and all of its eigenvalues are real and positive.

Because of the following theorem one can give an alternative and perhaps more intuitive definition of irreducibility.

2.8 Theorem. The matrix $A = (a_{i,j})$ of order N is irreducible if and only if

---

* See remarks following Definition 12-(18.16).

$N = 1$ or if $N > 1$ and for any i and j such that $1 \leq i \leq N$, $1 \leq j \leq N$, and $i \neq j$ either $a_{i,j} \neq 0$ or there exist $i_1, i_2, \ldots, i_s$ such that

2.9
$$a_{i,i_1} a_{i_1,i_2} \ldots a_{i_s,j} \neq 0.$$

Proof. If $N = 1$ the matrix is irreducible by definition. If $N > 1$ and if (2.9) holds for any $i \neq j$ and $i \in W, j \in W$, where $W = \{1, 2, \ldots, N\}$, let S and T be any two disjoint non-empty subsets of W such that $S + T = W$. Let i be any element of S and let j be any element of T. Then there exist $i_1, i_2, \ldots, i_s$ such that (2.9) holds. Clearly, at least one of the factors of the left member of (2.9) has the first subscript in S and the second in T. Thus the irreducibility follows.

Next, suppose $N > 1$ and that A is irreducible. Let i and j be any two distinct elements of W such that $a_{i,j} = 0$. Let $S_1 = \{i\}$ and $T_1 = W - S_1$. By the irreducibility there exists $i_1 \in T_1$ such that $a_{i,i_1} \neq 0$. Let $S_2 = \{i, i_1\}$, $T_2 = W - S_2$. Evidently $j \neq i_1$ and hence T_2 is nonempty. If $a_{i_1,j} \neq 0$, then (2.9) holds with $s = 1$. Otherwise, let i_2 be any element of T_2 such that $a_{i,i_2} \neq 0$ or $a_{i_1,i_2} \neq 0$. (Clearly i_2 exists by irreducibility.) Let $S_3 = \{i, i_1, i_2\}$ and $T_3 = W - S_3$. Evidently $j \neq i_2$ and hence T_3 is nonempty. If $a_{i_2,j} \neq 0$, then (2.9) holds with $s = 1$ or $s = 2$. Otherwise we continue this process until we find that (2.9) holds for some s.

The concept of irreducibility has a very simple interpretation in terms of graph theory. Given a matrix $A = (a_{i,j})$ of order N we can construct the associated *directed graph* as follows. Consider any N points in the plane and label them $1, 2, \ldots, N$. For each i and j such that $a_{i,j} \neq 0$ draw an arrow from the point i to the point j. Note that if $a_{i,j} \neq 0$ and $a_{j,i} \neq 0$ there will be one arrow from i to j and one from j to i. (If $a_{i,i} \neq 0$ draw a small loop containing the point i; however, this part of the graph has no effect on the irreducibility.) The matrix is irreducible if and only if the graph is connected in the following sense: given any i and j such that $1 \leq i \leq N$, $1 \leq j \leq N$ and $i \neq j$, there exist $i_1, i_2, \ldots, i_s$ such that there is an arrow from i to i_1, from i_1 to $i_2, \ldots$, from i_s to j.

As an example, consider the matrix

$$A = \begin{pmatrix} 0 & 1 & 0 & 1 \\ 1 & 0 & 1 & 0 \\ 0 & 1 & 0 & 1 \\ 1 & 0 & 0 & 1 \end{pmatrix}.$$

The directed graph is

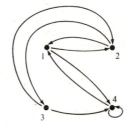

One can easily verify that the graph is connected. Indeed, one can touch all points on the graph by the path 1, 2, 3, 4.

The following matrix, on the other hand, is not irreducible.

$$A = \begin{pmatrix} 1 & 1 & 1 \\ 0 & 0 & 1 \\ 0 & 0 & 1 \end{pmatrix}.$$

Indeed, we have the directed graph

Clearly, we cannot reach 1 from 3. As a matter of fact, there are no paths starting at 3 except the loop containing 3 itself.

As in Chapter 15, we say that the set R_h is *connected* if any two points of R_h can be joined by a continuous path consisting of horizontal and vertical segments joining properly adjacent points of R_h. It is easy to show that if R_h is connected and if each $a_i(x, y)$ appearing in (1.5) is different from zero, then A is irreducible. Let N be the number of points of R_h. If $N = 1$, then by definition A is irreducible. Otherwise, let i and j be any two distinct integers such that $1 \le i \le N, 1 \le j \le N$. Let P and Q be the points of R_h corresponding to i and j, respectively. If P and Q are adjacent, then $a_{i,j} \ne 0$. Otherwise consider a path joining P and Q. Let the mesh points of the path be $P_1, P_2, \ldots, P_s$ and let these points correspond to the i_1-th, i_2-th, $\ldots, i_s$-th equations, respectively. Evidently, since the $a_i \ne 0$ then $a_{i,i_1}, a_{i_1,i_2}, \ldots, a_{i_s,j}$ are all different from zero. Hence (2.9) holds and by Theorem (2.8) the matrix is irreducible.

While the matrix A corresponding to (1.5) is not in general symmetric if the differential equation is self-adjoint and has the form 15-(2.24), we can use the analysis of Section 15.2 and obtain

2.10
$$\begin{cases} a_1 = A(x + \dfrac{h}{2}, y), \qquad a_3 = A(x - \dfrac{h}{2}, y) \\[2ex] a_2 = C(x, y + \dfrac{h}{2}), \qquad a_4 = C(x, y - \dfrac{h}{2}) \\[2ex] a_0 = a_1 + a_2 + a_3 + a_4 - h^2 F \\[1ex] t(x, y) = -h^2 G \end{cases}$$

which corresponds to a real symmetric matrix. As we show in the next section, if A is a real symmetric matrix which is irreducible and has weak diagonal dominance, then A is positive definite.

Exercises 16.2

1. For each of the following matrices

$$A_1 = \begin{pmatrix} 2 & -1 & -1 \\ -1 & 2 & -1 \\ -1 & -1 & 2 \end{pmatrix}, \qquad A_2 = \begin{pmatrix} 2 & -1 & 0 \\ -1 & 2 & -1 \\ 0 & -1 & 2 \end{pmatrix},$$

$$A_3 = \begin{pmatrix} 4 & -1 & -1 & 0 \\ -1 & 4 & 0 & -1 \\ -1 & 0 & 4 & -1 \\ 0 & -1 & -1 & 4 \end{pmatrix}, \qquad A_4 = (0),$$

$$A_5 = (1), \qquad A_6 = \begin{pmatrix} 2 & -1 & 0 \\ -1 & 2 & 0 \\ 0 & 0 & 2 \end{pmatrix},$$

$$A_7 = \begin{pmatrix} 2 & 0 \\ 0 & 1 \end{pmatrix}, \qquad A_8 = \begin{pmatrix} 1 & 0.9 & 0.9 \\ 0.9 & 1 & 0.9 \\ 0.9 & 0.9 & 1 \end{pmatrix},$$

$$A_9 = \begin{pmatrix} 1 & 2 & 2 \\ 2 & 1 & 2 \\ 2 & 2 & 1 \end{pmatrix}, \qquad A_{10} = \begin{pmatrix} 2 & -2 \\ 1 & -1 \end{pmatrix},$$

test whether the matrix is (has)
a) irreducible (construct the directed graph)
b) weak diagonal dominance
c) L-matrix
d) positive definite.

2. Prove Theorem 2.7.

16.3 EXISTENCE OF A UNIQUE SOLUTION

In order to prove that there exists a unique solution of (1.3) it is clearly sufficient to show that $\det A \neq 0$. We prove the following basic result.

3.1 Theorem. If A is an irreducible matrix of order N with weak diagonal dominance, then $\det A \neq 0$.

Proof. We first seek to show that $a_{i,i} \neq 0$ for all i. If $N = 1$, then $|a_{1,1}| > 0$ and $a_{1,1} \neq 0$ by the weak diagonal dominance. Hence det $A \neq 0$. Suppose $N > 1$ and $a_{i,i} = 0$ for some i. Then by the weak diagonal dominance $a_{i,j} = 0$ for all j. Therefore, if $j \neq i$ it is not possible to find $i_1, i_2, \ldots, i_s$ such that (2.9) holds. Hence by Theorem 2.8 the matrix is not irreducible. This contradiction shows that $a_{i,i} \neq 0$ for all i.

If det $A = 0$, then by Theorem A-7.7 there exists a nontrivial solution u of the homogeneous system

$$3.2 \qquad\qquad\qquad Au = 0.$$

Since $a_{i,i} \neq 0$ for all i, we can solve the i-th equation for u_i obtaining

$$3.3 \qquad\qquad u_i = \sum_{j=1}^{N} b_{i,j} u_j, \qquad i = 1, 2, \ldots, N$$

where

$$3.4 \qquad\qquad \begin{cases} b_{i,i} = 0, & \text{if } i = j, \\[2mm] b_{i,j} = -\dfrac{a_{i,j}}{a_{i,i}}, & \text{if } i \neq j. \end{cases}$$

Moreover, by the weak diagonal dominance we have

$$3.5 \qquad\qquad\qquad \sum_{j=1}^{N} |b_{i,j}| \leq 1$$

for all i, and for some i the strict inequality holds.

Since u is a nontrivial solution of (3.2), there exists an M such that $M = \max_i |u_i| > 0$. Let k be any value of i such that $|u_k| = M$. By (3.3) we have

$$3.6 \qquad\qquad\qquad u_k = \sum_{j=1}^{N} b_{k,j} u_j$$

and, by (3.5),

$$3.7 \qquad \sum_{j=1}^{N} |b_{k,j}| \, |u_j| \geq |u_k| \geq \sum_{j=1}^{N} |b_{k,j}| \, |u_k| = \sum_{j=1}^{N} |b_{k,j}| M.$$

Therefore,

$$3.8 \qquad\qquad \sum_{j=1}^{N} |b_{k,j}|(|u_j| - |u_k|) \geq 0.$$

Since $|u_k| \geq |u_j|$ for all j, the above inequality can hold only if $|b_{k,j}|(|u_j| - |u_k|) = 0$ for all j. Thus for each j such that $b_{k,j} \neq 0$ we have $|u_j| = |u_k| = M$.

For any j such that $j \neq k$, by the irreducibility of A, there exist $i_1, i_2, \ldots, i_s$ such that (2.9) holds with $i = k$. Hence $b_{k,i_1}, b_{i_1,i_2}, \ldots, b_{i_s,j}$ are all different from zero. Therefore

3.9
$$M = |u_k| = |u_{i_1}| = |u_{i_2}| = \cdots = |u_j|.$$

Thus $|u_i| = M$ for all i.

However, let i^* be any value of i such that

3.10
$$\sum_{j=1}^{N} |b_{i*,j}| < 1.$$

Then by (3.3) we have

3.11
$$|u_{i*}| \leq \sum_{j=1}^{N} |b_{i*,j}| \, |u_j|.$$

Since $|u_1| = |u_2| = \cdots = |u_N| = M$ we have

3.12
$$M = |u_{i*}| \leq \sum_{j=1}^{N} |b_{i*,j}| M$$

or

3.13
$$1 \leq \sum_{j=1}^{N} |b_{i*,j}|$$

which contradicts (3.10). This completes the proof of Theorem 3.1.

We are now in a position to prove

3.14 Theorem. If A is a real symmetric matrix with non-negative diagonal elements which is irreducible and has weak diagonal dominance, then A is positive definite.

Proof. Since A is real and symmetric, the eigenvalues of A are real. Suppose $\lambda \leq 0$ is an eigenvalue of A. Evidently $A - \lambda I$ is irreducible since A is. Moreover, $A - \lambda I$ has weak diagonal dominance since $\lambda \leq 0$, since A has non-negative diagonal elements and weak diagonal dominance. It therefore follows by Theorem 3.1 that $\det(A - \lambda I) \neq 0$ and hence λ is not an eigenvalue of A. Therefore all eigenvalues of A are positive.

We remark that the assumption that R_h is connected is no real restriction. If R_h is not connected, then we have one independent linear system for each connected component of R_h. A *connected component* of R_h is the set of all points of R_h which can be reached from a given point P by a chain of horizontal and vertical segments of length h connecting adjacent points of R_h. Clearly, the connected components corresponding to any pair of points P and Q are either disjoint or identical. Thus R_h can be divided up into connected components. For each corresponding system, the associated matrix has weak diagonal dominance (i.e., the inequality (2.6) is strict for at least one equation of the form (1.5) associated with each connected component). This is clear since each connected component has at least one point adjacent to a point of S_h.

Exercises 16.3

1. Show that the following matrices are nonsingular

$$A_1 = \begin{pmatrix} 2 & 0 & -2 & 0 \\ 0 & 2 & 0 & -2 \\ -1 & 0 & 2 & 0 \\ 0 & -1 & 0 & 2 \end{pmatrix}, \qquad A_2 = \begin{pmatrix} 3+4i & -3i & -2i \\ -i & 3 & -2i \\ -1 & 0 & 2 \end{pmatrix}.$$

Use Theorem 3.1 where possible. Also compute the determinant of each matrix.

2. Compute the determinant of each of the matrices of Exercise 1, Section 16.2, and verify Theorem 3.1.

3. For each of the symmetric matrices of Exercise 1, Section 16.2, compute the eigenvalues and verify Theorem 3.14.

16.4 ITERATIVE METHODS: THE JACOBI METHOD

Perhaps the simplest iterative method is the Jacobi method. In order that the method can be formally applied we require only that no diagonal element of A should vanish. We have already seen in the proof of Theorem 3.1 that if A is irreducible and has weak diagonal dominance, then this assumption holds. It follows immediately from the definition given in Chapter 11 (see Definition 11-(20.3)) that any positive definite matrix has positive diagonal elements. The condition $a_{i,i} \neq 0$ also holds, of course, when A is an L-matrix.

In order to define the Jacobi method and, in fact, several other methods treated later in this chapter, it is convenient to rewrite (1.1) in the form

4.1
$$u_i = \sum_{j=1}^{N} b_{i,j}u_j + c_i$$

or

4.2
$$u = Bu + c.$$

Here $B = (b_{i,j})$ where

4.3
$$b_{i,j} = \begin{cases} -\dfrac{a_{i,j}}{a_{i,i}}, & \text{if } i \neq j \\ 0, & \text{if } i = j \end{cases}$$

and $c = (c_1, c_2, \ldots, c_N)^T$ where

4.4
$$c_i = \frac{b_i}{a_{i,i}}.$$

Here we have, of course, assumed that none of the $a_{i,i}$ vanishes. In the case $N = 3$ the alternative system becomes

4.5
$$\begin{cases} u_1 = \quad\quad\quad b_{1,2}u_2 + b_{1,3}u_3 + c_1 \\ u_2 = b_{2,1}u_1 \quad\quad\quad + b_{2,3}u_3 + c_2 \\ u_3 = b_{3,1}u_1 + b_{3,2}u_2 \quad\quad\quad + c_3. \end{cases}$$

Evidently, (4.1) and (4.5) are obtained from solving (1.1) and (1.2), respectively, by solving the i-th equation for u_i in terms of the other unknowns.

One may also write (4.1) in the matrix form

4.6
$$u = Bu + c$$

where

4.7
$$B = D^{-1}C, \quad c = D^{-1}b$$

and where

4.8
$$C = D - A, \quad D = \operatorname{diag} A.$$

Here, in general, we let diag A be the diagonal matrix with the same diagonal elements as A. In the case $N = 3$ we have

4.9
$$C = \begin{pmatrix} 0 & -a_{1,2} & -a_{1,3} \\ -a_{2,1} & 0 & -a_{2,3} \\ -a_{3,1} & -a_{3,2} & 0 \end{pmatrix}, \quad D = \begin{pmatrix} a_{1,1} & 0 & 0 \\ 0 & a_{2,2} & 0 \\ 0 & 0 & a_{3,3} \end{pmatrix},$$

4.10
$$B = \begin{pmatrix} 0 & -a_{1,2}/a_{1,1} & -a_{1,3}/a_{1,1} \\ -a_{2,1}/a_{2,2} & 0 & -a_{2,3}/a_{2,2} \\ -a_{3,1}/a_{3,3} & -a_{3,2}/a_{3,3} & 0 \end{pmatrix}, \quad c = \begin{pmatrix} b_1/a_{1,1} \\ b_2/a_{2,2} \\ b_3/a_{3,3} \end{pmatrix},$$

and

4.11
$$\begin{pmatrix} u_1 \\ u_2 \\ u_3 \end{pmatrix} = \begin{pmatrix} 0 & b_{1,2} & b_{1,3} \\ b_{2,1} & 0 & b_{2,3} \\ b_{3,1} & b_{3,2} & 0 \end{pmatrix} \begin{pmatrix} u_1 \\ u_2 \\ u_3 \end{pmatrix} + \begin{pmatrix} c_1 \\ c_2 \\ c_3 \end{pmatrix}.$$

If we write (1.5) in a form corresponding to (4.1) we obtain

4.12 $u(x, y) = \beta_1 u(x + h, y) + \beta_2 u(x, y + h) + \beta_3 u(x - h, y) + \beta_4 u(x, y - h) + \tau(x, y)$

where

4.13
$$\beta_i = \frac{a_i}{a_0}, \qquad i = 1, 2, 3, 4$$

4.14
$$\tau(x, y) = \frac{t(x, y)}{a_0}.$$

If we write (1.8) in the form corresponding to (4.6) we obtain

4.15
$$\begin{pmatrix} u_1 \\ u_2 \\ u_3 \\ u_4 \end{pmatrix} = \begin{pmatrix} 0 & \frac{1}{4} & \frac{1}{4} & 0 \\ \frac{1}{4} & 0 & 0 & \frac{1}{4} \\ \frac{1}{4} & 0 & 0 & \frac{1}{4} \\ 0 & \frac{1}{4} & \frac{1}{4} & 0 \end{pmatrix} \begin{pmatrix} u_1 \\ u_2 \\ u_3 \\ u_4 \end{pmatrix} + \frac{1}{4} \begin{pmatrix} -h^2 G_1 + u_6 + u_{16} \\ -h^2 G_2 + u_7 + u_9 \\ -h^2 G_3 + u_{13} + u_{15} \\ -h^2 G_4 + u_{10} + u_{12} \end{pmatrix}.$$

We are now ready to define the Jacobi method. Starting with arbitrary initial values $u_1^{(0)}, u_2^{(0)}, \ldots, u_N^{(0)}$ we compute $u_1^{(1)}, u_2^{(1)}, \ldots, u_N^{(1)}, u_1^{(2)}, u_2^{(2)}, \ldots, u_N^{(2)}$, etc., using the formula

4.16
$$u_i^{(n+1)} = \sum_{j=1}^{N} b_{i,j} u_j^{(n)} + c_i, \qquad i = 1, 2, \ldots, N.$$

In the case $N = 3$, this becomes

4.17
$$\begin{cases} u_1^{(n+1)} = & b_{1,2} u_2^{(n)} + b_{1,3} u_3^{(n)} + c_1 \\ u_2^{(n+1)} = b_{2,1} u_1^{(n)} & + b_{2,3} u_3^{(n)} + c_2 \\ u_3^{(n+1)} = b_{3,1} u_1^{(n)} + b_{3,2} u_2^{(n)} & + c_3. \end{cases}$$

We can write the Jacobi method in the matrix form

4.18
$$u^{(n+1)} = B u^{(n)} + c$$

which in the case $N = 3$ becomes

4.19
$$\begin{pmatrix} u_1^{(n+1)} \\ u_2^{(n+1)} \\ u_3^{(n+1)} \end{pmatrix} = \begin{pmatrix} 0 & b_{1,2} & b_{1,3} \\ b_{2,1} & 0 & b_{2,3} \\ b_{3,1} & b_{3,2} & 0 \end{pmatrix} \begin{pmatrix} u_1^{(n)} \\ u_2^{(n)} \\ u_3^{(n)} \end{pmatrix} + \begin{pmatrix} c_1 \\ c_2 \\ c_3 \end{pmatrix}.$$

For the difference equation (4.12) we have

4.20 $u^{(n+1)}(x, y) = \beta_1 u^{(n)}(x + h, y) + \beta_2 u^{(n)}(x, y + h) + \beta_3 u^{(n)}(x - h, y)$
$$+ \beta_4 u^{(n)}(x, y - h) + \tau(x, y).$$

Let us now apply the Jacobi method to the system (1.8) with $G_1 = G_2 = G_3 = G_4 = 0$ and the boundary conditions $u_{12} = u_{13} = 1000$ with all other boundary values zero. Then (1.8) becomes

4.21
$$\begin{pmatrix} 4 & -1 & -1 & 0 \\ -1 & 4 & 0 & -1 \\ -1 & 0 & 4 & -1 \\ 0 & -1 & -1 & 4 \end{pmatrix} \begin{pmatrix} u_1 \\ u_2 \\ u_3 \\ u_4 \end{pmatrix} = \begin{pmatrix} 0 \\ 0 \\ 1000 \\ 1000 \end{pmatrix}.$$

Moreover, (4.15) becomes

4.22
$$\begin{pmatrix} u_1 \\ u_2 \\ u_3 \\ u_4 \end{pmatrix} = \begin{pmatrix} 0 & \frac{1}{4} & \frac{1}{4} & 0 \\ \frac{1}{4} & 0 & 0 & \frac{1}{4} \\ \frac{1}{4} & 0 & 0 & \frac{1}{4} \\ 0 & \frac{1}{4} & \frac{1}{4} & 0 \end{pmatrix} \begin{pmatrix} u_1 \\ u_2 \\ u_3 \\ u_4 \end{pmatrix} + \begin{pmatrix} 0 \\ 0 \\ 250 \\ 250 \end{pmatrix},$$

and from (4.18) we have

4.23
$$\begin{cases} u_1^{(n+1)} = & \frac{1}{4} u_2^{(n)} + \frac{1}{4} u_3^{(n)} \\ u_2^{(n+1)} = \frac{1}{4} u_1^{(n)} & + \frac{1}{4} u_4^{(n)} \\ u_3^{(n+1)} = \frac{1}{4} u_1^{(n)} & + \frac{1}{4} u_4^{(n)} + 250 \\ u_4^{(n+1)} = & \frac{1}{4} u_2^{(n)} + \frac{1}{4} u_3^{(n)} + 250. \end{cases}$$

If we let $u_1^{(0)} = u_2^{(0)} = u_3^{(0)} = u_4^{(0)} = 0$, then we get the following values.

n	$u_1^{(n)}$	$e_1^{(n)}$	$u_2^{(n)}$	$e_2^{(n)}$	$u_3^{(n)}$	$e_3^{(n)}$	$u_4^{(n)}$	$e_4^{(n)}$
0	0.	-125.0000	0.	-125.0000	0.	-375.0000	0.0000	-375.0000
1	0.	-125.0000	0.	-125.0000	250.0000	-125.0000	250.0000	-125.0000
2	62.5000	-62.5000	62.5000	-62.5000	312.5000	-62.5000	312.5000	-62.5000
3	93.7500	-31.2500	93.7500	-31.2500	343.7500	-31.2500	343.7500	-31.2500
4	109.3750	-15.6250	109.3750	-15.6250	359.3750	-15.6250	359.3750	-15.6250
5	117.1875	-7.8125	117.1875	-7.8125	367.1875	-7.8125	367.1875	-7.8125
6	121.0937	-3.9063	121.0937	-3.9063	371.0937	-3.9063	371.0937	-3.9063
7	123.0469	-1.9531	123.0469	-1.9531	373.0469	-1.9531	373.0469	-1.9531
8	124.0234	$-.9766$	124.0234	$-.9766$	374.0234	$-.9766$	374.0234	$-.9766$
9	124.5117	$-.4883$	124.5117	$-.4883$	374.5117	$-.4883$	374.5117	$-.4883$
10	124.7559	$-.2441$	124.7559	$-.2441$	374.7559	$-.2441$	374.7559	$-.2441$
11	124.8779	$-.1221$	124.8779	$-.1221$	374.8779	$-.1221$	374.8779	$-.1221$
12	124.9390	$-.0610$	124.9390	$-.0610$	374.9390	$-.0610$	374.9390	$-.0610$
13	124.9695	$-.0305$	124.9695	$-.0305$	374.9695	$-.0305$	374.9695	$-.0305$
14	124.9847	$-.0153$	124.9847	$-.0153$	374.9847	$-.0153$	374.9847	$-.0153$
15	124.9924	$-.0076$	124.9924	$-.0076$	374.9924	$-.0076$	374.9924	$-.0076$
16	124.9962	$-.0038$	124.9962	$-.0038$	374.9962	$-.0038$	374.9962	$-.0038$
17	124.9981	$-.0019$	124.9981	$-.0019$	374.9981	$-.0019$	374.9981	$-.0019$
18	124.9990	$-.0010$	124.9990	$-.0010$	374.9990	$-.0010$	374.9990	$-.0010$
19	124.9995	$-.0005$	124.9995	$-.0005$	374.9995	$-.0005$	374.9995	$-.0005$
20	124.9998	$-.0002$	124.9998	$-.0002$	374.9998	$-.0002$	374.9998	$-.0002$
21	124.9999	$-.0001$	124.9999	$-.0001$	374.9999	$-.0001$	374.9999	$-.0001$
22	124.9999	$-.0001$	124.9999	$-.0001$	374.9999	$-.0001$	374.9999	$-.0001$
23	125.0000	$-.0000$	125.0000	$-.0000$	375.0000	$-.0000$	375.0000	$-.0000$

Here we have given values of $e_i^{(n)} = u_i^{(n)} - \bar{u}$, where $\bar{u}$ is the true solution of (4.22), as well as values of $u_i^{(n)}$. Evidently the method is converging quite rapidly. As a matter of fact, the errors are being reduced by a factor of approximately $\frac{1}{2}$ after each iteration.

We also note that the largest eigenvalue of the matrix B corresponding to the Jacobi method is $\frac{1}{2}$. Thus, one can verify directly that the eigenvalues of

4.24
$$
B = \begin{pmatrix}
0 & \frac{1}{4} & \frac{1}{4} & 0 \\
\frac{1}{4} & 0 & 0 & \frac{1}{4} \\
\frac{1}{4} & 0 & 0 & \frac{1}{4} \\
0 & \frac{1}{4} & \frac{1}{4} & 0
\end{pmatrix}
$$

are $\frac{1}{2}, -\frac{1}{2}, 0, 0$. Thus the *spectral radius* $S(B)$ of B, that is, the modulus of the eigenvalue of largest modulus, is $\frac{1}{2}$.

We study the convergence of the Jacobi method as a special case of the somewhat more general iterative method

4.25 $$u^{(n+1)} = Gu^{(n)} + k$$

where G is a matrix and k is a vector such that

4.26 $$k = (I - G)A^{-1}b.$$

(Here we assume that A is nonsingular.)

4.27 Definition. The method (4.25) is *convergent* if for some $\hat{u}$ the sequence $u^{(0)}, u^{(1)}, u^{(2)}, \ldots$ generated by (4.25) converges to $\hat{u}$ for all $u^{(0)}$.

4.28 Theorem. If the method (4.25) is convergent, then

4.29 $$S(G) < 1$$

and for all $u^{(0)}$ the sequence $u^{(0)}, u^{(1)}, \ldots$ generated by (4.25) converges to the unique solution $\bar{u}$ of (1.3). On the other hand, if (4.29) holds then the method is convergent and the sequence $u^{(0)}, u^{(1)}, \ldots$ generated by (4.25) converges to $\bar{u}$ for all $u^{(0)}$.

Proof. If the method is convergent, then by taking limits* of both sides of (4.25) we have

4.30 $$\hat{u} = G\hat{u} + k$$

where $\hat{u}$ is the vector such that the sequence $u^{(0)}, u^{(1)}, \ldots$ generated by (4.25) converges to $\hat{u}$ for all $u^{(0)}$. Evidently

4.31 $$u^{(n+1)} - \hat{u} = G(u^{(n)} - \hat{u})$$

and

4.32 $$u^{(n)} - \hat{u} = G^n(u^{(0)} - \hat{u}).$$

We now prove

4.33 Lemma. For a given matrix G we have

4.34 $$\lim_{n \to \infty} G^n v = 0$$

for all v if and only if (4.29) holds.

Proof. If (4.29) holds, then for some natural matrix norm $\|\cdot\|_\beta$ we have $\|G\|_\beta < 1$ by Theorem 11–29.2. (See the footnote immediately preceding 11-(25.6) for the definition of a natural matrix norm.) For the vector norm $\|\cdot\|_\alpha$ corresponding to β we thus have

---

* It is easy to show using the methods of Chapter 11 that if $u^{(n)} \to \hat{u}$ then $Gu^{(n)} \to G\hat{u}$.

4.35
$$\|G^n v\|_\alpha \leq \|G^n\|_\beta \|v\|_\alpha$$
$$\leq \|G\|_\beta^n \|v\|_\alpha$$

which converges to zero as $n \to \infty$. Thus (4.34) holds.

If, on the other hand, (4.34) holds for all v then, as one can easily show,

4.36
$$\lim_{n \to \infty} G^n = 0.$$

But by Theorem 11-32.8 this implies (4.29), and the lemma follows.

If the method is convergent, then by (4.32) and Lemma 4.33 we must have (4.29). In that case, by (4.26) and (4.30) we have $\hat{u} = \bar{u}$.

On the other hand, if (4.29) holds, then $I - G$ is nonsingular and the equation

4.37
$$u = Gu + k$$

has a unique solution, say $\hat{u}$. Hence (4.32) holds. By Lemma (4.33), it follows that $u^{(n)} \to \hat{u}$ for any $u^{(0)}$. By (4.26) it follows that $\hat{u} = \bar{u}$ and the proof of Theorem 4.28 is complete.

Suppose we have an iterative method of the form (4.25) such that (4.29) holds. It can be shown* that

4.38
$$\lim_{n \to \infty} \|G^n\|_2^{1/n} = S(G).$$

Hence we have

4.39
$$\|u^{(n)} - \bar{u}\|_2 = \|G^n(u^{(0)} - \bar{u})\|_2$$
$$\leq \|G^n\|_2 \|u^{(0)} - \bar{u}\|_2$$
$$\sim S(G)^n \|u^{(0)} - \bar{u}\|_2.$$

Thus in some sense the spectral radius of G gives a measure of the rapidity of the convergence.

It is easy to show that if A is irreducible and has weak diagonal dominance, then the Jacobi method converges. If μ is an eigenvalue of B then

4.40
$$\det(B - \mu I) = \det(D^{-1} C - \mu I) = 0.$$

If $|\mu| \geq 1$, then this implies that

4.41
$$\det(D - \frac{1}{\mu} C) = 0.$$

---

* See, for instance, Varga [1962, Chapter 3] or Young [1971a, Chapter 3].

Evidently, if A has weak diagonal dominance, then so does $D - \mu^{-1}C$. Moreover, $D - \mu^{-1}C$ is irreducible if A is. Hence $\det(D - \mu^{-1}C) \neq 0$ by Theorem 3.1. This contradiction shows that if $|\mu| \geq 1$, then μ is not an eigenvalue of B. Hence $S(B) < 1$ and the Jacobi method converges.

We remark that even though A is positive definite, the Jacobi method need not converge. However, if A is also an L-matrix, then $S(B) < 1$. (See, for instance, Varga [1962] or Young [1971a].)

Let us now seek to estimate the spectral radius of the Jacobi method for the problem of solving the usual 5-point discrete analog of the Dirichlet problem for the unit square. Following Varga [1962] we shall refer to this as the *model problem*. The iteration equation is

4.42 $u^{(n+1)}(x, y) = \frac{1}{4}[u^{(n)}(x + h, y) + u^{(n)}(x, y + h) + u^{(n)}(x - h, y) + u^{(n)}(x, y - h)]$

where $u^{(n)}(x, y) = \bar{u}(x, y)$ on S_h and $\bar{u}(x, y)$ is the true solution of the model problem. Letting $e^{(n)}(x, y) = u^{(n)}(x, y) - \bar{u}(x, y)$ we have

4.43 $e^{(n+1)}(x, y) = \frac{1}{4}[e^{(n)}(x + h, y) + e^{(n)}(x, y + h) + e^{(n)}(x - h, y) + e^{(n)}(x, y - h)]$

where $e^{(n)}(x, y) = 0$ on S_h. Evidently an eigenvector of B associated with an eigenvalue μ corresponds to a function $v(x, y)$ vanishing on S_h such that on R_h we have

4.44 $\mu v(x, y) = \frac{1}{4}[v(x + h, y) + v(x, y + h) + v(x - h, y) + v(x, y - h)].$

If we assume a solution of the form

$$v(x, y) = \sin p\pi x \sin q\pi y$$

where p and q are positive integers, we have

4.45 $\mu = \mu_{p,q} = \frac{1}{2}(\cos p\pi h + \cos q\pi h).$

Thus for $p, q = 1, 2, \ldots, M - 1$, where $h = M^{-1}$, we obtain eigenvalues $\mu = \mu_{p,q}$ as given by (4.45). Moreover,* one can show that the $(M - 1)^2$ vectors $v_{p,q}$ corresponding to the functions

4.46 $v_{p,q}(x, y) = \sin p\pi x \sin q\pi y$

---

*It is easy to show that $\Sigma_{i=1}^{M-1} \sin p\pi i h \sin p'\pi i h = 0$ if $p \neq p'$ and if $h = M^{-1}$ and $1 \leq p \leq M - 1, 1 \leq p' \leq M - 1$. Hence $\sigma = \Sigma_{R_h} \sin p\pi x \sin q\pi y \sin p'\pi x \sin q'\pi y = 0$ unless $p = p'$ and $q = q'$. On the other hand, σ does not vanish if $p = p'$ and $q = q'$. If the functions $v_{p,q}$ were linearly dependent, there would exist constants $\{c_{p,q}\}$ not all zero such that $\Sigma_{p,q=1}^{M-1} c_{p,q} v_{p,q} = 0$. Multiplying by $\sin p'\pi x \sin q'\pi y$ and summing over R_h we get $c_{p',q'} = 0$ for all p' and q'. This contradiction shows that the $v_{p,q}$ are linearly independent.

are linearly independent.   Hence the $v_{p,q}$ form a basis for the vector space of functions defined on R_h and vanishing on S_h.

From (4.45) the largest eigenvalue, say $\bar{\mu}$ of B is given by

4.47
$$\bar{\mu} = S(B) = \cos \pi h.$$

We are now able, using (4.39), to estimate the number of iterations required to reduce $\|e^{(n)}\|_2$ to a specified fraction, say ρ, of $\|e^{(0)}\|_2$. By (4.39) and (4.47) we have, approximately

4.48
$$S(B)^n = \rho$$

or

4.49
$$n = \frac{-\log \rho}{R(B)} = \frac{-\log \rho}{-\log \cos \pi h} \doteq \frac{-2 \log \rho}{\pi^2 h^2}$$

for small h. Here, for convenience, we let $R(B)$ denote the (*asymptotic*) *rate of convergence* of the Jacobi method defined by

4.50
$$R(B) = -\log S(B).$$

Thus the number of iterations required for convergence increases as h^{-2}. For example, if $\rho = 10^{-6}$ we have

$$n \doteq \frac{27.6}{\pi^2 h^2} \doteq \frac{2.80}{h^2}.$$

If $h = \frac{1}{20}$, we have $n = 1120$, while if $h = \frac{1}{40}$, then $n = 4480$. Dividing h by 2, multiplies n by 4.

Exercises 16.4

1. Carry out the Jacobi method for solving

$$\begin{pmatrix} 4 & -1 & 0 \\ -1 & 4 & -1 \\ 0 & -1 & 4 \end{pmatrix} \begin{pmatrix} u_1 \\ u_2 \\ u_3 \end{pmatrix} = \begin{pmatrix} 2000 \\ 0 \\ 2000 \end{pmatrix}.$$

Carry out the iteration process until all of the values $|u_i^{(n+1)} - u_i^{(n)}|$ are less than 0.5. Use zero starting values.

2. Write down the matrix B corresponding to the Jacobi method of iteration for the model problem with $h = \frac{1}{4}$. Find the largest eigenvalue, $\bar{\mu}$, and an associated eigenvector v. Verify that $Bv = \bar{\mu}v$.

3. Show that the Jacobi method converges when applied to the linear system $Au = b$ where

$$A = \begin{pmatrix} 4i & -2 & -i & -1 \\ -3 & 4 & -i & 0 \\ 0 & -2i & 4 & -2 \\ 1+i & 0 & 1 & 4i \end{pmatrix}.$$

4. For each of the matrices of Exercise 1, Section 16.2, compute $S(B)$. Is it true that $S(B) < 1$ for every one of the matrices which is positive definite?

5. Show that the Jacobi method for solving $Au = b$ converges if A and $2D - A$ are positive definite, where $D = \text{diag } A$. Show that the eigenvalues of B are real and less than unity if A is positive definite, even though $2D - A$ is not necessarily positive definite.

6. Estimate the number of iterations required using the Jacobi method to solve the model problem with $h = \frac{1}{10}$. Assume that the initial error is reduced by a factor of 10^{-6}.

7. Show that

$$\sum_{i=1}^{M-1} \sin p\pi ih \sin p'\pi ih = 0$$

for $p \neq p'$ if $h = M^{-1}$ and $1 \leqq p \leqq M - 1$, $1 \leqq p' \leqq M - 1$.
Hint: Use the identity $\sin \alpha \sin \beta = \frac{1}{2}\cos(\alpha - \beta) - \frac{1}{2}\cos(\alpha + \beta)$.

8. Prove that if $v^{(n)} \to v$ as $n \to \infty$ then $Gv^{(n)} \to Gv$ for any matrix G.

16.5 THE GAUSS-SEIDEL METHOD

The convergence of the Jacobi method is frequently very slow. For the model problem, for example, the number of iterations required for convergence increases as h^{-2} as h decreases. We now consider a simple modification which yields somewhat faster convergence but not an order-of-magnitude improvement. In the next section we shall show how a further modification will lead to an order-of-magnitude improvement in the convergence rate.

The Gauss-Seidel method is the same as the Jacobi method except that in determining $u_i^{(n+1)}$ one uses $u_{i-1}^{(n+1)}, u_{i-2}^{(n+1)}, \ldots, u_1^{(n+1)}$ instead of $u_{i-1}^{(n)}, u_{i-2}^{(n)}, \ldots, u_1^{(n)}$, respectively, in (4.16). Thus, we have

5.1
$$u_i^{(n+1)} = \sum_{j=1}^{i-1} b_{i,j} u_j^{(n+1)} + \sum_{j=i+1}^{N} b_{i,j} u_j^{(n)} + c_i.$$

In the case $N = 3$, this becomes

5.2
$$\begin{cases} u_1^{(n+1)} = \qquad\qquad\quad b_{1,2}u_2^{(n)} + b_{1,3}u_3^{(n)} + c_1 \\ u_2^{(n+1)} = b_{2,1}u_1^{(n+1)} \qquad\qquad + b_{2,3}u_3^{(n)} + c_2 \\ u_3^{(n+1)} = b_{3,1}u_1^{(n+1)} + b_{3,2}u_2^{(n+1)} \qquad\quad + c_3. \end{cases}$$

For the difference equation (4.12) if we take the points in the "natural ordering" where (x, y) follows (x', y') if $y > y'$ or else $y = y'$ and $x > x'$ we have

5.3 $u^{(n+1)}(x, y) = \beta_1 u^{(n)}(x + h, y) + \beta_2 u^{(n)}(x, y + h) + \beta_3 u^{(n+1)}(x - h, y)$

$$+ \beta_4 u^{(n+1)}(x, y - h) + \tau(x, y).$$

If we apply the Gauss-Seidel method to the system (4.22) we have

5.4 $\begin{cases} u_1^{(n+1)} = & \frac{1}{4}u_2^{(n)} + \frac{1}{4}u_3^{(n)} \\[4pt] u_2^{(n+1)} = \frac{1}{4}u_1^{(n+1)} & + \frac{1}{4}u_4^{(n)} \\[4pt] u_3^{(n+1)} = \frac{1}{4}u_1^{(n+1)} & + \frac{1}{4}u_4^{(n)} + 250 \\[4pt] u_4^{(n+1)} = & \frac{1}{4}u_2^{(n+1)} + \frac{1}{4}u_3^{(n+1)} + 250. \end{cases}$

Letting $u_1^{(0)} = u_2^{(0)} = u_3^{(0)} = u_4^{(0)} = 0$ we get the following results:

n	$u_1^{(n)}$	$e_1^{(n)}$	$u_2^{(n)}$	$e_2^{(n)}$	$u_3^{(n)}$	$e_3^{(n)}$	$u_4^{(n)}$	$e_4^{(n)}$
0	0.	−125.0000	0.	−125.0000	0.	−375.0000	0.	−375.0000
1	0.	−125.0000	0.	−125.0000	250.0000	−125.0000	312.5000	−62.5000
2	62.5000	−62.5000	93.7500	−31.2500	343.7500	−31.2500	359.3750	−15.6250
3	109.3750	−15.6250	117.1875	−7.8125	367.1875	−7.8125	371.0937	−3.9063
4	121.0937	−3.9063	123.0469	−1.9531	373.0469	−1.9531	374.0234	−.9766
5	124.0234	−.9766	124.5117	−.4883	374.5117	−.4883	374.7559	−.2441
6	124.7559	−.2441	124.8779	−.1221	374.8779	−.1221	374.9390	−.0610
7	124.9390	−.0610	124.9695	−.0305	374.9695	−.0305	374.9847	−.0153
8	124.9847	−.0153	124.9924	−.0076	374.9924	−.0076	374.9962	−.0038
9	124.9962	−.0038	124.9981	−.0019	374.9981	−.0019	374.9990	−.0010
10	124.9990	−.0010	124.9995	−.0005	374.9995	−.0005	374.9998	−.0002
11	124.9998	−.0002	124.9999	−.0001	374.9999	−.0001	374.9999	−.0001
12	124.9999	−.0001	125.0000	.0000	375.0000	.0000	375.0000	.0000
13	125.0000	.0000	125.0000	.0000	375.0000	.0000	375.0000	.0000

Evidently the Gauss-Seidel method is converging much more rapidly than the Jacobi method. In fact, the errors are being reduced by a factor of approximately 4 after each iteration, as compared with a factor of approximately 2 with the Jacobi method.

To study the convergence properties of the Gauss-Seidel method we write the method in the matrix form

5.5 $u^{(n+1)} = Lu^{(n+1)} + Uu^{(n)} + c$

where L and U are strictly lower triangular and strictly upper triangular matrices such that

5.6 $L + U = B.$

Thus in the case $N = 3$ we have

5.7 $L = \begin{pmatrix} 0 & 0 & 0 \\ b_{2,1} & 0 & 0 \\ b_{3,1} & b_{3,2} & 0 \end{pmatrix}, \qquad U = \begin{pmatrix} 0 & b_{1,2} & b_{1,3} \\ 0 & 0 & b_{2,3} \\ 0 & 0 & 0 \end{pmatrix}.$

Hence in this case (5.5) is equivalent to

$$\begin{pmatrix} u_1^{(n+1)} \\ u_2^{(n+1)} \\ u_3^{(n+1)} \end{pmatrix} = \begin{pmatrix} 0 & 0 & 0 \\ b_{2,1} & 0 & 0 \\ b_{3,1} & b_{3,2} & 0 \end{pmatrix}\begin{pmatrix} u_1^{(n+1)} \\ u_2^{(n+1)} \\ u_3^{(n+1)} \end{pmatrix} + \begin{pmatrix} 0 & b_{1,2} & b_{1,3} \\ 0 & 0 & b_{2,3} \\ 0 & 0 & 0 \end{pmatrix}\begin{pmatrix} u_1^{(n)} \\ u_2^{(n)} \\ u_3^{(n)} \end{pmatrix}$$

$$+ \begin{pmatrix} c_1 \\ c_2 \\ c_3 \end{pmatrix} = \begin{pmatrix} b_{1,2}u_2^{(n)} + b_{1,3}u_3^{(n)} + c_1 \\ b_{2,1}u_1^{(n+1)} + b_{2,3}u_3^{(n)} + c_2 \\ b_{3,1}u_1^{(n+1)} + b_{3,2}u_2^{(n+1)} + c_3 \end{pmatrix},$$

which agrees with (5.2).

We note that $I - L$ is a lower triangular matrix with ones in the main diagonal. Hence, $\det(I - L) = 1$ and $I - L$ is nonsingular. Thus, we can solve (5.5) for $u^{(n+1)}$ obtaining

5.8 $u^{(n+1)} = \mathscr{L}u^{(n)} + (I - L)^{-1}c$

where

5.9 $\mathscr{L} = (I - L)^{-1}U.$

Since by (4.2) the true solution $\bar{u}$ of the original system (1.3) satisfies

5.10 $\bar{u} = B\bar{u} + c = L\bar{u} + U\bar{u} + c$

we have

5.11 $\bar{u} = (I - L)^{-1}U\bar{u} + (I - L)^{-1}c = \mathscr{L}\bar{u} + (I - L)^{-1}c.$

Hence, if $e^{(n)} = u^{(n)} - \bar{u}$ we obtain

5.12 $e^{(n+1)} = \mathscr{L}e^{(n)}, \qquad e^{(n)} = \mathscr{L}^n e^{(0)}.$

Thus the convergence of the Gauss-Seidel method depends on the eigenvalues of $\mathscr{L}$

in the same way that the convergence of the Jacobi method depends on the eigen-values of B.

We now prove

5.13 Theorem. If A is irreducible and has weak diagonal dominance then $S(\mathscr{L}) < 1$.

Proof. If λ is an eigenvalue of $\mathscr{L}$ then

5.14
$$
\begin{aligned}
\det(\mathscr{L} - \lambda I) &= \det((I - L)^{-1}U - \lambda I) \\
&= \det((I - L)^{-1}[U - \lambda(I - L)]) \\
&= \det[(I - L)^{-1}]\det(U - \lambda(I - L)) \\
&= \det(\lambda L + U - \lambda I) = 0.
\end{aligned}
$$

If $|\lambda| \geq 1$, then $\lambda \neq 0$ and

5.15
$$
\det(I - \frac{1}{\lambda}U - L) = 0.
$$

Evidently the matrix $I - \lambda^{-1}U - L$ has weak diagonal dominance since the sum of the absolute values of the off-diagonal elements in the i-th row does not exceed $\sum_{j=1, j\neq i}^{N}|b_{i,j}|$. Since the matrix $I - \lambda^{-1}U - L$ is irreducible since A is, it follows by Theorem 3.1 that it is nonsingular and $\det(\mathscr{L} - \lambda I)$ cannot vanish. Hence λ is not an eigenvalue of $\mathscr{L}$. This contradiction proves that if λ is an eigenvalue of $\mathscr{L}$ then $|\lambda| < 1$.

An alternative sufficient condition for the convergence of the Gauss-Seidel method is given by the following theorem.

5.16 Theorem. If A is a real positive definite matrix, then $S(\mathscr{L}) < 1$.

Proof. Suppose λ is an eigenvalue of $\mathscr{L}$. Then for some vector $v \neq 0$ we have $\mathscr{L}v = \lambda v$ and $(I - L)^{-1}Uv = \lambda v$, or,

5.17
$$
(\lambda L + U)v = \lambda v.
$$

Taking the inner product of both sides of the above equation with Dv, where $D = \text{diag } A$, and solving for λ we have

5.18
$$
\lambda = \frac{(Uv, Dv)}{(v, Dv) - (Lv, Dv)}.
$$

Since A is symmetric, then $A = D - C_L - C_U$ where C_L and C_U are strictly lower and strictly upper triangular matrices, respectively, such that $C_L^T = C_U$. Moreover, since $L = D^{-1}C_L$, $U = D^{-1}C_U$ we have

5.19 $(Lv, Dv) = (v, L^T Dv) = (v, C_L^T v) = (v, C_U v) = (\overline{C_U v, v}) = (\overline{DUv, v}) = (\overline{Uv, Dv})$

so that

5.20
$$\lambda = \frac{z}{1 - \bar{z}}$$

where

5.21
$$z = \frac{(Uv, Dv)}{(v, Dv)}.$$

Evidently,

5.22
$$|\lambda|^2 = \lambda\bar{\lambda} = \frac{|z|^2}{1 + |z|^2 - (z + \bar{z})}$$

and

5.23 $1 - (z + \bar{z}) = 1 - \dfrac{((L + U)v, Dv)}{(v, Dv)} = \dfrac{((I - (L + U))v, Dv)}{(v, Dv)}$

$$= \frac{((D - DL - DU)v, v)}{(v, Dv)} = \frac{(Av, v)}{(v, Dv)} > 0$$

since A and D are both positive definite and, hence, $(Av, v) > 0$ and $(v, Dv) = (Dv, v) > 0$ for all $v \neq 0$. From this it follows that $1 - (z + \bar{z}) > 0$ and $|\lambda|^2 < 1$. Therefore, $S(\mathscr{L}) < 1$ and the proof of Theorem 5.16 is complete.

If P is any nonsingular matrix then, as can be easily verified, the function

5.24
$$\|PAP^{-1}\|_2$$

has all the properties required of a matrix norm, and we let

5.25
$$\|A\|_P = \|PAP^{-1}\|_2.$$

Similarly we can define the vector norm

5.26
$$\|v\|_P = \|Pv\|_2.$$

Moreover

5.27
$$\|A\|_P = \sup_{v \neq 0} \frac{\|Av\|_P}{\|v\|_P}.$$

An alternative proof of the convergence of the Gauss-Seidel method for the case where A is a real positive definite matrix can be given in terms of the $A^{\frac{1}{2}}$-norm of $\mathscr{L}$ which is given by

5.28
$$\|\mathscr{L}\|_{A^{\frac{1}{2}}} = \|A^{\frac{1}{2}}\mathscr{L}A^{-\frac{1}{2}}\|_2.$$

Here $A^{\frac{1}{2}}$ is the unique positive definite matrix whose square is A. (For the existence and uniqueness of $A^{\frac{1}{2}}$ see Theorem 11-21.77.) Evidently, we have

5.29
$$\mathscr{L} = I - (I - L)^{-1}(I - L - U) = I - (I - L)^{-1}D^{-1}A$$
$$= I - (D - C_L)^{-1}A.$$

Therefore,

5.30
$$\mathscr{L}' = A^{\frac{1}{2}}\mathscr{L}A^{-\frac{1}{2}} = I - A^{\frac{1}{2}}(D - C_L)^{-1}A^{\frac{1}{2}}$$
$$(\mathscr{L}')^T = I - A^{\frac{1}{2}}(D - C_U)^{-1}A^{\frac{1}{2}}.$$

Hence

5.31 $\mathscr{L}'(\mathscr{L}')^T = I - A^{\frac{1}{2}}(D - C_L)^{-1}\{D - C_U + D - C_L - A\}(D - C_U)^{-1}A^{\frac{1}{2}}$
$$= I - A^{\frac{1}{2}}(D - C_L)^{-1}D(D - C_U)^{-1}A^{\frac{1}{2}}$$
$$= I - [A^{\frac{1}{2}}(D - C_L)^{-1}D^{\frac{1}{2}}][A^{\frac{1}{2}}(D - C_L)^{-1}D^{\frac{1}{2}}]^T.$$

Since $I - \mathscr{L}'(\mathscr{L}')^T$ is the product of a nonsingular matrix times its transpose, it follows that $I - \mathscr{L}'(\mathscr{L}')^T$ is positive definite and all eigenvalues of $\mathscr{L}'(\mathscr{L}')^T$ are less than unity. Since all such eigenvalues are non-negative it follows that

5.32
$$S(\mathscr{L}'(\mathscr{L}')^T) = \|\mathscr{L}\|_{A^{\frac{1}{2}}}^2 < 1$$

and hence $\|\mathscr{L}\|_{A^{\frac{1}{2}}} < 1$. Therefore, by Theorem 11-29.1 it follows that

5.33
$$S(\mathscr{L}) \leq \|\mathscr{L}\|_{A^{\frac{1}{2}}} < 1.$$

Stein and Rosenberg [1948] showed that if A is an L-matrix then the Gauss-Seidel method converges if and only if the Jacobi method converges, and if both converge, then the Gauss-Seidel method converges faster in the sense that $S(\mathscr{L}) \leq S(B)$. It can also be shown (see, for instance, Varga [1962]) that if A is an L-matrix then both methods converge if and only if A is an M-*matrix*, i.e., an L-matrix such that A is nonsingular and such that every element of A^{-1} is nonnegative.

Given L and U one can compute the eigenvalues of $\mathscr{L}$ by solving the equation

5.34
$$\det(\lambda L + U - \lambda I) = 0.$$

Thus in the case $N = 3$ we have

5.35
$$\det \begin{pmatrix} -\lambda & b_{1,2} & b_{1,3} \\ \lambda b_{2,1} & -\lambda & b_{2,3} \\ \lambda b_{3,1} & \lambda b_{3,2} & -\lambda \end{pmatrix} = 0.$$

In the example (1.8) we get

5.36
$$\det \begin{pmatrix} -\lambda & \frac{1}{4} & \frac{1}{4} & 0 \\ \frac{1}{4}\lambda & -\lambda & 0 & \frac{1}{4} \\ \frac{1}{4}\lambda & 0 & -\lambda & \frac{1}{4} \\ 0 & \frac{1}{4}\lambda & \frac{1}{4}\lambda & -\lambda \end{pmatrix} = \lambda^3(\lambda - \tfrac{1}{4}) = 0.$$

Hence the eigenvalues of $\mathscr{L}$ are $\frac{1}{4}, 0, 0, 0$ and $S(\mathscr{L}) = \frac{1}{4} = S(B)^2$. Thus the convergence of the Gauss-Seidel method is twice that of the Jacobi method.

We now prove

5.37 Theorem. Let the system (1.3) be derived from the five-point difference equation (1.5), where the points of R_h are taken in the natural ordering. Let $\mathscr{L}$ and B be the matrices corresponding to the Gauss-Seidel and Jacobi methods, respectively. If $\lambda \neq 0$ is an eigenvalue of $\mathscr{L}$ then $\sqrt{\lambda}$ and $-\sqrt{\lambda}$ are eigenvalues of B. Conversely, if μ is an eigenvalue of B then μ^2 is an eigenvalue of $\mathscr{L}$. Moreover,

5.38
$$S(\mathscr{L}) = S(B)^2.$$

Proof. If $\lambda \neq 0$ is an eigenvalue of $\mathscr{L}$ and if w is an associated eigenvector, then we have

5.39
$$\mathscr{L}w = (I - L)^{-1}Uw = \lambda w$$

or

5.40
$$(\lambda L + U)w = \lambda w.$$

Thus if we let $w(x, y)$ be the function defined on R_h and vanishing on S_h corresponding to w we have

5.41 $\beta_1 w(x + h, y) + \beta_2 w(x, y + h) + \lambda\beta_3 w(x - h, y) + \lambda\beta_4 w(x, y - h) = \lambda w(x, y),$

for all (x, y) in R_h. Let $v(x, y)$ be defined by

5.42
$$v(x, y) = \lambda^{-(x+y)/(2h)}w(x, y).$$

Thus we have

5.43 $\quad \beta_1 v(x + h, y) + \beta_2 v(x, y + h) + \beta_3 v(x - h, y) + \beta_4 v(x, y - h) = \lambda^{\frac{1}{2}} v(x, y).$

But this implies that $v(x, y)$ is an eigenvector of B with eigenvalue $\lambda^{\frac{1}{2}}$. Thus for some eigenvalue μ of B we have $\lambda^{\frac{1}{2}} = \mu$. Similarly, if we let $\hat{v}(x, y) = (-1)^{(x + y)/h} \lambda^{(x + y)/(2h)} w(x, y)$ we can show that $-\lambda^{\frac{1}{2}}$ is an eigenvalue of B.

Conversely, if μ is an eigenvalue of B and if v is an associated eigenvector, we let $v(x, y)$ be the function defined on R_h and vanishing on S_h which corresponds to v. If we let $\lambda = \mu^2$ and

5.44 $\qquad\qquad\qquad w(x, y) = \lambda^{(x + y)/(2h)} v(x, y)$

then we have

5.45 $\quad \beta_1 w(x + h, y) + \beta_2 w(x, y + h) + \lambda \beta_3 w(x - h, y) + \lambda \beta_4 w(x, y - h)$

$= \lambda^{(x + y + h)/(2h)} \{ \beta_1 v(x + h, y) + \beta_2 v(x, y + h) + \beta_3 v(x - h, y) + \beta_4 v(x, y - h) \}$

$= \lambda^{\frac{1}{2}} \mu \lambda^{(x + y)/(2h)} v(x, y) = \lambda^{\frac{1}{2}} \mu w(x, y).$

Hence, if $\lambda = \mu^2$, then $w(x, y)$ corresponds to an eigenvector of $\mathscr{L}$ with eigenvalue λ. Therefore, $\lambda = \mu^2$ is an eigenvalue of $\mathscr{L}$ and the proof of Theorem 5.37 is complete.

For the model problem it follows that

5.46 $\qquad\qquad\qquad S(\mathscr{L}) = \cos^2 \pi h = S(B)^2.$

Unfortunately, this implies that the convergence of the Gauss-Seidel method, while faster than that of the Jacobi method, nevertheless, is exceedingly slow. In the next section we shall show how a substantial, order-of-magnitude improvement can be obtained by a simple modification of the Gauss-Seidel method.

It can be shown (Young [1954]) that Theorem 5.37 holds for any *consistently ordered* matrix A, defined as follows.

The matrix $A = (a_{i,j})$ of order N is *consistently ordered* if there exist disjoint subsets $S_1, S_2, \ldots, S_t$ of $W = \{1, 2, \ldots, N\}$ such that $\sum_{k=1}^{t} S_k = W$ and such that if $a_{i,j} \neq 0$ or $a_{j,i} \neq 0$ then $j \in S_{k+1}$ if $j > i$ and $j \in S_{k-1}$ if $j < i$ where S_k is the subset containing i.

It can be verified (Young [1954]) that if the points of R_h are numbered in the natural ordering, then the matrix A corresponding to the 5-point difference equation (1.5) is consistently ordered with the S_k defined as follows. Let the mesh point of R_h labelled i be (x_i, y_i) and let $\alpha = \min_i (x_i + y_i)$. We let S_1 be the set of all i such that $x_i + y_i = \alpha$, S_2 be the set of all i such that $x_i + y_i = \alpha + h$, etc. In general, S_k is the set of all i such that $x_i + y_i = \alpha + (k - 1)h$. Clearly, the S_k have the desired properties.

For example, with the set of mesh points of R_h indicated in Fig. 5.47, we have $S_1 = \{3\}$, $S_2 = \{1, 4, 8\}$, $S_3 = \{2, 5, 9\}$, $S_4 = \{6\}$, $S_5 = \{7, 10\}$, $S_6 = \{11\}$.

5.47 Fig.

Labelling of the Mesh Points.

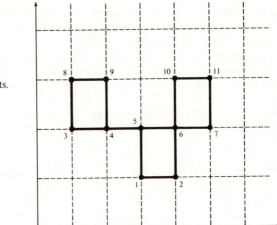

Another ordering of the mesh points which yields a consistently ordered matrix is the "red-black" ordering defined as follows. Let S_1 denote the set of all i such that $h^{-1}(x_i + y_i)$ is even (corresponding to "red" points) and let S_2 denote the set of all i such that $h^{-1}(x_i + y_i)$ is odd (corresponding to "black" points). Thus in the example we have $S_1 = \{3, 2, 5, 9, 7, 10\}$ and $S_2 = \{1, 4, 8, 6, 11\}$. Here, for convenience, we have assumed that $h^{-1}(x_3 + y_3) = h^{-1}\alpha$ is even. Otherwise we can interchange S_1 and S_2.

Given a 5-point difference equation defined on a set of mesh points R_h, the matrix may not actually be consistently ordered. Thus, for example, with the 4-points

the matrix is not consistently ordered. However, by relabelling the mesh points to obtain

or

we obtain a consistently ordered matrix. In the first relabelled case we have $S_1 = \{1\}$, $S_2 = \{2, 3\}$, $S_3 = \{4\}$, while in the second relabelled case $S_1 = \{1, 2\}$, $S_2 = \{3, 4\}$.

Evidently relabelling the mesh points yields a matrix, say A', which is the same as the original matrix A except that certain rows and the corresponding columns have been permuted. It can be shown (Young [1954]) that there exists a permutation of the rows and corresponding columns of a matrix A such that the resulting matrix A' is consistently ordered if and only if A has *Property A* defined as follows:

A matrix $A = (a_{i,j})$ of order N has *Property A* if there exist two disjoint subsets S_1 and S_2 of $W = \{1, 2, \ldots, N\}$ such that if $i \neq j$ and if either $a_{i,j} \neq 0$ or $a_{j,i} \neq 0$, then $i \in S_1$ and $j \in S_2$ or else $i \in S_2$ and $j \in S_1$.

One can test whether or not the matrix corresponding to a given numbering of the points of R_h is consistently ordered as follows (see, for instance, Young [1971a]). To each pair of adjacent mesh points of R_h draw an arrow in the direction of increasing index. The matrix is consistently ordered if for each closed path consisting of segments joining adjacent points of R_h the number of arrows in the direction of the path is the same as the number of arrows in the opposite direction.

Exercises 16.5

1. Carry out the Gauss-Seidel method for the problem of Exercise 1, Section 16.4. Carry out the iteration process until all of the values $|u_i^{(n+1)} - u_i^{(n)}|$ are less than 0.5. Use zero starting values.

2. Find all values of α such that for the matrix

$$A = \begin{pmatrix} 4 & -\alpha \\ -\alpha & 1 \end{pmatrix}$$

   a) the Gauss-Seidel method converges,
   b) A is positive definite.

3. Show that if P is a nonsingular matrix, then $\|PAP^{-1}\|_2$ defines a matrix norm, say $\|A\|_P$, and that $\|Pv\|_2$ defines a vector norm, say $\|v\|_P$, such that (5.27) holds.

4. Compute the eigenvalues of $\mathscr{L}$ for the matrices of Exercise 1, Section 16.2. Verify that $S(\mathscr{L}) < 1$ for each one of the positive definite matrices.

5. For the matrix

$$A = \begin{pmatrix} 2 & -1 \\ -1 & 2 \end{pmatrix}$$

compute $\|\mathscr{L}\|_{A^{\frac{1}{2}}}$ and $\|v\|_{A^{\frac{1}{2}}}$, where $v = (1, 2)^T$, and verify that

$$\|\mathscr{L}v\|_{A^{\frac{1}{2}}} \leq \|\mathscr{L}\|_{A^{\frac{1}{2}}} \|v\|_{A^{\frac{1}{2}}}.$$

Hint: Note that $\|\mathscr{L}\|_{A^{\frac{1}{2}}}^2 = S(\mathscr{L}A^{-1}\mathscr{L}^T A)$.

6. Find $S(\mathscr{L})$ and a corresponding eigenvector of $\mathscr{L}$ for the model problem with $h = \frac{1}{4}$

a) using the natural ordering,
b) using the red-black ordering.

7. Which of the following matrices are consistently ordered and which have Property A?

$$A_1 = \begin{pmatrix} 4 & -1 & 0 & -1 \\ -1 & 4 & -1 & 0 \\ 0 & -1 & 4 & -1 \\ -1 & 0 & -1 & 4 \end{pmatrix}, \quad A_2 = \begin{pmatrix} 4 & -1 & -1 & 0 \\ -1 & 4 & 0 & -1 \\ -1 & 0 & 4 & -1 \\ 0 & -1 & -1 & 4 \end{pmatrix},$$

$$A_3 = \begin{pmatrix} 4 & 0 & -1 & -1 \\ 0 & 4 & -1 & -1 \\ -1 & -1 & 4 & 0 \\ -1 & -1 & 0 & 4 \end{pmatrix}, \quad A_4 = \begin{pmatrix} 4 & -1 & -1 & -1 \\ -1 & 4 & -1 & -1 \\ -1 & -1 & 4 & -1 \\ -1 & 0 & -1 & 4 \end{pmatrix}.$$

In each case find the eigenvalues of B and of $\mathscr{L}$ and see whether $S(\mathscr{L}) = S(B)^2$.

8. Work out Exercise 6, Section 16.4 for the Gauss-Seidel method.

9. Show that the subsets S_k defined for the natural ordering in the paragraph following the definition of a consistently ordered matrix satisfy the conditions of that definition. Do the same for S_1 and S_2 corresponding to the red-black ordering.

10. For the set R_h considered below, show that the labelling indicated and a 5-point difference equation leads to a consistently ordered matrix.

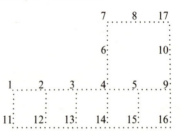

However, show that if the labelling of points 6 and 7 is interchanged, the corresponding matrix is not consistently ordered.

11. Show that if the red-black ordering is used, we may replace (5.44) by

$$w(x, y) = \lambda^{\gamma(x,y)} v(x, y)$$

where $\gamma(x, y) = 0$ if (x, y) is a red point, i.e., a point such that $h^{-1}(x + y)$ is even and $\gamma(x, y) = \frac{1}{2}$ if (x, y), is a black point.

16.6 THE SUCCESSIVE OVERRELAXATION METHOD

We now modify the Gauss-Seidel method slightly in order to improve the convergence. The *successive overrelaxation method* (SOR method) is defined by

6.1 $$u_i^{(n+1)} = \omega \left\{ \sum_{j=1}^{i-1} b_{i,j} u_j^{(n+1)} + \sum_{j=i+1}^{N} b_{i,j} u_j^{(n)} + c_i \right\} + (1 - \omega) u_i^{(n)}.$$

Here ω is a real parameter, known as the *relaxation factor*, which is to be chosen to make the convergence as fast as possible. If $\omega = 1$, then we have the Gauss-Seidel method. In the case $N = 3$ we have

6.2 $$\begin{cases} u_1^{(n+1)} = \omega \{ b_{1,2} u_2^{(n)} + b_{1,3} u_3^{(n)} + c_1 \} + (1 - \omega) u_1^{(n)} \\ u_2^{(n+1)} = \omega \{ b_{2,1} u_1^{(n+1)} + b_{2,3} u_3^{(n)} + c_2 \} + (1 - \omega) u_2^{(n)} \\ u_3^{(n+1)} = \omega \{ b_{3,1} u_1^{(n+1)} + b_{3,2} u_2^{(n+1)} + c_3 \} + (1 - \omega) u_3^{(n)}. \end{cases}$$

In effect, at each step one computes the Gauss-Seidel value, say $\hat{u}_i^{(n+1)}$ and then computes $u_i^{(n+1)}$ by

6.3 $$u_i^{(n+1)} = u_i^{(n)} + \omega [\hat{u}_i^{(n+1)} - u_i^{(n)}].$$

Thus, one is in effect using a kind of "extrapolation" or "overrelaxation" procedure.

The term "overrelaxation" is suggested by the terminology of Southwell [1946] and others. If one replaces $u_i^{(n)}$ by the Gauss-Seidel value, then one (temporarily) satisfies the i-th equation exactly. This process was considered in terms of "relaxing" a physical stress which is proportional to the *residual*, i.e., the amount by which the equation fails to be satisfied. It was found by Southwell that it was often better to overcorrect the value of $u_i^{(n)}$ so as to change the sign of the residual. Thus the term "overrelax" was introduced. The SOR method carries out the overrelaxation procedure in a fixed order, while with hand methods the residuals were usually relaxed in a varying order. Frequently the largest residual was relaxed at any stage.

For the difference equation (4.12) with the points taken in the natural ordering we have

6.4 $$\begin{aligned} u^{(n+1)}(x, y) = \omega \{ \beta_1 u^{(n)}(x + h, y) + \beta_2 u^{(n)}(x, y + h) + \beta_3 u^{(n+1)}(x - h, y) \\ + \beta_4 u^{(n+1)}(x, y - h) + \tau(x, y) \} + (1 - \omega) u^{(n)}(x, y). \end{aligned}$$

If we apply the SOR method to the system (4.22) we have

6.5 $$\begin{cases} u_1^{(n+1)} = \omega \{ \tfrac{1}{4} u_2^{(n)} + \tfrac{1}{4} u_3^{(n)} \} + (1 - \omega) u_1^{(n)} \\ u_2^{(n+1)} = \omega \{ \tfrac{1}{4} u_1^{(n+1)} + \tfrac{1}{4} u_4^{(n)} \} + (1 - \omega) u_2^{(n)} \\ u_3^{(n+1)} = \omega \{ \tfrac{1}{4} u_1^{(n+1)} + \tfrac{1}{4} u_4^{(n)} + 250 \} + (1 - \omega) u_3^{(n)} \\ u_4^{(n+1)} = \omega \{ \tfrac{1}{4} u_2^{(n+1)} + \tfrac{1}{4} u_3^{(n+1)} + 250 \} + (1 - \omega) u_4^{(n)}. \end{cases}$$

Letting $u_1^{(0)} = u_2^{(0)} = u_3^{(0)} = u_4^{(0)} = 0$ and letting $\omega = 1.072$ we get the following results:

n	$u_1^{(n)}$	$e_1^{(n)}$	$u_2^{(n)}$	$e_2^{(n)}$	$u_3^{(n)}$	$e_3^{(n)}$	$u_4^{(n)}$	$e_4^{(n)}$
0	0.	−125.0000	0.	−125.0000	0.	−375.0000	0.	−375.0000
1	0.	−125.0000	0.	−125.0000	268.0000	−107.0000	339.8240	−35.1760
2	71.8240	−53.1760	110.3217	−14.6783	359.0257	−15.9743	369.3178	−5.6822
3	120.6138	−4.3862	123.3585	−1.6415	373.4518	−1.5482	374.5543	−.4457
4	124.4610	−.5390	124.8543	−.1457	374.8476	−.1524	374.9522	−.0478
5	124.9589	−.0411	124.9867	−.0133	374.9871	−.0129	374.9964	−.0036
6	124.9959	−.0041	124.9989	−.0011	374.9989	−.0011	374.9997	−.0003
7	124.9997	−.0003	124.9999	−.0001	374.9999	−.0001	375.0000	.0000
8	125.0000	.0000	125.0000	.0000	375.0000	.0000	375.0000	.0000

Here the method is converging somewhat faster than the Gauss-Seidel method. As a matter of fact, in almost every case the error is reduced by a factor of 10 or more on each iteration. On the other hand, the ratios $e_i^{(n+1)}/e_i^{(n)}$ behave somewhat less regularly than in the case of the Gauss-Seidel method. This is because there are several eigenvalues of the associated matrix for the SOR method, each having the same modulus.

To study the convergence properties of the SOR method we write the method in the matrix form

6.6 $$u^{(n+1)} = \omega(Lu^{(n+1)} + Uu^{(n)} + c) + (1 - \omega)u^{(n)}.$$

Solving for $u^{(n+1)}$, as we can do since $I - \omega L$ is nonsingular, we have

6.7 $$u^{(n+1)} = \mathscr{L}_\omega u^{(n)} + (I - \omega L)^{-1}\omega c$$

where

6.8 $$\mathscr{L}_\omega = (I - \omega L)^{-1}(\omega U + (1 - \omega)I).$$

Since the true solution $\bar{u}$ of the original system (1.3) satisfies

6.9 $$\bar{u} = B\bar{u} + c = L\bar{u} + U\bar{u} + c$$

we have

6.10 $$\omega\bar{u} = \omega L\bar{u} + \omega U\bar{u} + \omega c$$

and

6.11 $$\bar{u} = \omega L\bar{u} + \omega U\bar{u} + \omega c + (1 - \omega)\bar{u}.$$

This is equivalent to

6.12 $$\bar{u} = \mathscr{L}_\omega \bar{u} + (I - \omega L)^{-1} \omega c.$$

Hence

6.13 $$e^{(n+1)} = \mathscr{L}_\omega e^{(n)}, \qquad e^{(n)} = \mathscr{L}_\omega^n e^{(0)}.$$

Thus the convergence of the SOR method depends on the eigenvalues of $\mathscr{L}_\omega$.

It can be shown (Varga [1962] or Young [1971a]) that if A is irreducible and has weak diagonal dominance and if $0 < \omega < 1$, then $S(\mathscr{L}_\omega) < 1$. This result is of little practical use since in order to obtain fast convergence we usually must let $\omega > 1$. However, the following result of Kahan [1958] shows that ω cannot lie outside of the range $0 < \omega < 2$ if we are to have convergence.

6.14 Theorem. If $S(\mathscr{L}_\omega) < 1$, then $0 < \omega < 2$.

Proof. The characteristic equation for $\mathscr{L}_\omega$ is

6.15 $$\begin{aligned} \det(\mathscr{L}_\omega - \lambda I) &= \det((I - \omega L)^{-1}(\omega U + (1 - \omega)I) - \lambda I) \\ &= \det(\omega U + (1 - \omega)I - \lambda(I - \omega L)) \\ &= \det(\omega U + \omega\lambda L - (\lambda + \omega - 1)I) \\ &= A_0\lambda^N + A_1\lambda^{N-1} + \cdots + A_N. \end{aligned}$$

But $A_0 = \pm 1$ and $A_N = \det \mathscr{L}_\omega = \det(\omega U - (\omega - 1)I)) = \pm(\omega - 1)^N$. Therefore, the product of the roots of the characteristic equation is $\pm(\omega - 1)^N$. Hence at least one of the eigenvalues must have modulus not less than $|\omega - 1|$. Thus, for convergence we must have $|\omega - 1| < 1$ and $0 < \omega < 2$.

It can be shown (see Ostrowski [1954]) that the SOR method converges if $0 < \omega < 2$ and if A is real and positive definite. Moreover, if A is symmetric and has positive diagonal elements, and if the SOR method converges, then A is positive definite and $0 < \omega < 2$. One can also show (see Varga [1962]) that if A is an L-matrix such that $S(B) < 1$, then $S(\mathscr{L}_\omega)$ converges for $0 < \omega \leq 1$ and $S(\mathscr{L}_\omega)$ is a monotone nonincreasing function of ω in that range. However, here again this general result is of little use since, as we shall see, we shall normally choose ω in the range $1 < \omega < 2$.

Given L and U one can compute the eigenvalues of $\mathscr{L}_\omega$ by solving the equation

6.16 $$\det\left(\lambda L + U - \frac{\lambda + \omega - 1}{\omega} I\right) = 0$$

provided $\omega \neq 0$. For the system (4.21) we get

6.17
$$\det \begin{vmatrix} -a & \frac{1}{4} & \frac{1}{4} & 0 \\ \frac{1}{4}\lambda & -a & 0 & \frac{1}{4} \\ \frac{1}{4}\lambda & 0 & -a & \frac{1}{4} \\ 0 & \frac{1}{4}\lambda & \frac{1}{4}\lambda & -a \end{vmatrix} = 0$$

where $a = (\lambda + \omega - 1)/\omega$. But the determinant can be shown to equal

6.18
$$\det \begin{vmatrix} -a & \frac{1}{4}\sqrt{\lambda} & \frac{1}{4}\sqrt{\lambda} & 0 \\ \frac{1}{4}\sqrt{\lambda} & -a & 0 & \frac{1}{4}\sqrt{\lambda} \\ \frac{1}{4}\sqrt{\lambda} & 0 & -a & \frac{1}{4}\sqrt{\lambda} \\ 0 & \frac{1}{4}\sqrt{\lambda} & \frac{1}{4}\sqrt{\lambda} & -a \end{vmatrix} = a^2(a^2 - \frac{1}{4}\lambda)$$

so that $(\lambda + \omega - 1)^2 = 0$ or else

6.19
$$(\lambda + \omega - 1)^2 = \frac{1}{4}\omega^2\lambda.$$

Thus if $\omega = 1.1$ we have for the eigenvalues of $\mathscr{L}_{1.1}$

6.20 $\lambda = -0.1, -0.1, 0.05125 + 0.08587i, 0.05125 - 0.085871i$

and hence all eigenvalues of $\mathscr{L}_{1.1}$ have modulus 0.1. Therefore

$$S(\mathscr{L}_{1.1}) = 0.1 < S(\mathscr{L}) = 0.25.$$

Similarly, if $\omega = 1.072$, the eigenvalues are

$$-0.072, -0.072, 0.072, 0.072$$

and

$$S(\mathscr{L}_{1.072}) = 0.072.$$

We shall see later that $\omega = 1.072$ minimizes $S(\mathscr{L}_\omega)$.

We now study the convergence properties of the SOR method as applied to the 5-point difference equation (1.5). We prove

6.21 Theorem. Under the hypotheses of Theorem 5.37 let $\mathscr{L}_\omega$ be the matrix corresponding to the SOR method. If $\lambda \neq 0$ is an eigenvalue of $\mathscr{L}_\omega$ then there exists an eigenvalue μ of B such that

6.22
$$(\lambda + \omega - 1)^2 = \omega^2 \mu^2 \lambda.$$

Conversely, if μ is an eigenvalue of B and if (6.22) holds, then λ is an eigenvalue of $\mathcal{L}_\omega$.

Proof. Suppose $\lambda \neq 0$ and λ is an eigenvalue of $\mathcal{L}_\omega$. Then for some $w \neq 0$ we have $\mathcal{L}_\omega w = \lambda w$ or

6.23
$$(I - \omega L)^{-1}(\omega U + (1 - \omega)I)w = \lambda w$$
$$(\omega U + \lambda \omega L)w = (\lambda + \omega - 1)w.$$

Let $w(x, y)$ be the function defined on R_h and vanishing on S_h corresponding to w, and let $v(x, y)$ be given by (5.42). Since $w(x, y)$ satisfies

6.24
$$\omega\{\beta_1 w(x + h, y) + \beta_2 w(x, y + h) + \lambda\beta_3 w(x - h, y)$$
$$+ \lambda\beta_4 w(x, y - h)\} = (\lambda + \omega - 1)w(x, y)$$

we have

6.25
$$\omega\{\beta_1 v(x + h, y) + \beta_2 v(x, y + h) + \beta_3 v(x - h, y) + \beta_4 v(x, y - h)\}$$
$$= \lambda^{-\frac{1}{2}}(\lambda + \omega - 1)v(x, y).$$

If $\omega = 0$, then all eigenvalues of $\mathcal{L}_\omega$ are unity. Hence any number μ satisfies (6.22). Thus there certainly exists an eigenvalue of B such that (6.22) holds. If $\omega \neq 0$, then (6.25) implies that

6.26
$$\frac{\lambda + \omega - 1}{\omega \lambda^{\frac{1}{2}}}$$

is an eigenvalue of B associated with the eigenvector $v(x, y)$, Hence (6.22) holds for some eigenvalue μ of B.

The converse is similar to that of Theorem 5.37 and we omit the proof.

We now consider the optimum choice of ω in the case where all eigenvalues of B are real and $S(B) < 1$. We remark that if A is a positive definite matrix which is consistently ordered, then the eigenvalues of B are real and $S(B) < 1$. For, by Theorem 5.16, since A is positive definite, then $S(\mathcal{L}) < 1$. Moreover, since A is consistently ordered, we have $S(B) = \sqrt{S(\mathcal{L})} < 1$. Hence $S(B) < 1$. Moreover, $B = D^{-1}C$ is similar to $\tilde{B} = D^{\frac{1}{2}}BD^{-\frac{1}{2}} = D^{-\frac{1}{2}}CD^{-\frac{1}{2}}$ which is symmetric since C is. Thus the eigenvalues of B are real.

Evidently, all of the eigenvalues of $\mathscr{L}_\omega$ can be found by solving the quadratic equation (6.22) for λ for all eigenvalues μ of B. If we let $r(\omega, \mu)$ denote the maximum of the moduli of the roots of (6.22), we have

6.27
$$S(\mathscr{L}_\omega) = \max_\mu r(\omega, \mu)$$

where the maximum is taken over all eigenvalues μ of B.

We first seek to show that if $\omega > 0$, then

6.28
$$\max_{-\bar\mu \leq \mu \leq \bar\mu} r(\omega, \mu) = r(\omega, \bar\mu),$$

where $\bar\mu = S(B)$. Evidently, $r(\omega, \mu)$ is given by

6.29
$$r(\omega, \mu) = \left| \frac{|\omega||\mu| + \sqrt{\omega^2 \mu^2 - 4(\omega - 1)}}{2} \right|^2.$$

If $\omega \leq 1$, then clearly $r(\omega, \mu)$ is an increasing function of $|\mu|$, and (6.28) holds. If $\omega > 1$, let us define μ_c by

6.30
$$\mu_c = \sqrt{\frac{4(\omega - 1)}{\omega^2}}.$$

If $|\mu| \leq \mu_c$ then $r(\omega, \mu) = \omega - 1$. If $|\mu| \geq \mu_c$ then $r(\omega, \mu)$ is an increasing function of $|\mu|$. Hence we obviously have (6.28).

From (6.28) and (6.27) it follows that

6.31
$$S(\mathscr{L}_\omega) = r(\omega, \bar\mu).$$

We now seek to show that if $\omega \neq \omega_b$, then

6.32
$$S(\mathscr{L}_\omega) > S(\mathscr{L}_{\omega_b}) = \omega_b - 1$$

where

6.33
$$\omega_b = \frac{2}{1 + \sqrt{1 - \bar\mu^2}}.$$

We first show that

6.34
$$S(\mathscr{L}_\omega) = \begin{cases} \omega - 1 & \text{if} \quad \omega_b \leq \omega < 2 \\ \left[\dfrac{\omega\bar\mu + \sqrt{\omega^2\bar\mu^2 - 4(\omega - 1)}}{2} \right]^2, & \text{if } 0 < \omega \leq \omega_b. \end{cases}$$

Since

$$\frac{d}{d\omega}\left\{\omega^2\bar{\mu}^2 - 4(\omega - 1)\right\} = 2(\omega\bar{\mu}^2 - 2)$$

it follows that $\omega^2\bar{\mu}^2 - 4(\omega - 1)$ is a decreasing function of ω for $0 < \omega < 2$. Moreover, $\omega^2\bar{\mu}^2 - 4(\omega - 1) = 4$ if $\omega = 0$, and $\omega^2\bar{\mu}^2 - 4(\omega - 1) = 4(\bar{\mu}^2 - 1) < 0$ if $\omega = 2$. Thus there is a unique value of ω in the range $0 < \omega < 2$ such that

6.35 $$\omega^2\bar{\mu}^2 = 4(\omega - 1).$$

Solving the quadratic

6.36 $$\left(\frac{1}{\omega}\right)^2 - \frac{1}{\omega} + \frac{\bar{\mu}^2}{4} = 0$$

for ω^{-1} and taking the root such that $\omega^{-1} > \frac{1}{2}$ we get (6.33). Thus, for $2 > \omega > \omega_b$ the expression $\omega^2\bar{\mu}^2 - 4(\omega - 1)$ is negative, while for $0 < \omega < \omega_b$ the expression is positive. In the former case we have $r(\omega, \bar{\mu}) = \omega - 1$, while in the latter case we have

6.37 $$r(\omega, \bar{\mu}) = \left[\frac{\omega\bar{\mu} + \sqrt{\omega^2\bar{\mu}^2 - 4(\omega - 1)}}{2}\right]^2.$$

Hence (6.34) holds.

We now seek to show that if $0 < \omega < \omega_b$ the function $S(\mathcal{L}_\omega)$ is a decreasing function of ω. But we have

6.38 $$\frac{d}{d\omega}\left[\omega\bar{\mu} + \sqrt{\omega^2\bar{\mu}^2 - 4(\omega - 1)}\right] = \bar{\mu} + \frac{\omega\bar{\mu}^2 - 2}{\sqrt{\omega^2\bar{\mu}^2 - 4(\omega - 1)}}$$

$$= \frac{\bar{\mu}\sqrt{\omega^2\bar{\mu}^2 - 4(\omega - 1)} + \omega\bar{\mu}^2 - 2}{\sqrt{\omega^2\bar{\mu}^2 - 4(\omega - 1)}}.$$

But the numerator of the above expression is negative since $\omega\bar{\mu}^2 - 2 < 0$ and

6.39 $$(\omega\bar{\mu}^2 - 2)^2 = \omega^2\bar{\mu}^4 - 4\omega\bar{\mu}^2 + 4$$

while

6.40 $$\left[\bar{\mu}\sqrt{\omega^2\bar{\mu}^2 - 4(\omega - 1)}\right]^2 = \omega^2\bar{\mu}^4 - 4\omega\bar{\mu}^2 + 4\bar{\mu}^2.$$

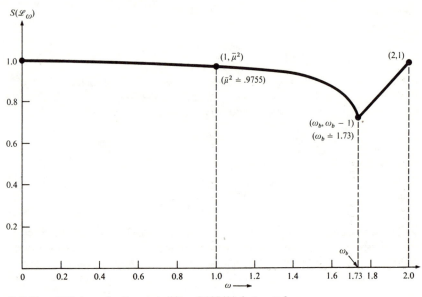

6.41 Fig. $S(\mathcal{L}_\omega)$ vs ω for $\bar{\mu} = \cos(\pi/20) \doteq 0.987688, 0 \leqq \omega \leqq 2$.

Therefore, we have the situation indicated in Fig. 6.41. Clearly, the minimum value of $S(\mathcal{L}_\omega)$ is assumed when $\omega = \omega_b$. Since $S(\mathcal{L}_{\omega_b}) = \omega_b - 1$ by (6.34) we have (6.32).

From (6.33) and (6.32) it follows that

6.42
$$S(\mathcal{L}_{\omega_b}) = \left(\frac{\bar{\mu}}{1 + \sqrt{1 - \bar{\mu}^2}} \right)^2.$$

Moreover, since $S(\mathcal{L}) = \bar{\mu}^2$ we can show using L'Hospital's rule that

6.43
$$\lim_{\mu \to 1-} \frac{R(\mathcal{L}_{\omega_b})}{2\sqrt{R(\mathcal{L})}} = 1;$$

hence for $\bar{\mu}$ close to unity we have

6.44
$$R(\mathcal{L}_{\omega_b}) \sim 2\sqrt{R(\mathcal{L})}.$$

For the model problem we have $\bar{\mu} = \cos \pi h$ and

6.45
$$\omega_b - 1 = \frac{1 - \sin \pi h}{1 + \sin \pi h} \sim 1 - 2\pi h$$

so that for small h

6.46
$$R(\mathscr{L}_{\omega_b}) \sim 2\pi h$$

whereas

6.47
$$R(\mathscr{L}) = -\log \bar{\mu}^2 \sim \pi^2 h^2.$$

These results are consistent with (6.44). We also note that in this case

6.48
$$\frac{R(\mathscr{L}_{\omega_b})}{R(\mathscr{L})} \sim \frac{2}{\pi h}$$

so that the factor of improvement of $\mathscr{L}_\omega$ over $\mathscr{L}$ is proportional to h^{-1}. Thus, the SOR method is superior to the Gauss-Seidel method by an order-of-magnitude in h^{-1}.

The following sample values of ω_b, $R(\mathscr{L}_{\omega_b})$, etc. are given for the model problem.

h^{-1}	ω_b	$\omega_b - 1$	$R(\mathscr{L}_{\omega_b})$	$R(\mathscr{L})$	$2\sqrt{R(\mathscr{L})}$	$\dfrac{R(\mathscr{L}_{\omega_b})}{R(\mathscr{L})}$
5	1.25962	.25962	1.34855	.42387	1.30211	3.18152
10	1.52786	.52786	.63892	.100363	.63360	6.36609
20	1.72945	.72945	.31546	.024776	.31481	12.73248
40	1.85450	.85450	.15724	.0061749	.15716	25.46438
80	1.92445	.92445	.078560	.0015425	.078550	50.93031
160	1.96149	.96149	.039272	.00038556	.039271	101.85704

It can be shown that Theorem 6.21 holds if A is a consistently ordered matrix with nonvanishing diagonal elements. Hence if B has real eigenvalues and $S(B) < 1$, which is certainly the case if A is positive definite and consistently ordered, then the results given above on the choice of ω and the comparison of the SOR method with the Gauss-Seidel method are valid.

A still further generalization is the following. If A is a positive definite L-matrix and if ω_b is determined by (6.33), then Kahan showed that

6.49
$$\omega_b - 1 \leqq S(\mathscr{L}_{\omega_b}) \leqq \sqrt{\omega_b - 1}.$$

Thus even if A is not consistently ordered the rate of convergence of $\mathscr{L}_{\omega_b}$ is not less than twice what it would be if the matrix were consistently ordered. It can also be shown that $\bar{\mu} \geqq S(\mathscr{L}) \geqq \bar{\mu}^2$. Hence one still obtains an order-of-magnitude improvement over the Gauss-Seidel method. We remark that ω_b may not give the smallest value of $S(\mathscr{L}_\omega)$.

Let us now consider the problem of choosing ω_b if $\bar{\mu}$ is not known. If one is planning to solve many cases with the same difference equation but with different boundary values and different functions $G(x, y)$, it may be appropriate to simply solve the various cases with different values of ω and to observe which value of ω gives the fastest convergence. This procedure is somewhat inefficient and can be improved by the following iterative procedure. We iterate using (6.7) with a value of ω which we expect will be less than ω_b. We observe the ratios

6.50 $$\delta_i^{(n)} = \frac{u_i^{(n+1)} - u_i^{(n)}}{u_i^{(n)} - u_i^{(n-1)}}$$

as n increases for each i, or for a sample of values of i.   We let $\bar{\lambda}_\omega$ be the average of the estimated limiting values of these ratios. (If the ratios do not appear to be converging to a limit independent of i it is probably a sign that we have used a value of ω greater than ω_b.) We then estimate $\bar{\mu}$ by the formula

6.51 $$\bar{\mu} = \frac{\bar{\lambda}_\omega + \omega - 1}{\omega \bar{\lambda}_\omega^{\frac{1}{2}}}.$$

We then determine a more accurate value of ω_b by (6.33) and perform subsequent iterations using this value.

For example, in the case of (4.21) one can verify from the table following (5.4) that the ratios $\delta_i^{(n)}$ obtained using the Gauss-Seidel method are approaching $\frac{1}{4}$. Hence, $\bar{\lambda} = \bar{\lambda}_1 = \frac{1}{4}$, and we can determine $\bar{\mu}$ by (6.51) obtaining

6.52 $$\bar{\mu} = \bar{\lambda}^{\frac{1}{2}} = \frac{1}{2}.$$

We then substitute in (6.33) and obtain

6.53 $$\omega_b = \frac{4}{2 + \sqrt{3}} = 1.072.$$

On the other hand, if we had let $\omega = 1.10$, then the values of $\delta_i^{(n)}$ as determined from the table following (6.5) would not converge to an obvious limit. In general, the use of $\omega > \omega_b$ does not usually lead to an improved value of ω.

The above scheme can be justified as follows. If $\omega < \omega_b$, then the eigenvalue of $\mathscr{L}_\omega$ of largest modulus, namely $\bar{\lambda}_\omega$, is real. Moreover, there are no other eigenvalues of $\mathscr{L}_\omega$ of the same modulus. In fact, all complex eigenvalues have modulus $\omega - 1$ and $\omega - 1 < \bar{\lambda}_\omega$. Hence repeated use of the SOR method with the value of ω given is equivalent to applying the direct power method (Section 14.3).

If $\omega < \omega_b$, the ratios $\delta_i^{(n)}$ will converge to $\bar{\lambda}_\omega$. Moreover, since $\bar{\lambda}_\omega$ corresponds to $\mu = \bar{\mu}$ by (6.31), we have

6.54 $$(\bar{\lambda}_\omega + \omega - 1)^2 = \omega^2 \bar{\mu}^2 \bar{\lambda}_\omega.$$

Solving for $\bar{\mu}$ we obtain (6.51).

Hageman and Kellogg [1968] developed a more rapidly converging scheme which involves starting with a vector $u^{(0)}$ which vanishes on S_h and computing $P_1(\mathcal{L}_\omega)u^{(0)}, P_2(\mathcal{L}_\omega)u^{(0)}, \ldots$, where $P_i(\mathcal{L}_\omega)$ are suitable polynomials in $\mathcal{L}_\omega$.

The following information should be borne in mind in considering methods for choosing ω_b. Because of the behavior of $S(\mathcal{L}_\omega)$ as a function of ω it is better to overestimate ω than to underestimate ω. Thus the derivative of $S(\mathcal{L}_\omega)$ with respect to ω becomes infinite as $\omega \to \omega_{b-}$ while the derivative is unity if $\omega > \omega_b$. On the other hand, using $\omega > \omega_b$ results in a less accurate determination of $\bar{\mu}$ than using $\omega < \omega_b$.

For the five-point discrete analog of the Dirichlet problem for a region R we can estimate $\bar{\mu}$ by computing $\bar{\mu}^*$, the spectral radius of B for a rectangle containing R. In fact, it can be shown (see, for instance, Varga [1962] or Young [1971a]) that $\bar{\mu} \leq \bar{\mu}^*$. Also, it is easy to show that if the sides of the rectangle are $a = Ih, b = Jh$, where I and J are integers, then

$$6.55 \qquad \bar{\mu}^* = \tfrac{1}{2}\left(\cos\frac{\pi}{I} + \cos\frac{\pi}{J}\right).$$

For other differential equations one can often obtain estimates of $\bar{\mu}^*$ (see, for example, Warlick [1955], Young and Shaw [1955], Henrici [1960], and Warlick and Young [1970]).

For instance, it can be shown that if A, C and F are constants such that $A > 0$, $C > 0$, and $F \leq 0$, then corresponding to the differential equation

$$6.56 \qquad (Au_x)_x + (Cu_y)_y + Fu = G$$

we have

$$6.57 \qquad \bar{\mu} = \left(\frac{A}{A + C - \tfrac{1}{2}Fh^2}\right)\cos\frac{\pi}{I} + \left(\frac{C}{A + C - \tfrac{1}{2}Fh^2}\right)\cos\frac{\pi}{J}.$$

Hence in the case of variable coefficients it would seem reasonable to use (6.57) with average values of the coefficients. A rigorous upper bound for $\bar{\mu}$, given by Young [1971c], is

$$6.58 \qquad \bar{\mu} \leq \frac{2(\bar{A} + \bar{C})}{2(\bar{A} + \bar{C}) + h^2(\underline{-F})}$$

$$\times \left\{ 1 - \frac{2\underline{A}\sin^2\dfrac{\pi}{2I} + 2\underline{C}\sin^2\dfrac{\pi}{2J}}{\left(\dfrac{\bar{A} + \underline{A}}{2}\right) + \left(\dfrac{\bar{C} + \underline{C}}{2}\right) + \left(\dfrac{\bar{A} - \underline{A}}{2}\right)\cos\dfrac{\pi}{I} + \left(\dfrac{\bar{C} - \underline{C}}{2}\right)\cos\dfrac{\pi}{J}} \right\}.$$

Here

$$\begin{cases} \underline{A} \leq A(x, y) \leq \bar{A} \\ \underline{C} \leq C(x, y) \leq \bar{C} \\ (-\underline{F}) \leq -F(x, y) \end{cases}$$

6.59

in $R + S$.

Exercises 16.6

1. Carry out the SOR method with $\omega = 1.05$ for the problem of Exercise 1, Section 16.4. Carry out the iteration process until all of the values $|u_i^{(n+1)} - u_i^{(n)}|$ are less than 0.5. Use zero starting values.

2. Carry out several iterations using the SOR method with $\omega = 1.1$ for the system (4.21) and observe the ratios $e_i^{(n+1)}/e_i^{(n)}$ and the $\delta_i^{(n)}$. Estimate ω_b from these ratios.

3. Prove that if A is positive definite and $0 < \omega < 2$, then the SOR method converges. Give two proofs: one similar to the proof of Theorem 5.16 and the other based on the $A^{\frac{1}{2}}$-norm.

4. For the model problem with $h = \frac{1}{4}$, find the largest (in modulus) eigenvalue of $S(\mathscr{L}_{1.1})$ and the corresponding eigenvector.

   a) with the natural ordering,
   b) with the red-black ordering.

   Draw a graph of $S(\mathscr{L}_{\omega})$ for this case, in the range $0 \leq \omega \leq 2$.

5. Work out the details of the proof of (6.43).

6. Compute the eigenvalues of B for the matrix

$$A = \begin{pmatrix} 4 & -1 & -1 \\ -1 & 4 & -1 \\ -1 & -1 & 4 \end{pmatrix}.$$

   Give bounds on $S(\mathscr{L}_{\omega_b})$ where $\omega_b = 2(1 + \sqrt{1 - \bar{\mu}^2})^{-1}$, $\bar{\mu} = S(B)$. Verify these bounds by direct calculation of the eigenvalues of $\mathscr{L}_{\omega_b}$.

7. Prove (6.55) and (6.57).

8. For the 5-point discrete analog of the Dirichlet problem for the region shown, estimate ω_b, the optimum ω to be used for the SOR method, if $h = \frac{1}{20}$.

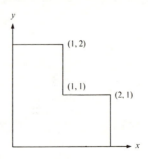

9. Work Exercise 8 if we replace Laplace's equation by

$$u_{xx} + 2u_{yy} - u = 1.$$

10. Give an upper bound for $\bar{\mu}$ for the differential equation

$$\frac{\partial}{\partial x}\left(\frac{1}{y}\frac{\partial u}{\partial x}\right) + \frac{\partial}{\partial y}\left(\frac{1}{y}\frac{\partial u}{\partial y}\right) - xu = 2$$

for the region $1 \le x \le 2, 1 \le y \le 2$. Also, give an estimate based on an average value of the coefficients. Assume $h = \frac{1}{20}$.

11. Prepare a computer program for solving the model problem with $u = 1$ on the side $y = 0$ and $u = 0$ on the other three sides. Use the SOR method and iterate until

$$\rho^{(n+1)} = \max_{R_h} |u^{(n+1)}(x, y) - u^{(n)}(x, y)| < \omega 10^{-6}$$

or until $n \ge 1000$. Let $h = \frac{1}{20}$ and consider the cases $\omega = 1, 1.50, 1.68, 1.73, 1.78$. Determine the number of iterations required for convergence and compare with the theoretical number of iterations, $-\log 10^6/\log S(\mathscr{L}_\omega)$, in each case. For each n, print out selected values of $u^{(n)}(x, y)$ as well as $\rho^{(n)}/\rho^{(n-1)}$. Let the initial values vanish and use the natural ordering.

12. In the previous example estimate ω_b by using the ratios $\rho^{(n)}/\rho^{(n-1)}$ in each of the cases $\omega = 1, \omega = 1.50, \omega = 1.68$.

13. Consider the model problem with the 9-point difference equation

$$20u(x, y) - 4u(x + h, y) - 4u(x, y + h) - 4u(x - h, y) - 4u(x, y - h) - u(x + h, y + h)$$

$$- u(x - h, y + h) - u(x + h, y - h) - u(x - h, y - h) = 0.$$

Determine $\bar{\mu} = S(B)$ by the method of separation of variables. Find upper and lower bounds on $S(\mathscr{L}_{\omega_b})$ where

$$\omega_b = \frac{2}{1 + \sqrt{1 - \bar{\mu}^2}}.$$

16.7 THE PEACEMAN-RACHFORD ALTERNATING DIRECTION IMPLICIT METHOD

We now consider a class of methods for solving the system (1.3) which is based on the "splitting" of the matrix A into the sum of three matrices

7.1 $$A = H_0 + V_0 + \Sigma$$

where Σ is a nonnegative diagonal matrix and where H_0, V_0, and Σ satisfy the following conditions:

a) $H_0 + \theta\Sigma + \rho I$ and $V_0 + \theta\Sigma + \rho I$ are nonsingular for any $\theta \ge 0, \rho > 0$;

b) for any vectors c and d and for any constants $\theta \ge 0$ and $\rho > 0$ it is "convenient" to solve the systems

7.2 $(H_0 + \theta\Sigma + \rho I)x = c,$ $(V_0 + \theta\Sigma + \rho I)y = d$

for x and y, respectively.

We shall be concerned here with the situation where H_0 and V_0 are either tri-diagonal matrices or can be made so by a suitable permutation of their rows and corresponding columns. Hence the procedure described in Section 10.4 can be used to solve the systems (7.2). This procedure is "convenient" in the sense that the work required is much less than would be required to solve the original system (1.3) directly.

For a system derived from a 5-point difference equation, a convenient splitting is to let H_0 correspond to the terms with derivatives in x and V_0 correspond to the terms with derivatives in y. Thus, for example, we could write the difference equation (1.5) in the form

7.3 $\mathscr{H}_0[u](x, y) + \mathscr{V}_0[u](x, y) - h^2 F u(x, y) = -h^2 G$

where

7.4 $\begin{cases} \mathscr{H}_0[u](x, y) = 2Au(x, y) - (A + \tfrac{1}{2}hD)u(x + h, y) - (A - \tfrac{1}{2}hD)u(x - h, y) \\ \mathscr{V}_0[u](x, y) = 2Cu(x, y) - (C + \tfrac{1}{2}hE)u(x, y + h) - (C - \tfrac{1}{2}hE)u(x, y - h). \end{cases}$

In the case of the Dirichlet problem we have

$$\mathscr{H}_0[u](x, y) = 2u(x, y) - u(x + h, y) - u(x - h, y)$$
$$\mathscr{V}_0[u](x, y) = 2u(x, y) - u(x, y + h) - u(x, y - h).$$

For the case of the unit square with $u = 0$ on all sides except $y = 1$ where $u = 1000$, if $h = \tfrac{1}{3}$ we get, using the natural ordering of the mesh points (see also (4.21)),

7.5 $(H + V)\begin{pmatrix} u_1 \\ u_2 \\ u_3 \\ u_4 \end{pmatrix} = \begin{pmatrix} 0 \\ 0 \\ 1000 \\ 1000 \end{pmatrix}$

where

7.6 $H = \begin{pmatrix} 2 & -1 & 0 & 0 \\ -1 & 2 & 0 & 0 \\ 0 & 0 & 2 & -1 \\ 0 & 0 & -1 & 2 \end{pmatrix}, \quad V = \begin{pmatrix} 2 & 0 & -1 & 0 \\ 0 & 2 & 0 & -1 \\ -1 & 0 & 2 & 0 \\ 0 & -1 & 0 & 2 \end{pmatrix}.$

We note that by interchanging the second and third rows and the second and third columns of V leads to the tridiagonal matrix

$$\begin{pmatrix} 2 & -1 & 0 & 0 \\ -1 & 2 & 0 & 0 \\ 0 & 0 & 2 & -1 \\ 0 & 0 & -1 & 2 \end{pmatrix}.$$

Let us write (1.3) in the form

7.7
$$(H_0 + V_0 + \Sigma)u = b$$

and let us consider two equivalent forms

7.8
$$\begin{cases} (H_0 + \theta\Sigma + \rho I)u = b - (V_0 + (1 - \theta)\Sigma - \rho I)u \\ (V_0 + \hat{\theta}\Sigma + \rho' I)u = b - (H_0 + (1 - \hat{\theta})\Sigma - \rho' I)u. \end{cases}$$

In the Peaceman-Rachford method [1955] one selects positive iteration parameters ρ and ρ' and determines $u^{(n+\frac{1}{2})}$ by

7.9
$$(H_0 + \theta\Sigma + \rho I)u^{(n+\frac{1}{2})} = b - (V_0 + (1 - \theta)\Sigma - \rho I)u^{(n)}.$$

Then one determines $u^{(n+1)}$ by

7.10 $$(V_0 + \hat{\theta}\Sigma + \rho' I)u^{(n+1)} = b - (H_0 + (1 - \hat{\theta})\Sigma - \rho' I)u^{(n+\frac{1}{2})}.$$

For simplicity, we shall consider here the special case where

7.11
$$\theta = \hat{\theta} = \tfrac{1}{2}, \qquad \rho = \rho',$$

and we let

7.12
$$H = H_0 + \tfrac{1}{2}\Sigma, \qquad V = V_0 + \tfrac{1}{2}\Sigma.$$

Evidently H and V satisfy the following conditions:

a) $H + \rho I$ and $V + \rho I$ are non-singular for any $\rho > 0$,

b) for any vectors c and d and for any $\rho > 0$ it is convenient to solve the systems

7.13
$$(H + \rho I)x = c, \qquad (V + \rho I)y = d.$$

Thus (7.7) becomes

7.14
$$(H + V)u = b$$

and (7.9)–(7.10) become, respectively,

7.15 $$(H + \rho I)u^{(n + \frac{1}{2})} = b - (V - \rho I)u^{(n)}$$

7.16 $$(V + \rho I)u^{(n + 1)} = b - (H - \rho I)u^{(n + \frac{1}{2})}.$$

For the difference equation (7.3) we have

7.17
$$(2A - \tfrac{1}{2}h^2 F + \rho I)u^{(n + \frac{1}{2})}(x, y) - (A + \tfrac{1}{2}hD)u^{(n + \frac{1}{2})}(x + h, y)$$
$$- (A - \tfrac{1}{2}hD)u^{(n + \frac{1}{2})}(x - h, y)$$
$$= -h^2 G(x, y) - \{(2C - \tfrac{1}{2}h^2 F - \rho I)u^{(n)}(x, y)$$
$$- (C + \tfrac{1}{2}hE)u^{(n)}(x, y + h) - (C - \tfrac{1}{2}hE)u^{(n)}(x, y - h)\}$$

for the horizontal sweep and

7.18
$$(2C - \tfrac{1}{2}h^2 F + \rho I)u^{(n + 1)}(x, y) - (C + \tfrac{1}{2}hE)u^{(n + 1)}(x, y + h)$$
$$- (C - \tfrac{1}{2}hE)u^{(n + 1)}(x, y - h)$$
$$= -h^2 G(x, y) - \{(2A - \tfrac{1}{2}h^2 F - \rho I)u^{(n + \frac{1}{2})}(x, y)$$
$$-(A + \tfrac{1}{2}hD)u^{(n + \frac{1}{2})}(x + h, y) - (A - \tfrac{1}{2}hD)u^{(n + \frac{1}{2})}(x - h, y)\}$$

for the vertical sweep. Evidently the determination of $u^{(n + \frac{1}{2})}(x, y)$ involves the solution of a linear system with a tridiagonal matrix. This is also true of the determination of $u^{(n + 1)}(x, y)$.

If we carry out the iterative process (7.15)–(7.16) for (7.5) with $\rho = \frac{1}{2}$, we obtain the following results using zero starting values.

n	$u_1^{(n)}$	$e_1^{(n)}$	$u_2^{(n)}$	$e_2^{(n)}$	$u_3^{(n)}$	$e_3^{(n)}$	$u_4^{(n)}$	$e_4^{(n)}$
0.	0.	−125.0000	0.	−125.0000	0.	−375.0000	0.	−375.0000
0.5	0.	−125.0000	0.	−125.0000	666.6667	291.6667	666.6667	291.6667
1.0	126.9841	1.9841	126.9841	1.9841	317.4603	−57.5397	317.4603	−57.5397
1.5	84.6561	−40.3439	84.6561	−40.3439	433.8624	58.8624	433.8624	58.8624
2.0	128.9997	3.9997	128.9997	3.9997	364.8274	−10.1726	364.8274	−10.1726
2.5	114.2185	−10.7815	114.2185	−10.7815	387.8391	12.8391	387.8391	12.8391
3.0	126.3442	1.3442	126.3442	1.3442	372.9699	−.0301	372.9699	−2.0301
3.5	122.3023	−2.6977	122.3023	−2.6977	377.9263	2.9263	377.9263	2.9263
4.0	125.3636	.3636	125.3636	.3636	374.5602	−.4398	374.5602	−.4398
4.5	124.3432	−.6568	124.3432	−.6568	375.6822	.6822	375.6822	.6822
5.0	125.0914	.0914	125.0914	.0914	374.9001	−.0999	374.9001	−.0999
5.5	124.8420	−.1580	124.8420	−.1580	375.1608	.1608	375.1608	.1608
6.0	125.0223	.0223	125.0223	.0223	374.9768	−.0232	374.9768	−.0232
6.5	124.9622	−.0378	124.9622	−.0378	375.0381	.0381	375.0381	.0381
7.0	125.0054	.0054	125.0054	.0054	374.9945	−.0055	374.9945	−.0055

Continued on next page

Continued from previous page

n	$u_1^{(n)}$	$e_1^{(n)}$	$u_2^{(n)}$	$e_2^{(n)}$	$u_3^{(n)}$	$e_3^{(n)}$	$u_4^{(n)}$	$e_4^{(n)}$
7.5	124.9910	−.0090	124.9910	−.0090	375.0091	.0091	375.0091	.0091
8.0	125.0013	.0013	125.0013	.0013	374.9987	−.0013	374.9987	−.0013
8.5	124.9979	−.0021	124.9979	−.0021	375.0022	.0022	375.0022	.0022
9.0	125.0003	.0003	125.0003	.0003	374.9997	−.0003	374.9997	−.0003
9.5	124.9995	−.0005	124.9995	−.0005	375.0005	.0005	375.0005	.0005
10.0	125.0001	.0001	125.0001	.0001	374.9999	−.0001	374.9999	−.0001
10.5	124.9999	−.0001	124.9999	−.0001	375.0001	.0001	375.0001	.0001
11.0	125.0000	.0000	125.0000	.0000	375.0000	.0000	375.0000	.0000

We observe that in each double iteration the error is reduced by a factor of approximately 4. Thus the convergence rate* in this case is about the same as that of the Gauss-Seidel method, though, of course, considerably more work is required per iteration.

We now seek to analyze the convergence properties of the Peaceman-Rachford method. If $\bar{u}$ is the true solution of (1.3), then

7.19
$$(H + V)\bar{u} = b$$

and

7.20
$$(H + \rho I)\bar{u} = b - (V - \rho I)\bar{u}.$$

Therefore by (7.15) we have

7.21
$$(H + \rho I)e^{(n + \frac{1}{2})} = -(V - \rho I)e^{(n)}.$$

Similarly

7.22
$$(V + \rho I)e^{(n + 1)} = -(H - \rho I)e^{(n + \frac{1}{2})}$$

and hence

7.23
$$e^{(n + 1)} = T_\rho e^{(n)}$$

where

7.24
$$T_\rho = (V + \rho I)^{-1}(H - \rho I)(H + \rho I)^{-1}(V - \rho I).$$

We now prove

---

* The eigenvalues of T_ρ are $\frac{1}{9}$, $\frac{5}{21}$, $\frac{5}{21}$, and $\frac{25}{49}$. The convergence is faster than anticipated because the component of the initial error vector, relative to the eigenvector associated with the eigenvalue $\frac{25}{49}$ vanishes. Note that $\frac{1}{9} = (1 - \frac{1}{2})^2(1 + \frac{1}{2})^{-2}$, $\frac{25}{49} = (3 - \frac{1}{2})^2(3 + \frac{1}{2})^{-2}$, $\frac{5}{21} = (3 - \frac{1}{2})$ $(3 + \frac{1}{2})^{-1}(1 - \frac{1}{2})(1 + \frac{1}{2})^{-1}$, and that the eigenvalues of H and V are 3 and 1.

7.25 Theorem. If H and V are real positive definite matrices and if $\rho > 0$ then

7.26
$$S(T_\rho) < 1.$$

Proof. Evidently T_ρ is similar to the matrix $\tilde{T}_\rho$ where

7.27
$$\tilde{T}_\rho = (V + \rho I) T_\rho (V + \rho I)^{-1}$$
$$= (H - \rho I)(H + \rho I)^{-1}(V - \rho I)(V + \rho I)^{-1}$$

and

7.28
$$\|\tilde{T}_\rho\|_2 \leq \|(H - \rho I)(H + \rho I)^{-1}\|_2 \|(V - \rho I)(V + \rho I)^{-1}\|_2.$$

But since H and V are symmetric and since $H - \rho I$ commutes with $(H + \rho I)^{-1}$ we have

7.29
$$\|(H - \rho I)(H + \rho I)^{-1}\|_2 = S((H - \rho I)(H + \rho I)^{-1})$$
$$= \max_\mu \left| \frac{\mu - \rho}{\mu + \rho} \right|$$

where μ ranges over all eigenvalues of H. But since H is positive definite, its eigenvalues are positive. Therefore

7.30
$$\|(H - \rho I)(H + \rho I)^{-1}\|_2 < 1.$$

Similarly $\|(V - \rho I)(V + \rho I)^{-1}\|_2 < 1$ and we have

7.31
$$S(T_\rho) = S(\tilde{T}_\rho) \leq \|\tilde{T}_\rho\|_2 < 1;$$

hence the convergence follows.

Even if H and V are not positive definite, convergence may still hold. Indeed we have

7.32 Theorem. If there exists a real nonsingular matrix P such that $\hat{H} = P^{-1}HP$ and $\hat{V} = P^{-1}VP$ are positive definite and if $\rho > 0$ then $S(T_\rho) < 1$.

Proof. As in the proof of Theorem 7.25, T_ρ is similar to $\tilde{T}_\rho$ given by (7.27); $\tilde{T}_\rho$ in turn is similar to

7.33
$$\hat{T}_\rho = P^{-1}\tilde{T}_\rho P = (\hat{H} - \rho I)(\hat{H} + \rho I)^{-1}(\hat{V} - \rho I)(\hat{V} + \rho I)^{-1}.$$

Therefore we have

7.34 $$S(T_\rho) = S(\hat{T}_\rho) \leq \|(\hat{H} - \rho I)(\hat{H} + \rho I)^{-1}\|_2 \|(\hat{V} - \rho I)(\hat{V} + \rho I)^{-1}\|_2 < 1$$

since $\hat{H}$ and $\hat{V}$ are positive definite.

As an application of the above result, suppose that for some real positive definite matrix Q the matrices QH and QV are positive definite. Then $S(T_\rho) < 1$. For if we let $P = Q^{-\frac{1}{2}}$ we have

7.35
$$\begin{cases} \hat{H} = Q^{\frac{1}{2}}HQ^{-\frac{1}{2}} = Q^{-\frac{1}{2}}(QH)Q^{-\frac{1}{2}} \\ \hat{V} = Q^{\frac{1}{2}}VQ^{-\frac{1}{2}} = Q^{-\frac{1}{2}}(QV)Q^{-\frac{1}{2}}. \end{cases}$$

Hence since QH and QV are positive definite, it follows that $\hat{H}$ and $\hat{V}$ are also*.

Let us now assume that H and V are real positive definite matrices and that the eigenvalues μ of H and ν of V lie in the ranges

7.36
$$0 < a \leqq \mu \leqq b, \qquad 0 < a \leqq \nu \leqq b.$$

Evidently, if $\rho > 0$ we have

7.37
$$S(T_\rho) \leqq S((H - \rho I)(H + \rho I)^{-1})S((V - \rho I)(V + \rho I)^{-1})$$

$$= \left(\max_{a \leqq \mu \leqq b} \left| \frac{\mu - \rho}{\mu + \rho} \right| \right) \left(\max_{a \leqq \nu \leqq b} \left| \frac{\nu - \rho}{\nu + \rho} \right| \right) = \left[\max_{a \leqq \gamma \leqq b} \left| \frac{\gamma - \rho}{\gamma + \rho} \right| \right]^2 = \phi(a, b; \rho).$$

Since $(\gamma - \rho)/(\gamma + \rho)$ is an increasing function of γ we have

7.38
$$\max_{a \leqq \gamma \leqq b} \left| \frac{\gamma - \rho}{\gamma + \rho} \right| = \max \left(\left| \frac{a - \rho}{a + \rho} \right|, \left| \frac{b - \rho}{b + \rho} \right| \right).$$

When $\rho = \sqrt{ab}$, then

7.39
$$\left| \frac{a - \rho}{a + \rho} \right| = \left| \frac{b - \rho}{b + \rho} \right| = \frac{\sqrt{b} - \sqrt{a}}{\sqrt{b} + \sqrt{a}}.$$

Moreover, if $0 < \rho < \sqrt{ab}$ we have

7.40
$$\left| \frac{b - \rho}{b + \rho} \right| - \frac{\sqrt{b} - \sqrt{a}}{\sqrt{b} + \sqrt{a}} = \frac{2\sqrt{b}\,(\sqrt{ab} - \rho)}{(b + \rho)(\sqrt{b} + \sqrt{a})} > 0$$

and if $\sqrt{ab} < \rho$, then

7.41
$$\left| \frac{a - \rho}{a + \rho} \right| - \frac{\sqrt{b} - \sqrt{a}}{\sqrt{b} + \sqrt{a}} = \frac{2\sqrt{b}\,(\rho - \sqrt{ab})}{(\rho + a)(\sqrt{b} + \sqrt{a})} > 0.$$

---

* Evidently for any $v \neq 0$ we have $(\hat{H}v, v) = (Q^{-\frac{1}{2}}(QH)Q^{-\frac{1}{2}}v, v) = ((QH)(Q^{-\frac{1}{2}}v), Q^{-\frac{1}{2}}v) > 0$ since $Q^{-\frac{1}{2}}v \neq 0$ and since QH and Q are positive definite.

Therefore $\phi(a, b; \rho)$ is minimized when $\rho = \sqrt{ab}$ and

7.42
$$S(T_{\sqrt{ab}}) \leq \phi(a, b; \sqrt{ab}) = \left(\frac{\sqrt{b} - \sqrt{a}}{\sqrt{b} + \sqrt{a}}\right)^2.$$

Thus $\rho = \sqrt{ab}$ is optimum in the sense that the bound $\phi(a, b; \rho)$ for $S(T_\rho)$ is minimized.

Let us now consider the model problem. Evidently H corresponds to the difference operator

7.43
$$\mathcal{H}[u](x, y) = 2u(x, y) - u(x + h, y) - u(x - h, y)$$

while V corresponds to the difference operator

7.44
$$\mathcal{V}[u](x, y) = 2u(x, y) - u(x, y + h) - u(x, y - h).$$

If $Hv = \mu v$, then there corresponds a function $v(x, y)$ vanishing on $x = 0$ and $x = 1$ such that

7.45
$$\mathcal{H}[v](x, y) = \mu v(x, y).$$

If we let

7.46
$$v(x, y) = \sin p\pi x$$

we see that (7.45) is satisfied with

7.47
$$\mu = \mu_p = 2 - 2 \cos p\pi h = 4 \sin^2 \frac{p\pi h}{2}.$$

Thus the eigenvalues of H are

7.48
$$\mu_p = 4 \sin^2 \frac{p\pi h}{2}, \qquad p = 1, 2, \ldots, M - 1,$$

where $M = h^{-1}$, and the eigenvalues of V are, similarly,

7.49
$$v_q = 4 \sin^2 \frac{q\pi h}{2}, \qquad q = 1, 2, \ldots M - 1.$$

Hence we have

7.50
$$a = 4 \sin^2 \frac{\pi h}{2}, \qquad b = 4 \cos^2 \frac{\pi h}{2}.$$

The optimum parameter is given by

7.51
$$\bar{\rho} = \sqrt{ab} = 4 \sin \frac{\pi h}{2} \cos \frac{\pi h}{2} = 2 \sin \pi h$$

and

7.52
$$S(T_{\bar{\rho}}) = \left(\frac{\cos \dfrac{\pi h}{2} - \sin \dfrac{\pi h}{2}}{\cos \dfrac{\pi h}{2} + \sin \dfrac{\pi h}{2}} \right)^2 = \frac{1 - \sin \pi h}{1 + \sin \pi h}.$$

But by (6.32) and (6.45) we have

7.53
$$S(\mathscr{L}_{\omega_b}) = S(T_{\bar{\rho}}).$$

Thus, based on a comparison of the spectral radii, the rate of convergence of the Peaceman-Rachford method with $\rho = \bar{\rho}$ is the *same* as that of the SOR method with $\omega = \omega_b$. On the other hand, considerably more work is required per iteration using the Peaceman-Rachford method. Not only is there a double sweep required but for each half-iteration one has to solve linear systems with tridiagonal matrices.

The convergence of the Peaceman-Rachford method is frequently very rapid if one allows ρ to vary from iteration to iteration. This rapid convergence can be proved to hold for an appropriate choice of the iteration parameters in the *commutative case*. Following Birkhoff, Varga, and Young [1962] we say that the commutative case holds if the matrices H_0, V_0 and Σ of (7.7) satisfy the following conditions:

7.54   a) $H_0 V_0 = V_0 H_0$;

7.55   b) $\Sigma = \sigma I$, where σ is a nonnegative constant;

7.56   c) H_0 and V_0 are similar to nonnegative diagonal matrices.

If these assumptions hold then the matrices $H = H_0 + \frac{1}{2}\Sigma$, $V = V_0 + \frac{1}{2}\Sigma$ satisfy the conditions

7.57   a') $HV = VH$

7.58   b') H and V are similar to nonnegative diagonal matrices.

We state without proof the following theorem of Frobenius*.

7.59 Theorem. If H and V are similar to diagonal matrices and if $HV = VH$,

---

* See Thrall and Tornheim [1957], page 190, exercise 1.

then there exists a nonsingular matrix W such that

7.60
$$W^{-1}HW = \Lambda_H, \qquad W^{-1}VW = \Lambda_V$$

where Λ_H and Λ_V are diagonal matrices.

It follows from (7.57) and (7.58) that there exists a set of linearly independent vectors $v_1, v_2, \ldots, v_N$, which correspond to the columns of W in Theorem 7.59, such that each v_i is an eigenvector both of H and of V.

We can verify the above result for the model problem with $h^{-1} = M$. Later in this section we shall show that $HV = VH$. Since H and V are positive definite, (7.58) holds. It is easy to verify that for any integer p and q such that $1 \leq p \leq M - 1$, $1 \leq q \leq M - 1$ the vector $v_{p,q}$ which corresponds to the function

7.61
$$v_{p,q}(x, y) = \sin p\pi x \sin q\pi y$$

is an eigenvector of H, corresponding to the eigenvalue μ_p given by (7.48), and is also an eigenvector of V corresponding to the eigenvalue v_q given by (7.49).

In the case of (7.5) we have

7.62 $\quad v_{1,1} = \frac{3}{4} \begin{pmatrix} 1 \\ 1 \\ 1 \\ 1 \end{pmatrix}, \qquad v_{1,2} = \frac{3}{4} \begin{pmatrix} 1 \\ 1 \\ -1 \\ -1 \end{pmatrix}, \qquad v_{2,1} = \frac{3}{4} \begin{pmatrix} 1 \\ -1 \\ 1 \\ -1 \end{pmatrix}, \qquad v_{2,2} = \frac{3}{4} \begin{pmatrix} 1 \\ -1 \\ -1 \\ 1 \end{pmatrix}$

and

7.63
$$W = \frac{3}{4} \begin{pmatrix} 1 & 1 & 1 & 1 \\ 1 & 1 & -1 & -1 \\ 1 & -1 & 1 & -1 \\ 1 & -1 & -1 & 1 \end{pmatrix}.$$

Evidently

7.64
$$HW = \frac{3}{4} \begin{pmatrix} 1 & 1 & 3 & 3 \\ 1 & 1 & -3 & -3 \\ 1 & -1 & 3 & -3 \\ 1 & -1 & -3 & 3 \end{pmatrix} = W \begin{pmatrix} 1 & 0 & 0 & 0 \\ 0 & 1 & 0 & 0 \\ 0 & 0 & 3 & 0 \\ 0 & 0 & 0 & 3 \end{pmatrix}$$

7.65
$$VW = \tfrac{3}{4} \begin{pmatrix} 1 & 3 & 1 & 3 \\ 1 & 3 & -1 & -3 \\ 1 & -3 & 1 & -3 \\ 1 & -3 & -1 & 3 \end{pmatrix} = W \begin{pmatrix} 1 & 0 & 0 & 0 \\ 0 & 3 & 0 & 0 \\ 0 & 0 & 1 & 0 \\ 0 & 0 & 0 & 3 \end{pmatrix}$$

so that, since W is nonsingular,

7.66
$$W^{-1}HW = \Lambda_H = \begin{pmatrix} 1 & 0 & 0 & 0 \\ 0 & 1 & 0 & 0 \\ 0 & 0 & 3 & 0 \\ 0 & 0 & 0 & 3 \end{pmatrix}$$

7.67
$$W^{-1}VW = \Lambda_V = \begin{pmatrix} 1 & 0 & 0 & 0 \\ 0 & 3 & 0 & 0 \\ 0 & 0 & 1 & 0 \\ 0 & 0 & 0 & 3 \end{pmatrix}.$$

For any column v of W we have

7.68
$$Hv = \mu v, \qquad Vv = vv$$

for some μ and v. Evidently

7.69
$$T_\rho v = (V + \rho I)^{-1}(H - \rho I)(H + \rho I)^{-1}(V - \rho I)v.$$

Since $(H + \rho I)^{-1}v = (\mu + \rho)^{-1}v, (V + \rho I)^{-1}v = (v + \rho)^{-1}v$ it follows that

7.70
$$T_\rho v = \left(\frac{(\mu - \rho)(v - \rho)}{(\mu + \rho)(v + \rho)} \right) v.$$

Thus v is an eigenvector of T_ρ for any ρ. Hence for any $\rho_1, \rho_2, \ldots, \rho_m$ we have

7.71
$$\left(\prod_{i=1}^{m} T_{\rho_i} \right) v = \left(\prod_{i=1}^{m} \frac{(\mu - \rho_i)(v - \rho_i)}{(\mu + \rho_i)(v + \rho_i)} \right) v.$$

Thus v is an eigenvector of the matrix

7.72
$$\prod_{i=1}^{m} T_{\rho_i}$$

which corresponds to the use of the Peaceman-Rachford method for m iterations using $\rho_1, \rho_2, \ldots, \rho_m$. If all eigenvalues μ of H and v of V lie in the ranges $0 < a \leq \mu \leq b, 0 < a \leq v \leq b$, we have

7.73
$$S\left(\prod_{i=1}^{m} T_{\rho_i}\right) \leq \max_{\substack{a \leq \mu \leq b \\ a \leq v \leq b}} \left(\prod_{i=1}^{m} \left|\frac{(\mu - \rho_i)(v - \rho_i)}{(\mu + \rho_i)(v + \rho_i)}\right|\right)$$

$$= \max_{a \leq \gamma \leq b} \left(\prod_{i=1}^{m} \left|\frac{\gamma - \rho_i}{\gamma + \rho_i}\right|\right)^2 = \phi(a, b; \rho_1, \ldots, \rho_m).$$

The problem of minimizing $\phi(a, b; \rho_1, \rho_2, \ldots, \rho_m)$ has been solved analytically in terms of elliptic functions by W. B. Jordan and E. L. Wachspress [1966].* Rather than describe this solution we consider instead parameters used by Peaceman and Rachford [1955], which yield results which are nearly as good as the theoretical optimum parameters. The Peaceman-Rachford parameters are

7.74
$$\rho_i^* = b\left(\frac{a}{b}\right)^{(2i-1)/(2m)}, \qquad i = 1, 2, \ldots, m.$$

With this set of parameters we can show that

7.75
$$S\left(\prod_{i=1}^{m} T_{\rho_i^*}\right) \leq \left(\frac{1 - \left(\frac{a}{b}\right)^{1/(2m)}}{1 + \left(\frac{a}{b}\right)^{1/(2m)}}\right)^2.$$

Since each factor of the product

7.76
$$\prod_{i=1}^{m} \left|\frac{\gamma - \rho_i^*}{\gamma + \rho_i^*}\right|$$

is less than unity it suffices to show that for any λ there is one factor which is less than $(1 - z)/(1 + z)$, where

7.77
$$z = \left(\frac{a}{b}\right)^{1/(2m)}.$$

Evidently by (7.74) we have

7.78
$$b > \rho_1^* > \rho_2^* > \cdots > \rho_m^* > a.$$

In the interval $a \leq \gamma \leq \rho_m^*$ we have

7.79
$$\left|\frac{\gamma - \rho_m^*}{\gamma + \rho_m^*}\right| \leq \left|\frac{a - \rho_m^*}{a + \rho_m^*}\right|$$

---

* Wachspress [1966] has also given a solution for the case $m = 2^p$, for any integer p.

since $(\gamma - \rho)(\gamma + \rho)^{-1}$ is an increasing function of γ. Therefore, since $\rho_m^* = b(a/b)^{1-(1/(2m))} = a/z$ we have

7.80
$$\left| \frac{\gamma - \rho_m^*}{\gamma + \rho_m^*} \right| \leqq \left| \frac{a - \frac{a}{z}}{a + \frac{a}{z}} \right| = \left| \frac{1 - z}{1 + z} \right|.$$

Similarly, in the interval $\rho_1^* \leqq \gamma \leqq b$ we have

7.81
$$\left| \frac{\gamma - \rho_1^*}{\gamma + \rho_1^*} \right| \leqq \left| \frac{1 - z}{1 + z} \right|.$$

In the interval $\rho_{i+1}^* \leqq \gamma \leqq \rho_i^*$ the function

7.82
$$\left(\frac{\gamma - \rho_{i+1}^*}{\gamma + \rho_{i+1}^*} \right) \left(\frac{\gamma - \rho_i^*}{\gamma + \rho_i^*} \right)$$

which vanishes at the end points ρ_i^* and ρ_{i+1}^*, has an extreme value at $\gamma = \sqrt{\rho_i^* \rho_{i+1}^*}$ since

7.83
$$\frac{d}{d\gamma} \left(\frac{\gamma - \rho_{i+1}^*}{\gamma + \rho_{i+1}^*} \right) \left(\frac{\gamma - \rho_i^*}{\gamma + \rho_i^*} \right) = \frac{2(\rho_i^* + \rho_{i+1}^*)(\gamma^2 - \rho_i^* \rho_{i+1}^*)}{(\gamma + \rho_{i+1}^*)^2 (\gamma + \rho_i^*)^2}.$$

The extreme value of (7.82) has modulus

7.84
$$\left(\frac{\sqrt{\rho_i^*} - \sqrt{\rho_{i+1}^*}}{\sqrt{\rho_i^*} + \sqrt{\rho_{i+1}^*}} \right)^2 = \left(\frac{1 - z}{1 + z} \right)^2 \leqq \frac{1 - z}{1 + z}.$$

Therefore

7.85
$$\max_{a \leqq \gamma \leqq b} \left(\prod_{i=1}^{m} \left| \frac{\gamma - \rho_i^*}{\gamma + \rho_i^*} \right| \right)^2 \leqq \left(\frac{1 - z}{1 + z} \right)^2$$

and, by (7.73), the result (7.75) follows.

For the model problem we have as before

7.86
$$a = 4 \sin^2 \frac{\pi h}{2}, \qquad b = 4 \cos^2 \frac{\pi h}{2}.$$

The Peaceman-Rachford parameters are given by

7.87
$$\rho_i^* = 4 \cos^2 \frac{\pi h}{2} \left(\tan^2 \frac{\pi h}{2} \right)^{(2i-1)/(2m)}, \qquad i = 1, 2, \ldots, m,$$

and

7.88
$$S\left(\prod_{i=1}^{m} T_{\rho_i^*}\right) \leqq \left(\frac{1 - \left(\tan\frac{\pi h}{2}\right)^{1/m}}{1 + \left(\tan\frac{\pi h}{2}\right)^{1/m}}\right)^2$$

$$\sim \left(\frac{1 - \left(\frac{\pi h}{2}\right)^{1/m}}{1 + \left(\frac{\pi h}{2}\right)^{1/m}}\right)^2 \sim 1 - 4\left(\frac{\pi h}{2}\right)^{1/m}.$$

Therefore for the rate of convergence we have

7.89
$$R\left(\prod_{i=1}^{m} T_{\rho_i^*}\right) \sim 4\left(\frac{\pi h}{2}\right)^{1/m}.$$

Thus the rate of convergence is asymptotically proportional to $h^{1/m}$, and the number of iterations needed to reduce the error by a certain fixed amount is proportional to $h^{-1/m}$. For $m > 1$ this represents an order of magnitude improvement over the SOR method where the rate of convergence is proportional to h.

As an example, let us consider the case $h = \frac{1}{20}$. Evidently we have

7.90
$$a = 4\sin^2\frac{\pi}{40} \doteq 0.024623, \qquad b \doteq 4\cos^2\frac{\pi}{40} = 3.97538.$$

The Peaceman-Rachford parameters for the cases $m = 1, 2, 3, 4$ are given in Table 7.91.

7.91 Table

i	$m = 1$	$m = 2$	$m = 3$	$m = 4$
1	.31287	.08777	.057458	.046489
2		1.11525	.31287	.16571
3			1.70362	.59070
4				2.10559

The bounds on $S\left(\prod_{i=1}^{m} T_{\rho_i^*}\right)$ and $\sqrt[m]{S\left(\prod_{i=1}^{m} T_{\rho_i^*}\right)}$ for $m = 1(1)10$ are given in Table 7.92. (Note that z_m is defined below.)

7.92 Table

m	z_m	$\left(\dfrac{1-z_m}{1+z_m}\right)^2$	$\sigma_m = \sqrt[m]{\left(\dfrac{1-z_m}{1+z_m}\right)^2}$
1	.07870	.72945	.72945
2	.28054	.31567	.56184
3	.42854	.16002	.54291
4	.52966	.09454	.55451
5	.60144	.06194	.57331
6	.65463	.04357	.59319
7	.69548	.03226	.61228
8	.72778	.02482	.63003
9	.75393	.01968	.64633
10	.77553	.01598	.66125

Evidently the bound on the average spectral radius $\left[S\left(\prod_{i=1}^{m} T_{\rho_i^*}\right)\right]^{1/m}$ initially decreases and then increases. Thus there is a value of m which is optimum in the sense that the bound on $\left[S\left(\prod_{i=1}^{m} T_{\rho_i^*}\right)\right]^{1/m}$ is minimized. (We remark that if one were to use the optimum parameters, the bound on the average spectral radius would be a decreasing function of m for all m.)

We seek to determine m so that

7.93
$$\sigma_m = \left(\frac{1-z_m}{1+z_m}\right)^{2/m}$$

is minimized, where

7.94
$$z_m = \left(\frac{a}{b}\right)^{1/(2m)}$$

Let

7.95
$$\delta = \frac{1-z_m}{1+z_m}.$$

Then

7.96
$$z_m = \frac{1-\delta}{1+\delta}$$

and

7.97
$$m = \frac{\log(a/b)}{2 \log[(1 - \delta)/(1 + \delta)]}.$$

Equating to zero the first derivative with respect to δ of

7.98
$$\zeta_m = -\log \sigma_m = -\frac{2}{m} \log\left(\frac{1 - z_m}{1 + z_m}\right) = -\frac{4 \log\left(\frac{1 - \delta}{1 + \delta}\right)}{\log\left(\frac{a}{b}\right)} \log \delta$$

we get

7.99
$$\frac{1 - \delta^2}{2} \log \frac{1 - \delta}{1 + \delta} = \delta \log \delta$$

which has the solution

7.100
$$\bar{\delta} = \sqrt{2} - 1 \doteq 0.414.$$

Thus, the optimum value of m is determined by finding the smallest integer such that

7.101
$$(0.414)^{2m} \leqq \frac{a}{b}.$$

The corresponding value of ζ_m is

7.102
$$\bar{\zeta}_m \doteq \frac{4(\log \bar{\delta})^2}{-\log(a/b)} \doteq \frac{3.11}{-\log(a/b)}.$$

For the model problem we have $a/b = \tan^2(\pi h/2)$ so that

7.103
$$\bar{\zeta}_m \doteq \frac{3.11}{-2 \log \tan \dfrac{\pi h}{2}} \sim \frac{1.55}{-\log \dfrac{\pi h}{2}}.$$

Thus the rate of convergence is proportional to $\log(\pi h/2)$, which decreases very slowly with h, as compared to the rate of convergence of the SOR method which is proportional to h.

In the case $h = \frac{1}{20}$ (7.101) becomes

7.104
$$(0.414)^{2m} \leq \tan^2 \frac{\pi h}{2} \doteq (0.0787)^2$$

and

7.105 $m = 3.$

This choice of m agrees with Table 7.92 where the smallest value of σ_m is assumed when $m = 3$. We remember that $S(\mathscr{L}_{\omega_b}) = S(\mathscr{L}_{1.73}) = 0.73$ for this problem. Since $S(\mathscr{L}_{\omega_b})^2 \doteq 0.533$ it follows that with the SOR method two iterations are slightly better than one double iteration of the Peaceman-Rachford method. However, two things should be pointed out. First, the Peaceman-Rachford method actually converges faster than indicated by the above analysis. Second, for smaller h the Peaceman-Rachford method becomes superior to the SOR method.

The results in Table 7.106 were obtained using the Peaceman-Rachford method and the SOR method for the model problem with various values of h. (See Birkhoff, Varga, and Young [1962] and Mouradoglou [1967].)

7.106 Table

SOR Method			Peaceman-Rachford Method						
h^{-1}	ω	n	$m = 1$	$m = 2$	$m = 3$	$m = 4$	$m = 5$	$m = 10$
5	1.27	12	12	10	9	8		·
10	1.54	28	23	16	15	15		
20	1.74	53	46	24	21	20		
40	1.86	117	91	36	27	27	25	27
80	1.93	236	183	49	37	31		
120		351*	274	61	44	38		
160		472*		71	47	39		

* *estimated values*

The boundary values were assumed to be zero and the initial values in R_h were assumed to be unity. The iteration was carried out until the largest value of $|u^{(n)}(x, y)|$ in R_h was less than 10^{-6}

Let us now discuss the circumstances under which the "commutative case" holds. We shall be primarily concerned with systems (7.7) derived from the self-adjoint differential equation

7.107 $$\frac{\partial}{\partial x}\left(A\frac{\partial u}{\partial x}\right) + \frac{\partial}{\partial y}\left(C\frac{\partial u}{\partial y}\right) + Fu = G.$$

We consider the difference equation

7.108 $\mathscr{H}_0[u](x, y) + \mathscr{V}_0[u](x, y) - h^2Fu(x, y) = -h^2G$

where

7.109 $\mathscr{H}_0[u](x, y) = [A(x + \frac{h}{2}, y) + A(x - \frac{h}{2}, y)]u(x, y)$

$$- A(x + \frac{h}{2}, y)u(x + h, y) - A(x - \frac{h}{2}, y)u(x - h, y)$$

7.110 $\mathscr{V}_0[u](x, y) = [C(x, y + \frac{h}{2}) + C(x, y - \frac{h}{2})]u(x, y)$

$$- C(x, y + \frac{h}{2})u(x, y + h) - C(x, y - \frac{h}{2})u(x, y - h).$$

Thus the matrix H_0 corresponds to the operator $\mathscr{H}_0[u]$, the matrix V_0 corresponds to the operator $\mathscr{V}_0[u]$, and the matrix Σ corresponds to the operator $-h^2 Fu$.

We seek to determine necessary and sufficient conditions for the existence of a diagonal matrix P with positive diagonal elements such that conditions (7.54), (7.55), and (7.56) hold with H_0, V_0, Σ replaced by $PH_0, PV_0, P\Sigma$, respectively, in other words, so that the commutative case holds for the system

7.111 $(PH_0 + PV_0 + P\Sigma)u = Pb.$

Thus we must have

7.112 $(PH_0)(PV_0) = (PV_0)(PH_0)$

7.113 $P\Sigma = \sigma I$, where σ is a nonnegative constant

7.114 PH_0 and PV_0 are similar to nonnegative diagonal matrices.

Evidently (7.114) is automatically satisfied for any diagonal matrix P with positive diagonal elements since PH_0 is similar to the positive definite matrix $P^{\frac{1}{2}}H_0 P^{\frac{1}{2}}$ and PV_0 is similar to $P^{\frac{1}{2}}V_0 P^{\frac{1}{2}}$.

Evidently, the matrices PH_0, PV_0 and $P\Sigma$ correspond to the operators

7.115 $\begin{cases} \mathscr{H}_0'[u](x, y) = A_0(x, y)u(x, y) - A_1(x, y)u(x + h, y) - A_3(x, y)u(x - h, y) \\ \mathscr{V}_0'[u](x, y) = C_0(x, y)u(x, y) - C_2(x, y)u(x, y + h) - C_4(x, y)u(x, y - h) \\ \mathscr{E}_0'[u](x, y) = -h^2 P(x, y)F(x, y)u(x, y) \end{cases}$

where

$$
\begin{cases}
A_0(x, y) = P(x, y)[A(x + \frac{h}{2}, y) + A(x - \frac{h}{2}, y)] \\[2mm]
A_1(x, y) = P(x, y)A(x + \frac{h}{2}, y) \\[2mm]
A_3(x, y) = P(x, y)A(x - \frac{h}{2}, y) \\[2mm]
C_0(x, y) = P(x, y)[C(x, y + \frac{h}{2}) + C(x, y - \frac{h}{2})] \\[2mm]
C_2(x, y) = P(x, y)C(x, y + \frac{h}{2}) \\[2mm]
C_4(x, y) = P(x, y)C(x, y - \frac{h}{2}).
\end{cases}
$$

7.116

Actually, the matrices PH_0 and PV_0 correspond more directly to the operators $\mathscr{H}_0'[u]$ and $\mathscr{V}_0'[u]$, respectively, which are defined for functions $u(x, y)$ vanishing on S_h and are such that $\mathscr{H}_0'[u](x, y)$ and $\mathscr{V}_0'[u](x, y)$ vanish on S_h. Thus, we have for any $u(x, y)$ vanishing on S_h

7.117
$$
\begin{cases}
\tilde{\mathscr{H}}_0'[u](x, y) = \Gamma(x, y)\mathscr{H}_0'[u](x, y) \\[2mm]
\tilde{\mathscr{V}}_0'[u](x, y) = \Gamma(x, y)\mathscr{V}_0'[u](x, y)
\end{cases}
$$

where

7.118
$$
\Gamma(x, y) =
\begin{cases}
1 & \text{if } (x, y) \in R_h \\[2mm]
0 & \text{if } (x, y) \notin R_h.
\end{cases}
$$

Evidently the product matrix $(PV_0)(PH_0)$ corresponds to the operator $\tilde{\mathscr{V}}_0'\tilde{\mathscr{H}}_0'[u](x, y)$ which for $(x, y) \in R_h$ can be determined by

7.119 $\quad \tilde{\mathscr{V}}_0'\tilde{\mathscr{H}}_0'[u](x, y) = C_0(x, y)\tilde{\mathscr{H}}_0'[u](x, y) - C_2(x, y)\tilde{\mathscr{H}}_0'[u](x, y + h)$

$$- C_4(x, y)\tilde{\mathscr{H}}_0'[u](x, y - h)$$

$$
\begin{aligned}
= \; & C_0(x, y)\mathscr{H}_0'[u](x, y) \\
& - C_2(x, y)\Gamma(x, y + h)\mathscr{H}_0'[u](x, y + h) \\
& - C_4(x, y)\Gamma(x, y - h)\mathscr{H}_0'[u](x, y - h) \\
= \; & A_0(x, y)C_0(x, y)u(x, y) - C_0(x, y)A_1(x, y)u(x + h, y) \\
& - \Gamma(x, y + h)C_2(x, y)A_0(x, y + h)\,u(x, y + h) \\
& \qquad\qquad - C_0(x, y)A_3(x, y)u(x - h, y) \\
& - \Gamma(x, y - h)C_4(x, y)A_0(x, y - h)u(x, y - h) \\
& + \Gamma(x, y + h)C_2(x, y)A_1(x, y + h)u(x + h, y + h) \\
& + \Gamma(x, y + h)C_2(x, y)A_3(x, y + h)u(x - h, y + h) \\
& + \Gamma(x, y - h)C_4(x, y)A_1(x, y - h)u(x + h, y - h) \\
& + \Gamma(x, y - h)C_4(x, y)A_3(x, y - h)u(x - h, y - h).
\end{aligned}
$$

A similar expression can be obtained for $\mathscr{H}_0'\mathscr{V}_0'[u](x, y)$. Indeed, we can summarize the results in Table 7.120.

7.120 Table

	Coefficient in $\mathscr{V}_0'\mathscr{H}_0'[u](x, y)$	Coefficient in $\mathscr{H}_0'\mathscr{V}_0'[u](x, y)$
$u(x, y)$	$A_0(x, y)C_0(x, y)$	$A_0(x, y)C_0(x, y)$
$u(x + h, y)$	$-C_0(x, y)A_1(x, y)$	$-\Gamma(x + h, y)A_1(x, y)C_0(x + h, y)$
$u(x, y + h)$	$-\Gamma(x, y + h)C_2(x, y)A_0(x, y + h)$	$-A_0(x, y)C_2(x, y)$
$u(x - h, y)$	$-C_0(x, y)A_3(x, y)$	$-\Gamma(x - h, y)A_3(x, y)C_0(x - h, y)$
$u(x, y - h)$	$-\Gamma(x, y - h)C_4(x, y)A_0(x, y - h)$	$-A_0(x, y)C_4(x, y)$
$u(x + h, y + h)$	$\Gamma(x, y + h)C_2(x, y)A_1(x, y + h)$	$\Gamma(x + h, y)A_1(x, y)C_2(x + h, y)$
$u(x - h, y + h)$	$\Gamma(x, y + h)C_2(x, y)A_3(x, y + h)$	$\Gamma(x - h, y)A_3(x, y)C_2(x - h, y)$
$u(x + h, y - h)$	$\Gamma(x, y - h)C_4(x, y)A_1(x, y - h)$	$\Gamma(x + h, y)A_1(x, y)C_4(x + h, y)$
$u(x - h, y - h)$	$\Gamma(x, y - h)C_4(x, y)A_3(x, y - h)$	$\Gamma(x - h, y)A_3(x, y)C_4(x - h, y)$

We first show that for commutativity we must have a rectangular region, assuming that R_h is connected and that $A > 0, C > 0, F \leq 0$. Suppose $(x + h, y + h)$ and (x, y) are in R_h. Then by equating the coefficients of $u(x + h, y + h)$ we see that $\Gamma(x, y + h) = \Gamma(x + h, y)$. Thus, $(x + h, y)$ belongs to R_h if and only if $(x, y + h)$ does. In this way one can show that if *any* three of the four points $(x, y), (x + h, y), (x, y + h), (x + h, y + h)$ lie in R_h the fourth does also. From this and the fact that R_h is connected, one can show that R_h is a rectangular set of points or else R_h is a line of points.

Suppose now that there exist positive functions $E_1(x), E_2(x), F_1(y), F_2(y)$ and a constant $c \geqq 0$ such that

7.121 $A(x, y) = E_1(x)F_1(y), C(x, y) = E_2(x)F_2(y), F(x, y) = -cE_2(x)F_1(y).$

If we let

7.122
$$P(x, y) = \frac{1}{E_2(x)F_1(y)}$$

then we have, by (7.116)

7.123
$$
\begin{cases}
A_0(x, y) = \dfrac{E_1\left(x + \dfrac{h}{2}\right) + E_1\left(x - \dfrac{h}{2}\right)}{E_2(x)} \\[4mm]
A_1(x, y) = \dfrac{E_1\left(x + \dfrac{h}{2}\right)}{E_2(x)}, \qquad A_3(x, y) = \dfrac{E_1\left(x - \dfrac{h}{2}\right)}{E_2(x)} \\[4mm]
C_0(x, y) = \dfrac{F_2\left(y + \dfrac{h}{2}\right) + F_2\left(y - \dfrac{h}{2}\right)}{F_1(y)} \\[4mm]
C_2(x, y) = \dfrac{F_2\left(y + \dfrac{h}{2}\right)}{F_1(y)}, \qquad C_4(x, y) = \dfrac{F_2\left(y - \dfrac{h}{2}\right)}{F_1(y)}.
\end{cases}
$$

Since the $A_i(x, y)$ are independent of y and since the $C_i(x, y)$ are independent of x it follows from Table 7.120 that $\mathcal{V}_0' \mathcal{H}_0'[u](x, y) = \mathcal{H}_0' \mathcal{V}_0'[u](x, y)$.

As an example, let us consider the equation

7.124
$$\frac{\partial}{\partial x}\left(\frac{1}{y}\frac{\partial u}{\partial x}\right) + \frac{\partial}{\partial y}\left(\frac{1}{y}\frac{\partial u}{\partial y}\right) - \frac{2}{y}u = e^x.$$

Evidently, the conditions (7.121) are satisfied with

7.125 $E_1(x) = 1, \quad F_1(y) = \dfrac{1}{y}, \quad E_2(x) = 1, \quad F_2(y) = \dfrac{1}{y}, \quad c = 2.$

Letting

7.126
$$P(x, y) = y$$

we have, by (7.116)

7.127
$$
\left\{ 2u(x, y) - u(x + h, y) - u(x - h, y) \right\} + \left\{ \left[\frac{y}{y + \dfrac{h}{2}} + \frac{y}{y - \dfrac{h}{2}} \right] u(x, y) \right.
$$

$$
\left. - \frac{y}{y + \dfrac{h}{2}} u(x, y + h) - \frac{y}{y - \dfrac{h}{2}} u(x, y - h) \right\}
$$

$$
+ 2h^2 u(x, y) = -h^2 y e^x.
$$

This leads to the linear system

7.128
$$
(H_0'' + V_0'' + \Sigma'')u = b''
$$

where $H_0'' V_0'' = V_0'' H_0''$, Σ'' is a constant times the identity matrix, and H_0'' and V_0'' are similar to positive definite matrices.

It can also be shown (see Birkhoff, Varga, and Young [1962]) that the conditions (7.121) are *necessary* in order that (7.112), (7.113), and (7.114) hold for some P and for some family of values of h tending to zero. (Of course, we also require that the region be a rectangle.)

On the other hand, if we have an equation which is not in the self-adjoint form (7.107) then we may satisfy the conditions (7.112), (7.113), and (7.114), even though conditions (7.121) are not satisfied. Thus consider the equation

7.129
$$
\frac{\partial^2 u}{\partial x^2} + \frac{\partial^2 u}{\partial y^2} + \frac{2}{x + y} \frac{\partial u}{\partial x} + \frac{2}{x + y} \frac{\partial u}{\partial y} = 0
$$

in the unit square. If we consider the difference equation

7.130
$$
\mathscr{H}[u](x, y) + \mathscr{V}[u](x, y) = 0
$$

where

7.131
$$
\left\{
\begin{array}{l}
\mathscr{H}[u](x, y) = 2u(x, y) - \left(1 + \dfrac{h}{x + y} \right) u(x + h, y) - \left(1 - \dfrac{h}{x + y} \right) u(x - h, y) \\[3mm]
\mathscr{V}[u](x, y) = 2u(x, y) - \left(1 + \dfrac{h}{x + y} \right) u(x, y + h) - \left(1 - \dfrac{h}{x + y} \right) u(x, y - h)
\end{array}
\right.
$$

one can easily show from Table 7.120 that $\mathscr{H}\mathscr{V}[u](x, y) = \mathscr{V}\mathscr{H}[u](x, y)$. Hence the corresponding matrices H and V commute. Moreover, it can be shown that H and V are similar to positive definite matrices. Hence the commutative case holds.

Even though the commutative case holds only for a very limited class of cases, nevertheless, numerical experience indicates that the Peaceman-Rachford method converges rapidly in many other cases. Some numerical experiments involving the model problem are described in papers by Young and Ehrlich [1960], Birkhoff, Varga, and Young [1962], and Mouradoglou [1967]. On the other hand, there do exist cases where H and V are positive definite and there exist $\rho_1 > 0, \rho_2 > 0$ such that

7.132 $$S(T_{\rho_1} T_{\rho_2}) > 1$$

(see Price and Varga [1962]). Guilinger [1965] has shown that if R is a convex region then for the discrete analog of the Dirichlet problem we have $S\left(\prod_{i=1}^{m} T_{\rho_i}\right) < 1$ for any set of positive parameters $\rho_1, \rho_2, \ldots, \rho_m$. So far no case involving the discrete analog of the Dirichlet problem has been observed where divergence occurs for any set of positive parameters. There are many problems involving nonconvex regions where neither convergence nor nonconvergence has been proved. Considerable research is still needed to determine precisely under what conditions the Peaceman-Rachford method will be effective.

Exercises 16.7

1. Perform one complete iteration (one double sweep) of the Peaceman-Rachford method for solving the model problem for the unit square with $u = 1000$ on the side $y = 1$ and $u = 0$ elsewhere on the boundary. Use zero starting values. Let $h = \frac{1}{3}$ and $\rho = 2$. Find the eigenvalues of T_ρ for $\rho = 2$.

2. In the previous exercise find the value of ρ which minimizes T_ρ and determine the corresponding value of $S(T_\rho)$.

3. Verify that W as given by (7.63) is nonsingular.

4. For the model problem with $h = \frac{1}{10}$, find the two Peaceman-Rachford parameters corresponding to $m = 2$. Determine

$$S\left(\prod_{i=1}^{2} T_{\rho_i}\right).$$

Estimate the number of double sweeps needed to reduce an initial error by a factor of 10^{-6}.

5. For the model problem with $h = \frac{1}{80}$, determine the optimum number of Peaceman-Rachford parameters and the values of the parameters. How many complete iterations would be required using these parameters to reduce an initial error by a factor of 10^{-4}? Make a similar estimate for the SOR method with the optimum ω.

6. Verify the statements following (7.131). Find H and V corresponding to (7.129) for the unit square with $h = \frac{1}{3}$ and verify that $HV = VH$.

7. Consider the generalized Dirichlet problem for the differential equation

$$(y^{-1}u_x)_x + (y^{-1}u_y)_y - (2/y)u = x^3$$

in the region $0 < x < 1, 1 < y < 2$. Show how one can construct a difference equation leading to a linear system

$$(H_0 + V_0 + \Sigma)u = b$$

where H_0, V_0, and Σ satisfy the conditions (7.54), (7.55), and (7.56). Obtain H_0, V_0, and Σ explicitly for the case $h = \frac{1}{3}$.

8. Show that the problem of finding ρ_1 and ρ_2 so as to minimize

$$\psi(\rho_1, \rho_2) = \max_{0 < a \leq \gamma \leq b} \left| \frac{(\rho_1 - \gamma)(\rho_2 - \gamma)}{(\rho_1 + \gamma)(\rho_2 + \gamma)} \right|$$

is equivalent to finding ρ to minimize

$$\max_{0 < \alpha \leq \gamma \leq \beta} \left| \frac{\rho - \gamma}{\rho + \gamma} \right|$$

for suitable α and β. (Assume the existence of a unique pair, $(\bar{\rho}_1, \bar{\rho}_2)$ which minimizes $\psi(\rho_1, \rho_2)$.) This is the basis for Wachspress' solution of the optimum parameter problem for the case $m = 2^p$. Find the optimum ρ_1 and ρ_2 if $a = 1$, $b = 3$.

16.8 OTHER METHODS

In this section we discuss briefly a number of other iterative methods, many of which are related to those already considered.

Group Iterative Methods

With group iterative methods one divides the set $W = \{1, 2, \ldots, N\}$ into disjoint subsets, or "groups," $R_1, R_2, \ldots, R_q$ such that $R_1 + R_2 + \cdots + R_q = W$. Given a matrix A of order N we let $D^{(\pi)}$ be the matrix formed from A by replacing all elements of A by zero except for those elements $a_{i,j}$ where i and j belong to the same group. Here the symbol π is used to indicate the division of the set W into subsets. As a special case, if π_0 is a division of W into the subsets $R_1 = \{1\}$, $R_2 = \{2\}, \ldots, R_N = \{N\}$, then $D^{(\pi_0)} = \text{diag } A$.

In general we let

8.1
$$C^{(\pi)} = D^{(\pi)} - A = C_L^{(\pi)} + C_U^{(\pi)}$$

where $C_L^{(\pi)}$ and $C_U^{(\pi)}$ are formed from A by replacing all elements of A by zero except those $a_{i,j}$ such that i and j belong to different groups and such that the group containing i comes before and after, respectively, the group containing j. We now define the group Jacobi method by

8.2
$$D^{(\pi)}u^{(n+1)} = C^{(\pi)}u^{(n)} + b,$$

the group Gauss-Seidel method by

8.3
$$D^{(\pi)}u^{(n+1)} = C_L^{(\pi)}u^{(n+1)} + C_U^{(\pi)}u^{(n)} + b,$$

and the group SOR method by

8.4 $D^{(\pi)}u^{(n+1)} = \omega[C_L^{(\pi)}u^{(n+1)} + C_U^{(\pi)}u^{(n)} + b] + (1 - \omega)D^{(\pi)}u^{(n)}.$

In order to carry out the above methods it is necessary to solve a system of equations whose matrix is $D^{(\pi)}$. In many cases, the larger $D^{(\pi)}$ the faster the convergence, but on the other hand, with a larger $D^{(\pi)}$ the more difficult each step of the iteration process will be. In practice, one usually chooses $D^{(\pi)}$ so that with some permutation of its rows and columns, it is a tridiagonal matrix.

An important special cause of group iteration is *block iteration* where π is a *partitioning* of the form

8.5
$$\begin{cases} R_1 = \{1, 2, \ldots, n_1\} \\ R_2 = \{n_1 + 1, n_1 + 2, \ldots, n_1 + n_2\} \\ \cdots \\ R_q = \left\{\sum_{k=1}^{q-1} n_k + 1, \quad \sum_{k=1}^{q-1} n_k + 2, \ldots, \quad \sum_{k=1}^{q} n_k\right\}. \end{cases}$$

In this case $D^{(\pi)}$ will be a block diagonal matrix and $C_L^{(\pi)}$ and $C_U^{(\pi)}$ will be strictly lower and strictly upper triangular matrices, respectively.

An important application of group iterative methods is *line iteration*. Let us illustrate for a linear system derived from the 5-point difference equation representation of Laplace's equation given by

8.6 $4u(x, y) - u(x + h, y) - u(x, y + h) - u(x - h, y) - u(x, y - h) = 0.$

The line Jacobi method is defined by

8.7 $4u^{(n+1)}(x, y) - u^{(n+1)}(x + h, y) - u^{(n+1)}(x - h, y)$
$$= u^{(n)}(x, y + h) + u^{(n)}(x, y - h).$$

For the line Gauss-Seidel method we have (using the natural ordering)

8.8 $4u^{(n+1)}(x, y) - u^{(n+1)}(x + h, y) - u^{(n+1)}(x - h, y)$
$$= u^{(n)}(x, y + h) + u^{(n+1)}(x, y - h).$$

For the line SOR method we have

8.9 $4u^{(n+1)}(x, y) - u^{(n+1)}(x + h, y) - u^{(n+1)}(x - h, y)$
$$= \omega(u^{(n)}(x, y + h) + u^{(n+1)}(x, y - h))$$
$$+ (1 - \omega)[4u^{(n)}(x, y) - u^{(n)}(x + h, y) - u^{(n)}(x - h, y)].$$

One can show that the same relation holds between the eigenvalues $\mu^{(\pi)}$ of the line Jacobi method and the eigenvalues $\lambda^{(\pi)}$ of the line SOR method as between the eigenvalues μ of B and λ of $\mathscr{L}_\omega$. Moreover, for the model problem it can be shown that

8.10
$$\bar{\mu}^{(\pi)} = \frac{\cos \pi h}{2 - \cos \pi h}.$$

Using a value of $\omega_b^{(\pi)}$ based on $\bar{\mu}^{(\pi)}$ it can further be shown that the rate of convergence of $\mathscr{L}_{\omega_b^{(\pi)}}^{(\pi)}$ is approximately $\sqrt{2}$ times that of $\mathscr{L}_{\omega_b}$ corresponding to point SOR.

It should be noted, of course, that each step of the line SOR method requires the solution of a linear system with a tridiagonal matrix. This additional work would at first sight seem to make the use of line SOR impractical. However, Cuthill and Varga [1959] have devised a "normalized block iteration" scheme which enables one to carry out line SOR with virtually no more work per iteration than for point SOR.

Varga [1960] showed that by the use of 2-line SOR one can achieve a factor of 2 advantage over point SOR without any appreciable increase in the work per iteration.

One of the advantages of line SOR and 2-line SOR is that the basic relation between the eigenvalues of $B^{(\pi)}$ and those of $\mathscr{L}_\omega^{(\pi)}$ will often hold even when such a relation does not hold for the point methods. For example, with the nine-point difference equation representation of Laplace's equation,

8.11 $20u(x, y) - 4u(x + h, y) - 4u(x, y + h) - 4u(x - h, y) - 4u(x, y - h)$

$- u(x + h, y + h) - u(x - h, y + h) - u(x + h, y - h) - u(x - h, y - h) = 0$

the eigenvalue relation does not hold for the point methods but does hold for the line methods.

Extensive use is made of group SOR methods for large problems arising in industrial applications. Even though the alternating direction methods converge more rapidly in certain cases, nevertheless, the class of problems where group SOR methods are effective is much larger.

Semi-iterative Methods

Let us consider a *linear stationary iterative method** defined by

8.12 $u^{(n+1)} = Gu^{(n)} + k$

for solving (1.3) where A is nonsingular. We assume that the method defined by (8.12) is *completely consistent* with (1.3) in the sense that every solution of (1.3) is a solution of the *related equation*

---

* See also the discussion of the Jacobi method in Section 16.4.

8.13 $u = Gu + k$

and vice versa. Since A is nonsingular, the complete consistency condition implies that $I - G$ is nonsingular and that $k = (I - G)A^{-1}b$, or equivalently, that $b = A(I - G)^{-1}k$.

Given a completely consistent linear stationary iterative method, one can often accelerate the convergence by means of a *semi-iterative method* based on the given method. For each nonnegative integer n one chooses $\alpha_{n,0}, \alpha_{n,1}, \ldots, \alpha_{n,n}$ such that

8.14 $\displaystyle\sum_{k=0}^{n} \alpha_{n,k} = 1, \qquad n = 0, 1, 2, \ldots$

and determines $v^{(0)}, v^{(1)}, \ldots$ by

8.15 $\displaystyle v^{(n)} = \sum_{k=0}^{n} \alpha_{n,k} u^{(k)}.$

If $\alpha_{n,n} = 1$ for all n and $\alpha_{n,k} = 0, n \neq k$, then the $v^{(k)}$ are the same as the $u^{(k)}$, and we are back to the original method. However, in many cases, we are able, by suitable choices of the $\alpha_{n,k}$, to substantially improve the convergence.

For example, suppose A is a positive definite matrix but does not have Property A and hence is not consistently ordered, nor can it be made so by a permutation of the rows and columns of A. Suppose further that A is not an L-matrix. In this case the SOR theory does not apply. As a matter of fact, the matrix

8.16 $A = \begin{pmatrix} 1 & a & a \\ a & 1 & a \\ a & a & 1 \end{pmatrix}$

is positive definite for $-\frac{1}{2} < a < 1$, but for a close to unity, the Jacobi method does not converge, and the convergence of the SOR method for any ω is not appreciably better than that of the Gauss-Seidel method. For example, the following results were obtained by direct calculation.

a	.90	.92	.94	.96	.98
$S(\mathscr{L})$	.85381	.88243	.91136	.94060	.97015
ω^*	1.22344	1.23203	1.24062	1.24961	1.25859
$S(\mathscr{L}_{\omega^*})$	.83569	.86717	.89932	.93217	.96572
$R(\mathscr{L})$	.15804	.12507	.09281	.06123	.03030
$R(\mathscr{L}_{\omega^*})$	.17950	.14252	.10611	.07024	.03488
$R(\mathscr{L}_{\omega^*})/R(\mathscr{L})$	1.13577	1.13951	1.14327	1.14705	1.15086

Here ω^* is the value of ω which minimizes $S(\mathscr{L}_\omega)$, correct to within $\pm.01$. Evidently, the SOR method is not in any sense an order-of-magnitude better than the Gauss-Seidel method.

On the other hand, using the optimum semi-iterative method based on the Jacobi method (4.2) we obtain the following results.

a	.90	.92	.94	.96	.98	.99
$\sigma = \dfrac{1}{z}$	.93103	.94520	.95918	.97297	.98658	.99331
τ	.68211	.71257	.74774	.79044	.84809	.89049
$-\log \tau$	.38256	.33888	.29070	.23516	.16477	.11599
R_1	.07146	.05635	.04167	.02740	.01351	.00671
R_5	.24824	.20688	.16271	.11472	.06135	.03190
R_{10}	.31329	.26968	.22169	.16675	.09909	.05605
R_{20}	.34790	.30422	.25605	.20051	.13018	.08181
$R(\mathscr{L})$	.15804	.12507	.09281	.06123	.03030	.01508
$-\log \tau / R(\mathscr{L})$	2.42	2.71	3.14	3.84	5.44	7.70

Here R_n is the average rate of convergence after n iterations, which equals $-n^{-1} \log S(\mathscr{B}_n)$, where

8.17
$$\mathscr{B}_n = \sum_{h=0}^{n} \alpha_{n,k} B^k.$$

The quantity $-\log \tau$ is the asymptotic average rate of convergence and is the limiting value of R_n. We note that the ratio $-\log \tau / R(\mathscr{L})$ increases rapidly as a increases. Thus, in a sense, we have an order-of-magnitude gain in the convergence rate.

The choice of the $\alpha_{n,k}$ which yields the convergence rate described above is based on the use of Chebyshev polynomials as described by Golub and Varga [1961]. Actually, from the standpoint of practical computation it is convenient to determine $v^{(1)}, v^{(2)}, \ldots$ by the formulas

8.18
$$v^{(n+1)} = \frac{\rho_{n+1}}{2 - (\alpha + \beta)} \{[2B - (\beta + \alpha)I]v^{(n)} + 2c\}$$
$$+ (1 - \rho_{n+1})v^{(n-1)}, n = 0, 1, 2, \ldots$$

where

8.19
$$\begin{cases} \rho_1 = 1, \quad \rho_2 = \dfrac{2z^2}{2z^2 - 1} \\ \rho_{n+1} = \left(1 - \dfrac{1}{4z^2}\rho_n\right)^{-1}, \quad n = 2, 3, \ldots \end{cases}$$

and where

8.20
$$z = \frac{2 - (\alpha + \beta)}{\beta - \alpha}, \quad \text{if} \quad \alpha \neq \beta.$$

Here α and β are, respectively, lower and upper bounds for the eigenvalues of B. (If $\alpha = \beta$, then we let all $\rho_i = 1$.)

For this method we have

8.21
$$S(\mathscr{B}_n) = \frac{2r^{n/2}}{1 + r^n}$$

where

8.22
$$r = \omega_b - 1 = \tau^2$$

and

8.23
$$\omega_b = \frac{2}{1 + \sqrt{1 - \bar{\mu}^2}}.$$

Therefore,

8.24
$$R_n = -\frac{1}{n} \log S(\mathscr{B}_n)$$

and

8.25
$$\lim_{n \to \infty} R_n = -\log \tau.$$

In the particular case at hand, we have $\alpha = -2a$, $\beta = a$ if $a > 0$, and hence

8.26
$$z = \frac{2 + a}{3a}.$$

We remark that the method described above is closely related to the method of Richardson [1910]. Richardson's method is defined by

8.27
$$u^{(n+1)} = u^{(n)} + \beta_{n+1}(Au^{(n)} - b)$$

where $\beta_1, \beta_2, \ldots$, are parameters. While Richardson did not specify the β_i except in a general way, later work on the method considered the determination of the β_i in terms of Chebyshev polynomials. Also, it was found that to prevent the excessive growth of rounding errors as observed by Young and Warlick [1953], one should use a formula, similar to (8.18) which gives $u^{(n+1)}$ in terms of $u^{(n)}$ and $u^{(n-1)}$ (see Frank [1960]).

Semi-iterative methods are primarily useful when all of the eigenvalues of the matrix corresponding to the basic iterative method are real. Thus, they are useful when the basic method is the Jacobi method or the Gauss-Seidel method, but are not particularly effective with the SOR method, since $\mathscr{L}_\omega$ normally has complex eigenvalues for $\omega > 1$ (see Varga [1957]). On the other hand, semi-iterative methods are sometimes effective for the symmetric successive overrelaxation method (SSOR method) defined by

8.28
$$\begin{cases} u^{(n+\frac{1}{2})} = (I - \omega L)^{-1}(\omega U + (1 - \omega)I)u^{(n)} + (I - \omega L)^{-1}\omega c \\ \\ u^{(n+1)} = (I - \omega U)^{-1}(\omega L + (1 - \omega)I)u^{(n+\frac{1}{2})} + (I - \omega U)^{-1}\omega c \end{cases}$$

wherein each iteration consists of a forward and a backward sweep. For the discrete Dirichlet problem with the natural ordering, the number of iterations required for convergence can be shown to increase like $h^{-\frac{1}{2}}$ as $h \to 0$ using a suitable semi-iterative method. (See Sheldon [1955], Habetler and Wachspress [1961], Ehrlich [1964], and Young [1971b].) Thus there is an order-of-magnitude improvement over the SOR method. If one uses the "red-black" ordering, on the other hand, the semi-iterative method based on the SSOR method is relatively ineffective.

The SOR Method with Variable ω

In applying the SOR method one can, of course, let ω vary from equation to equation and from iteration to iteration. Assuming that A is positive definite and ω varies only from iteration to iteration, Ostrowski [1954] has proved that if $0 < \varepsilon \leqq \omega_i \leqq 2 - \varepsilon, i = 1, 2, \ldots$, then the SOR method with $\omega_1, \omega_2, \ldots$, converges for any starting vector. McDowell [1967] has considered the situation where ω varies from equation to equation.

The case where A has the form

8.29
$$A = \begin{pmatrix} D_1 & H \\ K & D_2 \end{pmatrix}$$

where D_1 and D_2 are square diagonal matrices is of particular interest. DeVogelaere [1958] considered the SOR method with ω used for the equations corresponding to D_1 and ω' used for the equations corresponding to D_2. We call this the *modified SOR method with fixed parameters* (MSOR method with fixed parameters). Young, Wheeler, and Downing [1965] considered the more general form of the

MSOR method where ω and ω' vary from iteration to iteration. It was shown that for any choice $\omega_1, \omega'_1, \omega_2, \omega'_2, \ldots$

8.30
$$S\left(\prod_{k=1}^{m} \mathscr{L}_{\omega_k, \omega'_k}\right) \geqq S(\mathscr{L}_{\omega_b}^m) = (\omega_b - 1)^m.$$

Here $\mathscr{L}_{\omega,\omega'}$ is the matrix corresponding to the MSOR method using ω and ω'. The result (8.30) is valid under the assumption that the eigenvalues of B are real and $\bar{\mu} = S(B) < 1$. Thus, as far as the spectral radius is concerned, there is no advantage in letting ω vary. However, Sheldon [1959] and Golub and Varga [1961] showed that if one uses a different convergence measure, based on a norm of $\prod_{k=1}^{m} \mathscr{L}_{\omega_k,\omega'_k}$ rather than on the spectral radius, then other choices of ω are better. For example, Sheldon proposed the choice

8.31
$$\omega_1 = \omega'_1 = 1, \qquad \omega_2 = \omega'_2 = \omega_3 = \omega'_3 = \cdots = \omega_b.$$

Golub [1959] considered the choice

8.32
$$\omega_1 = 1, \qquad \omega'_1 = \omega_2 = \omega'_2 = \cdots = \omega_b.$$

Golub and Varga [1961] introduced the cyclic Chebyshev semi-iterative method where

8.33
$$\begin{cases} \omega_1 = 1, \qquad \omega'_1 = \dfrac{2}{2 - \bar{\mu}^2} \\[2mm] \omega_k = \left(1 - \dfrac{\omega'_{k-1}}{4}\bar{\mu}^2\right)^{-1}, \qquad \omega'_k = \left(1 - \dfrac{\omega_k}{4}\bar{\mu}^2\right)^{-1}, \qquad k = 2, 3, \ldots . \end{cases}$$

The cyclic Chebyshev semi-iterative method can also be derived from the semi-iterative method based on the Jacobi method by taking advantage of the special form of A.

All of the above methods yield a smaller $D^{\frac{1}{2}}$-norm than does the SOR method with $\omega_k = \omega'_k = \omega_b$ corresponding to $\mathscr{L}_{\omega_b}^m$. Here $D = \text{diag } A$ and we assume that the diagonal elements of A are positive. As in Section 16.5, the $D^{\frac{1}{2}}$-norm of a matrix G is defined by

8.34
$$\|G\|_{D^{\frac{1}{2}}} = \|D^{\frac{1}{2}}GD^{-\frac{1}{2}}\|_2.$$

Young and Kincaid [1969] showed that if A is positive definite then based on the $A^{\frac{1}{2}}$-norm, the SOR method with $\omega'_k = \omega_k = \omega_b$ is just about as good as the choices (8.31) and (8.32). Moreover, while it is not as good as the cyclic Chebyshev semi-iterative method, it compares much more favorably than with the $D^{\frac{1}{2}}$-norm.

Gradient Methods

We now describe briefly a class of nonlinear methods, known as *gradient methods*, for solving (1.3). Assuming that A is a real positive definite matrix, let us consider the quadratic form

8.35
$$Q(u) = \tfrac{1}{2}(u, Au) - (b, u).$$

One can show that the problem of minimizing (8.35) is equivalent to that of solving (1.3). Thus, suppose $\bar{u}$ is the solution of (1.3), and let $u = \bar{u} + v$. Then

8.36
$$
\begin{aligned}
Q(\bar{u} + v) &= \tfrac{1}{2}((\bar{u} + v), A(\bar{u} + v)) - (b, \bar{u} + v) \\
&= \tfrac{1}{2}(\bar{u}, A\bar{u}) + \tfrac{1}{2}(v, A\bar{u}) + \tfrac{1}{2}(\bar{u}, Av) + \tfrac{1}{2}(v, Av) - (b, \bar{u}) - (b, v) \\
&= Q(\bar{u}) + \tfrac{1}{2}(v, Av) + (v, A\bar{u} - b) = Q(\bar{u}) + \tfrac{1}{2}(v, Av).
\end{aligned}
$$

Thus, since A is positive definite, we have $(v, Av) > 0$ unless $v = 0$. Hence, if $w \neq \bar{u}$, then $Q(w) > Q(\bar{u})$. Thus $\bar{u}$ is the unique absolute minimum of $Q(u)$.

In the *method of steepest descent* one determines $u^{(n+1)}$ from $u^{(n)}$ by letting

8.37
$$u^{(n+1)} = u^{(n)} + \varepsilon_n p^{(n)}$$

where $-p^{(n)}$ is the "gradient" vector corresponding to Q and $u^{(n)}$ and where ε_n is a constant. The vector $p^{(n)}$ is defined by

8.38
$$
-p^{(n)} = \begin{pmatrix} \dfrac{\partial Q(u^{(n)})}{\partial u_1} \\[2ex] \dfrac{\partial Q(u^{(n)})}{\partial u_2} \\[1ex] \vdots \\[1ex] \dfrac{\partial Q(u^{(n)})}{\partial u_n} \end{pmatrix} = Au^{(n)} - b = r^{(n)}.
$$

We choose ε_n so as to minimize $Q(u^{(n)} + \varepsilon_n p^{(n)})$. Thus, one can show that

8.39
$$\varepsilon_n = -\frac{(p^{(n)}, r^{(n)})}{(p^{(n)}, Ap^{(n)})} = \frac{(r^{(n)}, r^{(n)})}{(r^{(n)}, Ar^{(n)})}.$$

One can also show that

8.40
$$Q(u^{(n+1)}) - Q(u^{(n)}) = -\frac{(r^{(n)}, r^{(n)})^2}{2(r^{(n)}, Ar^{(n)})}.$$

Thus the sequence $Q(u^{(0)}), Q(u^{(1)}), \ldots$ is a decreasing sequence of numbers bounded below by $Q(\bar{u})$, where $\bar{u}$ is the solution of (1.3). Hence the sequence $Q(u^{(0)}), Q(u^{(1)}), \ldots$ converges. One can then show that $Q(u^{(n)})$ actually converges to $Q(\bar{u})$ and that $u^{(n)} \to \bar{u}$.

Hestenes and Stiefel [1952] and Stiefel [1952] presented a modification of the method of steepest descent which, in the absence of rounding errors, will converge in no more than N iterations. This method is known as the *conjugate gradient method*. Here one seeks to determine vectors $p^{(0)}, p^{(1)}, \ldots, p^{(N-1)}$ which are pairwise *conjugate* in the sense that

8.41
$$(p^{(i)}, Ap^{(j)}) = 0$$

for $i \neq j$. The formulas for determining the $p^{(n)}$ and the $u^{(n)}$ for a given $u^{(0)}$ are as follows:

8.42
$$
\begin{cases}
u^{(n+1)} = u^{(n)} + \varepsilon_n p^{(n)}, & n = 0, 1, \ldots, N-1 \\[2mm]
r^{(n)} = Au^{(n)} - b, & n = 0, 1, \ldots, N \\[2mm]
p^{(n)} = \begin{cases} -r^{(0)}, & n = 0 \\ -r^{(n)} + \alpha_{n-1} p^{(n-1)}, & n = 1, 2, \ldots, N-1 \end{cases} \\[4mm]
\varepsilon_n = -\dfrac{(p^{(n)}, r^{(n)})}{(p^{(n)}, Ap^{(n)})}, & n = 0, 1, \ldots, N-1 \\[4mm]
\alpha_{n-1} = \dfrac{(r^{(n)}, Ap^{(n-1)})}{(p^{(n-1)}, Ap^{(n-1)})}, & n = 1, 2, \ldots, N-1.
\end{cases}
$$

One can verify that (8.41) holds and, moreover, that

8.43
$$(r^{(i)}, r^{(j)}) = 0$$

for $i \neq j$. It therefore follows that for some $m \leq N$ we must have

8.44
$$r^{(m)} = 0.$$

Hence the method converges in at most N iterations.

Numerical experiments were carried out by Poole [1965] based on the use of the conjugate gradient method and on a combination of the conjugate gradient method and a semi-iterative method based on the Jacobi method. The linear systems considered were derived from the discrete analog of the Dirichlet problem for various regions. For the unit square with mesh size h the following results were obtained.

	J—SI		CG		SOR		CCSI	
	n	t	n	t	n	t	n	t
$h^{-1} = 20$:	97	37.8	32	21.4	59	11.8	57	11.4
$h^{-1} = 40$:	194	283.5	65	165.5	117	90.1	114	85.5
$h^{-1} = 80$:	392	2331.2	132	1320.2	236	744.3	227	684.7

J—SI   the optimum semi-iterative method based on the Jacobi method
CG   conjugate gradient method
CCSI   cyclic Chebyshev semi-iterative method
n number of iterations
t time in seconds required to reduce the maximum absolute value of the
        error by a factor of 10^{-6} (program run on the Control Data 1604
        computer).

From these results it can be seen that while the number of iterations required using the conjugate gradient method was considerably less than with the other methods, because of the additional computations required for each iteration the overall time was considerably greater. Thus, it would appear that the conjugate gradient method is not competitive for this class of problems. However, it is quite possible that the algorithm for carrying out the conjugate gradient method could be modified to reduce the number of operations required, and, if so, the use of the method would be appropriate.

Exercises 16.8

1. For the model problem with $h = \frac{1}{4}$, determine $D^{(\pi)}$, $C^{(\pi)}$, $C_L^{(\pi)}$, and $C_U^{(\pi)}$ with the following groupings:

$$R_1 : \{1, 2, 3\}, \quad R_2 : \{4, 5, 6\}, \quad R_3 : \{7, 8, 9\};$$

$$R_1 : \{1, 4, 7\}, \quad R_2 : \{2, 5, 8\}, \quad R_3 : \{3, 6, 9\}.$$

Assume the points are labelled in the natural ordering. Show that in each case $D^{(\pi)}$ is either tridiagonal or can be made so by permuting its rows and corresponding columns.

2. Carry out two iterations of the line Jacobi method, of the Gauss-Seidel method, and of the line SOR method (with $\omega = 1.072$) for the model problem with $h = \frac{1}{3}$ and $u = 1000$ on the side with $y = 1$, and $u = 0$ elsewhere on the boundary. Use zero starting values.

3. Verify the basic relation between the eigenvalues of the matrix corresponding to the line SOR method and the eigenvalues of the matrix corresponding to the line Jacobi method for the five-point difference equation

$$a_0 u(x, y) - a_1 u(x + h, y) - a_2 u(x, y + h) - a_3 u(x - h, y) - a_4 u(x, y - h) = t(x, y)$$

where $a_0(x, y) > 0$ in R_h.

4. Verify the formula (8.10) for the spectral radius of $B^{(\pi)}$, corresponding to the line Jacobi method for the model problem. Compute $\omega_b^{(\pi)}$ for $h = \frac{1}{20}$.

5. Show that for the model problem $R(\mathcal{L}_{\omega_b}^{(\pi)})$ corresponding to line SOR is approximately $\sqrt{2} R(\mathcal{L}_{\omega_b})$ where $R(\mathcal{L}_{\omega_b})$ corresponds to point SOR.

6. Show that the basic relation between the eigenvalues of the line Jacobi method and those of the line SOR method holds for the nine-point difference equation (8.11), even though it does not hold for the corresponding point methods. Find the eigenvalues of the line Jacobi method for the case of the unit square with mesh size h.

7. Verify that if the SOR method is written in the form (8.12), then the matrix $I - G$ is non-singular for $0 < \omega < 2$ and the consistency condition

$$k = (I - G)A^{-1}b$$

holds, given that A is nonsingular.

8. Compute $S(B)$, $S(\mathcal{L})$, and $S(\mathcal{L}_\omega)$ for the matrix A given by (8.16) and for $\omega = 1(0.2)2$.

9. For the model problem with $u = 1$ for $y = 1$, and $u = 0$ elsewhere on the boundary, carry out 4 iterations of the cyclic Chebyshev semi-iterative method. (Use the red-black ordering.) Also carry out 4 iterations of Sheldon's method, and 4 iterations of the SOR method with $\omega = \omega_b$. In each case use zero starting values. Let $h = \frac{1}{3}$.

10. Show that ε_n as given by (8.39) minimizes $Q(u^{(n)} + \varepsilon_n p^{(n)})$. Also verify (8.40).

11. Show that with the method of steepest descent $Q(u^{(n)})$ converges to $Q(\bar{u})$ and $u^{(n)}$ converges to $\bar{u}$.

12. For the problem of Exercise 9, carry out four iterations of the method of steepest descent and also four iterations of the conjugate gradient method. In each case use zero starting values.

13. Verify (8.41) and (8.43).

SUPPLEMENTARY DISCUSSION

For additional information on iterative methods for solving large linear systems the reader is referred to the books of Forsythe and Wasow [1960], Varga [1962], Wachspress [1966], and Young [1971a].

Section 16.2.

Geiringer [1949] studied the convergence properties of various iterative methods for irreducible matrices with weak diagonal dominance. Stein and Rosenberg [1948] considered L-matrices.

Section 16.3.

As noted by Taussky [1949], many independent proofs of Theorem 3.1 have appeared in the literature.

Section 16.4.

In studying the literature it will be helpful to note that a positive definite L-matrix is often referred to as a *Stieltjes matrix*.

Section 16.5.

It was proved by E. Reich [1949] that if A is a real symmetric matrix with positive diagonal elements, then the Gauss-Seidel method converges if and only if A is positive definite.

The convergence proof involving the $A^{\frac{1}{2}}$-norm is based on that given by Wachspress [1966].

The determination of the eigenvalues and eigenvectors of $\mathcal{L}$ in terms of those of B is a special case of the analysis given by Young [1950] and [1954] (see also Young [1971a]).

The test for consistent ordering described at the end of Section 16.5 involves the construction of a compatible ordering vector (see Young [1950], [1954], [1971a]).

Section 16.6.

Ostrowski [1954] has shown that if A is a symmetric matrix with positive diagonal elements, then the SOR method converges if and only if $0 < \omega < 2$ and A is positive definite.

The analysis of the relation between the eigenvalues and eigenvectors of $\mathcal{L}_\omega$ and those of

B is a special case of the results of Young [1950].

Work on the analysis of the convergence properties of the SOR method for matrices which are not consistently ordered was done by Kahan [1958] and Varga [1959]. For a proof of (6.49) the reader is referred to Varga [1962]. An alternative proof of (6.49) is given by Wachspress [1966].

Section 16.7.

Besides the Peaceman-Rachford parameters and the optimum parameters, one can use the Wachspress parameters, given by

$$\rho_i = b\left(\frac{a}{b}\right)^{(i-1)/(m-1)}, \qquad m \geqq 2, i = 1, 2, \ldots, m$$

which were introduced by Wachspress [1957]. While not as effective as the optimum parameters, these parameters are better than the Peaceman-Rachford parameters in many cases (see Birkhoff, Varga, and Young [1962]).

Wachspress [1966] describes how one can by choosing ρ_i' different from ρ_i effectively handle the problem of finding the optimum iteration parameters in the commutative case where the ranges for the eigenvalues μ of H and ν of V are different.

Widlund [1966] has shown that by letting the iteration parameters vary over R_h, then one can achieve an order-of-magnitude improvement by using the Peaceman-Rachford method in the non-commutative case provided R is a rectangle.

Section 16.8.

Early theoretical work on group iterative methods was done by Arms, Gates, and Zondek [1956] and by Friedman [1957]. Subsequent work was done by Varga [1960], Parter [1959], [1961], [1965], and others.

For further discussion of semi-iterative methods the reader is referred to the works of Varga [1957], Blair, *et al.* [1959], and Golub and Varga [1961]. Actually, the idea of using Chebyshev polynomials to accelerate the convergence of methods for solving large linear systems seems to have been originated by Flanders and Shortley [1950] (see also Shortley [1953], and Young [1954a]).

A very readable account of the conjugate gradient method is given by Beckman [1960].

CHAPTER 17

PARTIAL DIFFERENTIAL EQUATIONS: INITIAL-VALUE PROBLEMS

17.1 INTRODUCTION

In Chapter 15 we considered the solution of the following problem: given a bounded region R with boundary S, find a function $u(x, y)$ continuous in $R + S$ twice differentiable in R and satisfying in R the elliptic partial differential equation

1.1
$$L[u] = G.$$

The function $u(x, y)$ was required to satisfy certain conditions on S. We had previously, in Chapter 10, considered a special case involving one independent variable where R is an interval and S is the set consisting of the two end points of the interval.

In this chapter we consider the introduction of a new variable, say t, and we seek to solve the differential equation

1.2
$$\psi[u] = u_t - L[u] = -G$$

for all (x, y, t) such that $(x, y) \in R$ and $t > t_0$, where t_0 is given. We require that for $(x, y) \in S$ and $t \geq t_0$, the solution u satisfy certain boundary conditions which may depend on t, and which may involve u and certain derivatives of u. Moreover, we assume that $u(x, y, t_0)$ is given for all $(x, y) \in R$. We designate the variables x and y as "space variables" to distinguish them from the variable t which in many problems denotes the time.

In Section 17.2 we consider the case of one space variable and show how by the use of semi-discretization the problem can be reduced to the solution of a system of ordinary differential equations. Various methods for solving this system lead to such standard methods for solving (1.2) as the forward difference method, Richardson's method, and the Crank-Nicolson method. These methods are studied in more detail in Sections 17.3–17.5. In Section 17.6 we consider more general discrete representations of (1.2) and study consistency and absolute and relative degrees of approximation.

A brief discussion of methods for solving problems with two space variables is given in Section 17.7. For such problems, alternating direction methods can be used as well as some of the standard methods which were considered for problems with one space variable.

17.2 PROBLEMS WITH ONE SPACE VARIABLE: SEMI-DISCRETIZATION

Let us consider the problem of finding a function $u(x, t)$ which satisfies the differential equation

2.1
$$u_t - L[u] = -G, \qquad 0 < x < 1, t > 0$$

together with the boundary conditions

2.2
$$\begin{cases} u(0, t) = g_1(t) \\ u(1, t) = g_2(t) \end{cases}$$

for $t \geq 0$ and the initial condition

2.3
$$u(x, 0) = f(x), \qquad 0 < x < 1.$$

Here $L[u]$ is the differential operator

2.4
$$L[u] = Au_{xx} + Du_x + Fu$$

where for $0 \leq x \leq 1$, the functions $A = A(x)$ and $F = F(x)$ satisfy

2.5
$$\begin{cases} A(x) > 0 \\ F(x) \leq 0. \end{cases}$$

The function $u(x, t)$ is required to be continuous for $x \in [0, 1]$, $t \geq 0$; also, $u_{xx}(x, t)$ and $u_t(x, t)$ are required to exist for $0 < x < 1$, $t > 0$.

For any positive integer M let*

2.6
$$u_i(t) = u(ih, t), \qquad i = 0, 1, \ldots, M$$

where

2.7
$$h = \Delta x = \frac{1}{M}.$$

The semi-discretization involves representing $L[u]$ at the point (ih, t) in terms of $u_i(t)$, $u_{i-1}(t)$, and $u_{i+1}(t)$. Substituting in (2.1) one obtains a system of ordinary differential equations involving the dependent variables $u_1(t), u_2(t), \ldots, u_{M-1}(t)$. Thus, for example, if

2.8
$$L[u] = u_{xx}$$

and if we let

2.9
$$u_{xx}(ih, t) \sim \frac{u_{i+1}(t) + u_{i-1}(t) - 2u_i(t)}{h^2},$$

---

*For the balance of this chapter we shall use $u_i(t)$ interchangeably with $u(x, t)$ where $x = ih$.

then we get, upon substitution in (2.1), the system of ordinary differential equations

2.10 $\dfrac{du_i(t)}{dt} = \dfrac{u_{i+1}(t) + u_{i-1}(t) - 2u_i(t)}{h^2} - G(ih, t), \qquad i = 1, 2, \ldots, M - 1.$

From (2.2) and (2.3) we have

2.11 $\begin{cases} u_0(t) = g_1(t) \\ u_M(t) = g_2(t) \end{cases}$

and

2.12 $u_i(0) = f(ih), \qquad i = 1, 2, \ldots, M - 1.$

The system (2.10)–(2.12) can be solved numerically by any of the methods considered in Chapter 8. Corresponding to the *Euler method* we get the *forward difference method* defined by

2.13 $u_i(t + k) = u_i(t) + r[u_{i+1}(t) + u_{i-1}(t) - 2u_i(t)] - kG(ih, t), \qquad i = 1, 2, \ldots, M - 1$

where we let

2.14 $k = \Delta t$

and the so-called "mesh ratio," r, is given by

2.15 $r = \dfrac{k}{h^2}.$

Since $u_0(t)$ and $u_M(t)$ are given for all t as well as $u_1(0), u_2(0), \ldots, u_{M-1}(0)$ one can proceed to compute $u_1(k), u_2(k), \ldots, u_{M-1}(k)$ then $u_1(2k), u_2(2k)$, etc.

Corresponding to the *midpoint method* we have the *method of Richardson* [1910] defined by

2.16 $u_i(t + k) = u_i(t - k) + 2r[u_{i+1}(t) + u_{i-1}(t) - 2u_i(t)] - 2kG(ih, t),$

$i = 1, 2, \ldots, M - 1.$

Here $u_1(k), u_2(k), \ldots, u_{M-1}(k)$ must be determined by some other method (e.g., by the forward difference method). Once this has been done, the calculation can proceed explicitly.

Corresponding to the *modified Euler method* we have the *method of Crank and Nicolson* [1947] defined by

2.17 $u_i(t + k) = u_i(t) + \dfrac{r}{2}\,[u_{i+1}(t + k) + u_{i-1}(t + k) - 2u_i(t + k) + u_{i+1}(t)$

$$+ u_{i-1}(t) - 2u_i(t)] - \frac{k}{2}[G(ih, t + k) + G(ih, t)],$$

$$i = 1, 2, \ldots, M - 1.$$

As in the case of the modified Euler method, the $u_i(t + k)$ are defined implicitly by (2.17). However, as we shall see, the solution of the implicit equations can be carried out with relatively little computational effort.

We remark that the above methods can be defined even if G is a function of u as well as a function of x and t. For the Crank-Nicolson method, instead of solving (2.17) exactly one might use the analog of the Heun or the modified Heun method (see Chapter 8).

EXERCISES 17.2

1. Consider the problem of solving $u_t = u_{xx}$, $0 < x < 1$, $t > 0$, with the boundary conditions $u(0, t) = 1$, $u(1, t) = 2$, $t \geq 0$, and the initial condition $u(x, 0) = x$, $0 < x < 1$. With $h = \frac{1}{3}$, use semi-discretization to obtain a pair of ordinary differential equations. Find the exact solution of these equations.

2. In the preceding problem use the Euler method, the midpoint method (with the Euler method for the first time step), and the modified Euler method to find approximate values of u for $t = 0.2$ with $k = \Delta t = 0.1$.

17.3 THE FORWARD DIFFERENCE METHOD

We now show that the forward difference method yields a satisfactory numerical procedure if the mesh ratio r given by (2.15) satisfies the condition

3.1 $$r \leq \tfrac{1}{2}$$

but that it is unsatisfactory if $r > \frac{1}{2}$. To simplify the analysis we consider the heat equation, or diffusion equation,

3.2 $$u_t = u_{xx}$$

with the boundary conditions

3.3 $$\begin{cases} u(0, t) = g_1(t) \\ u(1, t) = g_2(t), \quad t \geq 0 \end{cases}$$

and the initial conditions

3.4 $$u(x, 0) = f(x), \qquad 0 < x < 1.$$

Let $\bar{u}_i(t)$ be the exact solution of

3.5 $$u_i(t + k) = u_i(t) + r[u_{i+1}(t) + u_{i-1}(t) - 2u_i(t)], \qquad i = 1, 2, \ldots, M - 1$$

subject to the conditions (3.3) and (3.4) and let $\hat{u}_i(t)$ satisfy (3.5) and (3.3) for $t = nk, (n + 1)k, \ldots$, where n is a nonnegative integer. We seek to show that if (3.1) holds, then for some integer m with $m \geq n$ we have

3.6 $$E(m) \leq E(n)$$

where

3.7 $$E(n) = \max_{i=1,2,\ldots,M-1} |\hat{u}_i(nk) - \bar{u}_i(nk)|.$$

But if we let

3.8 $$e_i(t) = \hat{u}_i(t) - \bar{u}_i(t),$$

then we have, by (3.5), since $e_0(t) = e_M(t) = 0$,

3.9 $$|e_i(t + k)| \leq (1 - 2r)|e_i(t)| + r|e_{i+1}(t)| + r|e_{i-1}(t)|$$

so that

3.10 $$E(n + 1) \leq E(n)$$

and (3.6) follows. Thus the procedure is computationally stable in the sense that the effect of an error introduced at any point (ih, t) does not increase as the computation proceeds.

To analyze the case $r > \frac{1}{2}$ we use a more refined procedure described by O'Brien, Hyman, and Kaplan [1951] which was based on unpublished work of J. von Neumann. If $f(x)$ is continuous except for a finite number of finite jumps,* then the function

3.11 $$u(x, t) = \sum_{n=1}^{\infty} a_n \sin n\pi x \, e^{-n^2\pi^2 t}$$

---

* The function $f(x)$ has a *jump* at $x = a$ if $\lim_{x \to a+} f(x)$ and $\lim_{x \to a-} f(x)$ (as defined in A-(3.5)) exist but are different.

where

3.12
$$a_n = 2 \int_0^1 f(z) \sin n\pi z \, dz$$

satisfies (3.2), the boundary conditions

3.13
$$u(0, t) = u(1, t) = 0, \qquad t \geq 0,$$

and, for values of x such that $f(x)$ is continuous, the condition

3.14
$$\lim_{t \to 0+} u(x, t) = f(x), \qquad 0 < x < 1.$$

Whether or not the Fourier (sine) series for $f(x)$ converges, the function

3.15
$$u(x, t) = \begin{cases} \sum_{n=1}^{\infty} a_n \sin n\pi x \, e^{-n^2\pi^2 t}, & t > 0 \\ f(x), & 0 < x < 1, \quad t = 0 \\ 0, & x = 0, t = 0, \text{ and } x = 1, \quad t = 0 \end{cases}$$

is continuous* for $0 \leq x \leq 1$ and $t \geq 0$ except for points $(x, 0)$ where $f(x)$ is not continuous and except for the points $(0, 0)$ and $(1, 0)$ unless $f(0) = 0$ and $f(1) = 0$, respectively.

The result (3.11) can be derived by the method of separation of variables and by the use of the Fourier (sine) series representation for $f(x)$. Similarly, the exact solution of the problem defined by the forward difference method can also be obtained by letting

3.16
$$u_i(t) = X(ih)T(t).$$

Substituting in (3.5) we get

3.17
$$\frac{T(t + k) - T(t)}{T(t)} = r \left\{ \frac{X((i + 1)h) + X((i - 1)h) - 2X(ih)}{X(ih)} \right\} = -\alpha^2$$

where α is a constant. Evidently, for any integer m the above equation and the conditions

3.18
$$X(0) = X(1) = 0$$

---

* For a proof of the continuity of $u(x, t)$ for $0 \leq x \leq 1, t \geq 0$ (except at isolated points), see, for instance, Juncosa and Young [1954].

are satisfied if

3.19 $X_m(ih) = \sin m\pi ih$

3.20 $T_m(r) = \left(1 - 4r \sin^2 \dfrac{m\pi h}{2}\right)^{t/k}.$

Therefore for any $b_1, b_2, \ldots, b_{M-1}$ the function

3.21 $u_i(t) = \displaystyle\sum_{m=1}^{M-1} b_m \sin m\pi ih \left(1 - 4r \sin^2 \dfrac{m\pi h}{2}\right)^{t/k}$

satisfies (3.5) and (3.13). From the analysis of Section 6.13 it can be verified that if the b_m are given by

3.22 $b_m = \dfrac{2}{M} \displaystyle\sum_{j=1}^{M-1} f(jh) \sin m\pi jh, \qquad m = 1, 2, \ldots, M - 1,$

then

3.23 $f(ih) = \displaystyle\sum_{m=1}^{M-1} b_m \sin m\pi ih, \qquad i = 1, 2, \ldots, M - 1.$

Hence if the b_m are given by (3.22), then (3.21) satisfies the difference equation (3.5) as well as the conditions (3.13) and (3.4) for $x = 0, h, \ldots, Mh$, $t = 0, k, 2k, \ldots$.
If $r \leq \frac{1}{2}$, then we have

3.24 $\left|1 - 4r \sin^2 \dfrac{m\pi h}{2}\right| < 1, \qquad m = 1, 2, \ldots, M - 1,$

and hence each term of (3.21) converges to zero as $t \to \infty$. For this reason the procedure is said to be *stable*. On the other hand, if r is appreciably* greater than $\frac{1}{2}$, then at least some of the terms of (3.21) increase as $t \to \infty$. The procedure is then said to be *unstable*. It should be noted that this terminology is not the same as that used in the analysis given in Chapter 9 of methods for solving ordinary differential equations. Thus Euler's method for solving the semi-discretized problem is always stable in the sense of Chapter 9. However, even with a stable method one must choose the mesh size small in order to get useful solutions.

It can be shown that for $r \leq \frac{1}{2}$, the solution of the difference equation converges to the solution of the differential equation as $h \to 0$. For a simple proof of this

---

* More precisely, the condition (3.24) fails to be satisfied if

$$r > \dfrac{1}{2 \cos^2 \dfrac{\pi h}{2}}.$$

result under the assumption that the solution of the differential equation is sufficiently well-behaved, see, for instance, Isaacson and Keller [1966]. It can also be shown that the difference between the approximate solution and the exact solution is $O(h^2)$.*

Evidently, if $f(x) \geq 0$ and if $r \leq \frac{1}{2}$, the solution of the difference equation is nonnegative. The convergence theorem then implies that the solution of the differential equation is nonnegative.

As an illustration let us apply the forward difference method for the problem

3.25
$$\begin{cases} u_t = u_{xx}, & 0 < x < 1 \\ u(0, t) = u(1, t) = 0, & 0 \leq t \\ u(x, 0) = 1, & 0 < x < 1. \end{cases}$$

If $h = \frac{1}{4}$ and $r = \frac{1}{2}$, then we have the following results:

$j = t/k$ \ $i = x/h$	0	1	2	3	4
0	0	1.000	1.000	1.000	0
1	0	0.500	1.000	0.500	0
2	0	0.500	0.500($+\varepsilon$)	0.500	0
3	0	0.250($+\varepsilon/2$)	0.500	0.250($+\varepsilon/2$)	0
4	0	0.250	0.250($+\varepsilon/2$)	0.250	0
5	0	0.125($+\varepsilon/4$)	0.250	0.125($+\varepsilon/4$)	0
6	0	0.125	0.125($+\varepsilon/4$)	0.125	0

If we temporarily ignore the quantities in parentheses, we see that the values obtained are not smooth, but nevertheless decrease, for fixed x, as t increases. The quantities in parentheses show the effect of the introduction of an error ε for $i = 2, j = 2$. It can be seen that the effect of this error decreases as t increases.

Let us next consider the case $r = 1$. In this case we get

$j = t/k$ \ $i = x/h$	0	1	2	3	4
0	0	1	1	1	0
1	0	0	1($+\varepsilon$)	0	0
2	0	1($+\varepsilon$)	$-1(-\varepsilon)$	1($+\varepsilon$)	0
3	0	$-2(-2\varepsilon)$	3($+3\varepsilon$)	$-2(-2\varepsilon)$	0
4	0	5($+5\varepsilon$)	$-7(-7\varepsilon)$	5($+5\varepsilon$)	0
5	0	$-12(-12\varepsilon)$	17($+17\varepsilon$)	$-12(-12\varepsilon)$	0
6	0	29($+29\varepsilon$)	$-41(-41\varepsilon)$	29($+29\varepsilon$)	0

---

*Similar results can be proved under weaker hypotheses; see the Supplementary Discussion.

Since, as we have seen, the solution of the differential equation is nonnegative, the numbers obtained are completely worthless. That the process is unstable can be seen by considering the effect of an error of ε for $i = 2, j = 1$. After six time steps, the error has been magnified by a factor of 41.

It may happen that under certain conditions reasonable solutions of the difference equation may exist even if $r > \frac{1}{2}$. For example, if $u(x, 0) = f(x) = \sin \pi x$ then the exact solution of the difference equation with $r = 1$, namely

3.26
$$u_i(t) = \sin \pi i h \left(1 - 4 \sin^2 \frac{\pi h}{2} \right)^{t/k},$$

converges to the exact solution

3.27
$$u(x, t) = \sin \pi x e^{-\pi^2 t}$$

of the differential equation. With $h = \frac{1}{4}$, one obtains the following results.

$j = t/k$ \ $i = x/h$	0	1	2	3	4
0	0	0.707	1.000	0.707	0
1	0	0.293	0.414	0.293	0
2	0	0.121	0.172	0.121	0
3	0	0.051	0.070	0.051	0
4	0	0.019	0.032	0.019	0
5	0	0.013	0.006	0.013	0
6	0	-0.007	0.020	-0.007	0

Evidently, the values are quite well behaved until we reach $j = 5$ where, because of rounding errors, they begin to oscillate. If more decimal places had been carried in the calculation, the occurrence of the oscillation would have been delayed. If we had used exact values throughout, for example, $\sqrt{2}/2$ instead of 0.707, the oscillations would never have occurred. Of course, these considerations are largely academic, but they do tend to illustrate a distinction between stability and convergence. On the other hand, the above case would not be convergent in the sense of Richtmyer and Morton [1967, pp. 44–45] since for a method to be convergent when applied to a given problem, it should be convergent for all problems which are "close" in some sense. Clearly, if we change the initial function $f(x) = \sin \pi x$ even slightly, we shall spoil the convergence.

The requirement that k be chosen so small that $r = k/h^2 \leq \frac{1}{2}$ means that an excessively large number of time steps are required for a reasonably small value of h. Later we shall see that by the use of the Crank-Nicolson method we can overcome this limitation.

EXERCISES 17.3

1. Consider the problem of solving $u_t = u_{xx}$, $0 < x < 1$, $t > 0$ with the boundary conditions $u(0, t) = u(1, t) = 0$, $t \geq 0$ and the initial conditions $u(x, 0) = x$, $0 < x < 1$. Find an approximate value of $u(\frac{1}{2}, \frac{1}{8})$ by the forward difference method with $h = \frac{1}{4}$ and with: $r = \frac{1}{4}$; $r = 1$. Also, obtain the exact value of the solution.

2. In the preceding problem find the exact solution of the difference equation in each case.

3. Show that for fixed r, m, and t we have

$$\lim_{h \to 0} \left\{ \left(1 - 4r \sin^2 \frac{m\pi h}{2} \right)^{t/k} - e^{-m^2\pi t} \right\} = 0$$

where $k = rh^2$.

4. Give the largest value of r for which the forward difference method is stable when applied to the problem of Exercise 1.

17.4 RICHARDSON'S METHOD AND THE DUFORT-FRANKEL METHOD

In this section we show that Richardson's method is completely unsatisfactory for numerical computation but that by a simple modification one obtains the DuFort-Frankel method which is comparable with the forward difference method with respect to stability and accuracy.

We again consider the problem defined by (3.2), (3.13), and (3.4). As in Section 17.3, one can show that the exact solution of the Richardson difference equation

4.1 $u_i(t + k) = u_i(t - k) + 2r[u_{i+1}(t) + u_{i-1}(t) - 2u_i(t)]$, $i = 1, 2, \ldots, M - 1$

(see (2.16)) subject to (3.13) and (3.4) is given by

4.2 $$u_i(t) = \sum_{m=1}^{M-1} \sin m\pi i h (c_m \lambda_{m,1}^{t/k} + d_m \lambda_{m,2}^{t/k})$$

where

4.3 $$\lambda_{m,1} = -\alpha_m^2 + \sqrt{\alpha_m^4 + 4}, \qquad \lambda_{m,2} = -\alpha_m^2 - \sqrt{\alpha_m^4 + 4}$$

and

4.4 $$\alpha_m^2 = 4r \sin^2 \frac{m\pi h}{2}.$$

The coefficients c_m and d_m depend on the particular choice of $u_i(k)$, $i = 1, 2, \ldots, M - 1$. If we assume that

4.5 $$u_i(k) = \hat{f}(x) = \sum_{m=1}^{M-1} \hat{b}_m \sin m\pi i h$$

where

4.6
$$\hat{b}_m = \frac{2}{M} \sum_{j=1}^{M-1} \hat{f}(jh) \sin m\pi jh$$

then c_m and d_m are determined by the conditions

4.7
$$\begin{cases} c_m + d_m = b_m \\ c_m \lambda_{m,1} + d_m \lambda_{m,2} = \hat{b}_m. \end{cases}$$

Since $|\lambda_{m,2}| > 1$, it follows that unless $d_m = 0$ for all m the method will be unstable in the sense of Section 17.3. It should be pointed out that since the midpoint method for solving a system of ordinary differential equations is (weakly) stable and convergent, the same is true of Richardson's method. However, as shown by Eidson [1969], the time step size, k, must be chosen smaller than h by an order of magnitude (a condition much more severe than $k/h^2 \leq \frac{1}{2}$) in order to have convergence. Thus, while it is not true, as is frequently implied, that Richardson's method can *under no circumstances* give useful answers, it is certainly true that in many cases the work required to obtain useful answers is prohibitive.

By a slight modification of Richardson's method we can obtain a method with much better stability properties. If we replace the term $u_i(t)$ in (4.1) by

4.8
$$\tfrac{1}{2}(u_i(t + k) + u_i(t - k))$$

we obtain the *DuFort-Frankel method* [1953] given by

4.9
$$u_i(t + k) = \frac{1 - 2r}{1 + 2r} u_i(t - k) + \frac{2r}{1 + 2r} [u_{i+1}(t) + u_{i-1}(t)].$$

The exact solution of the difference equation is given by

4.10
$$u_i(t) = \sum_{m=1}^{M-1} \sin m\pi ih (\tilde{c}_m \tilde{\lambda}_{m,1}^{t/k} + \tilde{d}_m \tilde{\lambda}_{m,2}^{t/k})$$

where

4.11
$$\begin{cases} \tilde{\lambda}_{m,1} = \dfrac{2r \cos m\pi h - \sqrt{1 - 4r^2 \sin^2 m\pi h}}{1 + 2r} \\[4mm] \tilde{\lambda}_{m,2} = \dfrac{2r \cos m\pi h + \sqrt{1 - 4r^2 \sin^2 m\pi h}}{1 + 2r} \end{cases}$$

and $\tilde{c}_m$ and $\tilde{d}_m$ are determined by (4.7) with $\lambda_{m,1}$ and $\lambda_{m,2}$ replaced by $\tilde{\lambda}_{m,1}$ and $\tilde{\lambda}_{m,2}$, respectively. We note that $|\tilde{\lambda}_{m,1}|$ and $|\tilde{\lambda}_{m,2}|$ are less than unity. For if $1 - 4r \sin^2 m\pi h < 0$, then

4.12
$$|\tilde{\lambda}_{m,1}| = |\tilde{\lambda}_{m,2}| = \sqrt{\left|\frac{1-2r}{1+2r}\right|} < 1$$

while if $1 - 4r\sin^2 m\pi h > 0$, then

4.13 $\max(|\tilde{\lambda}_{m,1}|, |\tilde{\lambda}_{m,2}|) \leq \dfrac{1 + 2r|\cos m\pi h|}{1 + 2r} < 1, \qquad i = 1, 2, \ldots, M - 1.$

Thus, for $m = 1, 2, \ldots, M - 1$ each term of (4.10) approaches zero as $t \to \infty$ so that the method is stable for all $h > 0$, $k > 0$. However, as we shall see later, in Section 17.6 we cannot expect convergence if k is chosen too large relative to h.

EXERCISES 17.4

1. For the problem of Exercise 1, Section 17.3, determine $u(\frac{1}{2}, \frac{1}{8})$ by Richardson's method with $r = \frac{1}{4}$, $h = \frac{1}{4}$ and by the DuFort-Frankel method with $r = 1$, $h = \frac{1}{4}$. In each case use the forward difference method for the first time step.

2. In the preceding problem find the exact solution of the difference equation in each case.

17.5 THE CRANK-NICOLSON METHOD

Even though the Crank-Nicolson method as defined by (2.17) is *implicit*, nevertheless, the method can be carried out with relatively little computational effort since the determination of $u_1(t + k), u_2(t + k), \ldots, u_{M-1}(t + k)$ involves the solution of a linear system with a tri-diagonal matrix. To see this we write (2.17) in the form

5.1 $(1 + r)u_i(t + k) - \dfrac{r}{2}u_{i+1}(t + k) - \dfrac{r}{2}u_{i-1}(t + k)$

$$= (1 - r)u_i(t) + \frac{r}{2}[u_{i-1}(t) + u_{i+1}(t)] - \frac{k}{2}[G(ih, t + k) + G(ih, t)],$$

$$i = 1, 2, \ldots, M - 1.$$

The matrix of the system has $1 + r$ on the main diagonal and $-r/2$ on the two diagonals adjacent to the main diagonal. The algorithm described in Section 10.4 can then be used to obtain $u_1(t + k)$, $u_2(t + k), \ldots, u_{M-1}(t + k)$ from $u_1(t)$, $u_2(t), \ldots, u_{M-1}(t)$. Thus, we have

5.2 $\begin{cases} u_{M-1}(t + k) = q_{M-1} \\ u_i(t + k) = q_i - b_i u_{i+1}(t + k), \qquad i = M - 2, M - 3, \ldots, 1 \end{cases}$

where

5.3 $b_1 = \dfrac{-r}{2(1 + r)}, \quad b_i = -\dfrac{r}{2(1 + r)} \Big/ \Big[1 + \dfrac{r}{2(1 + r)} b_{i-1}\Big],$

$$i = 2, 3, \ldots, M - 2,$$

5.4 $q_1 = D_1, \quad q_i = \Big[D_i + \dfrac{r}{2(1 + r)} q_{i-1}\Big] \Big/ \Big[1 + \dfrac{r}{2(1 + r)} b_{i-1}\Big],$

$$i = 2, 3, \ldots, M - 1,$$

5.5 $D_i = \dfrac{r}{2(1 + r)} (u_{i-1}(t) + u_{i+1}(t)) + \dfrac{1 - r}{1 + r} u_i(t)$

$$\qquad - \dfrac{rh^2}{2(1 + r)} [G(ih, t + k) + G(ih, t)].$$

As an illustration, let us consider the example (3.25). The reader should verify that with $h = \frac{1}{4}$ and $r = 1$, one obtains the following values

$i = x/h$		0	1	2	3	4
$t/k = 0$	$u_i(jk)$	0	1.000	1.000	1.000	0
$t/k = 1$	D_i	$\cdots$	0.250	0.500	0.250	$\cdots$
	b_i	$\cdots$	−0.250	−0.267	$\cdots$	$\cdots$
	q_i	$\cdots$	0.250	0.600	0.428	$\cdots$
	u_i	0	0.428	0.714	0.428	0
Exact Values $\bar{u}_i$		0	0.488	0.686	0.488	0

The exact values were obtained from (3.11) and (3.12).

If the function G in (2.1) depends on u as well as on x and t one can use an iterative scheme based on the above process. Thus one can use $u_i(t)$ in place of $u_i(t + k)$ in the initial evaluation of G. For subsequent iterations one can use improved values. If $|\partial G/\partial u|$ is small and if k is small, then the process can be expected to converge. As in the modified Euler method, it may be best to carry out a fixed number of iterations, say one or two as in the Heun and modified Heun methods, respectively.

As in Sections 17.3 and 17.4 we can give a formula for the exact solution of (5.1), (3.13), and (3.4), in the case $G \equiv 0$. Indeed we have

5.6 $$u_i(t) = \sum_{m=1}^{M-1} b_m \sin m\pi ih \left(\frac{1 - 2r \sin^2 \dfrac{m\pi h}{2}}{1 + 2r \sin^2 \dfrac{m\pi h}{2}} \right)^{t/k}$$

where the b_m are given by (3.22). Since each term of (5.6) converges to zero as $t \to \infty$, the method is stable for any $h > 0, k > 0$.

Isaacson and Keller [1966] gave a proof of the convergence of the Crank-Nicolson method as $\Delta x \to 0$, if $\Delta t / \Delta x$ is a constant, under the assumption that the exact solution $u(x, t)$ is sufficiently well behaved. They also show that under this assumption the error is $O((\Delta x)^2)$. Juncosa and Young [1957] proved that convergence holds even if $f(x)$ is only piecewise continuous provided

5.7
$$k = O\left(\frac{h}{|\log h|}\right).$$

EXERCISES 17.5

1. For the problem of Exercise 1, Section 17.3, determine $u(\frac{1}{2}, \frac{1}{8})$ by the Crank-Nicolson method with $r = 1, h = \frac{1}{4}$.

2. In the preceding problem find the exact solution of the difference equation.

3. Apply the Crank-Nicolson method to solve

$$u_t - u_{xx} = \frac{1}{20} e^u, \qquad 0 < x < 1, t > 0$$

$$u(0, t) = u(1, t) = 0, \qquad t \geq 0$$

$$u(x, 0) = 0, \qquad 0 < x < 1.$$

Let $h = \frac{1}{8}, r = \frac{1}{2}$, and carry out the calculation for two time steps. For each time step perform two iterations.

17.6 ACCURACY OF THE DIFFERENCE EQUATIONS

In this section we carry out an analysis of the accuracy of the discrete operators corresponding to the various methods which have been considered so far. Each such method can be derived from a representation of the differential operator

6.1
$$\psi[u] = u_t - u_{xx}$$

of the form

6.2
$$\psi_h[u] = \sum_{i=0}^{6} \alpha_i u(x + s_i h, t + t_i k)$$

where we let

6.3
$$\begin{cases} s_0 = s_2 = s_4 = 0, & s_1 = s_5 = 1, & s_3 = s_6 = -1 \\ t_0 = t_1 = t_3 = 0, & t_2 = t_5 = t_6 = 1, & t_4 = -1. \end{cases}$$

As in Section 10.6 we require that ψ_h be *consistent* with ψ in the sense that for all sufficiently differentiable functions u we have

6.4
$$\lim_{h \to 0} [\psi_h[u] - \psi[u]] = 0.$$

Here we assume that

6.5
$$k = \theta(h)$$

for some function $\theta(h)$ such that

6.6
$$\lim_{h \to 0} \theta(h) = 0.$$

The *absolute degree of approximation* is the largest integer q such that

6.7
$$\psi_h[u] - \psi[u] = O(h^{q+1})$$

as $h \to 0$ for all sufficiently differentiable functions u. The *relative degree of approximation* of a consistent discrete operator is the largest integer p such that

6.8
$$\psi_h[u] - \psi[u] = O(h^{p+1})$$

for all sufficiently differentiable solutions of

6.9
$$\psi[u] = 0.$$

As in Theorem 15-5.2 we can show that if the relative degree of approximation is p, then for some $\beta_{i,j}$ tending to zero with h, only a finite number of which do not vanish, we have

6.10
$$\psi_h[u] - \psi[u] = \sum_{i,j=0}^{\infty} \beta_{i,j} \psi^{(i,j)}[u] + O(h^{p+1}).$$

We replace the differential equation

6.11
$$\psi[u] = -G$$

by the difference equation

6.12
$$\psi_h[u] = -G - \sum_{i,j=0}^{\infty} \beta_{i,j} G^{(i,j)}.$$

If u_h satisfies (6.12) and if u satisfies (6.11), then we have

6.13
$$\psi_h[e] = O(h^{p+1}).$$

where $e = u_h - u$. Under normal circumstances we would expect that if the relative degree of approximation is p, then

6.14
$$e = O(h^{p+1})$$

We now show that the forward difference method corresponds to a consistent operator which has an absolute degree of approximation of one if $\theta(h) = rh^2$ and a relative degree of approximation of one unless $r = \frac{1}{6}$ in which case the relative degree of approximation is three. For, we have

6.15 $\quad \psi_h[u] - \psi[u] = \dfrac{u(x, t+k) - u(x, t)}{k} - \dfrac{u(x+h, t) + u(x-h, t) - 2u(x, t)}{h^2} - [u_t - u_{xx}]$

$$= \left(u_{0,1} + \frac{k}{2} u_{0,2} + \frac{k^2}{6} u_{0,3} + \cdots\right) - \left(u_{2,0} + \frac{h^2}{12} u_{4,0} + \cdots\right)$$

$$- (u_{0,1} - u_{2,0})$$

$$= \frac{h^2}{2}\{(r - \tfrac{1}{6})u_{4,0} + r(\psi^{(0,1)}[u] + \psi^{(2,0)}[u])\} + O(h^4).$$

Here we let

6.16
$$u_{i,j} = \frac{\partial^{i+j} u}{\partial x^i \partial t^j}, \qquad i, j = 0, 1, 2, \dots .$$

If $r = \frac{1}{6}$, we evidently have a relative degree of approximation of three provided we use the difference equation

6.17
$$\psi_h[u] = -G - \frac{k}{2}[G^{(0,1)} + G^{(2,0)}].$$

We now show that Richardson's method corresponds to a consistent discrete operator which has an absolute degree of approximation of one if $\theta(h) = rh^2$ for some constant r. Moreover, the relative degree of approximation is also one. For we have

$$\psi_h[u] - \psi[u] = \frac{u(x, t+k) - u(x, t-k)}{2k} - \frac{u(x+h, t) + u(x-h, t) - 2u(x, t)}{h^2} - (u_t - u_{xx})$$

6.18

$$= \frac{k^2}{6} u_{0,3} - \frac{h^2}{12} u_{4,0} + O(h^4).$$

Even if $k = ch$ for some constant c, the absolute degree of approximation and the relative degree of approximation are not changed, as can easily be verified.

For the DuFort-Frankel method, the corresponding discrete operator is consistent and has an absolute degree of approximation of one if $\theta(h) = rh^2$ for some r.

In this case the relative degree of approximation is one unless $r = \sqrt{1/12}$ in which case it is three. For, we have

6.19

$$\psi_h[u] - \psi[u] = \frac{u(x, t+k) - u(x, t-k)}{2k} - \frac{u(x+h, t) + u(x-h, t) - u(x, t+k) - u(x, t-k)}{h^2}$$

$$- [u_t - u_{xx}]$$

$$= \frac{k^2}{6} u_{0,3} - \frac{h^2}{12} u_{4,0} + \frac{k^2}{h^2} u_{0,2} + \frac{k^4}{12h^2} u_{0,4} + O(k^4) + O(h^4)$$

and, if $k = rh^2$

6.20 $$\psi_h[u] - \psi[u] = h^2 \left[\left(r^2 - \frac{1}{12} \right) u_{0,2} + \frac{1}{12} (\psi^{(0,1)} + \psi^{(2,0)}) \right] + O(h^4).$$

Thus if $r = \sqrt{1/12}$ we obtain a relative degree of approximation of three provided we use the difference equation

6.21 $$\psi_h[u] = -G - \frac{h^2}{12} [G^{(0,1)} + G^{(2,0)}] = -G - \frac{k}{\sqrt{12}} [G^{(0,1)} + G^{(2,0)}].$$

Evidently the DuFort-Frankel method is consistent even if k decreases more slowly than rh^2. However, we cannot let $k = ch$, for some constant c; otherwise, the method is not consistent since

6.22 $$\lim_{h \to 0} (\psi_h[u] - \psi[u]) = c^2 u_{0,2}.$$

Actually, then, $\psi_h[u]$ is a representation, not of $\psi[u]$ but rather of

$$\psi[u] + c^2 u_{0,2}.$$

The following very crude analysis indicates that to let $k = ch^\alpha$ for some constant c and for $\alpha < 2$ is less advantageous than letting $\alpha = 2$. If we assume that

6.23 $$E_\alpha(h) \sim C_\alpha h^{2\alpha - 2}$$

is the error obtained using $k = ch^\alpha$ where $1 < \alpha \leq 2$, then to reduce the error by a factor of two we would use $\hat{h}$ where

6.24 $$\frac{\hat{h}}{h} = (\tfrac{1}{2})^{1/(2\alpha - 2)}.$$

Since the number of mesh points and hence the work is proportional to $h^{-(1+\alpha)}$, the factor of increase in the work is $(\hat{h}/h)^{-(1+\alpha)}$ in going from h to $\hat{h}$. Thus, to reduce the error by a factor of two, the work is multiplied by

6.25 $2^{(1+\alpha)/(2\alpha-2)}$.

This quantity is minimized in the interval $1 \leq \alpha \leq 2$ by letting $\alpha = 2$.

Next, we show that the Crank-Nicolson method corresponds to a consistent discrete operator which, if $\theta(h) = ch$ for some constant c has an absolute degree of approximation of zero and a relative degree of approximation of one. Indeed we have (assuming $k = ch$)

6.26 $\psi_h[u] - \psi[u] = \dfrac{u(x, t+k) - u(x, t)}{k} - \dfrac{1}{2}\left[\dfrac{u(x+h, t) + u(x-h, t) - 2u(x, t)}{h^2}\right.$

$$+ \left.\dfrac{u(x+h, t+k) + u(x-h, t+k) - 2u(x, t+k)}{h^2}\right] - [u_t - u_{xx}]$$

$$= \dfrac{k}{2}\psi^{(0,1)}[u] + O(h^2).$$

Since $\psi^{(0,1)}[u] = 0$ if $\psi[u] = 0$, it follows that the relative degree of approximation is one. Moreover, one should use the difference equation

6.27 $\psi_h[u] = -G(x, t) - \dfrac{k}{2} G^{(0,1)}(x, t) \sim -\frac{1}{2}[G(x, t) + G(x, t+k)]$.

Actually, there is no loss of accuracy in using the difference equation

6.28 $\psi_h[u] = -\frac{1}{2}[G(x, t) + G(x, t+k)]$

which is equivalent to (2.17).

EXERCISES 17.6

1. Consider the backward difference method for solving $u_t = u_{xx}$, $0 < x < 1$, $t > 0$, $u(0, t) = u(1, t) = 0$, $t \geq 0$, and $u(x, 0) = x(1 - x)$, $0 < x < 1$. The difference equation is

$$\dfrac{u(x, t+k) - u(x, t)}{k} = \dfrac{u(x+h, t+k) + u(x-h, t+k) - 2u(x, t+k)}{h^2}.$$

   a) With $h = \frac{1}{4}$, $k = h^2$, compute $u(\frac{1}{2}, \frac{1}{8})$.
   b) Obtain a formula for the exact solution of the difference equation and use it to compute $u(\frac{1}{2}, \frac{1}{8})$.
   c) Discuss the stability of the method.
   d) Show that the corresponding discrete operator is consistent with $\psi[u] = u_t - u_{xx}$ and find the absolute and the relative degrees of approximation.

2. Use the methods of Section 15.5 to develop a discrete operator of the form

$$\psi_h[u] = \alpha_0 u(x, t) + \alpha_1 u(x + h, t) + \alpha_2 u(x, t + k) + \alpha_3 u(x - h, t)$$
$$+ \alpha_4 u(x + h, t + k) + \alpha_5 u(x - h, t + k)$$

which is consistent with

$$\psi[u] = u_t - u_{xx}$$

and which has as high a relative degree of approximation as possible. Assume that $k = ch$ for some constant c. What difference equation would you use to solve the differential equation

$$\psi[u] = x^3 + t^3?$$

3. For the problem of Exercise 1, Section 17.3, with the differential equation $u_t - u_{xx} = x^3 + t^3$ what difference equation should one use with the forward difference method and with $h = \frac{1}{4}$ and $r = \frac{1}{6}$ to achieve maximum accuracy?

4. Verify (6.26).

5. Find a discrete operator of the form

$$\psi_h[u] = \sum_{i=0}^{8} \alpha_i u(x + s_i h, t + t_i k)$$

where

$$s_0 = s_2 = s_4 = 0, \qquad s_5 = s_1 = s_7 = 1, \qquad s_6 = s_3 = s_8 = -1,$$

and

$$t_0 = t_1 = t_3 = 0, \qquad t_5 = t_2 = t_6 = 1, \qquad t_7 = t_4 = t_8 = -1$$

such that $\psi_h[u]$ is consistent with

$$\psi[u] = u_t - u_{xx}$$

and such that the relative degree of approximation is as high as possible. What is the absolute degree of approximation? What difference equation would be used to solve

$$\psi[u] = -[x^8 + t^8 + x^4 t^4]?$$

Discuss the stability of the method.

17.7 PROBLEMS WITH TWO SPACE VARIABLES

For the solution of (1.2) by finite difference methods we first replace the region R and the boundary S by sets of mesh points. For simplicity, we assume $\Delta x = \Delta y = h$ and, as in Chapter 15, we construct R_h and S_h. We could then use a semi-discretization to construct a system of ordinary differential equations with one equation for each point of R_h. Evidently the forward difference method, Richardson's method, the DuFort-Frankel method, and the Crank-Nicolson method could be used.

Let us now consider the equation

7.1
$$u_t - u_{xx} - u_{yy} = -G(x, y, t)$$

where R is the rectangle with length $a = Mh$ and width $b = M'h$, where M and M' are integers. If the boundary conditions are

7.2
$$u(x, y, t) = 0, \qquad (x, y) \in S, t \geq 0,$$

and if the initial conditions are

7.3
$$u(x, y, 0) = f(x, y), \qquad (x, y) \in R,$$

where $f(x, y)$ is a given piecewise continuous function, then in the case $G \equiv 0$ one can express both the solution of the differential equation and the solution of the various difference equations in terms of infinite or finite series. For example, if $a = b = 1$, the solution of the differential equation is

7.4
$$u(x, y, t) = \sum_{n,n'=1}^{\infty} a_{n,n'} \sin n\pi x \sin n'\pi y \, e^{-\pi^2(n^2 + (n')^2)t}$$

where

7.5
$$a_{n,n'} = 4 \int_0^1 \int_0^1 f(x, y)\sin n\pi x \sin n'\pi y \, dx \, dy.$$

For the *forward difference method* we have

7.6
$$\frac{u(x, y, t + k) - u(x, y, t)}{k} = L_h[u(x, y, t)] - G(x, y, t)$$

where

7.7
$$L_h[u] = \frac{u(x + h, y, t) + u(x, y + h, t) + u(x - h, y, t) + u(x, y - h, t) - 4u(x, y, t)}{h^2}.$$

The solution of the above difference equation corresponding to the case $G \equiv 0$ and to the conditions (7.2) and (7.3) with $a = b = 1$ is

7.8
$$u(x, y, t) = \sum_{n=1}^{M-1} \sum_{n'=1}^{M-1} b_{n,n'} \sin n\pi x \sin n'\pi y \left[1 - 4r \left[\sin^2 \frac{n\pi h}{2} + \sin^2 \frac{n'\pi h}{2} \right] \right]^{t/k}$$

where

7.9
$$b_{n,n'} = \frac{4}{M^2} \sum_{j=1}^{M-1} \sum_{l=1}^{M-1} f(jh, lh)\sin n\pi jh \sin n'\pi lh.$$

Evidently for stability we must have

7.10
$$r = \frac{k}{h^2} \leq \tfrac{1}{4}.$$

As in the case of one space variable the condition (7.10) imposes a very serious limitation on k which results in an excessive amount of calculation for a given accuracy. In searching for a better method we reject Richardson's method, which is unsatisfactory. One could consider the DuFort-Frankel method. However, as in the case of one space variable, one should let $k = rh^2$ for some constant r, and hence many time steps are required if h is small. A more satisfactory method is the Crank-Nicolson method given by

7.11
$$\frac{u(x, y, t + k) - u(x, y, t)}{k} = \tfrac{1}{2}(L_h[u(x, y, t)] + L_h[u(x, y, t + k)])$$
$$- \tfrac{1}{2}[G(x, y, t) + G(x, y, t + k)].$$

The solution of the above difference equation corresponding to the case $G \equiv 0$ and to the conditions (7.2) and (7.3) with $a = b = 1$ is

7.12
$$u(x, y, t) = \sum_{n=1}^{M-1} \sum_{n'=1}^{M-1} b_{n,n'} \sin n\pi x \sin n'\pi y \left(\frac{1 - 2r\left[\sin^2 \frac{n\pi h}{2} + \sin^2 \frac{n'\pi h}{2} \right]}{1 + 2r\left[\sin^2 \frac{n\pi h}{2} + \sin^2 \frac{n'\pi h}{2} \right]} \right)^{t/k}$$

The method is stable for all r since the last factor is always less than unity. The reader should show that if $k = ch$, for some constant c, then the discrete operator for the Crank-Nicolson method is consistent with $\psi[u] = u_t - u_{xx} - u_{yy}$ and has a relative degree of approximation of one. For this reason we expect that, as in the case of one space variable, the overall accuracy is $O(h^2)$ at least if $f(x, y)$ is sufficiently well behaved.

In order to obtain $u(x, y, t + k)$ from $u(x, y, t)$ one can solve an elliptic boundary value problem by an iterative method. Thus, letting $u^{(0)}(x, y, t + k) = u(x, y, t)$, one can use the *SOR* method. It is easy to show that the spectral radius $\bar{\mu}$ of the Jacobi method is bounded by

7.13
$$\bar{\mu} \leq \frac{2r}{1 + 2r}$$

since the matrix of the Jacobi method has at most four nonzero elements per row, each of which is $(r/2)/(1 + 2r)$. Since $r = k/h^2$ and $k = ch$, we have $r = c/h$ and, for small h,

7.14
$$\bar{\mu} \sim 1 - \frac{h}{2c}.$$

If we use the *SOR* method with the relaxation factor

7.15
$$\omega_b = \frac{2}{1 + \sqrt{1 - \bar{\mu}^2}},$$

then, as in Chapter 16, the rate of convergence is given by

7.16
$$R(\mathscr{L}_{\omega_b}) \sim 2\sqrt{\frac{h}{c}}.$$

Thus the amount of work per time step is proportional to $h^{-1/2}$. Hence the amount of work needed to reach a given value of t is proportional to $h^{-5/2}$ as compared to h^{-3} for the forward difference method. This follows since the number of mesh points is proportional to h^{-2} for the Crank-Nicolson method and to h^{-3} for the forward difference method.

In order to carry out a time step with the Crank-Nicolson method one could, of course, use an iterative method other than the *SOR* method. For example, one could use the Peaceman-Rachford alternating direction implicit method. In some cases, for example for the equation (7.1) in the rectangle, it can be proved that this procedure will result in much less computational effort. As a matter of fact, one can actually replace the Crank-Nicolson method by one iteration of the Peaceman-Rachford method and a suitable value of the iteration parameter. This corresponds to using the following scheme

7.17
$$\frac{u\left(x, y, t+\frac{k}{2}\right) - u(x, y, t)}{k/2} - \frac{u\left(x+h, y, t+\frac{k}{2}\right) + u\left(x-h, y, t+\frac{k}{2}\right) - 2u\left(x, y, t+\frac{k}{2}\right)}{h^2}$$
$$-\frac{u(x, y+h, t) + u(x, y-h, t) - 2u(x, y, t)}{h^2} = -\frac{1}{2}\left[G(x, y, t) + G\left(x, y, t+\frac{k}{2}\right)\right]$$

7.18
$$\frac{u(x, y, t+k) - u\left(x, y, t+\frac{k}{2}\right)}{k/2} - \frac{u\left(x+h, y, t+\frac{k}{2}\right) + u\left(x-h, y, t+\frac{k}{2}\right) - 2u\left(x, y, t+\frac{k}{2}\right)}{h^2}$$
$$-\frac{u(x, y+h, t+k) + u(x, y-h, t+k) - 2u(x, y, t+k)}{h^2}$$
$$= -\frac{1}{2}\left[G\left(x, y, t+\frac{k}{2}\right) + G(x, y, t+k)\right].$$

The determination of $u(x, y, t + k/2)$ by (7.17) can be accomplished by solving a linear system with a tridiagonal matrix. Similarly, $u(x, y, t + k)$ can be found from (7.18).

It can be shown (see Douglas [1955]) that this scheme has accuracy comparable to that of the Crank-Nicolson method in the case of (7.1) with $G \equiv 0$ and the rectangle. One can, of course, carry out the procedure for more general problems even though the stability and accuracy are not guaranteed. It would seem safer to use the Crank-Nicolson method with the Peaceman-Rachford iteration procedure, since the worst that could happen would be that the convergence would be slower than predicted or else divergence would occur. In the latter case one could then use the SOR method. On the other hand, if one were to use the Peaceman-Rachford scheme by itself it would probably be desirable to develop some independent method for monitoring the results.

EXERCISES 17.7

1. Consider the problem of solving $u_t = u_{xx} + u_{yy}$, $x^2 + y^2 < 1, t > 0$, with the boundary conditions $u(x, y, t) = 0$, $t \geq 0$ and $x^2 + y^2 = 1$, and the initial conditions $u(x, y, 0) = 1$, $x^2 + y^2 < 1$. Carry out two time steps with the forward difference method with $h = \frac{1}{2}$, $r = \frac{1}{4}$. Carry out one time step with the Crank-Nicolson method with $h = \frac{1}{2}, r = 1$. Also carry out one time step with Peaceman-Rachford method with $h = \frac{1}{2}$ and $r = 1$ (for each half step).

2. Solve the differential equation $u_t = u_{xx} + u_{yy} - (x^3 + y^3 + t^3)$, in the square $0 < x < 1$, $0 < y < 1, t > 0$ subject to the condition $u(x, y, t) = 0$ on the boundary of the square for $t \geq 0$ and with the initial condition $u(x, y, 0) = 1, 0 < x < 1, 0 < y < 1$. Use $h = \frac{1}{4}, k = \frac{1}{4}$, and the Crank-Nicolson method with the SOR method used for each time step. Carry out the solution until $t = 1$. Also carry out the solution of the same problem for the Peaceman-Rachford method.

3. Develop a formula for the DuFort-Frankel method for the case of two space variables and the unit square. Show that the method is stable if $k = rh^2$ where $h = \Delta x = \Delta y$ and $k = \Delta t$. On the other hand, show that it is not consistent if $k = ch$ for some constant c.

4. Find a solution analogous to (7.8) and (7.12) for the Peaceman-Rachford method (7.17)–(7.18). Show that for given n and n' the "t-factor" is as good an approximation to $e^{-\pi^2(n^2 + (n')^2)t}$ as is the "t-factor" in (7.8) provided h is sufficiently small.

5. Consider the solution by the Crank-Nicolson method of the problem defined by

$$u_t = u_{xx} + u_{yy}, \qquad (x, y) \in R$$

$$u(x, y, t) = 0, \qquad (x, y) \in S, \qquad t \geq 0$$

$$u(x, y, 0) = xy, \qquad (x, y) \in R$$

where R is the region $0 < x < 1, 0 < y < 1$ and S is the boundary of R. Suppose $\Delta x = \Delta y = h = \frac{1}{20}$ and $\Delta t = k = h$. How many iterations would be required per time step using the SOR method (assuming one wished to reduce the error by a factor of 10^{-6})? How many iterations would be required using the Peaceman-Rachford method with two parameters?

6. In the previous example, with the differential equation replaced by

$$u_t = u_{xx} + 2u_{yy},$$

how many *SOR* iterations would be required for each time step?

SUPPLEMENTARY DISCUSSION

For more detailed accounts of finite difference methods for solving parabolic partial differential equations, see Richtmyer and Morton [1967], Saul'yev [1964], Douglas [1961], and Forsythe and Wasow [1960].

For recent work on the convergence of various finite difference methods, see Widlund [1968] and [1971], and Peetre and Thomée [1967].

Section 17.3

Richtmyer and Morton [1967] give a convergence proof for the forward difference method, and for other methods, using techniques of functional analysis, see also Widlund [1968, 1971]. Early proofs of convergence under relatively weak hypotheses were given by O'Brien, Hyman, and Kaplan [1951], by Leutert [1952], and by Juncosa and Young [1954]. Juncosa and Young [1953] also showed that under rather weak hypotheses the difference between the exact solution and the approximate solution is $O(h^2)$.

Section 17.5

For other work on alternating direction implicit methods see Douglas and Gunn [1962, 1964].

Section 17.7

A number of authors have studied relations between numerical methods for solving initial-value problems (usually associated with parabolic partial differential equations) and iterative methods for solving elliptic equations. See, for instance, Peaceman and Rachford [1955], Douglas and Rachford [1956], Garabedian [1956], Young [1962], Varga [1962], Saul'yev [1964], and many others.

In many cases the solution of a boundary value problem involving the elliptic differential equation

(*) $$L[u] = G$$

is the limit as $t \to \infty$ of the initial-value problem involving the differential equation

(**) $$u_t - L[u] = -G$$

with boundary conditions independent of t and with arbitrary initial conditions. A similar connection holds between the discrete analogs of the two problems. One can interpret the process of solving the elliptic problem by an iterative method with a given set of starting values as the solution of a time-dependent problem with initial values corresponding to these starting values. For example, the forward difference method for solving the time-dependent problem is analogous to a variant of the *Jacobi method* for solving the linear system $Au = b$. This variant which is referred to as the *JOR method* by Young [1971a], is defined by

$$u^{(n+1)} = (\omega B + (1 - \omega)I)u^{(n)} + \omega c.$$

(Here we use the notation of Chapter 16.) The limitation on k for stability for the time-dependent problem corresponds to the fact that if the matrix A is positive definite, for the convergence of the JOR method we must have

$$\omega \leqq \frac{2}{\bar{v}}$$

where $\bar{v}$ is the largest eigenvalue of A. This limitation can be largely overcome by using variable ω in a manner analogous to the use of varying step sizes for the time-dependent problem. We remark that the use of variable ω is closely related to the use of semi-iterative methods.

The correspondence between iterative methods for elliptic problems and methods for solving time-dependent problems is particularly striking in the case of alternating direction implicit methods including the Peaceman-Rachford method [1955] and the Douglas-Rachford method [1956]. The latter method is somewhat slower for solving elliptic problems in certain cases but can be used for solving problems involving more than two space variables.

As shown by Garabedian [1956] (see also Young [1962]), the Gauss-Seidel and SOR methods can be interpreted in terms of a time-dependent problem, but with a hyperbolic rather than a parabolic equation.

MATHEMATICAL PRELIMINARIES

A.1 INTRODUCTION

In this appendix we give many of the mathematical facts which are used in the text. Most of the material can be found in standard texts. However, the exact statement of a given theorem frequently varies from one book to another; hence it may be useful to give the exact form which we use. In some cases, proofs are given in the text.

A.2 SETS, SEQUENCES, AND SERIES

Sets of Real Numbers

Let $\mathbf{R}$ denote the set of real numbers, $\mathbf{R}^+$ the set of positive real numbers and $\mathbf{R}^-$ the set of negative real numbers.

2.1 Theorem. Given a set $S \subseteq \mathbf{R}$ such that for some α each element $x \in S$ does not exceed α, then there exists $\beta \in \mathbf{R}$ such that $\beta \geqq x$ for all $x \in S$ and such that if $\beta' \geqq x$ for all $x \in S$ then $\beta' \geqq \beta$. The number β is called the *least upper bound*, or *supremum*, of S, and we write

2.2
$$\beta = \sup_{x \in S} x.$$

Similarly, if for some γ we have $\gamma \leqq x$ for all $x \in S$, then there exists a *greatest lower bound* or *infimum*

2.3
$$\delta = \inf_{x \in S} x$$

of S which is the number δ such that $\delta \leqq x$ for all $x \in S$ and if $\delta' \leqq x$ for all $x \in S$ then $\delta' \leqq \delta$.

2.4 Definition. Given a set $S \subseteq \mathbf{R}$, if $\sup_{x \in S} x$ exists and belongs to S, then we say that $\max_{x \in S} x$ exists and

2.5
$$\max_{x \in S} x = \sup_{x \in S} x.$$

Similarly, if $\inf_{x \in S} x$ exists and belongs to S, then we say that $\min_{x \in S} x$ exists and

2.6
$$\min_{x \in S} x = \inf_{x \in S} x.$$

2.7 Definition. Given $x_0 \in \mathbf{R}$ and $\delta > 0$ the set of all x such that

$$|x - x_0| < \delta$$

is a *neighborhood* of x_0.

We use the following notation to describe various kinds of intervals in $\mathbf{R}$.

2.8
$$
\begin{cases}
[a, b] = \{x : a \leqq x \leqq b \text{ or } b \leqq x \leqq a\} \\
[a, b) = \{x : a \leqq x < b \text{ or } b < x \leqq a\} \\
(a, b] = \{x : a < x \leqq b \text{ or } b \leqq x < a\} \\
(a, b) = \{x : a < x < b \text{ or } b < x < a\}.
\end{cases}
$$

(Here the notation $\{x : a \leqq x \leqq b\}$ means the set of all x such that $a \leqq x \leqq b$.)

Sequences

2.9 Definition. The sequence $\{x_i\} = x_1, x_2, \ldots$ of real numbers *converges* (to a limit x) if there exists a number $x \in \mathbf{R}$ such that given any $\varepsilon > 0$ there exists N such that for all $n > N$

$$|x - x_n| < \varepsilon.$$

2.10 Theorem. If the sequence $\{x_i\}$ is a non-decreasing (non-increasing) sequence which has a finite upper (lower) bound, then the sequence converges.

2.11 Definition. A sequence $\{x_i\}$ is a *Cauchy sequence* if, given any $\varepsilon > 0$, there exists N such that for all m and n such that $m > N$ and $n > N$ we have

$$|x_n - x_m| < \varepsilon.$$

2.12 Theorem. If the sequence $\{x_i\}$ is a Cauchy sequence, then it converges.

2.13 Theorem. If for some a and b each element of the sequence $\{x_i\}$ satisfies the condition $x_i \in [a, b]$, then there exists a convergent subsequence. That is, there exist positive integers $n_1, n_2, \ldots$ such that

$$1 \leqq n_1 < n_2 < \cdots$$

and such that the sequence $\{y_i\}$ converges, where $y_1 = x_{n_1}, y_2 = x_{n_2}, \ldots$.

Series of Real Numbers

2.14 Definition. The series

$$\sum_{i=1}^{\infty} u_i$$

converges to a limit s if the sequence of partial sums

$$s_1 = u_1$$

$$s_2 = u_1 + u_2$$

$$s_3 = u_1 + u_2 + u_3$$

$$\vdots$$

converges to s.

2.15 Theorem. If the series $\sum_{i=1}^{\infty} u_i$ converges and if

$$0 \leqq v_i \leqq u_i, \qquad i = 1, 2, \ldots$$

then the series $\sum_{i=1}^{\infty} v_i$ converges.

2.16 Theorem. Let $u_1, u_2, \ldots$ be nonnegative numbers such that for some N

2.17 $$u_{n+1} \leqq u_n, \qquad n = N, N + 1, \ldots$$

and such that

$$\lim_{n \to \infty} u_n = 0.$$

Then the series

2.18 $$\sum_{i=1}^{\infty} (-1)^{n+1} u_i = u_1 - u_2 + u_3 - u_4 + \cdots$$

converges to a limit s. Moreover,

2.19 $$\left| s - \sum_{i=1}^{n} u_i \right| \leqq u_{n+1}$$

for $n \geqq N$.

A.3 FUNCTIONS OF A REAL VARIABLE
Limits

3.1 Definition. Given a function $f(x)$ defined in a neighborhood of x_0 (except possibly at x_0) we say that $\lim_{x \to x_0} f(x)$ exists and

3.2 $$\lim_{x \to x_0} f(x) = A$$

if given any $\varepsilon > 0$, there exists $\delta > 0$ such that for any x satisfying

3.3 $$0 < |x - x_0| < \delta$$

we have

3.4 $$|f(x) - A| < \varepsilon.$$

In the next definition $x \to x_0^+$ means "x approaches x_0 from the right" and $x \to x_0^-$ means "x approaches x_0 from the left." This terminology is not to be confused with $x \in \mathbf{R}^+$ (or $x \in \mathbf{R}^-$) which means x is a positive real number (or negative real number).

3.5 Definition. If $f(x)$ is defined for $I:(x_0, x_0 + \delta]$ for some $\delta > 0$, then $\lim_{x \to x_0+} f(x)$ exists and

3.6 $$\lim_{x \to x_0+} f(x) = A$$

if, given any $\varepsilon > 0$, there exists $\delta > 0$ such that for all $x \in I$ we have

3.7 $$|f(x) - A| < \varepsilon.$$

A similar definition can be given for

3.8 $$\lim_{x \to x_0-} f(x).$$

3.9 Definition. If $f(x)$ is defined in a neighborhood of x_0 (except possibly at x_0), then

3.10 $$\overline{\lim_{x \to x_0}} f(x) = \lim_{\delta \to 0} \left\{ \sup_{0 < |x - x_0| < \delta} f(x) \right\}$$

if the limit exists. The limit exists if, for some $\delta > 0$, the set of all $f(x)$ such that $0 < |x - x_0| < \delta$ has an upper bound. Similarly, we define

3.11 $$\underline{\lim_{x \to x_0}} f(x) = \lim_{\delta \to 0} \left\{ \inf_{0 < |x - x_0| < \delta} f(x) \right\}$$

if the limit exists.

3.12 Theorem. If

3.13 $$\overline{\lim_{x \to x_0}} f(x) = \underline{\lim_{x \to x_0}} f(x),$$

then $\lim_{x \to x_0} f(x)$ exists and

3.14 $$\lim_{x \to x_0} f(x) = \overline{\lim_{x \to x_0}} f(x) = \underline{\lim_{x \to x_0}} f(x).$$

Continuity

3.15 Definition. Given a function $f(x)$ defined in a neighborhood of x_0 we say that $f(x)$ is *continuous* at x_0 if $\lim_{x \to x_0} f(x)$ exists and

3.16
$$\lim_{x \to x_0} f(x) = f(x_0).$$

If $f(x)$ is continuous at each point of the interval $[a, b]$, we say that

3.17
$$f(x) \in C[a, b].$$

Similarly, $f(x) \in C[a, b)$, $C(a, b]$, or $C(a, b)$ if $f(x)$ is continuous in $[a, b)$, $(a, b]$, or (a, b), respectively.

3.18 Definition. The function $f(x)$, defined on the interval $I = [a, b]$, is *uniformly continuous* on I if, for every $\varepsilon > 0$, there exists a $\delta > 0$ such that for any two points x and x' in I satisfying

3.19
$$|x - x'| < \delta$$

we have

3.20
$$|f(x) - f(x')| < \varepsilon.$$

(It should be noted that δ is independent of x and x'.)

3.21 Theorem. If $f(x)$ is a continuous function defined on the interval $I = [a, b]$ then $f(x)$ is uniformly continuous on I.

3.22 Theorem. If $f(x)$ is a continuous function defined on the interval $I = [a, b]$, then for some ξ and η in I we have

3.23
$$f(\xi) = \max_{a \le x \le b} f(x), \qquad f(\eta) = \min_{a \le x \le b} f(x).$$

3.24 Theorem. If $f(x) \in C[a, b]$ and if $f(a)f(b) < 0$, then for some $\xi \in (a, b)$ we have

3.25
$$f(\xi) = 0.$$

3.26 Definition. A function $f(x)$ is *piecewise continuous* on the interval $I = [a, b]$ if $f(x)$ is defined and continuous for all but a finite number of points $x_1, x_2, \ldots, x_n$ in I. At each such point the limit

3.27
$$\lim_{x \to x_i+} f(x)$$

exists unless $x_i = b$ and the limit

3.28
$$\lim_{x \to x_i -} f(x)$$

exists unless $x_i = a$.

Integration

3.29 Definition. Given a function $f(x)$ defined on the interval $I = [a, b]$ we say that $f(x)$ is *integrable* over I, that $\int_a^b f(x)dx$ exists, and

3.30
$$\int_a^b f(x)dx = A$$

if, given any $\varepsilon > 0$, there exists $\delta > 0$ such that for any $x_1, x_2, \ldots, x_n$ satisfying

3.31
$$a = x_1 < x_2 < x_3 < \cdots < x_n = b$$

with

3.32
$$|x_{i+1} - x_i| < \delta, \qquad i = 1, 2, \ldots, n - 1$$

and for any $\xi_1, \xi_2, \ldots, \xi_{n-1}$ satisfying

3.33
$$\xi_i \in [x_i, x_{i+1}], \qquad i = 1, 2, \ldots, n - 1$$

we have

3.34
$$\left| A - \sum_{i=1}^{n-1} (x_{i+1} - x_i)f(\xi_i) \right| < \varepsilon.$$

3.35 Definition. If $f(x)$ is piecewise continuous on $[a, b]$, then $f(x)$ is integrable.

3.36 Theorem (Mean-value theorem for integrals). If $f(x)$ is continuous and $g(x)$ is piecewise continuous on $[a, b]$ and if $g(x) \geq 0$ (or $g(x) \leq 0$) on $[a, b]$, then

3.37
$$\int_a^b f(x)g(x)dx = f(\xi) \int_a^b g(x)dx$$

for some $\xi \in [a, b]$.

3.38 Corollary. If $f(x)$ is continuous on $[a, b]$, then

3.39
$$\int_a^b f(x)dx = f(\xi)(b - a)$$

for some $\xi \in [a, b]$.

3.40 Theorem. Let $g(x)$ be piecewise continuous function on the interval $I = [a, b]$. Then the relation

3.41
$$\int_a^b f(x)g(x)dx = f(\xi) \int_a^b g(x)dx$$

holds for all $f(x) \in C[a, b]$ and for some $\xi \in I$ if and only if $g(x)$ is one-signed in I (i.e., if $g(x) \geq 0$ for all $x \in I$ or $g(x) \leq 0$ for all $x \in I$).

Differentiation

3.42 Definition. Given a function $f(x)$ defined in a neighborhood of x_0, we say that $f(x)$ is *differentiable* at x_0 if

3.43
$$\lim_{x \to x_0} \frac{f(x) - f(x_0)}{x - x_0}$$

exists. If the limit exists, we let

3.44
$$f'(x_0) = \lim_{x \to x_0} \frac{f(x) - f(x_0)}{x - x_0}.$$

If $f(x)$ is differentiable (and hence continuous) at all points of the interval $[a, b]$, we say that

3.45
$$f(x) \in D^{(1)}[a, b].$$

If $f'(x)$ is continuous on the interval, then we say that

3.46
$$f(x) \in C^{(1)}[a, b].$$

Similarly, we say that $f(x) \in D^{(1)}[a, b)$ if $f'(x)$ exists on $[a, b)$, that $f(x) \in D^{(1)}(a, b]$ if $f'(x)$ exists on $(a, b]$, etc.

3.47 Theorem (Rolle's theorem). If $f(x) \in C[a, b]$ and $f(x) \in D^{(1)}(a, b)$ and if

3.48
$$f(a) = f(b) = 0,$$

then for some $\xi \in (a, b)$ we have

3.49
$$f'(\xi) = 0.$$

3.50 Theorem (Mean-value theorem). If $f(x) \in C[a, b]$ and $f(x) \in D^{(1)}(a, b)$, then for some $\xi \in (a, b)$ we have

3.51 $$f(b) - f(a) = (b - a)f'(\xi).$$

3.52 Theorem (L'Hospital's rule). Let $f(x) \in C^{(1)}[a, b]$ and $g(x) \in C^{(1)}[a, b]$ and let $c \in (a, b)$. If

3.53 $$g'(x) \neq 0, \qquad a \leq x \leq b, \qquad x \neq c$$

and if

3.54 $$f(c) = g(c) = 0,$$

then

3.55 $$\lim_{x \to c} \frac{f(x)}{g(x)} = \lim_{x \to c} \frac{f'(c)}{g'(c)},$$

provided the latter limit exists.

3.56 Theorem (Fundamental theorem of integral calculus). Let $f(x) \in C[a, b]$ and let $\phi(x)$ be defined for $x \in [a, b]$ by

3.57 $$\phi(x) = \int_a^x f(x)dx.$$

Then $\phi(x) \in C^{(1)}[a, b]$ and

3.58 $$\phi'(x) = f(x).$$

3.59 Theorem. If $f(x) \in C^{(1)}[a, b]$, then

3.60 $$\int_a^b f'(x)dx = f(b) - f(a).$$

3.61 Definition. If $f(x)$ is continuous and has continuous derivatives of all orders up to and including the n-th on the interval $I = [a, b]$, then we say that

3.62 $$f(x) \in C^{(n)}[a, b].$$

If $f(x) \in C^{(n-1)}[a, b]$ and $f^{(n)}(x)$ exists on I, then we say that

3.63 $$f(x) \in D^{(n)}[a, b].$$

3.64 Theorem (Taylor's theorem with the Lagrange form of the remainder). If $f(x) \in C^{(n)}[a, b]$ and if $f(x) \in D^{(n+1)}(a, b)$, then

3.65 $\quad f(b) = f(a) + (b - a)f'(a) + \cdots + \dfrac{(b - a)^n}{n!} f^{(n)}(a) + \dfrac{(b - a)^{n+1}}{(n + 1)!} f^{(n+1)}(\xi)$

for some $\xi \in (a, b)$.

3.66 Theorem (Taylor's theorem (with the integral form of remainder)). If $f(x) \in C^{(n+1)}(a, b)$, then

3.67 $\quad f(b) = f(a) + (b - a)f'(a) + \cdots + \dfrac{(b - a)^n}{n!} f^{(n)}(a) + \displaystyle\int_a^b \dfrac{(b - t)^n}{n!} f^{(n+1)}(t) dt.$

3.68 Definition. The function $f(x)$ is *analytic* at x_0 if it is defined in a neighborhood N of x_0 and has continuous derivatives of all orders in N, and if for all $x \in N$ we have

3.69 $\quad f(x) = f(x_0) + (x - x_0)f'(x_0) + \dfrac{(x - x_0)^2}{2!} f''(x_0) + \cdots.$

The function is analytic on an interval I if it is analytic at each point of I. The series (3.69) is the *Taylor series for $f(x)$.*

O and o Notation

3.70 Definition. If for some constant K we have

3.71 $\quad\quad\quad\quad\quad\quad\quad\quad |f(x)| \leq K|g(x)|$

for $|x - a|$ sufficiently small, then we say that

3.72 $\quad\quad\quad\quad\quad\quad\quad\quad f(x) = O(g(x))$

as $x \to a$. On the other hand, if

3.73 $\quad\quad\quad\quad\quad\quad\quad\quad \overline{\lim_{x \to a}} \left| \dfrac{f(x)}{g(x)} \right| = 0,$

then we say that

3.74 $\quad\quad\quad\quad\quad\quad\quad\quad f(x) = o(g(x))$

as $x \to a$.

Sequences and Series of Functions

3.75 Definition. The sequence of functions $s_1(x), s_2(x), \ldots$ converges to $s(x)$ if

3.76 $\quad\quad\quad\quad\quad\quad\quad\quad \lim_{n \to \infty} s_n(x) = s(x).$

The sequence *converges uniformly* on the interval $I = [a, b]$ if, given any $\varepsilon > 0$, there exists N such that for all $n > N$ we have

3.77 $$|s_n(x) - s(x)| < \varepsilon$$

for all $x \in I$. (It should be noted that N is independent of x.)

3.78 Definition. The series

3.79 $$\sum_{i=1}^{\infty} u_i(x)$$

converges to $s(x)$ if the sequence of partial sums $\{s_n(x)\}$, where

3.80 $$s_n(x) = \sum_{i=1}^{n} u_i(x),$$

converges to $s(x)$. The convergence of the series is uniform on the interval $I = [a, b]$ if the convergence of the sequence is uniform.

3.81 Theorem. If the series $\sum_{i=1}^{\infty} u_i(x)$ converges uniformly to $s(x)$ on the interval $I = [a, b]$, and if each $u_i(x)$ is continuous on I then $s(x)$ is continuous on I.

3.82 Definition. A *power series* is a series of the form

3.83 $$\sum_{k=0}^{\infty} a_k x^k.$$

3.84 Theorem. If a power series converges for $\hat{x} \neq 0$, then it converges for all x such that

$$|x| < |\hat{x}|.$$

Moreover, for any positive δ such that $\delta < |\hat{x}|$, the convergence is uniform in the interval

$$|x| \leq |\hat{x}| - \delta.$$

3.85 Definition. The *radius of convergence* r of the power series (3.83) is defined by

3.86 $$r = \frac{1}{\lim_{k \to \infty} \sqrt[k]{|a_k|}}.$$

3.87 Theorem. If r is the radius of convergence of the power series (3.83), then (3.83) converges for $|x| < r$ and diverges for $|x| > r$.

3.88 Theorem. If r is the radius of convergence of the power series (3.83), then

3.89
$$f(x) = \sum_{k=0}^{\infty} a_k x^k$$

defines a function which is analytic for $|x| < r$. Moreover, if $|x| < r$, then

3.90
$$\begin{cases} f'(x) = \sum_{k=1}^{\infty} k a_k x^{k-1} \\ \\ f''(x) = \sum_{k=2}^{\infty} k(k-1) a_k x^{k-2}, \end{cases}$$

etc., and if $|a| < r$, $|b| < r$, then

3.91
$$\int_a^b f(x)dx = \sum_{k=0}^{\infty} a_k \left(\frac{b^{k+1} - a^{k+1}}{k+1} \right).$$

A.4 FUNCTIONS OF TWO REAL VARIABLES

4.1 Definition. Given a point (x_0, y_0), for each $r > 0$ the circle

$$(x - x_0)^2 + (y - y_0)^2 < r^2$$

is a *neighborhood* of (x_0, y_0).

4.2 Definition. A *domain D* is a connected set of points such that for each point (x_0, y_0) of D there is a neighborhood of (x_0, y_0) contained in D.

4.3 Definition. A set S of points is *open* if for any point (x_0, y_0) of S there is a neighborhood of (x_0, y_0) each of whose points belongs to S.

4.4 Definition. A point (x_0, y_0) is a *limit point* of a set S if, given any $\varepsilon > 0$, there exists a point (x, y) (which may be (x_0, y_0) if $(x_0, y_0) \in S$) such that

$$(x - x_0)^2 + (y - y_0)^2 < \varepsilon^2.$$

4.5 Definition. A set S of points is *closed* if every limit point of S belongs to S.

4.6 Theorem (Taylor's theorem in two variables). If the function $f(x, y)$ is defined in a neighborhood N of (x_0, y_0) and has continuous partial derivatives of all orders up to and including the n-th in N, and if the partial derivatives of order $n + 1$ exist in N, then for any (x, y) in N we have

4.7 $f(x, y) = \displaystyle\sum_{i=0}^{n} \sum_{j=0}^{i} \frac{(x - x_0)^j(y - y_0)^{i-j}}{j!(i - j)!} \frac{\partial^i f(x_0, y_0)}{\partial x^j \partial y^{i-j}}$

$\qquad\qquad + \displaystyle\sum_{j=0}^{n+1} \frac{(x - x_0)^j(y - y_0)^{n+1-j}}{j!(n + 1 - j)!} \frac{\partial^{n+1} f(x_0 + \theta(x - x_0), y_0 + \theta(y - y_0))}{\partial x^j \partial y^{n+1-j}}$

for some θ such that $0 < \theta < 1$.

4.8 Definition. The function $f(x, y)$ is *analytic* at a point (x_0, y_0) if it is defined in a neighborhood N of (x_0, y_0) and has continuous partial derivatives of all orders in N, and if for all $(x, y) \in N$ we have

4.9 $$f(x, y) = \sum_{i=0}^{\infty} \sum_{j=0}^{i} \frac{(x - x_0)^i(y - y_0)^{i-j}}{j!(i - j)!} \frac{\partial^i f(x_0, y_0)}{\partial x^i \partial y^{i-j}}.$$

The function is analytic in a domain D if it is analytic at each point of D. The series (4.9) in the *Taylor series for* $f(x, y)$.

A.5 FUNCTIONS OF A COMPLEX VARIABLE

Let **C** denote the set of complex numbers z of the form

5.1 $z = x + iy$

where $x, y \in \mathbf{R}$ and where

5.2 $i^2 = -1.$

If (5.1) holds then we let

5.3 $\mathrm{Re}\, z = x, \quad \mathrm{Im}\, z = y.$

The *modulus*, $|z|$, of $z \in \mathbf{C}$ is given by

5.4 $|z| = \sqrt{(\mathrm{Re}\, z)^2 + (\mathrm{Im}\, z)^2}.$

If $z_1, z_2 \in \mathbf{C}$ then $z_1 = z_2$ if and only if

5.5 $\begin{cases} \mathrm{Re}\, z_1 = \mathrm{Re}\, z_2 \\ \mathrm{Im}\, z_1 = \mathrm{Im}\, z_2. \end{cases}$

Elementary Arithmetic Operations with Complex Numbers

If $z_1 = x_1 + iy_1$ and $z_2 = x_2 + iy_2$ where $x_1, y_1, x_2, y_2 \in \mathbf{R}$, then

5.6 $z_1 \pm z_2 = (x_1 \pm x_2) + i(y_1 \pm y_2).$

5.7
$$z_1 z_2 = (x_1 x_2 - y_1 y_2) + i(x_1 y_2 + x_2 y_1)$$

and

5.8
$$\frac{z_1}{z_2} = \frac{(x_1 x_2 + y_1 y_2) + i(-x_1 y_2 + x_2 y_1)}{x_2^2 + y_2^2}, \qquad \text{if } z_2 \neq 0.$$

Sets, Sequences, and Series

5.9 Definition. Given $z_0 \in \mathbf{C}$ and $\delta > 0$ the set of all z such that

$$|z - z_0| < \delta$$

is a *neighborhood* of z_0.

Definitions 2.7 and 2.9 and Theorems 2.12 and 2.13 concerning sequences of real numbers and Definition 2.14 concerning series of real numbers apply also to sequences and series of complex numbers. In the hypothesis of Theorem 2.13 the condition that each element of the sequence lie in $[a, b]$ should be replaced by the condition that each element lie in the circle

$$|z - z_0| < r$$

for some $z_0 \in \mathbf{C}$ and for some $r \in \mathbf{R}^+$.

5.10 Definition. Given a function $f(z)$ of a complex variable defined in a neighborhood of z_0 (except possibly at z_0) we say that $\lim_{z \to z_0} f(z)$ exists and

5.11
$$\lim_{z \to z_0} f(z) = A$$

if, given any $\varepsilon > 0$, there exists $\delta > 0$ such that for any z satisfying

$$0 < |z - z_0| < \delta$$

we have

$$|f(z) - A| < \varepsilon.$$

5.12 Definition. Given a function $f(z)$ of a complex variable defined in a neighborhood of z_0 we say that $f(z)$ is *continuous* at z_0 if $\lim_{z \to z_0} f(z)$ exists and

5.13
$$\lim_{z \to z_0} f(z) = f(z_0).$$

If $f(z)$ is continuous at each point of a domain D, we say that $f(z)$ is continuous in D.

5.14 Definition. Given a function $f(z)$ of a complex variable defined in a neighborhood of z_0, we say that $f(z)$ is *differentiable* at z_0 if

5.15
$$\lim_{z \to z_0} \frac{f(z) - f(z_0)}{z - z_0}.$$

If $f(z)$ is differentiable at each point of a domain D we say that $f(z)$ is differentiable in D.

5.17 Definition. The function $f(z)$ of a complex variable is *analytic* at z_0 if $f(z)$ is defined and differentiable at each point in a neighborhood of z_0. The function $f(z)$ is analytic in a domain D if it is analytic at each point of D.

5.18 Theorem. If $f(z)$ is analytic at z_0 then, for some neighborhood N of z_0, $f(z)$ has continuous derivatives of all orders in N and

5.19
$$f(z) = f(z_0) + (z - z_0)f'(z_0) + \frac{(z - z_0)^2}{2!}f''(z_0) + \cdots$$

for all z such that $|z - z_0| < \delta$, for some $\delta > 0$.

5.20 Theorem Cauchy-Riemann equations). Let $f(z) = u(x, y) + iv(x, y)$, where $u(x, y)$ and $v(x, y)$ are real-valued functions of the real variables x and y, and where $z = x + iy$. If $f(z)$ is analytic at $z_0 = x_0 + iy_0$, where $x_0, y_0 \in \mathbf{R}$, then for each point (x, y) in some neighborhood of (x_0, y_0) we have

5.21
$$\begin{cases} \dfrac{\partial u(x, y)}{\partial x} = \dfrac{\partial v(x, y)}{\partial y} \\[2mm] \dfrac{\partial v(x, y)}{\partial x} = -\dfrac{\partial u(x, y)}{\partial y}. \end{cases}$$

5.22 Theorem. Let $f(z)$ be analytic in a domain R and let S be the boundary of R. Then

5.23
$$\max_{z \in R + S} |f(z)| \leq \max_{z \in S} |f(z)|.$$

Moreover, if $f(z)$ does not vanish in R we have

5.24
$$\min_{z \in R + S} |f(z)| \geq \min_{z \in S} |f(z)|.$$

5.25 Definition. A *power series* is a series of the form

5.26
$$\sum_{k=0}^{\infty} a_k z^k.$$

5.27 Theorem. If a power series converges for $\hat{z} \neq 0$ then it converges for all z such that

$$|z| < |\hat{z}|.$$

Moreover, for any positive δ such that $\delta < |\hat{z}|$ the convergence is uniform for

$$|z| \leqq |\hat{z}| - \delta.$$

5.28 Definition. The *radius of convergence* r of the power series (5.26) is defined by

5.29
$$r = \frac{1}{\lim\limits_{k \to \infty} \sqrt[k]{|a_k|}}.$$

5.30 Theorem. If r is the radius of convergence of the power series (5.26) then (5.26) converges for $|z| < r$ and diverges for $|z| > r$.

5.31 Theorem. If r is the radius of convergence of the power series (5.26) then

5.32
$$f(z) = \sum_{k=0}^{\infty} a_k z^k$$

defines a function which is analytic for $|z| < r$. Moreover, if $|z| < r$ then

5.33
$$\begin{cases} f'(z) = \sum_{k=1}^{\infty} k a_k z^{k-1} \\ f''(z) = \sum_{k=2}^{\infty} k(k-1) a_k z^{k-2}, \end{cases}$$

etc.

A.6 EXTREMA OF A FUNCTION OF SEVERAL VARIABLES

Let $f(x_1, x_2, \ldots, x_n)$ be continuous and have continuous partial derivatives of orders one and two in a neighborhood of $(x_1^{(0)}, x_2^{(0)}, \ldots, x_n^{(0)})$. Then a *necessary* condition that $f(x_1, x_2, \ldots, x_n)$ have a relative extremum (a relative minimum or a relative maximum) at $(x_1^{(0)}, x_2^{(0)}, \ldots, x_n^{(0)})$ is that

6.1
$$\frac{\partial f(x_1^{(0)}, x_2^{(0)}, \ldots, x_n^{(0)})}{\partial x_i} = 0, \qquad i = 1, 2, \ldots, n.$$

A *sufficient* condition for a relative minimum (maximum) is that the matrix

6.2
$$A = \begin{bmatrix} \alpha_{11} & \alpha_{12} & \cdots & \alpha_{1n} \\ \alpha_{21} & \alpha_{22} & \cdots & \alpha_{2n} \\ \vdots & & & \\ \alpha_{n1} & \alpha_{n2} & \cdots & \alpha_{nn} \end{bmatrix}$$

where

6.3
$$\alpha_{ij} = \frac{\partial^2 f(x_1^{(0)}, x_2^{(0)}, \ldots, x_n^{(0)})}{\partial x_i \, \partial x_j}, \qquad i, j = 1, 2, \ldots, n$$

is positive definite (negative definite). See Chapter 11 for the definition of a positive definite matrix.

A.7 LINEAR ALGEBRA

Consider the system of n linear algebraic equations

7.1
$$\begin{cases} a_{11}u_1 + a_{12}u_2 + \cdots + a_{1n}u_n = b_1 \\ a_{21}u_1 + a_{22}u_2 + \cdots + a_{2n}u_n = b_2 \\ \vdots \\ a_{n1}u_1 + a_{n2}u_2 + \cdots + a_{nn}u_n = b_n \end{cases}$$

with n unknowns $u_1, u_2, \ldots, u_n$.

7.2 Theorem (Cramer's rule). If the determinant‡

7.3
$$\Delta = \det A$$

$$= \det \begin{bmatrix} a_{11} & a_{12} & \cdots & a_{1n} \\ a_{21} & a_{22} & \cdots & a_{2n} \\ \vdots & & & \\ a_{n1} & a_{n2} & \cdots & a_{nn} \end{bmatrix}$$

does not vanish then the system (7.1) has a unique solution which is given by

7.4
$$u_i = \frac{\Delta_i}{\Delta}, \qquad i = 1, 2, \ldots, n.$$

Here, for each i, Δ_i is the determinant of the matrix formed from A by replacing a_{1i} by b_1, a_{2i} by $b_2, \ldots, a_{ni}$ by b_n.

The *homogeneous system* corresponding to (7.1) is obtained from (7.1) by letting

7.5
$$b_1 = b_2 = \cdots = b_n = 0.$$

The homogeneous system always has the *trivial solution*.

---

‡ We assume that the reader is familiar with the definition of a determinant.

7.6
$$u_1 = u_2 = \cdots = u_n = 0.$$

Any other solution is said to be *nontrivial*.

7.7 Theorem. The homogeneous system, corresponding to (7.1) has a nontrivial solution if and only if

7.8
$$\Delta = 0.$$

7.9 Theorem. The system (7.1) has a unique solution if and only if the corresponding homogeneous system has no solution except the trivial solution.

A.8 POLYNOMIALS

8.1 Definition. Given the real or complex numbers $a_0, a_1, \ldots, a_n$, where $a_0 \neq 0$, the function

8.2
$$P(x) = a_0 x^n + a_1 x^{n-1} + \cdots + a_n$$

is a *polynomial of degree n*. (If $n = 0$, the function

8.3
$$P(x) = a_0$$

is a polynomial of degree zero whether or not $a_0 = 0$.)

8.4 Definition. The number α is a *zero* of *multiplicity m* of the polynomial $P(x)$ of degree n if

8.5
$$\begin{cases} P(\alpha) = P'(\alpha) = \cdots = P^{(m-1)}(\alpha) = 0 \\ P^{(m)}(\alpha) \neq 0. \end{cases}$$

8.6 Theorem (The fundamental theorem of algebra). Any polynomial of degree one or more has at least one zero.

8.7 Theorem. A polynomial of degree $n \geq 0$ has exactly n zeros, provided that a zero of multiplicity m is counted as m zeros.

A.9 MISCELLANEOUS

9.1 (*Principle of Mathematical Induction*). If a certain proposition involving the integer variable n is true for $n = 1$ and can be shown to be true for $n + 1$ provided it is true for all integers less than $n + 1$, then it is true for all n.

As an example, consider the proposition that

$$S_n = 1 + 2 + \cdots n = \frac{n(n+1)}{2}$$

for $n = 1, 2, \dots$. By direct verification the proposition is true for $n = 1$ since $S_1 = 1$. If the proposition is true for n then

$$S_{n+1} = S_n + n + 1$$

$$= \frac{n(n+1)}{2} + n + 1$$

$$= \frac{(n+1)(n+2)}{2}$$

$$= \frac{(n+1)((n+1)+1)}{2}$$

so that the proposition is true for $n + 1$. Hence the proposition is true in general.

9.2 (The symbol "$\sim$"). The symbol $\sim$ is used rather loosely throughout this book in the following three senses:

a) Asymptotic equality. For example, we write

$$f(x) \sim g(x) \qquad \text{as } x \to a,$$

if and only if

$$\lim_{x \to a} \frac{f(x)}{g(x)} = 1.$$

b) Approximately equal to (in some sense). The symbol "$\doteq$" is used to mean "is approximately equal to," also. For example, we might write

$$e \doteq 2.71828$$

(a numerical approximation) and

$$e^x \sim 1 + x + \frac{x^2}{2!} + \frac{x^3}{3!}$$

(an algebraic approximation).

c) Is represented by. For example, in 2-(4.4) we write

$$a \sim a_{59} a_{58} a_{57} \cdots a_2 a_1 a_0.$$

BIBLIOGRAPHY

Pages on which a reference is cited are listed in italic numerals at the end of each reference.

Achieser, N. I. [1956], *Theory of Approximation*, Frederick Ungar Pub. Co., New York; *328*.

Adams, Duane, A. [1967], "A stopping criterion for polynomial root finding," *Comm. Assoc. Comput. Mach.* 10, 655–658; *212*.

Ahlberg, J. H., E. N. Nilson, and J. L. Walsh [1967], *The Theory of Splines and Their Applications*, Academic Press, Inc., New York; *296*.

Ahlfors, L. V. [1966], *Complex Analysis* (second edition), McGraw-Hill Book Co., New York; *169*.

Aitken, A. C. [1932], "An interpolation by iteration of proportional parts, without the use of differences," *Proc. Edinburgh Math. Soc.* 3, Series 2, 56–76; *264*.

Allen, D. N. de G. [1954], *Relaxation Methods*, McGraw-Hill Book Co., New York; *997, 998*.

Antosiewicz, Henry A., and Walter Gautschi [1962], "Numerical methods in ordinary differential equations," Chapter 9 in *A Survey of Numerical Analysis*, edited by John Todd, McGraw-Hill Book Co., New York.

Arms, R. J., L. D. Gates, and B. Zondek [1956], "A method of block iteration," *J. Soc. Indust. Appl. Math.* 4, 220–229; *1074*.

Bareiss, E. H. [1960], "Resultant procedure and the mechanization of the Graeffe process," *J. Assoc. Comput. Mach.* 7, 346–386; *242*.

Bareiss, E. H. [1967], "The numerical solution of polynomial equations and the resultant procedure," Chapter 10 of Ralston and Wilf [1967]; *242*.

Batschelet, von Eduard [1952], "Über die numerische Auflösung von Randwertproblemen bei elliptischen partiellen Differentialgleichungen," *ZAMP* III, 165–193; *998*.

Bauer, F. L. [1963], "Optimally scaled matrices," *Numer. Math.* 5, 73–87; *815*.

Bauer, F. L., and C. T. Fike [1960], "Norms and exclusion theorems," *Numer. Math.* 2, 137–141; *947*.

Bauer, F. L., H. Rutishauser, and E. Stiefel [1963], "New aspects of numerical quadrature," *Proc. of Symposia in Applied Math.*, Vol. XV, Amer. Math. Soc., Providence, Rhode Island; *383, 384*.

Beckman, F. S. [1960], "The solution of linear equations by the conjugate gradient method," Chapter 4 of Ralston and Wilf [1960]; *1074*.

Bellman, R. [1960], *Introduction to Matrix Analysis*, McGraw-Hill Book Co., New York; *754*.

Bernstein, S. [1937], "Sur les formules de quadrature de Cotes et de Tchebycheff," *C. R. de l'Academie des Sciences de l'URSS* 14, 323–326; *414*.

Bickley, W. G. [1941], "Formulae for numerical differentiation," *Math. Gazette* 25, 19–27; *420*.

Bieberbach, L. [1930], *Theorie der Differentialgleichungen*, Springer, Berlin; *442*.

Birkhoff, G., C. de Boor, B. Swartz, and B. Wendroff [1966], "Rayleigh-Ritz approximation by piecewise cubic polynomials," *SIAM J. Numer. Anal.* 3, 188–203; *666*.

Birkhoff, Garrett, and Gian-Carlo Rota [1962], *Ordinary Differential Equations*, Ginn and Co., Boston; *429, 430, 486*.

Birkhoff, Garrett, and Saunders MacLane [1953], *A Survey of Modern Algebra*, The Mac-Millan Co., New York.

Birkhoff, Garrett, Richard S. Varga, and David Young [1962], "Alternating direction implicit methods," in *Advances in Computers*, Vol. 3, edited by F. Alt and M. Rubinoff, Academic Press, New York, 189–273; *1047, 1055, 1060, 1061, 1074*.

Birkhoff, G., David M. Young, and E. H. Zarantenello [1951], "Effective conformal transformation of smooth, simply connected domains," *Proc. Nat. Acad. Sci.* 37, 411–414.

Blair, A., N. Metropolis, J. von Neumann, A. H. Taub, and M. Tsingori [1959], "A study of a numerical solution of a two-dimensional hydrodynamical problem," *Math. Tables Aids Comput.* 13, 145–184; *1074*.

Blanch, G. [1964], "Numerical evaluation of continued fractions," *SIAM Review* 6, 383–421; *329*.

Blum, E. K. [1957], "A modification of the Runge-Kutta fourth-order method," Numerical Note NN–80, Ramo-Wooldridge Corp., Los Angeles; *492*.

Borosh, I., and A. S. Fraenkel [1966], "Exact solutions of linear equations with rational coefficients by congruence techniques," *Math. of Comp.* 20, 107–112; *68, 885, 888*.

Bramble, J. H., and B. E. Hubbard [1963], "A theorem on error estimation for finite difference analogues of the Dirichlet problem for elliptic equations," in *Contributions to Differential Equations*, Vol. II, John Wiley and Sons, New York; *992*.

Bramble, J. H., and B. E. Hubbard [1964], "On a finite difference analogue of an elliptic boundary problem which is neither diagonally dominant nor of nonnegative type," *Jour. of Math. and Phys.* 43, 117–132; *614*.

Bronson, R. [1969], *Matrix Methods, an Introduction*, Academic Press, New York; *717, 731, 753, 755*.

Businger, P. A. [1968], "Matrices which can be optimally scaled," *Numer. Math.* 12, 346–348; *817*.

Businger, P. A. [1969], "Reducing a matrix to Hessenberg form," *Math. of Comp.* 23, 819–822; *924*.

Businger, P. A. [1971], "Monitoring the numerical stability of Gaussian elimination," *Numer. Math.* 16, 360–361; *806*.

Butcher, J. C. [1965], "A modified multistep method for the numerical integration of ordinary differential equations," *J. Assoc. Comput. Mach.* 12, 124–135.

Champagne, W. P. [1964], "On finding roots of polynomials by hook or by crook," Master's thesis, The University of Texas at Austin, August 1964; also TNN–37, Computation Center, The University of Texas at Austin, 1964; *177, 216, 217, 235, 237, 238*.

Chartres, B. A. [1966], "Automatic controlled precision calculations," *J. Assoc. Comput. Mach.* 13, 386–403; *68*.

Chase, P. E. [1962], "Stability properties of predictor-corrector methods for ordinary differential equations," *J. Assoc. Comput. Mach.* 9, 457–468.

Cheney, E. W. [1966], *Introduction to Approximation Theory*, McGraw-Hill Book Co., New York.

Ciarlet, P., M. Schultz, and R. Varga [1967], "Numerical methods of high-order accuracy for nonlinear boundary value problems," *Numer. Math.* 9, 394–430; *665, 666*.

Clenshaw, C. W. [1962], "Chebyshev series for mathematical functions," *Mathematical Tables* 5, Nat. Physical Lab., London; *343*.

Cody, W. J. [1967], "The influence of machine design on numerical algorithms," *Proceedings Spring Joint Computer Conference 1967*, 305–309; *69*.

Cohen, Abraham [1931], *An Introduction to the Lie Theory of One-Parameter Groups*, Stechert and Co., New York; *431.*

Collatz, L. [1933], "Bemerkungen zur Fehlerabschätzung für das Differenzenverfahren bei partiellen Differentialgleichungen," *Z. Angew. Math. Mech.* 13, 56–57; *576, 612, 960.*

Collatz, L. [1960], *The Numerical Treatment of Differential Equations*, 3rd ed., Springer-Verlag, Berlin; *665, 666.*

Collatz, L. [1966], *Functional Analysis and Numerical Mathematics*, Academic Press, New York; *761.*

Conte, S. D. [1962], "The computation of satellite orbit trajectories," in *Advances in Computers*, Vol. 3, edited by Franz L. Alt and Morris Rubinoff, Academic Press: New York, 1–76; *488.*

Conte, S. D., and David M. Young [1957], "Eigenvalues in modern industry III: Problems involving differential operators," *Proceedings of the Joint N.Y.U.-I.B.M. Symposium on Digital Computing in the Aircraft Industry, New York, Jan. 31, 1957–Feb. 28, 1957; 659, 661.*

Control Data Corp. [1969], "*Control Data 6400/6500/6600 Computer Systems Reference Manual*, Publication 60100000, Control Data Corp., Minneapolis, Minn.; *45, 52.*

Courant, R., and D. Hilbert [1962], *Methods of Mathematical Physics*, Vol. II, Interscience Publishers, New York; *952.*

Courant, R., and F. John [1965], *Introduction to Calculus and Analysis*, Interscience Publishers, New York.

Crank, J., and P. Nicolson [1947], "A practical method for numerical evaluation of solutions of partial differential equations of the heat-conduction type," *Proc. Cambridge Philos. Soc.* 43, 50–67; *1078.*

Cullen, C. G. [1966], *Matrices and Linear Transformations*, Addison-Wesley, Reading, Mass.; *681.*

Cuthill, E. H., and R. S. Varga [1959], "A method of normalized block iteration," *J. Assoc. Comput. Mach.* 6, 236–244; *1064.*

Dahlquist, G. [1956], "Convergence and stability in the numerical integration of ordinary differential equations," *Math. Scand.* 4, 33–53; *492, 502, 513, 516, 522, 523, 574.*

Davis, Phillip J. [1963], *Interpolation and Approximation*, Blaisdell Publishing Company, New York; *288.*

Davis, Phillip J., and Phillip Rabinowitz [1967], *Numerical Integration*, Blaisdell Publishing Company, Waltham, Mass.; *420, 421.*

Dejon, Bruno, and Peter Henrici [1969], *Constructive Aspects of the Fundamental Theorem of Algebra.* (Proceedings of a Symposium conducted by the IBM Research Laboratory, Zurich-Ruschlikon, Switzerland, June 5–7, 1967), Wiley-Interscience, New York; *241.*

Delves, L. M., and J. N. Lyness [1967], "A numerical method for locating the zeros of an analytic function," *Math. Comp.* 21, 543–560; *245.*

DeVogelaere, R. [1958], "Over-relaxations," Abstract No. 539–53, *Amer. Math. Soc. Notices* 5, 147; *1068.*

Douglas, Jim, Jr. [1955], "On the numerical integration of $\partial^2 u/\partial x^2 + \partial^2 u/\partial y^2 = \partial u/\partial t$ by implicit methods," *J. Soc. Indust. Appl. Math.* 3, 42–65; *1097.*

Douglas, Jim, Jr. [1961], "A survey of numerical methods for parabolic differential equations," in *Advances in Computers*, 2, edited by F. L. Alt, Academic Press, New York, 1–54; *1098.*

Douglas, J., Jr., and J. E. Gunn [1962], "Alternating direction methods for parabolic systems in *m* space variables," *J. Assoc. Comput. Mach.* 9, 450–456; *1098.*

Douglas, J., Jr., and J. E. Gunn [1964], "A general formulation of alternating direction methods," *Numer. Math.* 6, 428–453; *1098.*

Douglas, J., Jr., and H. Rachford [1956], "On the numerical solution of heat conduction problems in two and three space variables," *Trans. Amer. Math. Soc.* 82, 421–439; *1098*.

Downing, J. A. [1966], "The automatic construction of contour plots with applications to numerical analysis," Master's thesis, The University of Texas at Austin; also TNN–58, Computation Center, The University of Texas at Austin; *162, 174*.

DuFort, E. C., and S. P. Frankel [1953], "Stability conditions in the numerical treatment of parabolic differential equations," *Math. of Comp.* (formerly *Math. Tables Aids Comput.*) 7, 135–152; *1085*.

Ehrlich, Louis W. [1964], "The block symmetric successive overrelaxation method," *SIAM J.* 12, 807–826; *1068*.

Eidson, Harold D., Jr. [1969], "The convergence of Richardson's finite-difference analogue for the heat equation," Master's thesis, The University of Texas at Austin; also TNN–90, Computation Center, The University of Texas at Austin; *1085*.

Engeli, Max E. [1969], "User's manual for the formula manipulation language SYMBAL," Report TRM-8.01, Computation Center, The University of Texas at Austin; *492*.

Faddeev, D. K., and V. N. Faddeeva [1963], *Computational Methods of Linear Algebra*, W. H. Freeman and Co., San Francisco; *721, 725, 747, 767, 823*.

Fanett, Mary [1963], "Application of the Remes algorithm to a problem in rational approximation," Master's thesis, The University of Texas at Austin; also TNN–23, Computation Center, The University of Texas at Austin; *317*.

Fejér, L. [1907], "Untersuchungen über Fouriersche Reihen," *Math. Annalen* 58, 51–69.

Fike, C. T. [1968], *Computer Evaluation of Mathematical Functions*, Prentice-Hall, Englewood Cliffs, N.J.; *68, 343*.

Flanders, D., and G. Shortley [1950], "Numerical determination of fundamental modes," *J. Appl. Physics* 21, 1326–1332; *1074*.

Ford, Lester R. [1933], *Differential Equations*, McGraw-Hill Book Co., New York; *389, 429, 431, 436, 643*.

Forsythe, G. E. [1957], "Generation and use of orthogonal polynomials for data-fitting with a digital computer," *J. Soc. Indust. Appl. Math.* 5, 74–88; *323, 324*.

Forsythe, G. E. [1966], "How do you solve a quadratic equation?" Tech. Report CS40, Stanford University; *92*.

Forsythe, G. E. [1967], "What is a satisfactory quadratic equation solver?" Tech. Report CS74, Stamford University; *92*.

Forsythe, G. E. [1969], "Solving a quadratic equation on a computer," *The Mathematical Sciences—A Collection of Essays*, edited by the National Research Council's Committee on Support of Research in the Mathematical Sciences, published for the National Academy of Sciences—National Research Council by the M.I.T. Press, Cambridge, Mass.; *92*.

Forsythe, G. E. [1970], "Pitfalls in computation," *Amer. Math. Monthly* 77, 931–956.

Forsythe, G. E., and C. Moler [1967], *Computer Solution of Linear Algebraic Equations*, Prentice-Hall, Englewood Cliffs, N.J.; *810, 812, 813, 814, 815, 816, 827, 833*.

Forsythe, G. E., and W. R. Wasow [1960], *Finite Difference Methods for Partial Differential Equations*, John Wiley and Sons, Inc., New York; *998, 1073, 1098*.

Fox, L. [1944], "Solution by relaxation methods of plane potential problems with mixed boundary conditions," *Quart. Appl. Math.* 2, 251–257; *993*.

Fox, L. [1947], "Some improvements in the use of relaxation methods for the solution of ordinary and partial differential equations," *Proc. Roy. Soc.*, London, Ser. A, Vol. 190, 31–59; *577, 666*.

Fox, L. [1957], *Numerical Solution of Two-point Boundary Value Problems in Ordinary Differential Equations*, Clarendon Press, Oxford; *665.*

Fox, L. [1962], *Numerical Solution of Ordinary and Partial Differential Equations*, Addison-Wesley Pub. Co., Reading, Mass.; *993, 998.*

Fox, L. [1965], *An Introduction to Numerical Linear Algebra*, Oxford Univ. Press, New York.

Fox, L., and D. F. Mayers [1968], *Computing Methods for Scientists and Engineers*, Clarendon Press, Oxford; *826, 834.*

Francis, J. G. F. [1961–1962], "The QR transformation, Parts I and II," *Computer Jour.* 4, 265–271 and 332–345; *923, 932, 934.*

Frank, W. L. [1958], "Computing eigenvalues of complex matrices by determinant evaluation and by methods of Danilewski and Wielandt," *J. Soc. Indust. Appl. Math.* 6, 378–392; *946.*

Frank, Werner [1960], "Solution of linear systems by Richardson's method," *J. Assoc. Comput. Mach.* 7, 274–286; *1068.*

Franklin, J. N. [1968], *Matrix Theory*, Prentice-Hall, Englewood Cliffs, N.J.; *773, 786.*

Fraser, M., and N. Metropolis [1968], "Algorithms in unnormalized arithmetic III. Matrix inversion," *Numer. Math.* 12, 416–428; *68.*

Friedman, B. [1956], *Principles and Techniques of Applied Mathematics*, John Yiley and Sons, Inc., New York; *753.*

Friedman, B. [1957], "The iterative solution of elliptic difference equations," A.E.C. Research and Development Report NYO-7698, Institute of Mathematical Sciences, New York University; *1074.*

Gantmacher, F. R. [1960], *The Theory of Matrices*, Vols. I and II (translated by K. A. Hirsch), Chelsea, New York; *739, 744, 747, 911.*

Garabedian, P. R. [1956], "Estimation of the relaxation factor for small mesh size," *Math. of Comp.* (formerly *Math. Tables Aids Comput.*); 10, 183–185; *1098.*

Geiringer, H. [1949], "On the solution of systems of linear equations by certain iterative methods," *Reissner Anniversary Volume*, University of Michigan Press, Ann Arbor, Mich., 365–393; *1073.*

Gerschgorin, S. [1930], "Fehlerabschätzung für das Differenzenverfahren zur Lösung partieller Differentialgleichungen," *Z. Agnew Math. Mech.* 10, 373–383; *576, 666, 970, 998.*

Gerschgorin, S. [1931], "Über die Abgrenzung der Eigenwerte einer Matrix," *Izv. Akad. Nauk SSSR*, Ser. fiz.-mat. 6, 749–754; *891.*

Gilbert, J. D. [1970], *Elements of Linear Algebra*, International Textbook Co., Scranton, Pa.

Gill, S. [1951], "A process for the step-by-step integration of differential equations in an automatic digital computing machine," *Proc. Cambridge Phil. Soc.* 47, 96–108; *492.*

Givens, Wallace [1953], "A method of computing eigenvalues and eigenvectors suggested by classical results on symmetric matrices," Chapter 17 of *Simultaneous Linear Equations and the Determination of Eigenvalues*, edited by L. J. Paige and Olga Taussky, National Bureau of Standards, Applied Mathematics Series No. 29, Washington, D.C.; *245.*

Givens, J. W. [1954], "Numerical computation of the characteristic values of a real symmetric matrix," Oak Ridge National Laboratory Report ORNL-1574; *10, 900, 909.*

Goldstine, H. H., F. J. Murray, and J. von Neumann [1959], "The Jacobi method for real symmetric matrices," *J. Assoc. Comput. Mach.* 6, 59–96; *896, 899.*

Golub, Gene H. [1959], "The use of Chebyshev matrix polynomials in the iterative solution of linear systems compared with the method of successive overrelaxation," doctoral thesis, University of Illinois.

Golub, G. H., and R. S. Varga [1961], "Chebyshev semi-iterative methods, successive over-relaxation iterative methods, and second order Richardson iterative methods," Parts I and II, *Numer. Math.* 3, 147–168; *1066, 1069, 1074.*

Gragg, William B., and Hans J. Stetter [1964], "Generalized multistep predictor-corrector methods," *J. Assoc. Comput. Mach.* 11, 188–209; *574.*

Greenstadt, J. [1960], "The determination of the characteristic roots of a matrix by the Jacobi method," Chapter 7 of Ralston and Wilf [1960]; *893, 896.*

Gregory, R. T. [1953], "Computing eigenvalues and eigenvectors of a symmetric matrix on the ILLIAC," *Math. of Comp.* 7, 215–220; *896.*

Gregory, R. T. [1957], "A method of deriving numerical differentiation formulas," *Amer. Math. Monthly* 64, 79–82.

Gregory, R. T. [1960], "Defective and derogatory matrices," *SIAM Review* 2, 134–139; *748.*

Gregory, R. T. [1963], *Numeral Systems,* Wm. C. Brown Book Co., Dubuque, Iowa (now Kendall/Hunt Publishing Company).

Gregory, R. T. [1966], "On the design of the arithmetic unit of a fixed-word-length computer from the standpoint of computational accuracy," *I.E.E.E. Trans. on Electronic Computers,* EC–15, 255–257; *69.*

Gregory, R. T., and D. L. Karney [1969], *A Collection of Matrices for Testing Computational Algorithms,* John Wiley and Sons, Inc., New York.

Griffith, H. W. [1971], "Preliminary investigations using interval arithmetic in the numerical evaluation of polynomials," doctoral dissertation, The University of Texas at Austin; also CNA–8, Center for Numerical Analysis, The University of Texas at Austin; *68.*

Guilinger, Willis H., Jr. [1965], "The Peaceman-Rachford method for small mesh increments," *J. of Math. Anal. and Appl.* 11, 261–277.

Habetler, G. J., and E. L. Wachspress [1961], "Symmetric successive overrelaxation in solving difference equations," *Math. of Comp.* 15, 356–362; *1068.*

Hageman, L. A., and R. B. Kellogg [1968], "Estimating optimum overrelaxation parameters," *Math. of Comp.* 22, 60–68; *1037.*

Hansen, Eldon [1969], "Cyclic composite multistep predictor-corrector methods," *Proc. of 24th National ACM Conference, New York,* 135–139; *574.*

Hansen, E. R. (editor) [1969], *Topics in Interval Analysis,* Clarendon Press, Oxford; *68, 574.*

Hart, J. F., E. W. Cheney, C. L. Lawson, H. J. Maehly, C. K. Mesztenyi, J. R. Rice, H. C. Thatcher, Jr., and C. Witzgall [1968], *Computer Approximations,* John Wiley and Sons, Inc., New York; *317, 326, 328, 329, 343.*

Hartree, D. R. [1952], *Numerical Analysis* (second edition 1958), Clarendon Press, Oxford; *11, 174, 343.*

Hayes, D. R., and L. Rubin [1970], "A proof of the Newton-Cotes quadrature formulas with error term," *Amer. Math. Monthly* 77, 1065–1072; *389, 392, 416.*

Henrici, Peter [1957], "Theoretical and experimental studies on the accumulation of error in the numerical solution of initial-value problems for systems of ordinary differential equations," *Proc. of the International Conference on Information Processing,* UNESCO, Paris, 15–20 June 1959 (published in 1960), 36–44.

Henrici, P. [1957a], unpublished lecture notes, Department of Mathematics, University of California, Los Angeles; *564, 569, 573.*

Henrici, Peter [1958], "On the speed of convergence of cyclic and quasicyclic Jacobi methods for computing eigenvalues of Hermitian matrices," *J. SIAM* 6, 144–162; *896.*

Henrici, P. [1960], "Estimating the best over-relaxation factor," Report NN-144, Ramo-Wooldridge Technical Memo, Los Angeles, Calif.; *1037*.

Henrici, Peter [1962], *Discrete Variable Methods in Ordinary Differential Equations*, John Wiley and Sons, Inc., New York; *389, 429, 442, 492, 498, 511, 513, 516, 520, 542, 543, 546, 563, 574*.

Henrici, P. [1963], *Error Propagation for Difference Methods*, John Wiley, New York; *486, 492*.

Henrici, Peter [1967], "Quotient-difference algorithms," Chapter 2 of Ralston and Wilf [1967]; *242*.

Hestenes, M. R., and E. Stiefel [1952], "Method of conjugate gradients for solving linear systems," *J. Res. Nat. Bur. Standards* 49, 409-436; *1071*.

Hildebrand, F. B. [1956], *Introduction to Numerical Analysis*, McGraw-Hill Book Co., Inc., New York; *416, 418, 420, 421*.

Hodge, W. V. D., and D. Pedoe [1947], *Methods of Algebraic Geometry*, Vol. I, Cambridge, at the University Press; *748*.

Hohn, F. E. [1958], *Elementary Matrix Algebra*, Macmillan, New York; *711, 734*.

Householder, Alston S. [1953], *Principles of Numerical Analysis*, McGraw-Hill Book Co., Inc., New York; *245*.

Householder, A. S. [1958], "The approximate solution of matrix problems," *J. Assoc. Comput. Mach.* 5, 205-243.

Householder, A. S. [1958a], "Unitary triangularization of a nonsymmetric matrix," *J. Assoc. Comput. Mach.* 5, 339-342; *901*.

Householder, A. S. [1964], *The Theory of Matrices in Numerical Analysis*, Blaisdell Publishing Co., New York; *202, 816, 901*.

Householder, A. S. [1970], *The Numerical Treatment of a Single Nonlinear Equation*, McGraw-Hill Book Co., New York; *173, 241, 245*.

Howell, J. A. [1971], "Algorithm 406. Exact solution of linear equations using residue arithmetic," *Comm. Assoc. Comput. Mach.* 14, 180-184; *888*.

Howell, J. A., and R. T. Gregory [1969a], "Solving systems of linear algebraic equations using residue arithmetic," Report TNN-82 (revised), Computation Center, The University Of Texas at Austin.

Howell, J. A., and R. T. Gregory [1969b], "An algorithm for solving linear algebraic equations using residue arithmetic, Parts I and II," *BIT* 9, 200-224 and 324-337; *888*.

Howell, J. A., and R. T. Gregory [1970], "Solving linear equations using residue arithmetic—Algorithm II," *BIT* 10, 23-37; *68, 874, 876, 888*.

Isaacson, Eugene, and Herbert B. Keller [1966], *Analysis of Numerical Methods*, John Wiley and Sons, New York; *390, 392, 1082, 1088*.

Jackson, Dunham [1930], *The Theory of Approximation*, American Math. Soc., Providence, R.I.

Jackson, Dunham [1941], *Fourier Series and Orthogonal Polynomials*, The Carus Mathematical Monographs, No. 6, Math. Assoc. of America; *330*.

Jenkins, M. A. [1969], "Three-stage variable-shift iterations for the solution of polynomial equations with *a posteriori* error bounds for the zeros," doctoral dissertation, Stanford University; also Tech. Report CS 138, Computer Science Department, Stanford University; *243*.

Jenkins, M. A., and J. F. Traub [1967], "An algorithm for an automatic general polynomial solver," Tech. Report No. CS 71, Computer Science Department, Stanford University; *243, 245*.

Jenkins, M. A., and J. F. Traub [1968, 1970], "A three-stage variable-shift iteration for polynomial zeros and its relation to generalized Rayleigh iteration," CS 107, Stanford Univ. (1968), *Numer. Math.* 14 (1970), 252–263; *243.*

John, F. [1952], "On integration of parabolic equations by difference methods," *Comm. Pure Appl. Math.* 5, 155–211.

John, F. [1967], *Lectures on Advanced Numerical Analysis*, Gordon and Breach, New York; *761, 769, 770.*

Juncosa, M. L., and David M. Young [1953], "On the order of convergence of solutions of a difference equation to a solution of the diffusion equation," *J. Soc. Indust. Appl. Math.* 1, 111–135; *1098.*

Juncosa, M. L., and David M. Young [1954], "On the convergence of a solution of difference equation to a solution of the equation of diffusion," *Proc. Amer. Math. Soc.* 5, 168–174; *1080, 1098.*

Juncosa, M. L., and David M. Young [1957], "On the Crank-Nicolson procedure for solving parabolic partial differential equations," *Proc. Cambridge Philos. Soc.* 53, part 2, 448–461; *1088.*

Kahan, W. [1958], "Gauss-Seidel methods of solving large systems of linear equations," doctoral thesis, University of Toronto; *1029, 1094.*

Kamke, E. [1948], *Differentialgleichungen Lösungsmethoden und Lösungen*, 1, Chelsea Pub. Co., New York; *431.*

Keller, Herbert B. [1968], *Numerical Methods for Two-point Boundary-value Problems*, Blaisdell, Waltham, Mass.; *665, 666.*

Kincaid, David [1969], "Solution of N simultaneous first-order differential equations," Program Writeup UTD2–01–CC07, The University of Texas Computation Center; *488.*

Kolman, B. [1970], *Elementary Linear Algebra*, Macmillan, New York.

Kopal, Zdeněk [1955], *Numerical Analysis*, John Wiley and Sons, Inc., New York; *420.*

Krause, E. F. [1970], *Introduction to Linear Algebra*, Holt, Rinehart and Winston, New York; *693, 694, 695, 699, 713, 715, 785.*

Kublanovskaya, V. N. [1961], "On some algorithms for the solution of the complete eigenvalue problem," *Zh. vych. mat.* 1, 555–570; *921.*

Kunz, K. S. [1957], *Numerical Analysis*, McGraw-Hill Book Co., New York; *276, 476.*

Kusmin, R. O. [1931], "Zur Theorie der mechanischen Quadraturen," *Nachr. Poly. Inst. Leningrad* 33, 5–14; *371.*

Lanczos, Cornelius [1956], *Applied Analysis*, Prentice-Hall, Englewood Cliffs, N.J.; *10.*

Lax, P. D., and R. D. Richtmyer [1956], "Survey of the stability of linear finite difference equations," *Comm. Pure Appl. Math.* 9, 267–293.

Lees, Milton [1966], "Discrete methods for nonlinear two-point boundary value problems," in *Numerical Solution of Partial Differential Equations*, edited by James Bramble, Academic Press, New York, 59–72; *665.*

Lehmer, D. H. [1961], "A machine method for solving polynomial equations," *J. Assoc. Comput. Mach.* 8, 151–161; *202.*

Leutert, Werner [1952], "On the convergence of unstable approximate solutions of the heat equation to the exact solution," *J. Math. Physics* 30, 245–251; *1098.*

Lindamood, G. E. [1964], "Numerical analysis in residue number systems," Univ. of Maryland Computer Science Report TR–64–7, College Park, Maryland; *68, 888.*

MacLane, S., and G. Birkhoff [1967], *Algebra*, 3rd edition, Macmillan, New York.

Maehly, H. J. [1959], "Rational approximations for transcendental functions," *Proc. of the International Conference on Information Processing,* UNESCO, Paris, 15–20 June 1959 (published in 1960), 57–62; *329.*

Marcus, M., and H. Minc [1964], *A Survey of Matrix Theory and Matrix Inequalities,* Allyn and Bacon, Inc., Boston; *816.*

Marden, Morris [1966], *Geometry of Polynomials,* Amer. Math. Soc., Providence, Rhode Island; *245.*

Martin, R. S., C. Reinsch, and J. H. Wilkinson [1968], "Householder's tridiagonalization of a symmetric matrix," *Numer. Math.* 11, 181–195.

Matula, D. W. [1969], "Towards an abstract mathematical theory of floating-point arithmetic," *Proc. Spring Joint Computer Conference,* IFIPS 34, 765–772; *68.*

McClellan, M. T. [1971], "The exact solution of systems of linear equations with polynomial coefficients." *Proc. of the Second Symposium on Symbolic and Algebraic Manipulation,* Assoc. for Comput. Mach., March 23–25, 1971, Los Angeles, Calif.; *68, 888.*

McDonald, A. E. [1970], "A multiplicity-independent, global iteration for meromorphic functions," doctoral dissertation. The University of Texas at Austin; also TNN–98, Computation Center, The University of Texas at Austin; *206, 210, 244, 245.*

McDowell, Leland K. [1967], "Variable successive overrelaxation," Report No. 244, Department of Computer Sciences, University of Illinois; *1068.*

McKeeman, W. M. [1962], "Algorithm 135, 'Crout with equilibration and iteration'," *Comm. Assoc. Comput. Mach.* 5, 553–555; *819.*

Milne, W. E. [1949], *Numerical Calculus,* Princeton University Press, Princeton, New Jersey.

Milne, W. E. [1953], *Numerical Solution of Differential Equations,* John Wiley and Sons, Inc., New York; *492.*

Milne, W. E., and R. R. Reynolds [1959], "Stability of a numerical solution of differential equations," *J. Assoc. Comput. Mach.* 6, 196–203; *574.*

Mitchell, B. E. [1953], "Normal and diagonalizable matrices," *Amer. Math. Monthly* 60, 94–96; *742.*

Montel, P. [1910], "Leçons sur les séries de polynomes à une variable complexe," Borel Monograph, Paris; *296,*

Moore, R. [1966], *Interval Analysis,* Prentice-Hall, Englewood Cliffs, N.J.; *68.*

Mouradoglou, A. J. [1967], "Numerical studies on the convergence of the Peaceman-Rachford alternating direction implicit method," Master's thesis, The University of Texas at Austin; also TNN–67, Computation Center, The University of Texas at Austin; *1035, 1061.*

Moursund, D. G. [1967], "Optimal starting values for Newton-Raphson calculation of $\sqrt{x}$," *Comm. Assoc. Comput. Mach.* 10, 430–432; *68.*

Muller, D. E. [1956], "A method for solving algebraic equations using an automatic computer," *Math. of Comp.* (formerly *Math. Tables Aids Comput.*) 10, 208–215; *195.*

Naiser, Lou Ann [1967], "The QR algorithm applied to Hessenberg matrices," Master's thesis, The University of Texas at Austin; also TNN–66, Computation Center, The University of Texas at Austin; *932, 934.*

Newman, M. [1967], "Solving equations exactly," *J. Res. Nat. Bur. Stds.* 17B, 171–179; *68, 868, 869, 881, 886, 888.*

Noble, Ben [1969], *Applied Linear Algebra,* Prentice-Hall, Inc., Englewood Cliffs, N.J.; *809, 817.*

O'Brien, George G., Morton A. Hyman, and Sidney Kaplan [1951], "A study of the numerical solution of partial differential equations," *J. Math. Phys.* 29, 223–252; *1079, 1098.*

Ortega, J. [1967], "The Givens-Householder method for symmetric matrices," Chapter 4 of Ralston and Wilf [1967]; *901, 938.*

Ortega, James M., and Werner C. Rheinboldt [1970], *Iterative Solution of Nonlinear Equations in Several Variables,* Academic Press, New York; *173, 174.*

Osborne, E. E. [1960], "On pre-conditioning of matrices," *J. Assoc. Comput. Mach.* 7, 338–335; *950.*

Ostrowski, A. M. [1954], "On the linear iteration procedures for symmetric matrices," *Rend. Mat. e Appl.* 13, 140–163; *1029, 1068, 1073.*

Ostrowski, A. M. [1958–1959], "On the convergence of the Rayleigh quotient iteration for the computation of the characteristic roots and vectors," (in six parts), *Arch. Rat. Mech. Anal:* **I,** vol. 1, pp. 233–241; **II,** vol. 2, pp.423–428; **III,** vol. 3, pp. 325–340; **IV,** vol. 3, pp. 341–347; **V,** vol. 3, pp. 472–481; **VI,** vol. 4, pp. 153–165; *245.*

Ostrowski, A. M. [1966], *Solution of Equations and Systems of Equations,* Academic Press, New York; *163, 173, 174, 549.*

Padé, H. [1892], "Sur la représentation approchée d'une fonction par des fractions rationnelles," thèse, Ann. de l'Ec. Nor. (3) 9; *326.*

Parlett, B. N. [1964], "Laguerre's method applied to the matrix eigenvalue problem," *Math. of Comp.* 18, 143–145; *242.*

Parlett, B. N. [1966], "Singular and invariant matrices under the QR transformation," *Math. of Comp.* 20, 611–615; *930.*

Parlett, B. N. [1967], "The LU and QR algorithms," Chapter 5 of Ralston and Wilf [1967]; *924, 930, 932, 936.*

Parlett, B. N. [1968], "Global convergence of the basic QR algorithm on Hessenberg matrices," *Math. of Comp.* 22, 803–817; *930.*

Parlett, B. N., and C. Reinsch [1969], "Balancing a matrix for calculation of eigenvalues and eigenvectors," *Numer. Math.* 13, 293–304; *950.*

Parter, Seymour V. [1959], "On 'two-line' iterative methods for the Laplace and biharmonic difference equations," *Numer. Math.* 1, 240–252; *1074.*

Parter, Seymour V. [1961], "'Multi-line' iterative methods for elliptic difference equations and fundamental frequencies," *Numer. Math.* 3, 305–319; *1074.*

Parter, Seymour V. [1965], "On estimating the 'rates of convergence' of iterative methods for elliptic difference equations," *Trans. Amer. Math. Soc.* 114, 320–354; *1074.*

Peaceman, D. W., and H. H. Rachford, Jr. [1955], "The numerical solution of parabolic and elliptic differential equations," *J. Soc. Indust. Appl. Math.* 3, 28–41; *1041, 1050, 1098.*

Peetre, J., and V. Thomee [1967], "On the rate of convergence for discrete initial-value problems," *Math. Scand.* 21, 159–176; *1098.*

Perlis, S. [1952], *Theory of Matrices,* Addison-Wesley, Reading, Mass.; *687, 713.*

Poole, William G. [1965], "Numerical experiments with several iterative methods for solving partial differential equations," Master's thesis, The University of Texas at Austin; also TNN–49, Computation Center, The University of Texas at Austin; *1071.*

Poole, W. G. [1970], "A geometric convergence theory for the QR, Rayleigh quotient, and power iterations," Computer Center Technical Report No. 41, Univ. of Calif., Berkeley; *941.*

Pope, D. A., and C. Tompkins [1957], "Maximizing functions of rotations," *J. Assoc. Comput. Mach.* 4, 459–466; *896.*

Price, H., and R. S. Varga [1962], "Recent numerical experiments comparing successive overrelaxation iterative methods with alternating direction implicit methods," Report No. 91, Gulf Research and Development Company, Pittsburgh, Pa.; *1061.*

Ralston, A. [1963], "On differentiating error terms," *Amer. Math. Monthly* 71, 187–189; *389*.

Ralston, Anthony [1965], *A First Course in Numerical Analysis*, McGraw-Hill Book Co., New York; *202, 242, 245, 327, 328, 329, 389, 421, 477*.

Ralston, Anthony, and Herbert S. Wilf [1960], *Mathematical Methods for Digital Computers*, Vol. I, John Wiley and Sons, Inc., New York.

Ralston, Anthony, and Herbert S. Wilf [1967], *Mathematical Methods for Digital Computers*, Vol. II, John Wiley and Sons, Inc., New York.

Raney, J. L. [1961], "Solution of N simultaneous differential equations by the Adams-Moulton method using a Runge-Kutta starter and partial double-precision arithmetic," Program Writeup UTD2-02-003, The University of Texas Computation Center; *488*.

Redish, K. A., and W. Ward [1971], "Environment enquires for numerical analysis," *SIGNUM Newsletter*, Assoc. for Comput. Mach. 6, 10–15; *92*.

Reich, E. [1949], "On the convergence of the classical iterative method of solving linear simultaneous equations," *Ann. Math. Statist.* 20, 448–451; *1073*.

Remes, E. [1934], "Sur un procédé convergent d'approximations successives pour déterminer les polynomes d'approximation," *Comptes Rendus*, 198, 2063–2065; *309*.

Remes, E. [1934a], "Sur le calcul effectif des polynomes d'approximation de Tchebichef," *Comptes Rendus* 199, 337–340; *309, 317*.

Rice, John R. [1964], *The Approximation of Functions I. Linear Theory*, Addison-Wesley, Boston; *317*.

Richardson, L. F. [1910], "The approximate arithmetical solution by finite differences of physical problems involving differential equations, with application to the stresses in a masonry dam," *Philos. Trans. Roy. Soc. London*, Ser. A, vol. 210, 307–357, and *Proc. Roy. Soc. London*, Ser. A, vol 83, 335–336; *577, 614, 1067, 1083*.

Richtmyer, Robert D., and K. W. Morton [1967], *Difference Methods for Initial-value Problems* (second edition), Interscience Publishers, New York, Second Edition; *1098*.

Romberg, W. [1955], "Vereinfachte numerische Integration," Det. Kong. Norske Videnskaber Selskab Forhandlinger, Band, 23, Nr. 7, Trondheim.

Ruhe, Axel [1970], "An algorithm for numerical determination of the structure of a general matrix," *BIT* 10, 196–216; *941*.

Runge, C. [1901], "Über empirische Funktionen und die Interpolation zwischen äquidistanten Ordinaten," *Zeit. für. Math. und Phys.*, XLVL, 229; *296*.

Rutishauser, H. [1957], *Der Quotienten-Differenzen-Algorithmus*, Birkhauser, Verlag, Basel; *242*.

Rutishauser, H. [1958], "Solution of eigenvalue problems with the *LR* transformation," in *Further Contributions to the Solution of Simultaneous Linear Equations and the Determination of Eigenvalues*, Nat. Bur. of Std. AMS 49; *921*.

Salzer, H. E. [1956], "Osculatory extrapolation and a new method for the numerical integration of differential equations," *J. Franklin Inst.* 262, 111–119; *574*.

Sard, A. [1963], "Linear approximation," *Mathematical Surveys, No. 9*, American Mathematical Society, Providence, R.I.; *421*.

Saul'yev, V. K. [1964], *Integration of Equations of Parabolic Type by the Method of Nets*, translated by G. J. Tee, Pergamon Press, New York; *1098*.

Schmidt, Jochen W., and Harmut Dressel [1967], "Fehlerabschätzungen bei Polynomgleichungen mit dem Fixpunktsatz von Brouwer," *Numer. Math.* 10, 42–50; *245*.

Schoenberg, I. J. [1964], "Spline interpolation and best quadrature formulas," *Bull. Amer. Math. Soc.* 70, 143–148; *421*.

Schröder, E. [1870], "Über unendlich viele Algorithmen zur Auflösung der Gleichungen," *Math. Ann.* 2, 317–365. English translation by G. W. Stewart, III, ORNL Translations No. 1851; *243, 245*.

Shanks, D. [1955], "Non-linear transformations of divergent and slowly convergent sequences," *J. Math. and Phys.* 34, 1–42; *174*.

Sheldon, J. [1955], "On the numerical solution of elliptic difference equations," *Math. of Comp.* (formerly *Math. Tables Aids Comput.*) 9, 101–112; *1068*.

Sheldon, J. W. [1959], "On the spectral norms of several iterative processes," *J. Assoc. for Comput. Mach.* 6, 494–505; *1069*.

Shortley. G. [1953], "Use of Tchebycheff polynomial operators in the numerical solution of boundary value problems," *J. Appl. Phys.* 32, 243–255; *1074*.

Simeunovic, D. M. [1967], "Les limites des modules des zéros des polynomes et des séries de Taylor," *Mat. Bech.* 4, 209–303; *245*.

Snyder, Martin A. [1966], *Chebyshev Methods in Numerical Approximation*, Prentice Hall, Inc., Englewood Cliffs, N.J.; *343*.

Southwell, R. V. [1946], *Relaxation Methods in Theoretical Physics*, Oxford University Press, New York; *1027*.

Steffensen, J. F. [1927], *Interpolation* (second edition 1950), Chelsea Publishing Co., New York; *296, 421*.

Stein, P., and R. Rosenberg [1948], "On the solution of linear simultaneous equations by iteration," *J. London Math. Soc.* 23, 111–118; *1021, 1073*.

Stewart, G. W. [1968], "Translation of the paper by E. Schröder [1870], 'On infinitely many algorithms for solving equations,'" Oak Ridge National Laboratory Report No. ORNL–tr–1851, Oak Ridge, Tennessee; *245*.

Stewart, G. W. [1969], "On some methods for solving equations related to Schröder's iterations," unpublished manuscript; *245*.

Stewart, G. W. [1970a], "Algorithm 384, Eigenvalues and eigenvectors of a real symmetric matrix," *Comm. Assoc. Comput. Mach.* 13, 369–371; *913*.

Stewart, G. W. [1970b], "Incorporating origin shifts into the QR algorithm for symmetric tridiagonal matrices," *Comm. Assoc. Comput. Mach.* 13, 365–367; *913*.

Stiefel, E. [1952], "Über einige Methoden der Relaxationrechnung," *Z. angew. Math. Phys.* 3, 1–33; *1071*.

Stiefel, E. L. [1959], "Numerical methods of Tchebycheff approximation," in *On Numerical Approximation*, edited by R. E. Langer, Univ. of Wisconsin Press, Madison, Wisconsin, 217–232; *309, 310, 316*.

Stiefel, E. L. [1959a], "Über diskrete und lineare Tschebyscheff-Approximationen," *Numer. Math.* 1, 1–28; *317*.

Strang, Gilbert [1970], "The finite element method and approximation theory," in *Numerical Solution of Partial Differential Equations*, II (SYNSPADE, 1970) edited by Bert Hubbard, Academic Press, New York; *998*.

Stroud, A. H., and Don Secrest [1966], *Gaussian Quadrature Formulas*, Prentice-Hall, Englewood Cliffs, N.J.; *421*.

Szabo, S., and R. Tanaka [1967]; *Residue Arithmetic and Its Applications to Computer Technology*, McGraw-Hill, New York; *68, 835, 837, 842*.

Takahasi, H., and Y. Ishibashi [1961], "A new method for 'exact calculation' by a digital computer," Information Processing in Japan 1, 28–42; *68, 888*.

Taussky, O. [1949], "A recurring theorem on determinants," *Amer. Math. Monthly* 56, 672–676; *1073*.

Thomas, L. H. [1949], "Elliptic problems in linear difference equations over a network," Watson Scientific Computing Laboratory, Columbia University, New York; *587*.

Thomason, John M. [1968], "Stabilizing averages for multistep methods of solving ordinary differential equations," Master's thesis, The University of Texas at Austin; also TNN–83, Computation Center, The University of Texas at Austin; *574*.

Thrall, R. M., and L. Tornheim [1957], *Vectors, Spaces and Matrices*, John Wiley & Sons, New York; *1047*.

Timlake, W. P. [1965], "On an algorithm of Milne and Reynolds," *BIT* 5, 276–281; *574*.

Todd, John [1963], *Introduction to the Constructive Theory of Functions*, Academic Press, Inc., New York; *309*.

Traub, J. F. [1964], *Iterative Methods for the Solution of Equations*, Prentice-Hall, Inc., Englewood Cliffs, N.J.; *120, 142, 146, 159, 173, 174*.

Traub, J. F. [1966a], "A class of globally convergent iteration functions for the solution of polynomial equations," *Math. Comp.* 20, 113–138; *243*.

Traub, J. F. [1966b], "Proof of global convergence of an iterative method for calculating complex zeros of a polynomial," *Notices Amer. Math. Soc.* 13, 117; *243*.

Traub, J. F. [1967], "The calculation of zeros of polynomials and analytic functions," Proceedings of Symposia in Applied Mathematics, Vol. 19, *Mathematical Aspects of Computer Science*, Amer. Math. Soc., Providence, Rhode Island, 138–152; *243*.

Tropper, A. M. [1969], *An Introduction to Linear Algebra*, American Elsevier, New York; *686, 704*.

Uspensky, J. V. [1948], *Theory of Equations*, McGraw-Hill, New York; *785, 787*.

Varah, James M. [1967], "The computation of bounds for the invariant subspaces of a general matrix operator," Tech. Report No. C.S. 66, Computer Science Dept., Stanford University; *941*.

Varga, Richard S. [1957], "A comparison of the successive overrelaxation method and semi-iterative methods using Chebyshev polynomials," *J. Soc. Indust. Appl. Math.* 5, 39–46; *1068, 1074*.

Varga, R. S. [1959], "Orderings of the successive overrelaxation scheme," *Pacific Jour. Math.* 9, 925–939; *1074*.

Varga, Richard S. [1960], "Factorization and normalized iterative methods," in *Boundary Problems in Differential Equations*, edited by R. E. Langer, University of Wisconsin Press, Madison, 121–142; *1064, 1074*.

Varga, Richard S. [1962], *Matrix Iterative Analysis*, Prentice-Hall, Inc., Englewood Cliffs, New Jersey; *816, 997, 998, 1013, 1014, 1021, 1029, 1037, 1073, 1074, 1098*.

Varga, R. S. [1970], *Functional Analysis and Approximation Theory in Numerical Analysis*, Proceedings of the Regional Conference Sponsored by the National Science Foundation at Boston University, Boston, Mass.; *666*.

Viswanathan, R. V. [1957], "Solution of Poisson's equation by relaxation method—normal gradient specified on curved boundaries," *Math. of Comp.* 11, 67–78; *993*.

Wachspress, E. L. [1957], "CURE: A generalized two-space-dimension multigroup coding for the IBM-704," Report KAPL-1724, Knolls Atomic Power Laboratory, Schenectady, New York; *1074*.

Wachspress, B. L. [1966], *Iterative Solution of Elliptic Systems and Applications to the Neutron Diffusion Equations of Reactor Physics*, Prentice-Hall, Inc., Englewood Cliffs, N.J.; *735, 1050, 1073, 1074*.

Walsh, J. L. [1931], "The existence of rational functions of best approximation," *Trans. Amer. Math. Soc.* 33, 668–689; *328*.

Warlick, Charles H. [1955], "Convergence rates of numerical methods for solving $\partial^2 u/\partial x^2 + (k/\rho)\partial u/\partial \rho + \partial^2 u/\partial \rho^2 = 0$," Master's thesis, The University of Maryland; *1037*.

Warlick, Charles H., and David M. Young [1970], "*A priori* methods for the determination of the optimum relaxation factor for the successive overrelaxation method," TNN–105, Computation Center, The University of Texas at Austin; *1037*.

Wedderburn, J. H. M. [1934], *Lectures on Matrices*, Amer. Math. Soc. Colloquium Publications, Vol. 27; *753*.

Wendroff, B. [1966], *Theoretical Numerical Analysis*, Academic Press, New York; *826, 828*.

Werner, H. [1912], "Die konstruktive Ermittlung der Tschebyscheff—Approximierenden im Bereich der rationalen Funktionen," *Arch. Rational Mech. Anal.* 11, 368–384; *328*.

Widder, David V. [1947], *Advanced Calculus*, Prentice-Hall, Englewood Cliffs, N.J.; *995*.

Widlund, O. B. [1966], "On the rate of convergence of an alternating direction implicit method in a noncommutative case," *Math. Comp.* 20, 500–515; *1074*.

Widlund, O. B. [1968], "On the rate of convergence for parabolic difference schemes, I," *Proc. Amer. Math. Soc. Symposium for Applied Mathematics 21*, Durham, North Carolina; *1098*.

Widlund, O. B. [1971], "On the rate of convergence for parabolic difference schemes, II," *Comm. Pure Appl. Math.* 23, 79–96; *1098*.

Wilf, Herbert S. [1960], "The numerical solution of polynomial equations," Chapter 21 of Ralston and Wilf [1960]; *210*.

Wilkinson, Belinda M. [1969], "A polyalgorithm for finding roots of polynomial equations," Master's thesis, The University of Texas at Austin; also TNN–93, Computation Center, The University of Texas at Austin; *177, 218, 229, 235*.

Wilkinson, J. H. [1954], "The calculation of the latent roots and vectors of matrices on the pilot model of the A.C.E.," *Proc. Camb. Phil. Soc.* 50, 536–566; *915*.

Wilkinson, J. H. [1960], "Householder's method for the solution of the algebraic eigenproblem," *Comp. Jour.* 3, 23–27; *901*.

Wilkinson, J. H. [1961], "Error analysis of direct methods of matrix inversion," *J. Assoc. Comput. Mach.* 8, 281–330; *813*.

Wilkinson, J. H. [1962], "Householder's method for symmetric matrices," *Numer. Math.* 4, 354–361; *901*.

Wilkinson, J. H. [1963], *Rounding Errors in Algebraic Processes*, Prentice-Hall, Inc., Englewood Cliffs, N.J.; *45, 46, 52, 53, 191, 213, 216, 230, 833, 889, 938, 939, 944*.

Wilkinson, J. H. [1965], *The Algebraic Eigenvalue Problem*, Clarendon Press, Oxford; *53, 745, 804, 808, 813, 826, 832, 891, 896, 915, 919, 923, 924, 932, 941, 946, 948, 950*.

Wilkinson, J. H. [1965a], "Convergence of the *LR, QR*, and related algorithms," *The Computer Jour.* 8, 77–84.

Wilkinson, J. H. [1967], "The solution of ill-conditioned linear equations," Chapter 3 of Ralston and Wilf [1967]; *806, 826, 828, 832, 834*.

Wilkinson, J. H. [1968], see Martin, Reinsch, and Wilkinson [1968]; *901*.

WPA [1944], "Tables of Lagrangian interpolation coefficients," *Mathematical Tables Project*, Works Progress Administration, Federal Works Agency; *271*.

Young, David M. [1950], "Iterative methods for solving partial difference equations of elliptic type," doctoral thesis, Harvard University; *1073, 1074*.

Young, David M. [1954], "Iterative methods for solving partial difference equations of elliptic type," *Trans. Amer. Math. Soc.* 76, 92–111; *1023, 1025, 1073*.

Young, David M. [1954a], "On Richardson's method for solving linear systems with positive definite matrices," *J. Math. Phys.* XXXII, 243–255; *1074*.

Young, David M. [1955], "Gill's method for solving ordinary differential equations," Numerical Note NN–4, Ramo-Wooldridge Corp., Los Angeles, Calif.

Young, David M. [1961], "The numerical solution of elliptic and parabolic partial differential equations," in *Numerical Analysis*, Vol. VI, edited by E. F. Beckenbach, McGraw-Hill Book Co., New York, 283–298.

Young, David M. [1962], "The numerical solution of elliptic and parabolic partial differential equations," in *Survey of Numerical Analysis*, edited by John Todd, McGraw-Hill Book Co., New York, 380–438; *1098, 1099*.

Young, David M. [1971], "On the consistency of linear stationary iterative methods," *SIAM Jour. of Numer. Analysis* 9, 89–96; *174*.

Young, David M. [1971a], *Iterative Solution of Large Linear Systems*, Academic Press, New York; *174, 755, 1013, 1014, 1025, 1029, 1037, 1073, 1098*.

Young, David M. [1971b], "Second-degree iterative methods for the solution of large linear systems," *Jour. of Approx. Theory* 5, 137–148; *1068*.

Young, David M. [1971c], "A bound for the optimum relaxation factor for the successive overrelaxation method," *Numer. Math.* 16, 408–413; *1037*.

Young, David M., and John H. Dauwalder [1965], "Discrete representations of partial differential operators," in *Error in Digital Computation*, Vol. 2, edited by L. B. Rall, John Wiley and Sons, Inc., New York, 181–217; *666, 989*.

Young, David M., and John H. Dauwalder [1971], "Discrete representations of partial differential operators—II," unpublished manuscript; *990*.

Young, David M., and Louis Ehrlich [1956], "On the numerical solution of linear and nonlinear parabolic equations on the Ordvac," Interim Tech. Report No. 18, Office of Ordnance Research Contract DA–36–034–ORD–1486, University of Maryland.

Young, David M., and Louis Ehrlich [1960], "Some numerical studies of iterative methods for solving elliptic difference equations," in *Boundary Problems in Differential Equations*, edited by R. E. Langer, The University of Wisconsin Press, Madison, 143–162; *1061*.

Young, David M., and Thurman Frank [1962], "A survey of computer methods for solving elliptic and parabolic partial differential equations," *Bulletin of the International Computation Center* 2, Rome, Italy, 3–61; also TNN–20, Computation Center, The University of Texas at Austin.

Young David M., L. D. Gates, Jr., R. J. Arms, and D. F. Eliezer [1955], "The computation of an axially symmetric free boundary problem on NORC," U.S. Naval Proving Ground Report No. 1413, Dahlgren, Va.; *997*.

Young, David M., and David R. Kincaid [1969], "Norms of the successive overrelaxation method and related methods," TNN–94, Computation Center, The University of Texas at Austin.

Young, David M., and Alvis E. McDonald [1969], "On the surveillance and control of number range and accuracy in numerical computation," in *Information Processing 68*, North-Holland Publishing Company, Amsterdam, 145–152; *87, 92*.

Young, D. M., Jr., A. E. McDonald, H. E. Eidson, and B. M. Wilkinson [1971], "Matrix eigenvalue methods for solving polynomial equations," in *Progress*; *177, 228, 235, 238, 240*.

Young, David M., and Harry Shaw [1955], "Ordvac solutions of $\partial^2 u/\partial x^2 + \partial^2 u/\partial y^2 + (k/y)\partial u/\partial y = 0$ for boundary value problems and problems of mixed type," Interim Tech. Report No. 14, Office of Ordnance Research Contract DA–36–034–ORD–1486, Univ. of Maryland; *1037*.

Young, David M., and Charles H. Warlick [1953], "On the use of Richardson's method for the numerical solution of Laplace's equation on the ORDVAC," Ballistic Research Labs. Memorandum Report No. 707, Aberdeen Proving Ground, Maryland; *1067*.

Young, David M., Mary F. Wheeler, and James A. Downing [1965], "On the use of the modified successive overrelaxation method with several relaxation factors," in *Proc. of IFIP Congress 65*, edited by W. A. Kalenich, Spartan Books, Inc., Washington, D.C., 177–182; *1068*.

Zelinsky, D. [1968], *A First Course in Linear Algebra*, Academic Press, New York; *730*.

INDEX